17th Edition

Technology in Action

Alan Evans • Kendall Martin • Mary Anne Poatsy

Content Development: Shannon Le-May Finn
Content Management: Stephanie Kiel
Content Production: Rudrani Mukherjee
Product Management: Marcus Scherer
Product Marketing: Wayne Stevens
Rights and Permissions: Jenell Forschler

Please contact https://support.pearson.com/getsupport/s/ with any queries on this content

Cover Image by © Neale Cousland/Shutterstock

Library of Congress Cataloging-in-Publication Data

Names: Evans, Alan (Alan D.), author. | Martin, Kendall (Kendall E.), author. | Poatsy, Mary Anne, author.
Title: Technology in action : complete / Alan Evans, Kendall Martin, Mary Anne Poatsy.
Description: 17th edition. | [Hoboken] : Pearson, [2022] | Includes index.
Identifiers: LCCN 2020048317 | ISBN 9780136903666 (paperback) | ISBN 9780136903154
Subjects: LCSH: Microcomputers. | Computer science. | Computer networks. | Information technology.
Classification: LCC QA76.5 .E9195 2021 | DDC 004.16–dc23
LC record available at https://lccn.loc.gov/2020048317

1 2020

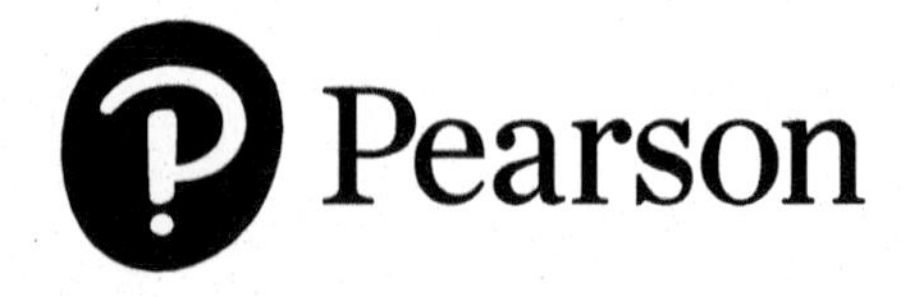

Rental
ISBN-10: 0-13-690366-5
ISBN-13: 978-013-690366-6

Contents at a Glance

Contents

Chapter 1

Chapter 2

Chapter 3

Chapter 9

Learn Technology by Using *Technology in Action* 17e

Technology in Action continues to be the best-selling title because it delivers an engaging approach to teaching the topics and skills students need to be digitally literate. The text provides an engaging approach to learning computer concepts in which students learn by using practical content, hands-on projects, and interactive activities.

To ensure every type of learner and instructor can choose the approach that best fits their style, *Technology in Action* is available in three formats:

1. **Print Rental for Technology in Action:** For courses focused on a more traditional, instructor-led approach, the print title is available as a rental option. The instructor resources provide everything needed to teach an in-person or online class in which students read the text and perform objective-based testing.
2. **MyLab IT for Technology in Action:** For courses that teach computer concepts *and* Microsoft Office skills, using the MyLab IT option is the right choice. This option provides students and instructors with an interactive etext and assignable activities that assess students' understanding in a variety of ways. By combining trusted author content with digital tools and a flexible platform, MyLab IT personalizes the learning experience and helps students absorb and retain key course concepts while developing skills that employers seek. All of the instruction, practice, review, and assessment resources are in one place, allowing instructors to arrange their course from an instructional perspective that gives students a consistent and measurable learning experience from chapter to chapter. New to this edition are Prebuilt Learning Modules that allow instructors to quickly assign author curated content in every chapter for quick course set up.
3. **Revel Technology in Action:** For courses focused on computer concepts only, Revel is an excellent, fully digital experience in which students read, practice, and study in one continuous experience. Students read a little and then interact with the content through videos, Check Your Understanding self-check questions, and a variety of interactives. Revel prepares students for meaningful class participation and provides the skills they need to be digitally literate in the workplace and their everyday lives. Revel replaces the traditional textbook by presenting an affordable, seamless blend of author-created digital text, media (videos, Helpdesks, Sound Bytes, and IT Simulations), and assessments based on learning science.

As with previous editions, we have thoroughly reviewed and updated the 17th edition to ensure coverage of the latest in technology. We have included topics such as Ransomware as a Service, CPUs that fight back against malware, and bandwidth throttling, as well as new Dig Deeper articles such as "Building Your Own Gaming Computer."

We made these updates with the instructor in mind—just know the content is timely and check out the transition guide, which provides the specifics on what we have updated or changed. We have retained popular 16e features, including the web-based survey activities that encourage students to engage with their peers and develop critical thinking skills through the *What Do You Think?* questions. These activities, along with the *Tech in the News* updates at the beginning of each chapter in the etext or Revel, videos, interactive Helpdesk activities, Sound Byte lessons, IT Simulations, and a variety of hands-on projects all help students learn the concepts and skills they need in order to be digitally literate in today's workplace. And, if they are using MyLab IT, they can earn the *Digital Competency* badge to demonstrate their skills to potential employers.

Hallmarks

- Engaging **question-and-answer writing style** that approaches topics as students do.
- **Ethics coverage** throughout, including in end-of-chapter activities, Point/Counterpoint ethical debate content found in relevant chapters, and a Sound Byte lesson on how to discuss and debate ethical issues.
- **Hands-on learning** with projects throughout each chapter:
  - **Try This** projects allow students to practice and demonstrate their proficiency with important topics. In MyLab IT and Revel, how-to videos accompany each project.

- **Solve This** projects put the concepts students are learning into action through real-world problem solving using Microsoft Office programs. MyLab IT provides Grader project versions of most of these projects.
- **Make This** projects provide activities where students build programs that run on their mobile devices. Twelve of the chapters have activities that build fully functional mobile apps, compatible with either Android or iOS. Each project includes instructions and a how-to video in MyLab IT and Revel.

- **Interactive activities** in MyLab or Revel engage students in active learning and *demonstration* of understanding:
  - **Helpdesk** interactive activities provide a review of chapter objectives by having students play the role of a helpdesk staffer assisting customers via a live chat using a decision-based simulation with a quiz.
  - **Sound Byte** lessons provide coverage of additional topics related to the chapter, including a brief quiz.
  - **IT Simulations** provide an in-depth chapter scenario that students work through in an active learning environment and complete with a brief quiz to demonstrate their understanding. These now include a "presentation mode" so instructors can walk through the simulation in class or with students.
- **Review and Quizzes**
  - **Check Your Understanding Quizzes** provide a self-check covering objectives in each part of the chapter so that students can see how well they are learning the content.
  - The **Chapter Quiz** provides a way for students to test that they have learned the material from the entire chapter.
  - **"Chew on This"** critical thinking questions require that students demonstrate their understanding through written answers that are manually graded.
  - **Testbank Exams** provide customizable prebuilt, autograded, objective-based questions covering the chapter objectives.
- **Videos**
  - **Chapter Overview Videos** provide an objective-based review of chapter content.
  - **Try This** and **Make This** project videos walk students through these projects.
- **Helpful Resources**
  - **PowerPoint and Audio Presentations** can be used in class for lecture or assigned to students, particularly online students, for instruction and review.
  - **Instructor Chapter Guides** provide teaching tips; homework and assessment suggestions; a brief overview of each chapter's *Try This, Make This*, and *Solve This* exercises; as well as select Sound Byte talking points and ethics debate starters.

What's New?

- **For the MyLab IT version, Prebuilt Learning Modules** have been created by the authors for each chapter. These will help instructors quickly set up their course to cover a range of resources to provide effective learning and assessment of the concepts. The modules are especially helpful when teaching fully online.
- **Key terms** have been scaled back by eliminating terms that the majority of today's students reported they already knew.
- The **Technology in the News** widget helps instructors keep their class current with weekly technology news. This currency widget is included in the in the MyLab IT interactive etext and Revel versions of *Technology in Action* throughout the semester.
- **Images and quizzes** have been updated throughout.
- **Team Time exercises** have been updated and designed for working in online environments like Microsoft Teams or Zoom.

Summary of Chapter Updates

All chapter Learning Outcomes and Learning Objectives have been revised as needed and throughout the text, figures and photos have been updated with new images, current topics, and state-of-the art technology coverage.

Chapter 1:

- We have added a new Bits&Bytes showing hands-on tools to let students explore AI applications in natural language processing and vision recognition.
- We have added a new Bits&Bytes taking students into the world of machine learning through sound synthesis.
- We have added a new discussion of digital forensics.
- We have added material focused on the challenges and successes of technology when faced with the global COVID-19 pandemic.
- We have added an ethics discussion on the impact of COVID-19 on the social equity issue of the digital divide.
- We have added a presentation on how to donate CPU time to the @home effort for vaccine discovery.

Chapter 2:

- We have streamlined the key terms to eliminate terms most students already know.
- We have added a new Dig Deeper: Building Your Own Gaming Computer.
- We have removed the Dig Deeper: How Touch Screens Work.
- We have added a new Bits&Bytes: Foldable Glass Is Making Foldable Phones Better!

- We have added a new Bits&Bytes: Teraflops in Your Living Room: Gaming on Overdrive.
- We have added a new Bits&Bytes: Smart Coral Reefs.
- We have deleted the Bits&Bytes: Distributed Computing: Putting Your Computer to Work While You Sleep.
- We have deleted the Bits&Bytes: Foldable Phones Are Here, and have added the content to the body of the text.
- We have deleted the Bits&Bytes: Coming Soon: USB 4 and DisplayPort 2 and have added the content to the body of the text.

Chapter 3:

- We have added a new Learning Objective to accommodate cloud computing, which we moved from the previous Dig Deeper into the body of the text.
- We have updated the content on online collaboration and file sharing tools.
- We have added a new Bits&Bytes on social media influencers.
- We have moved content from the Bits&Bytes: Secure Messaging Apps into the body of the text.
- We have added content in the E-Commerce Safeguards section about mobile wallets.
- We have added a new Dig Deeper about AI and cloud computing.
- We have updated content on web browsers.
- We have deleted content about Google specialized search tools.
- We have added content about predictive search results.
- We have moved the Bits&Bytes: Digital Assistants and Predictive Search Bits&Bytes into the body of the text.
- We have replaced the Digital Activism and Geolocation ethics topics with Deepfakes and Personalized Marketing.

Chapter 4:

- We have added a new Bits&Bytes: Portable Apps, which replaces the Bits&Bytes: Finding Alternative Software.
- We have added a new Trends In IT: Artificial Intelligence in Healthcare, which replaces Mobile Payment Apps: The Power of M-Commerce.
- We have updated Ethics: Can I Use Software on my Device When I Don't Own the License?
- We have added new content about Microsoft Teams to the Productivity Software section.
- We have moved the Bits&Bytes: Need to Work as a Team? Try These Collaboration Tools content into the body of the text.
- We have added a new Bits&Bytes: PDF: The Universal File Format.
- We have added a new Bits&Bytes: Scalable Vector Graphics: Text-Based Images.

Chapter 5:

- We have updated the text and screenshots to reflect updates to Windows 10 and its newest features.
- We have added a new Bits&Bytes on Quick Assist to control a computer remotely.
- We have added content on using Storage Sense to monitor space and free up storage automatically.
- We have added a new Bits&Bytes on finding the program appropriate to any file extension.
- We have added a new Dig Deeper outlining the usefulness of many Windows key shortcuts.

Chapter 6:

- We have retitled and expanded the Dig Deeper: How Storage Devices Work so that it is now How Storage Devices Work . . . and Fail.
- We have removed the Dig Deeper: The Machine Cycle (it is still discussed sufficiently in the text).
- We have added a new Bits&Bytes: Telepresence Robots Coming to a Workplace Near You.
- We have added a new Bits&Bytes: Surveillance Drives: Do You Need One?
- We have replaced the Trends in IT: USB 3.2 C Ports: One Port to Rule Them All with Trends in IT: Thunderbolt and USB Ports: Which Ones Do I Need? to cover the latest port standards.

Chapter 7:

- We have updated the Dig Deeper: P2P File Sharing.
- We have added a new Bits&Bytes: Watch Out for Wi-Fi 6E.
- We have updated the Bits&Bytes: Net Neutrality.
- We have added a new table comparing Wi-Fi standards.

Chapter 8:

- We have revised the sections on Digital Photography and Digital Video to be more phone camera centric.
- We have added a new Bits&Bytes: Professional Video Editing Tools . . . For Free!
- We have removed the Bits&Bytes: The Rise of Wearable Technology.
- We have added information on wearables into the body of the text.

Chapter 9:

- We have added discussion of ransomware and examples of its occurrence during the COVID-19 pandemic.
- We have added a new key term, Ransomware as a Service.
- We have added a new focus on employee monitoring software with emphasis as more employees are sent home to telecommute.
- We have updated references to operating system tools for security.
- We have added a coverage of new identity theft scams like using caller ID to pose as government agencies.

- We have updated coverage of audio captcha and more sophisticated analytical captcha techniques.
- We have updated coverage of password managers.
- We have added coverage of hardware keys like the Yubi key for multifactor authentication.
- We have added new information on Chrome extensions that enable you to work securely on public or private computers.
- We have updated the content on mobile alarm software/hardware solutions.
- We have updated our digital forensics coverage, including implications of IoT devices and cloud storage on data gathering for forensics.

Chapter 10:

- We have added a new Bits&Bytes: Max/MSP: Learn Programming with Patch Cords.
- We have added a new Bits&Bytes on Repl.IT.

Chapter 11:

- We have added a new Bits&Bytes: Use a Graph Database to Track Connected Data.

Chapter 12:

- We have moved the content in Trends In IT: Virtualization: Making Servers Work Harder into the body of the text.
- We have added a new Trends In IT: Protecting Privacy in Your Home Office.
- We have added a new Bits&Bytes: Encryption: Not Just For Businesses Any Longer.
- We have added a new Dig Deeper: Intent-Based Networking (IBN): Networks That Think for Themselves.
- We have moved the Dig Deeper: The OSI Model: Defining Protocol Standards content into the body of the text.
- We have deleted the Bits&Bytes: Guidance on Green Computing as the content is covered sufficiently in Chapter 2.

Chapter 13:

- We have added a new Bits&Bytes on packet analysis.
- We have added a new Bits&Bytes on recommended editors for web development.

Digital Approaches to *Technology in Action*

MyLab IT

To maximize student results in courses covering computer concepts and Microsoft 365 applications, we recommend using *Technology in Action* with **MyLab IT.** By combining trusted author content with digital tools and a flexible platform, MyLab IT personalizes the learning experience and will help students learn and retain key course concepts while developing skills that future employers seek.

With MyLab IT for *Technology in Action*, students have access to all of the instruction, practice, review, and assessment resources in one place. There are two ways instructors can set up their course:

1. Instructors can choose to use the new *Prebuilt learning modules* that allow them to create activities in the order they want students to complete them, providing a consistent, measurable learning experience from chapter to chapter.
2. Instructors can take a second approach for an interactive learning experience, where students use the interactive etext to read and learn actively with Helpdesk activities, Sound Bytes, IT Simulations, *Technology in the News* currency updates, *What do You Think?* surveys and critical thinking questions, hands-on projects, videos, accessible PowerPoint presentations, and more. Instructors assign the etext chapter, students engage in learning and practice, and then go back to their assignments to take the chapter quizzes.

MyLab IT Features

Technology in the News

Provide regular currency updates to the beginning of each chapter throughout the semester to deliver the latest technology news stories to use in the classroom. The update is live in the interactive etext and Revel, so no matter where students are in the content, instructors will have this weekly update to use for in-class discussion or as a reading assignment and can also use the discussion questions or activities included in most postings.

Solve This Projects

These exercises integrate and reinforce chapter concepts with Microsoft 365 skills.

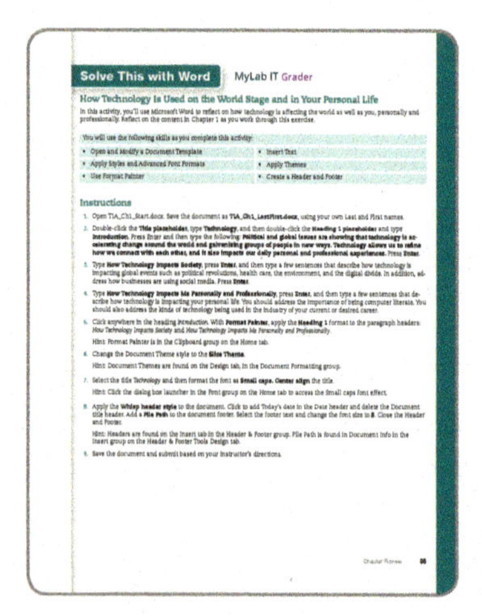
Solve This with Word MyLab IT Grader

How Technology Is Used on the World Stage and in Your Personal Life

In this activity, you'll use Microsoft Word to reflect on how technology is affecting the world as well as you, personally and professionally. Reflect on the content in Chapter 1 as you work through this exercise.

You will use the following skills as you complete this activity:

- Open and Modify a Document Template
- Apply Styles and Advanced Font Formats
- Use Format Painter
- Insert Text
- Apply Themes
- Create a Header and Footer

Instructions

1. Open TIA_Ch1_Start.docx. Save the document as **TIA_Ch1_LastFirst.docx**, using your own Last and First names.
2. Double-click the **Title placeholder**, type **Technology**, and then double-click the **Heading 1 placeholder** and type **Introduction**. Press Enter and then type the following: **Political and global issues are showing that technology is accelerating change around the world and galvanizing groups of people in new ways. Technology allows us to refine how we connect with each other, and it also impacts our daily personal and professional experiences.** Press Enter.
3. Type **How Technology Impacts Society**, press **Enter**, and then type a few sentences that describe how technology is impacting global events such as political revolutions, health care, the environment, and the digital divide. In addition, address how businesses are using social media. Press **Enter**.
4. Type **How Technology Impacts Me Personally and Professionally**, press **Enter**, and then type a few sentences that describe how technology is impacting your personal life. You should address the importance of being computer literate. You should also address the kinds of technology being used in the industry of your current or desired career.
5. Click anywhere in the heading *Introduction*. With **Format Painter**, apply the **Heading 1** format to the paragraph headers: *How Technology Impacts Society* and *How Technology Impacts Me Personally and Professionally*.
   Hint: Format Painter is in the Clipboard group on the Home tab.
6. Change the Document Theme style to the **Slice Theme**.
   Hint: Document Themes are found on the Design tab, in the Document Formatting group.
7. Select the title *Technology* and then format the font as **Small caps. Center align** the title.
   Hint: Click the dialog box launcher in the Font group on the Home tab to access the Small caps font effect.
8. Apply the **Whisp header style** to the document. Click to add Today's date in the Date header and delete the Document title header. Add a **File Path** to the document footer. Select the footer text and change the font size to **8**. Close the Header and Footer.
   Hint: Headers are found on the Insert tab in the Header & Footer group. File Path is found in Document Info in the Insert group on the Header & Footer Tools Design tab.
9. Save the document and submit based on your instructor's directions.

Chapter Review

Helpdesk Activities

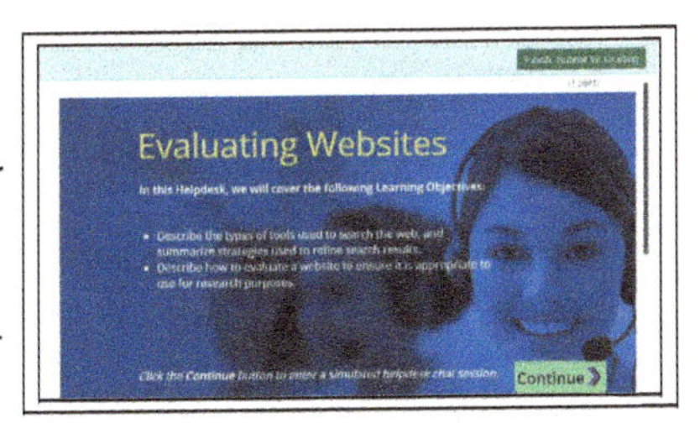

The Helpdesk training content, created specifically for *Technology in Action*, enables students to take on the role of a helpdesk staffer fielding questions posed by computer users so that students demonstrate their understanding in an active learning environment. Each Helpdesk ends with a quiz, ensuring students have grasped the content.

Sound Bytes

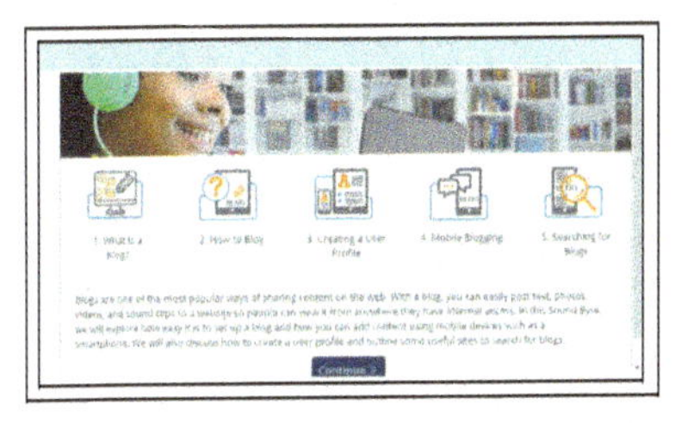

Sound Bytes expand student mastery of complex topics through engaging lessons with a brief quiz to check their understanding.

IT Simulations

These detailed interactive scenarios cover a core chapter topic in a hands-on environment where students can apply what they have learned and demonstrate understanding through active engagement.

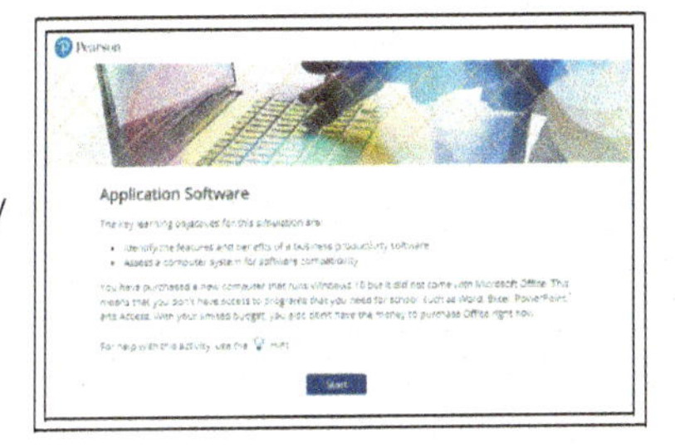

Revel

For courses focused on computer concepts only, Revel is an excellent, fully digital experience in which students read, practice, and study in one continuous experience. Students read a little and then interact with the content through videos, Check Your Understanding self-check questions, and a variety of interactives. Revel prepares students for meaningfully class participation and provides the skills they need to be digitally literate in the workplace and their everyday lives. Revel replaces the traditional textbook by presenting an affordable, seamless blend of author-created digital text, media (videos, Helpdesks, Sound Bytes, and IT Simulations), and assessments based on learning science.

Revel Functionality for Students

- **Full chapter audio** allows student to listen and learn in an alternative format.
- **Interactive exercises,** such as IT Simulations, allow students to engage with concepts and take an active role in learning.
- **Videos** in each chapter help explain key topics that can be hard to comprehend from a static photo or description.
- **Flashcards and study tools** provide practice with important key terms.
- **Chapter quizzes** test students' knowledge. These are graded instantly for immediate feedback and scores are tracked in the instructor dashboard.
- **Highlighting, note taking, and a glossary** let students read and study in a variety of ways.
- **The Revel mobile app** lets students read, practice, and study anywhere, anytime, on any device. Content is available both online and offline, and the app syncs work across all registered devices automatically, giving students great flexibility to toggle between phone, tablet and laptop.

Revel Functionality for Instructors

- **Adding notes** to precise locations within the narrative content allows instructors to include reminders, study tips, and links to additional resources for students.
- **Revel course management tools** allow instructors to indicate precisely which chapter readings must be completed on which dates. This clear, detailed schedule helps students stay on task by eliminating any ambiguity as to which material will be covered during each class.
- The **Performance dashboard** enables instructors to view assignment completion, time on task, points earned on chapter quizzes, how the class is trending, student performance by learning outcome, and more.
- **At a glance performance insights** enable instructors to access student progress at a glance with the new Educator Dashboard. Instructors can see how well students are preparing for class, the average grade distribution, who is struggling, and what's due next.
- **Enhanced Grades view** allows instructors to easily see how their students are doing, from performance on specific assignments to overall grades. New visual indicators help instructors identify struggling students.
- **Learning management system integration** provides easy access to Revel courses via Blackboard Learn™, Canvas™, Brightspace by D2L, and Moodle. With single sign-on, students can be ready to access Revel on their first day. Flexible, on-demand grade synchronization capabilities allow instructors to control exactly which grades should be transferred to their LMS Gradebook.

Instructor Teaching Resources

This program comes with the following teaching resources.

Supplements available to instructors at www.pearsonhighered.com/techinaction	Features of the Supplement
Accessible PowerPoint Presentation	PowerPoints meet accessibility standards for students with disabilities. Features include, but are not limited to: • Keyboard and Screen Reader Access • Alternative Text for Images • High Color Contrast between Background and Foreground Colors
Audio PowerPoint Presentation	This is a full audio version of the Accessible presentation.
End-of-Chapter Answer Key, Check Your Understanding Answer Key, Chapter Quiz Answer Key	Answers to all end-of-chapter questions.
Image Library	Every image in the book.
Instructor Chapter Guide	• Content Instruction • Student Preparation and Review • Active Learning Options • Chapter Assessment • End-of-Chapter Exercises • Currency Topics • Soft Skills and Team Work • Instructor Resources
***Make This* Projects**	Activities where students build programs that run on their mobile devices. Each project includes instructions and a how-to video.
Objectives Mapping	Outline of the objectives in every chapter.
***Solve This* Projects**	Real-world problem solving using Microsoft 365 programs. Grader versions of most of these projects are in MyLab IT.
Syllabus Template	Sample syllabus for help in setting up the course.
Test Bank (Textbook, Helpdesk, Sound Bytes)	Over 1,000 multiple-choice, true/false, short-answer, and matching questions with these annotations: • Difficulty level (1 for straight recall, 2 for some analysis, 3 for complex analysis) • Objective, which provides location in the text Provided for: • Textbook • Helpdesk • Sound Byte
Computerized TestGen	TestGen allows instructors to: • Customize, save, and generate classroom tests • Edit, add, or delete questions from the Test Item files • Analyze test results • Organize a database of tests and student results
Transition Guide	Detailed explanation of changes between the previous and current edition.
Web Projects	Discussion questions and additional projects that can be done on the Internet.

Letter from the Authors

Our 17th Edition—A Letter from the Authors

Why We Wrote This Book

The pace of technological change is ever increasing. In education, we have seen this impact us more than ever recently—the Maker movement, the Internet of Things, MOOCs, touch-screen mobile delivery, and Zoom streaming for online education are now fixed parts of our environment.

Even the most agile of learners and educators need support in keeping up with this pace of change. Our students have easier access to more information than any generation before them. We recognize the need for them to be able to think critically and investigate the data they see. We encourage instructors to use the chapter-opening features called *What do you think?* to promote students' critically thinking about a chapter topic. Students then follow up at the end of the chapter by answering additional related critical thinking questions in a *What do you think now?* feature. These make great topics for online course discussion forums.

We have also integrated material to help students develop skills for web application and mobile programming. We see the incredible value of these skills and their popularity with students and have included *Make This* exercises for each chapter. These exercises gently bring the concepts behind mobile app development to life. In addition, there is a *Solve This* exercise in each chapter that reinforces chapter content while also applying Microsoft 365 skills. These projects help to promote students' critical-thinking and problem-solving skills, which employers value highly.

The Helpdesk and Sound Byte training modules and IT Simulations continue to provide students with an active learning environment in which they can reinforce their learning of chapter objectives. In this edition, we continue to put the spotlight on critical thinking with integrated real-time surveys on important technology topics to foster classroom discussion and analytical skills. We have also included additional material on key challenges of a digital lifestyle, such as the impact of COVID-19 on the social equity issues of the digital divide.

We also continue to emphasize the many aspects of ethics in technology debates. Some of the Helpdesks and IT Simulations support instruction on how to conduct thoughtful and respectful discussion on complex ethical issues.

Our combined 73 years of teaching computer concepts have coincided with sweeping innovations in computing technology that have affected every facet of society. From setting up your Instacart order from your phone to attending your classes over Zoom, the use of digital technology is more than ever a fixture of our daily lives—and the lives of our students. But although today's students have a much greater comfort level with their digital environment than previous generations, their knowledge of the machines they use every day is still limited.

Part of the student-centered focus of our book has to do with making the material truly engaging to students. From the beginning, we have written *Technology in Action* to focus on what matters most to today's student. Instead of a history lesson on the microchip, we focus on tasks students can accomplish with their computing devices and skills they can apply immediately in the workplace, in the classroom, and at home.

We are constantly looking for the next technology trend or gadget. We have augmented the etext with weekly *Technology in the News* automatic updates. These updates are in each chapter, so regardless of where students are in the text, instructors will have current topics to talk about in class related to the latest breaking developments.

We also continue to include a number of multimedia components to enrich the classroom and student learning experience. The result is a learning system that sparks student interest by focusing on the material they want to learn (such as how to integrate devices into a home network) while teaching the material they need to learn (such as how networks work). The sequence of topics is carefully set up to mirror the typical student learning experience.

As they read this text, students will progress through stages and learning outcomes of increasing difficulty:

- Thinking about how technology offers them the power to change their society and their world and examining why it's important to be computer fluent
- Understanding the basic components of computing devices
- Connecting to and exploring the Internet
- Exploring application software
- Learning about the operating system and personalizing their computer
- Evaluating and upgrading computing devices
- Understanding home networking options
- Creating digital assets and understanding how to legally distribute them
- Keeping computing devices safe from hackers
- Going behind the scenes, looking at technology in greater detail

We strive to structure the book in a way that makes navigation easy and reinforces key concepts. We continue to design the text around learning outcomes and objectives, making them a prominent part of the chapter structure. Students will see the learning outcomes and objectives in the chapter opener, throughout the text itself, as well as in the summary so they understand just what they are expected to learn.

We also continue to structure the book in a progressive manner, intentionally introducing on a basic level in the earlier chapters concepts that students traditionally have trouble with and then later expanding on those concepts in more detail when students have become more comfortable with them. Thus, the focus of the early chapters is on practical uses for the computer, with real-world examples to help the students place computing in a familiar context. For example, we introduce basic hardware components in Chapter 2, and then we go into increasingly greater detail on some hardware components in Chapter 6. The Behind the Scenes chapters venture deeper into the realm of computing through in-depth explanations of how programming, networks, the Internet, and databases work. They are specifically designed to keep more experienced students engaged and to challenge them with interesting research assignments.

In addition to extensive review, practice, and assessment content, each chapter contains several problem-solving, hands-on activities that can be carried out in the classroom or as homework:

- The *Try This* exercises lead students to explore a particular computing feature related to the chapter.
- The *Make This* exercises are hands-on activities that lead students to explore mobile app development in both the Android and iOS environments.
- The *Solve This* exercises integrate and reinforce chapter concepts with Microsoft 365 skills.

Throughout the years we have also developed a comprehensive multimedia program to reinforce the material taught in the text and to support both classroom lectures and distance learning:

- Chapter-opening features called *What do you think?* allow students to critically think about a chapter topic. Students then follow up at the end of the chapter by answering additional related critical thinking questions in a *What do you think now?* feature.
- *Chew on This* critical-thinking questions require that students demonstrate their understanding through written answers that are manually graded.
- The Helpdesk training content, created specifically for *Technology in Action*, enables students to take on the role of a helpdesk staffer fielding questions posed by computer users so that students can demonstrate their understanding in an active learning environment.
- Sound Bytes expand student mastery of complex topics through engaging lessons with a brief quiz to check understanding.

- IT Simulations are detailed, interactive scenarios covering the core chapter topic. As students work through the simulation, they apply what they have learned and demonstrate understanding in an active learning environment.
- The *Technology in the News* (formerly *TechBytes Weekly*) is a weekly currency update that delivers the latest technology news stories to instructors for use in the classroom. In addition, the currency items have discussion points or activities included. The update is live in the etext chapters, so no matter where students are in the content, instructors will have this weekly update to use for an in-class discussion or reading assignment.

About the Authors

Alan Evans, MS, CPA

alevans@moore.edu

Alan is currently a faculty member at Moore College of Art and Design and Montgomery County Community College, teaching a variety of computer science and business courses. He holds a BS in accounting from Rider University and an MS in Information Systems from Drexel University, and he is a certified public accountant. After a successful career in business, Alan finally realized that his true calling is education. He has been teaching at the college level since 2000. He enjoys attending technical conferences and exploring new methods of engaging students.

Kendall Martin, PhD

kmartin@mc3.edu

Kendall is a full professor of Computer Science at Montgomery County Community College with teaching experience at both the undergraduate and graduate levels at a number of institutions, including Villanova University, DeSales University, Ursinus College, and Arcadia University. Her education includes a BS in electrical engineering from the University of Rochester and an MS and a PhD in engineering from the University of Pennsylvania. Kendall has industrial experience in research and development environments (AT&T Bell Laboratories) as well as experience with several start-up technology firms.

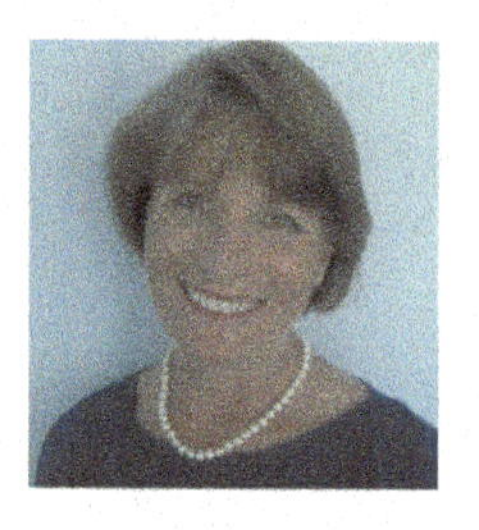

Mary Anne Poatsy, MBA

mpoatsy@mc3.edu

Mary Anne is a senior faculty member at Montgomery County Community College, teaching various computer application and concepts courses in face-to-face and online environments. She enjoys speaking at various professional conferences about innovative classroom strategies. Mary Anne holds a BA in psychology and education from Mount Holyoke College and an MBA in finance from Northwestern University's Kellogg Graduate School of Management. She has been in teaching since 1997, ranging from elementary and secondary education to Montgomery County Community College, Gwynedd-Mercy College, Muhlenberg College, and Bucks County Community College, as well as training in the professional environment. Before teaching, Mary Anne was a vice president at Shearson Lehman Hutton in the Municipal Bond Investment Banking Department.

Acknowledgments

For my wife, Patricia, whose patience, understanding, and support continue to make this work possible, especially when I stay up past midnight writing! And to my parents, Jackie and Dean, who taught me the best way to achieve your goals is to constantly strive to improve yourself through education.

—Alan Evans

For all the teachers, mentors, and gurus who have popped in and out of my life.

—Kendall Martin

For my husband, Ted, who unselfishly continues to take on more than his fair share to support me throughout this process, and for my children, Laura, Carolyn, and Teddy, whose encouragement and love have been inspiring.

—Mary Anne Poatsy

First, we would like to thank our students. We constantly learn from them while teaching, and they are a continual source of inspiration and new ideas.

We could not have written this book without the loving support of our families. Our spouses and children made sacrifices (mostly in time not spent with us) to permit us to make this dream into a reality.

Although working with the entire team at Pearson has been a truly enjoyable experience, a few individuals deserve special mention. The constant support and encouragement we receive from Jenifer Niles, Manager Higher Ed Content Strategy, Stephanie Kiel, Senior Content Analyst, and Marcus Scherer, Senior Product Manager, continually make this book grow and change. Our heartfelt thanks go to Shannon LeMay-Finn, our Developmental Editor. Her creativity, drive, and management skills helped make this book a reality. We also would like to extend our appreciation to Pearson Production partners, particularly Brian Hyland and his teams and the vendor teams, who work tirelessly to ensure that our book is published on time and looks fabulous. The timelines are always short, the art is complex, and there are many people with whom they have to coordinate tasks. But they make it look easy! We'd like to extend our thanks to the media and MyLab IT team—Becca Golden, Amanda Losonsky, and Heather Darby—for all of their hard work and dedication.

There are many people whom we do not meet at Pearson and elsewhere who make significant contributions by designing the book, illustrating, composing the pages, producing the media, and securing permissions. We thank them all.

And finally, we would like to thank the reviewers and the many others who contribute their time, ideas, and talents to this project. We appreciate their time and energy, as their comments help us turn out a better product each edition.

Don't just read about technology, interact with it.

HELPDESKS

These highly-interactive, almost game-like simulations let you take the role of a helpdesk staffer where you answer computer technology questions from customers. These simulations help reinforce the book content in a fun, engaging way.

CHAPTER 1
Technology Impacts
The Impact of Artificial Intelligence

CHAPTER 2
Understanding Bits and Bytes
Exploring Storage Devices and Ports

CHAPTER 3
Doing Business Online
Evaluating Websites

CHAPTER 4
Buying and Installing Software
Choosing Software

CHAPTER 5
Starting the Computer: The Boot Process
Organizing Your Computer: File Management

CHAPTER 6
Evaluating Your CPU and RAM
Evaluating Computer System Components

CHAPTER 7
Understanding Networking
Managing and Securing Wireless Networks

CHAPTER 8
Managing Digital Media
Understanding Intellectual Property and Copyright

CHAPTER 9
Threats to Your Digital Life
Understanding Firewalls

CHAPTER 10
Understanding Software Programming
A Variety of Programming Languages

CHAPTER 11
Using Databases
How Businesses Use Databases

CHAPTER 12
Using Servers
Transmission Media and Network Adapters

CHAPTER 13
Understanding IP Addresses, Domain Names, and Protocols
Keeping E-Mail Secure

SOUND BYTES

These multimedia lessons demystify complex computer concepts with short audio, animation, or video. The Sound Bytes now also include integrated learning objectives, a summary, and a quiz.

CHAPTER 1
Virtual Computer Tour
How to Debate Ethical Issues

CHAPTER 2
Binary Numbers Interactive
Smartphone Are Really Smart

CHAPTER 3
Blogging
Finding Information on the Web

CHAPTER 4
Where Does Binary Show Up?
Programming for End Users

CHAPTER 5
Using Windows Task Manager to Evaluate System Performance
Hard Disk Anatomy

CHAPTER 6
Installing RAM
Installing an SSD Drive

CHAPTER 7
Installing a Home Computer Network
Securing Wireless Networks

CHAPTER 8
Enhancing Photos with Image-Editing Software
Plagiarism and Intellectual Property

CHAPTER 9
Protecting Your Computer
Managing Computer Security with Windows Tools

CHAPTER 10
Using the Arduino Microcontroller
Programming with the Processing Language

CHAPTER 11
Creating and Querying an Access Database
Analyzing Data with Microsoft Power BI Suite

CHAPTER 12
Network Topology and Navigation Devices
A Day in the Life of a Network Technician

CHAPTER 13
Creating Web Pages with Squarespace
Client-Side Web Page Development

IT SIMULATIONS

IT Simulations are detailed, interactive scenarios covering the core chapter topic. Students work through the simulations to apply what they have learned and demonstrate understanding in an active learning environment.

CHAPTER 1
Technology and Ethics

CHAPTER 2
What Is a Computer?

CHAPTER 3
The Internet

CHAPTER 4
Application Software

CHAPTER 5
System Software

CHAPTER 6
Hardware

CHAPTER 7
Networks

CHAPTER 8
Digital Devices and Multimedia

CHAPTER 9
Security and Privacy

CHAPTER 10
Program Development

CHAPTER 11
Databases

CHAPTER 12
Client/Server Networks

CHAPTER 13
Communicating, Sharing on the Web

You will find the Helpdesks, Sound Bytes, and IT Simulations in **MyLab IT**.

Chapter 1

The Impact of Technology in a Changing World

 For a chapter overview, watch the **Chapter Overview videos**.

PART 1

Technology in Society

Learning Outcome 1.1 **You will be able to discuss the impact of the tools of modern technology on national and global issues.**

Technology in a Global Society 4

Objective 1.1 *Describe various technological tools being used to impact national and global issues.*

Objective 1.2 *Describe various global social issues that are being affected by technology.*

Technology Connects Us with Others 6

Objective 1.3 *Describe how technology is changing how and why we connect and collaborate with others.*

Objective 1.4 *Summarize how technology has impacted the way we choose and consume products and services.*

Helpdesk: Technology Impacts

The Importance of Computer Literacy 8

Objective 1.5 *Characterize computer literacy and explain why it is important to be computer literate.*

Sound Byte: Virtual Computer Tour

PART 2

Artificial Intelligence and Ethical Computing

Learning Outcome 1.2 **You will be able to describe emerging technologies, such as artificial intelligence, and how technology creates new ethical debates.**

Artificial Intelligence 13

Objective 1.6 *Describe artificial intelligence systems and explain their main goals.*

Helpdesk: The Impact of Artificial Intelligence

The Impact of Technology on Your Career 16

Objective 1.7 *Describe how artificial intelligence and other emerging technologies are important in many careers.*

Ethical Computing 21

Objective 1.8 *Define ethics and describe various ethical systems.*

Objective 1.9 *Describe influences on the development of your personal ethics.*

Objective 1.10 *Present examples of how technology creates ethical challenges.*

Sound Byte: How to Debate Ethical Issues

MyLab IT All media accompanying this chapter can be found here.

A Virtual Assistant on **page 12**

(John M Lund Photography Inc/Stone/Getty Images; Carlos Castilla/Shutterstock; Winui/Shutterstock; Ivan Trifonenko/123RF; Sergey Nivens/Shutterstock; Stuart Miles/123RF)

Journalism^CS

Robotics^CS

Education^CS

Psychology^CS

Medicine^CS

Literature^CS

Theater^CS

Biology^CS

Economics^CS

What do you think?

Having a background in technology and computer science enables you to be a powerful contributor in many career fields. More universities are realizing this and creating a new department, **CS + X**. What is CS + X? The CS stands for Computer Science, and the X stands for a **second area of study** that blends with computer technology. For example, the X might be **Music**; the combined degree would include study of the perception of music and use of digital programming to create new instruments. Or the X might be **Archaeology**, with a curriculum that includes courses on computer simulation of human behavior and creation of virtual reality models. Universities that are creating CS + X departments argue that combining computer science with other skills produces students who are better journalists, artists, scientists, and so on. The possibilities for collaboration between your **passions and technology** are endless.

Which field would you be most curious about combining with CS?

- *Literature*
- *Statistics*
- *Business*
- *Nursing*
- *Criminal Justice*
- *Anthropology*
- *Art*
- *Other*

See the end of the chapter for a follow-up question.

(Jacob Lund/Shutterstock)

Part 1

For an overview of this part of the chapter, watch **Chapter Overview Video 1.1.**

Technology in Society

Learning Outcome 1.1 You will be able to discuss the impact of the tools of modern technology on national and global issues.

Ask yourself: Why are you in this class? Maybe it's a requirement for your degree, or maybe you want to improve your computer skills. But let's step back and look at the bigger picture.

Technology is a tool that enables us all to make an impact beyond our own lives. We've all seen movies that dangle the dream in front of us of being the girl or guy who saves the world—and gets to drive a nice car while doing it! Technology can be your ticket to doing just that by enabling you to participate in projects that will change the world.

Technology in a Global Society

Recent national and global issues are showing that technology is accelerating change around the world and galvanizing groups of people in new ways. Let's look at a few examples.

Impact of Tools of Modern Technology

Objective 1.1 *Describe various technological tools being used to impact national and global issues.*

Social Media Tools

Social media platforms like Twitter, Facebook, and Instagram enable people to connect and exchange ideas. These platforms also bring together people facing similar problems to fight for social change. For example, the Twitter hashtag #MeToo began as a way of supporting women facing sexual harassment and assault but evolved to galvanize an international movement. A simple hashtag brought to light an important social issue and was a key means for revealing how widespread the problem was.

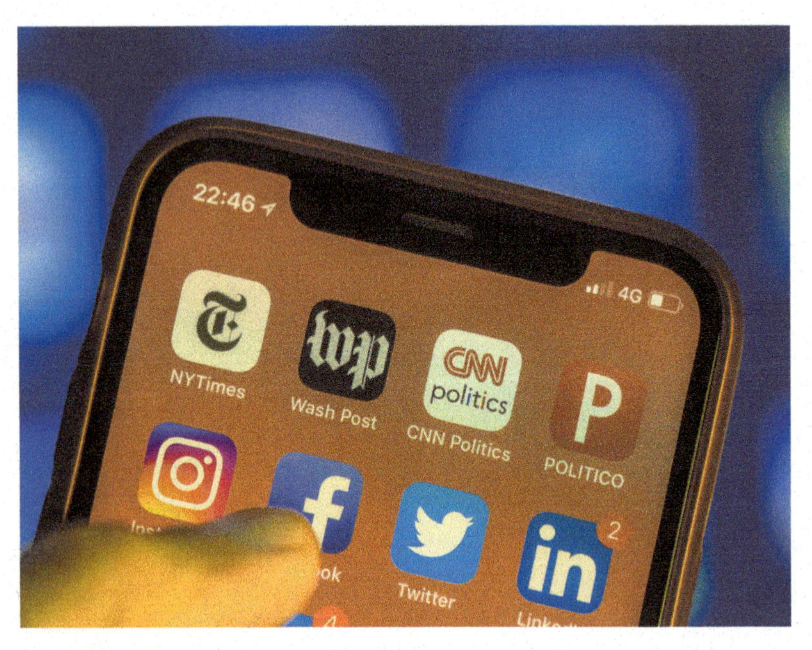

Figure 1.1 Social media has changed how we access, distribute, and evaluate information. *(Rudmer Zwerver/Shutterstock)*

How we conduct informed discussion in the age of social media is still developing, however. Bot accounts, automated programs retweeting news stories and quotes, have been used to create discord around controversial topics in many countries—inflaming the gun control debate in the United States or fanning prejudice into violence in Sri Lanka. For these and other reasons, knowing how to use and critically evaluate social media is an important skill needed by all (see Figure 1.1).

Crisis-Mapping Tool

Another example of the interaction of technology and society is the software tool Ushahidi. Following a disputed election in Kenya, violence broke out all over the country. Nairobi lawyer Ory Okolloh tried to get word of the violence out to the world through her blog, but she couldn't keep up with the volume of reports. However, two programmers saw her request for help and in a few days created Ushahidi. It is a crisis-mapping tool that collects information from e-mails, text messages, blog posts, and tweets and then maps them, instantly making the information publicly available. The developers then made Ushahidi a free platform anyone in the

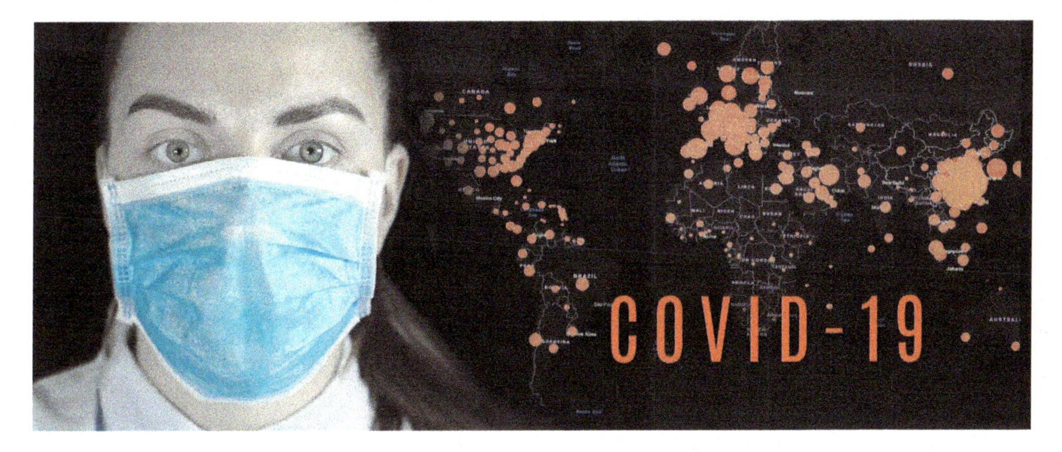

Figure 1.2 Crisis-mapping software connects resources with those in need in times of disaster. *(Carroti/Shutterstock)*

world can use (see Figure 1.2). It has since been used in several international disasters, including during the COVID-19 pandemic. In what other ways may technology help us face times of crisis?

Global Issues

Objective 1.2 *Describe various global social issues that are being affected by technology.*

Let's look at the different global social issues that technology affects.

Health Care

Infectious diseases account for about one-fifth of all deaths worldwide. The global pandemic brought by the COVID-19 virus infected millions, with catastrophic public health and economic consequences. The push for a vaccine that would allow a safe return to open public spaces was aided by technology. Scientific visualization tools help scientists develop tests for antibodies for specific viruses. Computationally intense modeling software helps researchers increase the pace of vaccine discovery. The Folding@Home software effort lets every person with a computer contribute. Folding@Home uses the time when your CPU is normally idling to run a vast number of simulations, increasing the chances of finding drugs that can work against COVID-19 and other diseases.

Technology is also playing a part in providing a solution for concussion injuries. Researchers now know that even without an actual concussion, athletes can sustain serious damage from repeated impacts of their brain against the skull. Computer programs have collected sensor data from impacts on the field, which scientists have analyzed and used to create a new kind of helmet. College programs and the NFL now use enhanced helmets, designed to distribute the impact of collisions better.

The Environment

What if every cell phone in the world had built-in atmospheric sensors? Then millions of points of data measuring air and water quality from around the world could be acquired. The data could be tagged with geographical information, alerting scientists to new trends in our environment. Researchers around the world are exploring ideas like these.

Smart Internet-connected water sprinklers are another technology that is already saving water in California and other dry areas of the country. The sprinkler system checks the weather forecast so it won't use water when rain is coming the next day. The system is showing a 30% reduction in water usage.

The Digital Divide

There is a serious gap in the levels of Internet access and the availability of technical tools in different regions of our country and the world. The term for this difference in access to technology is the digital divide. As we face a future in which schools may be forced to close as we combat viral

Figure 1.3 The Next Einstein Initiative (NEI) is rallying the support of the world to identify and encourage mathematical genius. *(Alistair Cotton/123RF)*

pandemics, the online delivery of public education requires all students to have a computer and Internet access. The danger of a digital divide is that it prevents the education of all.

The digital divide also prevents us from using all the minds on the planet to solve problems. But this challenge created by technology is also being answered by it. The Next Einstein Initiative (NEI) is a plan to focus resources on mathematical minds in Africa (see Figure 1.3). Cambridge professor Neil Turok founded the African Institute of Mathematical Sciences (AIMS) to bring together the brightest young minds across Africa and the best lecturers in the world. By capturing the enthusiasm of the world with presentations distributed through TED (*ted.com*), corporate supporters such as Google are helping AIMS push to create additional centers across Africa.

Table 1.1 shows additional examples of people putting technology into action to impact the world. How will you join them?

Table 1.1 Technology in Action: Taking on Global Problems

Person/ Organization	Global Problem	Technology Used	Action	Find Out More . . .
Start Network	Corruption	Blockchain, a digitized public ledger for recording a series of transactions	Blockchain technology can help track humanitarian aid funds as they flow from donors to recipients.	Start Network: *startnetwork.org*
SolaRoad/ Netherlands	The need for a renewable, nonpolluting energy resource	Solar cells	Solar cells are integrated into the asphalt roadway. They collect solar energy and distribute electricity all day.	Netherlands SolaRoad: *solaroad.nl*
Tech Against Trafficking	Human sex trafficking and slavery	Geographic mapping and tracking, AI data analysis	Tackles modern-day slavery and trafficking by identifying and tracking victims.	Tech Against Trafficking: *techagainsttrafficking.org*
Global Water Program	Dwindling access to pure water globally	Desalination, smart water meters, fog harvesting	Technological solutions for conserving, purifying, and desalinating water.	Global Water: *globalwater.org*

Technology Connects Us with Others

Technology also allows us to redefine fundamental parts of our social makeup—how we connect and collaborate with each other and how we purchase and consume products.

Technology Impacts How and Why We Connect and Collaborate

Objective 1.3 *Describe how technology is changing how and why we connect and collaborate with others.*

Collaborating for the Benefit of Others

With the arrival of many web applications that allow individuals to become creators of the web, a newer kind of Internet has come into being. Nicknamed Web 2.0, the web now allows users to contribute content and connect with one another easily. Web 2.0 has fostered a dramatic shift for technology users across the world, from simply consuming to having the ability to volunteer and collaborate on projects. The term cognitive surplus was coined to reflect the combination of leisure time and the tools to be creative. The availability of media tools and the easy connectivity of Web 2.0, along with generosity and a need to share, also enable projects like Ushahidi to emerge.

Connecting Through Business

One of the most profound ways we can connect with one another is to support other people's dreams. Someone with an idea can build a first attempt at an area makerspace. Then by posting a launch video on a site such as Kickstarter, that person can reach people who believe in the idea. Donors agree to put money forward, knowing the product may not come to fruition, in exchange for special rewards such as a discounted price or a special color version. This means of generating capital to start a business is known as crowdfunding. Business ideas are not the only projects benefiting from crowdfunding. Sites like GoFundMe enable people to crowdfund to raise money for things such as medical bills or tuition.

Figure 1.4 Lego's Idea website allows users to submit a product design and vote to select the most popular. The winners are turned into products for public sale. *(Everything You Need/Shutterstock)*

Technology Impacts How We Consume

Objective 1.4 *Summarize how technology has impacted the way we choose and consume products and services.*

Technology is also changing how we decide what we'll purchase and how we actually buy goods and services.

Marketing

There are billions of views of videos each month on YouTube, and marketers are taking note. *Influencers* are YouTube personalities with huge followings and high levels of interaction with their followers. Placing a product for review on their channel or showing a popular YouTube celebrity using that product in their own life can be a huge marketing win.

Marketers also use crowdsourcing—checking in with the voice of the crowd. Frito-Lay's "Do Us a Flavor" campaign asked customers to vote online for what flavor chips to make next. The Lego Ideas website lets fans contribute projects they would like to see and the most popular are produced for sale (see Figure 1.4). Marketing has shifted to take advantage of our digital lifestyle.

Sharing Economy

Even the idea of ownership is evolving. Items like cars and bikes can now be subscriptions instead of purchases. Services like Uber and Lyft let you use your car to provide rides for others, and Zipcar allows you to use a shared car whenever you need it. Bicycles can be shared in most cities with programs like New York City's Citi Bike. These new sharing options have revolutionized the transportation industry (see Figure 1.5).

Such collaborative consumption implies that we are joining together as a group to use a specific product more efficiently. There are increasing opportunities to share the services a product provides instead of owning it outright. Mounting environmental concerns and global financial pressures are other forces pushing us toward collaborative consumption.

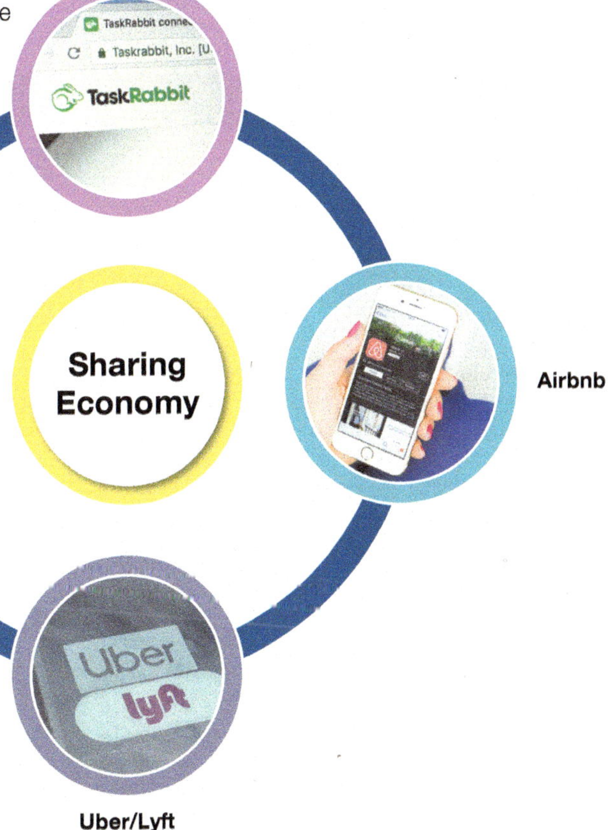

Figure 1.5 Smartphones and constant network connectivity have established a sharing economy, which has impacted many industries. *(Jarretera/Shutterstock, Kaspars Grinvalds/Shutterstock, Pisaphotography/Shutterstock, Jonathan Weiss/Shutterstock)*

The Importance of Computer Literacy

Everywhere you go, you see ads for computers and other devices. Do you know what all the words in the ads mean? What is a GPU? How fast do you need your computer to be, and how much memory should it have? If you're computer literate, you'll be a more informed consumer when it comes time to buy computers, peripherals, and technology services. Understanding computer terminology and keeping current with technology will help you better determine which computers and devices you need.

MyLab IT

Technology Impacts

In this Helpdesk, you'll play the role of a helpdesk staffer fielding questions about ways in which technology affects society.

Computer Literacy

Objective 1.5 *Characterize computer literacy and explain why it is important to be computer literate.*

Let's look at a few examples of what it means to be a savvy computer user and consumer.

Computer literacy. When you are computer literate (see Table 1.2), you understand the capabilities and limitations of computers and you know how to use them safely and efficiently. The topics listed in Table 1.2 and more are covered in detail in the remaining chapters.

Avoiding hackers and viruses. Hackers and viruses can threaten a computer's security. Being aware of how they operate and knowing the damage they can cause can help you avoid falling prey to them.

Protecting your privacy. If your identity is stolen, your credit rating can be quickly ruined. Do you know how to protect yourself from identity theft when you're online?

Table 1.2 What Does It Mean to Be Computer Literate?

 You can **avoid falling prey to hackers and viruses** because you are aware of how they operate.	 You know how to **protect yourself from identity theft**.	 You can **separate the real privacy and security risks from things you don't have to worry about**.
 You know how to find information and **use the web effectively**.	 You can **avoid being overwhelmed by spam, adware, and spyware**.	 You can **diagnose and fix problems** with your hardware and software.

(Peter Dazeley/The Image Bank/Getty Images; Yuri_Arcurs/E+/Getty Images; Zakai/DigitalVision Vectors/Getty Images; Justin Lewis/Stone/Getty Images; Argus/Shutterstock; Ivanastar/Istock/Getty Images)

Understanding the real risks. Being computer literate means being able to separate real privacy and security risks from things you don't have to worry about. Do you know whether cookies pose a privacy risk when you're on the Internet? Do you know how to configure a firewall for your needs?

Using the web wisely. People who are computer literate know how to find reliable, accurate information effectively. They also know how to use the web to work well with others. Are you effective in how you use the web?

Avoiding online annoyances. How can you avoid spam—unsolicited electronic junk mail? Do you know what adware and spyware are? Do you know the difference between viruses, worms, and Trojan horses? What software—the instructions that tell the computer what to do—should you install on your computer to avoid online annoyances?

Being able to maintain, upgrade, and troubleshoot your computer. Learning how to care for and maintain your computer and knowing how to diagnose and fix certain problems can save you time and money. Do you know how to upgrade your computer if you want more memory? Do you know which software and computer settings can keep your computer in top shape? Understanding the hardware of your computing system is a critical part of your computer literacy. (See the Virtual Computer Tour Sound Byte to get started.)

Keeping up to date. Finally, becoming computer literate means knowing about new technologies and how to integrate them into your life. Can you connect your TV to your wireless network? What is a media server, and do you need one? Can a USB type C connection carry HDMI video signals? Being able to stay up to date with technology is an important skill.

This book will help you become computer literate. In Chapter 3, you'll find out how to get the most from the web while staying free from the spam and clutter Internet surfing can leave on your computer. Chapter 6 shows you how to determine whether your hardware is limiting your computer's performance and how to upgrade or shop for a new device. Chapter 9 covers how to keep your computer and your digital life secure. You'll be able to save money, time, and frustration by understanding the basics of how computer systems operate.

Sound Byte MyLab IT

Virtual Computer Tour

In this Sound Byte, you'll take a video tour of the inside of a desktop system unit; locate the power supply, CPU, and memory; and learn more about what's inside a computer.

Before moving on to Part 2:

1. Watch Chapter Overview Video 1.1.
2. Take the Check Your Understanding quiz.

Check Your Understanding // Review & Practice

For a quick review to see what you've learned so far, answer the following questions.

multiple choice

1. Automated bot accounts use social media
   a. responsibly.
   b. to help humans who are busy.
   c. to manipulate opinion by posting news stories and quotes massive numbers of times.
   d. in some countries, but they are illegal in the United States.
2. The digital divide occurred because
   a. the United States has the fastest Internet access.
   b. not everyone has equal access to the Internet.
   c. crowdfunding is increasing.
   d. everyone now has a smartphone.
3. Cognitive surplus means that we now find many people with
   a. more money than free time.
   b. limited access to the Internet.
   c. mobile devices.
   d. excess time and free tools for collaboration.
4. Being computer literate means that you can
   a. read about computers.
   b. program in Java.
   c. use Twitter.
   d. use computers efficiently and safely.
5. Collaborative consumption is when people get together to
   a. find the best prices on products.
   b. increase the use of a single product by sharing access to it.
   c. fight diseases of the respiratory tract.
   d. exchange reviews on services and goods they have purchased.

chew on this

(3Dalia/Shutterstock)

Our thoughts are influenced by the information fed to our mind all day long. Web 2.0 has created numerous channels for people to offer their own work for free—open source software and free music, text, and artwork, to name a few. How has this affected your thinking, what you create, and what value you place on creative work?

Go to **MyLab IT** to take an autograded version of the *Check Your Understanding* review and to find all media resources for the chapter.

For the IT Simulation for this chapter, see MyLab IT.

Try This

What Does Facebook Know about You?

Social media sites like Facebook and Twitter do not charge you a fee. They make a profit by selling information about your behavior to marketers. By watching what groups you join and what posts you read, their algorithms make conclusions about what kind of person you are. In this exercise we'll show you how to check what information these sites have deduced about who you are. For more step-by-step instructions, watch the Chapter 1 Try This video on MyLab IT.

What You Need

A Facebook account

A Twitter account

(Rvlsoft/Shutterstock; Solomon7/Shutterstock)

From the Facebook website, download your Facebook data. On the top line of your Facebook page, click the dropdown arrow on the far right and select **Settings**. Next, select **Your Facebook Information** and click **Download Your Information**. Click the **Create File** button. Facebook will e-mail you when the file is ready.

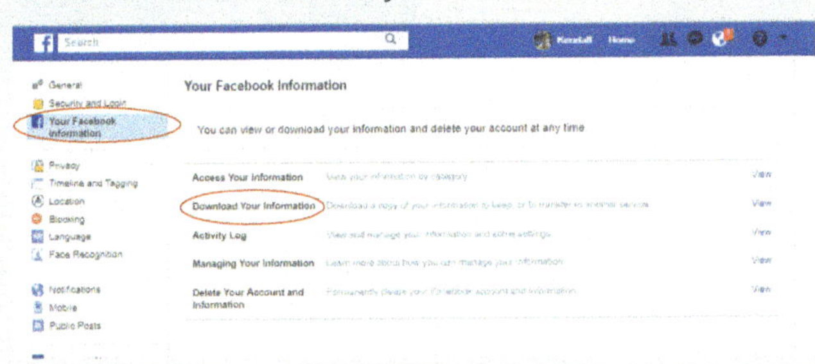

(Used with Permission from Kendall Martin)

From the Twitter website, download the information Twitter has created on you. From your Twitter home page, click **More** and select **Settings and privacy**. From the list on the right, select **Your Twitter data**. Locate **Download your Twitter data**, enter your password for access, and click the **Request archive** button. It will be e-mailed to you when the report is ready.

Demographics > Stay-at-home moms
Demographics > Trailing baby boomers
Demographics > Trendy moms
Demographics > Veteran in household
Demographics > Wife
Demographics > Working-class families
Demographics > Working-class moms
Demographics > Young adult age 18-25
Dining > Likely to dine at Chili's Grill & Bar
Dining > Likely to dine at Chipotle Mexican Grill
Dining > Likely to dine at Panera Bread Company

(Used with Permission from Kendall Martin)

Examine what these algorithms have decided about you. Check the list of interests and what items they think you will likely buy. How accurately have they guessed your household income, your politics, and your finances?

Make This

TOOL: IFTTT.com (If This Then That)

A Virtual Assistant

If This Then That (*IFTTT.com*) is an Internet-based tool that helps you automate tasks. By using recipes within this web-based tool, you can automate tasks you do during the day, such as:

- automatically silencing your phone when you go into class,
- automatically texting your manager when you're on your way to work, or
- notifying you when the president signs a new law.

In this exercise, you'll explore using IFTTT to create recipes like these.

Make the Internet work for you by knowing this one programming statement: IF THIS, THEN THAT.

You'll get the title of the law and a link to read more.

(IFTTT Inc.)

Receive an email when a NYT Technology article becomes popular

By kmartinphd

Stay up-to-date with the most popular NYTimes technology news from around the world by receiving an email when an article becomes popular. Want to stay on top of another topic? Choose the section you're interested in to customize.

(IFTTT Inc.)

For the instructions for this exercise, go to MyLab IT.

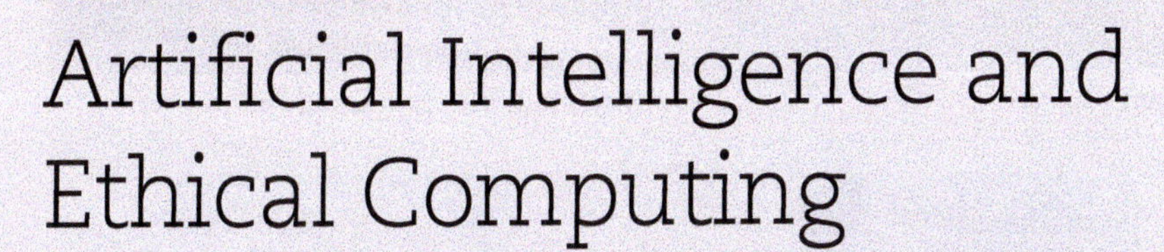

Part 2

For an overview of this part of the chapter, watch **Chapter Overview Video 1.2.**

Artificial Intelligence and Ethical Computing

Learning Outcome 1.2 **You will be able to describe emerging technologies, such as artificial intelligence, and how technology creates new ethical debates.**

Can computing devices really think? Are virtual assistants like Alexa intelligent, or do they just mimic thinking? Rapid developments in the field of artificial intelligence have forced us to consider many new ethical debates. It's important to learn about new advancements in technology, such as artificial intelligence, as well as to understand the ethical dilemmas technology presents.

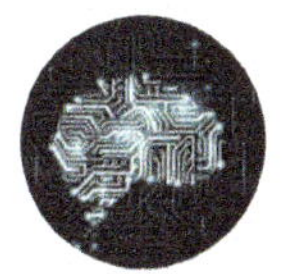

Artificial Intelligence

Let's explore what artificial intelligence is and how it impacts you.

Artificial Intelligence Basics

Objective 1.6 *Describe artificial intelligence systems and explain their main goals.*

What is artificial intelligence? Intelligence is the ability to acquire and apply knowledge and skills. Sociologists point to characteristics that make human beings intelligent, such as learning from experiences, using reasoning, solving problems, and using language. Artificial intelligence (AI) is a branch of computer science that focuses on creating computer systems able to perform tasks that are usually associated with human intelligence. By this definition, any computer-controlled device that accomplishes something thought of as intelligent by humans is considered AI.

Do computers "think" like human beings? In the 1950s, the goal was to create a machine that could think like a human. Early examples included expert systems that mimicked doctors in diagnosing illnesses, but this goal has shifted somewhat toward creating machines that generate intelligent output but that do not necessarily mimic the human thought process.

Consider a visit to the library, during which you could tell a human librarian your interests and the librarian could ask you some questions and recommend books you might like. The Amazon recommendation engine fulfills the same purpose. However, the Amazon recommendation engine doesn't mimic a human librarian's thought process but instead analyzes vast amounts of data about you and other shoppers to make its recommendations. It provides intelligent results, but it does not arrive at those results the same way a human would.

The Impact of Artificial Intelligence

In this Helpdesk, you'll play the role of a helpdesk staffer, fielding questions about artificial intelligence.

Bits&Bytes **Hands-On Artificial Intelligence**

None of our senses actually see or hear directly. When we view the world, our eyes do not see anything. All our eyes can do is report back to our brain a set of signals that indicate the pattern of brightness and colors. It is our brain or, more accurately, our mind that takes that information and constructs the world as we know it. That has never been the type of calculation that a computing machine can perform.

But AI advances have changed that. Visit the TryIt! demo site for Google Cloud Vision. You can drag any image file into the demo box (a photo, a screenshot), and software at the Google server will process it and report back what it sees as a series of percentages: 95% sure it is a puppy; 79% sure it is a stuffed toy (see Figure 1.6). AI has brought other human-like skills to computers. The Google Natural Language application processing interface (API) demo can even analyze a section of writing for emotion.

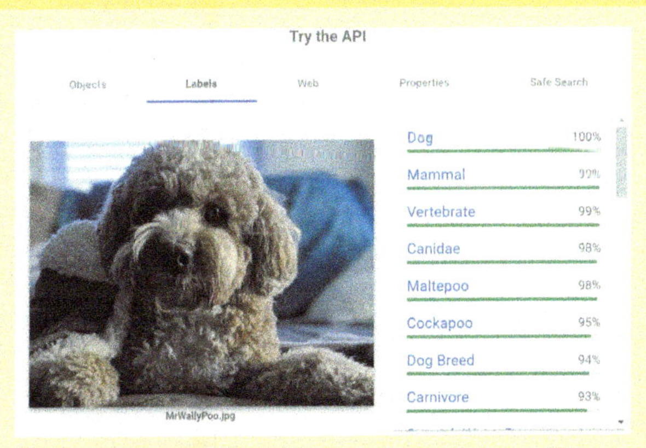

Figure 1.6 Google Cloud Vision is software that uses AI to enable a computer to "see" rather than just report pixels. *(https://cloud.google.com/vision/docs/drag-and-drop)*

Table 1.3 Main Areas of AI Research

Natural Language Processing

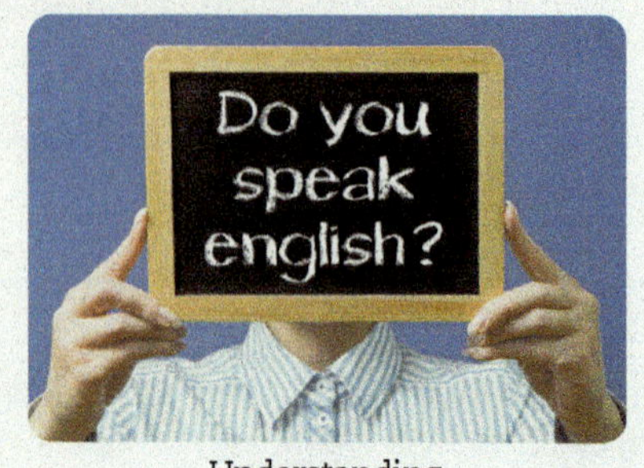

Understanding written and spoken words

Perception

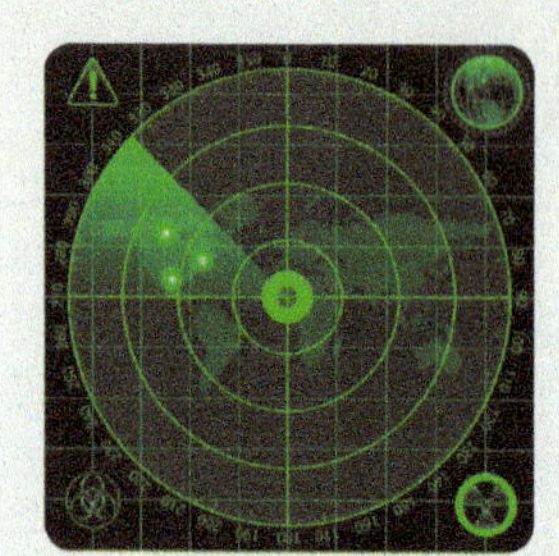

Using sensors to gather data

Knowledge Representation

Storing and accessing information about the world

Planning

Setting a goal and achieving it

Problem Solving

Using even incomplete information to achieve a solution

Learning

Improving through experience

(Convisum/123RF; Andrija Markovic/123RF; Galina Peshkova/123RF; Viktor Bondar/123RF; Tomertu/123RF; Sentavio/123RF)

What are the main areas of research for AI? AI's central goals can be grouped into a number of categories (see Table 1.3):

- Natural language processing (NLP): NLP works to develop AI systems that understand written and spoken words and can interact with humans, using language.
- **Perception**: AI systems have senses just as we do. AI systems use sonar, accelerometer, infrared, magnetic, and other electronic sensors to gather data. Being able to combine all the data from sensors and then construct information from it is a difficult challenge.
- Knowledge representation: Knowledge representation involves encoding information about the world into formats that the AI system can understand. Humans possess a vast collection of general knowledge based on their experiences. AI systems need to build knowledge bases to solve problems. Developing a knowledge base and using it efficiently have been demonstrated by IBM's AI computer Watson.
- **Planning**: AI systems need to set goals and then achieve them. An AI system might need to plan how to move a blue block out of the way to reach a red one or how to rotate a block as it moves to fit through a narrow opening.
- **Problem solving**: Humans tend to make intuitive judgments when solving a problem rather than performing a step-by-step analysis. AI programming combines a rules-based approach with trying to make judgments with incomplete information.
- **Learning**: Like humans, AI algorithms adapt and learn through experience. Supervised learning is used when the system can be trained with a huge number of examples. Unsupervised learning is when a system can look at data and build rules on its own to decide what it is seeing.

A project such as a self-driving (autonomous) car requires research in many of these areas. The car must have computer vision—the ability to interpret visual information the way humans do. The AI system must be able to tell the difference between trees and pedestrians, but it also needs to know many things about trees and people. Is the tree going to run into the path of the car? The car must scan the scene with sensors, recognize objects, and consult a knowledge base to create and execute a plan.

What has enabled us to deploy effective AI systems? Many information technology developments have contributed to the functional AI systems that we have today.

Artificial Neural Networks

Expert systems, computer programs that mimic the experience of human experts such as doctors or lawyers, were among the first attempts at producing AI. Rules-based systems, software that asks questions and responds based on preprogrammed algorithms, were the first expert systems designed. These systems asked questions ("Do you have a fever?") and initiated other questions or actions based on the answers ("How long have you had a fever?"). They worked adequately for some settings, but just a list of rules is not sophisticated enough to handle difficult tasks.

Artificial neural networks (ANNs) are designed based on the structure of the human brain, which is a network of loosely connected neurons. When a neuron receives a signal, it fires an electrical impulse and the signal travels to all the neurons connected to it. In ANNs, digital signals take the place of biological signals. Many modern ANNs feature layers of neurons that allow many degrees of complexity. ANNs have enabled researchers to tackle complex problems such as speech recognition.

Machine Learning

Machine learning (ML) is a type of AI that doesn't need to be specifically programmed. Instead, it analyzes patterns in data and then uses the patterns to draw conclusions and adjust the actions of the AI system accordingly. By learning, the AI system can adapt itself and constantly improve. You have interacted with these systems if you use Alexa, the Amazon voice-recognition device. Alexa gathers input from human speech and uses this information to become better at understanding language over time.

Deep learning (DL) is a subset of the ML field that describes systems capable of learning from mistakes, just as humans do. DL algorithms can learn from data that is not labeled as correct or incorrect. The algorithm adapts to improve its final result without being presented with a huge, labeled set of training data. This is known as *unsupervised learning.*

Bits&Bytes **Machine Learning to Create New Instruments**

The Google Creative Lab team is using machine learning to create new kinds of musical instruments. The team created a device called NSynth Super that uses machine learning tools to help musicians create new kinds of sounds (see Figure 1.7). NSynth pulls from each of four input signals, which might be sounds from other instruments or recordings of voices, the main characteristics that make up a sound. Using this data, it combines the four into a new kind of instrument. NSynth is not just mixing the sounds at different volumes; it is using sonic qualities from each source to create an instrument that never existed before.

You can find out more about this software because the entire project has been posted freely as open source. The instructions, parts list, and software to build your own NSynth Super are available at GitHub. If you want to get started by just exploring the NSynth in action, visit experiments.withgoogle.com/ai/sound-maker/view/.

Figure 1.7 Machine learning is creating new kinds of musical instruments. *(https://nsynthsuper.withgoogle.com/)*

Ethics in IT

Ethics in Computing

Here are some questions for you:

- Should there be rules for conduct in a virtual world? If so, what should they be?
- What does plagiarism mean in a world where people can easily copy, paste, and upload content of all kinds?
- Should workplaces be allowed to monitor their employees' computing activities without the employees' knowledge?
- Should websites be allowed to capture data related to what visitors do on their sites and analyze and sell that data?
- Should programmers be responsible if their software malfunctions and leads to personal injury or financial loss?
- Should Internet access be provided for free by communities to their citizens who cannot afford it?
- How should artificial intelligence be designed to behave ethically?

These are the sorts of ethical questions and challenges that technology poses—questions that did not even exist just a few years ago.

As the reach of technology continues to spread, these are questions that more and more societies must face and develop their own answers to. Because different societies have different ideas of what it means to behave ethically, there will be many solutions to ethical questions. How we navigate cultural responses to ethical challenges becomes more important as the pace of technology quickens.

How should U.S. companies respond to censorship of their websites in countries such as China? A state in the United States can declare that online gambling is illegal, but what does that mean when its citizens have access to foreign websites hosting gambling?

Answering challenging ethical questions related to technology is part of being a digitally literate citizen. This text will help you understand technology and the ethical issues it poses. Taking the time to think deeply about the connection between technology and ethics is one step in being a more knowledgeable and thoughtful global citizen.

Working with Artificial Intelligence and Other Information Technologies

Information technology (IT) is a field of study focused on the management and processing of information. Career opportunities in IT are on the rise, but no matter what career you choose, the workplace demands new skill levels in technology from employees. Understanding how AI systems and other technologies can be used in the workplace is an important skill for everyone.

The Impact of Technology on Your Career

Objective 1.7 *Describe how artificial intelligence and other emerging technologies are important in many careers.*

One of the benefits of being digitally literate is that you will most likely be able to perform your job more effectively. Your understanding of key concepts in technology can future-proof you so you easily and quickly react to the next round of new technologies.

Let's look at a whole range of industries and examine how current and emerging technologies are part of working efficiently.

Retail

The amount of data generated each second of the day is staggering (see Figure 1.8). AI systems deployed in the retail sector are responsible for managing huge amounts of data and performing

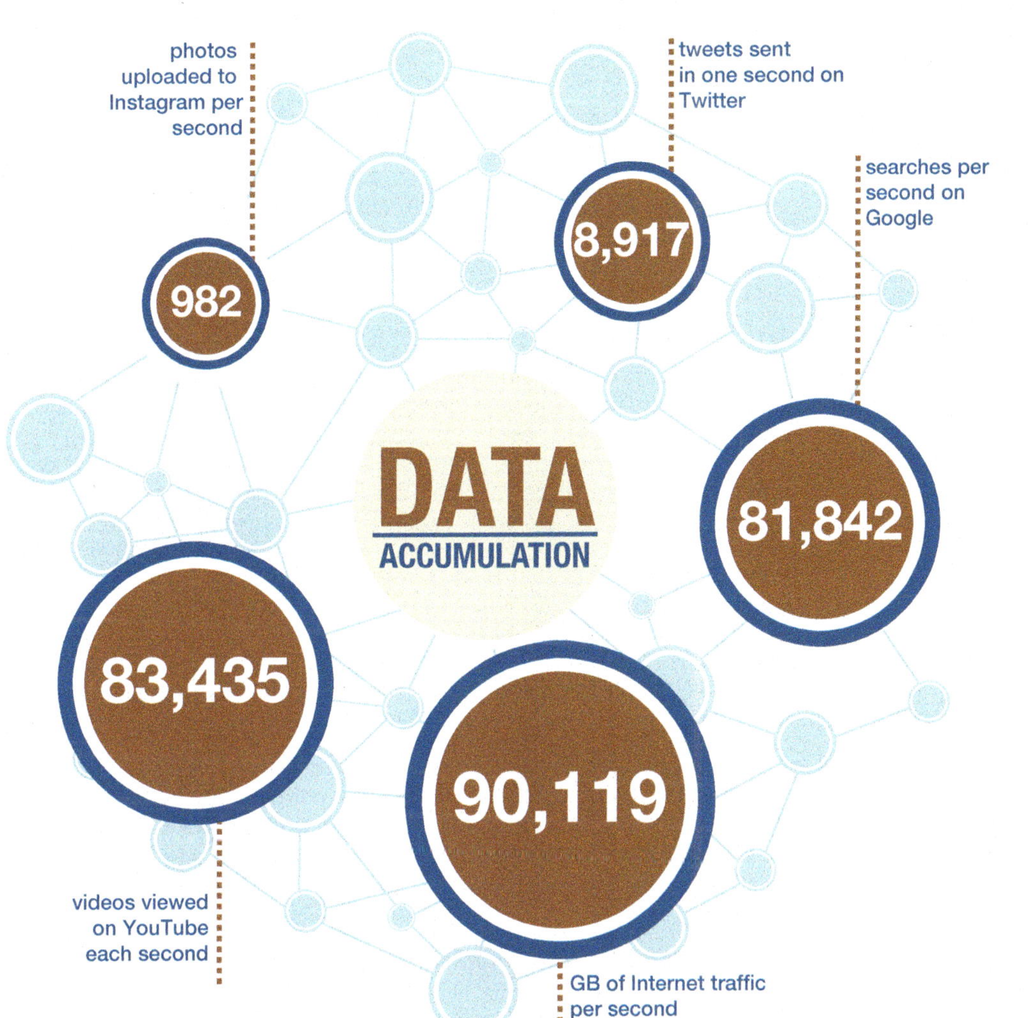

Figure 1.8 Enormous amounts of data are produced every second. Data mining is performed by sophisticated AI systems such as recommendation engines.
(Based on data from http://www.internetlivestats.com/)

data mining analysis for managers. For example, retailers often study the data gathered from register terminals to determine which products are selling on a given day and in a specific location. In addition to using inventory control systems, which helps managers determine how much merchandise they need to order to replace stock, managers can use mined data to determine that if they want a certain product to sell well, they must lower its price. Such data mining thus allows retailers to respond to consumer buying patterns.

Recommendation engines are AI systems that help people discover things they may like but are unlikely to discover on their own. The secret behind recommendation engines is crunching massive amounts of data effectively. Without the ability to collect this data and use data mining to understand it, recommendation engines wouldn't exist. Presenting effective recommendations and learning to make better recommendations are essential to Amazon's success. About 35% of Amazon's sales come from customers using the recommendation list.

Banking

Credit card processors and banks use AI systems to analyze huge volumes of transaction data to spot fraud. Banks use software to assess the risk of extending credit to customers by analyzing spending patterns, credit scores, and debt repayment. The same software also helps determine what interest rates and terms to offer on loans. Researchers at MIT determined that using AI to evaluate credit risk could help reduce an institution's loan losses by 25%.

Figure 1.9 Autonomous (self-driving) trucks are poised to revolutionize the transportation industry.
(Scharfsinn/Shutterstock)

Transportation

Most cars actually are moving for only 5 percent of their lifetime. Self-driving cars and fleets of trucks would optimize the use of vehicles and increase safety. The software that can anticipate actions of other drivers is proving more difficult to develop than predicted, but with all major car manufacturers working toward a solution, you can expect to see this area grow over the next several years. In the next few years, trucking companies will face major changes if self-driving trucks become common (see Figure 1.9).

Robots and Embodied Agents

Many robots are deployed in industrial settings doing hazardous or repetitive tasks, but it is difficult to design robots that can match human dexterity and mobility. This has led to designs that mimic how humans walk and grasp objects, producing robots called embodied agents (see Figure 1.10) that look and act like human beings. Still, there are many jobs in which human empathy and understanding are highly valued, such as in medicine, education, and counseling, but more robots will continue to appear in our everyday lives.

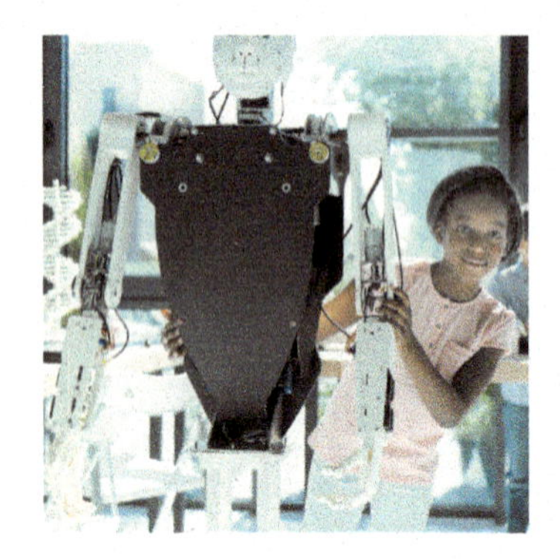

Figure 1.10 *Embodied agents* are robots that mimic humans in appearance.
(Dmytro Zinkevych/Shutterstock)

Education

In the education field, intelligent personal assistants are being used in a variety of ways to support learning. For example, virtual language instructors engage in conversations with students, correcting their grammar and offering new word choices to expand their vocabulary.

Plagiarism checkers such as Turnitin initially relied on brute-force comparisons of student work to databases of published material, looking for exact text matches, but with so much more data now being produced, this approach is impractical. Modern plagiarism agents use machine learning to spot similar patterns in writing and estimate the likelihood that it was plagiarized, so instructors need review only the papers flagged by the AI for suspected plagiarism, which increases the instructor's efficiency.

As an educator, being digitally literate will help you integrate computer technologies like these into your classroom.

Bits&Bytes Is It AI or Human? Take a Turing Test!

Alan Turing was an early pioneer in computer design and cryptography who inspired many designers of the first digital computers. He proposed a simple test to distinguish between a human and a computer system. The Turing test (see Figure 1.11) places a person in a room, asking written questions of two other people. One person they are questioning is a live human being; the other is a computer. If the questioner can't tell which one of the respondents is a computer, then Turing felt the computer had reached intelligence. The computer is permitted to be deceptive; however, the human is required to help the questioner reach the correct conclusion.

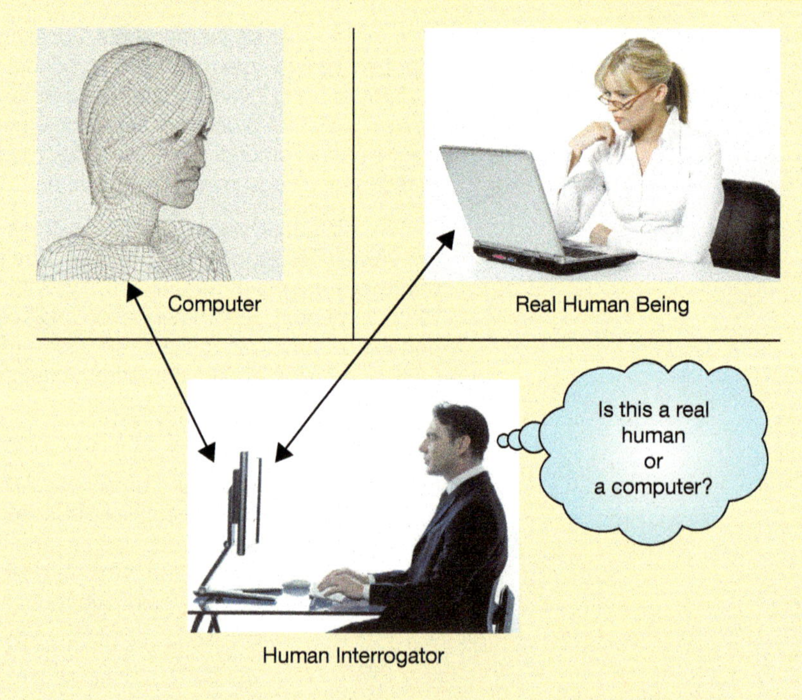

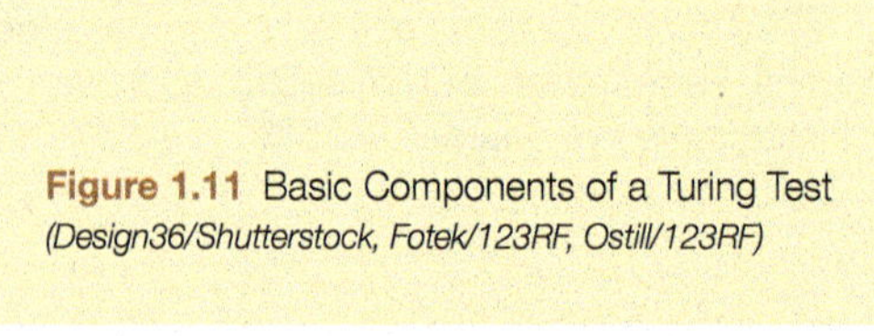

Figure 1.11 Basic Components of a Turing Test
(Design36/Shutterstock, Fotek/123RF, Ostill/123RF)

Law Enforcement

Today, AI systems are used in police cars and crime labs to solve an increasing number of crimes. For example, facial reconstruction systems like the one shown in Figure 1.12 can turn a skull into a digital image of a face, allowing investigators to proceed more quickly with identification.

Proprietary law enforcement databases can use AI-powered software to analyze for similarities between crimes to detect patterns that may reveal serial crimes. In fact, a law enforcement specialty called digital forensics is growing in importance to fight crime. Digital forensics analyzes computer devices with specific techniques to gather potential legal evidence. For example, the data from the pacemaker of a defendant in an arson case was recently used as evidence. The defendant insisted he was asleep during the crime, but the pacemaker data indicated he was under stress and heavy exertion. In many cases, other kinds of files, videos, and conversations conducted using a computer can be recovered by forensics specialists to use as evidence of criminal activity.

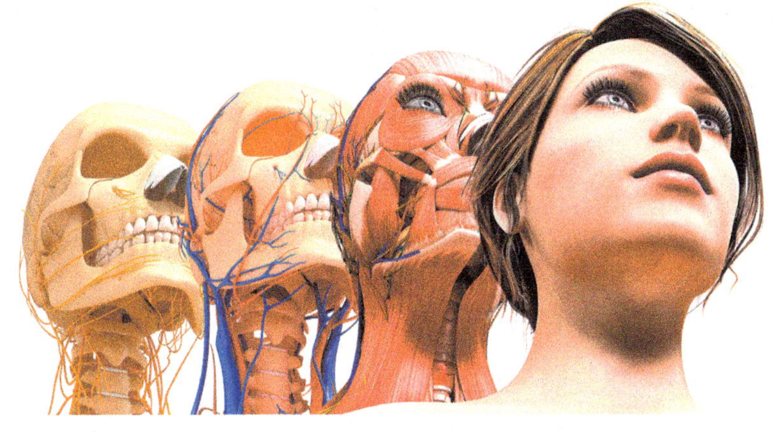

Figure 1.12 Tissue-rendering programs add layers of muscle, fat, and skin to create faces that can be used to identify victims. *(Pixologicstudio/Science Photo Library/Alamy Stock Photo)*

Medicine

A career in medicine will connect you to new ways of using technology to better people's lives. Websites such as Modernizing Medicine (*modmed.com*) use AI software to search through data on millions of patient visits and treatments provided by thousands of physicians. The website can help doctors quickly diagnose conditions with which they're not familiar and research alternative means of treatment for illnesses to reduce costs and side-effects for patients.

AI is also being integrated directly into patient information systems. Dashboards on the physician's computer screen can make recommendations about treatments to specific patients after the AI has analyzed their genetic traits and compared them to millions of other patients through records. The AI can recommend drugs for treatment as well as remind physicians about tests they may need to perform.

The design and construction of prosthetic devices is another area of medicine impacted by modern technology. MIT's biomechatronics lab has developed software that uses an array of pressure sensors to gauge the softness or stiffness of a patient's remaining tissue to create a better fit for a prosthetic to the limb. Meanwhile, 3D printing is allowing more inexpensive designs for prosthetic arms and legs, and more stylish artificial limbs as well (see Figure 1.13).

Figure 1.13 3D printing has become a tool for developing more inexpensive, and more stylish, prosthetic devices. *(Ilya Platonov/Shutterstock)*

Psychology

Fear of speaking in public is common, but for people with autism spectrum disorders, making proper eye contact and reacting to social cues is so difficult that it can severely limit their opportunities for relationships and jobs. Researchers at the MIT Media Lab have developed a system to help improve interpersonal skills for people who have autism.

MACH (My Automated Conversation coacH) is a computer system that generates an on-screen person that can, for example, conduct a job interview or appear ready for a first date. The computerized person (see Figure 1.14) nods and smiles in response to the user's speech and movement. This is an example of affective computing, developing systems that can recognize and simulate human emotions. MACH users can practice as many times as they wish in a safe environment. They receive an analysis that shows how well they modulated their voices, maintained eye contact, and smiled and how often they lapsed into umms and uhhhs.

Figure 1.14 My Automated Conversation coacH (MACH) generates an on-screen interviewer you can practice with over and over. *(Cultura Creative/Cultura Limited/SuperStock)*

Dig Deeper XR Extended Reality

What is *extended reality*? Extended reality, or XR, is the full set of ways we can manipulate reality with digital intervention, whether augmented reality (AR), new virtual realities (VR), or a combination of real-world and digitally created reality (mixed reality).

Augmented reality (AR) is the addition of digital information directly into our reality, either to add more detail or at times to remove unwanted visual effects. How does this happen?

AR combines our normal sense of the world around us with an additional layer of digital information. The extra information can be displayed on a separate device, such as in augmented reality apps for smartphones, or placed in the natural world, like a heads-up display on a car windshield to show car speed and location. Displays in stores that augment your image with the clothing you're interested in, creating a virtual fitting room (see Figure 1.15), are another example of AR.

Instead of adding information to the reality you perceive, virtual reality (VR) replaces it with a different world. It creates an artificial environment that is immersive and interactive. VR environments can be as simple as a pair of goggles or as elaborate as entire caves that you walk into (see Figure 1.16). VR has also come to the consumer gaming market with the sale of VR goggles, like the HTC Vive and the Occulus Rift (see Figure 1.17). These goggles have high pixel-count displays that wrap your full field of view.

Mixed reality is a mixture of both worlds. Digitally created objects are placed in the real world and can be interacted with directly. Surgeons might interact with a 3D hologram of their patient, and that information can be used to guide a robotic system to perform those steps (see Figure 1.18).

Thus, reality may be a bit less absolute than it once seemed. Whether we are adding, creating, or combining digital environments, XR technology is expanding our sense of what is real.

Figure 1.15 This high-tech fitting room uses augmented reality technology to allow shoppers to try on clothes virtually. *(Yoshikauz Tsuno/Staff/AFP/Getty Images)*

Figure 1.17 VR goggles like these wrap around your full field of view, creating a totally immersive environment. *(Philippe Plailly/Science Photo Library)*

Figure 1.16 Virtual reality caves replace our ordinary reality with a new, immersive environment. *(Idaho National Laboratory/US Department of Energy/Science Photo Library)*

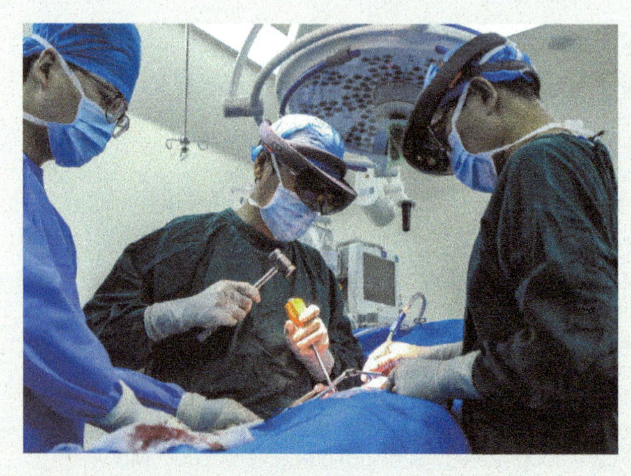

Figure 1.18 Surgeons use mixed reality. Their glasses project 3D scans of the structures they are manipulating in real time. *(ImaginechinaLimited/Alamy Stock Photo)*

Ethical Computing

As noted earlier, technology has brought about a new set of ethical challenges. Because technology often moves faster than rules can be formulated to govern it, how technology is used is often left up to the individual and the guidance of his or her personal ethics. Ethical considerations are complex, and reasonable people can have different, yet valid, views. In this section, we define ethics and then examine a set of ethical systems. Next, we discuss personal ethics and how technology affects them. Finally, we ask you to consider what your personal ethics are and how you make ethical choices.

Defining Ethics

Objective 1.8 *Define ethics and examine various ethical systems.*

Ethics is the study of the general nature of morals and of the specific moral choices individuals make. Morals involve conforming to established or accepted ideas of right and wrong (as generally dictated by society) and are usually viewed as being black or white. Ethical issues often involve subtle distinctions, such as the difference between fairness and equity. Ethical principles are the guidelines you use to make decisions each day, decisions about what you will say, do, and think.

There are many systems of ethical conduct (see Table 1.4). *Relativism* states that there is no universal truth and that moral principles are dictated by the culture and customs of each society. *Divine command theory* follows the principle that God is all knowing and that moral standards are perfectly stated by God's laws, such as the Ten Commandments. *Utilitarianism* judges actions to be right or wrong solely by their consequences. Actions that generate greater happiness are deemed to be better than those actions that lead to less happiness. *Virtue ethics* teaches that we each have an internal moral compass and should try to be a person who spontaneously follows that guide. Finally, the duty-based ethical system, or *deontology,* suggests we should all follow common moral codes and apply those to all humanity.

Systems of Ethics

What is the difference between laws and ethics? Laws are formal, written standards designed to apply to everyone. It's impossible to pass laws that cover every human behavior, though. Therefore, ethics provide a general set of unwritten guidelines for people to follow. Unethical behavior isn't necessarily illegal, however. Consider the death penalty. In many U.S. states, putting convicted criminals to death for certain crimes is legal. However, many people consider it unethical to execute a human being for any reason.

Not all illegal behavior is unethical, either. Civil disobedience, which is intentionally refusing to obey certain laws, is used as a form of protest to effect change. Gandhi's nonviolent resistance to the British rule of India, which led to India's establishment as an independent country, and Martin Luther King's protests and use of sit-ins in the civil rights movement are both examples of civil disobedience. What does civil disobedience look like in our hyper-connected age of social media and the Internet?

Note that there is also a difference between unethical behavior and amoral behavior. Unethical behavior is not conforming to a set of approved standards of behavior—cheating on an exam, for example. Amoral behavior occurs when a person has no sense of right and wrong and no interest in the moral consequences of his or her actions, such as when a murderer shows no remorse for his or her crime.

There is no universal agreement on which system of ethics is the best. Most societies use a blend of different systems. Regardless of the ethical system of the society in which you live, all ethical decisions are greatly influenced by personal ethics.

How to Debate Ethical Issues

This Sound Byte will help you consider important aspects of debating difficult issues arising from the ethical use of technology.

Table 1.4 Systems of Ethics

Ethical System	Basic Tenets	Examples
Relativism 	• There is no universal moral truth. • Moral principles are dictated by cultural tastes and customs.	Topless bathing is prevalent in Europe but generally banned on public beaches in the United States.
Divine Command Theory 	• God is all-knowing and sets moral standards. • Conforming to God's law is right; breaking it is wrong.	The Ten Commandments of Christian religions The Five Pillars of Islam
Utilitarianism 	• Actions are judged solely by consequences. • Actions that generate greater happiness are judged to be better than actions that lead to unhappiness. • Individual happiness is not important. Consider the greater good.	Using weapons of mass destruction ends a war sooner and therefore saves lives otherwise destroyed by conventional fighting.
Virtue Ethics 	• Morals are internal. • Strive to be a person who behaves well spontaneously.	A supervisor views the person who volunteers to clean up a park as a better person than the workers who are there because of court-ordered community service.
Deontology (Duty-Based) 	• Focus on adherence to moral duties and rights. • Morals should apply to everyone equally.	Human rights (like freedom of religion) should be respected for all people because human rights should be applied universally.

(Solomin Andrey/123RF; Sergey Galushko/123RF; Kheng Ho Toh/123RF; Alexmillos/123RF; Scanrail/123RF)

Personal Ethics

Objective 1.9 *Describe influences on the development of your personal ethics.*

Each day as you choose your words and actions, you're following a set of personal ethics—a set of formal or informal ethical principles you use to make decisions. Some people have a clear, well-defined set of principles they follow. Others' ethics are inconsistent or are applied differently in different situations.

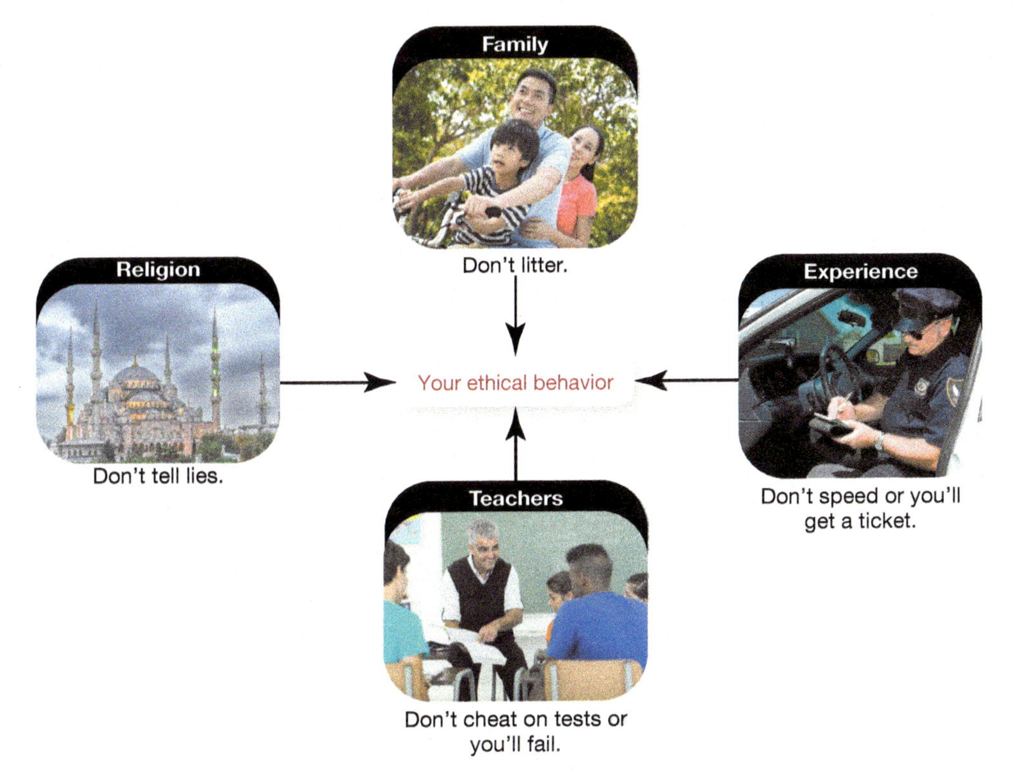

Figure 1.19 Many forces shape your personal ethics. *(Imtmphoto/123RF; Pisaphotography/Shutterstock; Lisa Young/123RF; Michaeljung/Shutterstock)*

A person's ethics develop in many ways. Naturally, your family plays a major role in establishing the values you cherish in your own life, and these might include a cultural bias toward certain ethical positions (see Figure 1.19). Your religious affiliation is another major influence on your ethics because most religions have established codes of ethical conduct. How these sets of ethics interact with the values of the larger culture is often challenging. Issues such as abortion, the death penalty, and war often create conflict between personal ethical systems and the larger society's established legal–ethical system.

As you mature, your life experiences also affect your personal ethics. Does the behavior you see around you make sense within the ethical principles that your family, your church, or your teachers taught you? Has your experience led you to abandon some ethical rules and adopt others? Have you modified how and when you apply these laws of conduct, depending on what's at stake?

Determining Your Personal Ethics

When you have a clear idea of what values are important to you, it may be easier to handle situations in your life that demand ethical action. You can follow these steps to help yourself define a list of personal values:

1. **Describe yourself.** Write down words that describe who you are, based on how others view you. Would a friend describe you as honest, helpful, or kind? These keywords will give you a hint as to the values and behaviors that are important to you.
2. **List the key principles you believe in.** Make a list of the key principles that influence your decisions. For example, would you be comfortable working in a lab that used animals for medical research? How important is it to you that you never tell a lie? List the key ideas you believe to be important in conducting your life. Do you always behave this way, or are there situations in which your answers might change?

3. **Identify external influences.** Where did your key principles come from—your parents? Your friends? Spiritual advisors? Television and movies? You may want to question some of your beliefs once you actually identify where they came from.
4. **Consider why.** After writing down your beliefs, think about why you believe them. Have you accepted them without investigation? Do they stand up in the context of your real-world experiences?
5. **Prepare a statement of values.** Distill what you have written into a short statement. Having a well-defined statement of the values you hold most important in your own life will make it easier for you to make ethical decisions in times of challenge.

Ethics and Technology

Objective 1.10 *Present examples of how technology creates ethical challenges.*

You'll encounter many ethical challenges relating to information technology. Your personal ethics—combined with the general ethical environment of society and ethical guidelines your company provides if your decision is related to the workplace—will help to guide your decisions.

Intellectual Property

One obvious area of ethical challenge in technology is intellectual property rights. Intellectual property is work that is the result of someone's creativity and knowledge (such as music, writing, and software) that is protected by copyright law and documented by copyrights, patents, and trademarks. Anti-piracy efforts by various organizations and governments aim to protect intellectual property rights—an issue that has been an ongoing and important one in the technology industry.

With increased use of electronic media, challenges to enforcing copyright laws have increased. Countries such as China and the Philippines are known for their high rates of pirated software use. Businesses and individuals that use such illegal software potentially gain an advantage in the international marketplace. Companies like Microsoft have taken several measures to combat illegal piracy but still haven't managed to reduce the amount of pirated software effectively in certain countries.

Piracy challenges have also placed significant pressures on existing business models for creative arts industries such as film and music. Time will tell whether copyright laws will need to be revised to adapt to the predominant digital culture, whether regulatory agencies will need to improve the methods of controlling piracy, or whether the industries themselves will need to change their business models.

Privacy

Technology is also posing new ethical issues related to privacy. Intelligent personal agents (like Alexa) are constantly listening to you and combing through your data (such as your e-mails and texts) to help determine your needs. Many people are concerned about the loss of privacy from such agents.

Social media sites such as Facebook and Twitter are designed to share information with others, which can be great for sharing graduation photos with family. Employers however now check social media sites to gather information on prospective employees. If you have privacy settings that allow them to see enough to infer your race, religion, or sexual orientation, they will have information they are not legally allowed to ask you directly. Facebook and Twitter also generate information about you based on your behavior on their sites. What is your household income? What is your political affiliation? Sophisticated algorithms make predictions about these matters from the choices you make using their sites. For example, what groups do you join? What links do you follow? The Try This exercise in this chapter will guide you in discerning exactly what information these companies have generated to describe you. The control and privacy of information will continue to be a fine balancing act for the foreseeable future.

Social Justice

Technological processing power has led to new issues that impact social justice. Consider predictive policing—gathering data from different sources, analyzing it, and then using the results to

Table 1.5 Point/Counterpoint: Social Justice

Issue: Predictive Policing	Ethical question: Should police agencies act on data trends that seem to predict crime?
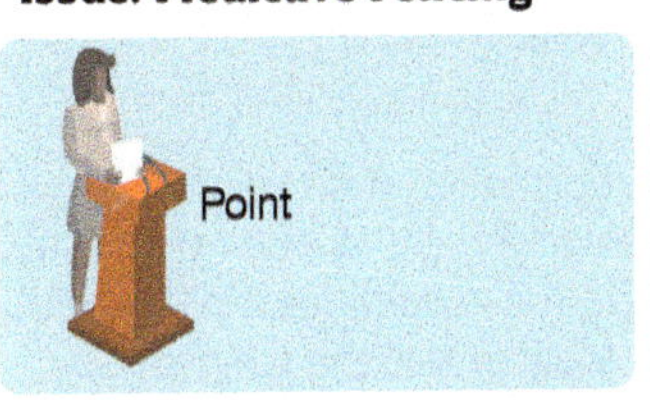	• This is just a use of modern technology to enhance law enforcement. • Data trends that show areas of high criminal activity allow the police to use their resources more effectively. • If law enforcement steps in before a crime has been committed, it helps both the potential victim and the potential criminal.
	• This is a high-tech form of racial and socioeconomic profiling and acts to reinforce existing stereotypes. • This decreases our privacy and encourages constant government surveillance of our activities. • It is impossible to create predictive algorithms in a way that is not harmful to law-abiding people living in high-crime areas.

Rvector/Shutterstock

prevent future crime. Data sources may include crime maps, traffic camera data, surveillance footage, and social media network analysis. But at what point does the possibility of a crime warrant intervention? Is predictive policing just another way to justify racial profiling? Table 1.5 lists different views on the issue of predictive policing. Where do your views fall?

Another social justice issue affected by technology is related to military and government secrets. There are military secrets and communication related to ongoing negotiations with foreign governments that the public does not have a right to see. Historically, tight controls have been in place over access to such information. With Web 2.0 tools, the entire model for the distribution of information has shifted, and now everyone can create and publish content on the Internet. This has altered the meaning of privacy in fundamental ways. WikiLeaks (*wikileaks.org*) makes private and secret documents available for viewing on its site. Different than the National Whistleblower website, WikiLeaks has published documents such as the script of a movie still in production or confidential reports detailing the names of foreign civilians passing information to the U.S. military. Does society have a right to see secret documents? And do we all have a responsibility to use technology to help achieve social justice? Where do the boundaries of those rights and responsibilities lie?

Liability

3D printing is revolutionizing the way items are manufactured, causing a range of issues involving intellectual property rights and liability. In traditional manufacturing, it is relatively easy to enforce safety laws. Consider the manufacture of bicycle helmets. Traditional manufacturers are held accountable for delivering safely designed products free from defects in workmanship. Products are subjected to safety testing prior to being sold to the public and must conform to legal safety guidelines. If you buy a helmet and are injured while riding your bike because of a defect in the design of or materials used in the helmet, you can sue the manufacturer for damages. Manufacturers protect their designs (intellectual property) by obtaining patents.

3D printers produce objects from design plans generated with various types of software. Anyone who is reasonably proficient with design software can generate a design for an object to be printed. However, a person could also use a 3D scanner to scan an existing object (say a helmet designed and sold by a traditional manufacturer) and use design software to create his or her own plan for how to print it. If the object is patented, this would be a violation of the law. However, it certainly makes it easier to steal someone's intellectual property.

Design plans for objects are often shared online. Even if the same design is used, the product quality can be vastly different, depending on the type of 3D printer used and the type of plastic selected for printing.

Consider the ethics of printing a helmet from a file you downloaded from a public website. You give it to a friend as a gift. While riding her bicycle, your friend is injured because the design of the helmet was flawed and the materials you used to make it were substandard. Who is responsible for the injuries? You? The manufacturer of the printer? The owner of the printer? The manufacturer of the raw materials used to make the helmet? The creator of the flawed design plans? The person who decided to use an untested product (i.e., your friend)? What if the design you used was created from a patented product? Table 1.6 lists different views on the issue of 3D printing regulation. Where do your views fall?

Censorship

Various countries have different answers to the question of what information their people should be allowed to see. Currently, India ranks first as the country whose government most often cuts off access to the Internet for its citizens. The government has argued at times that it is needed to stop cheating on exams or because there is political upheaval. Meanwhile, the Chinese government has demanded that search engine providers like Google self-censor their search engines, restricting access to foreign websites that express different views than the Chinese government and limiting information on the sensitive topics of Tibet, Taiwan, or the Tiananmen Square uprising. Table 1.7 lists different views on the issue of Internet censorship. What do you think?

Social Activism

Hacktivism (derived from *hack* and *activism*) involves using computers and computer networks in a subversive way to promote an agenda usually related to causes such as free speech,

Table 1.6 Point/Counterpoint: Liability

Issue: 3D Printing Regulation	**Ethical Question: Should 3D printing be regulated?**
Point	• The general public expects to buy products that are free from design and material defects. If governments do not regulate 3D printing materials and designs, the public is placed in unnecessary peril. • Because designs of existing products are easy to produce, there is an increased chance of intellectual property theft. • Users of 3D printers need to understand the risks, responsibilities, and liabilities of producing products on demand for their own use and for resale.
 Counterpoint	• Heavily regulating 3D printing will stifle creativity and ingenuity in the design and production of new products. • Manufacturers can seek legal remedies for violation of design patents under existing laws. • Consumers should be aware of the principle of caveat emptor (let the buyer beware) when using untested or unproven products. People should use untested products cautiously.

Table 1.7 Point/Counterpoint: Censorship

Issue: Censoring the Internet	**Ethical Question: Should an American company agree to restrict freedom of information to do business with a more restrictive country?**
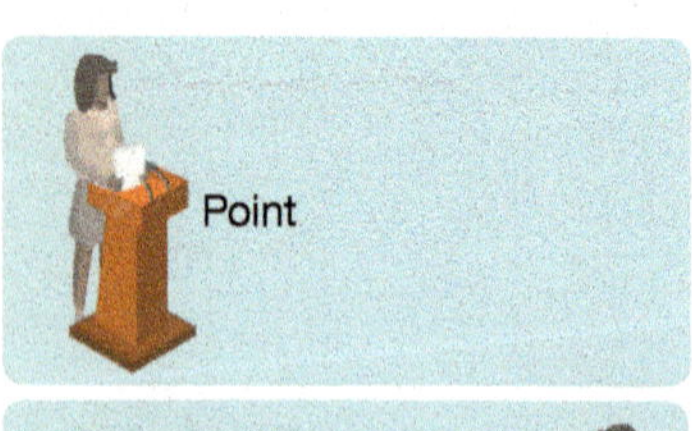 Point	• It is not the place of a company to try to change laws of foreign countries. Reform must come from within. • Working in China does not mean a company supports all of China's policies. • U.S. companies can ethically stay in China if they make an effort to improve human rights there. U.S. companies operating in China should agree on guidelines that respect human rights.
Counterpoint 	• Policies will never change unless there are financial and political incentives to do so. The departure of U.S. companies helps pressure foreign governments to reform. • If companies withdraw from China, the viability of many other resellers in China is threatened. • Countries cannot expect to compete in the global marketplace while refusing to have a global exchange of ideas.

Table 1.8 Point/Counterpoint: Social Activism

Issue: Hacktivism	Ethical Question: Is hacktivism a helpful form of civil disobedience?
	• Society as a whole benefits when injustices are exposed. • Civil disobedience is an inalienable right upon which the United States was founded. • Methods of civil disobedience need to keep pace with modern technology to have the greatest impact possible.
	• Computer technology provides many opportunities to reach a wide audience with a message without resorting to illegal activities such as hacking. • Personal data of individuals needs to be protected as a basic right of privacy. • Hacking into computer systems for alleged acts of civil disobedience are often hard to distinguish from acts of cyberterrorism.

human rights, or freedom of information. Essentially, it is using computer hacking to effect some sort of social change. An example of early hacktivism is Worms Against Nuclear Killers (WANK), a computer worm that anti-nuclear activists in Australia unleashed into the networks of NASA in 1989 to protest the launch of a shuttle that carried radioactive plutonium. As computer security has grown more sophisticated, attacks today are usually carried out by groups of hackers (such as the group Anonymous) or groups of computer scientists funded by nation states for the specific purpose of hacking.

Hacktivism often takes the form of denial-of-service attacks in which websites are bombarded with requests for information until they are overwhelmed and legitimate users can't access the site. At other times, hacktivism takes the form of cyberterrorism; the objective is to embarrass or harass a company by penetrating its computer networks and stealing (and often publishing) sensitive information. Certain countries are rumored to maintain groups of computer scientists whose main goal is to spy on other countries by hacking into their computer networks. Table 1.8 lists different views on the issue of hacktivism. What do you think?

Automated Robotic Machinery

Automobiles now contain sophisticated AI systems to exercise control over the vehicle and respond faster than humans can. For example, by using a wide set of sensors, accident-avoidance systems can apply the brakes and even change lanes to help drivers avoid accidents. In the coming years, these systems may exercise even more control of your vehicle, such as the self-driving cars that are under development by Google and major auto manufacturers.

But who controls the ethical constraints by which automated robotic machinery like this operates? Consider a selection between a set of bad choices: You suddenly find that you have to brake, knowing you will still hit the school bus ahead of you, swerve into oncoming traffic, or swerve to the other side into a tree. Which choice would you make? Which choice would you want your automated car to make? Should owners have a choice of overriding the programming in their vehicles and adjusting the ethical parameters of their robotic systems? What should manufacturers do until laws are passed regarding roboethics? Table 1.9 lists different views on the issue of robotic machinery ethics. Where do your views fall?

Ethical discussions are important in many fields but especially with technology. The pace of change is forcing us to consider implications for our families and countries as well as for our personal ideas of justice and proper conduct. Being able to analyze and then discuss your ethical positions with others are important skills to develop.

Table 1.9 Point/Counterpoint: Automated Robotic Machinery

Issue: Ethics of Robotic Machinery	Ethical Question: Should owners be able to set the ethical rules of behavior for their own robots?
 Point	• As long as choices over robotic ethics don't violate laws, individual owners should be able to choose ethical rules of behavior. • Individuals should have the right to set ethical parameters for their robots because they may wish to be even more selfless than society dictates. • Individuals make ethical decisions in times of crisis. Robotic systems should be an extension of the individual's ethical values.
 Counterpoint	• If all robots do not contain ethical constraints that prevent harm to human beings, human lives may be lost. • Allowing individuals to adjust the ethics of robotic devices exposes society to risks from careless, thoughtless, or psychotic individuals. • Every robot must have ethical programming that allows it to make decisions regarding the best possible outcomes in life-and-death situations.

Before moving on to the Chapter Review:

1.  Watch Chapter Overview Video 1.2.
2. Take the Check Your Understanding quiz.

Check Your Understanding // Review & Practice

For a quick review to see what you've learned so far, answer the following questions.

multiple choice

1. __________ is a branch of computer science that focuses on creating computer systems or computer-controlled machines that have an ability to perform tasks usually associated with human intelligence.
   a. Computer literacy
   b. Artificial intelligence
   c. Natural language
   d. Information technology
2. AI systems that help people discover things they may like but may not discover on their own are known as
   a. crowdsourcing systems.
   b. Turing testers.
   c. recommendation engines.
   d. intelligent personal assistants.
3. A working definition of *ethics* is
   a. impossible because it means different things to different people.
   b. the laws of a given society.
   c. the study of the general nature of morals and of the specific moral choices individuals make.
   d. the rules of conduct presented in Christian religions.
4. The set of formal or informal ethical principles you use to make decisions is your
   a. mindset.
   b. instinct.
   c. personal ethics.
   d. personality.
5. Digital forensics is the study of
   a. digital art.
   b. digital data that is evidence of a crime.
   c. using data to predict stocks.
   d. writings on the nature of the digital divide.

chew on this

(3Dalia/Shutterstock)

This part of the chapter lists many ways in which becoming computer literate is beneficial. Think about what your life will be like once you're started in your career. How would an understanding of computer hardware, software, and artificial intelligence help you in working from home, working with groups in other countries, and contributing your talents?

MyLab IT

Go to **MyLab** IT to take an autograded version of the *Check Your Understanding* review and to find all media resources for the chapter.

For the IT Simulation for this chapter, see MyLab IT.

1 Chapter Review

Summary

Part 1
Technology in Society

Learning Outcome 1.1 **You will be able to discuss the impact of the tools of modern technology on national and global issues.**

Technology in a Global Society

Objective 1.1 *Describe various technological tools being used to impact national and global issues.*

- Technology can be the means by which you find your voice in the world and impact others in meaningful ways.
- Social media is impacting elections and other issues worldwide.
- Crisis-mapping tools are an example of technology helping in different kinds of global conflicts and disasters.

Objective 1.2 *Describe various global social issues that are being affected by technology.*

- Global health care issues, like the spread of disease, require international cooperation and technological solutions.
- Environmental issues are global in nature and will require technology to address.
- The digital divide, an uneven distribution of access to computer technology, will make solving global problems difficult for us.

Technology Connects Us with Others

Objective 1.3 *Describe how technology is changing how and why we connect and collaborate with others.*

- Web 2.0 is a set of features and functionality that allows Internet users to contribute content easily and to be easily connected to one another.
- Cognitive surplus is the combination of leisure time and access to tools to work on problems and be creative.
- New collaborative tools available on the Internet allow us to work together on projects with much larger groups.
- Crowdfunding is a method of people connecting through the Internet to fund projects by donation.

Objective 1.4 *Summarize how technology has impacted the way we choose and consume products and services.*

- Marketing is changing because most consumers now shop with Internet access on their phones and can therefore check competing prices and online reviews.
- The idea of ownership is changing because technology is allowing subscription services for products like cars and bikes to be available for sharing. Such collaborative consumption means that we are joining together as a group to use specific products more efficiently.

The Importance of Computer Literacy

Objective 1.5 *Characterize computer literacy and explain why it is important to be computer literate.*

- If you're computer literate, you understand the capabilities and limitations of computers and know how to use them wisely.
- By understanding how a computer is constructed and how its various parts function, you'll be able to get the most out of your computer.
- You'll be able to avoid hackers, viruses, and Internet headaches; protect your privacy; and separate the real risks of privacy and security from things you don't have to worry about.
- You'll also be better able to maintain, upgrade, and troubleshoot your computer; make good purchasing decisions; and incorporate the latest technologies into your existing equipment.
- Being computer literate also enables you to understand the many ethical, legal, and societal implications of technology today.

Part 2

Emerging Technologies and Ethical Computing

Learning Outcome 1.2 **You will be able to describe emerging technologies, such as artificial intelligence, and how technology creates new ethical debates.**

Artificial Intelligence

Objective 1.6 ***Describe artificial intelligence systems and explain their main goals.***

- Artificial intelligence (AI) focuses on creating computer systems that can perform tasks associated with human intelligence.
- Machines using AI do not necessarily mimic the human thought process.
- Current AI research is focused in the following areas: natural language processing, perception, knowledge representation, planning, problem solving, and learning (both supervised and unsupervised).
- A Turing test can be used to determine whether a system is a computer AI or a human being.

Working with Artificial Intelligence and Other Information Technologies

Objective 1.7 ***Describe how artificial intelligence and other emerging technologies are important in many careers.***

- Artificial intelligence impacts the full range of careers, from retail and psychology to robotics and medicine.
- Understanding how to use software, how to use and maintain computer hardware, and how to take advantage of Internet resources will help you be a more productive and valuable employee, no matter which profession you choose.

Ethical Computing

Objective 1.8 ***Define ethics and describe various ethical systems.***

- Ethics is the study of moral choices.
- There are several ethical systems. There is no universal agreement on which system is best.
- Unethical behavior is not necessarily unlawful behavior.
- Amoral behavior is exhibited by a person who has no sense of right and wrong.

Objective 1.9 ***Describe influences on the development of your personal ethics.***

- Personal ethics are a set of formal or informal ethical principles you use to make decisions in your life.
- Your personal ethics develop from a number of sources: your family values, your religion, the values of the larger culture, and your life experiences.

Objective 1.10 ***Present examples of how technology creates ethical challenges.***

- Technology is posing new ethical challenges with regard to intellectual property, privacy, social justice, liability, censorship, social activism, and automated robotic machinery, among other concerns.

Be sure to check out **MyLab IT** for additional materials to help you review and learn. And don't forget to watch the Chapter Overview videos.

Key Terms

affective computing **19**
amoral behavior **22**
artificial intelligence (AI) **13**
artificial neural networks (ANNs) **15**
augmented reality (AR) **20**
bot accounts **4**
cognitive surplus **6**
collaborative consumption **7**
computer literate **8**
computer vision **14**
crisis-mapping tool **4**
crowdfunding **7**
crowdsourcing **7**
data mining **17**
deep learning (DL) **15**
digital divide **5**
digital forensics **19**
embodied agents **18**
ethics **22**
expert systems **15**
extended reality (XR) **20**
hacktivism **27**
information technology (IT) **16**
intellectual property **25**
intelligence **13**
knowledge representation **14**
machine learning (ML) **15**
natural language processing (NLP) **14**
personal ethics **22**
predictive policing **25**
recommendation engines **17**
rules-based systems **15**
social media **4**
software **9**
spam **9**
supervised learning **14**
Turing test **18**
unethical behavior **22**
unsupervised learning **14**
virtual reality (VR) **20**
Web 2.0 **6**

Chapter Quiz // Assessment

For a quick review to see what you've learned, answer the following questions. Submit the quiz as requested by your instructor. If you are using **MyLab IT**, the quiz is also available there.

multiple choice

1. The crisis-mapping tool Ushahidi
   - **a.** prevents two-way dialog between people.
   - **b.** is only available in the United States.
   - **c.** shows that a free software product can have great value.
   - **d.** relies on people using their desktop computing power to help others.
2. Sophisticated modeling software is helping international researchers
   - **a.** create more intricate screenplays and movie scripts.
   - **b.** analyze computer systems to gather potential legal evidence.
   - **c.** market new types of products to a wider audience.
   - **d.** increase the pace of research in finding and producing vaccines.
3. Which of the following would NOT impact the digital divide?
   - **a.** A decline in the price of Internet access
   - **b.** Less sharing of the work we produce
   - **c.** Expansion of free Wi-Fi connectivity in major cities
   - **d.** Legislation allowing communications companies to charge per website for access
4. Crowdfunding sites include
   - **a.** Citi Bike and Zipcar.
   - **b.** Bing and Google.
   - **c.** Kickstarter and GoFundMe.
   - **d.** Amazon and Netflix.
5. The sharing economy is exemplified by
   - **a.** Citi Bike and Zipcar.
   - **b.** Bing and Google.
   - **c.** Kickstarter and GoFundMe.
   - **d.** Amazon and Netflix.
6. Being computer literate includes being able to
   - **a.** avoid spam, adware, and spyware.
   - **b.** use the web effectively.
   - **c.** diagnose and fix hardware and software problems.
   - **d.** all of the above.
7. Unsupervised learning in an artificial training system improves
   - **a.** without being given specific data examples.
   - **b.** by using training data.
   - **c.** by using virtual reality scenarios.
   - **d.** by using augmented reality data.

8. Which of the following statements is FALSE?
   a. Unethical behavior is always illegal.
   b. Ethical decisions are usually influenced by personal ethics.
   c. Individuals who have no sense of right or wrong exhibit amoral behavior.
   d. Life experience affects an individual's personal ethics.
9. Which of the following actions would NOT help identify your personal ethics?
   a. Describe yourself.
   b. Conduct a genetic study of your extended family.
   c. Identify the influences of your work environment.
   d. Prepare a list of values that are most important to you.
10. Intellectual property
   a. is the result of someone's creative work and design.
   b. can be protected legally by the patent system.
   c. can include works of music, film, or engineering design.
   d. all of the above.

true/false

_____ 1. Intellectual property can be legally protected even though it is not a physical object.
_____ 2. The move toward access instead of ownership is a sign of cognitive surplus.
_____ 3. Machine learning systems analyze patterns in data to improve their performance automatically.
_____ 4. Web-based databases cannot be used to help investigators solve criminal cases.
_____ 5. Hacktivism is different from cyberterrorism.

What do you think now?

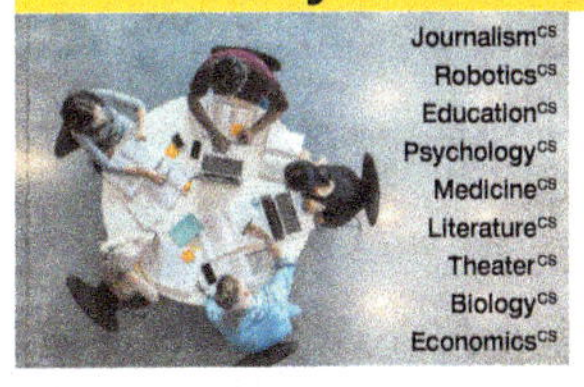

1. Examine your answer to the What Do You Think question from the beginning of the chapter. Describe how the topics of this chapter have influenced you. Have they pushed you to change your answer? Reinforced your thinking? Why?
2. Using the Reilly Top 10 list (*reillytop10.com*), investigate three areas where technology is producing conflict for society. What kind of CS + X combination majors would produce people who are well equipped to solve these challenges?

Team Time

A Culture of Sharing

problem

As more and more peer-to-peer music-sharing services appeared, such as BitTorrent, many felt a culture of theft was developing. Some argued there was a mind-set that property rights for intellectual works need not be respected and that people should be able to download, for free, any music, movies, or other digital content they wanted.

But there is another view of the phenomenon. Some are suggesting that the amount of constant access to other people—through texting, e-mail, blogging, and the easy exchange of digital content—has created a culture of trust and sharing. This Team Time will explore both sides of this debate as it affects three parts of our lives—finance, travel, and consumerism.

task

Each of three groups will select a different area to examine—finance, travel, or consumerism. The groups will find evidence to support or refute the idea that a culture of sharing is developing. The finance group will want to explore projects like Kickstarter (*kickstarter.com*) and Kiva (*kiva.org*). The travel group should examine what is happening with Airbnb (*airbnb.com*) to start their research. The team investigating consumerism will want to look at goods-exchange programs like Freecycle (*freecycle.org*).

process

1. Divide the class into three teams.
2. Discuss the different views of a culture of sharing. With the other members of your team, use the Internet to research up-and-coming technologies and projects that would support your position. People use social media tools to connect into groups to exchange ideas. Does that promote trust? Does easy access to digital content promote theft, or has the value of content changed? Are there other forces, like the economy and the environmental state of the world, that play a role in promoting a culture of sharing? What evidence can you find to support your ideas?
3. Present your group's findings to the class for debate and discussion.
4. Write a strategy paper that summarizes your position and outlines your predictions for the future. Will the pace of technology promote a change in the future from the position you are describing?

conclusion

The future of technology is unknown, but we do know that it will impact the way our society progresses. To be part of the developments that technology will bring will take good planning and attention, no matter what area of the culture you're examining. Begin now—learn how to stay on top of technology.

Ethics Project

Can Data Predict Your Grade?

background

As you move through your academic career, you leave an enormous swath of data: which courses you chose to register for; which ones you looked at but didn't pick; and how you did on each homework assignment, test, and project. Could this massive amount of data be analyzed to predict your grade in a course? Could the data suggest which courses you should take next? Should those predictions be public to your instructor, to you, and to financial aid officers?

research topics to consider

- Civitas Learning
- Tableau Higher Education Analytics

process

1. Divide the class into teams.
2. Team members should each think of a situation when a person would benefit from the type of results data mining can bring to a campus and a situation when it might be undesirable.
3. Team members should select the most powerful and best-constructed arguments and develop a summary conclusion.
4. Team members should present their findings to the class or submit a PowerPoint presentation for review by the rest of the class, along with the summary conclusion they developed.

conclusion

As technology becomes ever more prevalent and integrated with our lives, we will increasingly face ethical dilemmas. Weighing the value of whether personal data should be kept private or used publicly to help in decision making will return again and again. Being able to understand and evaluate both sides of the argument, and responding in a personally or socially ethical manner, will be an important skill.

Solve This with Word

MyLab IT Grader

How Technology Is Used on the World Stage and in Your Personal Life

In this activity, you'll use Microsoft Word to reflect on how technology is affecting the world as well as you, personally and professionally. Reflect on the content in Chapter 1 as you work through this exercise.

You will use the following skills as you complete this activity:

- Open and Modify a Document Template
- Apply Styles and Advanced Font Formats
- Use Format Painter
- Insert Text
- Apply Themes
- Create a Header and Footer

Instructions

1. Open TIA_Ch1_Start.docx. Save the document as **TIA_Ch1_LastFirst.docx**, using your own Last and First names.
2. Double-click the **Title placeholder**, type **Technology**, and then double-click the **Heading 1 placeholder** and type **Introduction**. Press Enter and then type the following: **Political and global issues are showing that technology is accelerating change around the world and galvanizing groups of people in new ways. Technology allows us to refine how we connect with each other, and it also impacts our daily personal and professional experiences.** Press **Enter**.
3. Type **How Technology Impacts Society**, press **Enter**, and then type a few sentences that describe how technology is impacting global events such as political revolutions, health care, the environment, and the digital divide. In addition, address how businesses are using social media. Press **Enter**.
4. Type **How Technology Impacts Me Personally and Professionally**, press **Enter**, and then type a few sentences that describe how technology is impacting your personal life. You should address the importance of being computer literate. You should also address the kinds of technology being used in the industry of your current or desired career.
5. Click anywhere in the heading *Introduction*. With **Format Painter**, apply the **Heading 1** format to the paragraph headers: *How Technology Impacts Society* and *How Technology Impacts Me Personally and Professionally*.

   Hint: Format Painter is in the Clipboard group on the Home tab.
6. Change the Document Theme style to the **Slice Theme**.

   Hint: Document Themes are found on the Design tab, in the Document Formatting group.
7. Select the title *Technology* and then format the font as **Small caps. Center align** the title.

   Hint: Click the dialog box launcher in the Font group on the Home tab to access the Small caps font effect.
8. Apply the **Whisp header style** to the document. Click to add Today's date in the Date header and delete the Document title header. Add a **File Path** to the document footer. Select the footer text and change the font size to **8**. Close the Header and Footer.

   Hint: Headers are found on the Insert tab in the Header & Footer group. File Path is found in Document Info in the Insert group on the Header & Footer Tools Design tab.
9. Save the document and submit based on your instructor's directions.

Chapter

2 Looking at Computers: Understanding the Parts

For a chapter overview, watch the Chapter Overview Videos.

PART 1

Understanding Digital Components

Learning Outcome 2.1 **You will be able to describe the devices that make up a computer system.**

PART 2

Processing, Storage, and Connectivity

Learning Outcome 2.2 **You will be able to describe how computers process and store data and how devices connect to a computer system.**

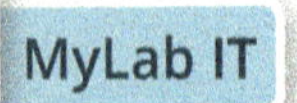

MyLab IT All media accompanying this chapter can be found here.

Make This A Mobile App on **page 58**

(Bennymarty/123RF; Aberration/123RF; Destinacigdem/123RF; Karyna Navrotska/123RF; Ruslan Danyliuk/123RF; Ale-kup/123RF; SciePro/Shutterstock)

What do you think?

Computer technology is making great strides toward **enabling individuals to overcome injuries and disabilities**. Implants that can restore sight are now being researched. Exo-skeletons, like the one shown here, are restoring the ability to walk. Cochlear implants can restore hearing. People with technology implanted in or attached to their bodies are referred to as ***augmented humans***. It is anticipated that augmented technologies may someday improve to the point that someone who uses such technology might have **superior abilities** to someone who doesn't use it. Perhaps in the future, we won't need a regular Olympics and a Special Olympics. But **is it fair to let augmented humans compete against non-augmented humans**?

?

Should augmented humans be allowed to compete in the Olympics and professional sports alongside non-augmented athletes?

- Augmented humans should only be allowed to compete against other augmented humans.
- Only augmented humans who have used technology to correct a deficiency should be allowed to compete with non-augmented humans.
- Any human being (augmented or not) should be allowed to compete.

See the end of the chapter for a follow-up question

Part 1

For an overview of this part of the chapter, watch **Chapter Overview Video 2.1**.

Understanding Digital Components

Learning Outcome 2.1 You will be able to describe the devices that make up a computer system.

After reading Chapter 1, you can see why becoming computer literate is important. But where do you start? You've no doubt gleaned some knowledge about computers just from being a member of society. However, even if you've used a computer before, do you really understand how it works, what all its parts are, and what those parts do?

Understanding Your Computer

Let's start our look at computers by discussing what a computer does and how its functions make it such a useful machine.

Computers Are Data Processing Devices

Objective 2.1 *Describe the four main functions of a computer system and how they interact with data and information.*

What exactly does a computer do? Strictly defined, a computer is a data processing device that performs four major functions:

1. **Input:** It gathers data or allows users to enter data.
2. **Process:** It manipulates, calculates, or organizes that data into information.
3. **Output:** It displays data and information in a form suitable for the user.
4. **Storage:** It saves data and information for later use.

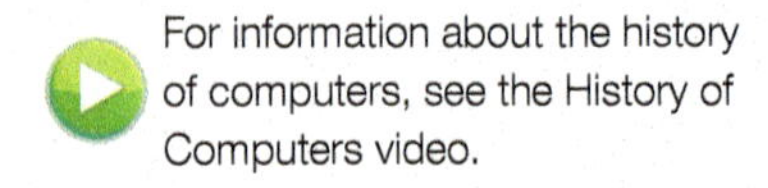

For information about the history of computers, see the History of Computers video.

What's the difference between data and information? In casual conversations, people often use the terms *data* and *information* interchangeably. However, for computers, the distinction between data and information is an important one. In computer terms, data is a representation of a fact, a figure, or an idea. Data can be a number, a word, a picture, or even a recording of sound. For example, the number 7135553297 and the names Zoe and Richardson are pieces of data. Alone, these pieces of data probably mean little to you. Information is data that has been organized or presented in a meaningful fashion. When your computer provides you with a contact listing that indicates that Zoe Richardson can be reached at (713) 555-3297, the data becomes useful—that is, it becomes information.

How do computers interact with data and information? Computers are excellent at processing (manipulating, calculating, or organizing) data into information. When you first arrived on campus, you probably were directed to a place where you could get an ID card. You most likely provided a clerk with personal data that was entered into a computer. The clerk then took your picture with a digital camera (collecting more data). All of the data was then processed so that it could be printed on your ID card (see Figure 2.1). This organized output of data on your ID card is useful information.

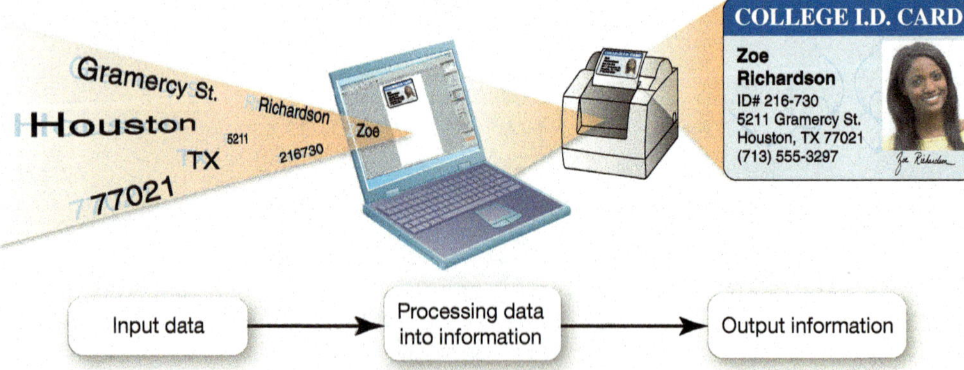

Figure 2.1 Computers process data into information. *(Mocker/123RF)*

Binary: The Language of Computers

Objective 2.2 *Define bits and bytes, and describe how they are measured, used, and processed.*

Binary Numbers Interactive

This Sound Byte helps remove the mystery surrounding binary numbers. You'll learn about base conversion among decimal, binary, and hexadecimal numbers using colors, sounds, and images.

How do computers process data into information? Unlike humans, computers work exclusively with numbers (not words). To process data into information, computers need to work in a language they understand. This language, called binary language, consists of just two digits: 0 and 1. Everything a computer does, such as processing data, printing a file, or editing a photo, is broken down into a series of 0s and 1s. Each 0 and 1 is a binary digit, or bit for short. Eight binary digits (or bits) combine to create one byte. In computers, each letter of the alphabet, each number, and each special character (such as @) consists of a unique combination of eight bits, or a string of eight 0s and 1s. So, for example, in binary language, the letter *K* is represented as 01001011. This is eight bits or one byte.

How does a computer keep track of bits and bytes? Computers understand only two states of existence: on and off. Inside a computer, these two states are defined using the numbers 1 and 0. Electrical switches are the devices inside the computer that are flipped between the two states of 1 and 0, signifying on and off. In fact, a computer system can be viewed as an enormous collection of on/off switches. These on/off switches are combined in different ways to perform addition and subtraction and to move data around the system.

You use various forms of switches every day. A water faucet is either on, allowing water to flow, or off. As shown in Figure 2.2, shutting off the faucet could represent the value 0, whereas turning it on could represent the value 1.

Figure 2.2 Water faucets can be used to illustrate binary switches.

What types of switches does a computer use? Early computers used transistors. Transistors are electrical switches built out of layers of a special type of material called a semiconductor. A semiconductor is any material that can be controlled either to conduct electricity or to act as an insulator (to prohibit electricity from passing through). Silicon, which is found in sand, is the semiconductor material used to make transistors.

By itself, silicon doesn't conduct electricity particularly well, but if specific chemicals are added in a controlled way to the silicon, it behaves like an on/off switch. The silicon allows electric current to flow when a certain voltage is applied; otherwise, it prevents electric current from flowing.

Advances in technology began to require more transistors than circuit boards could handle. Something was needed to pack more transistor capacity into a smaller space. Thus, integrated circuits were developed.

What are integrated circuits? Integrated circuits (or chips) are tiny regions of semiconductor material that support a huge number of transistors (see Figure 2.3). Most integrated circuits are no more than a quarter inch in size yet can hold billions of transistors. This advancement has enabled computer designers to create small yet powerful microprocessors, which are the chips that contain a central processing unit (CPU, or processor). The CPU can be considered the brains of the computer, since this is where the processing of data into information takes place. Today, more than 10 billion transistors can be manufactured in a space as tiny as your little fingernail!

Figure 2.3 An integrated circuit is packaged in a small case but holds billions of transistors. *(Volodymyr Krasyuk/Shutterstock)*

What else can bits and bytes be used for? Bits and bytes are not only used as the language that tells the computer what to do. They are also used to represent the quantity of data and information that the computer inputs and outputs. Word files, digital pictures, and software are represented inside computing devices as a series of bits and bytes. These files and applications can be quite large, containing billions of bytes.

To make it easier to measure the size of such files, we need units of measure larger than a byte. Kilobytes, megabytes, and gigabytes are therefore simply larger numbers of bytes. As shown in Figure 2.4, a kilobyte (KB) is approximately 1,000 bytes, a megabyte (MB) is about one million bytes, and a gigabyte (GB) is around one billion bytes. Today, personal computers can store terabytes (TB) (around one trillion bytes) of data, and many business computers can store up to a petabyte (PB) (1,000 terabytes) of data. The Google search engine processes more than one PB of user-generated data per hour!

Figure 2.4 How Much Is a Byte?

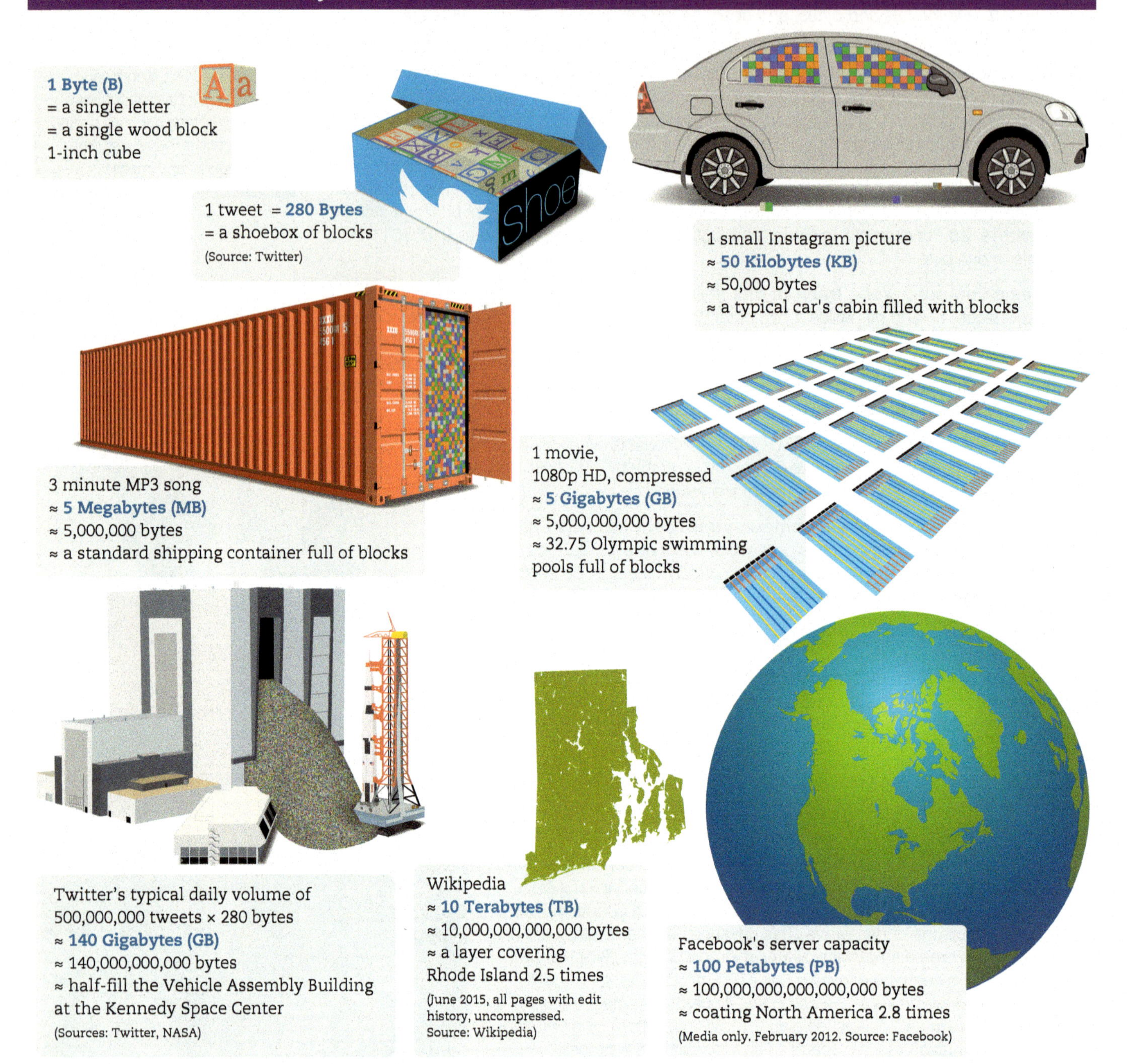

How does a computer process bits and bytes? A computer uses hardware and software to process data into information so you can complete tasks such as writing a letter or playing a game. Hardware is any part of the computer you can physically touch. Software is the set of computer programs that enables the hardware to perform different tasks, such as processing data into information. The *operating system* (*OS*)—the program that controls how your computer functions—is a common type of software. Most likely, your computer's operating system is a version of Microsoft Windows or the Apple macOS.

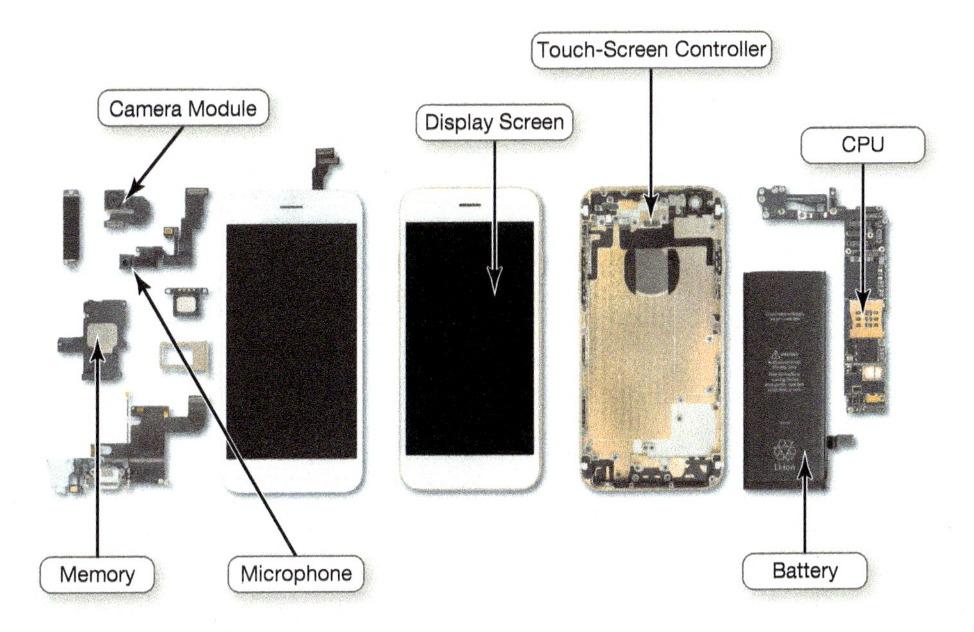

Figure 2.5 Inside your cell phone, you'll find a CPU, a memory chip, input devices such as a microphone and a touch screen, and output devices such as a display screen. *(iFixit)*

Types of Computers

Objective 2.3 *List common types of computers, and discuss their main features.*

What types of computers are popular for personal use? There are two basic designs of computers: portable and stationary. For portable computers, three main categories exist: cell phones, tablets, and laptops.

Cell Phones

Are cell phones really computers? All cell phones have the same components as any computer: a processor (central processing unit, or CPU), memory, and input and output devices, as shown in Figure 2.5. Cell phones also require their own software. So, in effect, all cell phones are indeed computers.

What makes a cell phone a smartphone? Smartphones use a CPU and an interface so powerful that they can take on many of the same tasks as much more expensive computers: videoconferencing, recording and editing high-definition (HD) video, and broadcasting live-streaming video. Most providers, like AT&T or Verizon, label a smartphone as one that has sufficient power for you to use the Internet features easily. You often must purchase a data plan with a smartphone. Smartphones like the iPhone and tablets like the iPad illustrate the true power of digital convergence—they incorporate a range of features that used to be available only in separate, dedicated devices (see Figure 2.6):

- Internet access
- Personal information management (PIM) features
- Voice recording features
- The ability to play and organize music files
- GPS services
- Digital image and video capture
- Computing power to run programs like word processors and video-editing software
- Control of other devices in your "smart" home

However, the drawback for phones as a single go-to convergence device has always been their screen size. Smartphones, though powerful, have less than optimal screen size for long periods of work. Foldable phones (see Figure 2.7), like the Samsung Galaxy Fold and the Huawei Mate X, unfold to provide you with 7- to 8-inch screens rivaling small tablets. Yet when folded, these devices still fit in your pocket. Perhaps folding phones will bring us closer to the ultimate goal of one device for all our computing needs.

Active Helpdesk

Understanding Bits and Bytes

In this Active Helpdesk, you'll play the role of a helpdesk staffer, fielding questions about the difference between data and information, what bits and bytes are, and how bytes are measured.

Music player

Digital camera

GPS

Remote control

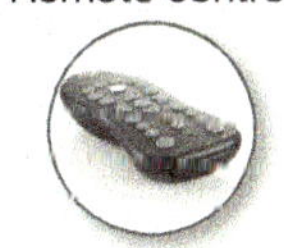

Figure 2.6 A single device like a tablet can play the role of many separate devices, illustrating the concept of digital convergence. *(Rostislavsedlacek/123RF; Science Photo Library/Shutterstock; OnBlast/Shutterstock; Andrey_Popov/Shutterstock; Oleksandr Makarenko/123RF)*

Figure 2.7 Foldable phones provide you with a larger screen but keep the phone small enough to put in your pocket. *(Wit Olszewski/ Shutterstock)*

Tablets

A tablet computer, such as the Apple iPad or Amazon Fire, is a portable computer integrated into a multi-touch sensitive screen. Tablet computers use an on-screen keyboard, but you can also connect external keyboards to tablets.

How do tablets compare with smartphones? Tablets are very light, portable devices, although not as light as smartphones. The top-selling tablets include the Apple iPad and the Samsung Galaxy, but there are more than 100 tablets on the market. One difference between tablets and smartphones is that you can't make cellular phone calls from most tablets. Traditional smartphones had screens of less than 5 inches, but many phones now have screen sizes over 6 inches. New folding phones further increase screen size to over 7 inches but still less than 10-inch tablets.

In most other regards, smartphones and tablets are similar. They have the following features:

- **Similar operating systems:** Common operating systems, such as iOS, Android, or Windows, operate on smartphones and tablets.
- **Similar processors:** The processing power of smartphones and tablets is often the same.
- **Touch-screen interfaces:** Smartphones and tablets are both equipped with touch-screen interfaces.
- **Long battery life:** Most tablets and smartphones run at least 10 hours on a single charge.
- **Similar software applications:** Most apps available for one device are available in a compatible version for the other type of device.
- **Similar Internet connectivity:** Both can offer cellular and wireless access to the Internet.
- **Bluetooth:** Most smartphones and tablets on the market today are Bluetooth-enabled, meaning they include a small Bluetooth chip that allows them to transfer data wirelessly to any other Bluetooth-enabled device. Bluetooth is a wireless transmission standard that enables you to connect devices such as smartphones, tablets, and laptops to devices such as keyboards and headsets. Bluetooth technology uses radio waves to transmit data signals over distances up to approximately 1,000 m.

Can tablets function as a communication device? Although tablets are currently unable to make cell phone calls, they can easily place audio or video phone calls if connected to a Wi-Fi network. An application like Skype that makes Internet phone calls is required for that. Apps are also available that allow tablets to handle texting. WhatsApp, for example, supports free national and international texting from a range of devices, including tablets.

Laptops and Their Variants

A laptop (or notebook) computer is a portable computer that has a keyboard, monitor, and other devices integrated into a single compact case. This was the first type of portable computer, but as our demand for lighter and more portable computers accelerated, tablets and smartphones were developed. There are several variations of laptops, including 2-in-1 PCs, ultrabooks, and Chromebooks.

Figure 2.8 Some 2-in-1 PCs have a hinge that allows them to fold the keyboard back so the device resembles a tablet. *(Andrey_Popov/ Shutterstock)*

What is a 2-in-1 PC? A 2-in-1 PC is a laptop computer that can convert into a tablet-like device (see Figure 2.8). In laptop mode, there is a physical keyboard, whereas in tablet mode you use an on-screen keyboard. On some models, such as Microsoft Surface devices, the touch screen is detachable from the keyboard so it can be carried and used as an independent tablet. On others, such as the Lenovo Yoga, there is a hinge so that the keyboard can be folded behind the screen.

Why would I want a 2-in-1 instead of a tablet? Although more expensive than a tablet, 2-in-1s have a more powerful CPU so they can function as a laptop. Also, whereas a tablet runs a mobile OS, like Android or iOS, a 2-in-1 uses a full traditional OS, like Windows, and can therefore run the same software applications as a laptop. In addition, a 2-in-1 has a physical keyboard when you need it. Many people find a 2-in-1 can replace the need to carry both a tablet and a laptop.

How are ultrabooks different from laptops? Ultrabooks are a category of full-featured computers that focus on offering a very thin, lightweight computing solution. Examples include the Apple MacBook Air and the Dell XPS 13. Ultrabooks don't offer optical drives, so they facilitate a very thin profile. Some do not even offer traditional USB ports, instead using the more compact USB-C connector so that they are only about 13 mm at the thickest point. Most ultrabooks offer SSD drives and so have very fast response times on startup and restoring from a low power state. They weigh in at fewer than three pounds even though they feature the same operating systems and CPUs as heavier, larger laptops. They also include full-size keyboards and 13- to 15-inch screens.

What is a Chromebook? A Chromebook is a special type of laptop that uses the Google Chrome OS and is designed to be connected to the Internet always. Documents and apps are stored primarily in the cloud as opposed to locally. Because the Chrome OS places less demand on computing hardware than a conventional operating system does (such as Windows or macOS), it can run the Chrome browser much more efficiently. If you do most of your work within a browser and need a lightweight, inexpensive computer, a Chromebook might be what you are seeking. However, remember that to use a Chromebook most effectively, you should have an active Internet connection.

Choosing a Portable Device

With all these choices, how do I know which device is best for me? Use these guidelines to determine which device best fits your personal needs:

- **Power:** How much computational power do you need?
- **Screen size and resolution:** These cannot be changed later, so make sure the quality and size of screen will fit your needs for the years you'll keep the device.
- **Style of keyboard:** Do you want a touch-based interface? Is a physical keyboard important, or is an on-screen keyboard sufficient? Does the feel of the keyboard work for you?
- **Battery life:** Some devices can operate for 15 hours continuously, others fewer than five. Investigate whether the battery can be upgraded and how much weight that would add.
- **Weight:** Does an additional two pounds matter? (Lighter devices usually cost more.) Remember to include the weight of any charging device you would need to carry when you travel as you consider the trade-off in price for a lighter device.
- **Number of devices:** Is this your only computing device? As technology prices fall, you may be able to have more than one device. You might find an affordable solution that includes both a very mobile device and a second, more powerful one.

Figure 2.9 summarizes the various mobile device categories.

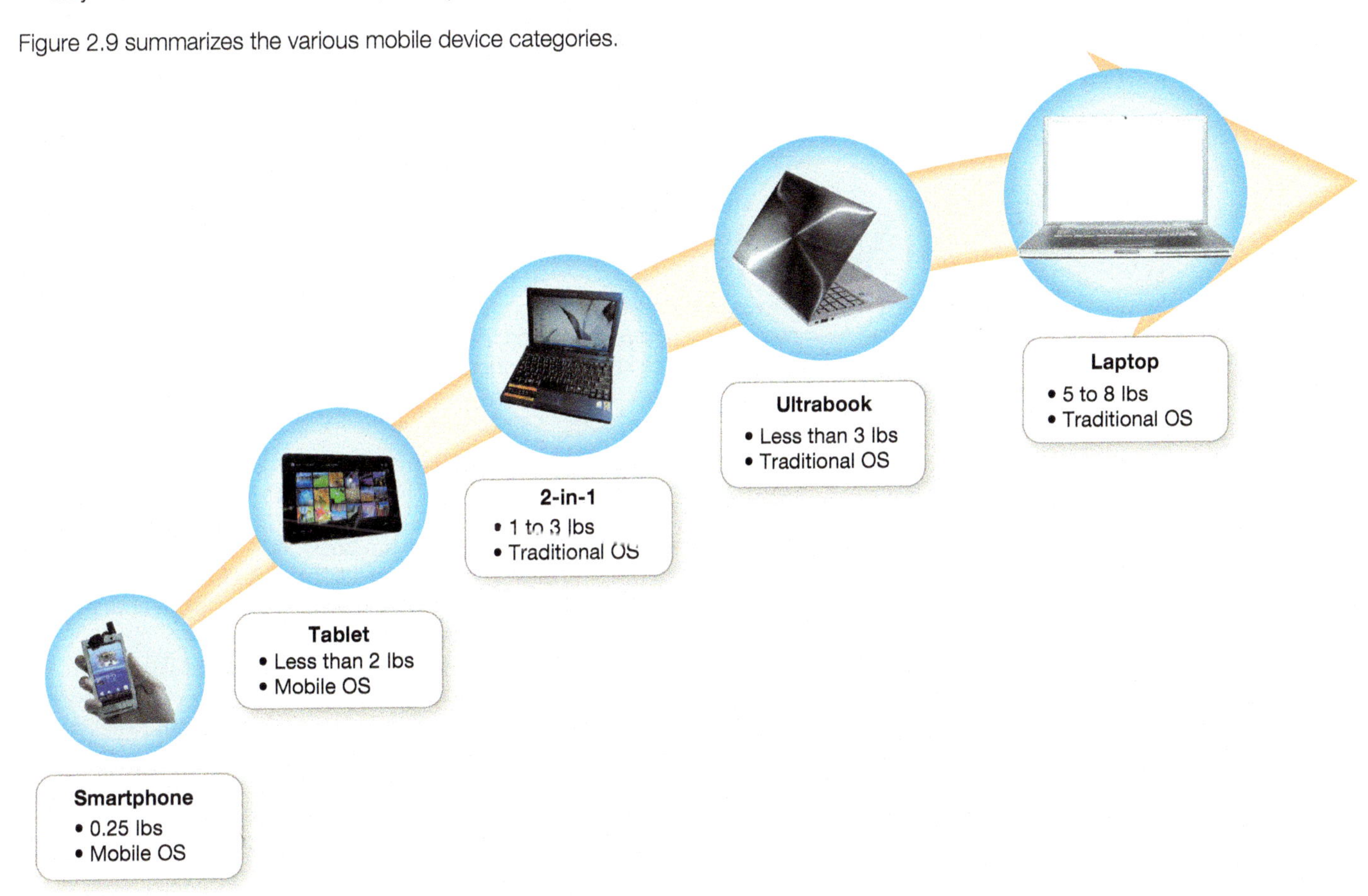

Figure 2.9 A full spectrum of mobile devices is available. *(Sean Gallup/Getty Images News/Getty Images; Josep Lago/Stringer/AFP/Getty Images; Ethan Miller/Getty Images News/Getty Images; Chris Tzou/Bloomberg/Getty Images; Peter Dazeley/The Image Bank/Getty Images)*

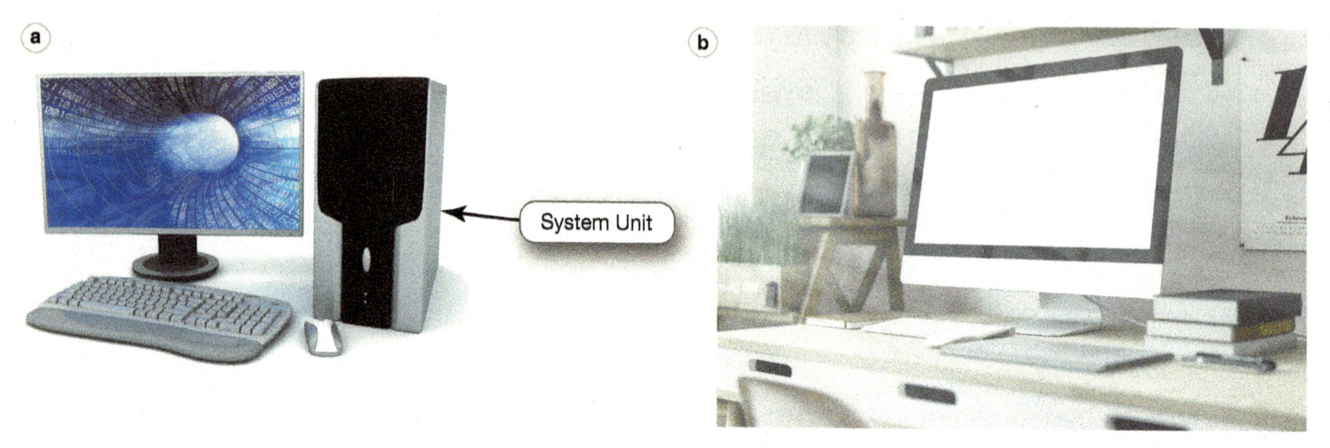

Figure 2.10 (a) A desktop computer has a system unit that holds the main components. (b) An all-in-one computer does not need a separate system unit. *(Maxim Kazmin/123RF; Best pixels/Shutterstock)*

Stationary Computers

What are the main stationary computers available to consumers? A desktop computer is intended for use at a single location, so it's stationary (see Figure 2.10a). Most desktop computers consist of a separate case or tower (called the system unit) that houses the main components of the computer and to which peripheral devices are attached. A peripheral device is a component, such as a monitor or keyboard that connects to the computer.

An all-in-one computer, such as the Apple iMac, eliminates the need for a separate tower because these computers house the computer's processor and memory in the monitor (see Figure 2.10b). Many all-in-one models also incorporate touch-screen technology.

What are the advantages of a desktop computer? Desktop computers are easier to upgrade than laptops or other portable computers (which often cannot be upgraded at all). Upgrades to components such as video cards are often important for gamers and graphic artists. Also, large screens are beneficial in many working environments (such as graphic design), and if a computer doesn't need portability, you can use a larger monitor without a problem. Finally, miniaturization of components is usually costly. Therefore, you can often get more computing power for your dollar when you buy a desktop computer.

Are there any other types of computers? Although you may never come into direct contact with the following types of computers, they are still very important and do a lot of work behind the scenes of daily life:

- A mainframe is a large, expensive computer that supports many users simultaneously. Mainframes are often used in businesses that manage large amounts of data, such as insurance companies, where many people are working at the same time on similar operations, such as claims processing. Mainframes excel at executing many computer programs at the same time.
- A supercomputer is a specially designed computer that can perform complex calculations extremely rapidly. Supercomputers are used when complex models requiring intensive mathematical calculations are needed (such as weather forecasting). Supercomputers are designed to execute a few programs as quickly as possible, whereas mainframes are designed to handle many programs running at the same time but at a slower pace.
- An embedded computer is a specially designed computer chip that resides in another device, such as your car, a drone, or the electronic thermostat in your home. Embedded computers are self-contained computer devices that have their own programming and that typically don't receive input from you or interact with other systems.

Each part of your computer has a specific purpose that coordinates with one of the functions of the computer—input, processing, output, or storage (see Figure 2.11).

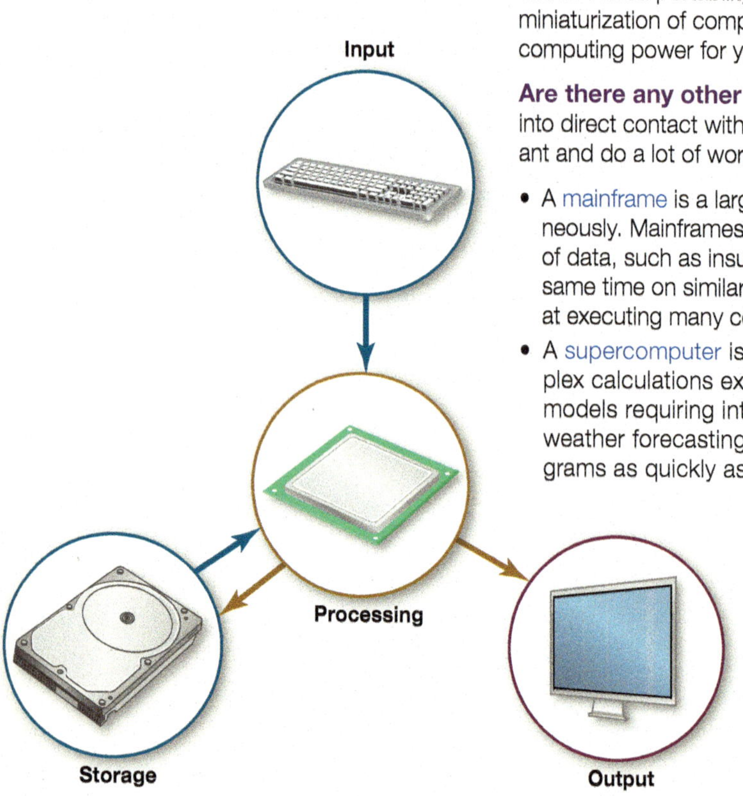

Figure 2.11 Each part of the computer serves a special function.

Bits&Bytes Today's Supercomputers: Faster Than Ever

Supercomputers are the biggest and most powerful type of computer. Scientists and engineers use these computers to solve complex problems or to perform massive computations. Some supercomputers are single computers with multiple processors, whereas others consist of multiple computers that work together. The IBM Summit (see Figure 2.12) is the fastest supercomputer deployed at the US Department of Energy. It features a whopping 4,356 nodes, each with two 22-core CPUs. Summit runs at a blazing 148.6 petaflops. How long do you think it will hold the top spot?

Figure 2.12 Deployed at the Oak Ridge National Laboratory in Tennessee, the IBM Summit is the world's fastest supercomputer—for now! Check *TOP500.org* for the latest rankings. *(https://www.ornl.gov/news/ornl-launches-summit-supercomputer)*

Additional devices, such as Wi-Fi adapters and routers, help a computer communicate with the Internet and other computers to facilitate the sharing of documents and other resources. Let's begin our exploration of hardware by looking at your computer's input devices.

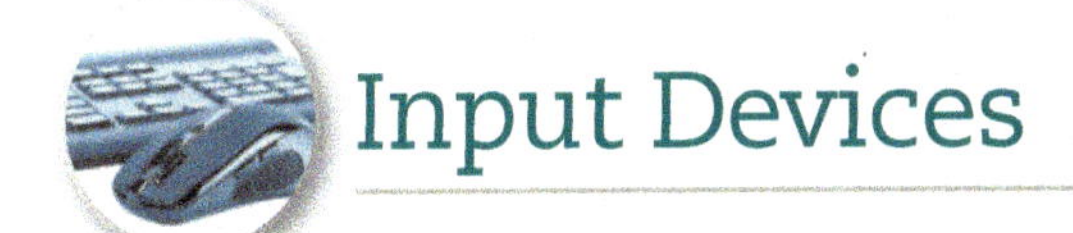

Input Devices

An input device enables you to enter data (text, images, and sounds) and instructions (user responses and commands) into your computer. Let's look at some of the most popular input devices used today.

Physical Keyboards and Touch Screens

Objective 2.4 *Identify the main types of keyboards and touch screens.*

What is the most common way to input data and commands? A keyboard is an input device you use to enter typed data and commands. However, as discussed earlier, computing devices such as smartphones, tablets, and many laptops now respond to touch. Touch screens are display screens that respond to commands initiated by touching them with your finger or a stylus—a device that looks like a pen and that you use to tap commands or draw on a screen. Touch-screen devices use a virtual keyboard that displays on screen when text input is required (see Figure 2.13). These keyboards show basic keyboard configurations but allow you to switch to other special keys. Virtual keyboards can also support dozens of languages and different character sets.

Figure 2.13 Touch-screen devices use virtual keyboards that appear on screen when needed. *(Lourens Smak/Alamy Stock Photo)*

Are all keyboards the same? Whether virtual or physical, the most common keyboard layout is a standard QWERTY keyboard. This keyboard layout gets its name from the first six letters in the top-left row of alphabetic keys and is the standard English-language keyboard layout. However, various languages in use throughout the world make use of different character sets and accent marks. Fortunately, operating systems allow for changing the keyboard layout on your device to the appropriate keyboard layout for the language you are using (see Figure 2.14).

How can I use my keyboard most efficiently? All keyboards have the standard set of alphabetic and numeric keys that you regularly use when typing. As shown in

Figure 2.14 Learning Korean? No problem! You can customize the layout of your keyboard using the Windows operating system. *(Courtesy of Microsoft Corporation)*

Figure 2.15, many keyboards for laptop and desktop computers have additional keys that perform special functions. Knowing how to use the special keys shown in Figure 2.15 will help you improve your efficiency.

What alternatives are there to a virtual keyboard? Virtual keyboards are not always convenient when a great deal of typing is required. Most computing devices can accept physical keyboards as an add-on accessory. Wired keyboards plug into a data port on the computing device. Wireless keyboards send data to the computer by using a form of wireless technology such as Bluetooth. Often, Bluetooth keyboards for tablets are integrated with a case to protect your tablet (see Figure 2.16a).

Figure 2.15 Keyboards have a variety of keys that help you work more efficiently. (Note that on Macs, function keys are slightly different: the Control function is the Apple key or Command key, and the Alt function is the Option key.) *(Artur Synenko/123RF)*

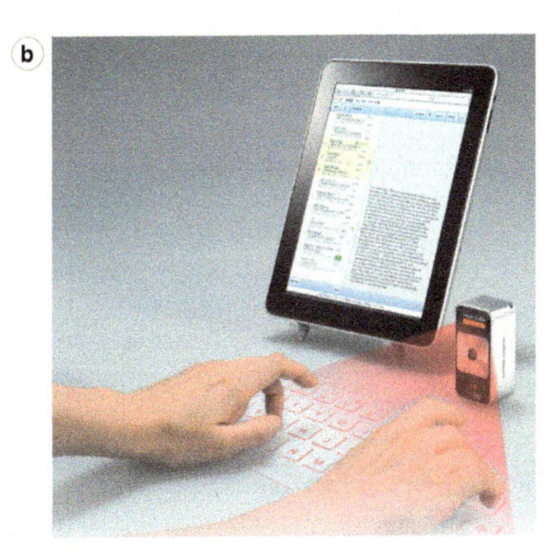

Figure 2.16 (a) Cases with integrated physical keyboards make tablets more typing-friendly. (b) Laser-projection keyboard devices project the image of a QWERTY keyboard on any surface. Sensors detect typing motions, and data is transmitted to your device via Bluetooth.
(Guteksk7/Shutterstock; Splash News/Hammacher Schlemmer/Newscom)

Flexible keyboards are a terrific alternative if you want a full-sized keyboard for your laptop or tablet. You can roll one up, fit it in your backpack, and plug it into a USB port when you need it. Another compact keyboard alternative is a laser-projection keyboard (see Figure 2.16b), which is about the size of a matchbox. They project an image of a keyboard onto any flat surface, and sensors detect the motion of your fingers as you type. Data is transmitted to the device via Bluetooth. These keyboards work with the latest smartphones and tablets.

Mice and Other Pointing Devices

Objective 2.5 *Describe the main types of mice and pointing devices.*

What input is a mouse responsible for? A mouse is the most common pointing device used to enter user responses and commands. One of the mouse's most important functions is to position the cursor. The cursor is an on-screen icon (often shown by a vertical bar or an arrow) that helps the user keep track of exactly what is active on the display screen. For instance, in word processing software, the cursor shows you exactly where the next character will be typed in a sentence based on its position. To move the cursor to another point on a touch screen, you just tap the screen with your finger. Or, to select an item on the screen, just tap it with your finger. However, laptops and desktop computers that are not touch-enabled need other types of devices (like mice) for positioning the cursor and selecting icons (by clicking the mouse).

What kinds of mice are there? The most common mouse type is the optical mouse, which uses an internal sensor or laser to detect the mouse's movement. The sensor sends signals to the computer, telling it where to move the pointer on the screen. Optical mice don't require a mouse pad, although you can use one to enhance the movement of the mouse on an uneven surface or to protect your work surface from being scratched.

Most mice have two or three buttons (mice for Macs sometimes have only one button) that let you execute commands and open shortcut menus. Many customizable mice have additional programmable buttons and wheels so you can quickly maneuver through web pages or games (see Figure 2.17). These mice are also customizable to fit any size hand and grip style by allowing for length and width adjustments. Aside from gamers, many people use customizable mice to reduce susceptibility to repetitive strain injuries or if they suffer from physical limitations that prevent them from using standard mice.

How do wireless mice connect to a computing device? Wireless mice usually connect the same way that wireless keyboards do—either through an integrated Bluetooth chip or a Bluetooth receiver that plugs into a USB port.

Figure 2.17 The R.A.T. Pro^X Precision Gaming Mouse offers the ultimate in customizability for serious gamers or users with special needs.
(Mad Catz, Inc.)

Why would I want to use a mouse with a touch-screen device? If you're using a physical keyboard with your touch-screen device, it's often easier to perform actions with a mouse rather than reaching to touch the screen. In addition, there are new kinds of mice, called *touch mice*, which are designed with touch-screen computers in mind. Unlike traditional mice, there are no

Bits&Bytes Foldable Glass Is Making Foldable Phones Better!

Foldable phones are now on the market and may become the ultimate digital convergence device, but they have one major drawback: they have plastic screens. Glass screens are the gold standard for delivering quality images, but you can't fold glass, can you?

Actually, folding glass has been around for a while. Manufacturers of fiber optic cable (primarily made from glass fibers) needed to make it more flexible without having the glass strands break when bent. So glass was developed that was more bendable (foldable). Some fiber optic cables can be bent 90 degrees without breaking. This glass, though, requires a certain thickness to be foldable, which is too thick for phone screens that must be thin to save weight. Fortunately, Corning and other glass manufacturers are working on developing new forms of foldable glass (see Figure 2.18) that can be made thin enough for phone screens. Once this is perfected, future foldable phones should have displays that are as sharp and crisp as today's smartphone screens.

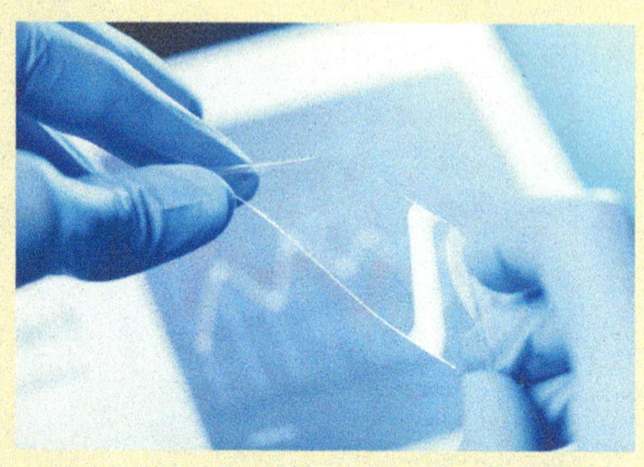

Figure 2.18 Development of thin, foldable glass is the key to enhancing screens on foldable phones. *(Nevodka/Shutterstock)*

specifically defined buttons. The top surface of a touch mouse is the button. You tap one, two, or three fingers to perform touch-screen tasks such as scrolling, switching through open apps, and zooming. Touch mice also enable you to perform traditional mouse tasks, such as moving the cursor when you move the mouse.

What pointing devices do laptops use? Most laptops have an integrated pointing device, such as a touch pad (or trackpad)—a small, touch-sensitive area at the base of the keyboard. Mac laptops include multi-touch trackpads, which don't have buttons but are controlled by various one-, two-, three-, and four-finger actions. For example, scrolling is controlled by brushing two fingers along the trackpad in any direction. Most touch pads are sensitive to taps, interpreting them as mouse clicks. Some laptops also have buttons under or near the pad to record mouse clicks.

What input devices are used with games? Game controllers such as joysticks, game pads, and steering wheels are also considered input devices because they send data to computing devices. Game controllers, which are like the devices used on gaming consoles such as Xbox One, are also available for use with computers. They have buttons and miniature pointing devices that provide input to the computer.

Image, Sound, and Sensor Input

Objective 2.6 *Explain how images, sounds, and sensor data are input into computing devices.*

What are popular input devices for images? Digital cameras, camcorders, and mobile-device cameras capture pictures and video and are considered input devices. Stand-alone devices can connect to your computer with a cable, transmit data wirelessly, or transfer data automatically through the Internet.

Flatbed scanners also input images. They work similarly to a photocopy machine; however, instead of generating the image on paper, they create a digital image, which you can then print, save, or e-mail.

The digital cameras in cell phones are also often used to scan information. For example, many health and fitness apps, like MyFitnessPal, capture data by using the cell phone's camera to scan UPC codes on food products. The devices used at supermarkets and retail stores to scan the UPC codes on your purchases are also a type of scanner that inputs information into another computer: the point-of-sale terminal (cash register).

A webcam is a camera that attaches to a desktop computer or is built into a computing device. Although webcams can capture still images, they're used mostly for capturing and transmitting live video. Videoconferencing apps on a device equipped with a webcam and a microphone transmit video and audio across the Internet. Video apps such as Zoom and Google Hangouts make it easy to videoconference with multiple people (see Figure 2.19).

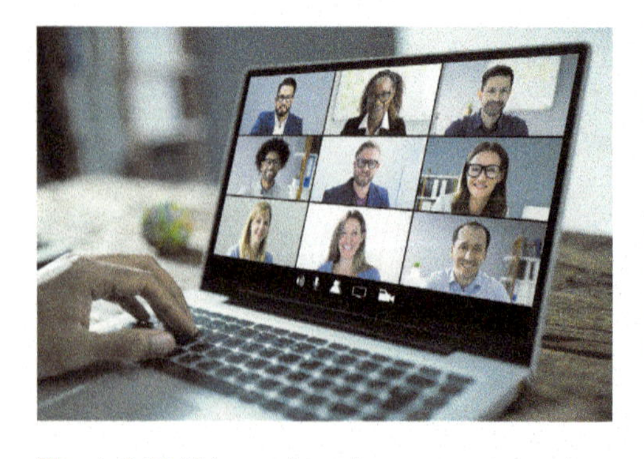

Figure 2.19 Videoconferencing apps saw massive increases in usage for working and learning from home during the coronavirus pandemic. *(Andrey_Popov/Shutterstock)*

How do my computing devices benefit from accepting sound input? Inputting sound to your computer requires using a microphone (or mic)—a device that captures sound waves (such as your voice) and transfers them to digital format on your computer. Laptops, tablets, and smartphones come with built-in microphones. In addition to enabling others to hear you in a videoconference, equipping your device to accept sound input enables you to conduct audio conferences, chat with people over the Internet, record podcasts, and even control computing devices with your voice.

How can I use my voice to control my computing device? Voice recognition software allows you to control your devices by speaking into a microphone instead of by using a keyboard or mouse. Apps like Dragon Naturally Speaking are available as stand-alone apps, but voice recognition features are built into the Windows and macOS operating systems as well.

Popular extensions of voice recognition software are intelligent personal assistants such as Apple's Siri, Google Assistant, Amazon's Alexa, and Microsoft's Cortana (see Figure 2.20). These so-called software agents respond to voice commands and then use your input, access to the Internet, and location-aware services to perform various tasks, such as finding the closest pizza parlor to your present location.

Figure 2.20 Just tap the microphone icon and ask Microsoft's intelligent personal assistant Cortana a question. She communicates using natural language processing techniques. *(Courtesy of Microsoft Corporation)*

What types of add-on microphones are available? For situations like videoconferencing, built-in microphones don't always provide the best performance. You may want to consider adding other types of microphones, such as those shown in Table 2.1, to your system for the best results.

Table 2.1 Types of Microphones

Microphone Type	Attributes	Best Used For
Close Talk	• Attached to a headset (allows for listening) • Leaves hands free	• Videoconferencing • Phone calls • Speech recognition software
Omnidirectional  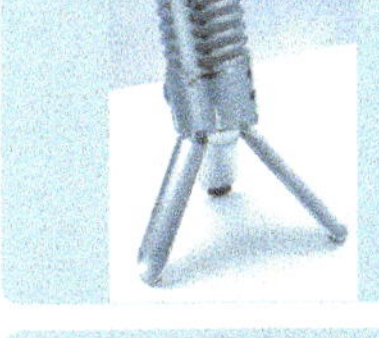	• Picks up sounds equally well from all directions	• Conference calls in meeting rooms
Unidirectional 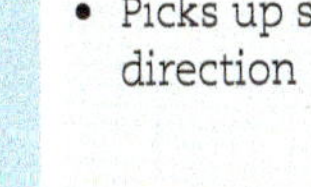 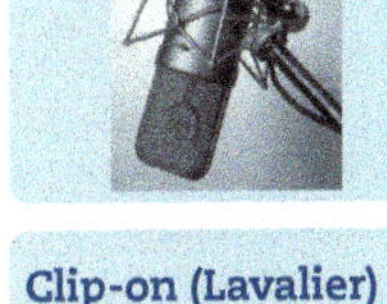	• Picks up sounds from only one direction	• Recordings with one voice (podcasts) • Videoconferencing (when high quality is required)
Clip-on (Lavalier)	• Clips to clothing • Available as wireless	• Presentations requiring freedom of movement • Leaves hands free for writing on whiteboards

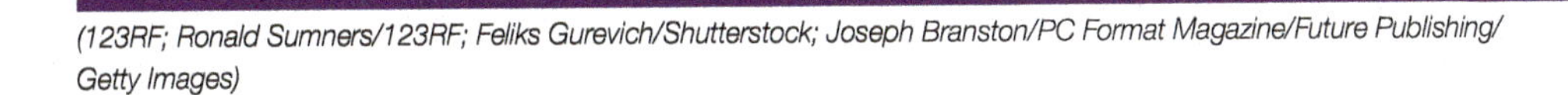

(123RF; Ronald Sumners/123RF; Feliks Gurevich/Shutterstock; Joseph Branston/PC Format Magazine/Future Publishing/Getty Images)

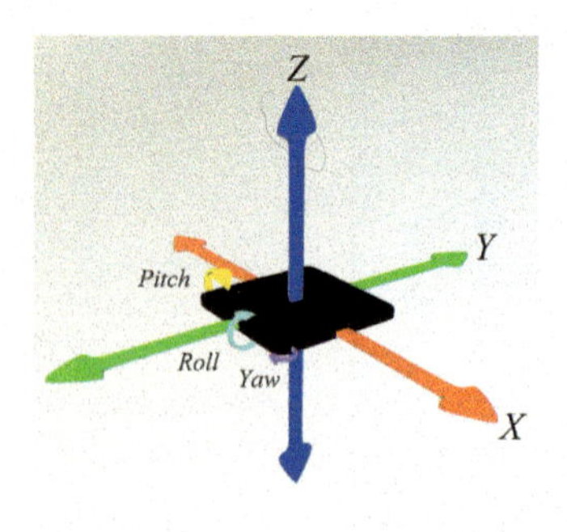

Figure 2.21 The accelerometer is a sensor in your devices that reports the acceleration in any direction. *(Chih Hsien Hang/123RF)*

What other types of input devices are found in computers? Smartphones and tablets are now arrayed with a variety of sensors. Broadly speaking, a sensor is any device that detects or measures something. Sensors feed data into the device and are used by the hardware or apps that rely on the input provided. The following sensors can be found in almost all smartphones and many tablets today (although your particular device may not contain every one):

- **Magnetometer:** detects magnetic fields. Compass and metal-detecting apps use this sensor.
- **Proximity sensor:** used in phones to determine when the phone is placed next to your ear so the phone knows to shut off the display while you are talking.
- **Light sensor:** measures ambient light brightness so the device can adjust the brightness of the display.
- **Accelerometer:** measures acceleration that the device is experiencing. It also helps determine the orientation of the device along three axes (see Figure 2.21), which allows the device to determine whether it is in landscape or portrait orientation.
- **Gyroscope:** provides orientation information but with greater precision than the accelerometer.
- **Barometer:** measures atmospheric pressure to determine how high the device is above sea level. This improves GPS accuracy and enables counting floors more accurately when step climbing.
- **Thermometer:** measures ambient temperature either inside or outside the device. This is mainly used to monitor overheating of delicate electronic components inside the device.
- **Pedometer:** used to record the number of steps taken while having the device in your possession.
- **Fingerprint sensor:** used to record and read fingerprint data to secure the device.
- **Heart rate monitor:** measures pulse rate by detecting pulsating blood vessels inside your finger.

Want to access the sensors in your smartphone? Try our Make This activity in the middle of the chapter to learn how to build your own apps that can gather information from the sensors in your smartphone.

Output Devices

An output device lets you send processed data out of your computer in the form of text, pictures (graphics), sounds (audio), or video. Let's look at some popular output devices.

Image and Audio Output

Objective 2.7 *Describe options for outputting images and audio from computing devices.*

Computing devices need options for outputting data so you can transfer it to other locations. Sometimes you need only soft copies of data, such as a graph displayed on your screen. Other times you need a hard copy, such as a printed page. Let's explore the most common output device: the display screen.

Image Output

What are the different types of display screens? A display screen (sometimes referred to as a monitor on desktops and laptops) displays text, graphics, and videos. The most common type of monitor for laptop and desktop computers is a liquid crystal display (LCD). An LCD monitor, also called a *flat-panel monitor*, is light and energy efficient. Some newer monitors use light-emitting diode (LED) technology, which is more energy efficient and may have better color accuracy and thinner panels than LCD monitors.

Organic light-emitting diode (OLED) displays use organic compounds that produce light when exposed to an electric current. Unlike LCDs and LEDs, OLEDs do not require a backlight to function and therefore draw less power and have a much thinner display, sometimes as thin as 3 mm. They are also brighter and more environmentally friendly than LCDs. Because of their lower power needs, OLED displays run longer on a single battery charge than do LEDs, which is why OLED technology is probably the technology used in your smartphone, tablet, and digital camera.

Companies like LG are now working on transparent and flexible OLED display screens (see Figure 2.22). These screens allow you to see what is behind the screen while still displaying information on the screen. These types of screens present interesting possibilities for augmentative reality. As described in Chapter 1, *augmented reality* (*AR*) is a view of a real-world environment whose elements are augmented (or supplemented) by some type of computer-generated sensory input such as video, graphics, or GPS data. For instance, if you had a transparent screen on your smartphone and held it up to view street signs that were in English, you could possibly have your phone display the signs in another language. Currently, applications like this exist (such as Pokémon Go) but require the use of a camera as well as your screen. But transparent screens will eliminate the need for the camera.

Figure 2.22 Because they don't need a backlight, OLED displays can be made transparent and flexible. *(Yonhap News/YNA/Newscom)*

How do display screens work? Display screens are grids made up of millions of tiny dots, called pixels. When these pixels are illuminated by the light waves generated by a fluorescent panel at the back of your screen, they create the images you see on the screen or monitor. Each pixel on 4K- and 8K-resolution TVs and monitors is made up of four yellow, red, blue, and green subpixels. Some newer TVs further split the subpixels into upper and lower, which can brighten and darken independently. LCD monitors are made of two or more sheets of material filled with a liquid crystal solution (see Figure 2.23). A fluorescent panel at the back of the LCD monitor generates light waves. When electric current passes through the liquid crystal solution, the crystals move around and either block the fluorescent light or let the light shine through. This blocking or passing of light by the crystals causes images to form on the screen. The various combinations of yellow, red, blue, and green make up the components of color we see on our monitors.

What factors affect the quality of a display screen? Some portable devices don't provide a choice of screens. The current iPhone has whatever screen Apple chooses to use. But when choosing a laptop or a monitor for a desktop, you have options to compare.

The most important factors to consider are aspect ratio and resolution. The aspect ratio is the width-to-height proportion of a monitor. Traditionally, aspect ratios have been 4:3, but newer monitors are available with an aspect ratio of 16:9 to accommodate HD video. The screen resolution, or the clearness or sharpness of the image, reflects the number of pixels on the screen. An LCD monitor may have a native (or maximum) resolution of 5120 × 2880, meaning it contains 5120 vertical columns with 2880 pixels in each column. The higher the resolution, the sharper and clearer the image, but generally, the resolution of an LCD monitor is dictated by the screen size and aspect ratio. Although you can change the resolution of an LCD monitor beyond its native resolution, the images may become distorted. Generally, you should buy a monitor with the highest resolution available for the screen size (measured in inches).

Is a bigger screen size always better? The bigger the screen, the more you can display, and depending on what you want to display, size may matter. In general, the larger the screen, the larger the number of pixels it can display. For example, a 27-inch monitor can display 5120 × 2880 pixels, whereas a 21.5-inch monitor may only be able to display 4096 × 2304 pixels. However, 4K monitors have at least the 3840 × 2160 resolution, which is sufficient for most users.

Larger screens can also display multiple documents or web pages at the same time. However, buying two smaller monitors might be cheaper than buying one large monitor. For either option—a big screen or two separate screens—check that your computer has the video hardware and/or software needed to support the display devices.

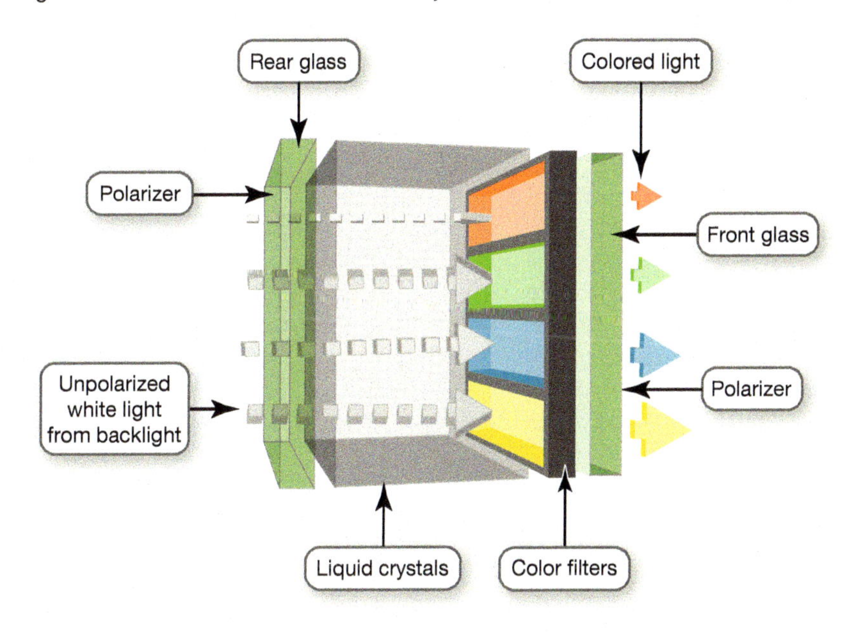

Figure 2.23 A Magnification of a Single Pixel in an LCD Monitor

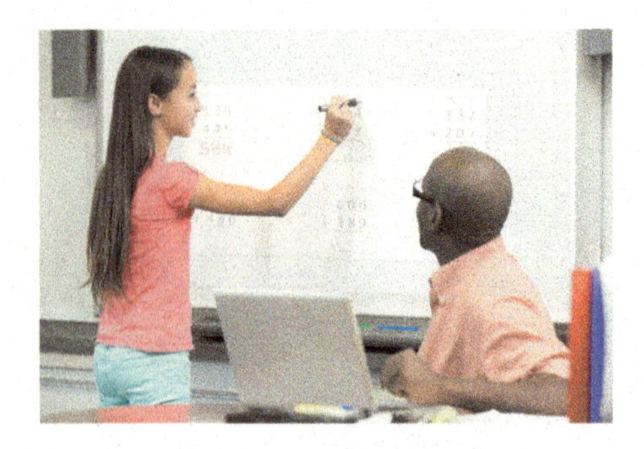

Figure 2.24 Interactive whiteboards are now common output/input devices found in classrooms and conference rooms. *(Shutterstock)*

What device provides the highest resolution? 8K monitors and TVs are now on the market and feature a whopping 7680 × 4320 resolution. The most cost-effective high-resolution devices are probably still Ultra HD (otherwise known as 4K) monitors and TVs. Ultra HD has a resolution of 3840 × 2160, which contains four times the pixels of the original HD devices. However, to take advantage of high-resolution devices, your media provider needs to stream your content at these higher resolutions, so make sure to check with your provider before running out and buying the newest high-resolution device the next time you need one.

How do I show output to a large group of people? You can always connect your computing device to an HDTV if you have one large enough to be seen by the entire group. Another option is a projector, which projects images from your device onto a wall or screen. Small and lightweight portable projectors are ideal for businesspeople who make presentations at client locations. Entertainment projectors include stereo speakers and multimedia connectors, making them a good option for use in the home to display TV programs, movies, or video games.

In classrooms and conference rooms, projectors are often combined with another output device, the interactive whiteboard (see Figure 2.24). A projector projects the computer's display onto the interactive whiteboard's surface. The touch-sensitive board doubles as an input device and enables users to control and provide input to the computer by using a pen, finger, or stylus. Notes and annotations can be captured and saved from the whiteboard to the attached computer.

Audio Output

What are the output devices for sound? All portable computing devices include integrated speakers, the output devices for sound. These speakers are sufficient for playing audio clips you find on the web and usually for making videoconference or phone calls over the Internet. However, if you plan to edit audio files digitally or are particular about how your music sounds, you may want a more sophisticated speaker system, such as one that includes subwoofers (special speakers that produce only low bass sounds) and surround-sound speakers. A surround-sound system is a set of speakers and audio processing equipment that envelops the listener in a 360-degree field of sound. Wireless speaker systems are widely available, enabling you to connect portable devices to quality speakers easily.

Headphones or earbuds connect wirelessly or plug into the same jack on your computing device to which external speakers connect. Studies of users of portable devices have shown that hearing might be damaged by excessive volume, especially when using earbuds, because they fit into the ear canals, so limit the volume when you use these devices.

Bits&Bytes Teraflops in Your Living Room: Gaming on Overdrive

When the Xbox Series X and the PlayStation 5 hit the market, both manufacturers bragged about the same main feature: blistering speed. The new Xbox clocks in at 12 teraflops and the PlayStation 5 delivers 10.28 teraflops. But what exactly are teraflops and how does the high speed impact gaming performance?

FLOPS is an acronym that stands for floating point operations per second. Floating-point arithmetic is the most efficient way to process numbers in computing devices and is much superior to its alternative, fixed-point arithmetic. Floating-point arithmetic offers much faster processing and more efficient code, which results in the superior graphics display necessary for today's complex games. The graphics processing unit (GPU) is the chip in computing devices that processes graphics data. GPU power is measured in FLOPS.

The original Xbox was capable of producing 20 gigaflops (20 billion FLOPS). Four years later, the PlayStation 3 clocked in at 230.4 gigaflops. Fast forward a decade and a half and the GPUs in the latest gaming consoles are delivering teraflop (trillions of FLOPS) speeds, which provide a vastly superior gaming experience (see Figure 2.25).

Figure 2.25 The latest games take advantage of GPUs that run at teraflop speeds. *(Gorodenkoff/Shutterstock)*

Printers

Objective 2.8 *Describe various types of printers, and explain when you would use them.*

What are the different types of printers? Another common output device is the printer, which creates hard copies (copies you can touch) of text and graphics. There are two primary categories of printers common in the home and office: inkjet and laser.

What are the advantages of inkjet printers? Inkjet printers (see Figure 2.26) are popular because they're affordable and produce high-quality printouts quickly and quietly. Inkjet printers work by spraying tiny drops of ink onto paper and are great for printing black-and-white text as well as color images. In fact, when loaded with the right paper, higher-end inkjet printers can print images that look like professional-quality photos. One thing to consider when buying an inkjet printer is the type and cost of the ink cartridges the printer needs. Some printers use two cartridges: black and color. Others use four or more cartridges, typically cyan, magenta, yellow, and black.

Figure 2.26 Inkjet printers are popular among home users, especially since high-end inkjet printers can print high-quality photographic images. *(Aleksey Mnogosmyslov/Shutterstock)*

Why would I want a laser printer? A laser printer uses laser beams and static electricity to deliver toner (similar to ink) onto the correct areas of the page (see Figure 2.27). Heat is used to fuse the toner to the page, making the image permanent. Laser printers are often used because

Figure 2.27 How Laser Printers Work

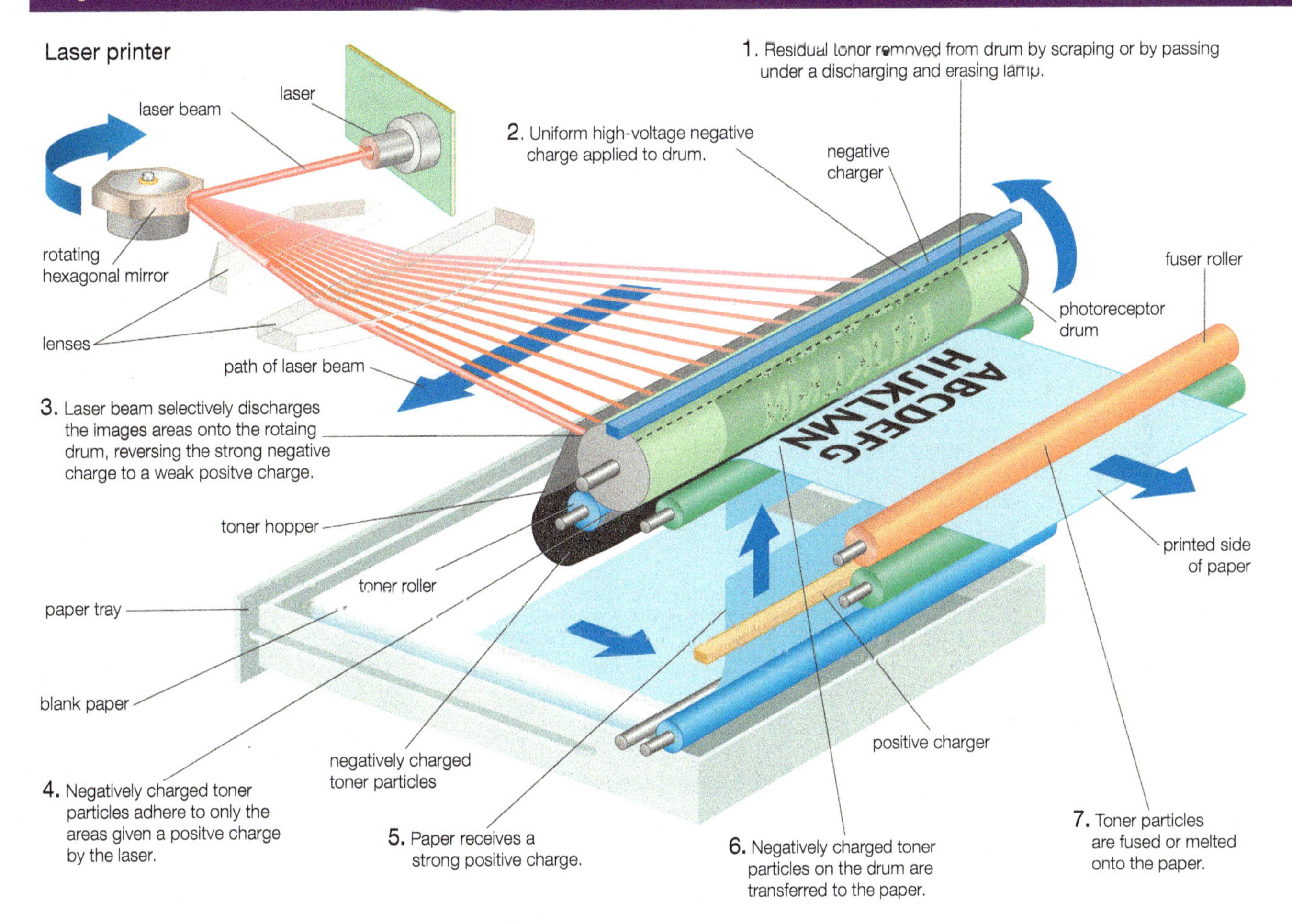

(Encyclopaedia Britannica/Universal Images Group North America LLC/Alamy Stock Photo)

they print faster than inkjet printers and produce higher-quality printouts. Black-and-white and color laser printers are both affordable for home use. If you print a high volume of pages, consider a laser printer. When you include the price of ink or toner in the overall cost, color laser printers can be more economical than inkjets.

What's the best way to print from portable devices such as tablets and smartphones? Wireless printers are great for printing from portable devices. Wireless printers are also a good option for home networks, because several people can print to the same printer from different devices and any location in the home. There are two types of wireless printers: Wi-Fi and Bluetooth. Both Wi-Fi and Bluetooth printers have a range of up to approximately 300 feet. Wi-Fi, however, sends data more quickly than Bluetooth does.

If you're using a device running the Apple iOS (such as an iPhone), AirPrint makes printing easy. AirPrint is a feature of iOS that facilitates printing to AirPrint-compatible wireless printers, and many printers produced today are AirPrint-compatible. For non-Apple devices, newer wireless printers come with their own app that enables printing from portable devices. After installing the app on your device, you can send print jobs to the wireless printer.

Are there any other types of specialty printers? Although you'll probably use laser or inkjet printers most often, you might encounter other types of printers (shown in Figure 2.28):

- An all-in-one printer combines the functions of a printer, scanner, copier, and fax into one machine. Popular for their space-saving convenience, all-in-one printers may use either inkjet or laser technology.
- A large-format printer generates oversize images such as banners, posters, and infographics that require more sophisticated color detail. Some of these printers use up to 12 different inks to achieve high-quality color images.
- A 3D printer prints three-dimensional objects. For example, parts can be built on the spot when needed, using a 3D printer. 3D printers build such objects one layer at a time from the bottom up. They begin by spreading a layer of powder on a platform. Then, depending on the technology, the printer uses nozzles to spray tiny drops of glue at specific places to solidify the powder, or the powder is solidified through a melting process. The printer repeats this process until the object is built. This technology is now used for manufacturing a variety of consumer goods, from toys to clothing.

How do I choose the best printer? If you're planning to print color photos and graphics, an inkjet printer or a color laser printer is a must, even though the cost per page will be higher. If you'll be printing mostly text-based documents or will be sharing your printer with others, a laser printer is best because of its speed and overall economy for volume printing. It's also important to determine whether you want just a printer or a device that prints and scans, copies, or faxes. In addition, you should decide whether you need to print from mobile devices.

Once you've narrowed down the type of printer you want, you can use the criteria listed in Table 2.2 to help you determine the best model to meet your needs.

Figure 2.28 Specialty printers: (a) all-in-one printer, (b) large-format printer, (c) 3D printer. *(Interior Design/Shutterstock; Koktaro/Shutterstock; Stockbroker/MBI/Alamy Stock Photo)*

Bits&Bytes CPUs That Fight Back

The University of Michigan has developed a new computer architecture called Morpheus. Unlike conventional CPUs, this one can defend itself against hackers (see Figure 2.29). The CPU randomly shuffles and encrypts parts of its own code 20 times per second, so even if a hacker can find a flaw in the code to exploit, the weakness vanishes milliseconds later. Given the volume of malicious attacks, this could be a key breakthrough in computer defense.

Figure 2.29 Researchers are working on CPUs that fight hacking without human intervention. *(Texelart/Shutterstock)*

Table 2.2 Major Printer Attributes

Attribute	Considerations
Speed 	• Print speed is measured in *pages per minute (PPM)*. • Black and-white documents print faster than color documents. • Laser printers often print faster than inkjets.

Resolution 	• Resolution refers to a printer's image clarity.
	• Resolution is measured in *dots per inch (dpi)*.
	• Higher dpi = greater level of detail and clarity.
	• Recommended dpi:
	  • Black-and-white text: 300
	  • General-purpose images: 1200
	  • Photos: 4800

Color Output  	• Printers with separate cartridges for each color produce the best quality output. • Inkjet and laser color printers generally have four cartridges (black, cyan, magenta, and yellow). • Higher-quality printers have six cartridges (the four above plus light cyan and light magenta). • With separate cartridges, you only need to replace the empty one.
Cost of Consumables 	• Consumables are printer cartridges and paper. • Printer cartridges can exceed the cost of some printers. • Consumer magazines such as *Consumer Reports* can help you research costs.

(Lightwise/123RF; Michaelnivelet/123RF; Thomas Amby Johansen/123RF; Tomislav Forgo/shutterstock)

Before moving on to Part 2:

1. Watch Chapter Overview Video 2.1.
2. Then take the Check Your Understanding quiz.

Check Your Understanding // Review & Practice

For a quick review to see what you've learned so far, answer the following questions.

multiple choice

1. Which of the following is NOT one of the four major functions of a computer?
   a. Input
   b. Processing
   c. Collating
   d. Output
2. Which of the following statements is TRUE?
   a. Smartphones do not contain memory.
   b. Smartphones generally have less computing power than desktop computers.
   c. Smartphones do not contain a CPU.
   d. Smartphones are not considered a type of computer.
3. Which of the following can be both an input device and an output device?
   a. Display screen
   b. Keyboard
   c. Mouse
   d. Laser printer
4. The number of pixels that are contained on a monitor is known as what?
   a. Viewing density
   b. Viewing plane
   c. Aspect ratio
   d. Screen resolution
5. Which type of printer works by using heat to fuse toner to the paper?
   a. Inkjet
   b. Laser
   c. 3D
   d. Resolution

(3Dalia/Shutterstock)

Embedded computers keep turning up in new places. They can be found in cars, household appliances, smoke detectors, and thermostats, too. Embedded computers enable us to interact with even more of these "smart" devices every day. What common objects do you think might benefit from an embedded computer? What capabilities can you envision?

Go to **MyLab IT** to take an autograded version of the *Check Your Understanding* review and to find all media resources for the chapter.

For the IT Simulation for this chapter, see MyLab IT.

Try This

What's Inside My Computer?

Understanding the capabilities your current computer has is one of the first steps toward computer literacy. In this exercise, you'll learn how to explore the components of your Windows computer. For step-by-step instructions, watch the Chapter 2 Try This video on MyLab IT.

Step 1 To gather information about the storage devices on your computer, click **File Explorer** from the Taskbar. Then in the navigation pane, click **This PC** to display information about your computer's storage devices and network locations.

(Courtesy of Microsoft Corporation)

Step 2 The **This PC** window displays information about internal storage devices (such as internal hard drives), and portable storage devices (such as flash drives and external hard drives). To display the System screen, click the **Computer tab** on the ribbon, and then click **System properties**.

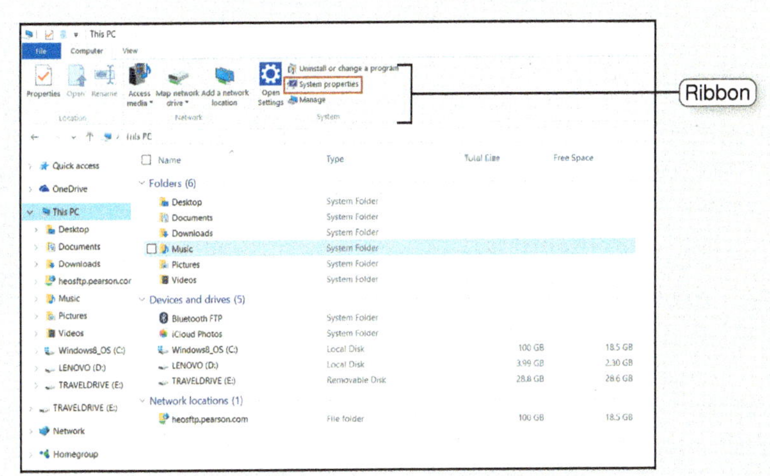

(Courtesy of Microsoft Corporation)

Step 3 Scroll down to gather quite a bit of information from the About screen, such as:

- Version of Windows
- Type of processor
- Speed of the processor
- Amount of RAM installed
- System type (32-bit or 64-bit)

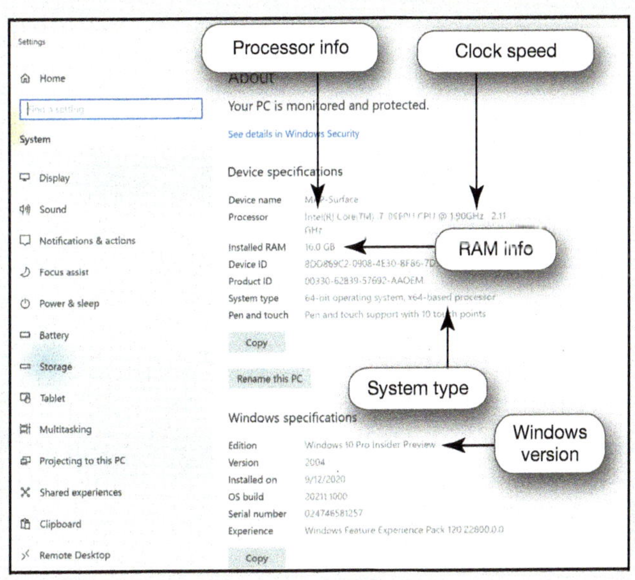

(Courtesy of Microsoft Corporation)

Make This  TOOL: App Inventor 2 or Thunkable

A Mobile App

Want to build your own Android app from scratch? You can, with a simple tool called **App Inventor**. To get started, have ready:

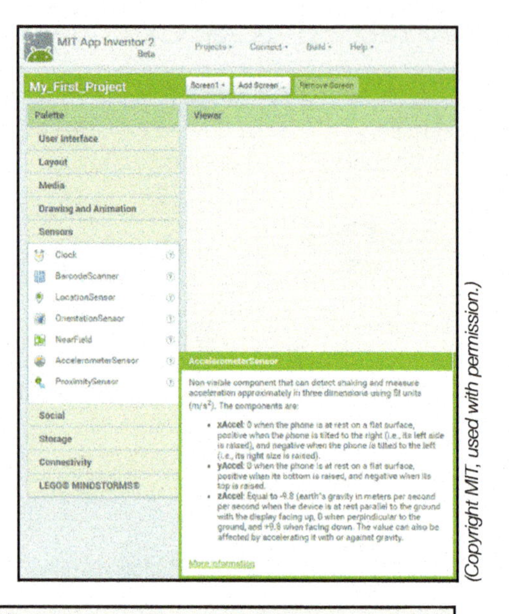

(Copyright MIT, used with permission.)

- A computer connected to a Wi-Fi network
- The Chrome browser
- A Google account
- The MIT AI2 Companion app (available in the Google Play Store)
- [optional] An Android device connected to the same Wi-Fi network

In this exercise, you'll explore the **App Inventor** tool and begin working with your first simple app. As you'll see, making your device work for you is as easy as drag and drop with **App Inventor**.

App Inventor is a programming platform used to create apps for Android devices. Using **App Inventor**, you can easily drag and drop components to design your app's interface and its behavior.

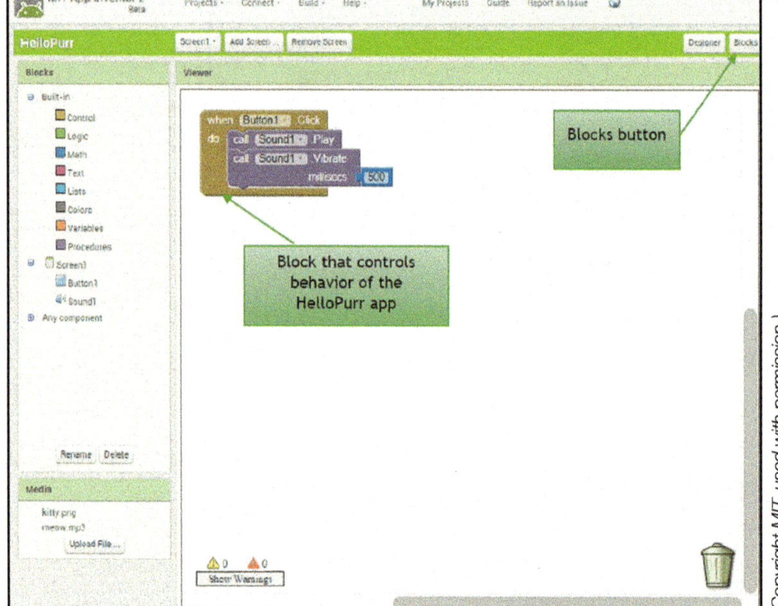

(Copyright MIT, used with permission.)

To create iOS apps, go to *Thunkable.com*, a programming platform based on App Inventor.

For the detailed instructions for this exercise, go to MyLab IT.

Part 2

For an overview of this part of the chapter, watch **Chapter Overview Video 2.2**.

Processing, Storage, and Connectivity

Learning Outcome 2.2 **You will be able to describe how computers process and store data and how devices connect to a computer system.**

So far, we have explored the components of your computer that you use to input and output data. But where does the processing take place, and where is the data stored? And how does your computer connect with peripherals and other computers?

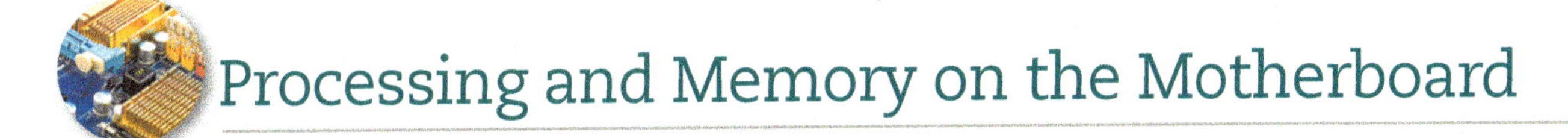

Processing and Memory on the Motherboard

The main processing functions of your computer take place in the CPU and memory, both of which reside on your computer's motherboard. In the following sections, we'll explore the components of the motherboard and how memory helps your computer process data.

The Motherboard and Memory

Objective 2.9 *Describe the functions of the motherboard and RAM.*

What exactly is a motherboard? The motherboard is the main circuit board that contains the central electronic components of the computing device, including the computer's processor (CPU), its memory, and the many circuit boards that help the computer function. On a desktop, the motherboard (see Figure 2.30) is located inside the system unit, the metal or plastic case that also houses the power source and all the storage devices (such as the hard drive). In a laptop, tablet, or phone, the system unit is combined with the monitor and the keyboard in a single package. Portable-device motherboards are smaller, flatter, and lack expansion slots (see the image of the cell phone motherboard in Figure 2.5).

What exactly is RAM? Random access memory (RAM) is the place in a computer where the programs and data that the computer is currently using are stored. RAM is much faster to read

Figure 2.30 A motherboard for a desktop computer contains the socket for the computer's processor (CPU), slots for memory (RAM) modules, ports, and slots for expansion cards. *(GIGA-BYTE Technology Co., Ltd)*

Smartphones Are Really Smart

In this Sound Byte, you'll learn how to use a smartphone as a powerful tool to communicate, calculate, and organize your workload.

from and write to than the hard drive and other forms of storage. The processor can request the RAM's contents, which can be located, opened, and delivered to the CPU for processing in a few nanoseconds (billionths of a second). If you look at a desktop motherboard, you'll see RAM as a series of small cards (called memory cards or memory modules) plugged into slots on the motherboard.

Because the entire contents of RAM are erased when you turn off the device, RAM is a temporary or volatile storage location. To save data permanently, you need to save it to your hard drive or to another permanent storage location, such as a flash drive or cloud storage.

Does the motherboard contain any other kinds of memory besides RAM? In addition to RAM, the motherboard contains a form of memory called read-only memory (ROM). ROM holds all the instructions the device needs to start up when it's powered on. Unlike data stored in RAM, which is volatile storage, the instructions stored in ROM are permanent, making ROM a nonvolatile storage location, which means the data isn't erased when the power is turned off.

What other functionality is contained on the motherboard? The motherboard on desktop computers also includes slots for expansion cards (or adapter cards), which are circuit boards that provide additional functionality. Typical expansion cards found in the system unit are sound and video cards. A sound card provides a connection for the speakers and microphone, whereas a video card provides a connection for the monitor. High-end gaming desktops use expansion cards to provide video and sound capabilities, making it easy for owners to upgrade to the latest video cards on the market. Laptops, tablets, and phones have video and sound capabilities integrated into their motherboards.

Other parts of the motherboard provide a means for connectivity to networks and peripheral devices. Wireless network interface modules provide connectivity to Wi-Fi networks. Connectivity ports such as USB and Thunderbolt ports (discussed later in this chapter) are also integrated into the motherboard.

On cell phone motherboards you'll also find a GPS (global positioning system) receiver, which is used to receive radio waves from satellites that help you pinpoint your location. The GPS receivers were initially installed in cell phones to allow the 911 emergency call operators to pinpoint the location of cell phones for people needing assistance. However, now many location-dependent apps, like Google Maps and Pokémon Go, also use the GPS information.

Processing

Objective 2.10 *Explain the main functions of the CPU.*

What is the CPU? As noted earlier, the *central processing unit* (*CPU* or *processor*) is sometimes referred to as the brains of the computer because it controls all the functions performed by the computer's other components and processes all the commands issued to it by software instructions.

How is processor speed measured? Processor speed is measured in units of hertz (Hz). Current systems run at speeds measured in gigahertz (GHz) or billions of operations per second. Therefore, a 3.8 GHz processor performs work at a rate of 3.8 billion operations per second. It's important to realize, however, that CPU processor speed alone doesn't determine the performance of the CPU.

What else determines processor performance? Although speed is an important consideration when determining processor performance, CPU performance is also affected by other factors. One factor is the number of cores, or processing paths, a processor has. Modern processors are designed so that they can have 2, 4, or even 10 paths, allowing them to process more than one instruction at a time (see Figure 2.31). Applications such as virus protection software and the operating system, which are always running behind the scenes, can have their own processor paths, freeing up the other paths to run other applications such as a web browser, Word, or iTunes more efficiently.

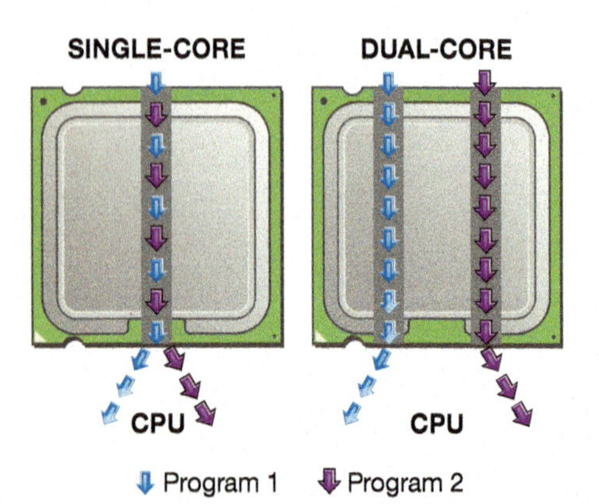

Figure 2.31 With multi-core processors, CPUs can work in parallel, processing two or more programs at the same time instead of switching back and forth between them.

How can I tell which processor is best for me? The best processor will depend on your particular needs and is not always the processor with the highest processor speed and the greatest number of cores. The website cpubenchmark.net provides a wealth of information that assists in comparing the performance of different models of CPUs.

Storing Data and Information

Because RAM is volatile storage, it can't be used to store information indefinitely. To save your data and information permanently, you need to save it to a nonvolatile storage device, such as a hard drive, cloud storage location, or flash drive.

Storage Options on Computing Devices

Objective 2.11 *Describe the various means of storing data and information with computing devices.*

Local Storage Devices

Are there different kinds of hard drives? The hard drive is your desktop's or laptop's primary device for permanent storage of software and documents. The hard drive is a nonvolatile storage device.

An internal hard drive resides within the desktop or laptop system unit and usually holds all permanently stored programs and data. Today's internal hard drives (see Figure 2.32) have capacities of as much as 30 TB or more. External hard drives offer similar storage capacities but reside outside the system unit and connect to the computer via a port.

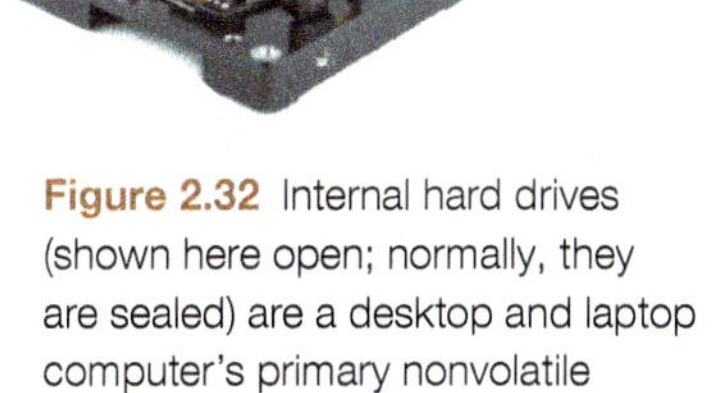

Figure 2.32 Internal hard drives (shown here open; normally, they are sealed) are a desktop and laptop computer's primary nonvolatile storage. *(Andrey Eremin/123RF)*

The most common type of hard drive—the hard disk drive—has moveable parts—spinning platters and a moving arm with a read/write head—that can fail and lead to disk failure. However, the solid-state drive (SSD) is now a popular option for ultrabooks and laptop storage. SSDs have no moving parts, so they're more efficient, run with no noise, emit little heat, and require little power. In addition, they're less likely to fail after being bumped or dropped. Replacing a conventional hard drive in an older laptop with an SSD can often significantly improve the computer's performance.

Permanent storage devices are located in your desktop or laptop computer in a space called a drive bay. On desktop computers, sometimes there are empty drive bays that can be used to install additional drives. Laptop computers generally do not give you the ability to add additional drives. Such expansion is done by attaching an external drive to the computer through a port.

Figure 2.33 Smaller, portable external hard drives enable you to take a significant amount of data and programs on the road with you. *(Inga Nielsen/Shutterstock)*

Portable Storage Options

How can I take my files with me? For large portable storage needs, there are portable external hard drives that are small enough to fit into your pocket and that have storage capacities of 16 TB (or more). These devices are lightweight and enclosed in a protective case. They attach to your computer by a USB port (see Figure 2.33).

A flash drive (sometimes referred to as a jump drive, USB drive, or thumb drive) uses solid-state flash memory, storing information on an internal memory chip. When you plug a flash drive into your computer's USB port, it appears in the operating system as another disk drive. You can write data to it or read data from it as you would a hard drive. Because a flash drive contains no moving parts, it's quite durable. It's also tiny enough to fit in your pocket. Despite their size, flash drives can have significant storage capacity—currently as much as 4 TB or more.

Another convenient means of portable storage is a flash memory card, such as an SD card. Like the flash drive, memory cards use solid-state flash memory. Most desktops and laptops include slots for flash memory cards. Flash memory cards enable you to transfer digital data between your computer and devices such as digital cameras, smartphones, tablets, video cameras, and printers. Table 2.3 compares the storage capacities of hard drives and flash drives.

How does my cell phone store data? Today's cell phones use various combinations of memory chips to store data. They do not contain hard drives.

Table 2.3 Hard Drive and Flash Drive Storage Capacity

Drive Type	Image	Capacity
Solid-state drive (SSD)		30 TB or more
External portable hard drive		16 TB or more
Mechanical internal hard drive		30 TB or more
Flash drive		4 TB or more
Flash memory card		1 TB or more

(Oleksiy Mark/Shutterstock; Ruslan Ivantsov/Shutterstock; D. Hurst/Alamy Stock Photo; Christophe Testi/123RF; Gene Blevins/ZUMA Press, Inc./Alamy Stock Photo)

Can I increase the storage capacity of my smartphone? Many smartphones let you add additional memory through micro SD flash cards (see Figure 2.34). Micro SD cards are easy to install inside a phone, and some models have external slots for an SD card. Not all smartphones allow memory upgrades in this way, however. For example, iPhones don't allow you to add memory.

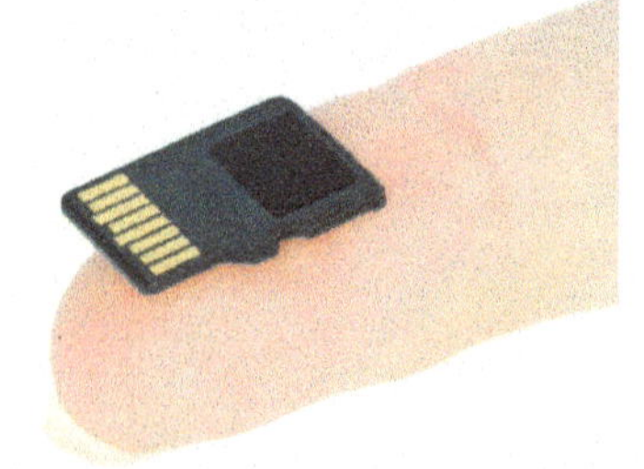

Figure 2.34 Micro SD flash cards add memory to some phones. *(Digitalr/123RF)*

Cloud Storage

How can I easily access my files from different devices? You might use multiple devices, such as a smartphone, laptop, and a tablet, at different times. Invariably, you'll need access to a file that is stored on a device other than the one you're using. If your devices are connected to the Internet, cloud storage provides a convenient option.

Cloud storage refers to using a service that keeps your files on the Internet (in the cloud) rather than storing your files solely on a device. Using a cloud storage service requires you to install software or an app on your device. Popular cloud storage options include Google Drive, Microsoft OneDrive, and Dropbox. For example, after installing the Dropbox software on your devices, any files you save in the Dropbox folder are accessible by all your other devices via the Internet. You can also share folders in Dropbox with other Dropbox users, making it ideal for group collaboration.

Most cloud storage providers give users an option of some amount of free storage, but the amount of space varies. All cloud storage providers offer additional storage space for a fee.

Ethics in IT

What Is Ethical Computing?

You've probably heard news stories about people using computers to unleash viruses or commit identity theft. You may also have read about students who were prosecuted for illegally sharing copyrighted material, such as songs and videos. These are both examples of *unethical* behavior while using a computer. However, what constitutes *ethical* behavior while using a computer?

As we discussed in Chapter 1, *ethics* is a system of moral principles, rules, and accepted standards of conduct (see Figure 2.35). So what are the accepted standards of conduct when using computers? The Computer Ethics Institute has developed the Ten Commandments of Computer Ethics, widely cited as a benchmark for companies developing computer usage and compliance policies for employees. These guidelines are applicable for schools and students as well. The following ethical computing guidelines are based on the Computer Ethics Institute's work.

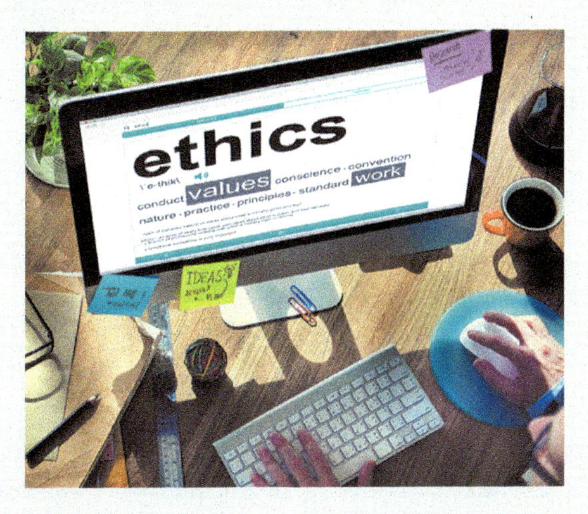

Figure 2.35 Striving to behave ethically isn't just for real-life interactions. It applies when using computing devices too. *(Rawpixel.com/Shutterstock)*

Ethical Computing Guidelines

1. Avoid causing harm to others when using computers.
2. Do not interfere with other people's efforts at accomplishing work with computers.
3. Resist the temptation to snoop in other people's computer files.
4. Do not use computers to commit theft.
5. Agree not to use computers to promote lies.
6. Do not use software (or make illegal copies for others) without paying the creator for it.
7. Avoid using other people's computer resources without appropriate authorization or proper compensation.
8. Do not claim other people's intellectual output as your own.
9. Consider the social consequences of the products of your computer labor.
10. Only use computers in ways that show consideration and respect for others.

The United States has enacted laws that support some of these guidelines, such as Guideline 6, the breaking of which would violate copyright laws, and Guideline 4, which is enforceable under numerous federal and state larceny laws. Other guidelines, however, require more subtle interpretation of what behavior is unethical because there are no laws designed to enforce them.

Consider Guideline 7, which covers unauthorized use of resources. The school you attend probably provides computer resources for you to use for coursework. But if your school gives you access to computers and the Internet, is it ethical for you to use those resources to run an online business on the weekends? Although it might not be technically illegal, you're tying up computer resources that other students could use for their intended purpose: learning and completing coursework. (This behavior also violates Guidelines 2 and 10.)

Throughout the chapters in this book, we touch on various ethical situations. You will likely encounter many topics related to these guidelines. So keep them in mind as you study, and think about how they relate to the actions you take as you use computers in your life.

Trends in IT Green Computing (Green IT)

"Going green" is a goal for many modern businesses. Green computing (or green IT) is a movement to encourage environmentally sustainable computing. The main goal is to reduce the overall carbon footprint of a company through the strategic use of computing resources and environmentally friendly computing devices. A business's *carbon footprint* is the total amount of greenhouse gases produced directly and indirectly to support the activities of the business. Carbon footprints are expressed in equivalent tons of carbon dioxide (CO_2), which is the main greenhouse gas contributing to global warming. Reduction of greenhouse gas emissions is critical to sustaining a healthy environment.

The main goals of green computing are as follows:

1. Reducing the use of hazardous processes or materials in the production of computing equipment
2. Promoting the use of recyclable or biodegradable materials to facilitate safe disposal of products
3. Buying products that use energy efficiently
4. Using technology to reduce employee travel
5. Reducing the use of energy and consumption of materials through shared computing resources

Sharing computing resources can make a vast difference in the consumption of resources and electricity. This is one of the reasons cloud storage is becoming so popular. Rather than having 20 individual companies, each maintaining a large group of computers to hold data, savings can be achieved by having one company maintain computer resources that are able to serve the 20 other companies. However, it's not only up to businesses to practice green computing. Table 2.4 lists a few ways you can participate.

Table 2.4 Green Computing Problems and Solutions

	Electricity	Commuting	Use Technology Longer

Problems	• Electricity is often generated using fossil fuels, which produce greenhouse gas emissions. • Devices are not energy efficient.	• Cars generate greenhouse gases. • Many people commute to work alone.	• Items are replaced before their useful life is over. • Old items are discarded instead of continuing to be used. • Technology is not disposed of or recycled properly.
Solutions	• Buy energy-efficient computing equipment with high Energy Star ratings. • Turn off computing devices when not in use. • Use appropriate power management settings to use less power when operating devices.	• Use technology to telecommute to your job. • Use public transportation to commute, which uses energy more efficiently than cars. • Use a green vehicle (bicycle, electric car) for your commute.	• Upgrade your technology only when absolutely necessary. • Donate your old technology to someone who will continue to use it (friends, family, charitable organization). • Dispose of electronic devices only at approved e-waste recycling facilities.

(Tomáš Koval?ík/123RF; MinDof/Fotolia; Marekuliasz/shutterstock)

Connecting Peripherals to the Computer

Throughout this chapter, we have discussed peripheral devices that input, store, and output data and information. We will now look at how these types of devices are connected to computers so they can exchange data.

Computer Ports

Objective 2.12 *Describe common types of ports used today.*

What is the fastest data transfer port available on today's computing devices? A port is a place through which a peripheral device attaches to the computer so that data can be exchanged between it and the operating system. Although peripherals may connect to devices wirelessly, ports are still often used for connections. Thunderbolt is the newest input/output technology on the market. Thunderbolt ports (see Figure 2.36) are very useful for laptops and ultrabooks because one Thunderbolt port can allow you to connect up to six peripherals to your computer. Thunderbolt 3 ports can achieve blazingly fast transfer speeds of up to 40 GBs (version 4 is expected to have the same speed).

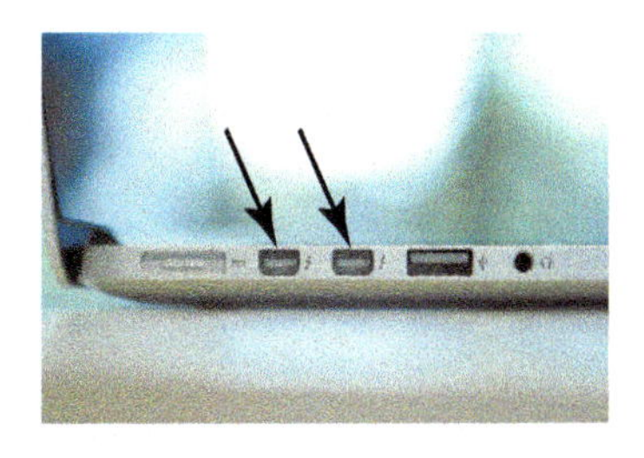

Figure 2.36 Thunderbolt ports are slim and speedy, making them popular on today's ultrabooks and laptops. *(David Paul Morris/Bloomberg/Getty Images)*

What is the most common port on digital devices? A universal serial bus (USB) port is the port type most commonly used to connect input and output devices to the computer. This is mainly because of the ready availability of USB-compatible peripherals. USB ports can connect a wide variety of peripherals to computing devices, including keyboards, printers, mice, smartphones, external hard drives, flash drives, and digital cameras. USB 4 is the newest USB standard. USB 4 uses the USB-C connector and has a transfer rate of 40 gigabits per second (Gbps) (twice the rate of USB 3.2) and backward compatibility with Thunderbolt 3. USB ports come in a variety of standard and proprietary configurations (see Figure 2.37), plus the new Type-C connector (and port), which is supplanting older connector types (see Figure 2.38).

Which ports help me connect with other computers and the Internet? A connectivity port can give you access to networks and the Internet. To find a connectivity port, look for a port that resembles a standard phone jack but is slightly larger. This port is called an Ethernet port. Ethernet ports transfer data at speeds up to 10,000 Mbps. You can use an Ethernet port to connect your computer to either a cable modem or a wired network.

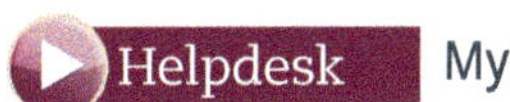

MyLab IT

Exploring Storage Devices and Ports

In this Helpdesk, you'll play the role of a helpdesk staffer, fielding questions about the computer's main storage devices and how to connect various peripheral devices to the computer.

Figure 2.37 USB connectors come in a wide variety of styles. *(TaraPatta/Shutterstock)*

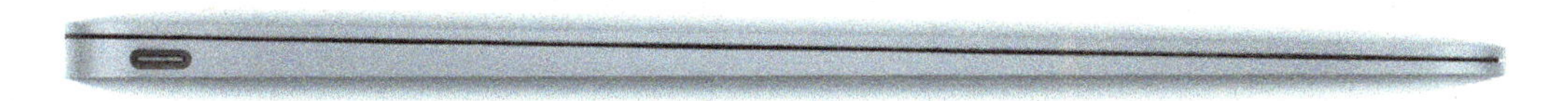

Figure 2.38 Many computers now feature the new USB-C port that supports data transfer, video output, and charging all in a single port. *(APPLE/HANDOUT/Epa european pressphoto agency b.v/Alamy Stock Photo)*

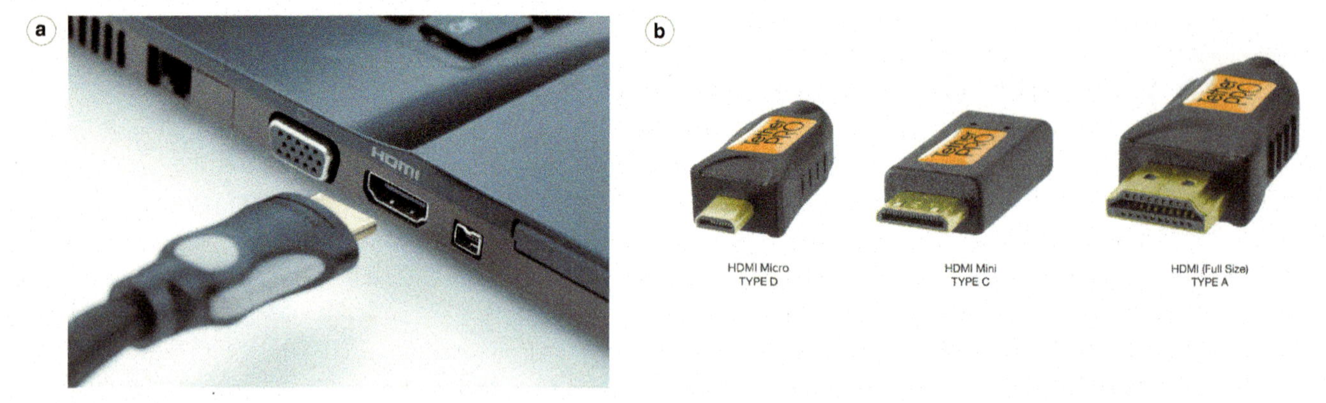

Figure 2.39 (a) HDMI is the industry standard digital connector type for HD monitors, TVs, and home theater equipment. (b) Cables with the appropriate connectors are available to facilitate connection of your mobile devices to your HDTV for enhanced viewing. *(Feng Yu/Shutterstock; Used with permission from Tether Tools.)*

How do I connect monitors and multimedia devices? Other ports on the back and sides of the computer include audio and video ports. Audio ports are where you connect headphones, microphones, and speakers to the computer. Whether you're attaching a monitor to a desktop computer or adding a second, larger display to a laptop computer, you'll use a video port.

HDMI ports are now the most common video port on computing devices. A high-definition multimedia interface (HDMI) port is a compact audio–video interface that allows both HD video and uncompressed digital audio to be carried on one cable. Because HDMI can transmit uncompressed audio and video, there's no need to convert the signal, which could ultimately reduce the quality of the sound or picture. All currently available monitors, Blu-ray players, TVs, and game consoles have at least one HDMI port (see Figure 2.39a).

If your computing device is equipped with an HDMI port, you can also choose to connect your computer directly to an HDTV, using an HDMI cable. The most common HDMI connector types are shown in Figure 2.39b. Full-size (or type A) connectors are found on most TVs and laptops. Tablets and phones are more likely to have a mini (type C) or micro (type D) HDMI port.

DisplayPort is a port that is similar to an HDMI video port. The original DisplayPort standard had capabilities comparable to HDMI but was usually only seen on high-end computing devices and TVs. However, the DisplayPort 2 standard supports up to 16K resolution video and data transfers of 77.37 Gbps. DisplayPort 2 ports are starting to appear on even lower-end devices to take advantage of this massive performance increase.

Power Management and Ergonomics

Conserving energy and setting up workspaces so they are comfortable are goals of many businesses. However, these are also goals to strive for at your home. In this section, we'll explore optimizing the power consumption of computing devices as well as the proper workspace setup to minimize injuries.

Power Controls and Power Management

Objective 2.13 *Describe how to manage power consumption on computing devices.*

Which components of a computing device drain the battery fastest? The number one culprit is the display. The backlights of LCD (and even OLED) displays draw a lot of power. You may love that new HD display in your phone, but your battery doesn't love it!

Wi-Fi and Bluetooth adapters also draw a lot of power because they're continuously scanning for compatible devices. In addition, any peripheral plugged into a port on your computer is most likely drawing power from the computer to function.

Certain apps are big culprits of battery drain also. Many apps continuously sync with the cloud and push alerts to you. So even though you may not be looking at an app (a *foreground* use), it is still drawing power (*background* use). Fortunately, you can determine which apps draw the most power:

- For Android devices, click Settings > Device > Battery or Settings > Power > Battery Use, and you'll see a list of apps and how much battery power they're consuming.
- For iOS devices, click Settings > Battery and wait for the battery usage list to populate. Click the clock icon to display information on foreground and background power use.
- For Windows, launch Settings, select System, select Battery, and then click the *See which apps are affecting your battery life* link.

Consider uninstalling apps that draw a lot of background power if you rarely use them. For apps you do use frequently, consider disabling or turning off sync and alert features in the app when your battery level is low.

How does my computer battery get its charge? The power supply comes in the form of a brick with a cord that plugs into your portable device to charge the battery. The power supply transforms the wall voltage to the voltages required by the battery or the computer chips (if you are operating your device on a/c power). A desktop computer's power supply is housed within the system unit.

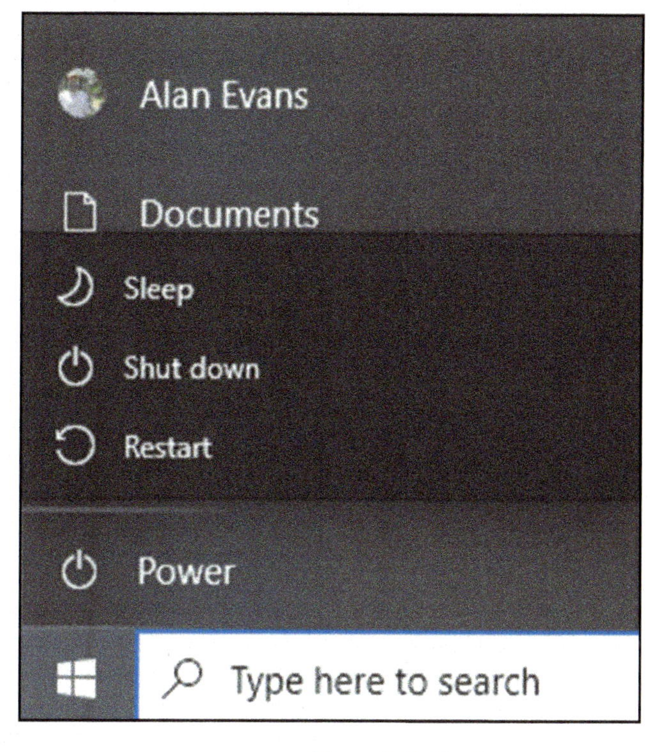

Figure 2.40 The Power icon on the Windows Start menu presents several power options. *(Courtesy of Microsoft Corporation)*

*> For a warm boot, choose **Restart**. To power down the computer completely, choose **Shut down**. To put your computer into a lower power mode, select **Sleep**.*

What is the best way to turn my computer off? In Windows and macOS, you can turn your computer completely off by displaying the power options and choosing the Shut down option (see Figure 2.40). If your computer's operating system is unresponsive, holding the power button down will force a shutdown to occur.

Can I rest my computer without turning it off completely? With many devices, an effective method of power conservation is Sleep. When your computer enters Sleep mode, all of the documents, applications, and data you were using remain in RAM (memory), where they're quickly accessible when you wake your computer.

In Sleep mode, the computer enters a state of greatly reduced power consumption, which saves energy and prolongs battery life. To put your computer into Sleep mode in Windows, open the Start menu, select Power, and then select the Sleep option. To wake up your computer, press a key or

Bits&Bytes Smart Coral Reefs

Monitoring the health of coral reefs is critical to determining the effects of climate change and pollution on oceans. The number of fish and the diversity of their species is a key indicator of the health of a coral reef. Monitoring coral reefs is currently a very labor intensive job involving human divers either observing or recording fish activity, but divers are limited to the amount of time they can spend underwater in one dive and the number of dives they can make in a day. Also, human activity can disrupt fish activity and change the results of observations.

Recently, Intel and Accenture teamed up with scientists in the Philippines to create a computer-based technology that records and analyzes fish activity in reefs without ongoing human intervention. The companies created an artificial reef (made of concrete), which is embedded with video cameras. Accenture's video analytics platform, which uses artificial intelligence, detects, classifies, and photographs fish as they go about their normal routines. Data is then transmitted to a buoy on the surface, which in turn transfers the data by Wi-Fi or 4G to onshore researchers. This allows for real-time analysis of the undisturbed reef environment to help scientists make better and faster interpretations of the data. In addition, the artificial reefs are used to repair damaged reefs since coral can readily adhere to the concrete and expand. The technology is expected to be implemented at other reefs around the world in the near future.

the physical power button. In a few seconds, the computer will resume with the same programs running and documents displayed as when you put it to sleep.

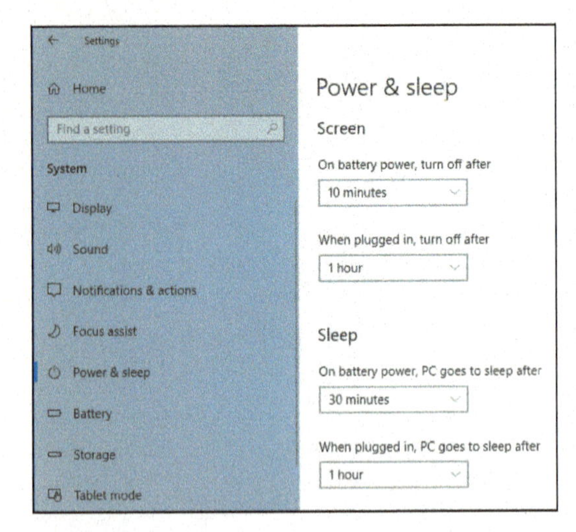

Figure 2.41 In Windows, you can set the options to your particular tastes to conserve power. *(Courtesy of Microsoft Corporation)*

*> To access Power & Sleep Settings, from the Start menu, select **Settings**, then **System**, then select **Power & sleep**.*

What's the Restart option for? Windows and macOS power options both provide a restart option. Restarting the system while it's powered on is called a warm boot. You might need to perform a warm boot if the operating system or other software application stops responding, if you've received operating system updates, or if you've installed new apps. It takes less time to perform a warm boot than to power down completely and then restart all your hardware. Powering on your computer from a completely turned-off state is called a cold boot.

Should I turn off my computer every time I'm finished using it? Some people argue that turning your computer on and off throughout the day subjects its components to stress because the heating and cooling process forces the components to expand and contract repeatedly. Other people say it's more environmentally friendly and less expensive to shut down your computer when you're not using it.

All modern operating systems include power-management settings that allow the most power-hungry components of the system (the hard drive and monitor) to shut down after a short idle period. The power-management options of Windows (see Figure 2.41), Android, iOS, and macOS provide you with flexibility to conserve power and battery life. Therefore, it's rarely necessary to shut your devices completely off, even when they're charging.

If you don't ever want to turn off your Windows laptop computer completely, you can change what happens when you press the power button or close the lid. By accessing the Power Options System Settings window (see Figure 2.42), you can decide whether you want your computer to Sleep, Hibernate, or Shut down when you press the power button. The Hibernate option is similar to Sleep except that your data is stored on your hard drive instead of in RAM, and your computer is powered off. This uses much less battery power than Sleep and is a good choice if you won't be using your laptop for a long time and won't have the opportunity to charge it. However, Sleep is still a good option if you just won't be using your computer for a short time. You may want to set your computer so it sleeps when you close the lid but hibernates when you press the power button, giving you quick access to either option.

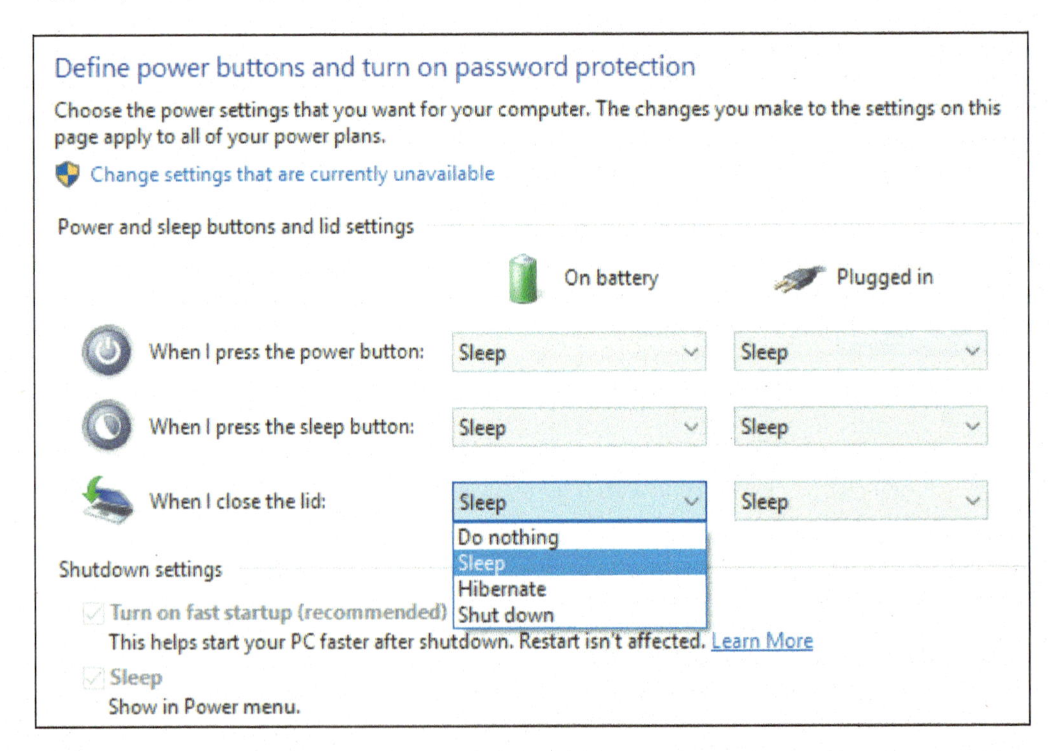

Figure 2.42 You can determine what happens when you press the power button on your computer or close the lid through the Power Options System Settings screen. *(Courtesy of Microsoft Corporation)*

*> To access Power Options System Settings, from the Start menu, select **Settings**, **System**, and **Power & sleep**, **and** then choose **Additional power settings**. Select **Choose what closing the lid does**.*

Dig Deeper Building Your Own Gaming Computer

Computers are complex devices and originally needed to be built by people with a solid background in electronics and engineering. But components are widely available today that are premanufactured and can be relatively easy for novices to put together to build functioning computers. Why would you want to build your own computer? Many gamers build computers to get exactly the components and features they want, often while saving money in the process. Fortunately, resources on the Internet can guide you through the process, including step-by-step videos and parts lists for computers that others have assembled. This Dig Deeper describes the basic hardware you need when building your own computer:

- **CPU:** Because the CPU is the brains of the computer (where all processing takes place), it is the most important (and expensive) component to consider. There are two primary manufacturers of CPUs, AMD and Intel, both of which offer a vast array of high-end and low-end choices. For a gaming PC, you will want to consider a higher-end choice from the Ryzen or Threadripper line from AMD or eighth or higher generation of the i9 for Intel.
- **Motherboard:** This is the circuit board to which all your components are attached. It also facilitates communication of your computer components. Your choice of motherboard will be driven somewhat by your choice of CPU, because not all CPUs are compatible with all motherboards. Motherboards have vast arrays of options (just like CPUs), ranging from inexpensive basics to expensive ones with high-end features (see Figure 2.43).
- **Graphics card (GPU):** Because a high-quality gaming experience depends on smooth processing of graphics, selecting a quality GPU will have a huge impact. AMD and Nvidia are the main graphics card manufacturers. If you want the best GPU on the market, its cost may exceed the cost of your CPU.

Figure 2.43 Motherboards are available with cool lighting that is visible if you have a see-through panel on your computer case *(FeelGoodLuck/Shutterstock)*

- **Memory (RAM):** RAM is important because your computer stores data in it that you use regularly, rather than storing the data on slower-to-access devices like a hard drive. There are many types of RAM, which vary in terms of speed (higher speed = higher cost). More RAM is usually better than less RAM, but 32 GB is sufficient for most gaming systems with solid GPUs.
- **Storage:** Your computer's operating system and games programs must be stored somewhere, so you'll need some type of nonvolatile storage. Conventional hard disk drives will usually save you money, but SSD drives (which are falling in price) will provide you with better data transfer rates and generate much less heat.
- **Power supply:** Electrical power is required for your computer components to function, and the power supply unit (PSU) fills this need. Make sure you select a PSU that has enough wattage for the components you have selected (wattage requirements are in the specs for the components).
- **System unit (case):** The components of your computer need a place to congregate. Cases come in a vast array of sizes and shapes but are pretty much all designed to hold the same set of computer components. Some may come with preinstalled fans to assist with cooling. If you're installing a high-end cooling system (like a liquid cooler), make sure it will fit in the case you select.
- **Cooling devices:** CPUs generate quite a bit of heat, as does the power supply. Therefore, cooling devices are essential (see Figure 2.44). Fans can be mounted in the case to promote airflow and reduce temperature. Usually, a dedicated CPU fan or heat sink is installed on or near the CPU. (Some CPUs are sold with their own cooling solutions.)
- **Monitor:** A basic high-resolution monitor (at least 1920 x 1080) is essential. Based on budget, you may want to purchase a higher-resolution 4K monitor with a higher refresh rate for increased performance.

Figure 2.44 Cooling fans with heat sinks are often placed on top of the CPU to dissipate heat. *(Ekipaj/Shutterstock)*

(continued)

- **Mouse:** Gaming mice are manufactured by companies such as Logitech, Corsair, and Razer. Getting one that matches your gaming style is essential. If you play many different games, you may want one with reconfigurable buttons.
- **Keyboard:** Keyboards range from basic to fancy lighted models. Testing some out is the key to getting one that has the right feel for you.

Although not hardware, you'll also need an operating system (like Windows or Linux). Linux, an open-source operating system, is free but has a slightly steeper learning curve than Windows.

Check out sites such as PC Gamer (pcgamer.com) and Tech Guided (techguided.com) for videos and guides. Talk to your friends who have built computers. And most of all, have fun!

Setting It All Up: Ergonomics

Objective 2.14 *Define ergonomics, and discuss the ideal physical setup for using computing devices.*

What is ergonomics? Ergonomics is the science that deals with the design and location of machines and furniture so that the people using them aren't subjected to an uncomfortable or unsafe experience. In terms of computing, ergonomics refers to how you set up your computer and other equipment to minimize your risk of injury or discomfort.

Why is ergonomics important? Studies suggest that teenagers, on average, spend 7.5 hours per day using computing devices. The repetitive nature of long-term computing activities can place stress on joints, tendons, and muscles, causing repetitive-stress injuries such as carpal tunnel syndrome and tendonitis. These injuries can take months or years to develop to a point that they become painful, but if you take precautionary measures now, you may prevent years of pain later.

How can I avoid injuries when I'm working at my computer? As Figure 2.45 illustrates, it's important to arrange your computer, chair, body, and keyboard in ways that will help you avoid injury, discomfort, and eyestrain. The following additional guidelines can help keep you comfortable and productive:

- **Position your monitor correctly.** Studies suggest it's best to place your monitor at least 25 inches from your eyes. Experts recommend that you position your monitor either at eye level or at an angle 15–20 degrees below your line of sight. If you have a laptop, this means placing the laptop on a stand (or stack of books) to achieve the correct height.
- **Purchase an adjustable chair.** Adjust the height of your chair (or use a footrest) so that your feet touch the floor. The back support needs to be adjustable so that it supports your lumbar (lower back) region. If your chair doesn't adjust, place a pillow behind your back to provide support.
- **Assume a proper position while typing.** Improperly positioned keyboards are one of the leading causes of repetitive stress injuries. Your wrists should be flat (not bent) with respect to the keyboard, and your forearms should be parallel to the floor. In addition, your wrists should not be resting on the keyboard while typing. When using a laptop on a stand, attach an external keyboard to achieve the proper typing position. You can adjust the height of your chair or install a height-adjustable keyboard tray to ensure a proper position. Specially designed ergonomic keyboards such as the one shown in Figure 2.46 can also help you achieve the proper wrist position.
- **Take breaks.** Remaining in the same position for long periods increases stress on your body. Shift your position in your chair and stretch your hands and fingers periodically. Likewise, staring at the screen for long periods can lead to eyestrain and dry eyes, so rest your eyes periodically by taking them off the screen and focusing them on an object at least eight feet away.

Figure 2.45 Using proper equipment that is adjusted correctly helps prevent repetitive-strain injuries while working at a computer. *(GARO/Phanie/Alamy Stock Photo)*

- **Ensure that the lighting is adequate.** Ensuring that you have proper lighting in your work area minimizes eyestrain. Eliminate sources of direct glare (light shining directly into your eyes) or reflected glare (light shining off the computer screen) and ensure that there is enough light to read comfortably. If you still can't eliminate glare from your screen, you can buy an antiglare screen to place over your monitor.

Is ergonomics important when using mobile devices? Working with mobile computing devices presents interesting challenges when it comes to injury prevention. For example, many users work with laptops resting on their laps, placing the monitor outside the optimal line of sight and thereby increasing neck strain. Table 2.5 provides guidelines on preventing injuries when computing on the go.

Figure 2.46 To reduce the risk of repetitive strain injuries, ergonomic keyboards have curved keyboards and wrist rests to help you maintain the proper hand position while typing. *(Dmitriy Melnikov/123RF)*

What devices are available for people with disabilities? Assistive (or adaptive) technologies are products, devices, equipment, or software that is used to maintain, increase, or improve the functional capabilities of individuals with disabilities. For users with visual impairment and individuals who can't type with their hand, voice recognition is a common option. For users whose visual limitations are less severe, keyboards with larger keys are available.

People with motor control issues may have difficulty with pointing devices. To aid such users, special trackballs are available that can be manipulated with one finger and can be attached to almost any surface, including a wheelchair. When arm motion is severely restrained, head-mounted pointing devices can be used. Generally, these involve a camera mounted on the computer monitor and a device attached to the head (often installed in a hat). When the user moves his or her head, the camera detects the movement and moves the cursor. In this case, mouse clicks are controlled by a switch that can be manipulated by the user's hands or feet or even by using an instrument that fits into the mouth and senses the user blowing into it.

Table 2.5 Preventing Injuries While on the Go

	Smartphone Repetitive-Strain Injuries	Portable Media Player Hearing Damage	Small-Screen Vision Issues	Lap Injuries	Tablet Repetitive-Strain Injuries
Malady	Repetitive-strain injuries (such as de Quervain's disease) from constant typing of instant messages	Hearing loss from high-decibel sound levels in earbuds	Blurriness and dryness caused by squinting to view tiny screens on mobile devices	Burns on legs from heat generated by laptop	Pain caused by using tablets for prolonged periods in uncomfortable positions
Preventive measures	Restrict length and frequency of messages, take breaks, and perform other motions with your thumbs and fingers during breaks to relieve tension.	Turn down volume (you should be able to hear external noises, such as people talking), use software that limits sound levels (not to exceed 60 decibels), and use external, over-ear style headphones instead of earbuds.	Blink frequently or use eye drops to maintain moisture in eyes, after 10 minutes take a break and focus on something at least eight feet away for five minutes, use adequate light, and increase the size of fonts.	Place a book, magazine, or laptop cooling pad between your legs and laptop.	Restrict the length of time you work at a tablet, especially when typing or gaming. Use the same ergonomic position you would use for a laptop when using a tablet.

Before moving on to the Chapter Review:

1. Watch Chapter Overview Video 2.2.
2. Then take the Check Your Understanding quiz.

Check Your Understanding // Review & Practice

For a quick review to see what you've learned so far, answer the following questions.

multiple choice

1. Which of the following is NOT found on a motherboard?
   a. RAM
   b. Hard drive
   c. ROM
   d. CPU
2. Which of these is considered the brains of the computer?
   a. CPU
   b. ROM
   c. RAM
   d. USB
3. Which of these statements about SSD drives is FALSE?
   a. SSD drives have no moving parts.
   b. SSD drives produce less heat than conventional hard drives.
   c. SSD drives have slower response times than conventional hard drives.
   d. SSD drives require less power than conventional hard drives.
4. Which of the following is NOT a port?
   a. HDMI
   b. USB
   c. ROM
   d. Thunderbolt
5. Which power control option leads to a warm boot?
   a. Sleep
   b. Restart
   c. Log off
   d. Shut down

chew on this

(3Dalia/Shutterstock)

The display screen of the future may be paper-thin, flexible, and transparent. Companies like Samsung and LG are already working on it. What uses can you envision for this new technology? What advantages and disadvantages do you foresee? Do you have any suggestions for manufacturers to help make this successful?

MyLab IT Go to **MyLab IT** to take an autograded version of the *Check Your Understanding* review and find all media resources for the chapter.

For the IT simulation for this chapter, see MyLab IT.

2 Chapter Review

Summary

Part 1
Understanding Digital Components

Learning Outcome 2.1 You will be able to describe the devices that make up a computer system.

Understanding Your Computer

Objective 2.1 *Describe the four main functions of a computer system and how they interact with data and information.*

- The computer's four major functions are: (1) input: gather data or allow users to enter data; (2) process: manipulate, calculate, or organize that data; (3) output: display data and information in a form suitable for the user; and (4) storage: save data and information for later use.
- Data is a representation of a fact or idea. The number 3 and the words *televisions* and *Samsung* are pieces of data.
- Information is data that has been organized or presented in a meaningful fashion. An inventory list that indicates that three Samsung televisions are in stock is processed information. It allows a retail clerk to answer a customer query about the availability of merchandise. Information is more powerful than raw data.

Objective 2.2 *Define bits and bytes, and describe how they are measured, used, and processed.*

- To process data into information, computers need to work in a language they understand. This language, called *binary language*, consists of two numbers: 0 and 1. Each 0 and each 1 is a binary digit or bit. Eight bits comprise one byte.
- In computers, each letter of the alphabet, each number, and each special character consists of a unique combination of eight bits (one byte)—a string of eight 0s and 1s.
- For describing large amounts of storage capacity, the terms *megabyte* (approximately one million bytes), *gigabyte* (approximately one billion bytes), *terabyte* (approximately one trillion bytes), and *petabyte* (1,000 terabytes) are used.

Objective 2.3 *List common types of computers, and discuss their main features.*

- A tablet computer is a portable computer integrated into a flat multi-touch sensitive screen.
- A laptop or notebook computer is a portable computer that has a keyboard, monitor, and other devices integrated into a single compact case.
- A 2-in-1 PC is a laptop computer that can convert into a tablet-like device.
- An ultrabook is a lightweight laptop computer featuring low-power processors and solid-state drives.
- Chromebook computers use the Google Chrome OS. Documents and apps are stored primarily in the cloud.
- Desktop computers consist of a separate case (called the system unit) that houses the main components of the computer plus peripheral devices.

Input Devices

Objective 2.4 *Identify the main types of keyboards and touch screens.*

- You use physical or virtual keyboards to enter typed data and commands. Most keyboards use the QWERTY layout.
- Touch screens are display screens that respond to commands initiated by a touch with a finger or a stylus.
- Wireless keyboards mainly use Bluetooth connectivity and provide alternatives to virtual keyboards on portable computing devices.

Objective 2.5 *Describe the main types of mice and pointing devices.*

- Mice are used to enter user responses and commands.
- Optical mice use a laser to detect mouse movement.
- Some mice can be adjusted to provide better ergonomics for users.
- Laptops have integrated pointing devices called touch pads (trackpads).

Objective 2.6 *Explain how images, sounds, and sensor data are input into computing devices.*

- Images are input into the computer with flatbed scanners, digital cameras, and mobile device cameras.
- Live video is captured with webcams and integrated cameras.
- Microphones capture sounds. There are many types of microphones, including desktop, headset, and clip-on models.
- Smartphones and tablets have a variety of sensors that detect or measure a variety of inputs (such as acceleration that the device is experiencing).

Output Devices

Objective 2.7 *Describe options for outputting images and audio from computing devices.*

- Monitors display soft copies of text, graphics, and video.
- Liquid crystal display (LCD) and light-emitting diode (LED) screens are the most common types of computer monitors.
- OLED displays use organic compounds to produce light and don't require a backlight, which saves energy.
- Aspect ratio and screen resolution are key aspects to consider when choosing a monitor.
- Speakers, headphones, and earbuds are the output devices for sound.
- More sophisticated systems include subwoofers and surround sound.

Objective 2.8 *Describe various types of printers, and explain when you would use them.*

- Printers create hard copies of text and graphics.
- The two primary categories of printers are inkjet and laser. Laser printers usually print faster and deliver output that is higher quality than that of inkjet printers. However, inkjet printers can be more economical for casual printing needs.
- Specialty printers are also available, such as all-in-one printers, large-format printers, and 3D printers.
- When choosing a printer, you should be aware of factors such as speed, resolution, color output, and cost of consumables.

Part 2

Processing, Storage, and Connectivity

Learning Outcome 2.2 **You will be able to describe how computers process and store data and how devices connect to a computer system.**

Processing and Memory on the Motherboard

Objective 2.9 *Describe the functions of the motherboard and RAM.*

- The motherboard, the main circuit board of the system, contains a computer's CPU, which coordinates the functions of all other devices on the computer.
- The motherboard also houses slots for expansion cards, which have specific functions that augment the computer's basic functions. Typical expansion cards are sound and video cards. In portable devices, sound and video are usually integrated directly into the motherboard.
- RAM, the computer's volatile memory, is also located on the motherboard. RAM is where data and instructions are held while the computer is running.

Objective 2.10 *Explain the main functions of the CPU.*

- The CPU controls all the functions performed by the computer's other components. The CPU also processes all commands issued to it by software instructions.
- The performance of a CPU is mainly affected by the speed of the processor (measured in GHz) and the number of processing cores.

Storing Data and Information

Objective 2.11 *Describe the various means of storing data and information with computing devices.*

- The internal hard drive is your computer's primary device for permanent storage of software and files. The hard drive is a nonvolatile storage device, meaning it permanently holds the data and instructions your computer needs, even after the computer is turned off.
- SSD drives have no moving parts, so they are more energy-efficient and less susceptible to damage.
- External hard drives are essentially internal hard drives that have been made portable by enclosing them in a protective case and making them small and lightweight.
- Cloud storage refers to nonvolatile storage locations that are maintained on the Internet (in the cloud). Examples are OneDrive, Google Drive, and Dropbox.

- Storing your data in the cloud allows you to access it from almost any computing device that is connected to the Internet.
- Flash drives are another portable means of storing data. Flash drives plug into USB ports.
- Flash memory cards enable you to transfer digital data between your computer and devices such as digital cameras, smartphones, and printers.

Connecting Peripherals to the Computer

Objective 2.12 *Describe common types of ports used today.*

- The fastest type of port used to connect devices to a computer is the Thunderbolt port.
- The most common type of port used to connect devices to a computer is the USB port.
- Connectivity ports, such as Ethernet ports, give you access to networks and the Internet.
- HDMI ports are the most common multimedia port, although DisplayPorts are becoming popular. They are used to connect monitors, TVs, and gaming consoles to computing devices and handle both audio and video data.
- Audio ports are used to connect headphones, microphones, and speakers to computing devices.

Power Management and Ergonomics

Objective 2.13 *Describe how to manage power consumption on computing devices.*

- Turning off your computer when you won't be using it for long periods of time saves energy. In Windows, you can turn your computer off by accessing the Power option on the Start menu and then selecting Shut down.
- If you are not using your computer for short periods of time, selecting the Sleep option will help your computer save energy but allows it to be quickly awakened for use.

Objective 2.14 *Define ergonomics, and discuss the ideal physical setup for using computing devices.*

- Ergonomics refers to how you arrange your computer and equipment to minimize your risk of injury or discomfort.
- Achieving proper ergonomics includes positioning your monitor correctly, buying an adjustable chair, assuming a proper position while typing, making sure the lighting is adequate, and not looking at the screen for long periods. Other good practices include taking frequent breaks and using specially designed equipment such as ergonomic keyboards.
- Ergonomics is also important to consider when using mobile devices.

Be sure to check out **MyLab IT** for additional materials to help you review and learn. And don't forget to watch the Chapter Overview videos.

Key Terms

Chapter Quiz // Assessment

For a quick review to see what you've learned, answer the following questions. Submit the quiz as requested by your instructor. If you are using **MyLab IT**, the quiz is also available there.

multiple choice

1. Which of the following functions of a computer is mostly responsible for turning data into information?
- **a.** Output
- **b.** Storage
- **c.** Input
- **d.** Processing

2. In a computer, each _________ can represent a single character such as "G".
- **a.** byte
- **b.** bit
- **c.** integrated circuit
- **d.** megabyte

3. A(n) _________ is a computing device focused on offering a very thin, lightweight computing solution.
- **a.** 2-in-1 PC
- **b.** Chromebook
- **c.** ultrabook
- **d.** all-in-one PC

4. Touch-screen devices usually feature _________ keyboards.
- **a.** physical
- **b.** laser-projection
- **c.** virtual
- **d.** optical

5. Which of the following is a NOT sensor found in certain smartphones?
- **a.** HDMI
- **b.** Accelerometer
- **c.** Gyroscope
- **d.** Barometer

6. Ergonomics is an important consideration
- **a.** for all computing devices.
- **b.** only for laptop computers.
- **c.** only for laptop and desktop computers, but never for mobile devices.
- **d.** only for desktop computers.

7. The most common output device for soft output is a
   a. printer.
   b. scanner.
   c. mouse.
   d. display screen.
8. _________ printers work by spraying tiny drops of ink onto paper.
   a. Inkjet
   b. 3D
   c. Laser
   d. Large format
9. The most common computer port is the _________ port.
   a. USB
   b. expansion
   c. Thunderbolt
   d. Wi-Fi
10. Which component of a computing device drains the battery the fastest?
   a. Hard drive
   b. Display screen
   c. Wi-Fi adapter
   d. Bluetooth adapter

true/false

_____ 1. Data and information are NOT interchangeable terms.

_____ 2. RAM is volatile storage.

_____ 3. Conventional disk drives are superior to SSD drives because they have no moving parts.

_____ 4. A touch pad is a pointing device usually found in laptops.

_____ 5. The "brain" of the computer is the ROM.

What do you think now?

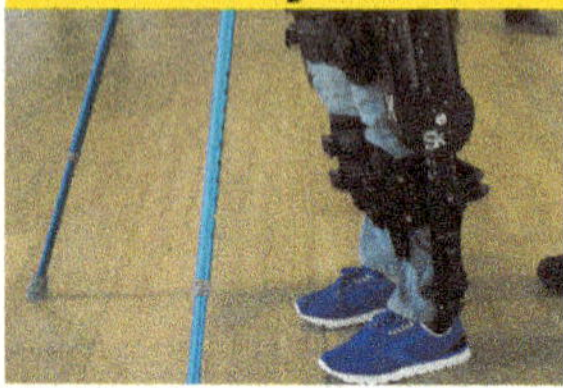

MIT is one of many educational institutions leading the way in researching devices for augmenting humans. Explore the Biomechatronics Group website at *biomech.media.mit.edu* and familiarize yourself with their current projects. Consider other avenues of research that you think have merit to explore. Explain what human augmentations you feel would provide the most benefits to humankind, and provide rationale for your opinions.

Team Time

Portable Computing Options

problem

You've joined a small business that's beginning to evaluate its technology setup. Because of the addition of several new sales representatives, customer service reps, and other administrative employees, the company needs to reconsider the various computing devices that it needs. Salespeople are out in the field, as are the customer service technicians. You've been asked to evaluate options for portable computing devices.

task

Split your class into teams of three and assign the following tasks:

- Member A explores the benefits and downfalls of smartphones.
- Member B explores the benefits and downfalls of tablets.
- Member C explores the benefits and downfalls of ultrabooks.

process

1. Think about what the portable computing needs are for the company and what information and resources you'll need to tackle this project.
2. Research and then discuss the components of each method you're recommending. Are any of these options better suited for the particular needs of certain types of employees (sales representatives versus administrative staff)? Consider the types of battery life needed and the appropriate screen size for common tasks. How significant a factor is device weight?

3. Consider the different types of employees in the company. Would a combination of devices be better than a single solution? If so, what kinds of employees would use which type of device?
4. As a team, write a summary position paper. Support your recommendation for the company. Each team member should include why his or her portable device will or will not be part of the solution.

conclusion

There are advantages and disadvantages to any portable computing device. Being aware of the pros and cons and knowing which device is best for a particular scenario or employee will help you to become a better consumer as well as a better computer user.

Ethics Project

Green Computing

Ethical conduct is a stream of decisions you make all day long. In this exercise, you'll research and then role-play a complicated ethical situation. The role you play may or may not match your own personal beliefs, but your research and use of logic will enable you to represent whichever view is assigned. An arbitrator will watch and comment on both sides of the arguments, and together, the team will agree on an ethical solution.

background

Green computing—conducting computing needs with the least possible amount of power and impact on the environment—is on everyone's minds. Although it's hard to argue with an environmentally conscious agenda, the pinch to our pocketbooks and the loss of some comforts sometimes make green computing difficult. Businesses, including colleges, need to consider a variety of issues and concerns before jumping into a complete green overhaul.

research areas to consider

- End-of-life management: e-waste and recycling
- Energy-efficient devices
- Renewable resources used in computer manufacturing
- Costs of green computing
- Government funding and incentives

process

1. Divide the class into teams.
2. Research the areas previously cited and devise a scenario in which your college is considering modifying its current technology setup to a greener information technology (IT) strategy.
3. Team members should take on the role of a character—for example, environmentalist, college IT administrator, or arbitrator—from the scenario you devise. Each team member should write a summary that provides background information for their character and that details their character's behaviors to set the stage for the role-playing event. Then, team members should create an outline to use during the role-playing event that outlines what will occur.
4. Team members should arrange a mutually convenient time to meet, using a virtual meeting tool or by meeting in person.
5. Team members should present their case to the class or submit a PowerPoint presentation for review by the rest of the class, along with the summary and resolution they developed.

conclusion

As technology becomes ever more prevalent and integrated into our lives, more and more ethical dilemmas will present themselves. Being able to understand and evaluate different sides of the argument, while responding in a personally or socially ethical manner, will be an important skill.

Solve This with Excel MyLab IT Grader

Technology Wish List

You are in need of a significant technology upgrade, and your parents have told you they will help you finance your purchases by loaning you the money. You will need to repay them at a modest 2.5% interest rate over two years. The only catch is that they want you to create a list of all the new devices that you need, note the cost, and provide a website for each device where they can find more information. Then, they want you to calculate how much you will need to give them each month to pay them back.

You will use the following skills as you complete this activity:

- Merge and Center
- Modify Workbook Themes
- Apply Number Formats
- Use the SUM, PMT, and COUNTA Functions
- Modify Column Widths
- Insert a Hyperlink
- Create a Formula
- Wrap Text

Instructions

1. Open *TIA_Ch2_Start.xlsx* and save as **TIA_Ch2_LastFirst.xlsx**.
2. Format the title in cell A1 with the **Title Cell Style**, and format the column headers in cells A3:F3 with the **Heading 3 Cell Style**.

   Hint: To format cell styles, on the Home tab, in the Styles group, click **Cell Styles**.
3. **Merge and Center** A1 across columns A through F, and **Center align** the column headers in cells A3:F3.

   Hint: To Merge and Center text, on the Home tab, in the Alignment group, click **Merge & Center**.
4. Modify column widths so that Column A is **25** and Column D is **45**.

   Hint: To modify column widths, on the Home tab, in the Cells group, click **Format**, and then select **Column Width**.
5. In **cells B4:E9**, fill in the table with the Brand and Model of the six devices that you would like to purchase. The device type is filled out for you. In the *Reason* column, write a brief note for why this device will help you. (You'll format the text so it all displays later.) Enter the cost of the device in the Cost column. Don't include tax and/or shipping.
6. Change the Workbook Theme to **Integral**.

   Hint: To apply the Theme, on the Page Layout tab, in the Themes group, click Themes.
7. In **cells F4:F9**, create a **Hyperlink** to a webpage that features each respective product so your parents can have access to more information if they need it. Ensure that each hyperlink includes the URL to the exact webpage for the device in Address, but displays the Make/Model of the device in the worksheet.

   Hint: To insert a hyperlink, on the Insert tab, in the Links group, click Hyperlink. In the Insert Hyperlink dialog box, enter the URL in the Address box and enter the Make/Model in the Text to display box.
8. **Wrap the text** in **cells C4:C9, D4:D9**, and **F4:F9** so all text appears.

   Hint: To wrap text, on the Home tab, in the Alignment group, click Wrap Text.
9. Format the values in **cells E4:E9** with the **Accounting Number format with two decimals**.

   Hint: To apply number and decimal formats, look on the Home tab, in the Number group.
10. In **cell A10**, type **Subtotal** and then, in **cell E10**, use a **SUM function** to calculate the total cost of all devices. Format the results in **Accounting Number format with two decimals**.

    Hint: To apply number and decimal formats, look on the Home tab, in the Number group.
11. In **cell A11**, type **Estimated Tax** and then, in **cell E11**, create a formula that references the subtotal in cell E10 and multiplies it by a tax of 6%. Format the results in **Accounting Number format with two decimals**.

    Hint: The formula will be =E10*0.06

12. In **cell A12**, type **Estimated Shipping** and then, in **cell E12**, create a formula to calculate the shipping charge by using the **COUNTA function** to determine the number of devices being purchased and then multiplying that by a $10 shipping charge. Format the results in **Accounting Number format with two decimals**.

    Hint: The formula will be =COUNTA(B4:B9)*10.

13. In **cell A13**, type **Total Cost** and then, in **cell E13**, use the **SUM function** to create a formula that adds up the *Subtotal, Estimated Tax*, and *Estimated Shipping* costs. Format the results in **Accounting Number format with two decimals**. Format cells A13:E13 with the **Total Cell Style**.

14. **Right-align** cells A10:A13.

15. In **cell D14**, type **Estimated Monthly Payment** and then, in **cell E14**, use the **PMT function** to calculate the monthly payment owed to your parents to pay back the total purchase amount in two years at a 2.5% annual interest rate.

    Hint: The formula will need to adjust the annual interest rate to a monthly rate and adjust the annual term of the loan to monthly payments.

16. Save the workbook and submit based on your instructor's directions.

Chapter

3 Using the Internet: Making the Most of the Web's Resources

For a chapter overview, watch the Chapter Overview Videos.

PART 1

Collaborating and Working on the Web

Learning Outcome 3.1 **You will be able to explain how the Internet works and how it is used for collaboration, communication, commerce, and entertainment purposes.**

PART 2

Using the Web Effectively

Learning Outcome 3.2 **You will be able to describe the tools and techniques required to navigate and search the web.**

MyLab IT All media accompanying this chapter can be found here.

(Smatch/Shutterstock; Marish/Shutterstock; Happy Art/Shutterstock; Rabiulalam/Shutterstock; Dinosoft/Shutterstock; Rashad Ashur/Shutterstock)

What do you think?

Ever seen a headline on your social media feed or in sponsored content on a website that's too tempting to ignore? You know it's probably not legit, but you click anyway. Such **sensationalized headlines** are called ***clickbait***, and they're designed to tempt you to click on them. Why? Because clicking generates ad revenue. Clickbait has been around for years. People create intentionally sensationalized headlines and photos that **bait users to click through** to read what ends up being a pointless story (hence the term clickbait). Similarly, sensationalized stories (referred to as fake news) are posted on the social web in the hopes of going viral. **Viral stories increase website traffic** and that in turn generates income. Given that sites such as Facebook and Twitter can reach billions, clickbait and fake news have become **social media plagues**. Facebook has taken strides to curtail the distribution of fake news by providing additional fact-checking information about potential fake news stories and has also cut back on the ability for clickbait sites to use Facebook advertising services. Google has also instituted a policy disallowing fake news sites from using AdSense. These policy changes may help to limit fake news and clickbait, but are we also **responsible for the proliferation of this content?** If we didn't click, they wouldn't post. Where do you stand on clickbait?

Have you been enticed to click a false story?

- Yes, I was tricked, but read the story anyway.
- Yes, I was tricked and was mad it wasted my time.
- No, I can recognize a trick headline and know not to click it.

See the end of the chapter for a follow-up questi[illegible]

(Prostock-studio/Shutterstock)

Part 1

For an overview of this part of the chapter, watch **Chapter Overview Video 3.1**.

Collaborating and Working on the Web

Learning Outcome 3.1 **You will be able to explain how the Internet works and how it is used for collaboration, communication, commerce, and entertainment purposes.**

You most likely know at least a bit about how to use the web's resources to communicate and collaborate with others and how business is conducted over the web. In this section, we'll explore these and related topics. But first, let's start with a brief overview of the Internet.

The Internet and How It Works

It's hard to imagine life without the Internet. The Internet is actually a network of networks that connects billions of computer users globally, but its beginnings were much more modest.

The Origin of the Internet

Objective 3.1 *Describe how the Internet got its start.*

Why was the Internet created? The concept of the Internet—the largest computer network in the world—was developed in the late 1950s while the United States was in the midst of the Cold War with the Soviet Union (see Figure 3.1). At that time, the US Department of Defense needed a computer network that wouldn't be disrupted easily in the event of an attack.

1958 The **Advanced Research Projects Agency (ARPA)** is established for the U.S. Department of Defense. This agency creates the ARPANET—the beginnings of the Internet.

ASCII Code - Character to Binary

0	0011 0000	I	0100 1001	b	0110 0010	v	0111 0110
1	0011 0001	J	0100 1010	c	0110 0011	w	0111 0111
2	0011 0010	K	0100 1011	d	0110 0100	x	0111 1000
3	0011 0011	L	0100 1100	e	0110 0101	y	0111 1001
4	0011 0100	M	0100 1101	f	0110 0110	z	0111 1010
				g	0110 0110		

1963 **ASCII code** is developed as the standard for computers from different manufacturers to exchange data.

1964 A **new network scheme** is developed with multiple paths so if one communication path was destroyed (from a potential Soviet Union attack), the rest of the network would still be able to communicate.

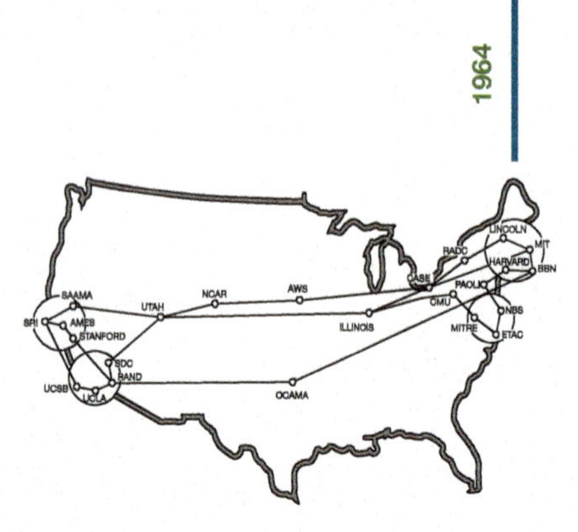

1969 Researchers at UCLA **send the first message via a networked computer system** to researchers at Stanford University.

Stanford
L
O
G
UCLA

Figure 3.1 How the Internet Began

At the same time, researchers for the Department of Defense were trying to get different computers to work with each other, using a common communication method that all computers could use. The Internet was created to respond to these two concerns: establishing a secure form of communications and creating a means by which *all* computers could communicate.

Who invented the Internet? The modern Internet evolved from an early US government-funded internetworking project called the Advanced Research Projects Agency Network (ARPANET). ARPANET began as a four-node network involving UCLA, Stanford Research Institute, the University of California at Santa Barbara, and the University of Utah in Salt Lake City. The first real communication occurred in 1969 between the computer at Stanford and the computer at UCLA. Although the system crashed after the third letter of "Login" was transmitted, it was the beginning of a revolution. Many people participated in the creation of the ARPANET, but two men, Vinton Cerf and Robert Kahn, are generally acknowledged as the fathers of the Internet. These men earned this honor because in the 1970s, they were primarily responsible for developing the communications protocols (standards) still in use on the Internet today.

So, are the web and the Internet the same thing? Because the World Wide Web (WWW or the web) is what we use the most, we sometimes think of the Internet and the web as being interchangeable. However, the web is only a subset of the Internet, dedicated to hosting HTML pages; it is the means by which we access information over the Internet. The web is based on the Hypertext Transfer Protocol (HTTP), which is why you see an http://at the beginning of web addresses. What distinguishes the web from the rest of the Internet is its use of the following:

- Common communications protocols that enable computers to talk to each other and display information in compatible formats
- Special links that enable users to navigate from one place to another on the web

Who created the web? The web began in 1991. It was based on a protocol developed by Tim Berners-Lee, a physicist at the European Organization for Nuclear Research (CERN), who wanted a method for linking his research documents so that other researchers could access them. In conjunction with Robert Cailliau, Berners-Lee developed the basic architecture of the web and

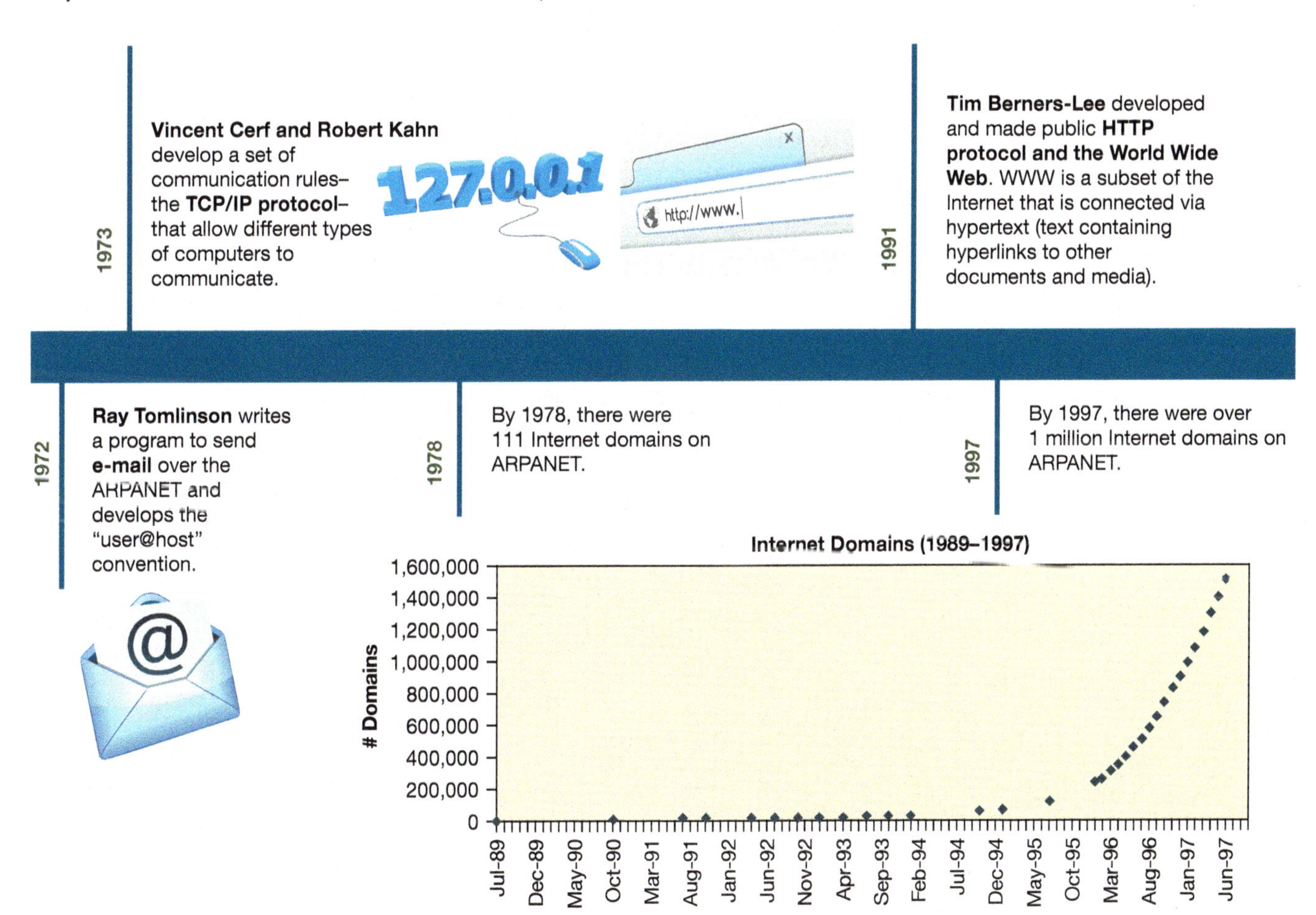

(Matthias Pahl/Shutterstock; LvNL/Shutterstock; Phonlamai Photo/Shutterstock)

created the first web browser, software that enables you to display and interact with text and other media on the web. The original browser could handle only text. Then, in 1993, the Mosaic browser, which could display graphics as well as text, was released. The once-popular Netscape Navigator browser evolved from Mosaic and heralded the beginning of the web's monumental growth.

How the Internet Works

Objective 3.2 *Explain how data travels on the Internet.*

How does the Internet work? All computers and other devices such as tablets and smartphones that are connected to the Internet create a network of networks. These Internet-connected devices communicate with each other in turns, just as we do when we ask a question or reply with an answer. Thus, a computer (or other device) connected to the Internet acts in one of two ways: Either it's a client, a computer that asks for data, or it's a server, a computer that receives the request and returns the data to the client. Because the Internet uses clients and servers to exchange data and information, it's referred to as a client/server network.

When a client computer puts out a request for information from the server computer, the request travels along transmission lines. These transmission lines are similar to our highway system of roads, with some roads having different speed limits. The transmission lines with the fastest speeds are referred to as Internet backbones.

How do computers talk to each other? Suppose you want to order something from Amazon. Figure 3.2 illustrates what happens when you type *www.amazon.com* in your web browser and when Amazon's home page appears on your computer monitor. As you can see, the data request from your computer (the client computer) is sent by Internet communication pathways to a server computer. The server computer (in this case, Amazon's server) processes the request and returns the requested data to your client computer, also by Internet communication pathways. The data reply most likely takes a different route than did the data request. The web browser on your client computer interprets the data and displays it on its monitor.

How is the data sent to the correct computer? Each time you connect to the Internet, your computer is assigned a unique identification number. This number, called an Internet Protocol (IP) address, is a set of four groups of numbers separated by periods, such as 123.45.245.91, and is commonly referred to as a dotted quad or dotted decimal. IP addresses are the means by which computers connected to the Internet identify each other. Similarly, each website is assigned a unique IP address. However, because the numbers that make up IP addresses are difficult

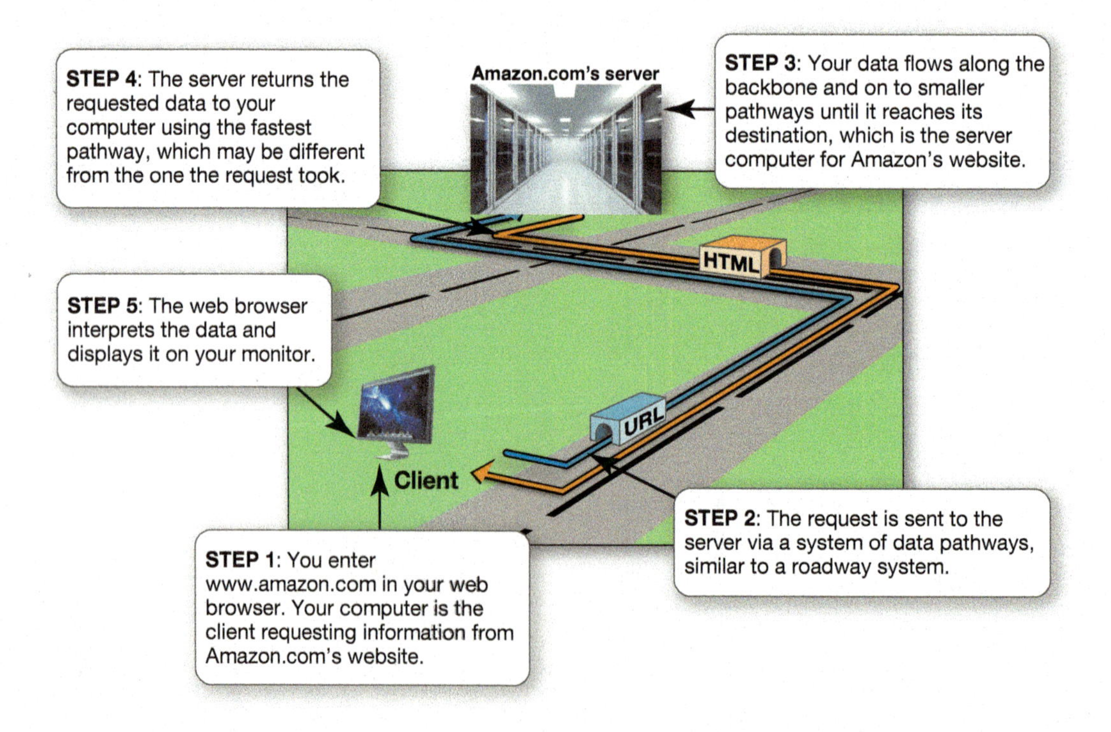

Figure 3.2 How the Internet's Client/Server Model Works *(Phonlamai Photo/Shutterstock; Cigdem/Shutterstock)*

for people to remember, websites are given text versions of their IP addresses. Suppose that Amazon's website has an IP address of 72.21.211.176 and a name of *www.amazon.com*. When you enter *www.amazon.com* into your browser, your computer (with its unique IP address) looks for Amazon's IP address (72.21.211.176). Data is exchanged between Amazon's server computer and your computer, using these unique IP addresses.

Collaborating and Communicating on the Web

Collaborating, or working together to achieve a common goal, is often done in the classroom, in business, and even at home. The advancement in web technologies has facilitated the collaboration process in many ways, especially enabling participants to collaborate from distant locations. In this section, we'll explore some popular means of collaborating on the web.

Collaborating with Web Technologies

Objective 3.3 *Evaluate the tools and technologies used to collaborate on the web.*

How has the web evolved? Over time, our use of the web has evolved from passively browsing web pages to actively creating our own web content and sharing and collaborating on it with others. This collaborative, user-created web is sometimes called the *social web*. Through the social web, we can collaborate on documents, using web applications such as Google Drive, collaborate on projects with tools such as Slack and Basecamp, rate and recommend products or services with Yelp, make connections on social networks such as LinkedIn, and share a video on YouTube or an image on Pinterest. These means of web-based communications are called social media and include social networking, online collaboration and file-sharing tools, blogs, podcasts, and webcasts.

Social Networking

Is there more to social networking than Facebook, Snapchat, and Twitter? Social networking refers to using the web to communicate and share information among your friends and others. Professional, business-oriented networks such as LinkedIn are helpful for members seeking clients, business opportunities, jobs, or job candidates. These sites can help you meet other professionals through connections with people you know. In addition, businesses use social networking for marketing to and communicating with their customers. For example, companies may post special deals on their Facebook page or solicit responses from LinkedIn followers that may help with product development. Table 3.1 lists social networking sites.

How has social networking changed how we relate to each other? When Facebook was created, its purpose was to connect people. As Facebook expanded, and other social networking sites popped up, they were often criticized as being an antisocial force, driving people to their individual devices and away from traditional personal connections. Over time, however, social media has enabled people from diverse locations to form communities of similar interests and causes. Hashtags, in particular, enabled focused communication around specific topics. During the recent coronavirus pandemic, when traditional means of socializing and gathering were prohibited, social networks and other means of online communication connected people. On the other hand, social networks have been flooded with hate speech and false information. Blatant criticisms and accusations have become prominent themes in many politically and racially charged posts, the extent of which ulimately has driven us apart rather than forming bonds between us. Although social networks allow for the freedom of thought and expression, how can they evolve to promote more productive societal conversations that bring about understanding rather than hate?

What are some dos and don'ts of social networking? Until recently, social media users believed that improved privacy policies of most social media sites offered enough privacy protection and protection from identity thieves. But frequently reported privacy violations are a reminder that personal vigilance and oversight are always necessary. Users must be cautious about the type of content they post on these sites. Consider these precautions as you use social networking sites:

- Keep your personal information personal. Don't broadcast the year you were born, your physical address, and the routines of your life (sports games, practices, work schedules) to the general public.

Table 3.1 Types of Social Networking Sites

Social Communication	Media Sharing	Business Networking and Collaboration	Information Sharing

• Facebook • Snapchat • Twitter • WhatsApp	• Instagram • Pinterest • Spotify • YouTube	• Kickstarter • LinkedIn • Slack • StartupNation	• Digg • Reddit • SlideShare • StumbleUpon

Social Commerce and Payment	Social Travel	Health and Fitness

• GoFundMe • Groupon • PayPal • Venmo	• Airbnb • TripAdvisor • Uber • Waze	• Fitocracy • Happier • MapMyFitness • MyFitnessPal

(Rvector/Shutterstock; Best4u/Shutterstock; Aha-Soft/Shutterstock; Blan-k/Shutterstock; ValentinT/Shutterstock; Kristinasavkov/Shutterstock; Sarahdesign/Shutterstock)

- Know who your friends are and know who can see the information you post. Review your privacy settings periodically, because sites change and update their privacy practices frequently.
- Do not post information you might use to set up security questions such as your pet's name.
- Use caution when posting images and know what images others are posting of you. Although privacy settings may offer some comfort, some images may be available for viewing through search engines and may not require site registration to be seen. Online images may become public property and subject to reproduction, and you might not want some—or any—of those images to be distributed.
- Be sure to research posts before sharing them to avoid spreading misinformation.

Many employers and colleges use social networks as a means of gaining information about potential applicants before granting an interview or extending admission or a job offer. In fact, people have been fired from their jobs and expelled from schools for using social media in a questionable way. Generally, questionable content on social media includes negative discussion about the poster's job, employer, or colleagues, or inappropriate content about the poster. The responsibility for your content rests with you. Even though you may have strong privacy settings, you can't control what those you allow to see your content do with it. Therefore, treat all information posted on the web as public and avoid posting damaging words and pictures. Bottom line: Make sure your profile, images, and site content project an image that positively and accurately represents you.

Online Collaboration and File Sharing Tools

What are the best tools to use when collaborating on a project? Rather than passing documents back and forth via e-mail and possibly losing track of which version is the most recent, web-based products such as Google Docs and Microsoft Office Online have features that promote online collaboration. These products offer capabilities that allow users to add, remove, or edit a document's content in real time.

Online collaboration has several benefits, the most obvious of which is that people who aren't in the same location can easily work together on a project. Another benefit of online collaboration is that it helps save time and money. Companies routinely hold meetings online especially given the rise of employees working from home. Online collaboration can also make team projects easier to manage.

Bits&Bytes Social Media Influencers

You may be aware of or even follow certain people online to see what products they're using, trends they're promoting, or advice they're giving. These individuals, known as *social media influencers*, seem like everyday people but actually have a lot of power when it comes to consumer behavior. How so? Because of the more casual way in which influencers share with the public, they're trusted more than information delivered through traditional advertisements.

Influencer marketing has grown into a multibillion-dollar industry. In the early days, brands would send influencers free products or offer to pay small commissions in exchange for a casual promotion. But as the practice has evolved into an invaluable marketing tool, influencers have begun to command significantly higher fees for each post, mention, or product placement.

Influencers generally pick a niche in which they're interested, such as fashion, travel, sports, or cooking. Many influencers use blogs in addition to a social media platform to deliver their content, and they structure their messages so they're relevant and engaging to their target audience. The content and delivery methods are usually casual, honest, and informative, similar to an exchange between friends. Once an influencer has built up a solid following, they can begin to collaborate with brands and be on the path to becoming a major influencer.

Full-featured business enterprise solutions are available, but for smaller collaborative projects, such as those you might encounter in a school group project, you have a variety of web-based options to choose from, many of which are free. Project management tools incorporate tasks and calendars so a project can stay on schedule. Trello, for example, is a free project management tool that organizes projects onto boards and specific tasks onto cards. Cards can contain checklists, images, labels, discussion notes, and the like. All cards are searchable and shareable and can have reminders. Slack is another popular project management tool that incorporates text messaging, file sharing, and other collaborative features.

Figure 3.3 Videoconferencing tools enable colleagues to work together from remote locations. *(Girts Ragelis/ Shutterstock)*

What tools are best to use when you want to work together but from remote locations? Screen-sharing or videoconferencing software can be helpful when team members are working from different locations but still need to collaborate and communicate in real time. Products such as Zoom, Microsoft Teams, and Google Hangouts offer free screen-sharing and videoconferencing (see Figure 3.3).

However, there are certain forms of etiquette and best practices that make videoconferences a good experience for everyone. Keep in mind the following when participating in a videoconference:

- As a host, ensure you send the meeting invite to specific individuals. Do not post on social media or other public sites.
- As a host, manage screen-sharing privileges as needed for each meeting. In some instances just the host needs to share the screen, such as giving a PowerPoint presentation. In other instances, it's necessary for the host and other participants to be able to share their screens.
- As a host, use poll questions periodically to engage the audience and keep them focused.
- As a participant, mute your microphone as a default. Unmute only when you need to speak. The host also has the ability to mute and unmute participants.
- As a participant, avoid multitasking with your video on. If necessary, turn off your video briefly while handling unexpected interruptions.
- As a host or participant, ensure all files and or links to files are ready to go before the meeting begins.
- As a host or participant, ensure your live background and lighting is appropriate and not distracting. If necessary, use a virtual background provided in most videoconferencing applications.

What is the best way to share music, pictures, and videos on the web? Social media sharing sites such as Instagram, YouTube, Flickr, and SoundCloud enable anyone to create and share multimedia. Instagram allows users to share photos and videos and to tag not only their images but others' images as well, thus providing a strong search capability to locate images with similar tags. YouTube enables users to upload videos, whereas SoundCloud is both a distribution platform and a music streaming service. Fledgling musicians can upload original music to share, and music enthusiasts can track and follow new and favorite artists.

Figure 3.4 Cooking classes are often taught using vlogs. *(Lordn/Shutterstock)*

MyLab IT

Blogging

In this Sound Byte, you'll see why blogs are one of today's most popular publishing mediums. You'll also learn how to create and publish your own blog.

Blogs

What is a blog? A blog (short for weblog) is a personal log or journal posted on the web. Anyone can create a blog, and there are millions of blogs available to read, follow, and comment on. Blogs are generally written by a single author and are arranged as a listing of entries on a single page, with the most recent blog entry appearing at the top of the list. In addition, blogs are public, and readers can post comments on a blog, often creating an engaging, interactive experience. Blogs have searchable and organized content, making them user friendly. They're accessible from anywhere using a web browser. BlogCatalog is a blog directory that can help you find blogs that fit your interests.

Many people use blogs as a means to share their story, experiences, or ideas. Some bloggers focus on a particular niche or demographic, at times becoming influencers of products or services because of their perceived knowledge and expertise. Some other blogs focus on a particular topic, such as Engadget, which discusses technogadgets. Bloggers are often attracted to this medium because of the community a blog creates through the comments and interaction of its readers. Additionally, some bloggers begin in hopes of making money by using tools such as Google AdSense and the Amazon Associates Program. Although not every blog is guaranteed to generate large sums, bloggers with strong followings can earn a good income over time.

Are all blogs text-based? The traditional form of a blog is primarily text-based but may also include images and audio. A vlog (short for video blog) is a blog that uses video as the primary content (although it can also contain text, images, and audio). Vlogs are a popular means of personal expression but are also great for online instruction (see Figure 3.4). You can find vlogs by searching YouTube.

How do I create a blog? Many websites provide the tools you need to create your own blog. Two sites that offer free blog hosting are Blogger and WordPress. Such tools also let you add features like pictures or subpages to your blog. You can also choose to host your blog yourself so that the URL reflects your name or the name of your business. If you choose this option, you need your own website and a URL so people can access it.

Are Twitter and Tumblr blogs? Twitter and Tumblr are examples of microblogs, where users post short text messages with frequent updates. Twitter and Tumblr also allow users to post multimedia and other content. Posts can be public or restricted to a certain audience.

Podcasts and Webcasts

How would I distribute a series of audio or video files over the Internet to a wide audience? A podcast is audio content delivered in a series over the Internet. There are podcasts for radio shows, audiobooks, magazines, and educational programs, which you can download to your device. For example, you can access lessons on yoga, foreign language classes, or do-it-yourself tips. Most podcasts are audio only, but there are great video podcasts for topics that require a visual aspect, such as fitness and cooking.

How are podcasts different from simple audio files? A podcast is a series of audio files. What makes a podcast different is that you can subscribe to the series of audio files (called episodes). When a new episode is available, if you've subscribed to the podcast, the new episode is pushed to you automatically. You don't need to search for an update. The mechanism that podcasts use to deliver their content is Really Simple Syndication (RSS). RSS is a format that sends the latest content of the podcast series automatically to an aggregator such as Stitcher or Apple Podcast. An aggregator locates all the RSS series to which you've subscribed and automatically downloads the new content to your device. If you have several favorite podcasts, rather than checking each one for updated content, aggregators collect all the podcast updates in one place and deliver them to your device

Where can I find podcasts? Several podcast directories and aggregators offer a wide variety of podcasts, as shown in Table 3.2.

What is a webcast (or live streaming)? A webcast is the (usually live) broadcast of streaming audio or video content over the Internet. Unlike podcasts, which are prerecorded and made available for download, most webcasts are distributed in real time, as the event is taking place. Live streaming is as simple as taking video with your phone or a GoPro camera and posting it to Facebook Live or Instagram Live. Some webcasts are archived so people can view them later. Unlike podcasts, webcasts don't use an aggregator to update new episodes automatically.

Table 3.2 Podcast Directories and Aggregators

Podcast Directories and Aggregators
Overcast (overcast.fm)
• Podcast player for Apple mobile devices
Pocket Casts (pocketcasts.com)
• Broad collection of podcasts
Stitcher (stitcher.com)
• Customize podcast playlists
Spotify (spotify.com)
• Good source for audio and video podcasts

Webcasts use continuous audio and video feeds, which let you view and download large audio and video files. Webcasts can include non-interactive content, such as a simultaneous broadcast of a radio or television program, but some webcasts invite interactive responses from the audience. Nugs.net provides access to live music and professionally recorded concerts, and BroadcastMed provides webcasts that demonstrate the latest surgical innovations and techniques. Webcasts are used in the corporate world to broadcast annual meetings and in the educational arena to transmit seminars.

Communicating over the Web

Objective 3.4 *Summarize the technologies used to communicate over the web.*

Why do I still need e-mail? Despite the popularity of social media, e-mail (short for *electronic mail*)—a written message sent or received over the Internet—still remains the most widely used form of communication on the Internet. E-mail is the primary method of electronic communication worldwide because it's fast and convenient, and because it's asynchronous, users do not need to be communicating at the same time. They can send and respond to messages at their own convenience. E-mail is also convenient for exchanging files via attachments.

How private is e-mail? Although e-mail is a more private exchange of information than public social networking sites are, it is not really private. Consider the following:

- Because e-mail can be printed or forwarded, you never know who may read it.
- Most e-mail content is not encrypted, so you should never use e-mail to send personal or sensitive information, such as bank account or Social Security numbers. Doing so could lead to your identity being stolen.
- Employers have the right to access to e-mail sent from the workplace, so use caution when putting negative or controversial content in e-mail from your workplace.
- Even after you've deleted a message, it doesn't really vanish. Many Internet service providers and companies archive e-mail, which can be accessed or subpoenaed in the event of a lawsuit or investigation.

What are some tips on e-mail etiquette? When you write a casual e-mail to friends, you obviously don't need to follow any specific e-mail guidelines. But when you send e-mail for professional reasons, you should use proper e-mail etiquette. The following are a few guidelines:

- Be concise and to the point.
- Use the spell-checker and proofread your e-mail before sending it.
- Avoid using abbreviations such as u, r, and LOL.
- Include a meaningful subject line to help recipients prioritize and organize e-mail messages.
- Add a signature line that includes your contact information.
- Include only those people on the e-mail who truly need to receive it.

Following such guidelines can promote a professional image and make your message more effective.

Are there different types of e-mail systems? There are two types of e-mail systems:

1. Web-based e-mail, such as Yahoo! Mail or Gmail, is managed with your web browser; you access your e-mail with an Internet connection.

Figure 3.5 Texting uses the same cellular network you use to send voice calls from your mobile device. *(Rfranca/Shutterstock)*

2. A desktop client requires a program, such as Microsoft Outlook, to be installed on your computer. When you open the program, your e-mail is downloaded to your computer.

The primary difference between the two is access. Web-based e-mail allows you to access e-mail from any Internet-connected device. This means you can't send or receive e-mail unless you're online. With a desktop client, you view your e-mail on the computer on which the desktop client software is installed. Desktop clients also often have integrated functionalities such as a calendar, although some web-based e-mail clients are beginning to offer some integrated functionalities as well. You can set up either e-mail system to display e-mail from other web-based or desktop client accounts. So if you have separate work, school, and personal e-mail accounts and use both desktop and web-based e-mail systems, you don't need to have multiple accounts and systems open at the same time.

How do texting and IM work? Texting (or text messaging) is the most popular means of mobile communication (see Figure 3.5). Texting uses the same cellular network you use to make voice calls from your mobile device. It uses Short Message Service (SMS) for text-based messages and Multimedia Message Service (MMS) to send images or videos. Snapchat is a popular messaging app that uses MMS so users can send brief videos or images. With Snapchat, once you open a message, the video or image is available for you to view for only a brief time and you generally can't save it.

Instant messaging (IM) is a way of communicating in real time over the Internet. People may use IM when they can't access a cellular network, such as when they travel abroad. Some messaging apps, such as WhatsApp and Facebook Messenger, let you conduct video chats using a webcam. Messaging apps are proprietary, meaning you can IM or chat only with those who share the same IM service. For example, you can't use Facebook Messenger to send a message to a Google Hangouts user or vice versa.

IM is also an important means of quick and efficient communication in the business world. In fact, some IM services, such as Cisco Jabber and Oracle Beehive, were developed primarily for business use. IM is also great for business communications because you can monitor who's allowed to contact you, and people can communicate with you only if they know your exact e-mail or IM address.

Figure 3.6 VoIP technology lets your computing device behave like a phone or video phone. *(David Malan/Stone/Getty Images)*

How secure are texting and IM? The messages you send by text or IM may not be secure and can be intercepted. Luckily, messaging apps exist that encrypt your messages before they leave your device, so if they're intercepted, they'll be meaningless. Encryption ensures that no one can read messages while they're being sent to the recipient. WhatsApp, Signal, and ChatSecure all use end-to-end encryption, meaning conversations can't be read by anyone but you and the person you're messaging.

What's the best way to make phone calls over the Internet? Traditional telephone communications use analog voice data and telephone connections. In contrast, VoIP (Voice over Internet Protocol) is a digital phone service that transmits calls over the Internet rather than over traditional phone lines or cellular networks (see Figure 3.6). You can place VoIP calls from anywhere you have Internet access. Major Internet service

Table 3.3 Methods of Online Collaboration and Communication

Social Networking	Online Collaboration and File Sharing	Blogs	Podcasts and Webcasts

• Lets you build an online network of friends • Lets you share media content	• Web-based document products such as Google Docs • Can include screen sharing, project management features, and videoconferencing • Images, videos, and music are shared via social media sharing sites	• Chronologic entries • Searchable content • May include images, audio, and video • Video logs are called *vlogs*	• Audio or video content delivered as a series, using RSS • New content collected with aggregator • Webcasts are usually live streamed broadcasts
E-Mail	**Texting**	**Instant Messaging**	**VoIP**
	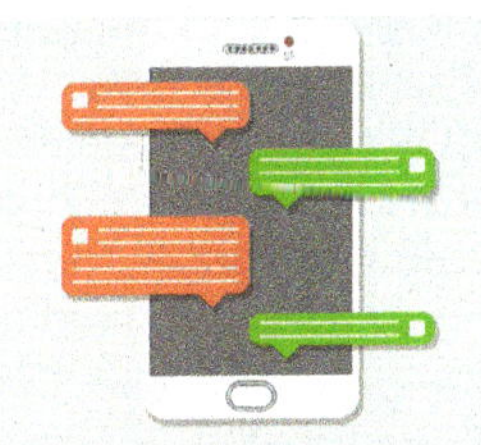		
• Most common form of online communication • Asynchronous (not having to be done at the same time)	• Needs a cellular network • Uses Short Messaging Service (SMS) for text messages • Uses Multimedia Message Service (MMS) for video and audio messages	• Real-time communication • Used with individuals or groups • Capable of doing video/audio chats	• Voice communication via the Internet • Free or low-cost calls

(Creative Stall/Shutterstock; Creative Stall/Shutterstock; VectorsMarket/Shutterstock; Icon99/Shutterstock; RedlineVector/Shutterstock; Marnikus/Shutterstock; Vlad Kochelaevskiy/Shutterstock; Jemastock/Shutterstock; Sky vectors/Shutterstock; Yuriy Vlasenko/Shutterstock)

providers (ISPs), like Comcast and Verizon, provide VoIP phone services as an option you can package with your Internet or cable plan. Table 3.3 lists the popular methods of online collaboration and communication that we've discussed.

Conducting Business on the Web

You can buy nearly anything on the web, including big-ticket items such as homes and cars. Now, with mobile apps and cloud technology, there is even greater ability to conduct business from virtually any location, at any time. In this section, we'll look at various ways of conducting business on the web.

Being Productive with Cloud Technologies

Objective 3.5 *Describe how cloud technologies are used to create, store, and share files.*

What exactly is cloud computing? Cloud computing refers to storing data, files, and applications on the Internet and being able to access and manipulate these files and applications from any Internet-connected device. When you work from the cloud, you don't need to have everything stored on your own computer's drives. Instead, you can access your files and programs from any device as long as you have access to the Internet. In addition, cloud computing makes it easier to collaborate and communicate with others.

How does cloud computing work? There are two sides to cloud computing (see Figure 3.7):

- The *front end* is the side we see as users. It involves a web browser like Google Chrome, Microsoft Edge, or Mozilla Firefox.
- The *back end* consists of data centers and server farms that house the files and programs you access on the cloud. These data centers and server farms are warehouses full of computers and servers located all over the world. The computers in the data centers or server farms are designed to work together, adjusting to the demand placed on them at any time.

What are examples of cloud computing? Google was one of the first true explorers in the cloud. The Chromebook is Google's cloud-based system where all the applications and files are stored on the web. Little is installed or saved to a Chromebook hard drive, not even the operating system! The web-based operating system enables users to sign in on any Internet-connected computer and display their computer setup (desktop configurations and images, programs, files, and other personalized settings). Microsoft Windows 10 and macOS operating systems, although installed on a computer's hard drive, host the user's account data in the cloud and thus also enable users to access their operating system setup on any Internet-connected computer by entering their username and password. Google offers other cloud-based services such as Google Drive, the G-Suite of software, Gmail, and the Chrome web browser.

Microsoft has an online version of Office that is accessed through its cloud-based storage system, OneDrive. Other online storage systems such as Dropbox, Box, and iCloud offer the ability to store files in a single cloud location and provide access to saved files by multiple users. This single-location access facilitates collaboration on a file by multiple users without passing files around by e-mail that generates multiple versions of the same file.

Are there things to watch out for with cloud computing? There are some considerations with cloud computing, such as security, backup, and access issues:

- **Security and privacy:** Cloud security is based on the passwords we set and the security systems that the data centers put in place to keep our information away from unauthorized users. OneDrive and Google Drive have separate vault storage folders that are intended to store important or sensitive files. These vaults use stronger authentication methods, such as requiring a fingerprint, face recognition, or other two-step authentication. Regardless, caution is always warranted, because nothing is completely safe and private.
- **Backup:** Because even the devices in these large data centers and server farms inevitably break down, the cloud computing systems have redundant systems to provide backup. However, for critical files that you must have, it might be a good idea not to rely completely on the cloud, but to have your own offline backup system as well.
- **Access issues:** With cloud computing, you access your files and programs only through the Internet. If you can't access the Internet due to a power failure or system failure with your Internet service provider, you won't be able to access your files. Storing your most critical files and programs offline can help reduce the inconvenience and loss of productivity while access to the Internet is being restored. Some services such as Dropbox and OneDrive store a synced version of your online files to the local storage on your computer.

THE CLOUD
amazon.com
Gmail
Google docs

Figure 3.7 There are two sides to cloud computing: the side we see as users and the banks of computers and servers that house the files and programs we access. *(Godshutter/Shutterstock; SFIO CRACHO/Shutterstock; Yakobchuk Viacheslav/Shutterstock Ferenc Szelepcsenyi/Shutterstock; Anatolii Babii/Alamy Stock Photo; Alamy; rvlsoft/Shutterstock; Digitallife/Alamy Stock Photo; Used with Permission from Dropbox Inc; Courtesy of Google Inc.; PhotoEdit/Alamy Stock Photo)*

Dig Deeper Artificial Intelligence and Cloud Computing

Our society has quickly become dependent on Internet-based devices, software, and services that generate enormous amounts of data, known as big data. Cloud computing is the compilation of large server centers that use the Internet, the cloud, to store, manage, and process all that big data. As more and more data goes into the cloud, businesses need more powerful ways to organize and understand their data. Artificial intelligence (AI), especially the branches of deep learning and machine learning, is being used to analyze all that data stored in the cloud and apply it to solve society's issues. The ability to expand the capacity of cloud computing further fuels the capabilities of AI to analyze huge amounts of data, enabling companies and industries to learn from the analyses and apply the results to create better outcomes.

For example, the aerospace industry uses AI to process large volumes of satellite data to help improve the accuracy of autonomous, self-driving vehicles. Law enforcement personnel use AI image recognition and processing to help prevent crime. Business call centers use AI-fueled language translators to facilitate global commerce and communication, including the translation of sign language into text and speech. AI cloud computing services help the medical community predict clinical outcomes based on data from scientific papers, electronic medical records, and physician notes. AI and cloud computing was used to analyze all collected COVID-19 data to help detect and diagnose carriers of the virus to curb the global pandemic. The applications of AI and cloud computing are seemingly as endless as the data our devices generate.

Conducting Business Online

Objective 3.6 *Describe how business is conducted using the Internet.*

What are the different types of e-commerce? E-commerce—short for electronic commerce—is the process of conducting business online. Typically, e-commerce is identified by whom the business is being conducted between:

1. Business-to-consumer (B2C) transactions take place between businesses and consumers. Such transactions include those between customers and completely online businesses, such as Zappos, and those between customers and stores that have both an online and a physical presence, such as Target. Businesses with an online and physical presence are referred to as click-and-brick businesses. Many click-and-bricks allow online purchases and in-store pickups and returns.
2. Business-to-business (B2B) transactions occur when businesses buy and sell goods and services to other businesses. An example is Omaha Paper Company, which distributes paper products to other companies.
3. Consumer-to-consumer (C2C) transactions occur when consumers sell to each other through sites such as Craigslist and Etsy.

Social commerce is a subset of e-commerce that uses social networks to assist in marketing and purchasing products. Consumers voice their opinions on social media sites, such as Yelp and Facebook, about products and services by providing ratings and reviews, and studies show that such peer recommendations have a major influence on buying behavior. When you see your friend recommend a product or service, you're more likely to click through to that retailer and check out the product.

Other peer-influenced e-commerce trends include group-buying and individual customization. Groupon and LivingSocial are two popular group purchase services that offer discounted deals. CafePress and Zazzle sell T-shirts and other items that are customized with your own graphic designs.

E-commerce encompasses more than just online shopping opportunities from your desktop or laptop computers. Mobile commerce (or m-commerce) is conducting commercial transactions through a smartphone, tablet, or other mobile device. Apple Pay, Venmo, and other digital payment systems are popular, as is making payments by using proprietary mobile apps such as at Starbucks. Many businesses push coupons and offers through their mobile apps to customers, using location-based services. Some fast-food places and restaurants such as Panera and Chipotle enable you to place an order through their mobile app for in-store pick-up or delivery. Mobile commerce also includes services that enable users to check their bank and credit card account balances, deposit checks, pay bills online, and manage investment portfolios. Tickets and boarding passes can also be sent to smartphones, adding another level of convenience.

Bits&Bytes Looking for Extra Money? Try a Side Hustle

The web offers many opportunities through which you can earn extra money by working part-time. These gigs are called *side hustles*. Side hustles come in many shapes and sizes, ranging from a hobbyist venture to a full-time career. The web has facilitated side hustles such as becoming an Uber driver, renting your home or apartment through Airbnb, being an errand-runner through TaskRabbit, writing a blog, creating websites, or offering child or adult care services through Care.com. You can also explore freelance opportunities through Upwork.

Figure 3.8 Mobile wallet payment systems make in-store purchases simple and safe.

E-Commerce Safeguards

Objective 3.7 *Summarize precautions you should take when doing business online.*

Should I trust using a mobile wallet? Paying for items by using a smartphone or smartwatch with a mobile wallet, an app that stores payment card information on your mobile device such as Apple Pay or Android Pay, is pretty standard when purchasing items in a brick and mortar store (see Figure 3.8). Assuming you've taken the basic precaution to protect your device from being used by anyone other than yourself, using a mobile payment is safer than using a physical credit card. Using a mobile payment generates an authentication code for that specific purchase that no one else can use again. In addition, mobile payments avoid being caught up in skimming scams that involve taking small amounts of cash off the transaction before it is formally entered in the system. Ensure that you turn on the Find Your Device setting in case you lose your phone or watch, and make sure you're able to wipe the data remotely to prevent unauthorized users from accessing your information.

Just how safe are online transactions? When you use a credit card to buy something online, money is exchanged directly between your credit card company and the online merchant's bank. However, take the following precautions, which are also summarized in Table 3.4, when shopping online to ensure a safe transaction:

- Avoid making online transactions when using public computers. Public computers may have spyware installed, which are programs that track and log your keystrokes and can retrieve your private information. Similarly, unless you have specific protection on your mobile device, avoid making wireless transactions on public hotspots.
- Look for visual indicators that the website is secure. Check that the beginning of the URL changes from "http://" to "https://"—with the *s* standing for *secure*, indicating that the Secure Sockets Layer protocol has been applied to manage the security of the website. Also, look for a small icon of a closed padlock in the address bar or a green-colored text or a green address bar—indications that the site may be secure. (However, note that even if a site has these indicators, it still might not be safe. Consider the validity of the site before making a purchase.)
- Shop at well-known, reputable sites. If you aren't familiar with a site, investigate it with the Better Business Bureau or at Bizrate. Make sure the company has a phone number and street address in addition to a website. You can also look for third-party verification such as that from TrustArc or the Better Business Bureau, but let common sense prevail. Online deals that seem too good to be true are generally just that—and may be pirated software or illegal distributions.
- Pay by credit card, not debit card. Federal laws protect credit card users, but debit card users don't have the same protection. If possible, reserve one credit card for Internet purchases only; even better, use a prepaid credit card that has a small credit limit. For an extra layer of security, find out whether your credit card company has a service that confirms your identity with an extra password or code that only you know to use when making an online transaction or that offers a one-time-use credit card number. In addition, consider using a third-party payment processor such as PayPal or a mobile wallet app. PayPal also offers a security key that provides additional security to your PayPal account.
- When you place an order, check the return policy, save a copy of the order, and make sure you receive a confirmation number. Make sure you read and understand warranties, return policies, and the retailer's privacy statements. If the site disappears overnight, this information may help you in filing a dispute or reporting a problem to a site such as the Better Business Bureau.

MyLab IT

Doing Business Online

In this Helpdesk, you will play the role of a helpdesk staffer, fielding questions about e-commerce and e-commerce safeguards.

Whether you're doing business or collaborating and communicating with friends or colleagues, the Internet makes these activities more accessible. The Internet can enrich these experiences and activities as well, although you must take precautions for the safest experiences.

Table 3.4 Online Shopping Precautions

When shopping at home, use a firewall and antivirus software for general computer protection.	Don't shop on public Wi-Fi networks, because they may contain spyware.	Check for visual indicators such as https://in the URL, a closed padlock icon, and green text or a green address bar.	Look for third-party verification from TrustArc or the Better Business Bureau symbol.
			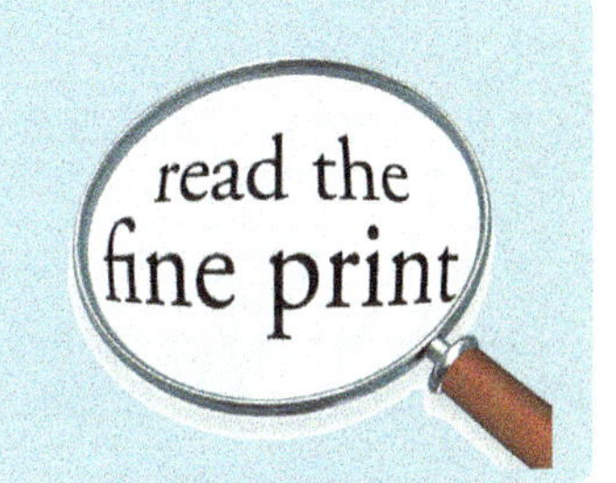
Use a credit card, not a debit card, to protect transactions, or use a third-party payer such as PayPal or a mobile wallet app.	Create a strong password for all online accounts (one that includes numbers and other symbols such as @).	Deals that are too good to be true are usually just that.	Read and understand the fine print on warranties, return policies, and the retailer's privacy statements.

(Beboy/Shutterstock; Photoinnovation/Shutterstock; Jozsef Bagota/Shutterstock; Piotr Swat/Shutterstock; Maxx-Studio/Shutterstock; Isak55/Shutterstock; Alhovik/Shutterstock; Iqoncept/123RF)

Bits&Bytes Bitcoin: A Form of Virtual Currency

Bitcoin is a form of virtual currency (or cryptocurrency) that eliminates intermediaries or banks to conduct transactions, thus enabling direct anonymous transactions between users (see Figure 3.9). Payments are sent peer to peer from the payee to the recipient with no transaction fees and without the need to supply personal information. The time and amount of every finalized bitcoin transaction are kept on a massive public online ledger known as the *blockchain*. The public nature of the blockchain makes it difficult for bitcoin transactions to be manipulated by anyone. However, the value of the currency is extremely volatile and can fluctuate widely within hours of a day.

If you want to hold and spend bitcoins, you must create a bitcoin wallet, which stores the information needed to complete bitcoin transactions. The security of bitcoin wallets is evolving, but there is a real risk of a wallet being hacked and the contents of the wallet stolen.

Although the anonymity of the currency initially encouraged illicit transactions, the use of bitcoins for legitimate purposes is taking hold. Mainstream retailers such as Home Depot, Microsoft, Expedia, and Overstock.com are accepting bitcoins, as are many small businesses. Although Bitcoin has the largest user base and market capitalization, it's not the only virtual currency. Alternatives include Litecoin, Ethereum, Zcash, Dash, Ripple, and Monero.

Figure 3.9 Bitcoin is a type of virtual currency. *(Zapp2Photo/Shutterstock)*

Before moving on to Part 2:

1. Watch Chapter Overview Video 3.1.
2. Then take the Check Your Understanding quiz.

Check Your Understanding // Review & Practice

For a quick review to see what you've learned so far, answer the following questions.

multiple choice

1. Which is NOT an event associated with the beginning of the Internet?
   a. Vinton Cerf helped develop the Internet communications protocol.
   b. *Amazon.com* was one of the first websites on the Internet.
   c. "Login" was intended to be the first network transmission
   d. The U.S. Department of Defense created ARPANET.
2. Which of the following best describes the transmission lines with the fastest speeds?
   a. Internet backbone
   b. Main Internet
   c. Main backbone
   d. Primary backbone
3. Which of the following is not considered a cloud computing application?
   a. Google Drive
   b. Microsoft Online Office apps
   c. Dropbox
   d. Windows 10
4. Which of the following does not describe a desktop e-mail client?
   a. The e-mail program is installed on your computer.
   b. E-mail is downloaded when the program is opened.
   c. Sending e-mail does not depend on an Internet connection.
   d. Can access e-mail from any Internet-connected device.
5. Which of the following best describes a mobile wallet?
   a. A case to hold credit cards that attaches to your smartphone
   b. An app that exchanges payments between friends
   c. An app that stores payment card information
   d. Another name for PayPal or Venmo

chew on this

(3Dalia/Shutterstock)

E-commerce businesses collect purchase data and web-browsing data from customers. They use it to give you more personalized services. You may receive special offers based on your interests, for example. Some businesses sell their data to other companies for a profit. Are these practices an invasion of privacy? Where do you stand?

MyLab IT Go to **MyLab IT** to take an autograded version of the *Check Your Understanding* review and to find all media resources for the chapter.

For the IT Simulation for this chapter, see MyLab IT.

Try This

Use OneDrive to Store and Share Your Files in the Cloud

You may use a USB drive or use e-mail to access files when away from your computer. However, if you've updated the file on multiple devices or at different times, there can be confusion about which is the most current version. You can also lose the file completely by misplacing your USB drive or deleting the e-mail attachment.

Instead, use a web-based storage and sharing service such as OneDrive, Dropbox, or Google Drive. In this Try This, we'll explore OneDrive. You can access OneDrive through Microsoft Office Online or with an Office 365 subscription. For more step-by-step instructions, watch the Chapter 3 Try This video on MyLab IT.

Step 1 **Sign in to OneDrive:** Go to **onedrive.com**. Sign in with your Microsoft account. If you don't have a Microsoft account, creating one is easy.

Step 2 **Create a Folder and Add Files:** Once in OneDrive, you can create a folder or begin to add files to OneDrive.

- To create a folder: Click **New** at the top of the page, click **Folder**, and then give your new folder a name. Click **Create** and then select the new folder to open it.
- To add a file: Click **Upload** at the top of the page, select **Files**, and then locate the file on your local computer and click **Open**. To upload more than one file, press and hold **Ctrl** while you select each file.

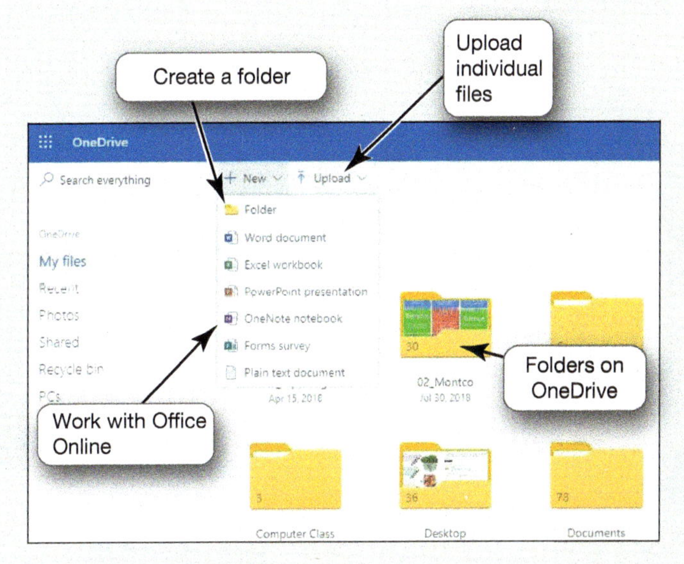

(Courtesy of Microsoft Corp.)

Step 3 **Share a File or Folder:** To share a file or folder, complete the following steps:

1. Right-click the file or folder that you want to share and click **Share** or click **Share** in the top menu after selecting the desired file or folder.
2. Enter a name or e-mail address to send a link to the shared file or folder. Add an optional message. Click **Send**.
3. Editing privileges are established by default, so if you want to restrict editing privileges, click to **clear** the **Allow editing** box.
4. To see which files have been shared with you, click **Shared** in the left menu.

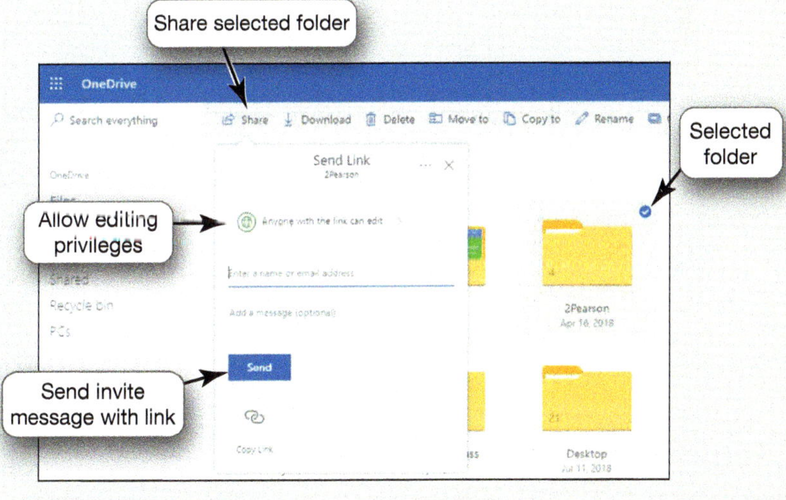

(Courtesy of Microsoft Corp.)

Make This  TOOL: App Inventor 2 or Thunkable

A Web-Capable App

Want your app to be able to display a web page?

In this exercise, you will continue your mobile app development by adding a new power to your apps: web browsing.

It is as easy as using the WebViewer component in **App Inventor**. Drag a WebViewer component onto your Designer screen and then control it with the Blocks for WebViewer. These allow you to go back, forward, to the home page, or to any specific URL.

The WebViewer component allows you to control a live browser inside your mobile app.

```
when ListPicker1 .BeforePicking
do  set ListPicker1 . Elements to  make a list  " beach "
                                                " island "
                                                " waterfall "
                                                " national park "

when ListPicker1 .AfterPicking
do  set TextBox1 . Text to  ListPicker1 . Selection
    call WebViewer1 .GoToUrl
                       url  join  " https://www.flickr.com/search/?q= "
                                  ListPicker1 . Selection
    call TextToSpeech1 .Speak
                   message  join  " My happy place is a "
                                  ListPicker1 . Selection
```

(Copyright MIT, used with permission.)

(Copyright MIT, used with permission.)

To create this app for iOS, go to Thunkable.com, a programming platform based on App Inventor.
For the instructions for this exercise, go to MyLab IT.

Part 2

 For an overview of this part of the chapter, watch **Chapter Overview Video 3.2**.

Using the Web Effectively

Learning Outcome 3.2 You will be able to describe the tools and techniques required to navigate and search the web.

You no doubt know how to use the web—to buy products, send e-mail, visit Facebook, and use Google—but do you know how to use it effectively? In this section, we'll look at ways to make your online experience more enjoyable, more productive, and more efficient.

Accessing and Moving Around the Web

Can you imagine a day without accessing the Internet? Although Internet usage is virtually ubiquitous and essential to some, others still don't have the ability or the desire to access the Internet. Regardless of how little or how often we access the Internet, none of our online activities could happen without an important software application: a web browser.

Web Browsers

Objective 3.8 *Explain what web browsers are and describe their common features.*

What is a web browser exactly? A web browser is software you can use to locate, view, and navigate the web. Most browsers today are graphical browsers, meaning they can display pictures (graphics) in addition to text and other forms of multimedia such as sound and video. A list of the most common browsers appears in Figure 3.10.

What features do web browsers offer? The most popular browsers share similar features that make the user experience more efficient (see Figure 3.11). For example, browsers include an omnibox, a combined search and address bar, so you can both type a website URL or search the web from the address bar. Other common features include the following:

- **Tabbed browsing:** Web pages are loaded in tabs within the same browser window. Rather than having to switch among web pages in several open windows, you can flip between the tabs in one window. You may also save a group of tabs as a Favorites group if there are several tabs you often open at the same time.
- **Pinned tabs:** You can pin tabs to lock your most used tabs to the left of the tab bar. Pinned tabs are smaller in size and can't be closed by accident. They will display automatically when the browser opens.
- **Tear-off tabs:** An opened tab can be dragged or moved away from its current window so that it's then opened in a new window.
- **Thumbnail previews:** Another convenient navigation tool that most browsers share is providing thumbnail previews of all web pages in open tabs. Microsoft Edge also enables you to display newsfeed items as well.
- **Tab isolation:** With this feature, tabs are independent of each other, so if one crashes, it does not affect the other tabs.
- **SmartScreen filter:** Most browsers offer built-in protection against phishing, malware, and other web-based threats.

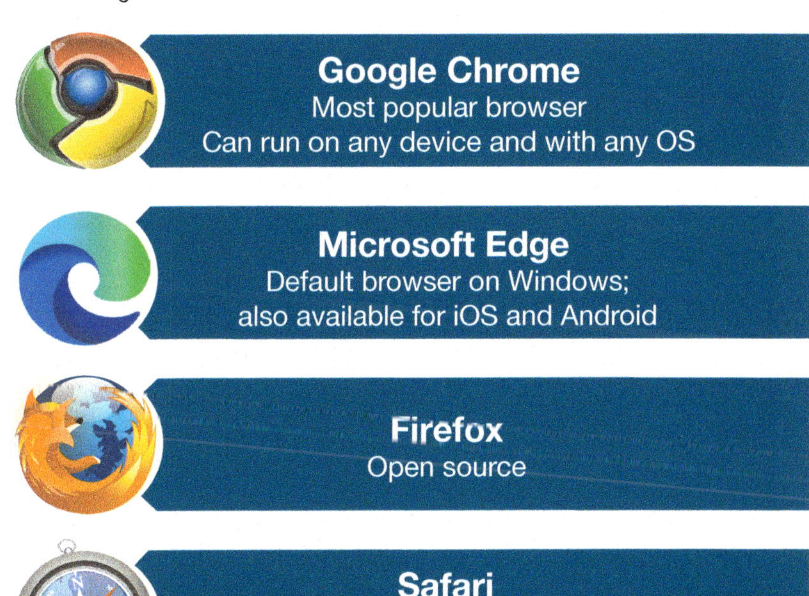

Figure 3.10 Common Web Browsers *(Tommy (Louth)/Alamy Stock Photo; Emil Mammadov/Shutterstock; Lucia Lanpur/Alamy Stock Photo; 2020WEB/Alamy Stock Photo)*

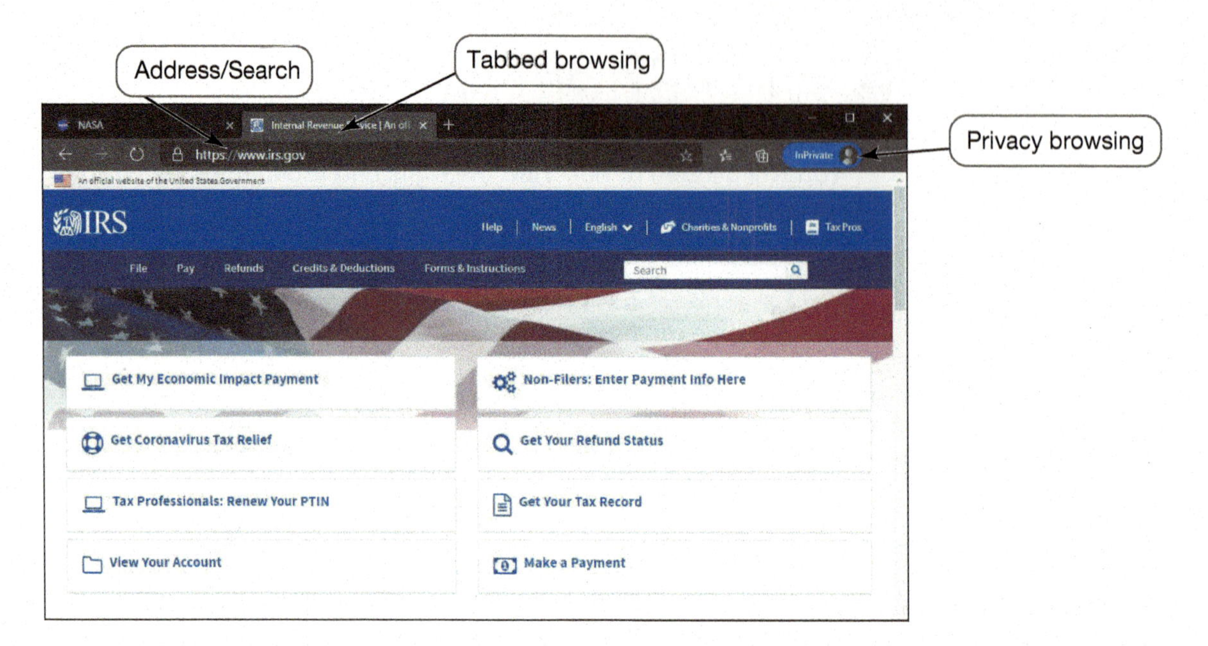

Figure 3.11 Tabbed and privacy browsing and an address bar that doubles as a search bar are common features in today's browsers. *(Homepage, Internal Revenue Service, Homepage, Aeronautics and Space Administration; Courtesy of Google Inc.)*

- **Privacy browsing:** Privacy features (such as InPrivate Browsing in Microsoft Edge or Incognito in Chrome) enable you to browse the web without retaining a history trail, temporary Internet files, cookies, or usernames and passwords. These features are especially helpful when you use public computers at school, for example.
- **Add-ons and extensions:** Add-ons (also known as extensions) are small programs that customize and increase the functionality of the browser. Examples include Video DownloadHelper, which converts web videos like those found on YouTube to files you can save.
- **Session Restore:** Brings back all your active web pages if the browser or system shuts down unexpectedly.
- **Multiple profiles:** Chrome and Edge allow for individual profiles if multiple users are on the same device.

Microsoft Edge has some unique features. You can make Web Notes by annotating a web page and then share them, using Onenote or Evernote or save them for later use. Edge is also integrated with Microsoft's personal assistant Cortana, which shares web content with you based on your indicated preferences and interests. You can also save online content to read later, and if the noise of a web page is distracting to you, you can use Reading View for a cleaner, simpler layout.

URLs, Protocols, and Domain Names

Objective 3.9 *Explain what a URL is and discuss its main parts.*

What do all the parts of a URL mean? You gain initial access to a particular website by typing its unique address, or Uniform Resource Locator (URL, pronounced "you-are-ell"), in your browser. A website is comprised of many web pages, each of which is a separate document with its own unique URL. Like a regular street address, a URL is made up of several parts that help identify the web document it stands for (see Figure 3.12):

- the *protocol* (set of rules) used to retrieve the document
- the domain name, which includes the *top-level domain*
- the *path* or subdirectory

What's the protocol? You're probably most familiar with URLs that begin with http, which is short for Hypertext Transfer Protocol (HTTP). HTTP is the protocol that transfers files from a web server—a computer that hosts the website you're requesting—so that you can see it on your computer. The web is based on the HTTP protocol. Because it is so universal, most current browsers no longer require you to enter the http://protocol or the www of an address.

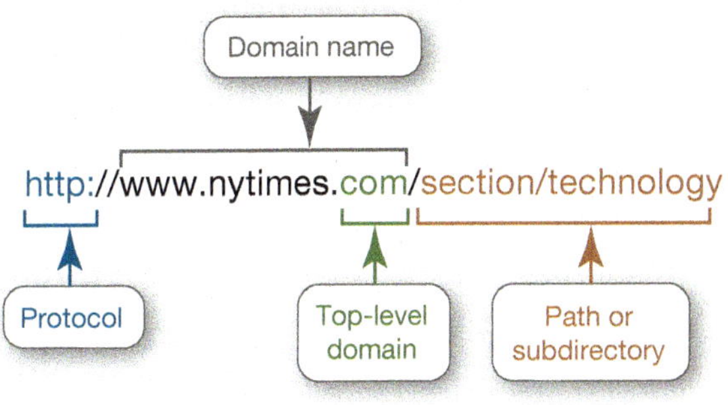

Figure 3.12 The Parts of a URL. *(The New York Times Company)*

Is HTTP the only protocol I need to use? HTTP is the most common protocol, but it's not the only one. HTTP is part of the Internet Protocol suite, a group of protocols that govern how information is exchanged on a network. Other important protocols are the communication protocols developed by Vincent Cerf and Bob Kahn. These protocols, Transmission Control Protocol (TCP) and Internet Protocol (IP), are usually referred together as TCP/IP. TCP/IP allows different computers to work together without issues and governs how information passes through the Internet. Another common protocol is the File Transfer Protocol (FTP). As its name implies, FTP was originally designed to transfer files from a computer to a web server. Today, FTP is often used when you have large files to upload or download. To connect to most FTP servers, you need a user ID and password. To upload and download files from FTP sites, you can use a browser or file transfer software such as FileZilla.

BitTorrent, like FTP, is a protocol used to transfer files, although it's not part of the Internet protocol suite. To use BitTorrent, you install a software client program. It uses a peer-to-peer networking system so that sharing occurs between connected computers on which the BitTorrent client is installed. BitTorrent is popular especially among users who want to share music, movies, and games. Use caution, however, when accessing BitTorrent content. Because it is a peer-to-peer system, it's possible for copyrighted material to be shared illegally.

What's in a domain name? The domain name identifies the site's host, the location that maintains the computers that store the website files. For example, www.berkeley.edu is the domain name for the University of California at Berkeley website. The suffix in the domain name after the dot (such as "com" or "edu") is called the top-level domain. This suffix indicates the kind of organization to which the host belongs. Some browsers don't require you to enter the top-level domain if it is a .com—the browser enters the domain automatically. For example, you can just type in Amazon in any browser and the browser automatically fills in the .com. Table 3.5 lists the most frequently used top-level domains.

Each country has its own top-level domain. These are two-letter designations such as .za for South Africa, and .de for Germany. A sampling of country codes is shown in Table 3.6. Within a country-specific domain, further subdivisions can be made for regions or states. For instance, the .us domain contains subdomains for each state, using the two-letter abbreviation of the state.

What's the information after the domain name that I sometimes see? When the URL is the domain name, such as *www.nytimes.com*, you're requesting a site's home page. However, sometimes a forward slash and additional text follow the domain name, such as in *www.nytimes.com/section/technology*. The information after each slash indicates a particular file or path (or

Table 3.5 Common Top-Level Domains and Their Authorized Users

Domain Name	Who Can Use It
.biz	Businesses
.com	Originally for commercial sites, but now can be used by anyone
.edu	Degree-granting institutions
.gov	Local, state, and federal U.S. governments
.info	Information service providers
.mil	US military
.net	Originally for networking organizations, but no longer restricted
.org	Organizations (often not-for-profits)

Table 3.6 Examples of Country Codes

Country Code	Country
.au	Australia
.ca	Canada
.jp	Japan
.uk	United Kingdom

Note: For a full listing of country codes, refer to *www.iana.org/domains/root/db/*.

subdirectory) within the website. If you use the URL in Figure 3.12, you would connect to the technology section on the *New York Times* site.

Navigating the Web

Objective 3.10 *Describe tools used to navigate the web.*

What's the best way to get around in a website? As its name implies, the web is a series of interconnected paths, or links. You've no doubt moved around the web by clicking hyperlinks, specially coded elements that take you from one web page to another within the same website or to another site altogether (see Figure 3.13). Generally, text that operates as a hyperlink appears in a different color (often blue) and is usually underlined, but sometimes images also act as hyperlinks. When you hover your cursor over a hyperlink, the cursor changes to a hand with a finger pointing upward.

What other tools can I use to navigate a website? To move back or forward one page at a time, you can use the browser's Back and Forward buttons (see Figure 3.13).

To help navigate more quickly through a website, some sites provide a breadcrumb trail—a navigation aid that shows users the path they have taken to get to a web page, or where the page is located within the website. It usually appears at the top of a page. Figure 3.13 shows an example of a breadcrumb trail. By clicking earlier links in a breadcrumb trail, you can go directly to a previously visited web page without having to use the Back button to navigate back through the website.

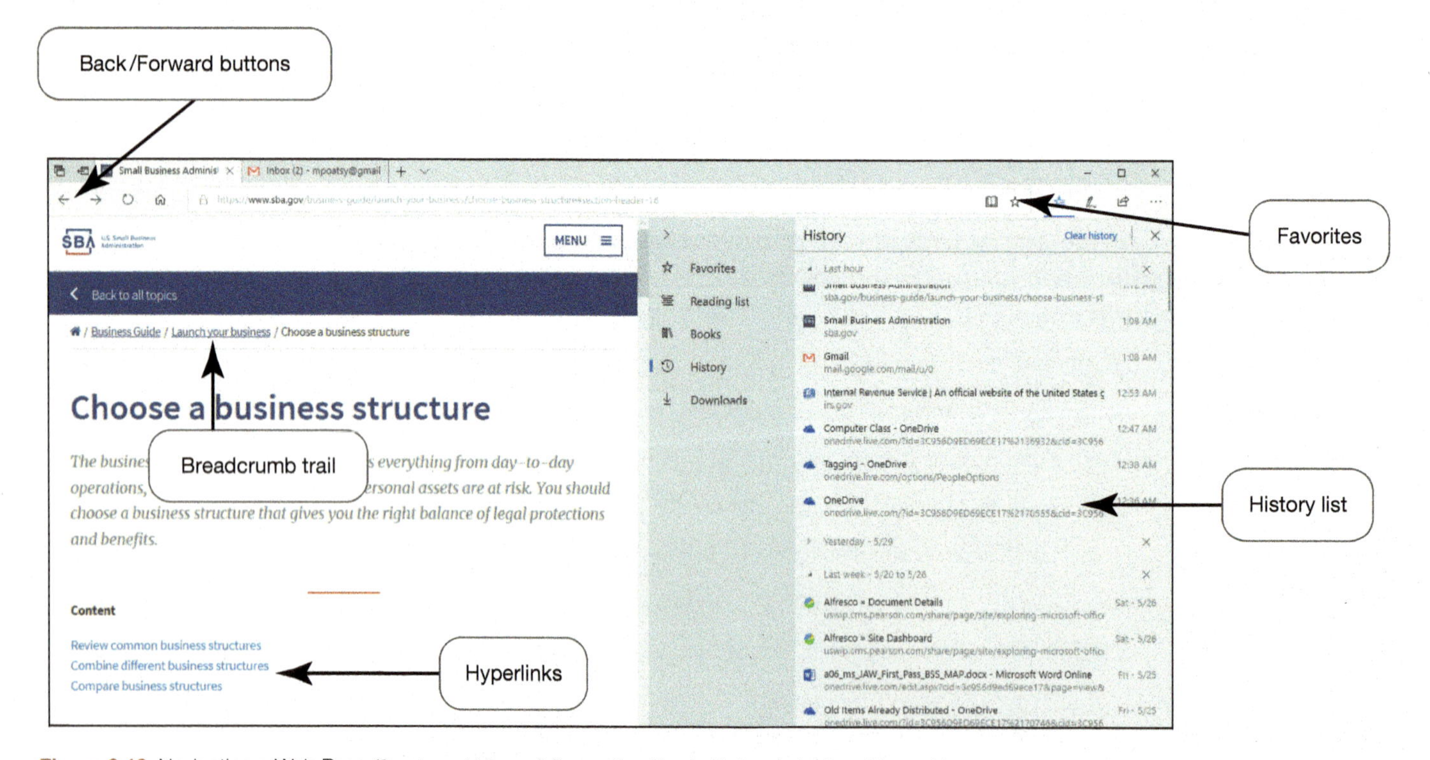

Figure 3.13 Navigating a Web Page *(Courtesy of Microsoft Corporation. Used with Permission from Microsoft.)*

Most browsers offer caret browsing; you use your keyboard to grab text instead of swiping across with a mouse. To activate caret browsing in Microsoft Edge, press F7 on your keyboard while you have the browser open. In Google Chrome you must add an extension, and in Firefox you must enable a preference before you can activate caret browsing. You can then use your mouse pointer as an insertion pointer (similar to working in Word) and the arrow keys to move around. Press Shift and an arrow key to select text. Press F7 again to turn off caret browsing. This method offers more precise text selection and is very handy when you frequently copy and paste text from web pages.

What's the best way to mark a site so I can return to it later? If you want an easy way to return to a specific web page, you can use your browser's Bookmarks feature (Microsoft Edge calls this feature Favorites). This feature places a marker of the site's URL in an easily retrievable list in your browser's toolbar. To organize the sites into categories, most browsers offer tools to create folders.

Favorites and Bookmarks are great for quickly locating those sites you use. Most browsers provide features by which to export the list of bookmarks to a file you can import to another computer or another browser. Another way to access your Bookmarks and Favorites from any computer is to use MyBookmarks—a free Internet service that stores your Bookmarks and Favorites online.

How can I get to a site if it hasn't been bookmarked? If you forget to add a site as a bookmark or favorite but would like to return to a site you've previously visited, try using the History list. A browser's History list shows all the websites and pages you have visited over a certain period of time. These sites are organized according to date and can go back several months, depending on your browsing activity. To access the History list, click the three horizontal lines in the upper right-hand corner of the browser window (often referred to as the hamburger) and then click History in Chrome, Search History in Microsoft Edge, or Library and then History in Firefox.

What is social bookmarking? Social bookmarking is like bookmarking your favorite website, but instead of saving it to your browser for only you to see, you're saving it to a social bookmarking site so that you can share it with others. A *social bookmark* or *tag* is a term that you assign to a web page, digital image, or video. A tag can be something you create to describe the digital content, or it can be a suggested term provided by the website. For example, if you came across a web page with an article on places to go for spring break, you might tag the article *vacations*. Others on the same social bookmarking site who are looking for websites about vacations may use *vacations* as the search term and find the article you tagged.

You can use sites like Pocket (*getpocket.com*) to save interesting web content such as articles, videos, and images to view or review at a more convenient time. Giving a tag to each article you save helps categorize your content in a meaningful way. When you're ready to access your saved content, you can do so from any device, online or offline. In addition to your content, you can also access trending articles and content based on others' tags.

Other social bookmarking sites include Reddit, StumbleUpon, and Pinterest. Reddit encourages its community to vote on links and stories, so the most popular rise to the top. StumbleUpon offers a Stumble! button that, when clicked, provides new web content that you can like or dislike, which eventually contours the Stumble results more to your interests. Unlike other social bookmarking sites, Pinterest enables you to share only images. You can post images that you find online or that you upload directly.

Bits&Bytes Maintain Your Privacy While Searching the Web

Privacy is important to many people, but creating a private online experience is not always an easy task. Although you can use the privacy or incognito mode in some web browsers, these only disable your browsing history and web cache so that data generated from your searches can't be retrieved later. DuckDuckGo is different—this search engine doesn't collect your personal browsing information to share with third parties for marketing purposes. And although most search engines sell your search term information to third parties, which then use that information to generate display ads and recommendations on pages you visit, DuckDuckGo doesn't sell your search terms to other sites. Moreover, DuckDuckGo doesn't save your search history. Using DuckDuckGo eliminates third-party sharing of your personal information.

Searching the Web Effectively

You've most likely googled something today, and if you did, your search is one of over five billion daily searches. Let's look at how to search the web effectively.

Using Search Engines

Objective 3.11 *Describe the types of tools used to search the web and summarize strategies used to refine search results.*

How do search engines work? Google is the world's most popular search engine—a set of programs that searches the web for keywords (specific words you wish to look for or *query*) and then returns a list of the sites on which those keywords are found. Search engines have three components:

1. The first component is a program called a *spider*, which constantly collects data on the web, following links in websites and reading web pages. Spiders get their name because they crawl over the web using multiple legs to visit many sites simultaneously.
2. As the spider collects data, the second component of the search engine, an *indexer program*, organizes the data into a large database.
3. When you use a search engine, you interact with the third component: the *search engine software*. This software searches the indexed data, pulling out relevant information according to your search.

The resulting list appears in your web browser as a list of hits—sites that match your search.

Why don't I get the same results from all search engines? Each search engine uses a unique program, or algorithm, to formulate the search and create the resulting index of related sites. In addition, search engines differ in how they rank the search results. Most search engines rank search results based on the frequency of the appearance and location of the queried keywords in websites. This means that sites that include the keywords in their URL or site name will most likely appear at the top of the results list. An important part of a company's marketing strategy is search engine optimization (SEO): designing the website to ensure that it ranks near the top of search results.

How do I search just for images and videos? With the increasing popularity of multimedia, search engines help you search the web for digital images as well as audio and video files. Click Images on the Google search page or enter google.com/images in your browser. Similarly, google.com/videos allows you to search Google for videos. Bing has an image icon built into the search box that you can click to search for images or click Images or the More button in Bing to access images and video search capabilities.

How else can I customize my searches? In addition to being able to search for videos and images, both search engines allow you to restrict search results by time. Google provides additional search customization capabilities, providing specific searches for Books and Flights. Google provides additional search functionality specific to the type of search being conducted through the Search Tools feature (see Figure 3.14a). For example, when searching for videos, you can continue to customize by Any time, Any quality, or Any duration; if searching for images, Search Tools enables you to restrict the search by Size, Color, Type, Time, or Usage rights. By clicking the Settings link on Google's search page, you can access an Advanced Search form (see Figure 3.14b) through which you can further narrow your search results by language, region, specific terms, usage rights, or file type. Although many of the features in Advanced Search can be found elsewhere in Google's search tools, it's often convenient to have them located in one spot.

Can I use voice commands to search the web? Voice-activated digital assistants such as Siri, Cortana, and Google Now have become standard on smartphones. In addition, Cortana and Siri are available on computers with Windows 10 and macOS Sierra, respectively. Each assistant offers voice-activated capabilities to search files and the web, get directions, and create reminders. Google Now and Cortana have predictive search capabilities; they can recognize a user's repeated actions and searches and, using stored data, offer relevant and predictive information. This might include alerts for upcoming flights, news stories relevant to your interests, travel time to your next appointment, and other information deemed relevant to you.

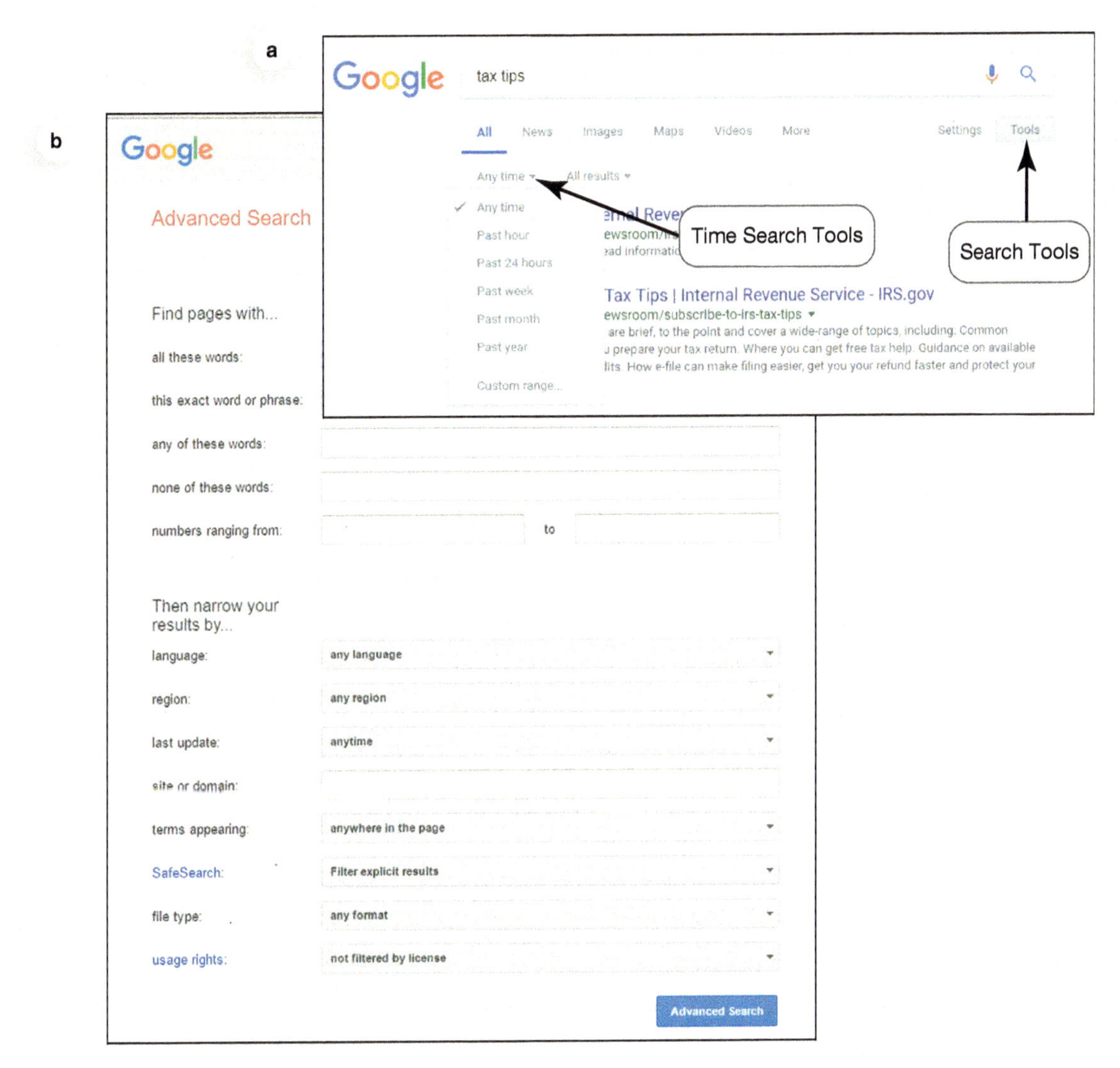

Figure 3.14 (a) Google Search tools and (b) Advanced Search form offer ways to narrow search results. *(Courtesy Google Inc.)*

Are there search engines that offer more specialized results? Search engines differ according to which sites they search. For instance, Google and Bing search nearly the entire web, whereas specialty search engines search only sites that are relevant to a particular topic or industry. Specialty search engines exist for almost every industry or interest. DailyStocks is a search engine, used primarily by investors, that searches for corporate information. Search Engine Watch has a list of many specialty search engines, organized by industry. If you can't decide which search engine is best, you may want to try a metasearch engine. Metasearch engines, such as Dogpile, search other search engines rather than individual websites.

Why do I get suggestions before finishing my search request? For years, Google and other search engines have been offering suggested searches as you type. For example, typing in the term *chocolate* will bring up search terms for chocolate chip cookies, chocolate cake recipe, chocolate mousse, chocolate martini, and chocolate-covered pretzels. These terms are based on your own previous searches as well as popular searches by other people. Search suggestions are also based on what is currently popular. In 2020, typing in *corona* brings up various search suggestions about the coronavirus, but prior to 2020, that same term might have brought up suggestions relating to Corona beer. Of course, you can ignore these suggestions and continue typing in your intended search term, but if one of the suggestions matches your own intentions, you can stop typing and click the suggested term.

Can I refine my key terms for better results? You've no doubt searched for something on Google and gotten a list of hits that includes thousands—even millions—of web pages that have no relevance to the topic you're interested in. Initially, Boolean operators were needed to help refine a search. Boolean operators are words such as *AND, NOT*, and *OR* that describe the relationships between keywords in a search. With the simple addition of a few words or constraints, you can

Finding Information on the Web

In this Sound Byte, you'll learn how and when to use search engines and subject directories. Through guided tours, you'll learn effective search techniques, including how to use Boolean operators and metasearch engines.

narrow your search results to a more manageable and more meaningful list. Other strategies can help refine your searches when entering your search keywords:

- **Search for a phrase.** To search for an exact phrase, place quotation marks around your keywords. The search engine will look for only those websites that contain the words in that exact order. For example, if you searched the phrase *ford was an explorer*, the results would reflect web pages about the Ford Motor Company and, in particular, the Ford Explorer model. However, if you put quotes around the phrase, the results are far fewer, and all include the exact phrase "ford was an explorer." The quotation marks guarantee that search results will include this exact phrase.
- **Search within a specific website.** To search just a specific website, you can use the search keyword, then site: followed by the website's URL. For example, searching with *processor site:* www.wired.com returns results about processors from the *Wired* magazine website. The same method works for entire classes of sites in a given top-level domain or country code.
- **Use a wild card.** The asterisk (*) is a wild card, or placeholder, feature that is helpful when you need to search with unknown terms. Another way to think about the wild card search feature is as a "fill in the blank." For example, searching with *Congress voted * on the * bill* might bring up an article about the members of Congress who voted no on the healthcare bill or a different article about the members of Congress who voted yes on the energy bill.

Evaluating Websites

Objective 3.12 *Describe how to evaluate a website to ensure it is appropriate to use for research purposes.*

How can I make sure a website is appropriate to use for research? When you're using the Internet for research, you shouldn't assume that every web article you find is accurate and appropriate to use. The following is a list of questions to consider before you use an Internet resource; the answers to these questions will help you decide whether you should consider a website to be a good source of information:

Evaluating Websites

In this Helpdesk, you'll play the role of a helpdesk staffer, fielding questions about how websites can be evaluated as appropriate to use for research.

- **Authority:** Who is the author of the article or the sponsor of the site? If the author is well known or the site is published by a reputable news source (such as *The New York Times*), you can feel more confident using it as a source than if you are unable to locate such information. Note: Some sites include a page with information about the author or the site's sponsor.
- **Bias:** Is the site biased? The purpose of many websites is to sell products or services or to persuade rather than inform. These sites, though useful in some situations, present a biased point of view. Look for sites that offer several sets of facts or consider opinions from several sources.
- **Relevance:** Is the information on the site current? Material can last a long time on the web. Some research projects (such as historical accounts) depend on older records. However, if you're writing about cutting-edge technologies, you need to look for the most recent sources. Therefore, look for a date on information to make sure it's current.
- **Audience:** For what audience is the site intended? Ensure that the content, tone, and style of the site match your needs. You probably wouldn't want to use information from a site geared toward teens if you're writing for adults, nor would you use a site that has a casual style and tone for serious research.
- **Links:** Are the links available and appropriate? Check out the links provided on the site to determine whether they're still working and appropriate for your needs. Don't assume that the links provided are the only additional sources of information. Investigate other sites on your topic as well. You should also be able to find the same information on at least three websites to help verify that the information is accurate.

If you're still unsure whether an article or website is legitimate, fact-check article details by using sites such as Factcheck.org, Snopes, and PolitiFact. If you want to check whether an image has been altered to represent something untrue, try using Google's reverse image search (images.google.com) as shown in Figure 3.15.

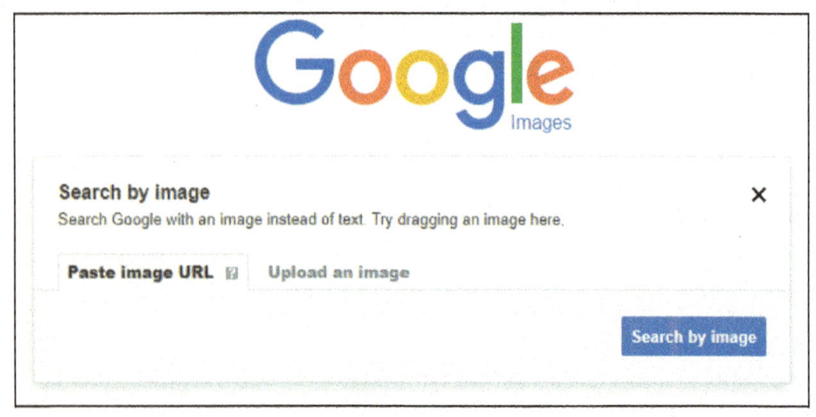

Figure 3.15 Google's Reverse Image Search *(Courtesy of Google Inc.)*

Trends in IT Linked Data and the Semantic Web

Think about all the types of data on the web that you access manually, such as appointment times, transportation and entertainment schedules, and store locations and hours. This type of data is referred to as *structured data*—data that can be defined by a particular category (or field) and a specific data type (such as number, text, or date). Structured data is easily entered, stored, queried, and analyzed and is usually found in databases.

However, the web currently is not set up to manipulate the structured data directly that resides on web pages. Initially, the web was set up to link documents and for people to read web pages. Right now, search engines function by recognizing keywords such as *office hours* and *dentist*, but they can't determine, for example, in which office and on which days Dr. Smith works and what his available appointment times are. The capacity to have linked data, data that is formally defined and that can be expressed in relationships, is beginning to be available, especially by using digital assistants as intermediaries.

The Linking Open Data project is an ongoing effort to convert existing data into a usable format and link it together. The process ensures that the same categorization structures are used so that similar information shares the same attributes, ensuring consistency of metadata throughout the web. The project began with formatting and linking the structured information from Wikipedia but has begun to broaden the sources of data as the project continues to evolve. This web of linked data has been coined the semantic web. The semantic web is an evolving extension of the web in which data is defined so that computers can process it more easily. Ultimately, each website would have text and pictures (for people to read) and metadata (for computers to read) describing the information on the web (see Figure 3.16).

With the assistance of a virtual agent, computers would find relevant metadata, coordinate it with your schedules and preferences, and then make appointments for you. Linked data will also assist you in comparing products, prices, and shipping options by finding the best product option based on specified criteria and then placing the order for you. Eventually, the agent could record the financial transaction in your personal bookkeeping software and arrange for a technician to help install the software, if needed.

Although some of the semantic web's functionalities are beginning to emerge in technologies such as the Siri, Google Now, and Cortana digital assistants, much of its functionality and implementation are still in development. The greatest challenge has been the recoding of all the data currently available on the web into the type of metadata that computers can recognize. However, as more data is converted, and the capabilities of digital assistants continue to be enhanced, we are closer than ever to realizing the full capabilities of the semantic web and linked data.

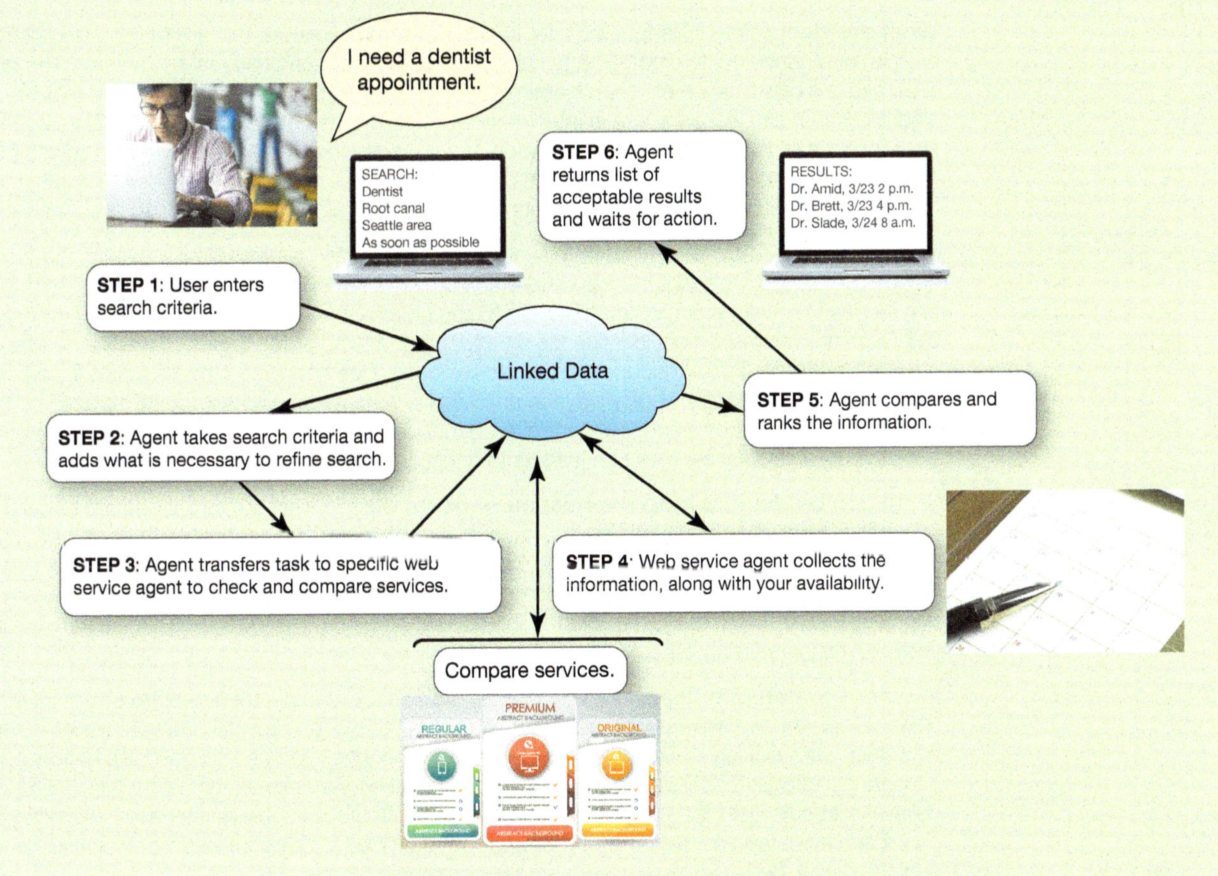

Figure 3.16 Linked Data and the Semantic Web *(Mangostar/Shutterstock; Kaitshot/Shutterstock; Lfor/Shutterstock)*

Bits&Bytes Why Isn't Wikipedia Good to Use as a Source for a Research Paper?

The idea behind content that is managed and edited by many users, such as that found on Wikipedia and other large public wikis, is that the group will keep the content current and valid. However, because wikis are publicly editable, they can't be trusted completely. If a user adds erroneous content, the community of users can catch and correct it, but if you used Wikipedia as a source, you may have referenced erroneous content before it was corrected. To help address these concerns, Wikipedia has implemented tighter access controls, requiring users who want editing privileges to register with the site, but the risks remain. Citizendium, another open wiki encyclopedia, requires contributors to provide real names and sign an ethics pledge, and all postings are monitored.

Using the Web Ethically

As you've read in this chapter, the web offers a vast array of opportunities for business, communication, and collaboration. It also poses some temptations for inappropriate behavior, often to seek gain at another's expense. In this section, we'll explore several ethical issues that are challenging how we use the web.

Deepfakes

Objective 3.13 *Demonstrate an understanding of Internet-related ethical issues such as deepfakes.*

What are deepfakes? Deepfakes refer to digital videos or images that are manipulated using artificial intelligence (AI) to create images and sounds that appear to be real but are not. The term *deepfake* is a combination of "deep learning" and "fake." Deep learning is a form of AI that uses algorithms that can learn and make intelligent decisions on their own. The deep-learning algorithms study photos and videos of the target person and then use the learned intelligence to mimic the target person's behavior and speech patterns. Eventually, those patterns are applied to another image or video, thus transforming one person into another.

What harm do deepfakes cause? Although good fun can be had with deepfakes, the technology can be used to make people believe something is real when it is not. Deepfake technology can be used to make a public figure appear to say or do something that they never actually said or did. Such misinformation could influence an election or affect the reputation of a public leader or other influential person. Although deepfakes have been used mostly to defame public figures such as politicians, professional sports players, and entertainers, eventually they could be used by businesses to fake a person's endorsement of products or services. Because you don't need expert skills to create deepfakes, they can lead us to question the validity of any image or video.

What can be done to stop the misuse of deepfakes? Given the threat deepfakes can have on society, many major tech companies such as Facebook and Microsoft are working hard to create methods to detect deepfakes. In addition, the Department of Defense has launched initiatives to spot and stop deepfakes and other automated disinformation. But the deepfake technology improves just as fast, if not faster, than any detection program, making the challenge to control the falsifications increasingly difficult.

Is it hard to detect a deepfake? It once was fairly easy to separate a deepfake from an actual video. In earlier deepfakes, erratic blinking or discolorations often provided clues to viewers that what they were looking at was a deepfake. But as the technology has improved, it's becoming increasingly difficult for casual viewers to detect deepfakes. As a result, it's best to view hard to believe videos as just that—unbelievable. Before accepting what is being portrayed in a video, consider the related information and who and why the video is being shared. Look closely for the original source, because these videos tend to be passed around virally.

Table 3.7 Point/Counterpoint: Deepfakes

Issue: Deepfakes	Ethical Question: Are deepfakes dangerous or just a form of entertainment?
 (Rvector/Shutterstock)	• Most deepfakes are unrealistic, so nobody will believe them anyway. • Deepfakes are a form of graphical art and should be allowed as a form of personal expression.
 (Rvector/Shutterstock)	• Deepfakes generate misinformation and can be used to persuade an uneducated audience. • It's the responsibility of the technology platform that deepfakes appear on to monitor and weed out deepfakes.

Table 3.7 lists different views on the issue of deepfakes. Where do your views fall?

Personalized Marketing

Objective 3.14 *Demonstrate an understanding of Internet-related ethical issues such as personalized marketing.*

Why does it seem like online ads know exactly what I'm looking for? A few hours ago, you were searching online for a present for your uncle who is really into gardening. Now, as you're researching for a political science paper and are on websites completely outside of your gift search, you see ads for exactly the same John Deere hat that you were considering buying for your uncle. It seems creepy, right? Indeed, companies have been targeting online marketing messages to consumers by using predictive analytics for many years. *Predictive analytics* is used to predict what a customer wants before the customer knows they actually want it. Amazon and Netflix, for example, use this technology to make recommendations for future viewing or purchases.

What's the harm in companies directing ads to products that are exactly what I want? Marketing is a fact of our consumer-driven lives. So if we're to be bombarded with advertisements, isn't it better to be shown ads for products or services we're interested in than in wasting our time on ads that are irrelevant to us? Indeed, data has shown that targeted personalized marketing strategies are effective in getting viewers to click the ads, whereas they would otherwise be ignored.

Most consumers are aware that their online behaviors create data that is used for marketing purposes, but are these tactics too intrusive? Although consumers have expressed concerns about the creepiness of personalized ads that seem to follow them from website to website as well as bombard their e-mail, consumers' online behaviors speak otherwise. People are still willing to give their e-mail address and other information to enter sweepstakes or product giveaways.

Are marketers responsible for protecting privacy? Studies about privacy and personalized marketing have shown that most consumers appreciate the convenience of marketing that matches their interests and needs. However, most consumers don't understand what's at stake when they share certain private data, and younger consumers, in particular, are more willing to forgo privacy for convenience. Major data breaches, such as with Cambridge Analytical and Equifax, have created an awareness of the risks when consumer information ends up in the wrong hands. Because marketers are in control of their advertising tactics, are they taking advantage of consumer naiveté about some of the risks? Will marketing tactics become more invasive, or should marketers be more responsible in how they use consumer data in their marketing strategies? To date, there is still much debate about who is responsible for protecting personal information and online consumer behaviors.

Table 3.8 lists different views on the issue of personalized marketing. What do you think?

Table 3.8 Point/Counterpoint: Personalized Marketing

Issue: Personalized Marketing	Ethical Question: Is personalized marketing a threat to privacy or another modern convenience?
 Point (Rvector/Shutterstock)	• Personalized marketing is helpful because it provides more relevant content. • Personalized marketing is not a threat to personal privacy.
 Counterpoint (Rvector/Shutterstock)	• Personalized marketing increases concerns about privacy. • Marketers should use other methods that don't threaten personal privacy to promote products or services.

Bits&Bytes Human-Implanted Data Chips: Protection or Invasive Nightmare?

Many of us give up quite a bit of our privacy voluntarily through the devices we carry and the apps we use. So why does the thought of implanting a data chip inside the human body cause so much uproar and dissent?

Human-implantable microchips (see Figure 3.17) were approved for use by the Food and Drug Administration in 2004. The basic advantages of implanted chips are that they can receive and transmit information and are extremely difficult to counterfeit. An implanted chip could positively identify you when you make a retail transaction, virtually eliminating fraud. A chip could provide instant access to medical records to facilitate treatment for unconscious accident victims or track lost or missing children or Alzheimer patients, and law enforcement could track the movement of convicted criminals out on parole.

Despite the benefits, people see great potential for privacy infringement and potential misuse of data. Would you want your parents (or the government) to know where you are at any time? Would someone be able to read the medical information in your chip without your permission and use it against you? In fact, there is so much resistance from the public to implants that some states such as Wisconsin, California, Georgia, and North Dakota have already enacted legislation to prohibit mandatory data chips implants in the future, even though it is not being considered now.

The debate will most likely rage on as technologies continue to be developed that will allow implanted chips to gather and provide even more data. Whether the public embraces this technology remains to be seen. Would you ever be willing to have a data chip implanted in your body?

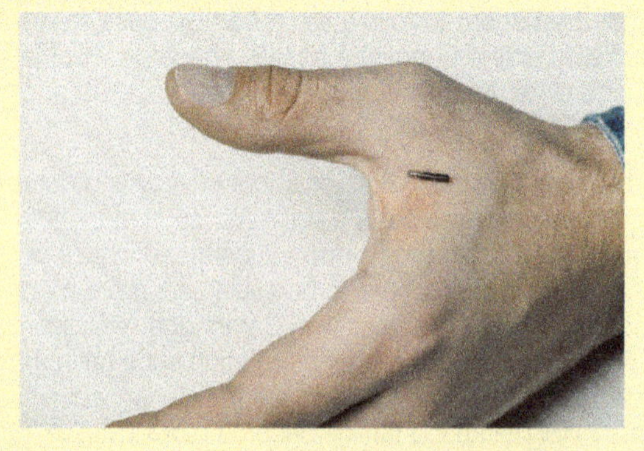

Figure 3.17 Human-implantable microchips can provide personal information, including credit card data or medical records, but there are differing opinions on this technology. *(LightField Studios/Shutterstock)*

Ethics in IT

Cyber Harassment

The Internet provides an array of benefits and advantages, but a darker side has emerged, with a growing epidemic of abusive online behaviors collectively referred to as cyber harassment. Cyber harassment is used to describe the pursuit of individual(s) by an individual with intention to frighten, embarrass, teach the victim a lesson, or seek revenge. Cyber harassment includes acts such as cyberbullying, cyberstalking, trolling, and catfishing.

Cyberbullying and *cyberstalking* involve the use of digital technologies such as the Internet, cell phones, or video to harass, stalk, or bully another. The distinction between cyberbullying and cyberstalking is primarily age. Cyberbullying is generally used when children and adolescents are involved (see Figure 3.18). Cyberbullying and cyberstalking can include actions such as bombarding a victim with harassing instant messages or text messages, spreading rumors or lies on social networking sites, posting embarrassing photos or videos of a victim on the web, or infecting the victim's computer with malware, usually to spy on that person.

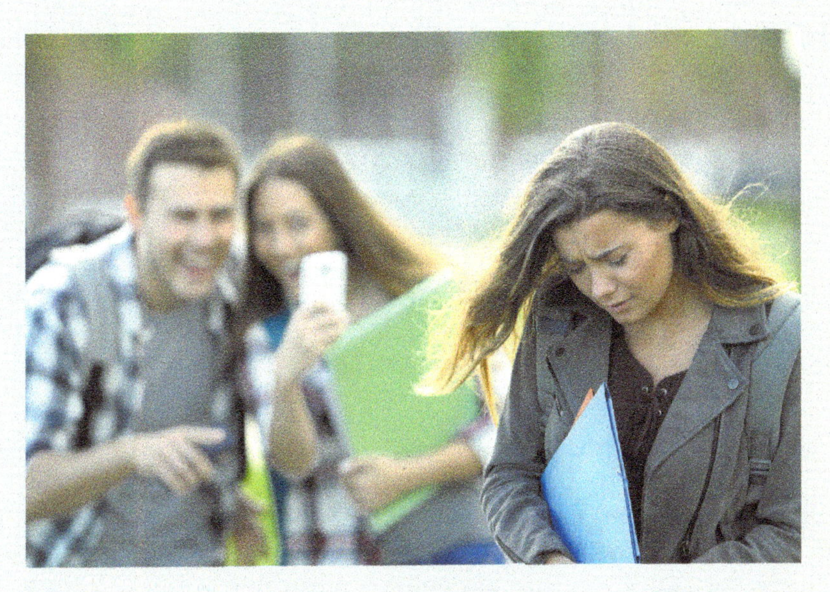

Figure 3.18 Cyberbullying involves the use of digital technologies both to bully and to disseminate acts of bullying. *(Antonio Guillem/Shutterstock)*

The effects of these violations can be devastating. Infamous cases of cyberbullying include Hannah Smith, the English girl who at 14 committed suicide after being repeatedly taunted on social networking sites, and Tyler Clementi, a Rutgers freshman who committed suicide after his roommate showed fellow students videos of him having sex. There are plentiful examples of workplace bullying and celebrity cyberstalking, but often these cases are not as publicized as those involving children.

Trolling is the act of posting inflammatory remarks online for the sheer pleasure of soliciting an angry or negative response. Although both trolling and cyberbullying are generally done with the intention of inflicting fear or anger, the main difference between trolling and cyberbullying is that a troll generally does not personally know his or her target. Trolls derive pleasure from annoying others, and when they are confronted about their behavior, they often shrug off the incident, claiming it was all in good fun. Trolls are more interested in the reaction than the personal confrontation that cyberbullies and cyberstalkers seek. When confronted by a troll, the best thing to do is to not react at all and to leave the conversation as quickly as possible.

Catfishing is another type of Internet harassment, in which some individual scams others into a false romantic relationship. Catfishers create fake profiles and trick others into thinking they are someone else entirely. The fabricated relationships are developed through online and phone interactions, but never in person. One of the most public catfish scams involved Notre Dame football player Manti Te'o, who fell in love with a fictitious girl. The resulting tragic realization of the scam played out in public for Te'o. *Catfish*, a documentary that highlights a true catfishing situation and a reality TV series of that same name, continues to bring attention to this type of scam.

There is currently no federal law prohibiting cyberbullying, but a recently passed law against cyberstalking may cover this area. According to the Cyberbullying Research Center (*cyberbullying.org*), currently all states have anti-bullying laws or policies on the books. However, only 43 states have criminal sanctions for cyberbullying or electronic harassment. Many legislatures are reluctant to pass laws that instruct parents on how to raise their children because this tends to raise issues about personal freedom. Therefore, anti-cyberbullying laws tend to place the burden of detection on the schools or the workplace. For instance, the Massachusetts law requires schools to provide age-appropriate education on bullying to students, to train school employees in detection and prevention of bullying, and to develop plans for detecting and reporting bullying.

Before moving on to the Chapter Review:

1. **Watch Chapter Overview Video 3.2.**
2. **Then take the Check Your Understanding quiz.**

Check Your Understanding // Review & Practice

For a quick review to see what you've learned so far, answer the following questions.

multiple choice

1. Which feature of a browser enables you to separate a browser tab into its own window?
   a. Tabbed browsing
   b. Pinned tabs
   c. Tear-off tabs
   d. Tab isolation
2. Which of the following is NOT an Internet protocol?
   a. HTTP
   b. FTP
   c. BitTorrent
   d. ARPANET
3. What browser feature is used to mark a website so it can be easily returned to in the future?
   a. Favorites
   b. Breadcrumb trail
   c. History
   d. Social bookmarks
4. Which search strategy should you use to search for a specific phrase?
   a. Use quotation marks around the phrase.
   b. Include Boolean operators in the phrase.
   c. Use asterisks around the phrase.
   d. Use a wild card around the phrase.
5. Which of the following is true when evaluating web articles for research?
   a. You should only use an article when an author is listed.
   b. You should evaluate web articles for bias and relevance.
   c. You can assume everything with an author listed is accurate and appropriate.
   d. All links in a web article are timely and relevant.

chew on this

(3Dalia/Shutterstock)

Google is the largest and most popular search engine. Because of its size and popularity, Google can have enormous power to influence a web user's search experience solely by its website-ranking processes. Read some articles about search engine optimization (SEO) and discuss how SEO works and why it can be so important to a company's success.

Go to **MyLab IT** to take an autograded version of the *Check Your Understanding* review and find all media resources for the chapter.

For the IT Simulation for this chapter, see MyLab IT.

3 Chapter Review

Summary

Part 1
Collaborating and Working on the Web

Learning Outcome 3.1 **You will be able to explain how the Internet works and how it is used for collaboration, communication, commerce, and entertainment purposes.**

The Internet and How It Works

Objective 3.1 *Describe how the Internet got its start.*

- The Internet is the largest computer network in the world, connecting millions of computers.
- Government and military officials developed the early Internet as a reliable way to communicate in the event of war. Eventually, scientists and educators used the Internet to exchange research.
- Today, we use the Internet and the web (which is a part of the Internet) to shop, research, communicate, and entertain ourselves.

Objective 3.2 *Explain how data travels on the Internet.*

- A computer (or other device) connected to the Internet acts as either a client (a computer that asks for information) or a server (a computer that receives the request and returns the information to the client).
- Data travels between clients and servers along a system of communication lines or pathways. The largest and fastest of these pathways form the Internet backbone.
- To ensure that data is sent to the correct computer along the pathways, IP addresses (unique ID numbers) are assigned to all computers connected to the Internet.

Collaborating and Communicating on the Web

Objective 3.3 *Evaluate the tools and technologies used to collaborate on the web.*

- Collaboration on the web uses social media and includes social networking, online collaboration and file sharing tools, blogs, podcasts, webcasts, and media sharing.
- Social networking enables you to communicate and share information with friends as well as meet and connect with others.
- File-sharing tools, such as web-based document products, promote online collaboration.
- Project management tools incorporate texting, tasks, and calendar features for individual team members as well as for the project.
- Other useful web tools for group collaboration online are screen sharing and videoconferencing applications.
- Blogs are journal entries posted to the web that are generally organized by a topic or area of interest and are publicly available.
- Video logs are personal journals that use video as the primary content in addition to text, images, and audio.
- Podcasts use mainly audio content and are distributed over the Internet. Users subscribe to receive updates to podcasts.
- Webcasts are live broadcasts of audio or video content over the Internet.

Objective 3.4 *Summarize the technologies used to communicate over the web.*

- E-mail allows users to communicate electronically without requiring the parties involved to be available at the same time.
- Texting sends text-based messages as well as images and videos, using the same cellular network you use to send voice calls from your mobile device.
- Instant-messaging services are programs that enable you to communicate in real time with others who are online.
- VoIP (Voice over Internet Protocol) is the technology used to place phone calls by using the Internet.

Conducting Business on the Web

Objective 3.5 *Describe how cloud technologies are used to create, store, and share files.*

- Cloud computing refers to storing data, files, and applications on the Internet and the ability to access and manipulate files and applications from any Internet-connected device.
- There are two sides to cloud computing: the side we see as users and the banks of computers and servers that house the files and programs we access.
- Productivity and collaboration are enhanced because files and programs are available on any Internet-connected device and stored in a single cloud-based location. The single location eliminates the confusion generated by e-mailing various versions of files, because all changes can be made to the same file by many users.
- Issues exist with cloud computing, such as security and privacy, backup, and access if the Internet is unavailable.

Objective 3.6 *Describe how business is conducted using the Internet.*

- E-commerce is the business of conducting business online.
- E-commerce includes transactions between businesses (B2B), between consumers (C2C), and between businesses and consumers (B2C).

Objective 3.7 *Summarize precautions you should take when doing business online.*

- Precautions include using firewalls and up-to-date antivirus software and employing strong passwords for online accounts.
- Check to see whether the website is secure, shop at reputable websites, pay by credit card not debit card, check the return policy and save a copy of the receipt, and avoid making online transactions on public computers.

Part 2
Using the Web Effectively

Learning Outcome 3.2 You will be able to describe the tools and techniques required to navigate and search the web.

Accessing and Moving Around the Web

Objective 3.8 *Explain what web browsers are and describe their common features.*

- Once you are connected to the Internet, to locate, navigate to, and view web pages, you need software called a web browser.
- The most common web browsers are Google Chrome, Microsoft Edge, Firefox, and Safari. Each browser offers tabbed browsing, pinned tabs, privacy browsing, and add-ons and extensions.

Objective 3.9 *Explain what a URL is and discuss its main parts.*

- You gain access to a website by typing in its address, called a Uniform Resource Locator (URL).
- A URL is composed of several parts, including the protocol, the domain, the top-level domain, and paths (or subdirectories).

Objective 3.10 *Describe tools used to navigate the web.*

- One unique aspect of the web is that you can jump from place to place by clicking specially formatted pieces of text or images called hyperlinks.
- You can also use the Back and Forward buttons, History lists, and breadcrumb trails to navigate the web.
- Favorites, live bookmarks, and social bookmarking help you return to specific web pages without having to type in the URL and help you organize the web content that is most important to you.

Searching the Web Effectively

Objective 3.11 *Describe the types of tools used to search the web and summarize strategies used to refine search results.*

- A search engine is a set of programs that searches the web using specific keywords you wish to query and then returns a list of the websites on which those keywords are found.
- Search engines can be used to search for images, podcasts, and videos in addition to traditional text-based web content.
- Metasearch engines search other search engines.

Objective 3.12 *Describe how to evaluate a website to ensure it is appropriate to use for research purposes.*

- To evaluate whether it is appropriate to use a website as a resource, determine whether the author of the site is reputable, whether the site is intended for your particular needs, that the site content is not biased, that the information on the site is current, and that all the links on the site are available and appropriate.

- Use a reverse image search to check whether an image has been falsified through alterations or tampering.

Using the Web Ethically

Objective 3.13 *Demonstrate an understanding of Internet-related ethical issues such as deepfakes.*

- Deepfakes refer to manipulated digital videos and images that use artificial intelligence to create images and sounds that appear to be real but are not.
- Deepfakes can be used to deceive and consequently alter the reputation of a public figure, affect the outcome of an election, or falsely endorse products.

Objective 3.14 *Demonstrate an understanding of Internet-related ethical issues such as personalized marketing.*

- Personalized marketing uses predictive analytics to target online messages and advertisements to influence consumer purchases and add convenience to online shopping experiences.
- Personalized marketing, when taken to extremes, can seem to invade personal privacy.

Be sure to check out **MyLab IT** for additional materials to help you review and learn. And don't forget to watch the Chapter Overview Videos.

Key Terms

aggregator **90**
blog (weblog) **90**
bookmarks **105**
Boolean operators **107**
breadcrumb trail **104**
business-to-business (B2B) **95**
business-to-consumer (B2C) **95**
catfishing **113**
client **86**
client/server network **86**
cloud computing **94**
consumer-to-consumer (C2C) **95**
cyber harassment **113**
deepfake **110**
desktop client **92**
domain name **102**
e-commerce (electronic commerce) **95**
e-mail **91**
Favorites **105**
File Transfer Protocol (FTP) **103**
host **103**
hyperlink **104**
Hypertext Transfer Protocol (HTTP) **102**
instant messaging (IM) **92**
Internet **84**
Internet backbone **86**
Internet Protocol (IP) address **86**
keyword **106**
linked data **109**
metasearch engine **107**
microblog **90**
mobile commerce (m-commerce) **95**
path (subdirectory) **103**
podcast **90**
Really Simple Syndication (RSS) **90**
search engine **106**
search engine optimization (SEO) **106**
secure sockets layer **96**
semantic web **109**
server **86**
social bookmarking **105**
social commerce **95**
social media **87**
social networking **87**
TCP/IP **103**
texting **92**
top-level domain **103**
trolling **113**
Uniform Resource Locator (URL) **102**
vlog **90**
VoIP (Voice over Internet Protocol) **92**
web-based e-mail **91**
web browser **101**
web server **103**
webcast **90**
World Wide Web (WWW or the web) **85**

Chapter Quiz // Assessment

For a quick review to see what you've learned, answer the following questions. Submit the quiz as requested by your instructor. If you are using **MyLab IT**, the quiz is also available there.

multiple choice

1. Which statement is not true about the web?
   - **a.** The web was based on a protocol developed by Tim Berners-Lee.
   - **b.** Special links are used to navigate from page to page on the web.
   - **c.** The web is based on the WWW protocol.
   - **d.** The web is a subset of the Internet.
2. Which of the following describes an IP address?
   - **a.** It is referred to as a dotted quad.
   - **b.** It identifies any computer connecting to the Internet.
   - **c.** It identifies a website.
   - **d.** All of the above.
3. Which type of social networking site would best describe LinkedIn?
   - **a.** business networking
   - **b.** social payment
   - **c.** social communication
   - **d.** media sharing
4. What web browser feature would be particularly useful when using public computers?
   - **a.** pinned tabs
   - **b.** privacy browsing
   - **c.** session restore
   - **d.** all of the above
5. In the URL *http://www.whitehouse.gov/blog*, what part of the URL is represented by www.whitehouse.gov?
   - **a.** protocol
   - **b.** domain name
   - **c.** top-level domain
   - **d.** path or subdirectory
6. Which of the following is NOT a component of a search engine?
   - **a.** spider
   - **b.** indexer program
   - **c.** search assistant
   - **d.** search engine software
7. Which of the following is NOT considered appropriate e-mail etiquette?
   - **a.** Avoiding text abbreviations such as LOL, u, and r
   - **b.** Adding a signature line that includes your contact information
   - **c.** Always using Reply to All when responding to e-mail that is sent to multiple people.
   - **d.** Including a meaningful subject in the subject line
8. Ensuring that website content is not slanted in its point of view is checking for
   - **a.** Audience
   - **b.** Authority
   - **c.** Bias
   - **d.** Relevance
9. Which of the following is a good precaution to take when making online purchases?
   - **a.** Use the same strong password for all online accounts.
   - **b.** Pay by debit card.
   - **c.** Make sure http: and an open padlock displays in the address bar.
   - **d.** Use a private browser when purchasing items from a public computer.
10. Which of the following would NOT be a concern with relying completely on cloud computing?
    - **a.** security and privacy of stored data and information
    - **b.** the collaborative nature of cloud-based applications
    - **c.** creating an offline back-up of critical documents
    - **d.** accessing files when Internet access is not available

true/false

_____ 1. Deepfakes are mirror websites that are stored in the deep web.

_____ 2. Podcasts can only be created by professional organizations.

_____ 3. Conducting business with stores that have an online and physical presence is referred to as C2C.

_____ 4. The History feature of a browser enables you to retrace your browsing history over a short period of time.

_____ 5. Google Docs is an example of cloud computing.

What do you think now?

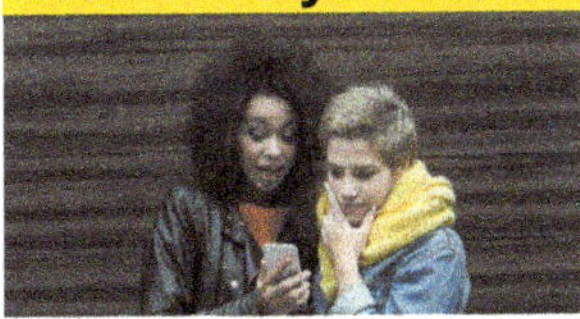

(Prostock-studio/Shutterstock)

Clickbait is used to generate interest and ultimately revenue. Many might argue that although clickbait is annoying, it is relatively harmless. But is it? Explore and discuss the dangers of clickbait.

Team Time

Collaborating with Technology

problem

Collaborating on projects with team members is a regular part of business and academia. Many great tools are available that facilitate online collaboration, and it's important to be familiar with them. In this Team Time, each team will create a group report on a specific topic, using online collaboration tools, and compare and rate the tools and the collaboration process.

process

Split your group into teams. To appreciate fully the benefits of online collaboration, each team should have at least five or six members. Each group will create a team report on a topic that is approved by your instructor. As part of the report, one group member should record the process the group took to create the report, including a review of the tools used and reflections on the difficulties encountered by the group.

1. Conduct a virtual meeting. Agree on an online meeting and video collaboration tool such as Microsoft Teams, Zoom, or Google Hangouts and conduct a group chat. In this phase, outline your group project strategy and delegate work responsibilities.
2. Share documents and collaborate online. Your group must create one document that is accessible to every member at all times. Explore document-sharing sites such as Google Drive, Evernote, OneDrive, or Dropbox and collaboratively create your group document. All members are responsible for reviewing the entire document.

conclusion

After all the team group reports have been completed and shared, discuss the following with your class: What is the benefit of using online collaboration technology to create group projects? How did collaboration technologies help or hinder the team process?

Ethics Project

Internet Privacy

In this exercise, you'll research a complicated ethical situation. The position you are asked to take may or may not match your own personal beliefs, but your research and use of logic will enable you to represent whichever view is assigned. Together the team will agree on an ethical solution.

problem

Our smartphones and other devices that connect to the Internet enable us to share our information, such as where we are, what we've eaten, who we're friends with, when we've exercised, and what we're interested in. In essence, we've become a very public society. And studies show that for the most part, we don't care about being so open. But should we? It's not just our friends and family that we're sharing this information with. Governments and companies are listening, too, and we've learned that often they can't be trusted in how they collect and use our data. In many cases, we agree to privacy terms when we want to use a device or software. We've come to accept that there is a trade-off between being able to use software or a device and giving up some level of privacy, but is this the way it should be? Can we have smart devices that offer us great conveniences while preserving our fundamental right to privacy? Who is ultimately responsible for controlling our privacy, us or the companies that make the devices?

research areas to consider

- Internet privacy
- Internet of Things
- Geolocation
- Social media and privacy

process

1. Divide the class into teams and assign each team a device that is connected to the Internet or an app that is used to share information. Each team is to research the aspect of privacy from the perspective of digital device or software owners who enjoy the social networking feature of the device/software, and from the perspective of business management that has developed the device/software and is collecting and using the owners' location, usage, and other data.
2. Team members should write a summary of the position their team assumes, outlining the pros and cons of their position.
3. Team members should present their case to the class or submit a PowerPoint presentation for review by the rest of the class, along with the summary and resolution they developed.

conclusion

As the use of social media and the Internet of Things becomes more prevalent, our definition of privacy and need for privacy may begin to erode. Being able to determine the benefits and potential sacrifices to our privacy is important in this Internet-connected society in which we live.

Solve This with Word

MyLab IT Grader

Create a Report: Conducting Research on the Web

You and a partner have been asked to write a report on alternatives to using the Google search engine, as well as how to evaluate a website for a research paper. The paper needs to cite references within the body of the text and include a list of your works cited at the end of the report. Your partner has begun the report. You will modify what has been started.

You will use the following skills as you complete this activity:

- Use Find and Replace
- Format Bulleted Lists
- Insert SmartArt
- Insert a Hyperlink
- Add Sources
- Insert Citations and Bibliography

Instructions

1. Start Word. Open *TIA_Ch03_Start.docx* and save it as **TIA_Ch03_LastFirst.docx**, using your last and first name.
2. Using Find and Replace, find all instances of *meta-search* and replace them with **metasearch**.
3. At the blank paragraph after the end of the second paragraph of text under *Alternative Search Engines*, insert a Vertical Box List SmartArt graphic. Open the text pane, if necessary. With the cursor active in the first bullet, type **Google Scholar**. (Do not include the period.) Press **Enter**, press **Tab**, and then type **Searches scholarly literature**. (Do not include the period.)

   Repeat these steps to add the following information for the next two bullets:

   Dogpile
   Metasearch engine that searches Google, Yahoo!, and Bing
   Specialty Search Engines
   Search only sites that are relevant to a topic or industry

   Hint: To insert a SmartArt graphic, on the Insert tab, in the Illustrations group, click SmartArt.
4. Change the SmartArt graphic colors to Colorful—Accent Colors. Move the Google Scholar box and bullet point to the bottom of the SmartArt. Change the SmartArt Style to Intense Effect.

   Hint: To change the colors, on the SmartArt Tools Design tab, in the SmartArt Styles group, click Colors. To move the box and bullet point, select the content and the, in the Create Graphic group, click Move Down. To change the style, in the SmartArt Styles group, click the More button for Styles.
5. In the second-to-last sentence of the first paragraph in the *Alternative Search Engines* section, select the text *specialty search engines*. Then, create a hyperlink to the http://bestonlineuniversities.com/2011/20-useful-specialty-search-engines-for-college-students/web page.
6. At the end of the first paragraph in the *Evaluating Websites* section, immediately to the left of the period, insert a citation to a new website. Before entering information, click the check box for Show All Bibliography Fields. Use the following information:

   Author: **Kapoun, Jim**
   Name of Web Page: **Five criteria for evaluating Web pages**
   Name of Web Site: **Olin & Uris Libraries, Cornell University**
   Year: **2010**
   Month: **May**
   Day: **10**
   URL: http://olinuris.library.cornell.edu/ref/research/webcrit.html
7. In the Evaluating Websites section, create a bulleted list with the five points, beginning with *Authority*, *Bias or Objectivity*, *Relevance*, *Audience*, and *Coverage*. Use a check mark as the bullet point.
8. Press Ctrl + End to go to the end of the document, press **Enter** twice, and then insert a References page. Change the citation style to **APA Sixth Edition**.
9. Save the document, and then close Word.
10. Submit the document as directed.

Chapter

Application Software: Programs That Let You Work and Play

 For a chapter overview, watch the **Chapter Overview Videos**.

PART 1

Accessing, Using, and Managing Software

Learning Outcome 4.1 **You will be able to explain the ways to access and use software and describe how best to manage your software.**

Software Basics 124

Objective 4.1 *Compare application software and system software.*

Objective 4.2 *Explain the differences between commercial software and open source software and describe models for software distribution.*

Managing Your Software 125

Objective 4.3 *Explain the different options for purchasing software.*

Objective 4.4 *Describe how to install and uninstall software.*

Objective 4.5 *Explain the considerations around the decision to upgrade your software.*

Objective 4.6 *Explain how software licenses function.*

Helpdesk: Buying and Installing Software

Sound Byte:: Where Does Binary Show Up?

PART 2

Application Software

Learning Outcome 4.2 **Describe the different types of application software used for productivity and multimedia.**

Productivity and Business Software 135

Objective 4.7 *Categorize the types of application software used to enhance productivity and describe their uses and features.*

Objective 4.8 *Summarize the types of software that small and large businesses use.*

Sound Byte: Programming for End Users

Multimedia and Educational Software 146

Objective 4.9 *Describe the uses and features of multimedia software.*

Objective 4.10 *Describe the uses and features of audio software.*

Objective 4.11 *Describe the features of app creation software.*

Objective 4.12 *Categorize educational and reference software and explain their features.*

Helpdesk: Choosing Software

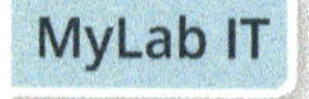

All media accompanying this chapter can be found here.

A More Powerful App on **page 134**

(Ronstik/123RF; Rawpixel.com/Shutterstock; Rawpixel.com/Shutterstock; Laurent Davoust/123RF)

What do you think?

Now that it's effortless to use technology from almost anywhere, more of us are becoming dependent on the **tiny computers** we carry. So it's no surprise that health experts are seeing a **rise in addictive tendencies** that involve technology. Some researchers say that people who use their phones or stay online for many hours a day experience a similar high as with other addictions. They also **feel withdrawal when cut off**. It's not just the amount of time spent with technology that defines addiction, but how excessive use adversely affects one's mental and physical health, daily life, relationships, and academic or job performance.

Apple is adding tools to help limit phone use in response to concerns about phone addiction. A new activity report will **track the time** you're spending on your mobile device. Apple is also enabling you to see which apps are sending the most notifications, so you can **set your own limits**, such as one hour a day for Instagram.

Keep a technology-use diary for one week and note the amount of time you spend using technology.

- *Less than 4 hours*
- *Between 4 and 8 hours*
- *Between 9 and 12 hours*
- *More than 12 hours*

Part 1

For an overview of this part of the chapter, watch **Chapter Overview Video 4.1**.

Accessing, Using, and Managing Software

Learning Outcome 4.1 **You will be able to explain the ways to access and use software and describe how best to manage your software.**

Although a computer's hardware is critical, a computer system does nothing without software. In this section, we'll discuss the way software operates and how you can manage your software to make sure it is current and being used legally.

Software Basics

It's important to understand what software is, how it's created, and the different models used to distribute it.

Application vs. System Software

Objective 4.1 *Compare application software and system software.*

What is software? Technically speaking, the term software refers to a set of instructions that tells the computer what to do. An instruction set, also called a program, provides a means for us to interact with and use the computer, even if we lack specialized programming skills.

What is the difference between application and system software? Your computer has two main types of software. Application software is the software you use to do tasks at home, school, and work. System software is software that helps run the computer and coordinate instructions between application software and the computer's hardware devices. System software includes the operating system (such as Windows and macOS) and utility programs (programs in the operating system that help manage system resources). We discuss system software in detail in Chapter 5.

Other types of application software, such as web browsers, virus protection, and backup and recovery software, are used every day. We'll discuss these types of software elsewhere in this book. In this chapter, we'll discuss productivity software and other types of software that you use to get things done.

Distributing Software

Objective 4.2 *Explain the differences between commercial software and open source software and describe models for software distribution.*

How is software created? Software is created in two main ways. Proprietary (or commercial) software is created by companies for profit and then sold to you. Adobe Photoshop is an example of proprietary software. When you use proprietary software, you agree to specific terms of use. Open source software is available free of charge and with few licensing and copyright restrictions. One advantage of open source software is that changes are made by community users, rather than having to go through a set process that is required by commercial software developers. This means that open source software is often more responsive to correcting bugs and implementing useful changes. However, unlike proprietary applications, open source applications offer little or no formal support. Instead, they're supported from their community of users across websites and newsgroups. Table 4.1 illustrates how some specific software products fit into each of these categories.

How is software distributed? Software is distributed in three main ways:

1. **Local installation:** With locally installed software (often referred to as *perpetual-license software*), you pay a one-time fee for either an electronic download or a disc of the software, which you then install on your device. Because the software resides on your hard drive, you can run the software anytime, whether you're connected to the Internet or not. You can only use the software on the number of devices permitted by license. You'll have the option to upgrade to a new version as it's released and you'll likely need to pay again to purchase the upgraded version. Desktop tax preparation software is an example of software distributed in this way.
2. **SaaS:** With Software as a Service (SaaS), the vendor hosts the software online and you access and use the software over the Internet without having to install it on your computer's

Table 4.1 Open Source vs. Commercial Software

	System Software		Application Software	
Open Source	Linux		Gimp	LibreOffice
Commercial	macOS	Windows	Microsoft Office	Adobe Photoshop CC

(Stanislaw Mikulski/Shutterstock; Retrieved from https://commons.wikimedia.org/wiki/File:GIMP_Icon_without_brush.svg; LibreOffice; Stanislaw Mikulski/Shutterstock; Courtesy of Microsoft Corp; Courtesy of Microsoft Corp; PixieMe/Shutterstock)

hard drive. Because the program resides on a company's server and you run the program through your browser, you can run the program only when you're connected to the Internet. However, because the manufacturer updates the program on its server, you don't have to worry about manually upgrading the software yourself. Google Docs and Microsoft Office Online are examples of programs that are distributed in this way. SaaS programs are often free of charge but not as full-featured as locally installed versions.

3. **Subscription:** With subscription software, you pay a monthly or annual fee to use the software. It is then available for you to download and install on your computer. As soon as updates are available, the manufacturer pushes these new features to you. You can use the software without being on the Internet but can only receive updates when you're connected to the Internet. Adobe Creative Cloud and Microsoft Office 365 are examples of software distributed in this way.

Bits&Bytes **Portable Apps**

Most often, you install software on your device or access it from the web. But sometimes you won't have access to your own device or an Internet connection. In these instances, consider using portable apps. Portable apps are run from a flash drive you insert into a USB port, so they are not installed. (Note, however, that they must be compatible with the operating system of the computer they're used on.) When you finish using the app, just carefully eject the flash drive. No personal data or files are stored on the host machine because everything is self-contained within the portable app. Portable apps are typically open source. You can find them at PortableApps.com.

Managing Your Software

It's important to know how to choose the best software for your computing device, how to get it onto your computer correctly, and how to remove it. We discuss all these topics next.

Purchasing Software

Objective 4.3 *Explain the different options for purchasing software.*

Where is the best place to buy software? Most commonly, software is purchased online and downloaded to your computer. Developer websites offer their full product line as digital downloads. If you have a machine with an optical drive (a DVD or Blu-ray reader), you may purchase the software in disc form. Software for gaming consoles is still sold on discs, for example. Software for mobile devices is sold electronically through online sites for each operating system. Apple runs the App Store, Microsoft has Microsoft Store, and Google administers the Google Play site (see Figure 4.1).

Is discounted software for students available? If you're a student, you can buy substantially discounted software that is no different from regularly priced software. Campus computer stores

Figure 4.1 Software for mobile devices is sold through online sites like Apple's App Store or Google Play. *(Bloomicon/Shutterstock)*

and college bookstores offer discounted prices to students who possess a valid student ID. Online software suppliers such as JourneyEd (*journeyed.com*) and Academic Superstore (*academicsuperstore.com*) also offer popular software to students at reduced prices. Software developers, such as Microsoft and Adobe, often offer their products to students at a discount, so it's always good to check a manufacturer's website before purchasing software or hardware. In addition, be sure to check with your college or school. Many times, they have agreements with large vendors like Microsoft for free or reduced pricing for students.

Can I get software for free legally? In addition to open source software, freeware is copyrighted software that you can use for free. Explore sites like FileHippo (*filehippo.com*) to see how many good freeware programs are available. If you use Amazon devices like the Fire tablet or Android smartphone, you can get free applications through Amazon Underground. Developers are paid by Amazon for every minute each app is used. However, although much legitimate freeware exists, some unscrupulous people use freeware to distribute viruses and malware. Be cautious when installing freeware, especially if you're unsure of the provider's legitimacy.

Can I try new software before it's released? Some software developers offer beta versions of their software free of charge. A beta version is an application that is still under development. By distributing free beta versions, developers hope users will report errors, or bugs, they find in their programs. Many beta versions are available for a limited trial period and are used to help developers respond to issues before they launch the software on the market.

Are there risks associated with installing beta versions? By their very nature, beta products are unlikely to be bug free, so you always run the risk of something going awry with your system by installing and using beta versions. Unless you're willing to deal with potential problems, it may be best to wait until the last beta version is released—often referred to as the *gold version*. By that time, most of the serious bugs have been worked out.

Trends in IT: Artificial Intelligence in Health Care

Doctors, nurses, scientists, and hospital administrators increasingly depend on artificial intelligence (AI) to help deliver more reliable and efficient health care. Health care–related data is being generated continuously from clinical, genetic, and social sources. Scientists and physicians are beginning to rely on AI to sift through and make sense of the phenomenal amounts of data to help diagnose conditions and create reliable solutions for common and not-so-common medical issues. AI is also used to analyze symptoms and deliver personalized health care recommendations and treatments with more efficiency and reliability than most humans can deliver. AI is also used to read scans and images and then suggest the most likely diagnosis by comparing the outcomes to similar images stored in patient databases. These technologies help reduce human-related errors that are often caused by fatigue, moods, or other human-related distractions.

Physicians are also beginning to use chatbots and digital personal assistants to review patients' symptoms and then recommend whether a face-to-face visit or a virtual check-in is required. This outcome helps relieve congestion in physician offices, keeps the sickest away from the healthy, and improves patient care.

Robots led by AI also assist surgeons with complex or routine surgical procedures with precision and control. Usually the patient recovers more quickly, with fewer complications and less pain (see Figure 4.2). AI is also being used to help with communication between doctors to lessen the risk of complications so patients can go home sooner and avoid being readmitted.

Hospital administrators are also beginning to use AI to streamline the patient experience and improve hospital efficiency. For example, such technology is being used to schedule surgeries and imaging tests by predicting the amount of time each will take, thus reducing patient wait times and improving staff workflow. In addition, discharge procedures are being managed with more precision and efficiency, so patients may leave the hospital earlier in the day.

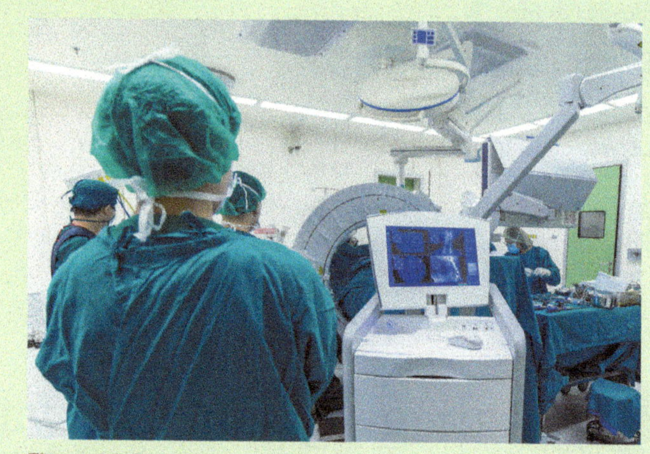

Figure 4.2 Doctors use artificial intelligence and robotic technologies to assist with surgical procedures.

Although the impact of AI has had immediate and dramatic effects on health care, it is still in its infancy. It's not expected that AI will replace doctors in the future, but continued improvements in AI and technology will be essential in helping doctors and hospitals provide better care.

As a precaution, you should be comfortable with the reliability of the source before downloading a beta version of software. If it's a reliable developer whose software you're familiar with, you can be more certain that a serious bug or virus isn't hiding in the software. Similarly, you should be sure that the software you're downloading is meant for your system and that your system has met all the necessary hardware and operating system requirements.

Helpdesk MyLab IT

Buying and Installing Software

In this Helpdesk, you'll play the role of a help desk staffer, fielding questions about how best to purchase software or to get it for free, how to install and uninstall software, and where you can go for help when you have a problem with your software.

Installing and Uninstalling Software

Objective 4.4 *Describe how to install and uninstall software.*

How do I know the software I buy will work on my computer? If you're buying software, it's your responsibility to check for compatibility with your system. Every software program has a set of system requirements that specify the minimum recommended standards for the operating system, processor, primary memory (random access memory, or RAM), and storage. These requirements are printed on the software packaging or are available at the manufacturer's website.

What steps should I take before installing a new program? Before installing any software, make sure your virus protection software is up to date. It's equally important to back up your system and create a restore point. A restore point doesn't affect your personal data files but saves all the apps, updates, drivers, and information needed to restore your computer system to the exact way it's configured at that time. That way, if something does go wrong, you can return your system to the way it was before the problem occurred. You can create a restore point by using Windows 10 System protection tools. To create a restore point, type *restore point* in the Cortana search box and select *Create a Restore Point Control Panel* to open the System Properties dialog box. On the System Protection tab, click *Configure* and then click *Turn on System Protection.* Type a description for the restore point (such as "Before installing [name of software]") and click *Create.* You will be notified when the restore point is created.

What's the difference between a custom installation and a full installation? The exact installation process will differ slightly, depending on whether you're installing the software from a disc, purchasing it from an app marketplace, or downloading it from the web. Depending on the type of software you're installing, you may need to decide between a full installation and a custom installation. A full installation (sometimes referred to as a typical installation) copies all the commonly used files and programs to your computer's hard drive. By selecting custom installation, you can decide which features you want installed, which saves space on your hard drive.

How do I uninstall a program? An application contains many files—library files, help files, and other text files—in addition to the main file that runs the program. By deleting only the main file, you're not ridding your system of all the related files. Windows 10 makes it easy to uninstall a program: On the Start menu, right-click the app and select *Uninstall.* Alternatively, you can use Apps & features from Settings. Going this route, you can choose whether you want to uninstall the program or simply modify or repair the program if it's not working correctly.

If my computer crashes, can I get the preinstalled software back? If you've installed your software through a digital download, chances are you have created an account that contains

Bits&Bytes Ridding Your Computer of Bloat

Manufacturers often include software on new computers that you don't want or need. Called *bloatware,* this software can slow down your computer and degrade its performance. If you purchase a computer with bloatware, you can run an application such as Should I Remove It or PC Decrapifier. These utilities help you decide which programs to remove. Or, if you'd rather do it yourself, consider some of these tips:

- *Uninstall preinstalled antivirus software:* If you have antivirus software on your old computer, you may be able to transfer the unexpired portion of your software license to your new computer. If this is the case, you can uninstall the preinstalled trial version on your new computer.
- *Remove manufacturer-specific software:* Some computer manufacturers install their own software. Some of these programs can be useful, but others are help features and update reminders that are also found in your operating system. You can remove these support applications and instead just check the manufacturer's website for updates or new information.

the product key along with other installation and product information. Assuming your account is up to date, you can then download the software again, using the same product key. This applies to application software as well as operating system software. However, it's always good practice to create a recovery drive of your operating system. A recovery drive contains all the information needed to reinstall your operating system if it should become corrupted. Often the manufacturer will have placed a utility on your system to create this. To create a Windows 10 recovery drive, type *Recovery* in the Cortana search box and then select *Create a recovery drive*. When the Recovery Drive dialog box opens, click Next. Then insert a blank flash drive into a USB port and follow the steps in the wizard. Once you've created the recovery drive, label the flash drive and put it in a safe place.

Upgrading Software

Objective 4.5 *Explain the considerations around the decision to upgrade your software.*

When do I need to upgrade software? Periodically, software developers improve the functionality of their software by releasing a software upgrade. When software is sold by subscription or using the cloud-based SaaS model, you don't need to worry about upgrading because it's handled automatically. Software purchased through mobile app stores also updates automatically. However, if you're using software that was installed locally, you'll need to make the upgrade decision yourself. Before upgrading, it's important to understand what's included in the upgrade and how stable the

Dig Deeper How Number Systems Work

As we discussed in Chapter 2, to process data into information, computers need to work in a language they understand. This language, called *binary language*, consists of two digits: 0 and 1. Everything a computer does, including running the applications we've been discussing, is broken down into a series of 0s and 1s. *Electrical switches* are the devices inside the computer that are flipped between the two states of 1 and 0, signifying on and off. How can simple switches be organized so that you can use a computer to pay your bills online or write an essay? Let's look at how number systems work.

A number system is an organized plan for representing a number. Although you may not realize it, you're already familiar with one number system. The base-10 number system, also known as decimal notation, is the system you use to represent the numeric values you use each day. It's called base 10 because it uses 10 digits—0 through 9—to represent any value.

To represent a number in base 10, you break the number down into groups of ones, tens, hundreds, thousands, and so on. Each digit has a place value, depending on where it appears in the number. For example, the whole number 6,954 contains 6 sets of thousands, 9 sets of hundreds, 5 sets of tens, and 4 sets of ones. Working from right to left, each place in a number represents an increasing power of 10, as follows:

$$6{,}954 = (6 * 1{,}000) + (9 * 100) + (5 * 10) + (4 * 1)$$
$$= (6 * 10^3) + (9 * 10^2) + (5 * 10^1) + (4 * 10^0)$$

Note that in this equation, the final digit 4 is represented as $4 * 10^0$ because any number raised to the zero power is equal to 1.

Anthropologists theorize that humans developed a base-10 number system because we have 10 fingers. However, computer systems are not well suited to thinking about numbers in groups of 10. Instead, computers describe a number in powers of 2 because each switch can be in one of two positions: on or off. This numbering system is referred to as the binary number system (or base-2 number system). In the base-10 number system, a whole number is represented as the sum of 1s, 10s, 100s, and 1,000s—that is, sums of powers of 10. The binary number system works in the same way but describes a value as the sum of groups of 1s, 2s, 4s, 8s, 16s, 32s, 64s, and so on—that is, powers of 2.

Let's look at the number 67. In base 10, the number 67 would be six sets of 10s and seven sets of 1s, as follows:

$$67_{\text{Base }10} = (6 * 10^1) + (7 * 10^0)$$

One way to figure out how 67 is represented in base 2 is to find the largest possible power of 2 that could be in the number 67. Two to the seventh power is 128, which is bigger than 67. Two to the sixth power is 64, which is less than 67, meaning there is a group of 64 inside a group of 67.

67 has	1	group of	64	that leaves 3 and
3 has	0	groups of	32	
	0	groups of	16	
	0	groups of	8	
	0	groups of	4	
	1	group of	2	that leaves 1 and
1 has	1	group of	1	Now nothing is left.

new software version is. Depending on the software, some upgrades may not be sufficiently different from the previous version to make it cost-effective for you to buy the newest version. Unless the upgrade adds features that are important to you, you may be better off waiting to upgrade. You also should consider whether you use the software frequently enough to justify an upgrade and whether your current system can handle the new system requirements of the upgraded version. In between upgrades, developers will make available software updates (sometimes referred to as software patches). Updates are usually downloaded automatically and provide smaller enhancements to the software or fix program bugs.

Software Licenses

Objective 4.6 *Explain how software licenses function.*

Don't I own the software I buy? Unlike most other items you purchase, the software you buy doesn't belong to you. What you're actually purchasing is a license that gives you the right to use the software for the specific purposes described in the license.

A software license, also known as an End User License Agreement (EULA), is an agreement between you, the user, and the software company (see Figure 4.3). You accept this agreement before installing the software on your machine. It's a legal contract that outlines the acceptable uses of the program and any actions that violate the agreement. Generally, the agreement states who the

So, the binary number for 67 is written as 1000011 in base 2:

$$67_{\text{Base 10}} = 64 + 0 + 0 + 0 + 0 + 2 + 1$$
$$= (1 * 2^6) + (0 * 2^5) + (0 * 2^4) + (0 * 2^3) + (0 * 2^2) + (1 * 2^1) + (1 * 2^0)$$
$$= 1000011_{\text{Base 2}}$$

A large integer value becomes a very long string of 1s and 0s in binary. For convenience, programmers often use hexadecimal notation to make these expressions easier to use. Hexadecimal is a base-16 number system, meaning it uses 16 digits to represent numbers instead of the 10 digits used in base 10 or the two digits used in base 2. The 16 digits it uses are the 10 numeric digits, 0 to 9, plus six extra symbols: A, B, C, D, E, and F. Each of the letters A through F corresponds to a numeric value, so that A equals 10, B equals 11, and so on. Therefore, the value 67 in binary notation is 1000011, or 43 (= 4 * 16 + 3 * 1) in hexadecimal notation. It is much easier for computer scientists to use the 2-digit 43 than the 7-digit string 1000011.

We've just been converting integers from base 10, which we understand, to base 2 (binary state), which the computer understands. Similarly, we need a system that converts letters and other symbols that we understand to a binary state the computer understands. To provide a consistent means for representing letters and other characters, certain codes dictate how to represent characters in binary format. Most of today's personal computers use the American National Standards Institute (ANSI, pronounced "AN-see") standard code, called the American Standard Code for Information Interchange (ASCII, pronounced "AS-key"), to represent each letter or character as an 8-bit (or 1-byte) binary code.

When you convert a base 10 number to binary format, the binary format has no standard length. For example, the binary format for the number 2 is two digits (10), whereas for the number 10 it is four digits (1010). Although binary numbers can have more or fewer than 8 bits, each single alphabetic or special character is 1 byte (or 8 bits) of data. The ASCII code uses 8 bits to represent each of the 26 uppercase letters and 26 lowercase letters used in the English language, along with many punctuation symbols and other special characters. Because the ASCII code represents letters and characters using only 8 bits, it can assign only 256 (or 2^8) codes. Although this is enough to represent English and many other characters found in the world's languages, the ASCII code can't represent all languages and symbols. Thus, an encoding scheme, called Unicode, was created. By using 16 bits instead of 8 bits, Unicode can represent nearly 1,115,000 code points and currently assigns more than 128,000 unique character symbols. As we continue to become a more global society, it is anticipated that Unicode will replace ASCII as the standard character formatting code.

MyLab IT

Where Does Binary Show Up?

This Sound Byte helps remove the mystery around binary numbers. You'll learn about base conversion among decimal, binary, and hexadecimal numbers by using colors, sounds, and images.

ultimate owner of the software is, under what circumstances copies of the software can be made, and whether the software can be installed on any other computing device. Finally, the license agreement states what, if any, warranty comes with the software.

Does a license only cover one installation? Some software comes with a single license that covers only one person's use on a single device. You can't share these licenses, and you can't extend the license to install the software on more than one of your devices. However, many manufacturers are allowing users to install software on several devices. For example, some Apple software and most versions of Microsoft Office come with a license that allows you to install the software on multiple computers.

Businesses and educational institutions often buy multiuser licenses that allow more than one person to use the software. Some multiuser licenses are per-seat and limit the number of users overall, whereas others, called *concurrent licenses*, limit the number of users accessing the software at any given time.

Does open source software require a license? Anyone using open source software has access to the program's code. Therefore, open source software programs can be tweaked by another user and redistributed. A free software license, the GNU General Public License, is required and grants the recipients the right to modify and redistribute the software. Without such a license, the recipient would be in violation of copyright laws. This concept of redistributing modified open source software under the same terms as the original software is known as copyleft. Thus, all enhancements, additions, and other changes to copyleft software must also be distributed as free software.

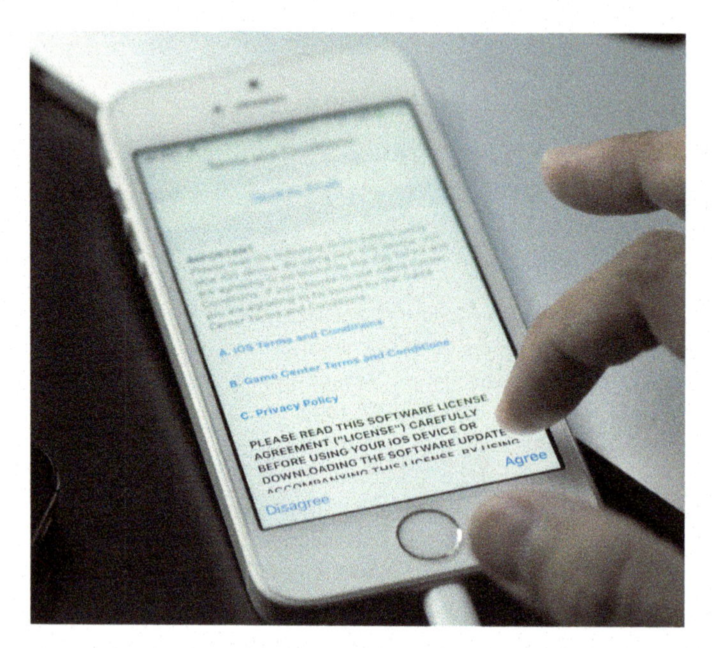

Figure 4.3 You must accept the terms of the software license before using the product. *(Karanik yimpat/Shutterstock)*

Ethics in IT

Can I Use Software on My Device When I Don't Own the License?

As noted earlier, when you purchase software, you're purchasing a license to use it rather than purchasing the actual software. That license indicates the rules of using the software. Not all licenses are the same, so it's important to read it. If you use, copy, or distribute more copies of the software than the license permits, you're participating in software piracy (see Figure 4.4).

There are several forms of software piracy. Some forms are more purposeful and aggressive, such as counterfeiting and hard-disk loading. *Counterfeiting* refers to illegally duplicating and selling software that appears to be authentic. *Hard-disk loading* involves someone installing an illegal copy of software onto a hard disk and installing the hard disk in a computer intended for resale. Other forms of software piracy may occur unknowingly, such as installing the software on more devices than the license allows or acquiring the use of software through peer-to-peer file sharing.

So how bad is it to use software when you don't own the license, or to use it beyond the agreed constraints in the license? As reported by the Business Software Alliance, over one third of all software used is pirated. Not only is pirating software unethical and illegal, the practice has financial impacts on all software consumers. The financial loss to the software industry decreases the amount of money available for further software research and development while increasing the upfront costs to legitimate consumers. Moreover, pirated software may come with unexpected viruses that could have dire consequences. In addition, it does not provide warranty, technical support, upgrades, or regular bug fixes.

So far, there's no such thing as an official software police force, but if you're caught with pirated software, severe penalties do exist. A company or individual can pay large fines for each software title copied. You can also be criminally prosecuted for copyright infringement, which involves fines or even jail time. Efforts to stop groups involved with counterfeit software are in full force.

Figure 4.4 Making more copies than the software license permits is pirating and is illegal. *(Alexskopje/123RF)*

In addition, software manufacturers also are becoming more aggressive in programming mechanisms into software to prevent illegal installations. For instance, with many products, installation requires you to activate the serial number of your software with a database maintained by the software manufacturer. Failure to activate your serial number or attempting to activate a serial number used previously results in the software going into a reduced-functionality mode after a certain number of uses. This usually precludes you from doing useful things like saving files created with the software.

Before moving on to Part 2:

1. Watch Chapter Overview Video 4.1.
2. Then take the Check Your Understanding quiz.

Check Your Understanding // Review & Practice

For a quick review to see what you've learned so far, answer the following questions.

multiple choice

1. Which type of software distribution would NOT be hosted and accessed online?
   a. Open source software
   b. Software as a Service (Saas)
   c. Locally installed software
   d. Subscription software
2. Beta software is never
   a. system software.
   b. the final finished version of the software.
   c. open source software.
   d. application software.
3. Choosing which features of software you want to install is which kind of installation?
   a. Custom installation
   b. Full installation
   c. Restore installation
   d. Single installation
4. Which type of software is not upgraded automatically?
   a. Software as a service
   b. Mobile apps
   c. Locally installed
   d. Subscription based
5. Which best describes copyleft?
   a. A single license covering a single user on a single device
   b. A license that grants users the right to modify and redistribute the software
   c. A license covering one user with multiple installations
   d. A pirated license

chew on this

(3Dalia/Shutterstock)

Because app marketplaces work so well with mobile devices, it is difficult to find alternate sources of software.

a. Will this end software piracy, because mobile apps are not distributed on discs and must be purchased at one specified place?

b. Should software for laptops and desktops be distributed only through app marketplaces to prevent piracy?

c. What else might be done to combat software piracy? Explain your answers.

MyLab IT

Go to **MyLab IT** to take an autograded version of the *Check Your Understanding* review and to find all media resources for the chapter.

For the IT Simulation for this chapter, see MyLab IT.

Try This

Citing Website Sources

You've been assigned a research paper, and your instructor requires citations. In the past, you might have used websites such as Son of Citation Machine (*citationmachine.net*) to create your citations. As you'll see in this Try This, tools that are built right into Microsoft Word do the same thing. (Note: For more step-by-step instructions and information on how to build a bibliography, watch the Try This video on MyLab IT.)

Step 1 Click at the end of the sentence or phrase that you want to cite. On the References tab, in the Citations & Bibliography group, set the Bibliography style to the format you wish to follow, for example, MLA or APA. Click **Insert Citation**. To add a source, click **Add New Source**.

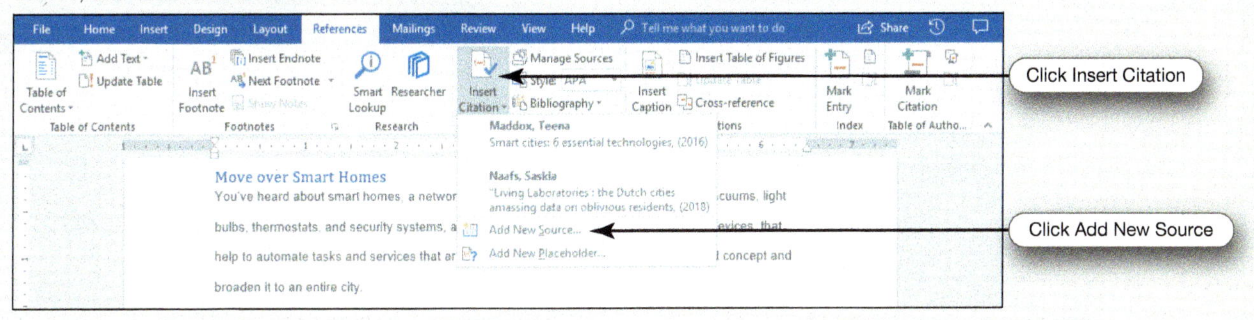

(Courtesy of Microsoft Corp.)

Note: To create a citation and fill in the source information later, click **Add New Placeholder**.

Step 2 Click the arrow next to Type of Source and begin to fill in the source information. Click OK when you're done.

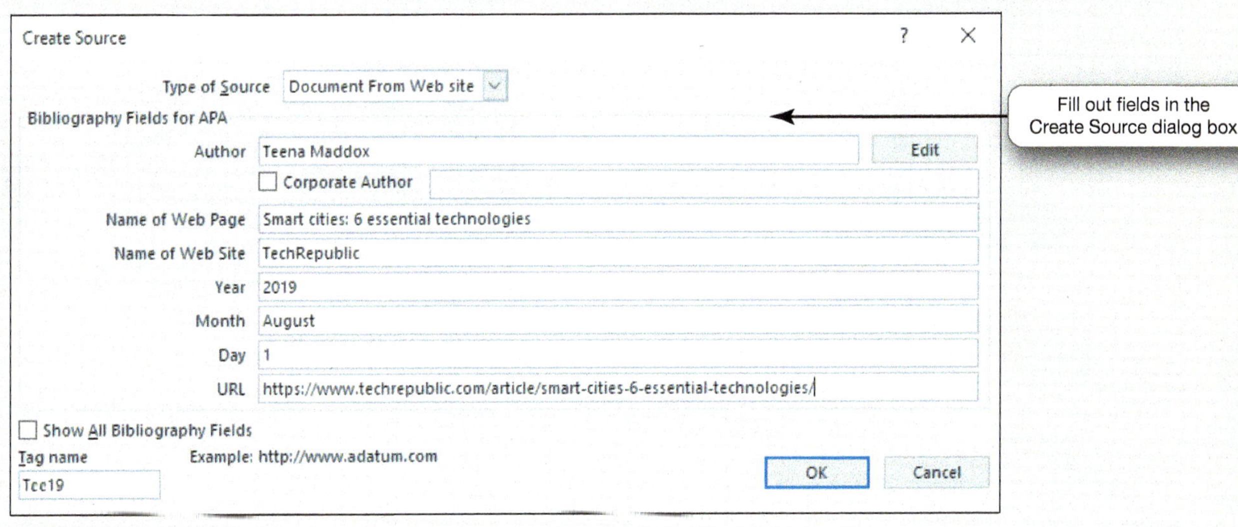

(Courtesy of Microsoft Corp.)

Step 3 The citation will appear in your document.

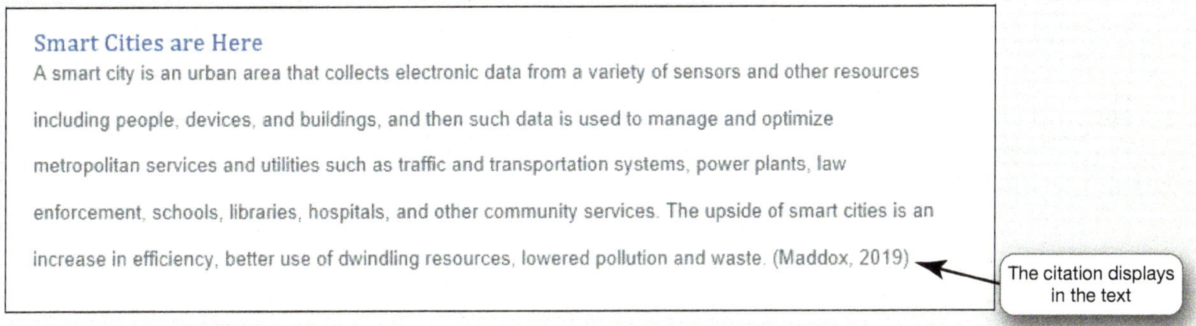

Smart Cities are Here

A smart city is an urban area that collects electronic data from a variety of sensors and other resources including people, devices, and buildings, and then such data is used to manage and optimize metropolitan services and utilities such as traffic and transportation systems, power plants, law enforcement, schools, libraries, hospitals, and other community services. The upside of smart cities is an increase in efficiency, better use of dwindling resources, lowered pollution and waste. (Maddox, 2019)

(Courtesy of Microsoft Corp.)

Make This  TOOL: App Inventor 2 or Thunkable

A More Powerful App

Want your app to be able to open a file from the SD card on your phone, fire up the YouTube app, or display a location in the Maps app?

In this exercise, you'll use ActivityStarter to incorporate the software already on your device into your mobile app.

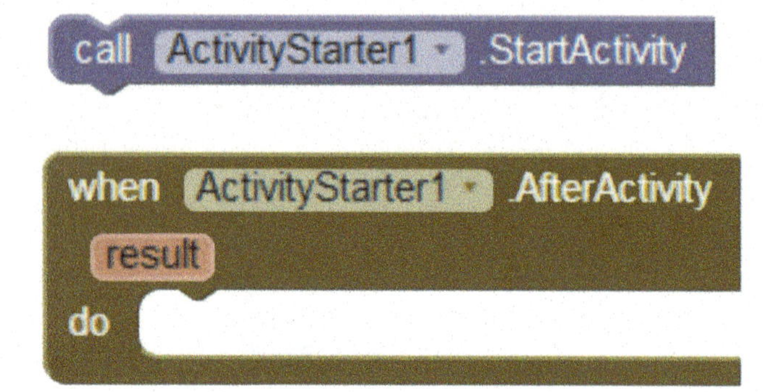

(Copyright MIT, used with permission.)

It's as easy as using the Connectivity drawer of the Designer. Drag the ActivityStarter component onto your Designer screen and then control it with the Blocks for ActivityStarter. Your apps now have the power of all the software on your device behind them!

The ActivityStarter component allows you to use the existing software on your device from within your mobile app.

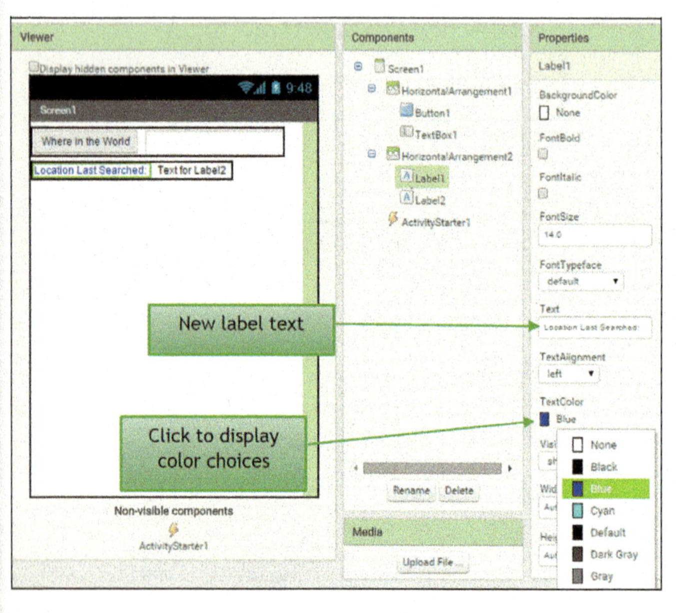

(Copyright MIT, used with permission.)

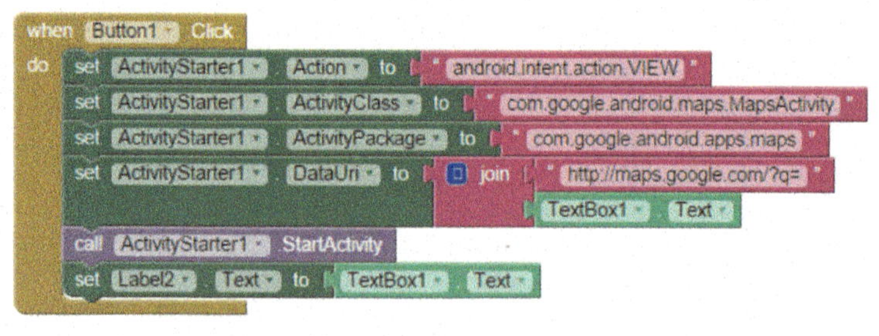

(Copyright MIT, used with permission.)

To create this app for iOS, go to *Thunkable.com*, a programming platform based on App Inventor.

For the instructions for this exercise, go to MyLab IT.

Part 2

For an overview of this part of the chapter, watch **Chapter Overview Video 4.2**.

Application Software

Learning Outcome 4.2 Describe the different types of application software used for productivity and multimedia.

One reason computers are invaluable is that they make it easier to complete daily tasks. Although many programs help you be more productive, there are also programs that entertain you with audio, video, and images, as well as games, animations, and movies.

In this section, we'll look at applications that help enhance your productivity, support your business projects, and let you be creative or entertained.

Productivity and Business Software

Let's start by looking at productivity and business software.

Productivity Software

Objective 4.7 *Categorize the types of application software used to enhance productivity and describe their uses and features.*

What is productivity software? Productivity software enables you to perform various tasks required at home, school, and business. The main types of productivity software include programs for word processing, spreadsheet analysis, presentations, database management, note-taking, and personal information management. In addition, financial planning software and tax preparation software are available for personal use.

How do I choose among the many available programs? When you're looking for a specific product, say, a word processing program, you have many choices. As discussed earlier, you can choose between commercial products and open source products or choose to install the program on your local device or use it purely online. Table 4.2 guides you through these many choices.

Microsoft Office is the most popular set of productivity programs and is offered as locally installed software (Office 2019) or as a subscription (Office 365). There are also online versions of most Office 365 applications, although they are not as full-featured as the premium, installed versions of Office 2019 or Office 365. Google offers its G-Suite of online productivity software such as Google Docs and Google Sheets. The open source LibreOffice is free to download and install and is supported by an active community.

Next, we'll examine each type of program to understand its key features.

Bits&Bytes **Productivity Software Tips and Tricks**

Looking for tips on how to make better use of your productivity software? Some websites send subscribers periodic e-mail full of tips, tricks, and shortcuts for their favorite software programs:

- The Microsoft website includes many tips and tricks for its Office applications (*support.office.com*).
- *MakeTechEasier.com* has tidbits for a variety of applications, including Windows and Mac products, LibreOffice, and cell phone apps.
- *GCFLearnFree.org* offers free instructional tutorials on a variety of technology topics, including Microsoft Office applications.
- You can also find tips as videos online from YouTube, TED, and other websites.

Table 4.2 Productivity Software

PROGRAM	WORD PROCESSING	SPREADSHEET	PRESENTATION	DATABASE	NOTE-TAKING	PIM/E-MAIL
Installed: Proprietary						
Office 365	Word	Excel	PowerPoint	Access	OneNote	Outlook
Apple iWork	Pages	Numbers	Keynote			
Installed: Open Source						
LibreOffice	Writer	Calc	Impress	Base		
Web-Based						
Office 365 Online	Word	Excel	PowerPoint		OneNote	Outlook
G-Suite	Docs	Sheets	Slides			Gmail
Zoho	Writer	Sheet, Books	Show	Creator	Notebook	
ThinkFree	Write	Calc	Show		Note	

Word Processing Software

What are the most common word processing applications? You've probably used word processing software to create and edit documents such as research papers, class notes, and résumés. Microsoft Word is the most popular word processing program. If you're looking for a more affordable alternative, you might want to try an open source program such as Writer from LibreOffice (*libreoffice.org*) or a free online word processing application such as Google Docs or Microsoft Word Online.

What are the special features of word processing programs? Word processors come with basic tools such as spelling and grammar checking, a thesaurus, and a language translator. There are tools to format bibliographical references and organize them into a database so you can reuse them in other papers quickly. You can also enhance the look of your document by creating an interesting background, inserting images, or by adding a theme of coordinated colors and styles (see Figure 4.5).

Bits&Bytes How to Open Unknown File Types

Normally, when you double-click a file to open it, the program associated with the file opens automatically. For example, when you double-click a file with a .doc or .docx extension, the file opens in Microsoft Word. However, if the file has no extension or Windows has no application associated with that file type, a *How do you want to open this file?* dialog box appears and asks what program you want to use to open the file. In other cases, a document may open with a program other than the one you want to use. This is because many applications can open several file types, and the program you expected the file to open in isn't the program currently associated with that file type. To assign a program to a file type or to change the program to open a particular file type, follow these instructions:

1. Use the search and navigation tools in File Explorer to locate the file you want to change. (For example, you can search for all Word files by searching for *.doc or *.docx.) Right-click the file and then point to **Open with**.
2. A list of programs installed on your computer appears. Click the program you want to use to open this file type. If you're sure the selected program is the one that should always be used for this file type, select **Choose another app,** which opens the **How do you want to open this file?** dialog box. Check the **Always use this app to open [extension] files** and click the default program from the list.

When you open the file in the future, the file will open in the program you selected.

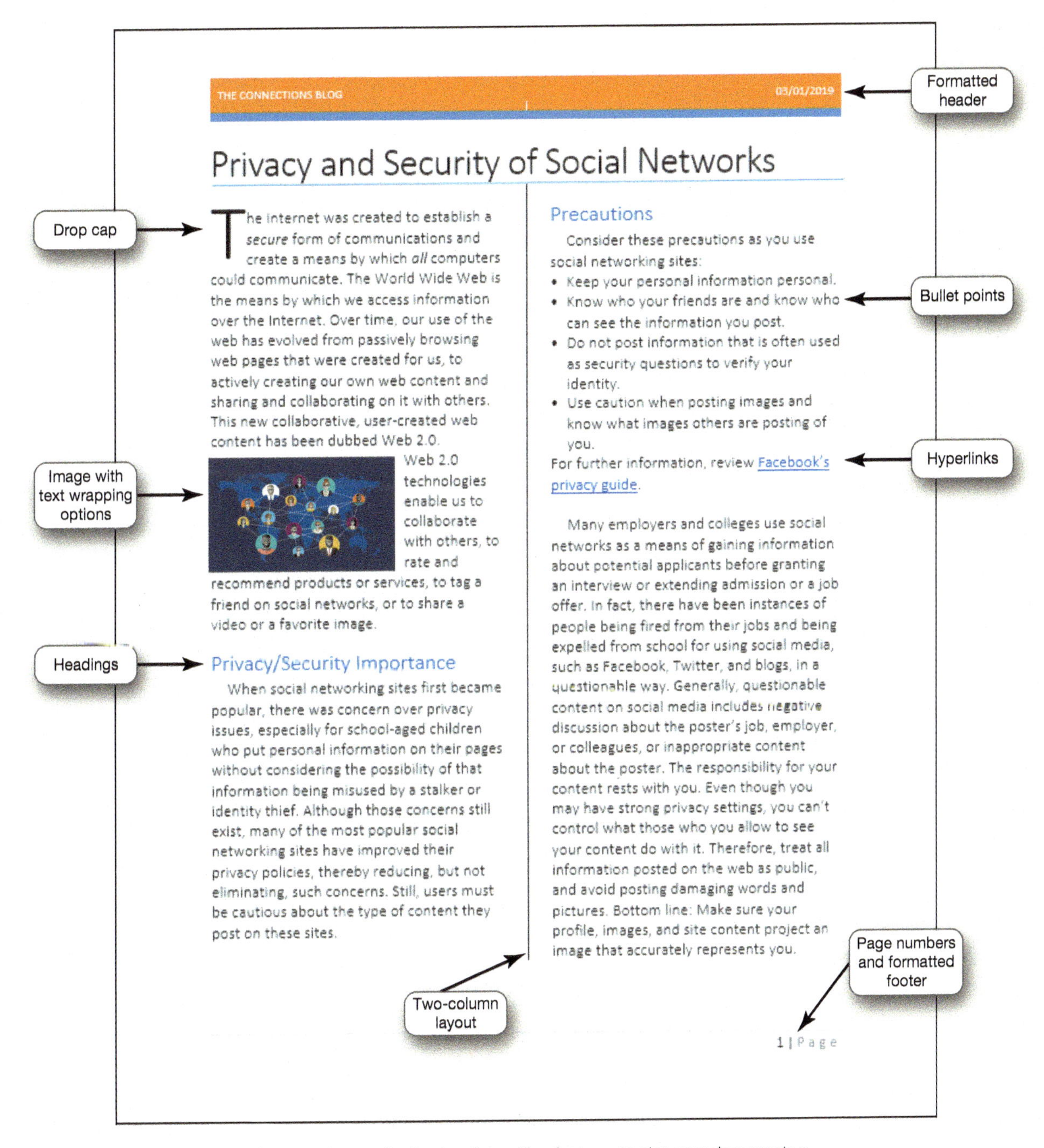

THE CONNECTIONS BLOG 03/01/2019

Privacy and Security of Social Networks

The internet was created to establish a *secure* form of communications and create a means by which *all* computers could communicate. The World Wide Web is the means by which we access information over the Internet. Over time, our use of the web has evolved from passively browsing web pages that were created for us, to actively creating our own web content and sharing and collaborating on it with others. This new collaborative, user-created web content has been dubbed Web 2.0.

Web 2.0 technologies enable us to collaborate with others, to rate and recommend products or services, to tag a friend on social networks, or to share a video or a favorite image.

Privacy/Security Importance

When social networking sites first became popular, there was concern over privacy issues, especially for school-aged children who put personal information on their pages without considering the possibility of that information being misused by a stalker or identity thief. Although those concerns still exist, many of the most popular social networking sites have improved their privacy policies, thereby reducing, but not eliminating, such concerns. Still, users must be cautious about the type of content they post on these sites.

Precautions

Consider these precautions as you use social networking sites:

- Keep your personal information personal.
- Know who your friends are and know who can see the information you post.
- Do not post information that is often used as security questions to verify your identity.
- Use caution when posting images and know what images others are posting of you.

For further information, review Facebook's privacy guide.

Many employers and colleges use social networks as a means of gaining information about potential applicants before granting an interview or extending admission or a job offer. In fact, there have been instances of people being fired from their jobs and being expelled from school for using social media, such as Facebook, Twitter, and blogs, in a questionable way. Generally, questionable content on social media includes negative discussion about the poster's job, employer, or colleagues, or inappropriate content about the poster. The responsibility for your content rests with you. Even though you may have strong privacy settings, you can't control what those who you allow to see your content do with it. Therefore, treat all information posted on the web as public, and avoid posting damaging words and pictures. Bottom line: Make sure your profile, images, and site content project an image that accurately represents you.

1 | Page

Figure 4.5 Nearly every word processing application has formatting features to give your documents a professional look. *(Picture Alliance; Courtesy of Microsoft Corp.)*

Spreadsheet Software

Why would I use spreadsheet software? Spreadsheet software calculates and performs numerical analyses. For example, you can use it to create a simple budget, as shown in Figure 4.6a. Microsoft Excel and LibreOffice Calc are two examples of spreadsheet software. Web-based options are available from Google and Microsoft Office Online. One benefit of spreadsheet software is that it can automatically recalculate all formulas and functions in a spreadsheet when values for some of the inputs change. For example, as shown in Figure 4.6b, you can insert an additional row in your budget (Membership) and change a value (for September Financial aid), and the results for Total Expenses and Net Income recalculate automatically.

Because automatic recalculation immediately shows you the effects different options have on your spreadsheet, you can quickly test different assumptions. This is called *what-if analysis*. Look again

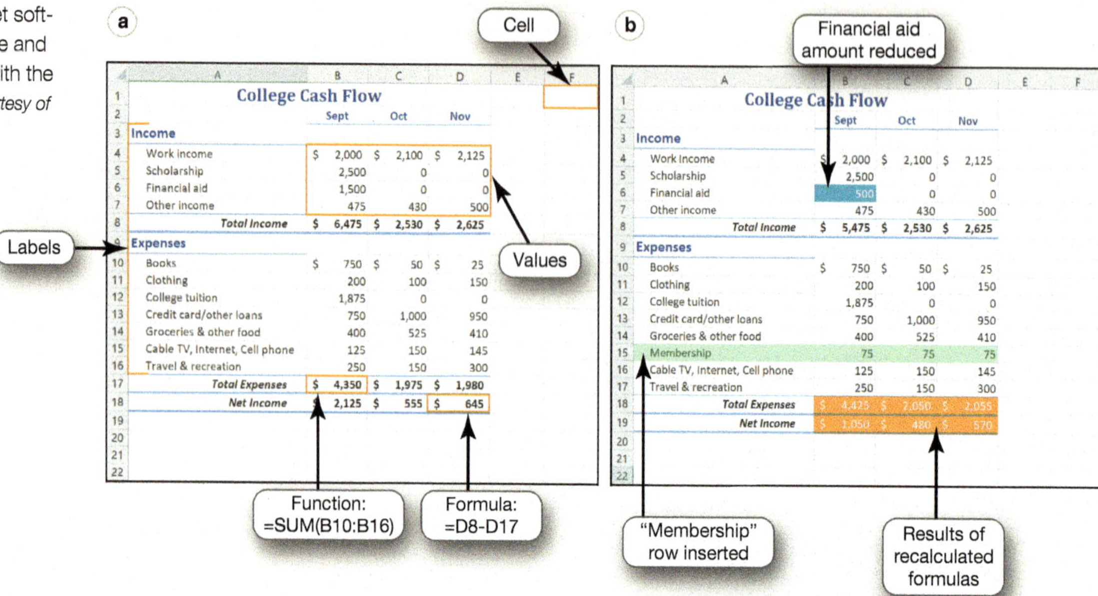

Figure 4.6 With spreadsheet software, you can easily calculate and manipulate numerical data with the use of built-in formulas. *(Courtesy of Microsoft Corp.)*

at Figure 4.6b and ask, "If I don't get as much financial aid next semester, what impact will that have on my total budget?" The recalculated cells in rows 18 and 19 help answer your question. In addition to financial analysis, many spreadsheet applications have limited database capabilities to chart, sort, filter, and group data.

How do I use spreadsheet software? The basic element in a spreadsheet program is the *worksheet*, which is a grid consisting of columns and rows. As shown in Figure 4.6a, the columns and rows form boxes called *cells*. Each cell can be identified according to its column and row position. For example, the cell in column A, row 1 is cell A1. You can enter several types of data in a cell:

- **Text:** Any combination of letters, numbers, symbols, and spaces. Text is often used as labels to identify the contents of a worksheet or chart.
- **Values and dates (numerical data):** These values can represent a quantity or a date/time and are often the basis for calculations.
- **Formulas:** Equations that use addition, subtraction, multiplication, and division operators as well as values and cell references. For example, in Figure 4.6a, you would use the formula =D8-D17 to calculate net income for November.
- **Functions:** Functions help develop formulas. Adding a group of numbers (SUM) or determining the monthly loan payment (PMT) can be done using built-in functions. Functions free you from needing to know the math behind the calculation. In Figure 4.6a, to calculate the total expenses for September, you could use the built-in summation function =SUM(B10:B16).

What kinds of graphs and charts can I create with spreadsheet software? Most spreadsheet applications enable you to create a variety of charts, including basic column charts, pie charts, and line charts, with or without 3D effects (see Figure 4.7). In addition, you can make stock charts (for investment analysis) and scatter charts (for statistical analysis) or create custom charts. Another feature in Excel is *sparklines*—charts that fit into a single cell and make it easy to show data trends (see Figure 4.8a). To analyze very large data sets, Excel has added newer chart types like treemaps, sunbursts, and waterfall charts (see Figure 4.8b).

Figure 4.7 (a) Column charts show comparisons. (b) Pie charts show how parts contribute to the whole. (c) Line charts show trends over time. *(Courtesy of Microsoft Corp.)*

Presentation Software

How can software help with my presentations? You've no doubt sat through presentations where the speaker used presentation software such as Microsoft PowerPoint (see Figure 4.9) or

Apple's Keynote or Google Slides to outline and illustrate their talking points. Because these applications are simple to use, you can produce high-quality presentations without a lot of training. You can further enhance presentations by embedding videos and adding animation and transition effects. You can also integrate quizzes and surveys into your PowerPoint presentations to make them more interactive.

What are some tips to make a great presentation? Some features in Microsoft PowerPoint can help you improve your slide layouts as well as how you deliver your presentation. Design Ideas provides suggestions for professionally designed layouts based on your own slide content. In addition, as you rehearse your presentation, the Presenter Coach feature uses the recorded audio and evaluates your presentation for pacing, pitch, informal speech, and culturally sensitive terms. In addition, it checks for filler words such as *ums, ahs*, and *you knows*, and it detects when you're being overly wordy or are simply reading the text on a slide. After each rehearsal, you're given a dashboard with statistics and tips for improvement.

As you create your presentation, keep some of these standard guidelines in mind:

- **Use images:** Images can convey a thought or illustrate a point. Make sure your audience can read any text over an image.
- **Be careful with color:** Choose dark text on a light background or light text on a dark background. Avoid using text and background colors that are too similar or that clash.
- **Use bullets for key points:** Use no more than four to six bulleted points per slide. Avoid full sentences and paragraphs.
- **Consider font size and style:** Keep the font size large enough to read from the back of the room. Avoid script or fancy font styles. Use only one or two font styles per presentation.
- **Keep animations and/or background audio to a minimum:** They can be distracting.

	Current	1 Year History Trend	High	Low
Microsoft	$ 25.79		30.54	20.26
Apple	$ 255.96		261.09	125.83
Intel	$ 20.95		22.84	15.72
Hewlett Packard	$ 46.05		53.15	34.35
Dell	$ 13.24		16.2	11.57

a Sparklines

b

Figure 4.8 (a) Sparklines and (b) sunbursts are two of many charting tools in Excel. *(Courtesy of Microsoft Corp.)*

Database Software

Why is database software useful? Database software such as Oracle, MySQL, and Microsoft Access can store and organize data for you. Spreadsheet applications, such as Excel, offer some database functionalities for simple tasks such as sorting, filtering, and organizing data. However, you need to use a more robust, fully featured database application to manage larger and more

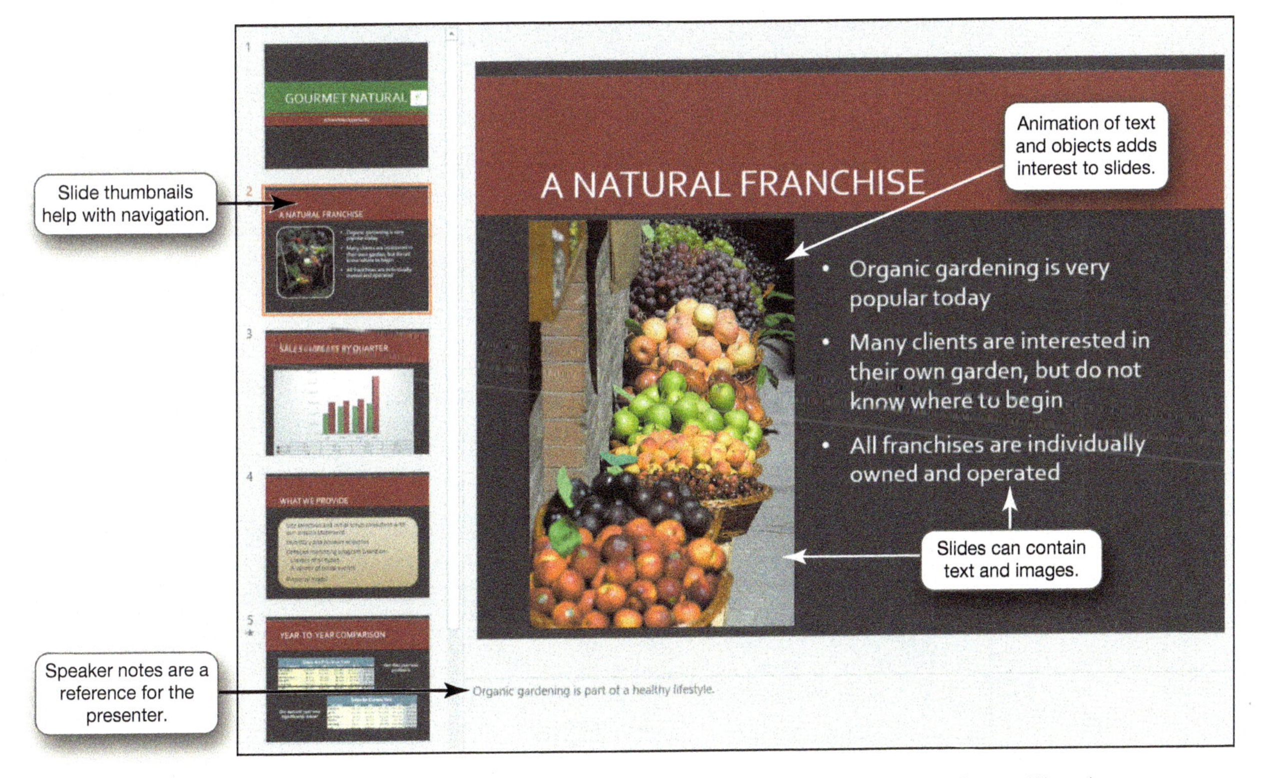

Figure 4.9 You can use presentation software to create dynamic slide shows. *(Courtesy of Microsoft Corp.; Picture Alliance)*

Bits&Bytes Going Beyond PowerPoint

Several applications offer an alternative to PowerPoint. One option is Prezi. Rather than using a set of slides, Prezi takes a less linear approach and enables you to connect ideas and pan and zoom on text or images for emphasis and interest. Other options include Canva, Visme, and Haiku Deck. Sway is an online app from Microsoft that allows you to create presentations that are stored in the cloud and linked to your Microsoft account. Sway facilitates importing content from a variety of sources to allow you to tell a story and is designed primarily for creating web-based presentations (see Figure 4.10). You can begin a Sway presentation with the content of a Word document, using the Transform to Web Page feature.

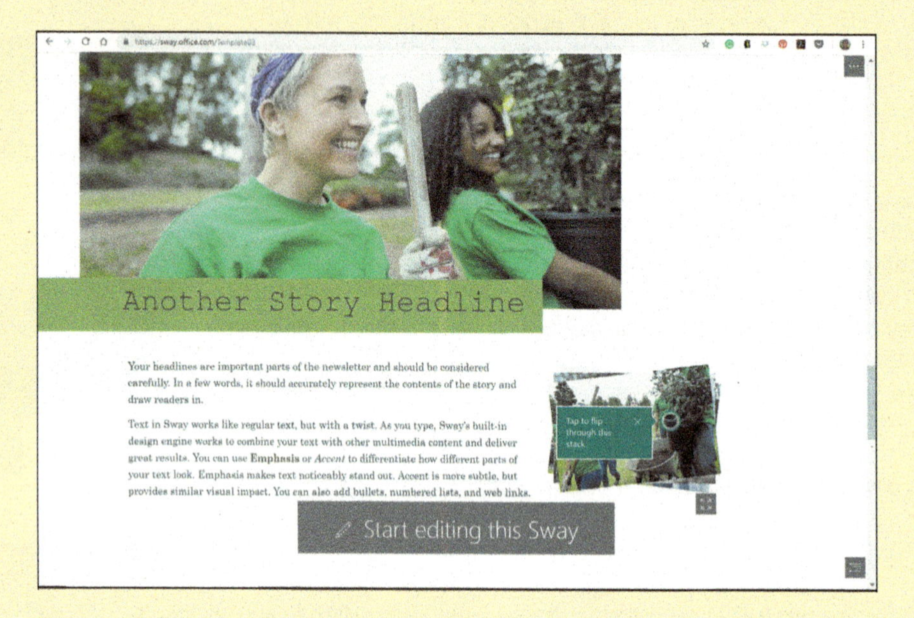

Figure 4.10 Sway is used to create web-based presentations. *(Courtesy of Microsoft Corporation; Sway.com/Template08)*

complicated data. Database programs organize data into one or more tables and facilitate the management of the data through grouping, sorting, and retrieving certain data based on specific criteria. Moreover, databases help to input data and generate reports. Traditional databases are organized into *fields*, *records*, and *tables*, as shown in Figure 4.11.

How do businesses use database software? Most websites rely on databases to keep track of products, clients, invoices, and personnel information. Often, some of that information is available to a home computer user. For example, at Amazon, you can access the history of all the purchases you've made on the site. Shipping companies let you search their online databases for tracking numbers, allowing you to get information on your package status.

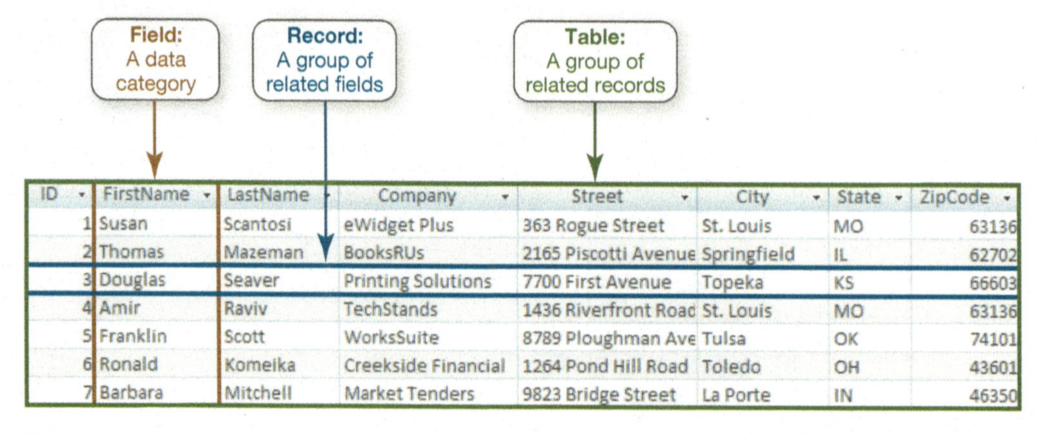

ID	FirstName	LastName	Company	Street	City	State	ZipCode
1	Susan	Scantosi	eWidget Plus	363 Rogue Street	St. Louis	MO	63136
2	Thomas	Mazeman	BooksRUs	2165 Piscotti Avenue	Springfield	IL	62702
3	Douglas	Seaver	Printing Solutions	7700 First Avenue	Topeka	KS	66603
4	Amir	Raviv	TechStands	1436 Riverfront Road	St. Louis	MO	63136
5	Franklin	Scott	WorksSuite	8789 Ploughman Ave	Tulsa	OK	74101
6	Ronald	Komeika	Creekside Financial	1264 Pond Hill Road	Toledo	OH	43601
7	Barbara	Mitchell	Market Tenders	9823 Bridge Street	La Porte	IN	46350

Figure 4.11 In databases, information is organized into fields, records, and tables. *(Courtesy of Microsoft Corp.)*

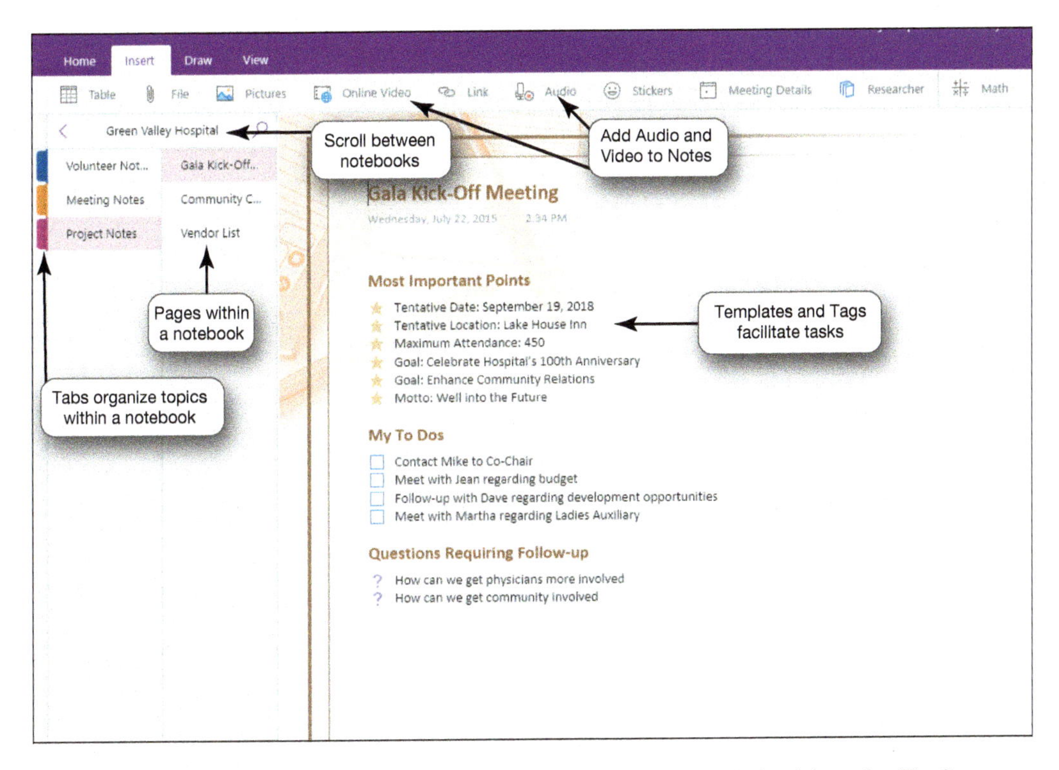

Figure 4.12 Microsoft OneNote is a great way to collect and organize notes and other information. The files are readily searchable and easy to share. *(Courtesy of Microsoft Corp.)*

Note-Taking Software

Is there software to help me take notes? Microsoft OneNote is a note-taking and organizational software tool you can use for research, brainstorming, and collaboration as well as for organizing information. You can organize your notes into tabbed sections with OneNote (see Figure 4.12).

You can add audio or video recordings of lectures to OneNote, and you can search for a term across all the digital notebooks you created during the semester to find common ideas, such as key points that might appear on a test. There is also a OneNote mobile app.

Several free online note-taking options are also available to help you take notes or jot down a quick reminder. Evernote, for example, lets you take notes via the web, your phone, or your computer and then syncs your notes between your devices. You can then share your notes with other Evernote users for easy collaboration. Google Keep is great for quick and simple notes that are similar to digital Post-It® notes. Notes can be paired with reminders and collaborated on in real time. Another simple note app is Sticky Notes, which is built into the Windows operating system. You can pin individual notes to the desktop or view all sticky notes in a list format that you can also search. Table 4.3 lists some other popular alternative note-taking applications.

Personal Information Manager Software

How can software help me manage my e-mail, time, contact lists, and tasks? Most people need some form of personal information manager (PIM) software to help manage e-mail, contacts, calendars, and tasks in one place. Microsoft Outlook (see Figure 4.13) is a widely used PIM program. If you share a network and are using the same PIM software as others on a common network, a PIM program simplifies sharing calendars and scheduling meetings.

Many web-based e-mail clients, such as Yahoo! and Google, also include coordinating calendar and contacts similar to those of Microsoft Outlook.

Table 4.3 Beyond Microsoft OneNote: Alternative Note-Taking Applications

Application	Features
Evernote (evernote.com)	• Web-based • Notes can be shared for easy collaboration • Syncs notes between all devices
AudioNote (luminantsoftware.com)	• Synchronized note taking and audio recording • Allows text or handwritten notes • Highlights notes during playback
Simplenote (simplenote.com)	• Web-based, open source • Notes organized by tags • Mobile apps available
Notability (gingerlabs.com)	• PDF annotations • Advanced word processing • Linked audio recordings to notes • Auto-sync notes between devices

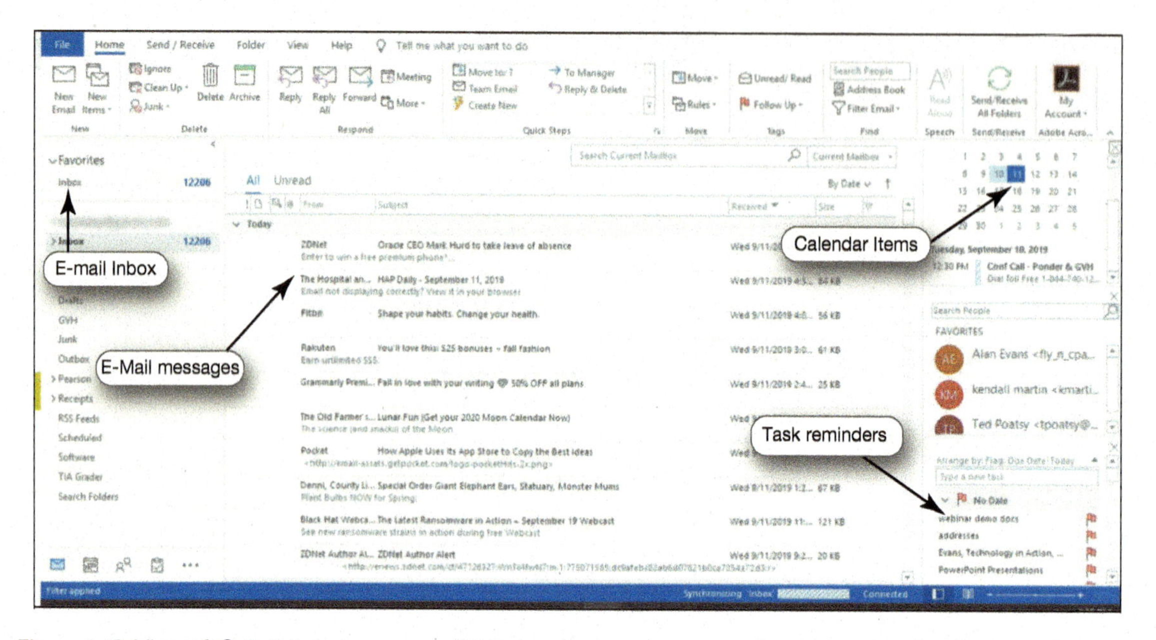

Figure 4.13 Microsoft Outlook includes common PIM features such as a summary of appointments, a list of tasks, and e-mail messages. *(Courtesy of Microsoft Corp.)*

Google's calendar and contacts sync with Outlook so you can access your Outlook calendar information by logging on to Google. This gives you access to your schedule anywhere you have access to a computer and an Internet connection.

A wide variety of other to-do lists and simple organizers work with mobile and computing devices. For example, Toodledo is a free program that coordinates well with Microsoft Outlook, and OmniFocus 3 is a more full-featured option for Mac devices, including a slick interface for Apple Watch.

Project and Collaboration Software

What software should I use when working remotely with team members? Microsoft Teams is a communication and collaboration application (see Figure 4.14). Microsoft Office products have allowed team members easily to share files created in Word, Excel, and PowerPoint for years. Now, using Teams as the hub for your group work, collaborating with others is even more efficient. Teams is available for free with a Microsoft 365 subscription as a mobile or installable app. In Teams, you can break down your projects into separate channels to coordinate the members, files, and communications. In each channel, you can share files, participate in chats, and create collaborative documents. You can add other apps to extend the capabilities of teams. For example, installing a whiteboard app makes virtual brainstorming easy. You can also hold video conferences and share screens with Teams, enabling you to stay productive when working remotely. Last, Microsoft Teams is exceedingly secure, with full data encryption.

What other collaboration tools are available? In addition to Microsoft Teams, other tools are good for group work, team coordination, and project collaboration:

- Zoom is an app that facilitates conference calls. You can share your screen with others, add another person during the conference, record the conference, and mute callers. You can see everyone on the call and even view their social networking profiles. The free version is limited to 45-minute sessions.
- Slack is a messaging app you can use to organize your conversations into channels. You can link to files stored in cloud services like Google Drive and Dropbox instead of attaching the files in an e-mail or sharing them outside of the Slack platform. The files you link into the conversation are searchable along with your conversations. Slack runs on mobile apps as well as in a web browser.
- Trello is a visually oriented app used for managing projects. Unlike traditional project management apps, it features a board filled with lists, and each list is filled with cards representing tasks to be completed. It is easy to add checklists and due dates to cards as well as upload files to them. You can transfer cards between lists to show progress on various aspects of a project. You can invite as many people to your board as you want. You can then assign people to cards

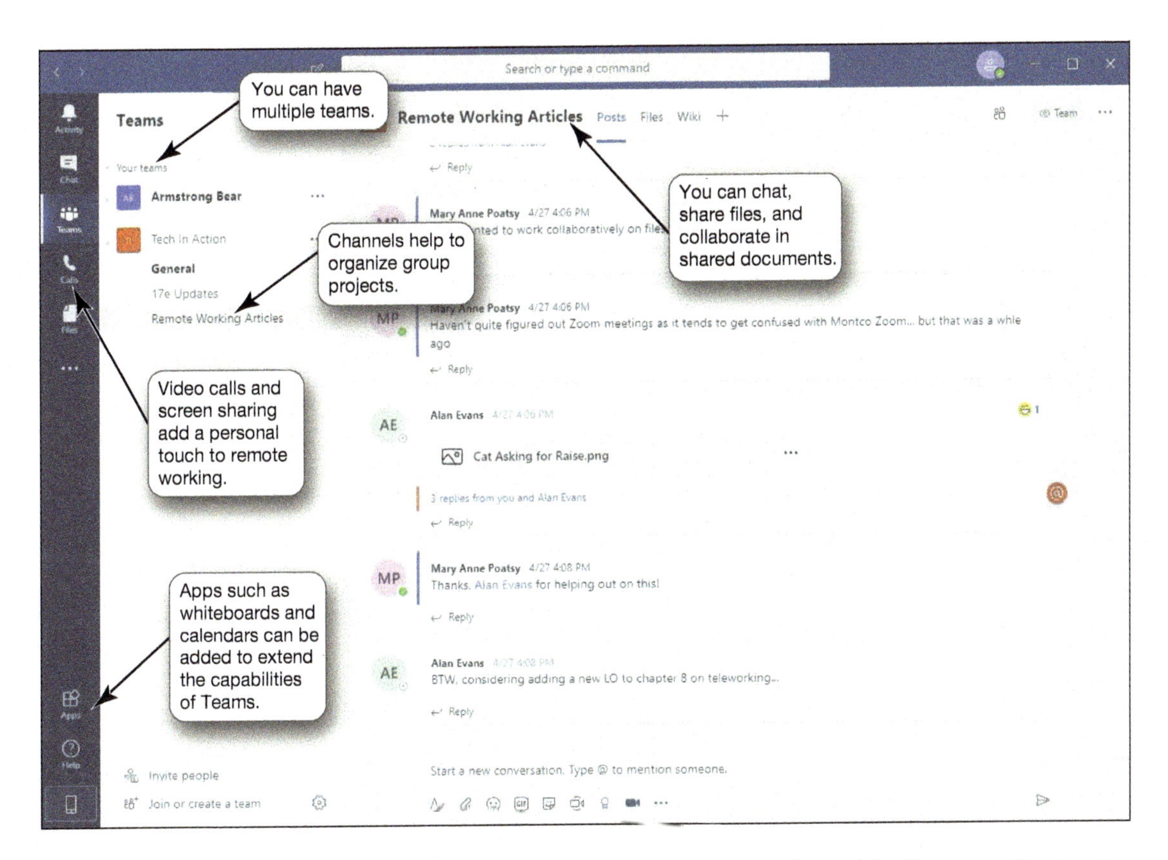

Figure 4.14 Collaboration software, such as Microsoft Teams, makes it easier to work remotely with a group. *(Courtesy of Microsoft Corporation. Used with Permission from Microsoft.)*

to divide up tasks. You can also see an overview of your project by glancing at your Trello board. The free version of the app offers a powerful feature set.

- Scribblar (*scribblar.com*) is a multiuser whiteboard with live audio chat, which is great for holding virtual brainstorm sessions.

Programming for End Users

In this Sound Byte, you'll be guided through the creation of a macro in Microsoft Office. You'll learn how Office enables you to program with macros to customize and extend the capabilities it offers.

Personal Financial Software

How can I use software to keep track of my finances? Financial planning software helps you manage your daily finances. Financial planning programs include electronic checkbook

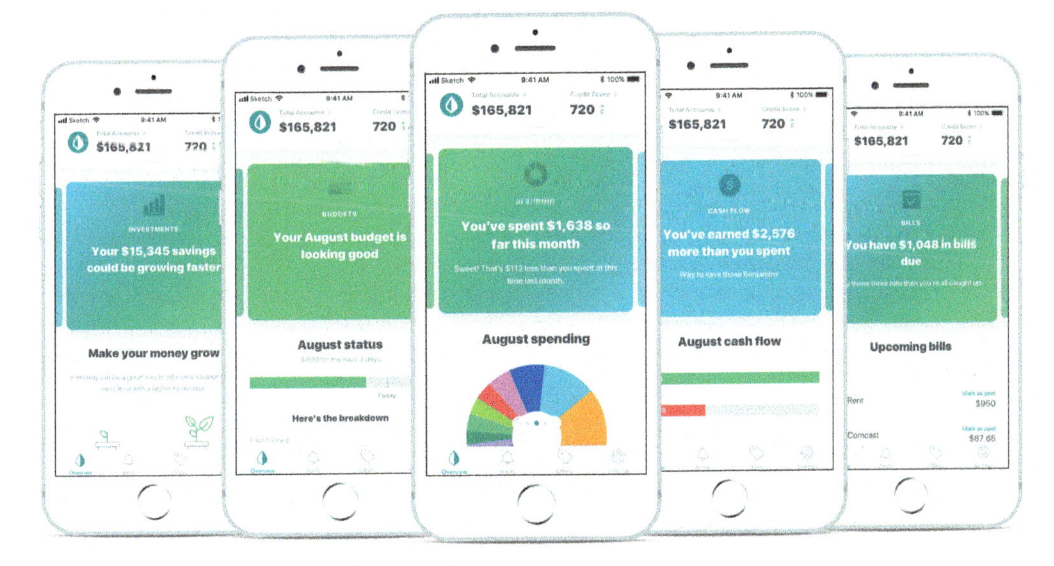

Figure 4.15 Mint (*mint.com*) is an online financial management tool. An extensive online community provides helpful tips and discussions with other people in similar situations. *(Courtesy of Intuit Inc.)*

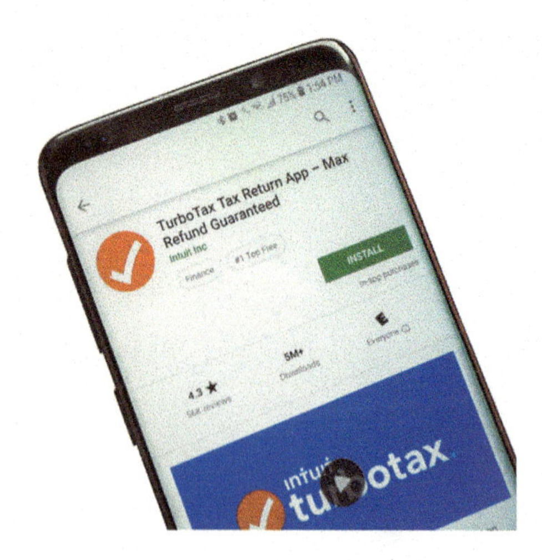

Figure 4.16 Tax preparation software lets you prepare and file your taxes on mobile devices. *(Ricky Of The World/Shutterstock)*

registers and bill payment tools. You can make recurring monthly payments, such as rent, with scheduled online payments, and the software records all transactions in the checkbook register. In addition, you can assign categories to each transaction and use these categories to create budgets and analyze your spending patterns.

Intuit's installed and web-based products, Quicken and Mint (*mint.com*), are the market leaders in financial planning software (see Figure 4.15). Both track and analyze your spending habits and offer advice on how to manage your finances better. With either, you can also track your investment portfolio. With Mint, you can monitor and update your finances from any computer with a private and secure setting. You can use the Mint app on a smartphone or tablet, so accessing your financial information is convenient. Mint also provides access to a network of other users with whom to exchange tips and advice.

What software can I use to prepare my taxes? Tax preparation software, such as Intuit TurboTax and H&R Block Tax Software, lets you prepare your state and federal taxes on your own. Both programs offer tax forms and instructions as well as videos with expert advice on how to complete each form. Each program offers free web-based versions of federal forms and instructions. In addition, error-checking features are built into the programs to catch mistakes as well as provide audit alerts. Both programs allow you to file your return electronically and offer financial planning guidance to help you plan and manage your financial resources in the following year (see Figure 4.16).

Some financial planning applications also coordinate with tax preparation software. Both Quicken and Mint, for example, integrate with TurboTax, so you never have to go through your debit card statements and bills to find tax deductions, tax-related income, or expenses. Many banks and credit card companies also offer online services that download a detailed monthly statement into Quicken and Mint. Remember, however, that the tax code changes annually, so you must obtain an updated version of the software each year.

Business Software

Objective 4.8 *Summarize the types of software that small and large businesses use.*

How does business software differ from personal software? Many businesses rely on the types of software we've discussed so far in the chapter. In addition, specialized software is used in both large and small businesses.

Small Business Software

What kinds of software are helpful for small business owners? If you have a small business or a hobby that produces income, you know the importance of keeping good records and tracking your expenses and income. Accounting software helps small business owners manage their finances more efficiently by providing tools for tracking accounts receivable and accounts payable. In addition, these applications offer inventory management, payroll, and billing tools. Intuit QuickBooks is an example of an accounting application. It includes templates for invoices, statements, and financial reports so that small business owners can create common forms and reports.

If your business produces newsletters, catalogs, annual reports, or other large publications, consider using desktop publishing (DTP) software. Although many word processing applications include many features that are hallmarks of desktop publishing, specialized DTP software, such as Adobe InDesign, allows professionals to design books and other publications that require complex layouts.

What software do I use to create a web page? Web authoring software allows even novices to design interactive web pages without knowing any HTML code. Web authoring applications often include wizards, templates, and reference materials to help novices complete most web authoring tasks. More experienced users can take advantage of these applications' advanced features to make the web content current, interactive, and interesting. Adobe Dreamweaver is an

Table 4.4 Common Types of Business-Related Software

Project Management	Customer Relationship Management (CRM)	Enterprise Resource Planning (ERP)
 Creates scheduling charts to plan and track specific tasks and coordinate resources	 Stores sales and client contact information in one central database	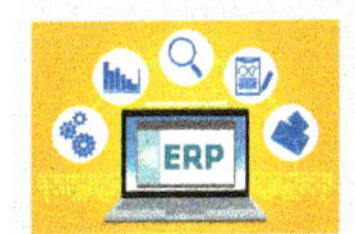 Controls many back office operations and processing functions such as billing, production, inventory management, and human resources management
E-Commerce	**Computer-Aided Design (CAD)**	**Vertical Market**
 Facilitates website creation and hosting services, shopping cart setup, and credit card–processing services	 Creates automated designs, technical drawings, and 3D model visualizations for architecture, automotive, aerospace, and medical engineering industries	 Addresses the needs of businesses in a specific industry or market such as the real estate, banking, and automotive industries

(Golden Sikorka/Shutterstock; Macrovector/Shutterstock; Aa Amie/Shutterstock; Khanisom Chalermchan/123RF; VENTRIS/Science Photo Library/Alamy Stock Photo; Nmedia/Shutterstock)

installed program that both professionals and casual web page designers use. Squarespace is an online-based option that is full featured and easy to use.

Note that if you need to produce only the occasional web page, you'll find that many applications include features that let you convert your document to a web page. For example, in some Microsoft Office applications, you can choose to save a file as a web page.

Software for Large and Specialized Businesses

What types of software do large businesses use? An application exists for almost every aspect of business. Table 4.4 lists many of the common types of business-related software. Some applications are tailored to the specific needs of a particular company or industry. Software designed for a specific industry, such as property management software for real estate professionals, is called vertical market software.

What software is used to make 3D models? Engineers use computer-aided design (CAD) programs such as Autodesk's AutoCAD to create automated designs, technical drawings, and 3D model visualizations. Here are some cool applications of CAD software:

- Architects use CAD software to build virtual models of their plans and readily visualize all aspects of design before actual construction.
- Engineers use CAD software to design everything from factory components to bridges. The 3D nature of these programs lets engineers rotate their models and adjust their designs if necessary, eliminating costly building errors.
- CAD software (and other 3D modeling software) is used to generate designs for objects that will be printed using 3D printers.
- The medical engineering community uses CAD software to create anatomically accurate solid models of the human body, developing medical implants quickly and accurately.

The list of CAD applications keeps growing as more and more industries realize the benefits CAD can bring to their product development and manufacturing processes.

Multimedia and Educational Software

Although many programs help you increase your productivity, there are also programs that entertain you with audio, video, and digital images and through games, animations, and movies. This section discusses multimedia and educational software, both for your mobile devices and for professional projects (see Figure 4.17).

Multimedia Software

Objective 4.9 *Describe the uses and features of multimedia software.*

What is multimedia software? Multimedia software includes image- and video-editing software, audio software, and other specialty software required to produce computer games, animations, and movies. Many mobile and desktop software programs allow everyone, from beginner to professional, to create animations, videos, audio clips, and images.

What will help improve my digital images? The camera app installed in your mobile device gives you control over standard elements like the use of flash and filtering options and useful modes like panorama or virtual 360 shots. In addition, each camera app has editing features that remove red eye and adjust qualities like saturation, contrast, and hue. Specialized apps like Snapseed or Enlight Photofox let you merge images, add artistic effects like a pencil sketch look, or apply special filters.

Professionals and advanced amateur photographers often use fully featured image-editing desktop software like Adobe Photoshop and Corel PaintShop Pro. A free alternative is Gimp (*gimp.org*), a download that has most of the features offered by the commercial applications. Desktop photo-editing

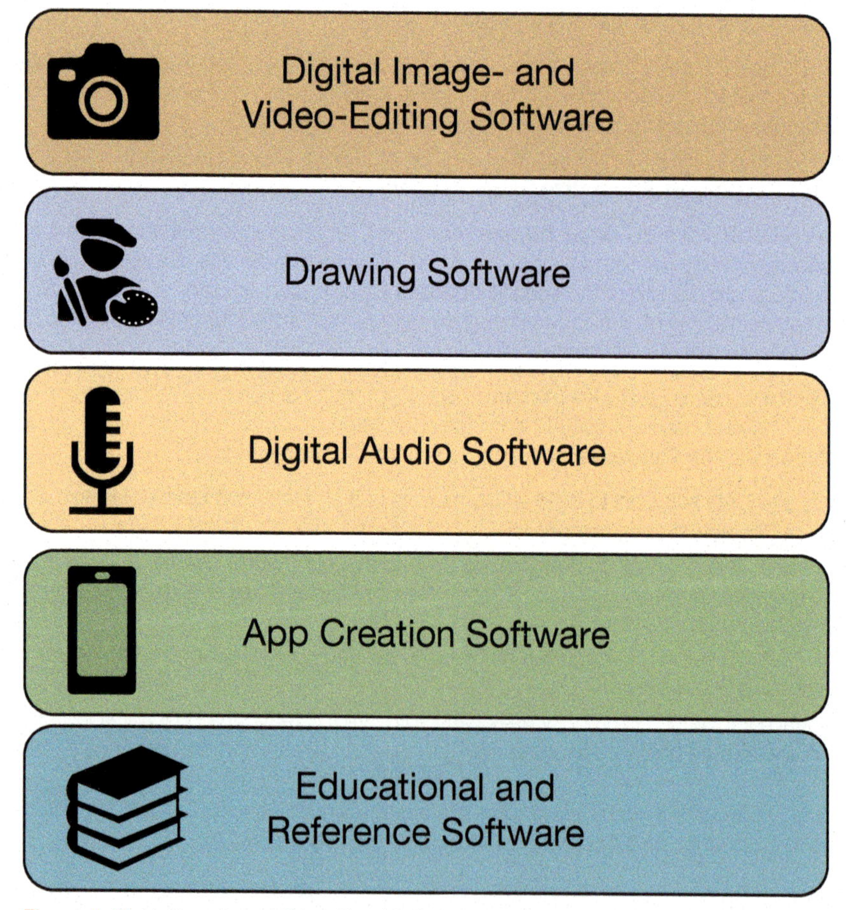

Figure 4.17 Multimedia and Educational Software

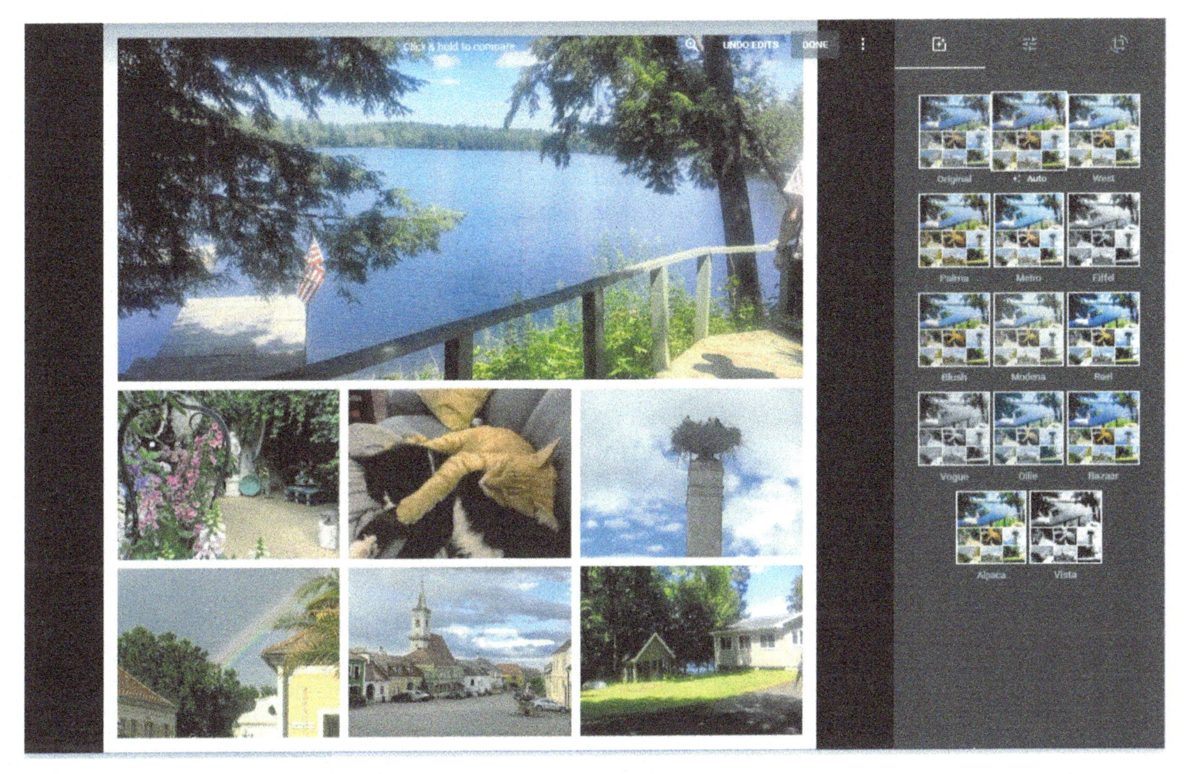

Figure 4.18 You can create collages of your favorite images by using Google Photos. *(Pearson Education, Inc)*

software offers sophisticated tools (see Figure 4.18) to create the enhanced digital images found commercially in logos and advertisements.

How do I organize all the photos I take? The photo app on your mobile device stores and sorts images by such things as date and location. These apps can be taught the faces of people you know and then automatically identify and group photos by people. Some apps go further and let you search your photo collection. For instance, searching "food" will let you pull together all photos of any meals you've photographed.

You may want to consider moving your photos and videos from your mobile device and storing them in a cloud storage system. Each mobile OS is designed to work with a specific cloud system (Apple devices use iCloud and Android devices automatically save to Google Photos), but you can use any cloud service you like. Tools like If This Then That (IFTTT) allow you to create recipes that automatically back up your images to Dropbox, for example. (To learn more about using If This Then That, see the Make This activity in Chapter 1.)

What software do I need to create and edit digital videos? The same camera apps that come preinstalled on mobile devices record video. They typically include features that let you record in slow motion, in fast motion, and in time lapse. Although it's easy to upload videos directly to YouTube or Facebook unedited, you can use digital video-editing software to refine your videos. Although the most expensive products (such as Adobe Premiere Pro and Apple's Final Cut Pro) offer the widest range of special effects and tools, more moderately priced video-editing programs often have enough features to keep the casual user happy. Apple iMovie and Camtasia, for example, both have intuitive drag-and-drop features and numerous templates and effects that make it simple to create professional-quality movies with little or no training (see Figure 4.19).

Figure 4.19 Video editing programs make it easy to create and edit movies to then share on social media. *(Andrey_Popov/Shutterstock)*

What kind of software should I use to create illustrations? Drawing (or illustration) software lets you create and edit 2D, line-based drawings. You can use it to create technical diagrams or original non-photographic drawings and illustrations, using software versions of traditional artistic tools such as pens, pencils, and paintbrushes. You also can drag geometric objects from a toolbar onto the canvas area to create images and can use paint bucket, eyedropper, and spray can tools to add color and special effects to the drawings.

Adobe Illustrator includes tools that let you create professional-quality creative and technical illustrations such as muscle structures in the human body. Its warping tool allows you to bend, stretch, and twist portions of your image or text. Because of its many tools and features, Illustrator is the preferred drawing software program of most graphic artists. Be sure also to explore Inkscape, an open source program similar to Illustrator. For touch-based devices, there is a great collection of software for creating vector art. Apps like Paper and Concepts turn your tablet into a terrific tool for creating illustrations.

Bits&Bytes **Scalable Vector Graphics: Text-Based Images**

If you've ever created a website and added an image, you've probably used a photograph that is based on pixels or a vector image that is based on points, lines, and shapes. The problem with using either type of image on a website is that they lose quality when the website is viewed on different sized screens. Scalable Vector Graphics (SVG) resolves this problem. SVG is a text-based graphics language that describes lines, curves, shapes, and colors in simple, plain language. Unlike some programming languages, SVG files are easily understood and modified.

Graphics created as SVG files are more convenient to use than other image file formats such as PNG, GIF, or JPG. Because text is used to describe the graphic, an SVG file is scalable. That means it can be resized without losing quality and viewed or printed with high quality at any resolution. SVG files are becoming the standard for web-based vector graphics.

Because the SVG format was developed under the W3C (World Wide Web Consortium), it is open standard and therefore works well with other open standard web languages and technologies such as JavaScript, cascading style sheets (CSS), and HTML.

Audio Software

Objective 4.10 *Describe the uses and features of audio software.*

What's the difference between all the audio file types on my computer? You probably have a variety of audio files stored on your computer, such as downloaded music files, audiobooks, and podcasts. These types of audio files have been compressed so they're more manageable to transfer to and from your computer and over the Internet. MP3, short for MPEG-1 Audio Layer 3, is the most common audio compression format, but there are other compressed audio formats, such as AAC and WMA.

You may also see uncompressed audio files on your computer, such as WAV or AIFF files. Uncompressed files—the files found on audio CDs, for example—have not had any data removed, so the quality is a perfect representation of the audio as it was recorded. Unfortunately, the file size is much larger than that of compressed files. Compressed formats remove data, such as high frequencies that the human ear does not hear, to make the files smaller and easier to download and store. MP3 format, for example, makes it possible to transfer and play back music on smartphones and other music players. Finite storage space on smartphones makes smaller file sizes more desirable. The smaller file size not only lets you store and play music in less space but also allows quick and easy distribution over the Internet. Ogg Vorbis (or just Ogg) is a free, open source audio compression format alternative to MP3. Some say that Ogg produces a better sound quality than MP3.

What do I use to create my own audio files? Many audio applications let you create and record your own audio files. Digital audio workstation software (DAWs) lets you create individual tracks to build songs or soundtracks with virtual instruments, voice recorders, synthesizers, and special audio effects, and these will end up as uncompressed MIDI (Musical Instrument Digital Interface) files. Examples of DAWs include Apple GarageBand and Ableton Live. You can use these programs to record audio from live sources as well, such as riffs from your electric guitar or vocals.

MyLab IT

Choosing Software

In this Helpdesk, you'll play the role of a help desk staffer, fielding questions about the different kinds of multimedia software, educational and reference software, and entertainment software.

How do I edit audio files? Audio-editing software includes tools that make editing audio files as easy as editing text files. Software such as the open source Audacity (*audacityteam.org*) lets you perform basic editing tasks like cutting dead air space from the beginning or end of a song or clipping a portion from the middle. You can also add special sound effects, such as reverb or bass boost, and remove static or hiss from MP3 files. These applications support recording sound files from a microphone or any source you can connect through the input line of a sound card.

App Creation Software

Objective 4.11 *Describe the features of app creation software.*

I'm not a programmer, but could I build my own mobile app? There are many types of app creation software that you can use to produce apps that run on various mobile devices. Many are approachable for beginners. MIT's App Inventor open source web application and Thunkable (which are used in the Make This exercises in this book) make it easy for beginners to create functional apps for Android and Apple mobile devices (see Figure 4.20). These applications feature a drag-and-drop interface and enable users to begin quickly developing powerful apps without actually knowing how to write program code.

Scratch (scratch.mit.edu) is another MIT programming environment that facilitates the creation of interactive stories, games, and animations that you can then share with an online community. The programming interface of Scratch is somewhat similar to that of App Inventor. Many elementary schools use Scratch to introduce young children to the ideas behind computer programming.

Corona (*coronalabs.com*) is a powerful, free programming environment that has been used to develop games and business apps. It comes with a large library of application programming interfaces (APIs) that can be used as building blocks to make writing computer code less time-consuming (although you will have to learn the Lua programming language). The Composer graphical user interface (GUI) is a visual editor for Corona, which makes it even easier to create apps without doing as much actual coding. Corona supports both iOS and Android.

Apple also has a powerful development environment for all its operating systems, called Swift. Swift uses more concise code than some other programming environments. It has a feature that allows you to see the results of each line of code visually as you write it. Many professional developers like using Swift because the code works side by side with Objective-C, which is a major programming language used in many businesses.

To get started in Swift, download the Swift Playgrounds app. It has a fun set of puzzles, games, and challenges that guide you into the world of programming for iOS. Despite the power of the Swift environment, it's still very approachable for beginners.

I just want to make video games. Is that a reasonable goal? Yes, video games represent an industry that generates billions in revenue each year, so designing and creating video games is a desirable career path. Professionally created video games involve artistic storytelling and design as well as sophisticated programming. Major production houses such as Electronic Arts use applications not easily available to the casual home enthusiast. However, you can use the editors and game engines available for games such as EverQuest and Unreal Tournament to create custom levels and characters to extend the games.

If you want to try your hand at creating your own video games right now at home, multimedia applications such as Unity and Unreal Engine provide the tools you need to explore game design and

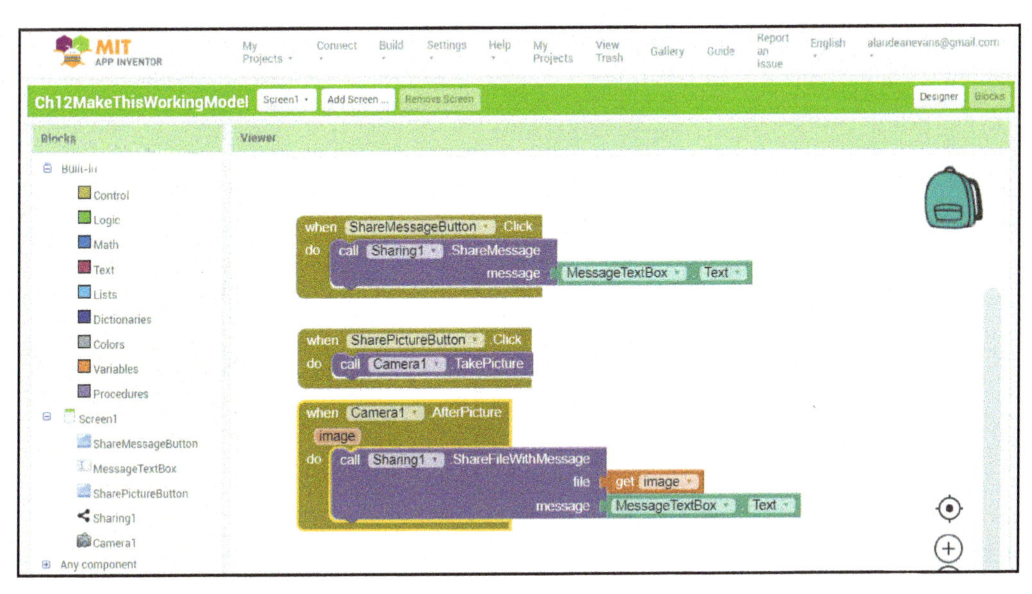

Figure 4.20 MIT's App Inventor is an open source web application that enables beginners to create apps. *(Copyright MIT, used with permission.)*

creation. The program GameMaker (*yoyogames.com*) is a free product that lets you build a game without any programming; you drag and drop key elements of the new game creation into place. Alice (*alice.org*) is another free environment to check out; use it to create 3D animations, virtual reality, and simple games easily.

Bits&Bytes Mirror, Mirror . . .

Sometimes when you download software, especially open source software, you must choose a download location. Each location in the list is a mirror site. A *mirror site* is a copy of a website or set of files hosted at a remote location. Software developers sometimes use geographically distributed mirror sites so that users can choose the mirror site closest to their location to expedite the download time. Mirror sites are also used as backups, so if a server goes down for some reason, users can access the software from a different site, ensuring that the software is always accessible. Developers also use mirror sites to help offset issues associated with a sudden influx of traffic that would otherwise overload a single server.

Educational and Reference Software

Objective 4.12 *Categorize educational and reference software and explain their features.*

What fun educational and reference software should I check out? If you want to learn more about a subject, you can turn to the web for many instructional videos and documents, but sometimes, it's best to use software for more complete or detailed instructions. Educational and reference software is available to help you master, study, design, create, or plan. As shown in Table 4.5, there are software products that teach new skills such as typing, languages, cooking, and learning how to play an instrument.

Table 4.5 Educational and Reference Software: A Sample of What's Available

Test Preparation	Simulation	Instructional	Health and Fitness

Designed to improve your performance on standardized tests	Allows you to experience a real situation through a virtual environment	Designed to teach you almost anything from playing a musical instrument to learning a language or cooking	Helps monitor physical activity, diet, and general health status
Home Design/Improvement	**Course Management**	**Brain Training**	**Genealogy**

Provides 2D or 3D templates and images so you can visualize indoor and outdoor remodeling projects and landscaping ideas better	Web-based software system that creates a virtual learning experience, including course materials, tests, and discussion boards	Features games and activities to exercise your brain to improve your memory, processing speed, attention, and multitasking capabilities	Helps chart the relationships between family members through multiple generations

(Aduchinootonosama/123RF; Greg Pease/The Image Bank/Getty Images; IQoncept/Shutterstock; ALMAGAMI/Shutterstock; Zack C/Shutterstock; Scott Maxwell LuMaxArt/Shutterstock; Vlad Kochelaevskiy/Shutterstock; Alfazet Chronicles/Shutterstock)

Students who will be taking standardized tests like the SAT often use test preparation software. In addition, many computer and online brain-training games and programs are designed to improve the health and function of your brain. Lumosity (*lumosity.com*) is one such site that has a specific workout program that you can play on your PC or smartphone. Brain Age (*brainage.com*) has software for the Nintendo and is designed for players of all ages.

What types of programs are available to train people to use software or special machines? Many programs provide tutorials for popular computer applications (you may even use one in your course provided with MyLab IT). These programs use illustrated systematic instructions, known as simulation programs, to guide users through unfamiliar skills in an environment that acts like the actual software, without the software actually being installed.

Simulation programs allow you to experience or control the software as if it were an actual event. Such simulation programs include commercial and military flight training, surgical instrument training, and machine operation training. One benefit of simulated training programs is that they safely allow you to experience potentially dangerous situations such as flying a helicopter during high winds. Consequently, users of these training programs are more likely to take risks and learn from their mistakes—something they could not afford to do in real life. Simulated training programs also help prevent costly errors. If something goes awry, the only cost of the error is restarting the simulation program.

Do I need special software to take courses online? Although some courses are run from an individually developed website, many online courses are run using course management software such as Blackboard, Moodle, and Canvas. In addition to traditional classroom tools such as calendars and grade books, these programs provide special areas for students and instructors to exchange ideas and information through chat rooms, discussion forums, and e-mail. Collaboration tools such as whiteboards and desktop sharing facilitate virtual office hour sessions. Depending on the content and course materials, you may need a password or special plug-ins to view certain videos or demos.

Before moving on to the Chapter Review:

1. Watch Chapter Overview Video 4.2.
2. Then take the Check Your Understanding quiz.

Check Your Understanding // Review & Practice

For a quick review to see what you've learned so far, answer the following questions.

multiple choice

1. Which type of productivity software would you most likely use to do calculations and what-if analyses?
   a. Database
   b. Tax preparation
   c. Spreadsheet
   d. Word processing
2. The software that allows you to edit digital images, videos, and digital audio is called
   a. multimedia software.
   b. encryption algorithms.
   c. distributed software.
   d. productivity software.
3. Which type of software is used to hold learning materials for your classes?
   a. Accounting software
   b. Desktop publishing software
   c. Web authoring software
   d. Course management software
4. Which type of software would be best to create individual audio tracks with virtual instruments?
   a. App creation software
   b. Audio-editing software
   c. Digital audio workstation software
   d. Simulation software
5. Which of the following is considered a benefit of using simulation programs?
   a. They allow users to experience potentially dangerous situations without risk.
   b. They help prevent costly errors.
   c. They allow users to experience or control the software as if it were an actual event.
   d. All of the above.

chew on this

(3Dalia/Shutterstock)

Cloud computing is becoming more popular, and many users are working from the cloud and not even realizing it. Open Google Docs and Office Online and compare these applications with an installed counterpart (e.g., Excel Online and Google Sheets versus Excel). What similarities and differences do you find between the online applications and the installed version? Envision a time when all software is web-based and describe how being totally on the cloud might be an advantage. What disadvantages might a cloud-based environment present?

MyLab IT Go to **MyLab IT** to take an autograded version of the *Check Your Understanding* review and to find all media resources for the chapter.

For the IT Simulation for this chapter, see MyLab IT.

4 Chapter Review

Summary

Part 1
Accessing, Using, and Managing Software

Learning Outcome 4.1 You will be able to explain the ways to access and use software and describe how to best manage your software.

Software Basics

Objective 4.1 *Compare application software and system software.*

- The term *software* refers to a set of instructions that tells the computer what to do.
- *Application software* is the software you use to do everyday tasks at home, school, and work. Microsoft Word and the Edge browser are examples of application software.
- *System software* is the software that helps run the computer and coordinates instructions between application software and the computer's hardware devices. System software includes the operating system and utility programs.

Objective 4.2 *Explain the differences between commercial software and open source software and describe models for software distribution.*

- There are two basic types of software that you can install on your computer:
  - *Proprietary (or commercial) software* is software you buy.
  - *Open source software* is program code that is free and publicly available with few licensing and copyright restrictions. The code can be copied, distributed, or changed without the stringent copyright protections of software products you purchase
- Software is delivered in three main ways:
  - *Local installation* means that software is installed on your computing device. These programs generally do not require an Internet connection to function.
  - *SaaS* (Software as a Service) is a model for delivery of software whereby the vendor hosts the software online and you access and use the software over the Internet without having to install it on your computer's hard drive.
  - *Subscription* is a model whereby the user pays a fee to use the software. The software is downloaded and installed locally but is routinely updated by connection to the manufacturer's server.

Managing Your Software

Objective 4.3 *Explain the different options for purchasing software.*

- Although most software today is downloaded from the web, it can also be purchased at retail stores or directly from software developers.
- *Freeware* is copyrighted software that you can use for free.
- *Beta software* comprises apps that are still under development but are released to the public to gather feedback on design features and errors.

Objective 4.4 *Describe how to install and uninstall software.*

- Before installing software on your computer, ensure that your system setup meets the *system requirements*. System requirements specify the minimum recommended standards for the operating system, processor, primary memory (RAM), and hard drive capacity.
- When installing software, you're often given the choice between a full (typical) or custom installation. Before installing any software, it's important to back up your system as well as create a *restore point*.
- When uninstalling software, it's best to use the uninstall feature that comes with the operating system. To uninstall an app in Windows 10, on the Start menu, right-click the app and select Uninstall.

Objective 4.5 *Explain the considerations around the decision to upgrade your software.*

- Software that is sold by subscription or that is cloud-based will receive periodic upgrades and updates automatically. Software that is acquired by a one-time purchase and installed locally will not be automatically upgraded. Therefore, the user must decide whether the changes are worth buying a new version of the software that contains all the latest features.

Objective 4.6 *Explain how software licenses function.*

- A *software license*, also known as an *End User License Agreement (EULA)*, is an agreement between you, the user, and the software company that owns the software. Software licenses permit installation on a specific number of devices.
- When you purchase software, you're actually purchasing the license to use it and therefore must abide by the terms of the licensing agreement you accept when installing the program.
- *Copyleft* refers to the ability to redistribute modified open source software under the same terms as the original software.

Part 2
Application Software

Learning Outcome 4.2 Describe the different types of application software used for productivity and multimedia.

Productivity and Business Software

Objective 4.7 *Categorize the types of application software used to enhance productivity and describe their uses and features.*

- *Productivity software* programs include the following:
  - Word processing: to create and edit written documents
  - Spreadsheet: to do calculations and numerical and what-if analyses easily
  - Presentation: to create slide presentations
  - Database: to store and organize data
  - Note-taking: to take notes and easily organize and search them
  - Personal information manager (PIM): to keep you organized by putting a calendar, address book, notepad, and to-do lists within your computer
  - Project and collaboration software: to help groups work collaboratively on a project.
- *Financial planning software* helps you manage your daily finances. Examples include Quicken and Mint.
- *Tax preparation software*, such as Intuit TurboTax and H&R Block Tax Software, help you prepare your state and federal taxes on your own instead of hiring a professional.

Objective 4.8 *Summarize the types of software that small and large businesses use.*

- *Accounting software* helps small business owners manage their finances more efficiently by providing tools for tracking accounts receivable and accounts payable.
- *Desktop publishing (DTP) software* is used to create newsletters, catalogs, annual reports, or other large, complicated publications.
- *Web authoring software* allows even novices to design interesting and interactive web pages without knowing any HTML code.
- There are also specialized programs for project management; customer relationship management (CRM); enterprise resource planning (ERP); e-commerce, marketing and sales; finance, point of sale, security, networking, and data management; and human resources, to name just a few.

Multimedia and Educational Software

Objective 4.9 *Describe the uses and features of digital multimedia software.*

- *Image-editing desktop software* includes tools for basic modifications to digital photos, such as removing red eye; modifying contrast, sharpness, and color casts; or removing scratches or rips from scanned images of old photos.
- *Digital video-editing software* is used to apply special effects, change the sequence of scenes, or combine separate video clips into one movie.
- *Drawing (or illustration) software* facilitates the creation and editing of 2D, line-based drawings. It is used for the creation of technical diagrams and original non-photographic drawings.

Objective 4.10 *Describe the uses and features of digital audio software.*

- *Digital audio workstation software (DAWs)* enables you to create individual tracks to build songs or soundtracks with virtual instruments, voice recorders, synthesizers, and special audio effects that will become uncompressed MIDI files.
- *Audio-editing software* includes tools that make editing audio files as easy as editing text files.

Objective 4.11 *Describe the features of app creation software.*

- *App creation software* provides professionals and novices alike with the ability to create their own apps.
- Although used widely in game development, app creation software can be used to create many other types of apps that have business applications.

Objective 4.12 *Categorize educational and reference software and explain their features.*

- *Test preparation software* is designed to improve your performance on standardized tests.
- *Simulation software* allows you to experience or control the software as if it were an actual event. Often this allows you to experience dangerous situations safely.
- *Course management software* creates a virtual learning experience for students and houses course materials, tests, and discussion boards.

Be sure to check out **MyLab IT** for additional materials to help you review and learn. And don't forget to watch the Chapter Overview Videos.

Key Terms

accounting software **144**
American Standard Code for Information Interchange **129**
app creation software **149**
application software **124**
audio-editing software **148**
base-10 number system (decimal notation) **128**
beta version **126**
binary number system (base-2 number system) **128**
computer-aided design (CAD) **145**
copyleft **130**
course management software **151**
custom installation **127**
database software **139**
desktop publishing (DTP) software **144**
digital audio workstation software (DAWs) **148**
digital video-editing software **147**
drawing (or illustration) software **148**
End User License Agreement (EULA) **129**
financial planning software **143**
freeware **126**
full installation **127**
hexadecimal notation **129**
image-editing desktop software **146**
locally installed software **124**
multimedia software **146**
number system **128**
open source software **124**
personal information manager (PIM) software **141**
presentation software **138**
productivity software **135**
program **124**
proprietary (commercial) software **124**
recovery drive **128**
restore point **127**
simulation programs **151**
software **124**
Software as a Service (SaaS) **124**
software license **129**
software piracy **131**
spreadsheet software **137**
subscription software **125**
system requirements **127**
system software **124**
tax preparation software **144**
Unicode **129**
vertical market software **145**
web authoring software **144**
word processing software **136**

Chapter Quiz // Assessment

For a quick review to see what you've learned, answer the following questions. Submit the quiz as requested by your instructor. If you are using **MyLab IT**, the quiz is also available there.

multiple choice

1. Which would NOT be considered application software?
   a. Microsoft Windows 10
   b. Microsoft PowerPoint
   c. Chrome browser
   d. Adobe Illustrator

2. LibreOffice is an example of which type of software?
   a. System software
   b. Open source software
   c. Beta software
   d. Software as a Service (SaaS)

3. Beta software is made available because
   a. new programmers have been hired.
   b. wireless networking must be added.
   c. a company is making an early release to test bugs.
   d. the company needs more profits.
4. Which type of software controls many back-office operations and processing functions such as billing, production, and inventory management?
   a. Course management software
   b. Enterprise resource planning software
   c. Customer relationship management software
   d. Data management software
5. The minimum recommended standards for the operating system, processor, primary memory (RAM), and storage capacity required for certain software applications are called
   a. redistributing standards.
   b. system requirements.
   c. hardware.
   d. minimum standards.
6. Ginny downloads and modifies an open source software program and then uploads the program with a different name. Which type of software license is she most likely to be working under?
   a. Copyleft
   b. GNU General Public License
   c. Open Source License
   d. No license is necessary
7. Thunkable is an example of which type of software?
   a. Test preparation software
   b. App creation software
   c. Productivity software
   d. Course management software
8. Which of the following is an example of financial management software?
   a. Lumosity
   b. AudioAcrobat
   c. Photos
   d. Mint
9. Image-editing software allows you to
   a. compute complex formulas.
   b. remove red eye.
   c. create drafting blueprints.
   d. conduct product simulations.
10. Which kind of software might be used to train airline pilots?
   a. Project management software
   b. Test preparation software
   c. Course management software
   d. Simulation software

true/false

_____ 1. Google Docs is an example of Software as a Service.
_____ 2. App creation software includes tools like Adobe Photoshop.
_____ 3. Simulation software creates a virtual learning experience for students and houses course materials, tests, and discussion boards.
_____ 4. When you subscribe to software, no license is necessary.
_____ 5. The ability to share computer screens is a feature in project and collaboration software.

What do you think now?

After reviewing your technology use diary, determine whether you are spending too much time with your devices. Describe the signs of technology addiction and a plan to help reduce tech dependency.

Team Time

Software for Startups

problem

You and your friends have decided to start Recycle Technology, a not-for-profit organization that will recycle and donate used computer equipment. In the first planning session, your group recognizes the need for certain software to help you with various parts of the business, such as tracking inventory, designing notices, mapping addresses for pickup and delivery, and soliciting residents by phone or e-mail about recycling events, to name a few.

task

Split your class into as many groups of four or five as possible. Make some groups responsible for locating free or web-based software solutions and other groups responsible for finding proprietary solutions. Another group could be responsible for finding mobile app solutions. The groups will present and compare results with each other at the end of the project.

process

1. Identify a team leader who will coordinate the project and record and present results.
2. Each team is to identify the various kinds of software that Recycle Technology needs. Consider software that will be needed for all the various tasks required to run the organization, such as communication, marketing, tracking, inventory management, and finance.
3. Create a detailed and organized list of required software applications. Depending on your team, you will specify either proprietary software or open source software.

conclusion

Most organizations require a variety of software applications to accomplish different tasks. Compare your results with those of other team members. Were there applications that you didn't think about, but that other members did? How expensive is it to ensure that even the smallest company has all the software required to carry out daily activities, or can the needs be met with free, open source products?

Ethics Project

Open Source Software

Ethical conduct is a stream of decisions you make all day long. In this exercise, you'll research and then role-play a complicated ethical situation. The role you play might or might not match your own personal beliefs, but your research and use of logic will enable you to represent the view assigned. An arbitrator will watch and comment on both sides of the arguments, and together, the team will agree on an ethical solution.

topic: proprietary software vs. open source software

Proprietary software has set restrictions on use and can be expensive, whereas open source software is freely available to use as is or to change, improve, and redistribute. Open source software has become acceptable as a cost-effective alternative to proprietary software—so much so that it's reported that the increased adoption of open source software has caused a drop in revenue to the proprietary software industry. However, determining which software to use involves more than just reviewing the IT budget.

research areas to consider

- Open source software
- Proprietary software
- Copyright licensing
- Open source development

process

1. Divide the class into teams.
2. Research the areas cited above and devise a scenario in which someone is a proponent for open source software but is being rebuffed by someone who feels that "you get what you pay for" and is a big proponent of using proprietary software.
3. Team members should write a summary that provides background information for their character—for example: open source proponent, proprietary developer, or arbitrator—and details their character's behaviors, to set the stage for the role-playing event. Then, team members should create an outline to use during the role-playing event.
4. Team members should present their case to the class or submit a PowerPoint presentation for review by the rest of the class, along with the summary and resolution they developed.

conclusion

As technology becomes ever more prevalent and integrated into our lives, more and more ethical dilemmas will present themselves. Being able to understand and evaluate both sides of the argument, while responding in a personally or socially ethical manner, will be an important skill.

Solve This with Excel

MyLab IT Grader

Analyzing Benchmark Data

You work for a design firm that uses many of the software applications in the Adobe Creative Suite, especially Photoshop, Illustrator, and InDesign. You have been asked to evaluate whether it would be worthwhile to upgrade the software to the latest version. In addition to reviewing any new or revised features, you also want to provide an analysis of any improvements in product efficiency and performance, so you have repeated using the same skills with the current and new data and have recorded the results. Now you just need to analyze it.

You will use the following skills as you complete this activity:

- AutoFill Data
- Insert AVERAGE Function
- Add Borders
- Align and Wrap Text
- Apply Conditional Formatting
- Create Bar Chart

Instructions

1. Open *TIA_Ch04_Start*.xlsx and save as **TIA_Ch04_LastFirst.xlsx**.
2. In **cell F1**, type **Current Average** and, in cell J1, type **New Average**.
3. In **cell F2**, use the **AVERAGE function** to compute the average of **range C2:E2**. In **cell J2**, use the **AVERAGE function** to compute the average of **range G2:I2**.
4. Select **cell F2** and then drag the Fill Handle to cell F13. Select **cell J2** and drag the Fill Handle to cell J13.
5. Select **cells C2:J13** and then format the cell contents with **Number format with one decimal**.
6. Select **cell A2** and **Merge and Center** across cells A2:A5. Adjust the orientation of the text to **Rotate Text Up (90 degrees). Middle Align** cell contents and **Bold** text.
7. Use Format Painter to copy these formats to cell A6 and cell A10.
8. Select range **A2:J5** and then apply **Thick Outside Borders**. Repeat with ranges **A6:J9** and **A10:J13**.
9. Select range **F1:F13** and then apply **Thick Outside Borders**. Repeat with range **J1:J13**.
10. Select **J2:J13** and then apply a **Conditional Format** that will format cells that are Less Than those cells in range F2:F13 with Green Fill and Dark Green Text.

    Hint: In the Format cells that are in the LESS THAN box, enter =F2 (*not* F2). The rest of the cells in the range will update automatically.
11. Create a **Clustered Bar Chart**, using ranges B1:B13, F1:F13, and J1:J13.

    Hint: Hold down the Ctrl button as you select ranges F1:F13 and J1:J13.
12. Add the title **Benchmark Comparison: New and Current Versions of CS Software** to the Clustered Bar Chart, add a Horizontal Axis title, and type **Seconds. Position and resize** the chart so it fills range A16:J34.
13. Save the workbook and submit, based on your instructor's directions.

Chapter 5

System Software: The Operating System, Utility Programs, and File Management

 For a chapter overview, watch the Chapter Overview videos.

PART 1

Understanding System Software

Learning Outcome 5.1 **You will be able to explain the types and functions of operating systems and explain the steps in the boot process.**

Operating System Fundamentals 162

Objective 5.1 *Discuss the functions of the operating system.*

Objective 5.2 *Explain the most popular operating systems for personal use.*

Objective 5.3 *Explain the different kinds of operating systems for machines, networks, and business.*

What the Operating System Does 167

Objective 5.4 *Explain how the operating system provides a means for users to interact with the computer.*

Objective 5.5 *Explain how the operating system helps manage hardware such as the processor, memory, storage, and peripheral devices.*

Objective 5.6 *Explain how the operating system interacts with application software.*

Sound Byte: Using Windows Task Manager to Evaluate System Performance

Starting Your Computer 172

Objective 5.7 *Discuss the process the operating system uses to start up the computer and how errors in the boot process are handled.*

Helpdesk: Starting the Computer: The Boot Process

PART 2

Using System Software

Learning Outcome 5.2 **You will be able to describe how to use system software, including the user interface, file management capabilities, and utility programs.**

The Windows Interface 178

Objective 5.8 *Describe the main features of the Windows interface.*

File Management 181

Objective 5.9 *Summarize how the operating system helps keep your computer organized and manages files and folders.*

Helpdesk: Organizing Your Computer: File Management

Utility Programs 186

Objective 5.10 *Outline the tools used to enhance system productivity, back up files, and provide accessibility.*

Sound Byte: Hard Disk Anatomy

All media accompanying this chapter can be found here.

A Notification Alert on **page 177**

(Dizain/shutterstock; Display intermaya/Shutterstock; Revers/Shutterstock; Courtesy of Microsoft Corporation; Stanislav Popov/Shutterstock; Alexey Bogatyrev/123RF)

What do you think?

Think of the **private information** on your smartphone—your texts, contact lists, and e-mails. If someone has your phone and can log on to it, could they read your information? Not if it were encrypted. *Encryption* changes the data stored on your device to a **secret key.** When you log on to an encrypted phone, it automatically unencrypts the data for you to read. Without the encryption key, the data is just garbled nonsense.

If your phone is stolen, even without the means to log on, the new owner could still get information off it. They could plug the SD card into a computer or could use software to read the data stored on the phone memory—unless your operating system encrypted it.

Almost all modern smartphones offer **encryption**. However, tech companies have faced legal action from government agencies when, for example, the FBI wants to break into the phone of a suspect to look for evidence of crime. **Law enforcement** wants tech companies to work on a way to defeat the encryption installed in the device or to add a backdoor that would allow the companies to defeat the encryption any time they are asked to. The debate over how much privacy you should expect is an important one.

?

We need to balance the privacy rights of individuals against the need for security. Which position should tech companies take?

- *Produce products that support full encryption that can't be broken by anyone—even the tech company.*
- *Produce products that support encryption but enable the tech company to defeat it if asked to by authorities.*
- *Produce products that do not support encryption so that law enforcement can access the data.*

See the end of the chapter for a follow-up question.

(Wutzkohphoto/Shutterstock)

Part 1 For an overview of this part of the chapter, watch **Chapter Overview Video 5.1**.

Understanding System Software

Learning Outcome 5.1 You will be able to explain the types and functions of operating systems and explain the steps in the boot process.

Your computer uses two basic types of software: application software and system software. *Application software* is the software you use to do everyday tasks at home and at work. *System software* is the set of programs that helps run the computer and coordinates instructions between application software and the computer's hardware devices. From the moment you turn on your computer to the time you shut it down, you're interacting with system software.

Operating System Fundamentals

Every computer, from smartphones to supercomputers, has an operating system. Even game consoles, cars, and some appliances have operating systems. The role of the operating system is critical; a computer can't operate without it.

Operating System Basics

Objective 5.1 *Discuss the functions of the operating system.*

What does the operating system do? System software consists of two primary types of programs: the *operating system* and *utility programs*. The operating system (OS) is a group of programs that controls how your computer functions. The operating system has three primary functions:

- *Managing hardware,* including the processor, memory, and storage devices as well as peripheral devices such as the printer.
- *Managing software,* which allows application software to work with the central processing unit (CPU).
- *Managing tasks,* such as scheduling and coordinating processes (like reading strokes from the keyboard) and managing network resources.

You interact with your OS through the user interface—the *desktop, icons*, and *menus* that enable you to communicate with your computer.

How many operating systems are there? There are several operating systems in the marketplace. Microsoft Windows and Apple's macOS are the most common on desktops and laptops. In terms of mobile devices, operating systems must be designed to manage power and use the limited screen space efficiently. Apple created iOS for use on its tablets, watchOS for use on its Apple Watch, and tvOS for use on its Apple TV product. Meanwhile, Windows 10 can run on desktops, laptops, and mobile devices. As devices continue to converge in functionality, developers are making mobile and desktop operating systems that have similar functionality (such as macOS and iOS) or single operating systems (such as Windows 10) that can run on a variety of devices. Table 5.1 lists some popular operating systems.

How are operating systems categorized? Early operating systems were designed for one person performing one task at a time. These are named *single-user, single-task operating systems*. Modern operating systems allow a single user to multitask—to perform more than one process at a time. And operating systems such as Windows and macOS provide networking capabilities as well, essentially making them *multiuser, multitasking operating systems*.

Operating systems can also be categorized by the type of device in which they are installed, such as mobile devices, personal computers, robots or mainframes, and network computers.

Operating Systems for Personal Use

Objective 5.2 *Explain the most popular operating systems for personal use.*

What are the most popular operating systems for personal computers? Microsoft Windows, Apple macOS, and Linux (an open source OS) are popular operating systems for personal computers. Google's Chrome OS is used on the Chromebook series of devices. Although each OS has unique features, they share many features as well.

Table 5.1 Popular Operating Systems

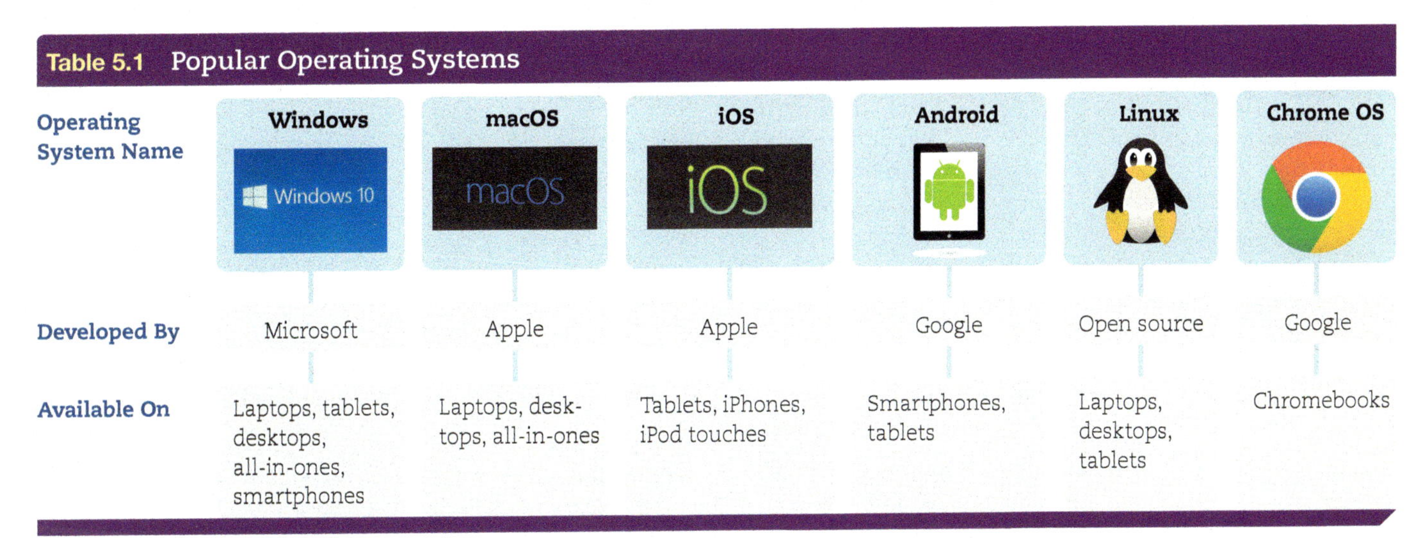

Operating System Name	Windows	macOS	iOS	Android	Linux	Chrome OS
	Windows 10	macOS	iOS			
Developed By	Microsoft	Apple	Apple	Google	Open source	Google
Available On	Laptops, tablets, desktops, all-in-ones, smartphones	Laptops, desktops, all-in-ones	Tablets, iPhones, iPod touches	Smartphones, tablets	Laptops, desktops, tablets	Chromebooks

(Top Photo Corporation/Alamy Stock Photo; David Paul Morris/Bloomberg/Getty Images; David Paul Morris/Bloomberg/Getty Images; Mtkang/YAY Media AS/Alamy Stock Photo; Asnia/Shutterstock; Rose Carson/Shutterstock)

What common features are found in personal computer operating systems? All personal computer operating systems include a window-based interface with icons and other graphics that facilitate point-and-click commands. They also include many utility programs. A utility program is a small program that performs many of the general housekeeping tasks for your computer, to help the computer run more efficiently. Utilities include such things as virus protection, backup and restore software, and system management tools. A set of utility programs is bundled with each OS, but you can also buy stand-alone utility programs that often provide more features.

Other features operating systems incorporate include natural language search capability as well as virtual desktops. They support Bluetooth and Wi-Fi connectivity. Although macOS doesn't have touchscreen capabilities like Windows, iOS offers this capability for iPhones and iPads.

What kind of OS do mobile devices use? Smartphones and tablets use a mobile operating system. Android is the mobile OS for devices designed by Google, Samsung, and other companies. The main OS for Apple mobile devices is iOS. Both iOS and Android support devices like cameras, the sensors built into mobile devices, touch screen displays, and multiple types of connectivity (Wi-Fi, Bluetooth, near-field communication [NFC]). They also support mobile payment systems, on-screen note taking, and voice recognition.

Are any operating systems web-based? Google Chrome OS (see Figure 5.1), is a web-based OS. With the Chrome OS, very few files are installed on your computing device. Rather, the main functionality of the OS is provided through a web browser. Chrome OS is only available on certain devices called *Chromebooks* from Google and Google's manufacturing partners. Chrome OS should not be confused with the Google Chrome browser. The browser is application software that can run on many operating systems.

How do other operating systems use the cloud? Operating systems have features that are tied to cloud computing. Windows 10 uses your Microsoft account to store your settings so you can see your familiar desktop and applications on any device you log on to. You can access and store files online from OneDrive, Windows' cloud-based storage system. Similarly, macOS allows you to sign in to any Apple device with your Apple ID, which provides access to Apple's iCloud system. Both systems store your content online and automatically push it out to all your associated devices. There are third-party products that do the same thing (for example, Dropbox) but a product built into the OS by the manufacturer might be more tightly integrated into the operating system.

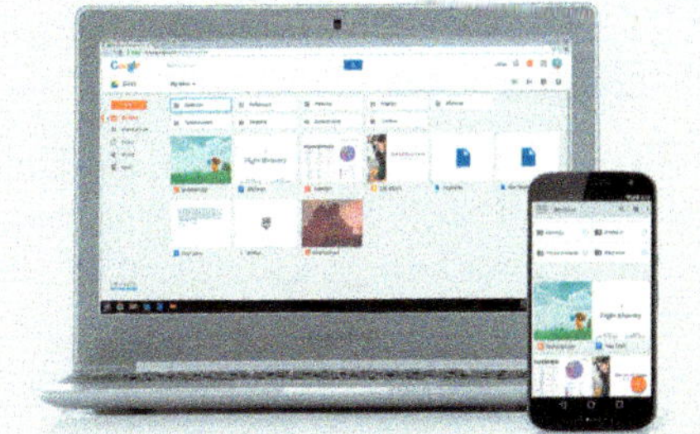

Figure 5.1 The Google Chrome OS is web-based and has a very minimalist look. *(Courtesy of Google, Inc.)*

What is Linux? Linux is a free, open source OS designed for use on personal computers and web servers. Open source software is available for anyone to use or modify. Linux began in 1991 as a project of Finnish university student Linus Torvalds. It has since been tweaked by scores of programmers as part of the Free Software Foundation GNU Project (*gnu.org*).

Linux has a reputation as a stable OS, one that is not subject to crashes. Because the code is available to anyone, Linux can be updated quickly by programmers around the world.

Where can I get Linux? Linux is available for download in various packages known as distributions or distros. Distros include the underlying Linux kernel (the code that provides an operating system's basic functionality) and special modifications to the OS,

Bits&Bytes Quick Assist

Have you ever had a family member or a friend who needed computer help but you didn't have time to drive over and support them? Quick Assist to the rescue! Locate the Windows Search bar next to the Start button, and type Quick Assist to find the tool. Select Assist *Another Person* and send them the code provided. They can now give you remote control of their system. You will see their screen in front of you and take any steps you would have if you were right in the room with them.

and may also include additional open source software (such as LibreOffice). A good place to start researching distros is *distrowatch.com*. This site tracks Linux distros and provides helpful tips for beginners on choosing one.

Does it matter what OS is on my computer? Most application software is OS dependent. You need to make sure you get the correct version of software, matched to your OS. Although you might find the same app in Google Play, the Windows store, or the Apple store, you must download and install the app from the specific store that works with your OS.

Can I have more than one OS on my computer? Yes, you can. You might want to test a program in both operating systems or you may want to use one specific application that only exists for a specific OS. Windows and Linux can run on most of the hardware being sold today. And although macOS and iOS run only on Apple equipment, Apple devices can support Windows and Linux as well. Your choice of an OS is mostly a matter of price and personal preference.

Note that Mac users might install Windows on their computers so they can install and run a program they like that is only available on Windows. A Mac computer can boot up into either Windows or macOS, using the utility included in macOS named Boot Camp. You can also run both macOS and Windows at the same time, using software such as Parallels or VMware Fusion.

In Windows, you can create a separate section of your hard drive (called a *partition*) and install Linux on it while leaving your original Windows installation untouched. After installing Linux, when your computer starts, you're offered a choice of which OS to use.

How do I update my OS? Microsoft Windows delivers automatic OS updates, including new features, apps, and patches, as necessary. You can manage your update settings through the Update & Security section of Settings. Advanced options enable you to choose to download updates automatically or defer a feature update for a certain amount of time.

Apple releases periodic new versions of its desktop and mobile operating systems. The newest versions as of this writing are iOS 14 and macOS Catalina. These versions for Apple mobile and desktop OSs are pushed to the user, and the user can decide whether to install the updates.

Updates to the mobile Android OS are pushed to users but at the discretion of the device manufacturer. For example, when a new version of Android is released, Samsung may release it immediately for the Galaxy smartphone but may not release it for another Samsung device, whereas LG may choose to delay the release for all its smartphones. If having the latest update on your Android device is important to you, consider purchasing Google hardware, because the updates come to Google devices first.

Bits&Bytes Operating Systems for the Smart Home

Have you thought about having an operating system for your home? Smart home devices are becoming more inexpensive and powerful. These devices include light switches, alarm systems, doorbell cameras, door locks, garage door openers, and smoke detectors—each connected to the Internet. Having a central controller for your home to manage them all and to provide a consistent interface is important.

Amazon's Echo and Google's Home are two candidates for organizing and controlling your smart home. These devices have operating systems that respond to voice commands and can run your smart devices as well as send text messages and place or receive phone calls. There are also open source solutions like openHAB (*openhab.org*) that can run on inexpensive hardware like the Raspberry Pi. Explore making your home smarter!

Operating Systems for Machinery, Networks, and Business

Objective 5.3 *Explain the different kinds of operating systems for machines, networks, and business.*

Why do machines with built-in computers need an OS? Machinery that performs a repetitive series of specific tasks in an exact amount of time requires a real-time operating system (RTOS). Also referred to as *embedded systems*, RTOSs require minimal user interaction. This type of OS is a program with a specific purpose, and it must guarantee certain response times for certain computing tasks; otherwise, the machine is useless. The programs are written specifically for the needs of the devices and their functions. Devices that must perform regimented tasks or record precise results require RTOSs. Examples include a pacemaker or the anti-lock braking system in your car.

You also encounter RTOSs every day in devices such as point-of-sale systems, fuel-injection systems in car engines, and some common appliances. RTOSs are also found in categories of devices like medical equipment or surveillance devices (see Figure 5.2).

What kind of operating systems do networks use? A multiuser operating system (or network operating system) allows more than one user access to the computer system at a time by handling and prioritizing requests from multiple users. Networks (groups of computers connected to each other so that they can communicate and share resources) need a multiuser OS because many users simultaneously access the server (the computer that manages network resources such as printing and communications). The latest versions of Windows and macOS can be considered network operating systems because they enable multiple computers in a home or small business to connect to one another and share resources.

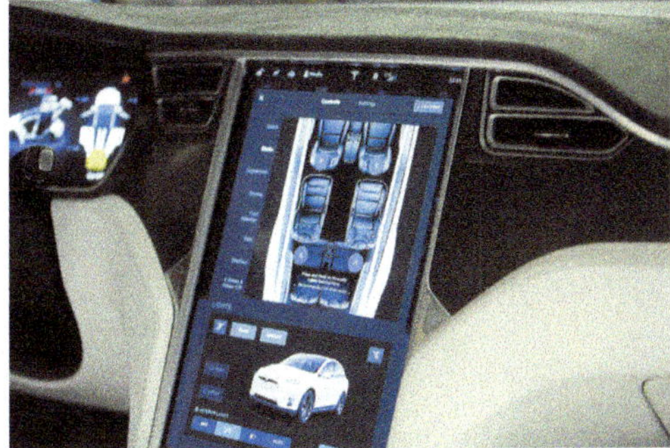

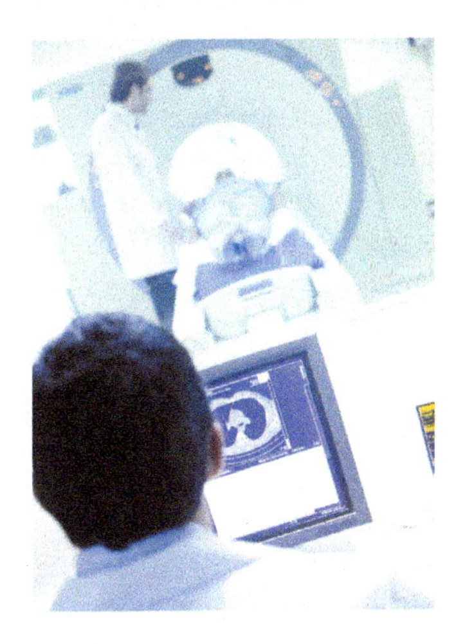

Figure 5.2 Real-time operating systems can be found in devices such as medical equipment, cars, surveillance systems, and point-of-sale machines. *(ESB Professional/Shutterstock; Aleksandra Suzi/Shutterstock; Henk Vrieselaar/Shutterstock; Vereshchagin Dmitry/Shutterstock)*

Ethics in IT

The Great Debate: Is macOS Safer Than Windows?

Many Mac users feel they're impervious to viruses and malware (software that can disable or interfere with the use of a device) because those are "just Windows problems." This means that Mac users often run their computers with only the basic protection provided by Apple in its operating system software. Are Mac users wild risk takers, or are they safe from hackers? As with many issues, it's a little bit of both.

Threats Are Out There

Windows users have been bombarded by malware and virus attacks for decades. When you bought your last Windows device, it came with a Windows Defender security solution built into Windows. Manufacturers also often include trial versions of third-party antivirus/anti-malware software. Running a Windows computer without antivirus/anti-malware software is just asking for trouble.

However, over the past several years, the attacks against Macs have increased. Why weren't Macs attacked frequently in the past? Most malware is designed to steal sensitive information such as credit card numbers. When thieves expend the time, money, and effort to develop malware, they want to ensure that it targets the largest population of potential victims. As macOS gains market share, Mac users are becoming a larger group of potential targets. In fact, in many affluent nations, Mac ownership has reached 20% of the market, and because Macs tend to cost more than Windows machines, it can be argued that Mac users may have more disposable income than other computer buyers. Wealthy people are always attractive targets for thieves.

But Isn't macOS Safer Than Windows by Design?

To a certain extent, this is true; macOS does have certain design features that tend to prevent the installation and spread of malware. Apple has also designed current versions of iOS and macOS to prevent the installation of unapproved software (i.e., software not available on Apple's approved online outlets like the App Store). In addition, apps sold on the App Store are required to be designed using the access control technology known as *App Sandbox*.

When most software programs or apps are running, they have broad latitude to interact with the OS. Usually, they have all the rights that the user has over the OS, so if hackers can design an exploit that takes advantage of a security flaw in an app, they can potentially gain extensive control over the computer, using the access that the user has to the OS. As noted, Apple requires all approved apps to be sandboxed. When an app is sandboxed, the developer defines what the app needs to do to interact with the OS. The OS then grants only those specific rights and privileges to the app and nothing else. By doing this, it severely limits what hackers can do in an OS if they breach the security of an app. It's like being in a high-walled sandbox as a child. You can play within the confines of your space, but you can't make mischief outside certain limits.

So I'll Buy a Mac and Be Safe Forever, Right?

Although it's more difficult to design exploits for macOS and iOS, it's not impossible, and a great deal of cybercrime relies on social engineering techniques like those used in scareware scams. *Scareware* is software designed to make it seem as if there is something wrong with your computer. The author of the scareware program then persuades you to buy a solution to the problem, acquiring your credit card number in the process. Scareware victims can be both Mac and PC users, so even if you own a Mac, you need to be aware of such scams and avoid falling prey to them (see Chapter 9).

The Solution: Extra Security Precautions

The current versions of macOS, iOS, and Windows all include security tools. Here are a few additional things you should do to protect yourself:

1. **Make sure your software is set to download and install updates automatically.** As OS developers discover holes in their software's security, they provide updates to repair these problems.
2. **Consider using third-party antivirus/anti-malware software (even on a Mac).** Although no product will detect 100% of malware, detecting some is better than detecting none.
3. **Be aware of social engineering techniques.** Use vigilance when surfing the Internet so you don't fall prey to scams.

So, no OS is 100% safe. But if you're informed and proceed with caution, you can avoid a lot of schemes perpetrated by hackers and thieves.

In larger networks, a more robust network OS is required to support workstations and the sharing of databases, applications, files, and printers among multiple computers in the network. This network OS is installed on servers and manages all user requests. For example, on a network users share for a printer, the network OS ensures that the printer prints only one document at a time in the order the requests were made. Examples of network operating systems include Windows Server, Linux, and UNIX.

What is UNIX? UNIX is a multiuser, multitasking OS that's used as a network OS, although it can also be found on PCs. Developed in 1969 by Ken Thompson and Dennis Ritchie of AT&T's Bell

Labs, the UNIX code was initially not proprietary—no one company owned it. Rather, any programmer could use the code and modify it to meet their needs. UNIX is now a brand that belongs to The Open Group, but any vendor that meets the testing requirements and pays a fee can use the UNIX name. Individual vendors then modify the UNIX code to run specifically on their hardware.

What other kinds of computers require a multiuser OS? Mainframes and supercomputers also require multiuser operating systems. Mainframes routinely support hundreds or thousands of users at a time, and supercomputers are often accessed by multiple people working on complex calculations. Examples of mainframe operating systems include UNIX, Linux on System z, and IBM's z/OS, whereas the clear majority of supercomputers use Linux.

What the Operating System Does

As shown in Figure 5.3, the OS is like an orchestra's conductor. It coordinates and directs the flow of data and information through the computer system. In this section, we explore the operations of the OS in detail.

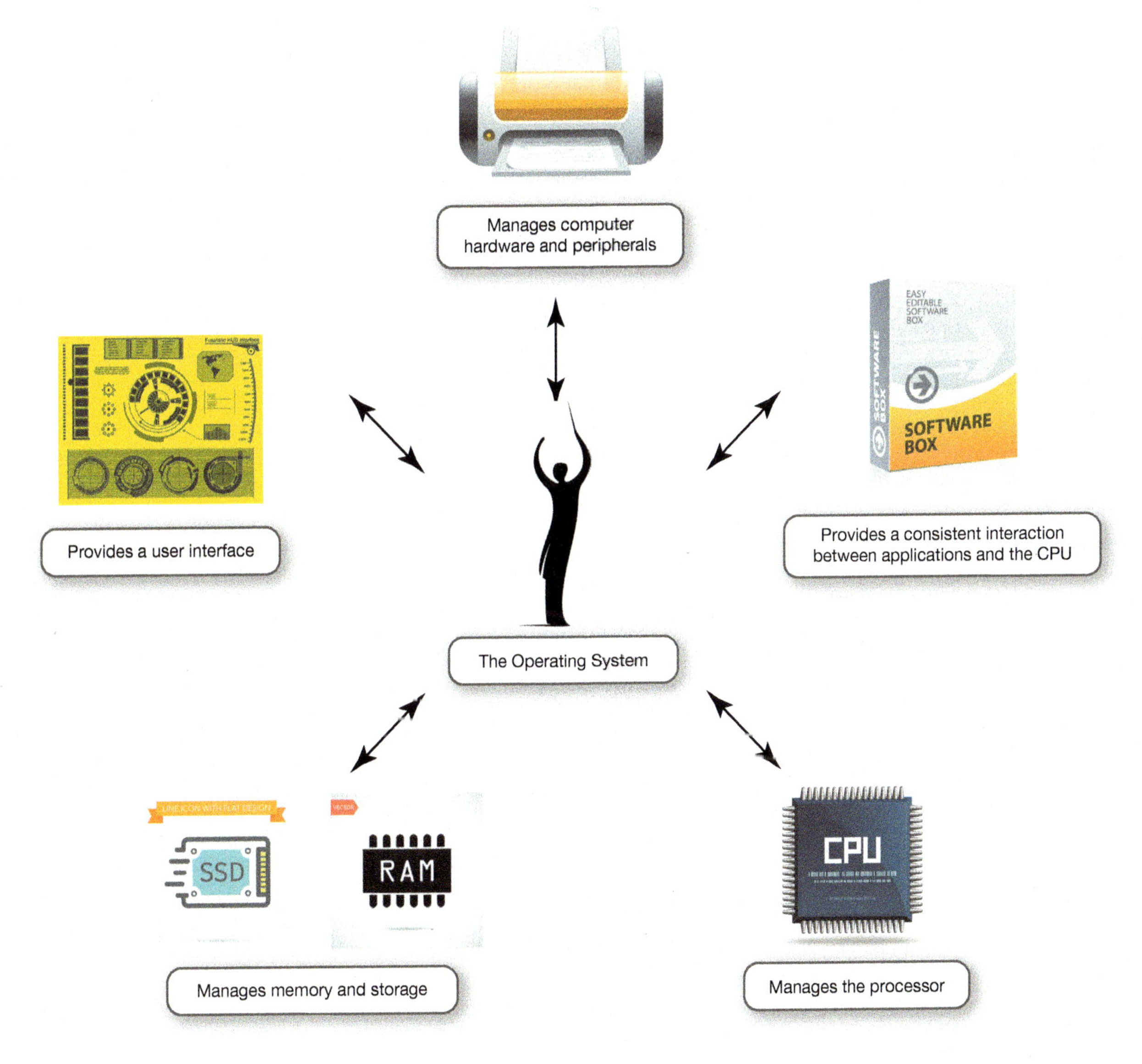

Figure 5.3 The OS is the orchestra conductor of your computer, coordinating its many activities and devices. *(White snow/Shutterstock; Andrew metto/Shutterstock; Dmitri Mikitenko/Shutterstock; Graphego/Shutterstock; Icon Craft Studio/Shutterstock; Jiunn/Shutterstock; MIKHAIL GRACHIKOV/Shutterstock)*

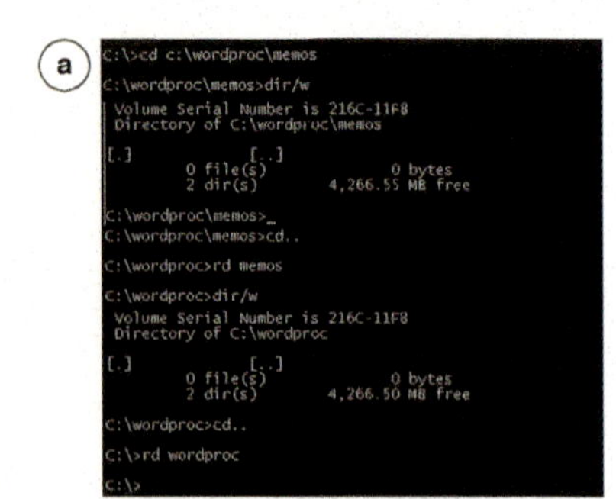

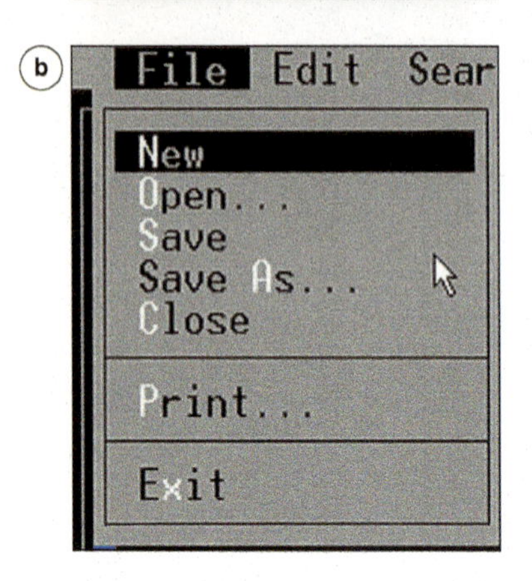

Figure 5.4 (a) A command-driven interface. (b) A menu-driven interface. *(Courtesy of Microsoft Corp.)*

Using Windows Task Manager to Evaluate System Performance

In this Sound Byte, you'll learn how to use the utilities provided by Windows to evaluate system performance. You'll also learn about shareware utilities that expand on the capabilities the Task Manager utility provides.

The User Interface

Objective 5.4 *Explain how the operating system provides a means for users to interact with the computer.*

How does the OS control how I interact with my computer? The OS provides a user interface by which you interact with your device. Early personal computers used the *Microsoft Disk Operating System* (*MS-DOS* or just *DOS*), which had a command-driven interface, as shown in Figure 5.4a. A command-driven interface is one in which you enter commands to communicate with the computer system. The DOS commands were not always easy to understand and the interface proved to be complicated.

The command-driven interface was later improved by incorporating a menu-driven interface, as shown in Figure 5.4b. A menu-driven interface is one in which you choose commands from menus displayed on the screen. Menu-driven interfaces eliminated the need for users to know every command because they could select from a menu. However, they still were not intuitive for most people.

What kind of interface do operating systems use today? Current computer and mobile operating systems such as Microsoft Windows and macOS use a graphical user interface, or GUI (pronounced "gooey"). Unlike command- and menu-driven interfaces, GUIs display graphics and use the point-and-click technology of the mouse and cursor (or human finger), making them much more user-friendly.

With Linux-based operating systems, users are free to choose among many commercially available and free interfaces, such as GNOME and KDE, each of which provides a different look and feel.

Hardware Coordination

Objective 5.5 *Explain how the operating system helps manage hardware such as the processor, memory, storage, and peripheral devices.*

Processor Management

Why does the OS have to manage the processor? When you use your computer, you're usually asking the processor, or CPU, to perform several tasks at once. You might be creating a PowerPoint presentation, have a Word document printing out, and be streaming music from Spotify—all at the same time, or at least what appears to be at the same time. Although the CPU is powerful, it needs the OS to arrange the execution of all these activities so it can work on them one at a time but so quickly that you think it is happening simultaneously.

How does the OS coordinate all the activities? The OS assigns a slice of its time to each activity that requires the processor's attention. The OS then switches between different processes billions of times per second to make it appear that all the processes are running at the same time.

Every keystroke and every mouse click creates an action, or event, in the respective device (keyboard, mouse, etc.) to which the OS responds. Sometimes these events occur one after another (such as when you type characters on the keyboard), but other events involve multiple devices working at the same time (such as the printer printing while you type and stream music). When you tell your computer to print your document, the printer generates a unique signal called an interrupt that tells the OS that it's in need of immediate attention. Every device has its own type of interrupt, which is associated with an interrupt handler, a special numerical code that prioritizes the requests. These requests are placed in the interrupt table in the computer's primary memory (RAM). The OS runs the task with a higher priority before running a task assigned a lower priority. This is called preemptive multitasking.

In our example, when the OS receives the interrupt from the printer, it suspends the CPU's typing activity and the next command Spotify needs to run and puts a "memo" in a special location in RAM called a *stack*. The memo is a reminder of what the CPU was doing before it started to work on the printer request. The CPU then retrieves the printer request from the interrupt table and begins to process it. On completion of the printer request, the CPU goes back to the stack, retrieves the memo it placed about the keystroke or Spotify, and returns to that task until it is interrupted again, in a very quick and seamless fashion, as shown in Figure 5.5.

What happens if many processes need the same resources? When the processor receives a request to send information to the printer, it first checks with the OS to ensure that the printer is not already in use. If it is, the OS puts the request in another temporary storage area in RAM, called the *buffer*. The request then waits in the buffer until the spooler, a program that helps coordinate all print jobs currently being sent to the printer, indicates the printer is available. If more than one print job is waiting, a line (or *queue*) is formed so that the printer can process the requests in order.

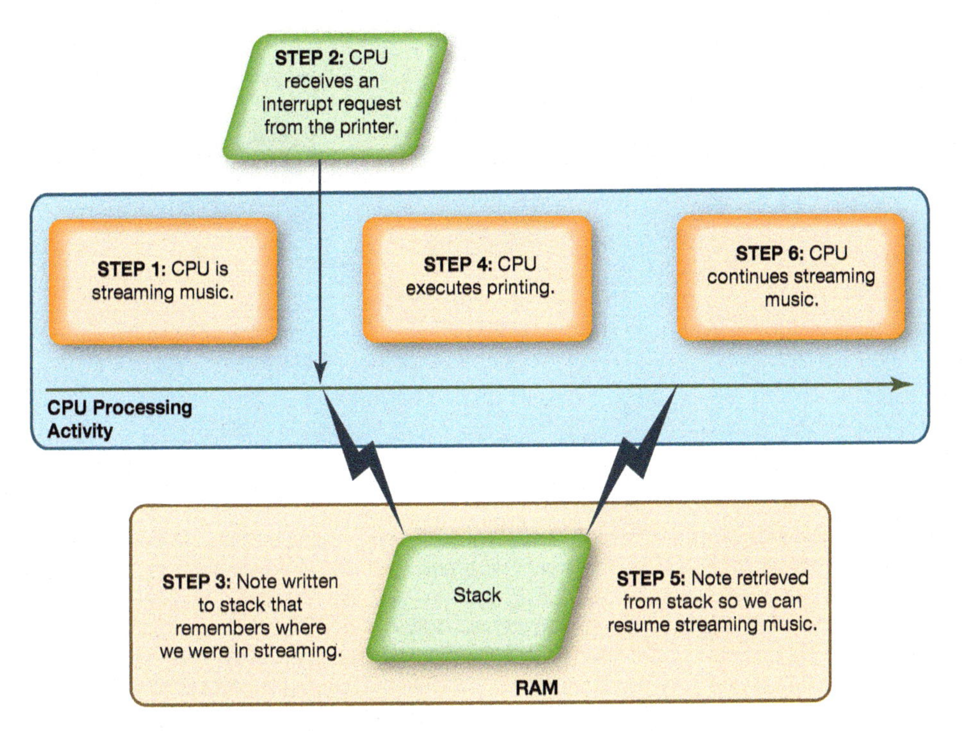

Figure 5.5 How Preemptive Multitasking Works

Memory and Storage Management

How does the OS manage the computer's memory? As the OS coordinates the activities of the processor, it uses RAM as a temporary storage area for instructions and data the processor needs. The processor then accesses these instructions and data from RAM when it's ready to process them. So the OS is responsible for coordinating the space allocations in RAM, making sure there is enough space for all the pending instructions and data. The OS then clears the items from RAM when the processor no longer needs them.

What happens if the applications I'm using require more RAM than what's installed? Any system has a fixed amount of RAM installed. When there isn't enough RAM for the OS to store the required data and instructions, the OS borrows from the more spacious hard drive. This process of optimizing RAM storage by borrowing hard drive space is called virtual memory. As shown in Figure 5.6, when more RAM is needed, the OS swaps out from RAM the data or instructions that haven't recently been used and moves them to a temporary storage area on the hard drive called the swap file (or page file). If the data or instructions in the swap file are needed later, the OS swaps them back into active RAM and replaces them in the hard drive's swap file with less active data or instructions. This process of swapping is known as paging.

Can I ever run out of virtual memory? Only a portion of the hard drive is allocated to virtual memory. You can manually change this setting to increase the amount of allocated hard drive space, but eventually your computer will become sluggish as it is forced to page more often. This condition of excessive paging is called thrashing. The solution to this problem is to increase the amount of RAM in your computer, if excess capacity is available. If you increase the RAM, it won't be necessary to send data and instructions to virtual memory.

How does the OS manage files? All of the details of how a file is stored, retrieved, named, and compressed—all of the things we do with files—are the responsibility of the operating system. Different operating systems allow different naming schemes and have different tools for searching and optimizing file access. One operating system, say Windows, can even support different file systems. Some Windows versions support file allocation table (FAT) file architecture and some New Technology File System (NTFS). Before you format a disk drive, make sure you know which file system you plan to use with it.

Hardware and Peripheral Management

How does the OS manage the hardware and peripheral devices? Each device attached to your computer comes with a special program called a device driver that facilitates communication between the device and the OS. Because the OS must be able to communicate with every

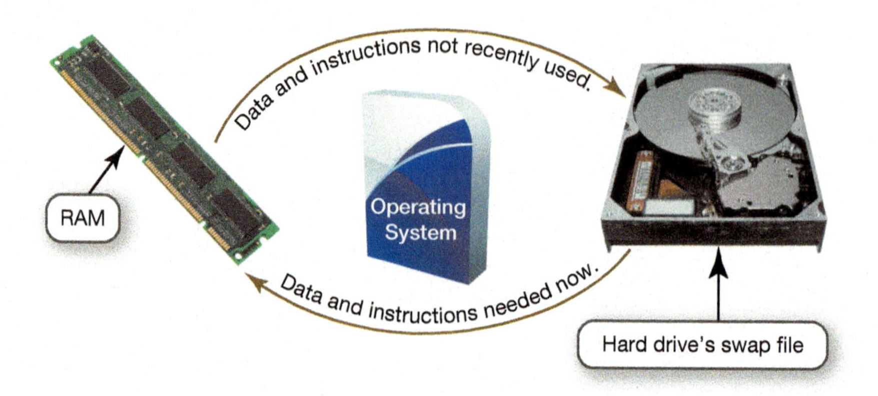

Figure 5.6 Virtual memory borrows excess storage capacity from the hard drive when there isn't enough capacity in RAM. *(Ragnarock/Shutterstock; Tim Dobbs/Shutterstock)*

device in the computer system, the device driver translates the device's specialized commands into commands the OS can understand, and vice versa. Devices wouldn't function without the proper device drivers because the OS wouldn't know how to communicate with them.

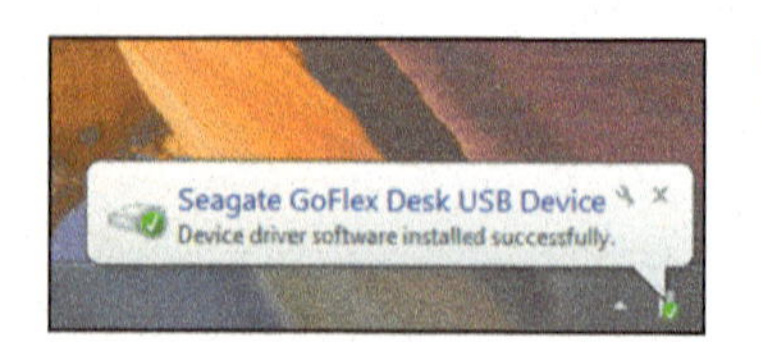

Figure 5.7 A Windows message showing a successful driver installation the first time an external hard drive was connected to the computer. *(Courtesy of Microsoft Corp.)*

Do I always need to install drivers? Today, most devices, such as flash drives, mice, keyboards, and digital cameras, come with the driver already installed in Windows. The devices whose drivers are included in Windows are called Plug and Play devices. Plug and Play (PnP) is a software and hardware standard designed to facilitate the installation of new hardware in PCs by including in the OS the drivers these devices need in order to run. Because the OS includes this software, incorporating a new device into your computer system seems automatic. With PnP, you plug a new device into your computer, turn it on, and immediately use the device (see Figure 5.7).

What happens if the device is not Plug and Play? Sometimes you may have a device, such as a printer, whose driver is not incorporated in Windows. You'll then be prompted to install or download, from the Internet, the driver required for that device. If you obtain a device secondhand without the device driver, or if you're required to update the device driver, you can often download the necessary driver from the manufacturer's website. You can also go to websites such as DriverZone (*driverzone.com*) to locate drivers.

Can I damage my system by installing a device driver? Occasionally, when you install a driver, your system may become unstable. Programs may stop responding, certain actions may cause a crash, or the device or the entire system may stop working. Although this is uncommon, it can happen. Fortunately, to remedy the problem, Windows has a Roll Back Driver feature that removes a newly installed driver and replaces it with the last one that worked (see Figure 5.8).

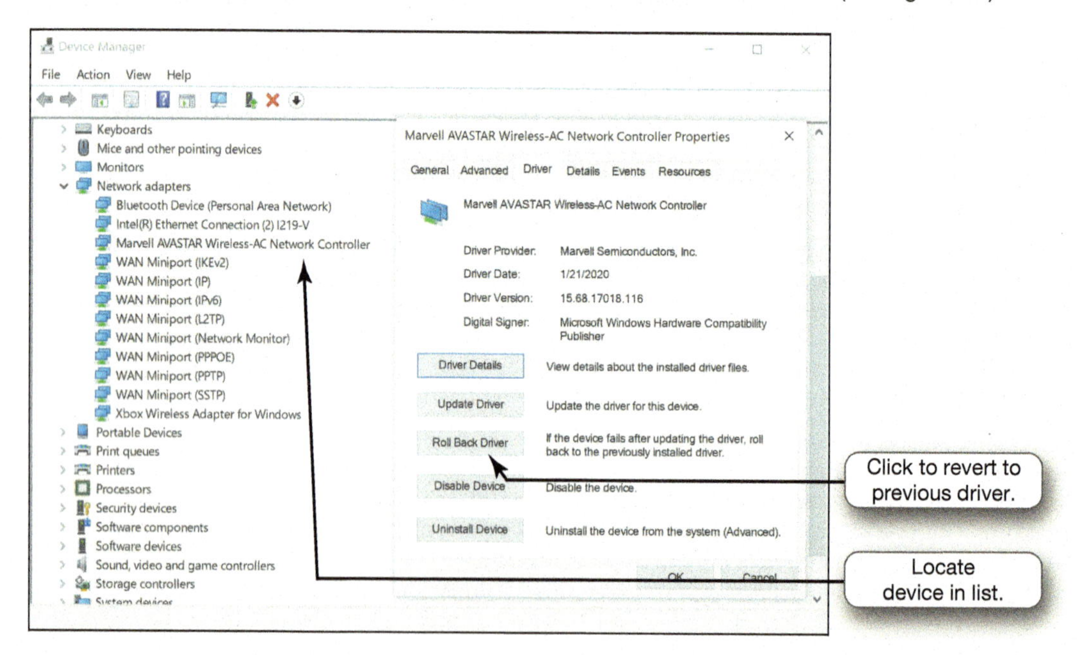

Figure 5.8 The Roll Back Driver feature in Windows removes a newly installed driver and replaces it with the last one that worked.

> *To access Device Manager, right-click the* ***Start*** *button and select* ***Device Manager****. To display the Properties dialog box, double-click a device. Click the* ***Driver*** *tab.* *(Courtesy of Microsoft Corp.)*

Software Application Coordination

Objective 5.6 *Explain how the operating system interacts with application software.*

How does the OS help application software run on the computer? Every computer program, no matter what its type or manufacturer, needs to interact with the CPU by using computer code. For programs to work with the CPU, they must contain code the CPU recognizes. Some operations that happen are common to so many programs that the operating system includes these blocks of code. These are called an application programming interface (API). Microsoft DirectX, for example, is a group of multimedia APIs built into the Windows OS that improves graphics and sounds when you're playing games or watching videos on your PC.

What are the advantages of APIs? To create applications that can communicate with the OS, software programmers need only refer to the API code when they write an application. This standardizes the code and saves developers time. APIs not only prevent redundancies in software code but also make it easier for software developers to respond to changes in the OS. Software companies also take advantage of APIs to ensure that applications in software suites (such as Microsoft Office) have a similar interface and functionality. Because these applications share common APIs, data exchange is facilitated between two programs, such as inserting a chart from Excel into a Word document.

Trends in IT — Are Personal Computers Becoming More Human?

Watch nearly any sci-fi movie or show and you'll see that making computers more human-like has long been a dream. Now, that dream is becoming a reality with advances in the field of artificial intelligence (AI). For example, IBM's AI supercomputer Watson was put to the test in a *Jeopardy* challenge and beat two of the show's greatest human champions. Although we've grown to expect great processing capabilities from computers, the *Jeopardy* challenge was different. Watson, like its human competitors, needed to discern specific pieces of data from complex questions in the *Jeopardy* game. Computers hadn't been capable of processing these types of complex situations previously.

What's behind Watson's abilities? The computer's ability to generate answers from complex questions is a matter of synthesizing large amounts of unstructured data with natural language processing and machine learning. Natural language processing is a field of AI that uses software to analyze, understand, and generate human languages naturally, enabling users to communicate with computers as if they were humans. Machine learning is another form of AI that provides computers with the ability to learn without being explicitly programmed. In other words, machine learning enables computer programs to grow and adapt when exposed to new data (much like human brains). Machine learning uses data to detect patterns and adjust programmed actions accordingly

But can computers cross the barrier of simply doing calculations into the kind of fuzzy decision-making we humans do and call intuition? The game of Go is famous for being complex—there are more possible board positions than the number of atoms in the universe. Strong players need to think creatively and use their intuition (see Figure 5.9). To develop a computer-based solution, the AlphaGO algorithm was shown a few games played by amateurs, and then it played itself in thousands of games. AlphaGO taught itself strategies that enabled it to defeat the world's champion Go player, Ke Jie. AI enthusiasts were stunned; they thought such an accomplishment was still 20 years in the future.

The constant flood of more and more data makes the need to have intelligent machine processing critical to new discoveries. Scientists are coming to rely on machine learning to help them focus on the most interesting subsets of data to examine, but with current algorithms, we do not actually know how the AI systems came to their conclusions. Their decision making is spread across neural networks and it is not possible for programmers to point to the specific logic the system used to arrive at its conclusions. That logic may be based on biases that the machine learned from its training data and which remain hidden in the system.

The future will reward people who have skills that are the most difficult for computer systems to mimic: higher-order critical thinking, creative and innovative thinking, and emotional intelligence—the ability to handle interpersonal relationships with empathy and compassion. Cultivate these along with your growing technical skill set, and you'll be positioned to work with computer systems of the future.

Figure 5.9 The world champion of the difficult game Go was defeated by an AI program. The program learned to improve its game by playing itself many thousands of times. *(Mopic/Alamy Stock Photo)*

Starting Your Computer

Many things happen quickly between the time you turn on your computer and the time it's ready for you to start using it. Data and instructions, including the OS, are stored in RAM while your computer is on. When you turn off your computer, RAM is wiped clean of all its data, including the OS. How does the computer know what to do when you turn it on if there is nothing in RAM? It runs through a *boot process,* a special start-up procedure that loads the OS into RAM.

The Boot Process

Objective 5.7 ***Discuss the process the operating system uses to start up the computer and how errors in the boot process are handled.***

What are the steps involved in the boot process? As illustrated in Figure 5.10, the boot process consists of four basic steps. The term *boot*, from *bootstrap loader* (a small program used to start a larger program), alludes to the straps of leather, called *bootstraps*, that people used to use to help them pull on their boots. This is also the source of the expression "pull yourself up by your bootstraps." Let's look at each step of the boot process in detail.

Step 1: Activating BIOS

What's the first thing that happens after I turn on my computer? In the first step of the boot process, the CPU activates the basic input/output system (BIOS). BIOS is a program that

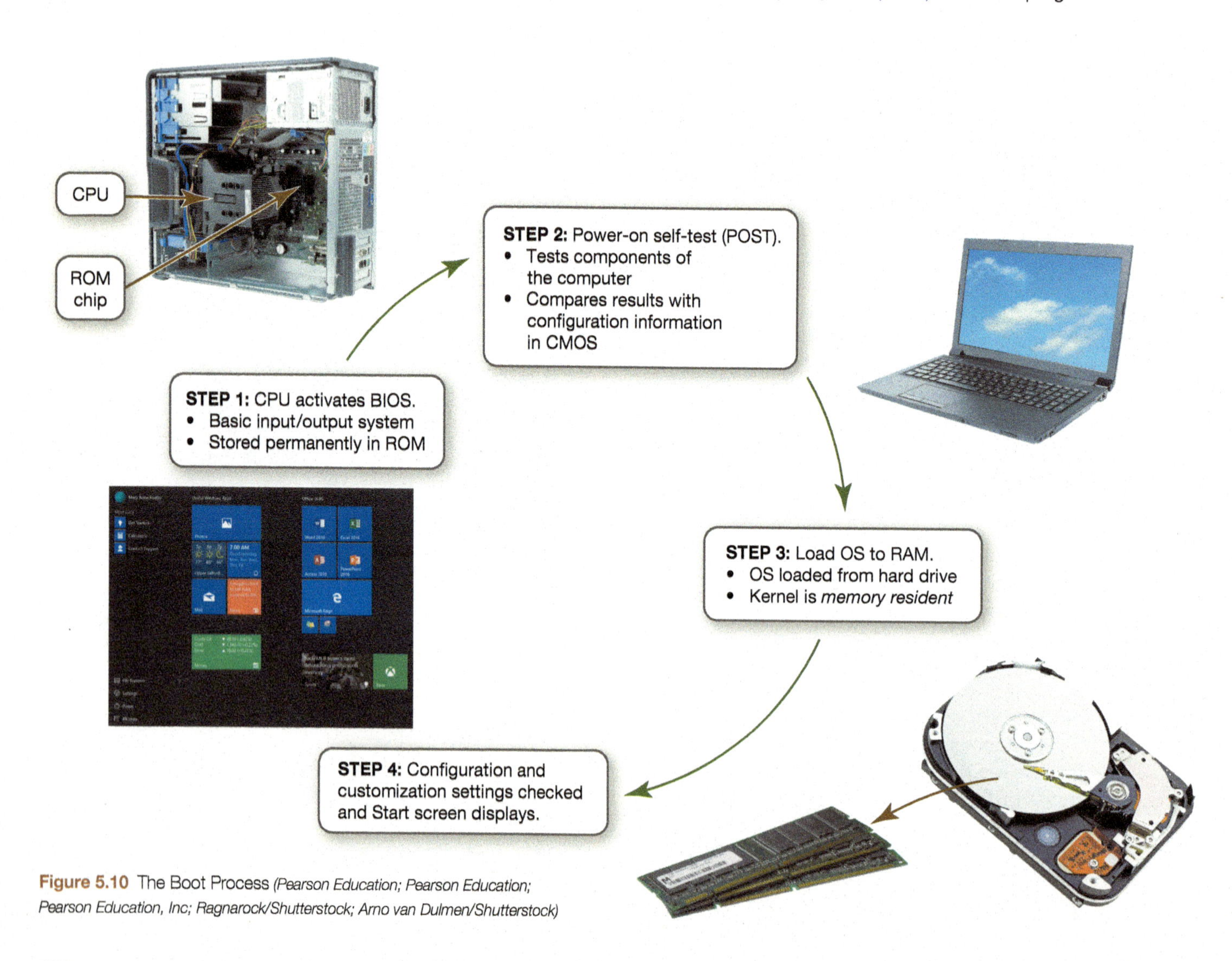

Figure 5.10 The Boot Process *(Pearson Education; Pearson Education; Pearson Education, Inc; Ragnarock/Shutterstock; Arno van Dulmen/Shutterstock)*

manages the exchange of data between the OS and all the input and output devices attached to the system. BIOS is also responsible for loading the OS into RAM from its permanent location on the hard drive. BIOS itself is stored on a read-only memory (ROM) chip on the motherboard. Unlike data stored in RAM, data stored in ROM is permanent and is not erased when the power is turned off.

Step 2: Performing the Power-On Self-Test

How does the computer determine whether the hardware is working properly? The first job BIOS performs is to ensure that essential peripheral devices are attached and operational—a process called the power-on self-test (POST). The BIOS compares the results of the POST with the various hardware configurations permanently stored in CMOS (pronounced "see-moss"). CMOS, which stands for *complementary metal-oxide semiconductor*, is a special kind of memory that uses almost no power. A small battery provides enough power so that the CMOS contents won't be lost after the computer is turned off. CMOS contains information about the system's memory, types of disk drives, and other essential input and output hardware components. If the results of the POST compare favorably with the hardware configurations stored in CMOS, the boot process continues.

Step 3: Loading the OS

How does the OS get loaded into RAM? Next, BIOS goes through a preconfigured list of devices in its search for the drive that contains the system files, the main files of the OS. When the system files are located, they load into RAM.

Once the system files are loaded into RAM, the kernel is loaded. The kernel is the essential component of the OS. It's responsible for managing the processor and all other components of the computer system. Because it stays in RAM the entire time your computer is powered on, the kernel is said to be *memory resident*. Other, less critical, parts of the OS stay on the hard drive and are copied over to RAM on an as-needed basis so that RAM is managed more efficiently. These programs are referred to as *nonresident*. Once the kernel is loaded, the OS takes over control of the computer's functions.

Step 4: Checking Further Configurations and Customizations

Is that it? Finally, the OS checks the registry for the configuration of other system components. The registry contains all the configurations (settings) the OS and other applications use. It contains the customized settings you put into place, such as mouse speed, as well as instructions as to which programs should be loaded first.

How do I know whether the boot process is successful? The entire boot process takes only a few minutes to complete. If the entire system is checked out and loaded properly, the process completes by displaying the lock screen. After logging on, the computer system is now ready to accept your first command.

Why do I sometimes need to enter a login name and password at the end of the boot process? The verification of your login name and password is called authentication. The authentication process blocks unauthorized users from entering the system. You may have your home computer set up for authentication, especially if multiple users are accessing it. All large networked environments, such as your college, require user authentication for access.

Starting the Computer: The Boot Process

In this Helpdesk, you'll play the role of a help desk staffer, fielding questions about how the operating system helps the computer start up.

Handling Errors in the Boot Process

What should I do if my computer doesn't boot properly? Sometimes during the boot process, BIOS skips a device (such as a keyboard) or improperly identifies it. Your only indication that this problem has occurred is that the device won't respond after the system has been booted. Try one of the following suggestions:

1. Power the computer off and on again.
2. If you've recently installed new software or hardware, try uninstalling it. Make sure you uninstall software using Windows *Apps & features*. From the Settings app, select Apps then select the program you wish to uninstall to remove the software.
3. Try accessing the Windows Advanced Options Menu (accessible by pressing the F8 key during the boot process). If Windows detects a problem in the boot process, it will add Last Known Good Configuration to the Windows Advanced Options Menu. Every time your

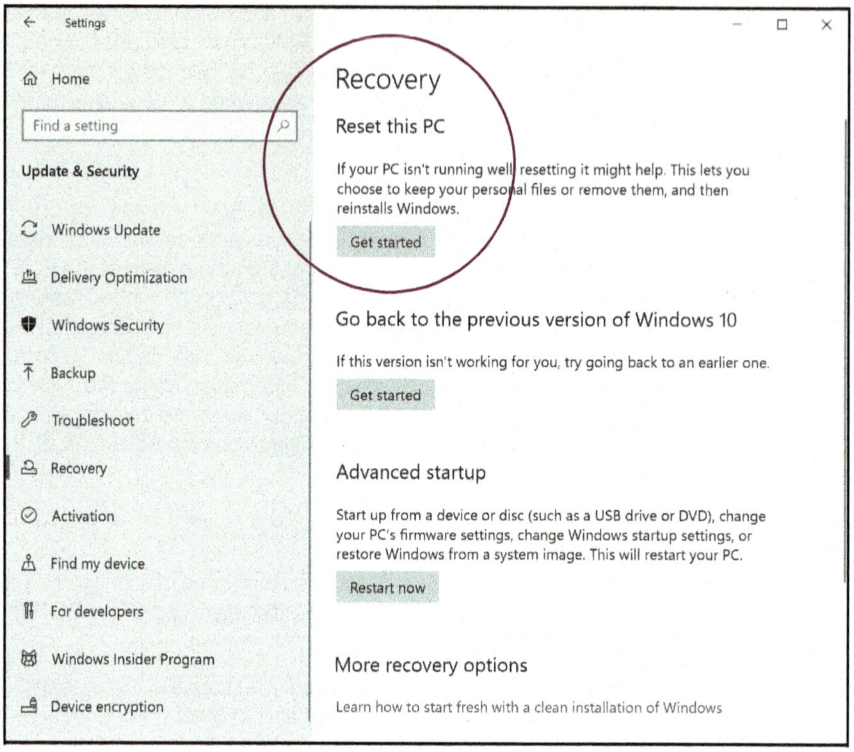

Figure 5.11 The Recovery control in Windows 10 provides access to the Reset this PC option.
> *To access the Recovery control, from the* ***Start*** *menu, click* ***Settings****, and then click* ***Update & Security****. On the Update & Security screen, select* ***Recovery****.* (Courtesy of Microsoft Corp.)

computer boots successfully, a configuration of the boot process is saved. When you choose to boot with the Last Known Good Configuration, the OS starts your computer by using the registry information that was saved during the last successful shut down.

4. Try resetting your computer from the Update & Security Settings screen. It's recommended that you back up your PC prior to resetting it as a precautionary measure. Reset this PC is a utility program in Windows 10 that attempts to diagnose and fix errors in your Windows system files that are causing your computer to behave improperly (see Figure 5.11). When a PC is reset, your data files and personalization settings are not changed. Apps that you have downloaded from the Windows Store are kept intact, but apps that you have downloaded from the Internet will be removed from your PC. Therefore, you'll need to reinstall them after the reset.
5. Finally, try *Go back to an earlier build* to revert to a past configuration. System Recovery is covered in more detail later in this chapter.

Before moving on to Part 2:

1. Watch Chapter Overview Video 5.1.
2. Then take the Check Your Understanding quiz.

Check Your Understanding // Review & Practice

For a quick review to see what you've learned so far, answer the following questions.

multiple choice

1. Which is an example of an operating system primarily for mobile devices?
   a. iOS
   b. Windows 10
   c. Linux
   d. macOS
2. You're most likely to find an RTOS
   a. on an iPad.
   b. in a robotic camera.
   c. on a supercomputer.
   d. on a mainframe.
3. Plug and Play is a software and hardware standard that
   a. allows you to play video games with better graphics.
   b. includes device drivers for lots of hardware as part of the OS.
   c. adds a new type of port (plug) to your system.
   d. optimizes the work of the GPU.
4. The OS can optimize RAM storage by using
   a. virtual memory.
   b. thrashing.
   c. an interrupt handler.
   d. a spooler.
5. BIOS stands for
   a. bundled input output software.
   b. billable input output services.
   c. basic input/output system.
   d. basic in order services.

(3Dalia/Shutterstock)

As more devices and appliances use RTOSs, protecting them from hackers becomes increasingly critical. Developers are working to improve the security of the software and safeguard communications between such devices. How concerned are you about the security of RTOSs in cars, smart homes, and wearable technology? Is enough being done to ensure the safety of these devices? What else can be done?

Go to **MyLab IT** to take an autograded version of the *Check Your Understanding* review and to find all media resources for the chapter.

For the IT Simulation for this chapter, see MyLab IT.

Try This

Using Virtual Desktops in Windows 10

Virtual desktops are a great way to organize your working space into different displays when you're multitasking between projects. This exercise will walk you through the process of organizing your files into virtual desktops. Have several applications open at the same time before you begin. For more detail, watch the Try This video.

Step 1 Click **Task View** on the taskbar to bring up the Task View interface. This consists of a display area that shows large thumbnails of all open windows running on your system. You haven't created any virtual desktops yet, so you will only see a button + New Desktop in the upper left corner. To organize open windows into specific groups, we will make a virtual desktop.

Step 2 While in Task View, click + **New desktop** in the upper left corner. This adds a virtual desktop, named Desktop 2. (A right click on the thumbnail gives you the option to rename it,)

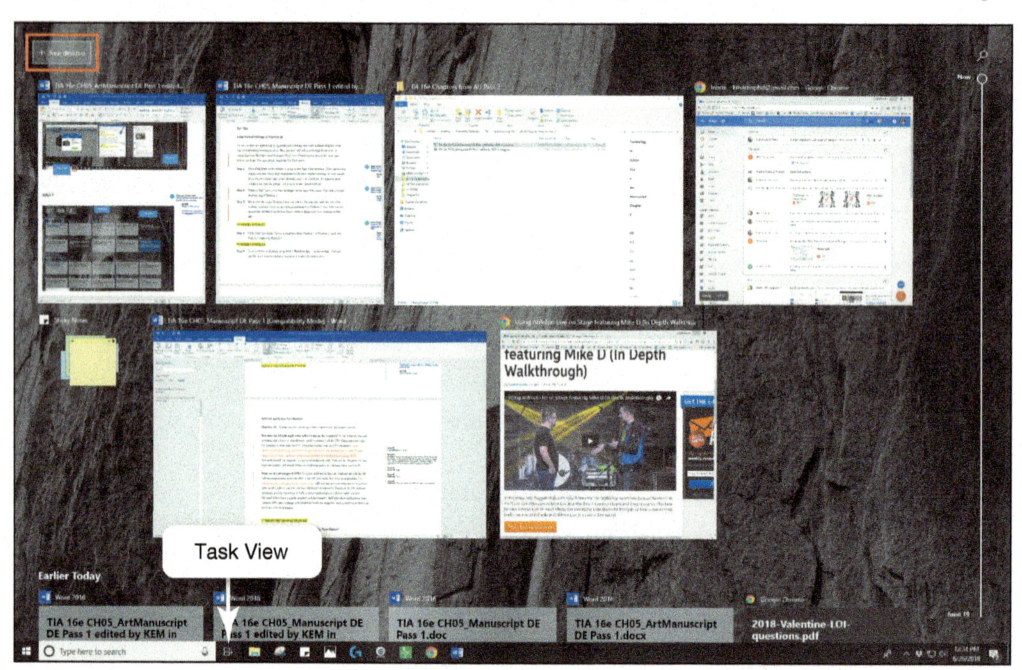

(Courtesy of Microsoft Corporation)

Step 3 Hover over Desktop 1 to see the programs and files you were working on earlier. Click any of them and drag one into Desktop 2.

Note: You can also use the timeline to go back in time to retrieve things you were working on days ago.

Step 4 Now to move between desktops, press **Ctrl + Windows key + an arrow key**. You can quickly move between desktops customized to specific projects now.

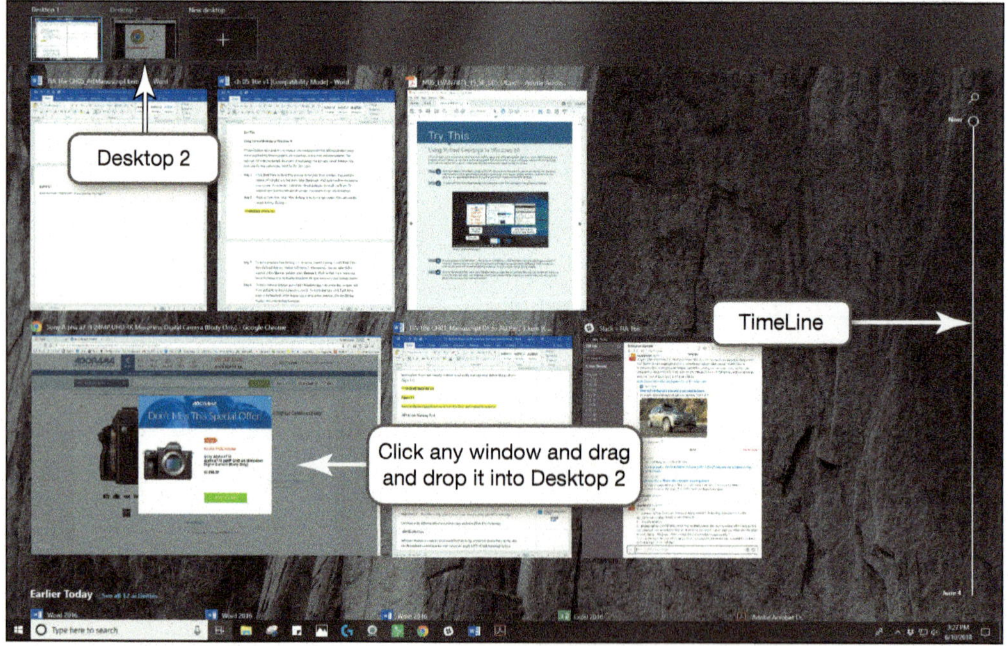

(Courtesy of Microsoft Corporation)

Make This  TOOL: App Inventor 2 or Thunkable

A Notification Alert

Does your app need to communicate with or provide feedback to the user? With the Notifier component of **App Inventor**, you can make your program generate message boxes or ask for a response from the user.

In this exercise, you'll use the Notifier component of **App Inventor** to show a message box, post a user choice with two buttons for response, or post an alert. Your apps can now use more of the features of the operating system to communicate!

The Notifier component allows you to communicate with alerts, message boxes, and text choice pop-ups within your mobile app.

```
when Button1 .Click
do  call Notifier1 .ShowAlert
                  notice  " This is how to Show an Alert notification. "

when Button2 .Click
do  call Notifier1 .ShowMessageDialog
                  message  " Force the user to click that they read this one! "
                  title  " MessageDialog "
                  buttonText  " Click Me "

when Button3 .Click
do  call Notifier1 .ShowChooseDialog
                  message  " or You can give the user a choice "
                  title  " Example of ChooseDialog "
                  button1Text  " I want it Faster "
                  button2Text  " I want it Cheaper "
                  cancelable  true

when Button4 .Click
do  call Notifier1 .ShowTextDialog
                  message  " or You can have the user tell you in their own words. "
                  title  " Text Dialog example "
                  cancelable  true
```

(Courtesy of MIT, used with permission.)

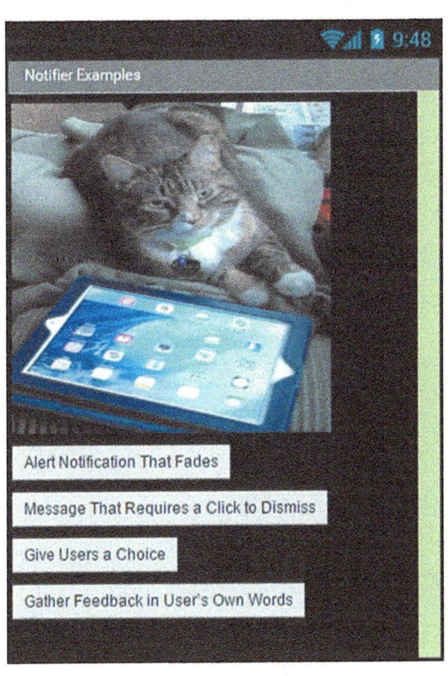

(Courtesy of MIT, used with permission.)

To create this app for iOS, go to *Thunkable.com*, a programming platform based on App Inventor.
For the detailed instructions for this exercise, go to MyLab IT.

Part 2

For an overview of this part of the chapter, watch **Chapter Overview Video 5.2.**

Using System Software

Learning Outcome 5.2 **You will be able to describe how to use system software, including the user interface, file management capabilities, and utility programs.**

Now that you know how system software works, let's explore how specific operating systems and their tools function.

The Windows Interface

One of the functions of the operating system is to provide a user interface that enables you to communicate with your computer. Today's operating systems use a graphical user interface. We describe the Windows 10 user interface in this section, but it's very similar to the user interfaces for the macOS and Linux operating systems.

Using Windows 10

Objective 5.8 *Describe the main features of the Windows interface.*

What are the main features of the Windows 10 desktop? After logging on to your Microsoft account by entering or confirming your e-mail address and entering a password, you're brought to the primary working area: the desktop. At the bottom of the Windows 10 desktop is the taskbar, which displays open and favorite applications for easy access. You can point to an icon to preview windows of open files or programs, or you can move your mouse over a thumbnail to preview a full-screen image. You can also right-click an icon to view a Jump List—a list of the most recently or commonly used files or commands for that application.

How do I find and open applications? The Start menu provides access to all applications and apps installed on your device as well as access to settings and power options. You open the Start menu by clicking the Windows icon (or Start button) on the far left side of the taskbar or by pressing the Windows key on the keyboard.

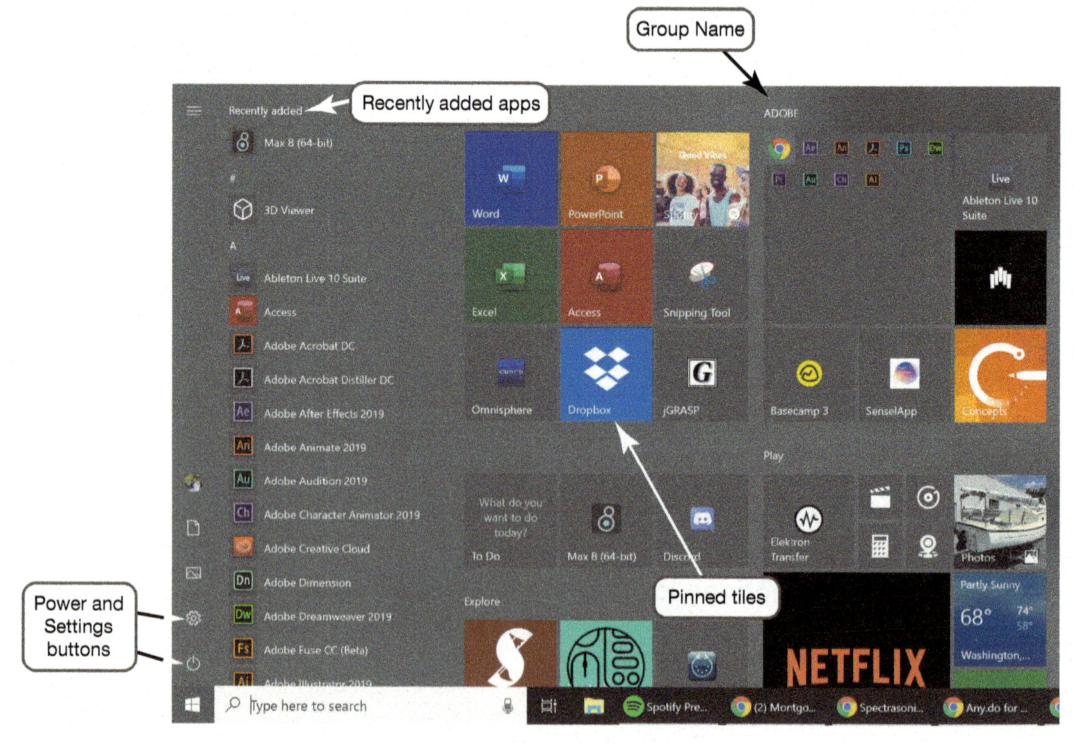

Figure 5.12 The Windows 10 Start menu provides access to your recently added apps. You can customize the Start menu by adding, resizing, and grouping tiles. *(Courtesy of Microsoft, Corp.)*

The Start menu is divided into two sections (see Figure 5.12). The right side has block tiles that provide easy access to your most frequently used software and Windows apps (such as Weather, Mail, and Photos). In addition to programs, tiles can represent files and folders. If there are more tiles on the Start menu than displayed, a scroll bar becomes available. The left side of the Start menu provides access to your Documents, Pictures, Settings, Power, and a list of all the installed apps and programs on your computer.

Can I modify the Start menu? You can customize the Start menu to meet your needs (refer to Figure 5.12). It's easy to add, remove, resize, move, and group application tiles on the Start menu. You can choose which applications are tiles on the Start menu through a process called pinning. Just right-click any application in the app list or in the Most used list, and choose Pin to Start.

Once on the Start menu, a tile can be resized by right-clicking the tile, pointing to Resize, and then selecting Small, Medium, Wide, or Large. You can rearrange apps on the Start menu by dragging them to a new location. You can also arrange several tiles near each other to create a group. Point to the top of the group to display a name box where you can type a group name. If there is a tile on the Start menu you don't need, just right-click it (or touch and hold), and select Unpin from Start.

You can also customize the left section of the Start menu. Right-click any app in the Most used section to get Pin to Start, More, and Uninstall options. If you click More, Pin to taskbar and other options that vary by app appear. Go to Settings, Personalization and click Start, and you can determine which folders appear on the left side of the Start menu. The default is Documents and Settings, but there are others to choose from, including Music, Pictures, or Downloads.

Because of the potential for customization, the Start menu on your computer will most likely be different from your friend's Start menu. However, because your Windows 10 settings and preferences are saved in your Microsoft account, you'll see your own personal settings, preferences, and applications reflected on the Start menu when you log on to any Windows 10 computer.

How can I change the appearance of my desktop and lock screen? You can also personalize the desktop and lock screen (the screen where you enter a password to resume using your computer) to suit your tastes. You can choose to have a picture, a slide show, or a single-color display on the desktop and lock screen. You can also determine whether you'd like other information to display on the lock screen, such as the weather, time, or your e-mail or calendar. Choosing a theme sets a color and font scheme for your entire system.

How can I organize more than one window on my screen at a time? You can "snap" windows—fixing open programs into place on either the left or the right side of the screen, to display two apps at the same time easily. Once you snap a window into place, you see thumbnails of all the other open windows. Clicking any thumbnail will snap that window into place. You can also snap windows to the corners, thus having up to four windows displayed at the same time.

How can I see open windows and quickly move between them? Task View allows you to view all the tasks you're working on in one glance. To see all open windows at once, click Task View on the taskbar. This is similar to using Alt + Tab.

Can I group my open programs and windows into task-specific groups? When faced with situations in which you're working with multiple documents for multiple projects, it might be easier to group the sets of documents for each project and just switch between projects. You can do that in Windows 10 with a feature called virtual desktops, which allows you to organize groups of windows into different displays (see Figure 5.13). For example, you can put an open Word document, web browser, and PowerPoint presentation that you need for your business class in one

Bits&Bytes Snip and Sketch

Snip and Sketch is a Windows tool that replaces the Snipping Tool and allows you to take screenshots. Snip and Sketch can be found in the Action Center (or use Windows key + Shift + S). It allows you to capture any part of your screen or a window and save it as an image file, pop it on the clipboard, or share it to others. You can also easily draw over the image or add writing. The delay timer even lets you pause capture for a few seconds so you can open dialog boxes or menus.

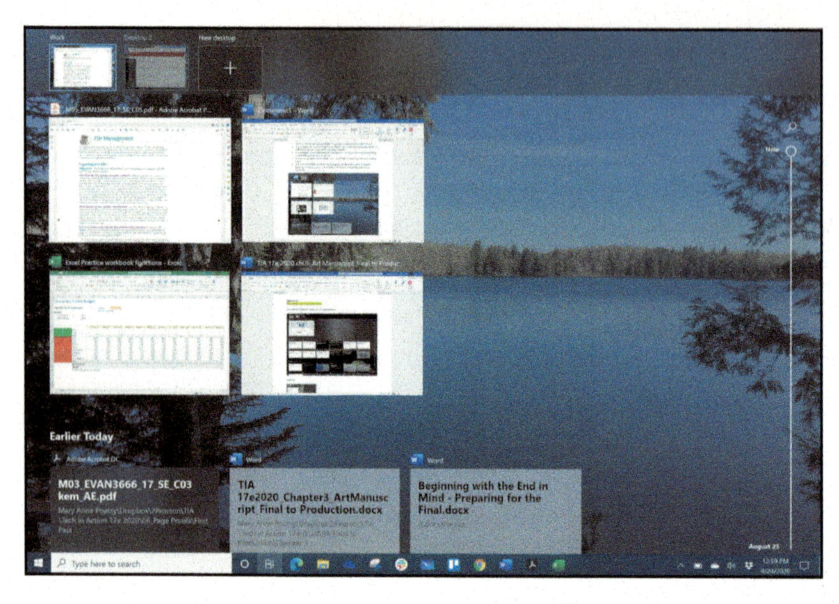

Figure 5.13 Virtual desktops enable you to organize your open files into separate working spaces. *(Courtesy of Microsoft, Corp.)*

desktop, and, in a second desktop, you can put a Word document and an Excel spreadsheet for a project you're working on for your job. There is no limit to the number of desktops you can create. You can see all virtual desktops by clicking Task View or using Windows key + Tab. For more information on how to work with virtual desktops in Windows 10, see the Try This on page 176.

User Interfaces for Other OSs

How does the Mac user interface compare with that of Windows? Although macOS and the Windows operating systems aren't compatible, they're extremely similar in terms of functionality; macOS programs appear in resizable windows and use menus and icons, just like Windows (see Figure 5.14). Although macOS doesn't have a Start menu, it features a Dock with icons for your most popular programs, very similar to the Windows taskbar.

Apple has developed different operating systems for different categories of devices: macOS for desktops and laptops, iOS for the iPhone and iPad, tvOS for Apple TV units, and watchOS for the Apple watch; macOS features a tight integration between the desktop and portable devices running iOS. With the macOS desktop, you can take and make phone calls, receive and send text messages, and easily transfer files with iOS devices.

Is a Linux user interface similar to that of macOS and Windows? Different distros of Linux feature different user interfaces, but most of them, like Ubuntu (see Figure 5.15), are based on familiar Windows and macOS paradigms, such as using icons to launch programs and having apps run in a window environment.

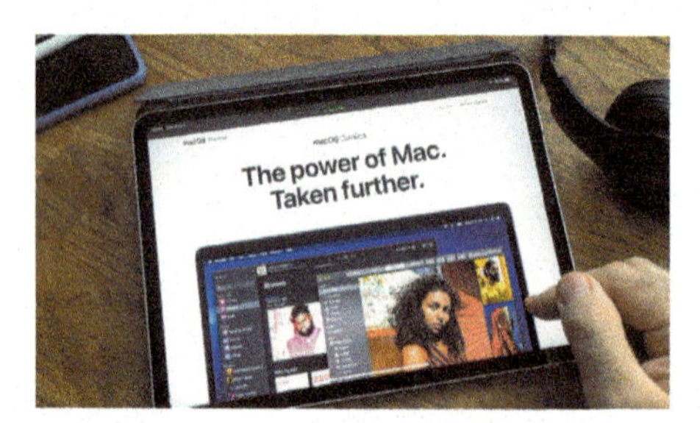

Figure 5.14 MacOS Catalina is Apple's latest operating system for desktop and laptop computers. *(Ifeelstock/Alamy Stock Photo)*

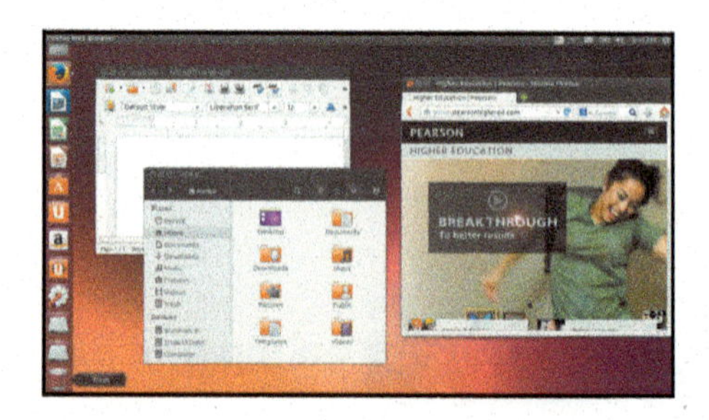

Figure 5.15 The Ubuntu Linux user interface resembles the Windows desktop. *(Courtesy of Canonical Limited.)*

File Management

So far we've discussed how the OS manages the processor, memory, storage, and devices, and how it provides a way for applications and users to interact with the computer. An additional function of the OS is to enable file management, which provides an organizational structure to your computer's contents. In this section, we discuss how you can use this feature to make your computer more organized and efficient.

Organizing Your Files

Objective 5.9 *Summarize how the operating system helps keep your computer organized and manages files and folders.*

How does the OS organize computer contents? Windows organizes the contents of your computer in a hierarchical directory structure composed of *drives, libraries, folders, subfolders*, and *files*. The hard drive, represented as the C: drive, is where you permanently store most of your files. Each additional storage drive, such as an optical drive, flash drive, or external hard drive, is given a unique letter (*D, E, F*, and so on). The C: drive is like a large filing cabinet in which all files are stored. The C:\ directory is the top of the filing structure of your computer and is referred to as the root directory. All other folders and files are organized within the root directory. There are areas in the root directory that the OS has filled with files and folders holding special OS files. The programs within these files help run the computer and generally shouldn't be accessed. In addition, sometimes the manufacturer will store files used to reinstate your computer back to the original factory state. Those files should not be accessed unless necessary.

What exactly are files, folders, and libraries? In an OS, a file is a collection of program instructions or data that is stored and treated as a single unit. Files can be generated from an application such as a Word document or an Excel workbook. In addition, files can represent an entire application, a web page, a set of sounds, an image, or a video. Files can be stored on the hard drive, a flash drive, online, or on any other permanent storage medium. As the number of files you save increases, it becomes more important to keep them organized in folders and libraries. A folder is a collection of files.

How can I easily locate and see the contents of my computer? In Windows, File Explorer is the main tool for finding, viewing, and managing the contents of your computer. It shows the location and contents of every drive, folder, and file. As illustrated in Figure 5.16, File Explorer is divided into multiple panes or sections:

1. The *Navigation pane* on the left shows the contents of your computer. It displays commonly accessed files and folders in the Quick Access area, files stored online in OneDrive, and other storage areas in This PC such as Documents, Music, Pictures, and Videos.
2. When you select a folder, drive, or library from the Navigation pane, the contents of that particular area are displayed in the *File list* on the right. The File list can be viewed in different ways: icons, tiles, or a detailed list.
3. A third pane can be added. From the View tab, the Pane group allows you to add either a Details pane (as in Figure 5.16) or a Preview pane. The Details pane displays the file or folder's properties. The Preview pane displays the first page of the selected document.

For folders that have been shared with others by using an online storage system such as OneDrive or Dropbox, Windows 10 displays a shared icon instead of the traditional yellow folder icon.

When you save a file for the first time, you give the file a name and designate where you want to save it. For easy reference, the OS includes libraries where files are saved unless you specify otherwise. In Windows, the default libraries are Documents for document files, Music for audio files, Pictures for graphics files, and Videos for video files. These are in the This PC section of File Explorer.

You can determine the location of a file by its file path. The file path starts with the drive in which the file is located and includes all folders, subfolders (if any), the file name, and the extension. For example, if you were saving a picture of Andrew Carnegie for a term paper for a US History course, the file path might be C:\Documents\HIS182\Term Paper\Illustrations\ACarnegie.jpg.

As shown in Figure 5.17, "C:" is the drive on which the file is stored (in this case, the hard drive), and "Documents" is the file's primary folder. "HIS182," "Term Paper," and "Illustrations" are successive subfolders within the Documents main folder. Last is the file name, "ACarnegie," separated from the file extension (in this case, "jpg") by a period. The backslash character (\), used by

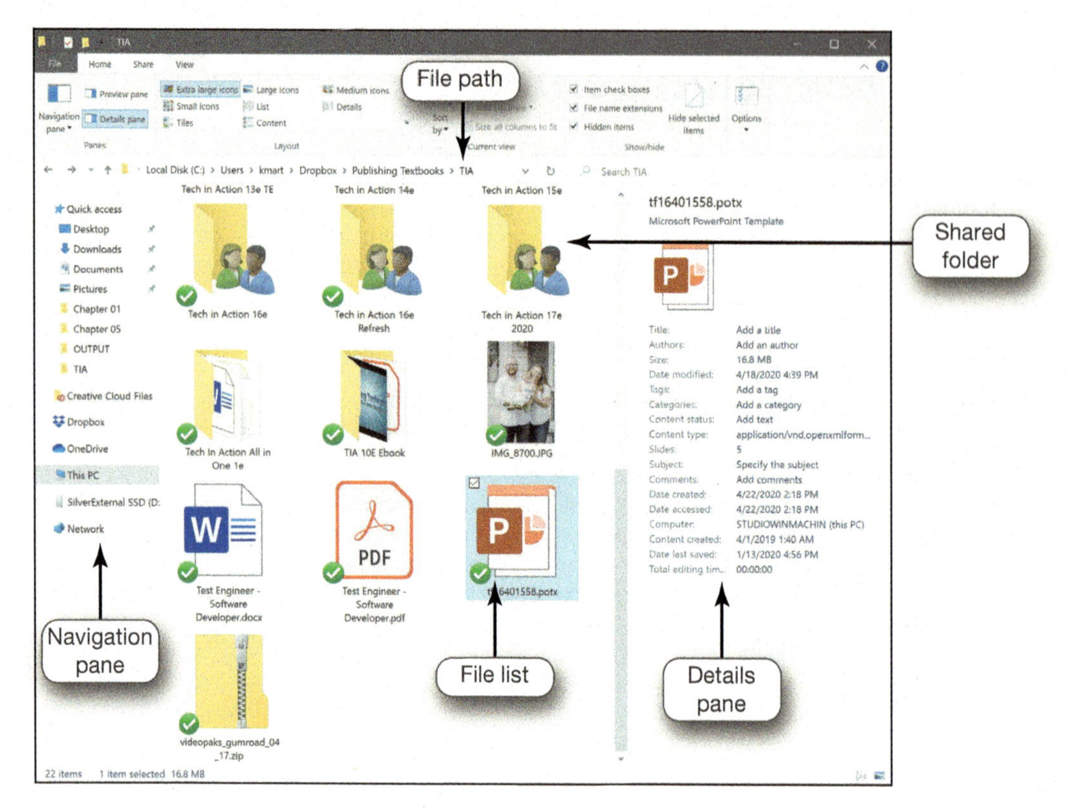

Figure 5.16 File Explorer shows you the contents of your computer.
> *Click **File Explorer** from the Start menu or from the taskbar. (Courtesy of Microsoft Corp.)*

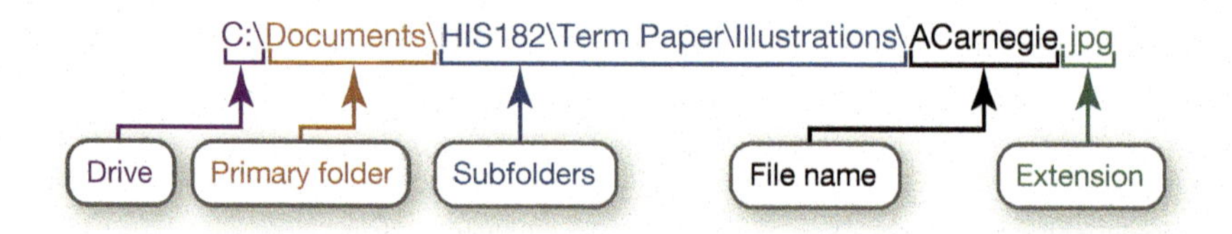

Figure 5.17 Understanding File Paths

Windows, is referred to as a path separator; macOS files use a colon (:), whereas UNIX and Linux files use the forward slash (/) as the path separator.

Are there different ways I can view and sort my files and folders? Clicking the View tab in File Explorer offers you different ways to view and sort the folders and files.

1. **Details view:** This is the most interactive view. Files and folders are displayed in list form, and the additional file information is displayed in columns alongside the name of the file. You can sort and display the contents of the folder by clicking any of the column headings, so you can sort the contents alphabetically by name or type or hierarchically by date last modified or file size. Right-click the column heading area to modify the display of columns.
2. **Icons view:** A number of icon views display files and folders in sizes ranging from small to extra-large. In most icon views, the folders are displayed as *Live Icons*, which allow you to preview the actual contents of a specific folder without opening the folder. Large Icons view, as shown in Figure 5.19, is the best view to use if your folder contains picture files, because you can see a bit of the actual images peeking out of the folder. It's also good to use if your folder contains PowerPoint presentations because the title slide of the presentation will be displayed, making it easier for you to distinguish among presentations.

You can change the views by using the controls in the Layout group on the View tab or by clicking between Details view and Large Icons view from the bottom right corner of File Explorer.

Bits&Bytes Save Files in the Cloud

If you're still using a flash drive, consider saving to the cloud instead. Cloud storage sites such as Microsoft OneDrive, Google Drive, and Dropbox provide easy solutions (see Figure 5.18). One advantage of saving files to the cloud is that your files are accessible from any Internet-connected device—including your smartphone. Using cloud storage will ensure the availability of your files when you need them. Moreover, files and folders stored in the cloud are easily shared with others to facilitate collaboration. Another advantage is that you have a backup. If your computer breaks or is lost, you simply download the files onto your new system from the cloud.

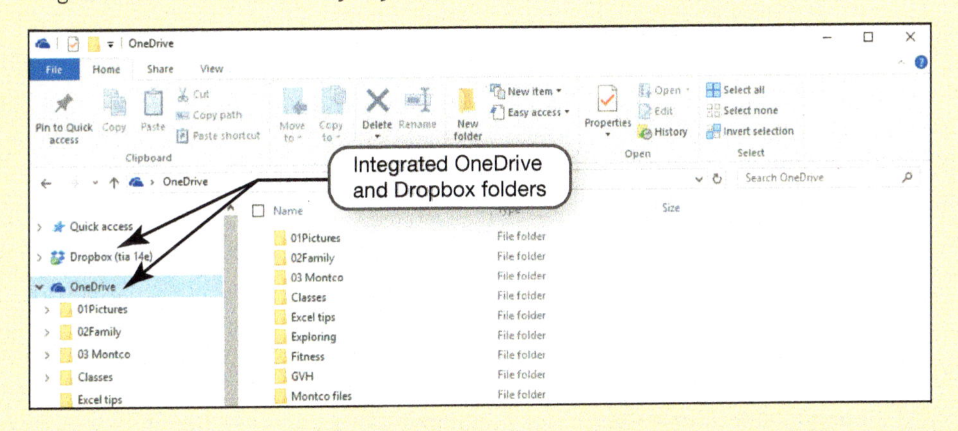

Figure 5.18 OneDrive and Dropbox are integrated in Windows File Explorer for easy, cloud-based file storage. *(Courtesy of Microsoft Corporation)*

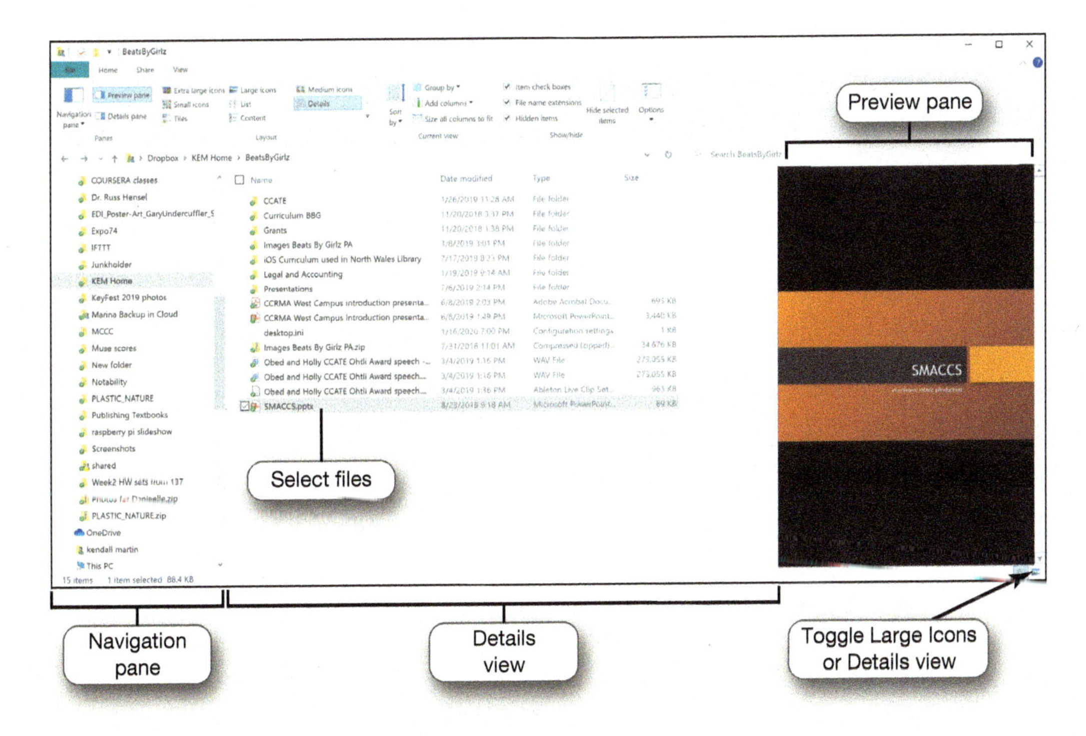

Figure 5.19 The View tab in File Explorer allows you to use different views to display information about files and folders and even show their contents. The bottom right icons switch between Large Icon and Details views. *(Courtesy of Microsoft Corp.)*

Naming Files

Are there special rules I have to follow when I name files? The first part of a file, or the file name, is generally the name you assign to the file when you save it. For example, "bioreport" may be the name you assign a report you have completed for a biology class.

In a Windows application, an extension, or file type, follows the file name and a period or dot (.). Like a last name, this extension identifies what kind of family of files the file belongs to, or which application should be used to read the file. For example, if "bioreport" is a spreadsheet created in Microsoft Excel, it has an .xlsx extension and its name is bioreport.xlsx. You can choose to display file extensions by checking or clearing the File name extensions box in the Show/hide group on the View tab in File Explorer.

Table 5.2 lists some common file extensions and the types of documents they indicate.

Why is it important to know the file extension? When you save a file created in most applications running under the Windows OS, you don't need to add the extension to the file name; it is added automatically for you. Mac and Linux operating systems don't require file extensions, because the information about the type of application the computer should use to open the file is stored inside the file itself. However, if you're using the Mac or Linux OS and will be sending files to Windows users, you should add an extension to your file name so that Windows can more easily open your files.

Table 5.2 Common File Name Extensions

Extension	Type of Document	Application
.docx	Word processing document	Microsoft Word 2007 and later
.xlsx	Workbook	Microsoft Excel 2007 and later
.accdb	Database	Microsoft Access 2007 and later
.pptx	Presentation	Microsoft PowerPoint 2007 and later
.pdf	Portable Document Format	Adobe Acrobat or Adobe Reader
.rtf	Text (Rich Text Format)	Any program that can read text documents
.txt	Text	Any program that can read text documents
.htm or .html	Hypertext Markup Language (HTML) for a web page	Any program that can read HTML
.jpg	Joint Photographic Experts Group (JPEG) image	Most programs capable of displaying images
.zip	Compressed file	Various file compression programs

Bits&Bytes File Extension Help

There are thousands of file extensions and more are created all the time. If you run into a file extension you're not familiar with, there is help. Visit FILExt at filext.com. This site serves as an encyclopedia of file extensions. It maintains a list of over 26,000 different file extensions so you're sure to be able to find out what kind of file you're looking at!

If you try to open a file with an unknown extension, Windows 10 displays a dialog box that offers several suggested programs that can be used (see Figure 5.20). If you're certain of the type of file you're opening and know the program you'll want to use in the future to open similar types of files, select the Always use the selected program option. However, if you're at all unsure of the program to use or don't want to set a default program for this type of file, make sure that option is not selected.

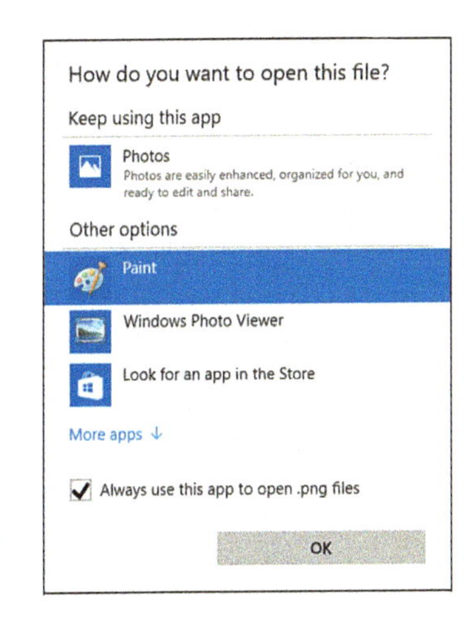

Figure 5.20 When you open a file with an unknown file type, you can choose the app you want to use. *(Courtesy of Microsoft Corp.)*

Are there things I shouldn't do when naming my files? It's important for you to name your files so that you can easily identify them. File names can have as many as 255 characters, so don't be afraid to use as many characters as you need. A file name such as "BIO 101 Research Paper First Draft.docx" makes it very clear what that file contains.

Keep in mind, however, that all files must be uniquely identified, unless you save them in different folders or in different locations. Therefore, although files may share the same file name (such as "bioreport.docx" or "bioreport.xlsx") or share the same extension ("bioreport.xlsx" or "budget.xlsx"), no two files stored in the same folder can share *both* the same filename and the same file extension. In addition, some characters can't be used in a file name, such as a quotation mark, and the file will not be saved until the file name is modified.

Copying, Moving, and Deleting Files and Folders

How can I move and copy files and folders? You can move or copy a file or folder to another location by using the Cut or Copy commands, respectively, on the ribbon or by right-clicking the file and selecting an option from the shortcut menu. After selecting Cut or Copy, select the new location and then select Paste from the ribbon or right-click the new location folder and select Paste. Alternatively, you can drag and drop a file or folder to a new location. One method is to open two File Explorer windows side by side and drag a file or folder from one window to the other. Another method is to drag a file or folder from the File list to a new location displayed in the Navigation pane. When dragging a file or folder, the default action is to move the file or folder. If you'd rather make a copy, press Ctrl while you drag the file or folder to its new location.

Where do deleted files go? The Recycle Bin is a folder on the desktop, represented by an icon that looks like a recycling bin, where files deleted from the hard drive reside until you permanently purge them from your system. Unfortunately, files deleted from other storage locations or files stored in the cloud don't go to the Recycle Bin but are deleted from the system immediately.

Mac systems have something similar to the Recycle Bin, called Trash, which is represented by a wastebasket icon. To delete files on a Mac, drag the files to the Trash icon.

How do I permanently delete files from the Recycle Bin? Files placed in the Recycle Bin or the Trash remain in the system until they're permanently deleted. To delete files from the Recycle Bin permanently, select Empty Recycle Bin after right-clicking the desktop icon. On Macs, select Empty Trash from the Finder menu in macOS.

What happens if I need to recover a deleted file? Getting a file back after the Recycle Bin has been emptied may still be possible:

- File History is a Windows 10 utility that automatically backs up files and saves previous versions of files to a designated drive (such as an external hard drive). If you're using File History, you can restore previously deleted files or even previous versions of files you've changed. (We'll discuss File History in more detail later in the chapter.)
- Cloud-based storage systems, such as Dropbox and Google Drive, maintain histories of file changes, and you can not only recover a deleted file but also roll back files edit by edit.

File Compression

When would I ever see a compressed file? File compression (or "zipping a file") makes a large file more compact, making it easier and faster to send large attachments by e-mail, to upload them to the web, or to save them onto a flash drive or other storage medium. Windows has a built-in file compression utility that you can access through File Explorer. You can unzip compressed folders that you receive, as shown in Figure 5.21. You can also create your own zipped file by selecting a group of files and then compressing them from inside File Explorer. Some people use stand-alone freeware and shareware programs for compression, such as WinZip (for Windows) and StuffIt (for Windows or Mac), because they like their features.

Organizing Your Computer: File Management

In this Helpdesk, you'll play the role of a help desk staffer, fielding questions about the desktop, window features, and how the OS helps keep the computer organized.

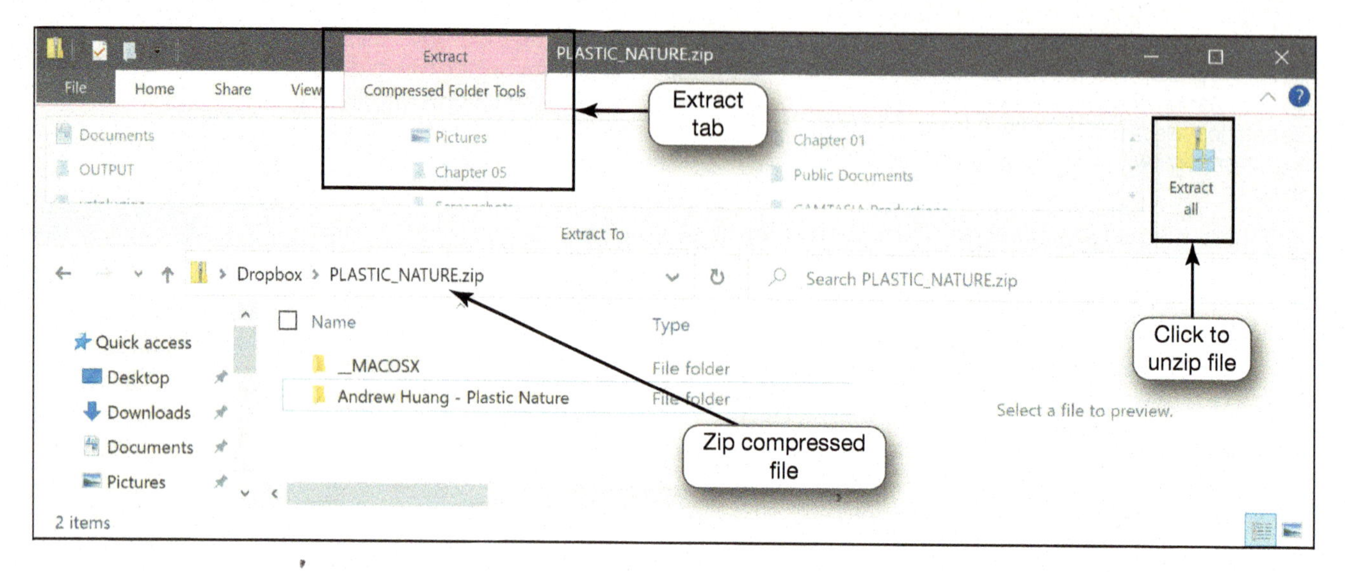

Figure 5.21 Compressed zip files can be expanded using tools built into File Explorer.
> To uncompress (extract) a zipped file, select the file, click the ***Extract*** *tab, and then click* ***Extract All****. (Courtesy of Microsoft Corporation)*

How does file compression work? Compression takes out redundancies in a file to reduce the file size. Most compression programs look for repeated patterns of letters and replace these patterns with a shorter placeholder. The repeated patterns and the associated placeholder are cataloged and stored temporarily in a separate file called the *dictionary*. For example, in the following sentence, you can easily see the repeated patterns of letters.

The rain in Spain falls mainly on the plain.

Although this example contains obvious repeated patterns (ain and the), in a large document, the repeated patterns will be more complex. The compression program's algorithm (a set of instructions designed to complete a solution in a step-by-step manner), therefore, runs through the file several times to determine the optimal repeated patterns needed to obtain the greatest compression.

How effective are file compression programs? Current compression programs can reduce text files by 50% or more, depending on the file. However, some files, such as PDF files, already contain a form of compression, so they don't need to be compressed further. Image files such as JPEG, GIF, and PNG files discard small variations in color that the human eye might not pick up. Likewise, MP3 files permanently discard sounds that the human ear can't hear. These graphic and audio files don't need further compression.

Utility Programs

The main component of system software is the operating system. However, *utility programs*—small applications that perform special functions on the computer—are also an essential part of system software. Utility programs come in three flavors:

1. Those included with the OS (such as System Restore)
2. Those sold as stand-alone programs (such as Norton AntiVirus)
3. Those offered as freeware (such as anti-malware software like Ad-Aware from Lavasoft)

Table 5.3 lists some of the various types of utility programs available within Windows as well as some alternatives available as stand-alone programs.

Table 5.3 Utility Programs Available Within Windows and as Stand-Alone Programs

Windows Utility Program	Stand-Alone Alternatives	What It Does
Disk Cleanup	CCleaner	Removes unnecessary files from your hard drive
Defragment and Optimize Drives	Norton Utilities, iDefrag	Rearranges files on your hard drive to allow for faster access of files
Task Manager and Resource Monitor	Process Explorer	Displays performance measures for processes; provides information on programs and processes running on your computer
File History, File Recovery	Acronis True Image, Norton Online Backup	Backs up important files, makes a complete mirror image of your current computer setup
System Restore	Acronis True Image	Restores your system to a previous, stable state

Windows Administrative Utilities

Objective 5.10 *Outline the tools used to enhance system productivity, back up files, and provide accessibility.*

What kind of utilities are in the operating system? In general, the basic utilities designed to manage and tune your computer hardware are incorporated into the OS. The stand-alone utility programs typically offer more features or an easier user interface for backup, security, diagnostic, or recovery functions.

System Performance Utilities

What utilities can make my system work faster? Disk Cleanup is a Windows utility that removes unnecessary files from your hard drive. These include files that have accumulated in the Recycle Bin as well as temporary files—files created by Windows to store data temporarily while a program is running. Windows usually deletes these temporary files when you exit the program, but sometimes it forgets to do this or doesn't have time because your system freezes or incurs a problem that prevents you from properly exiting a program.

Storage Sense is another utility built into Windows 10 for deleting unnecessary files. To turn on Storage Sense, click the Start button, click Settings, click System, then select Storage. Storage Sense runs automatically to empty your Recycle bin, delete temporary files, and clean out your Downloads folder. You can specify how often it should run or how long it should allow files to stay (see Figure 5.22).

How can I check on a program that's stopped running? If a program has stopped working, you can use the Windows Task Manager utility (see Figure 5.23) to check on the program or exit the unresponsive program. The Processes tab of Task Manager lists all the programs you're using and indicates their status. "Not responding" will be shown next to a program that stopped improperly. You can terminate programs that aren't responding by right-clicking the app name and selecting End task from the shortcut menu.

If you're working on a Mac, Force Quit can help you close out unresponsive applications. You can access Force Quit by clicking the Apple menu or simply pressing Command + Option + Esc. Alternatively, you can press and hold the Option and Ctrl keys and click the unresponsive program's icon on the dock.

How can I improve my computer drive's performance? Data is recorded on hard disks in concentric circles called tracks. Each track is further broken down into pie-shaped wedges, each called a sector. The data is further identified by clusters, which are the smallest segments within the sectors (see Figure 5.24). After you have used your computer for a while, uploading, creating, changing, and deleting content, a mechanical hard drive becomes fragmented. Parts of files get scattered (fragmented) around the hard disk platter, the thin circular plate of metal covered with

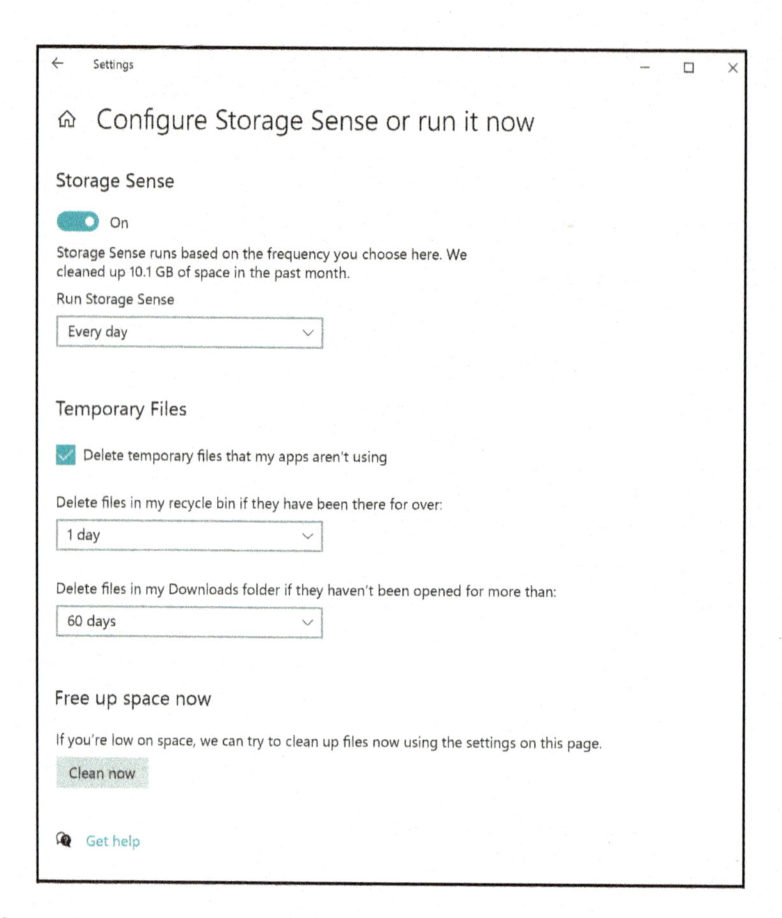

Figure 5.22 Storage Sense runs automatically to delete temporary files and items you have deleted or downloaded long ago.

> *To turn on Storage Sense, click the **Start** button, click **Settings**, click **System**, then select **Storage**.*

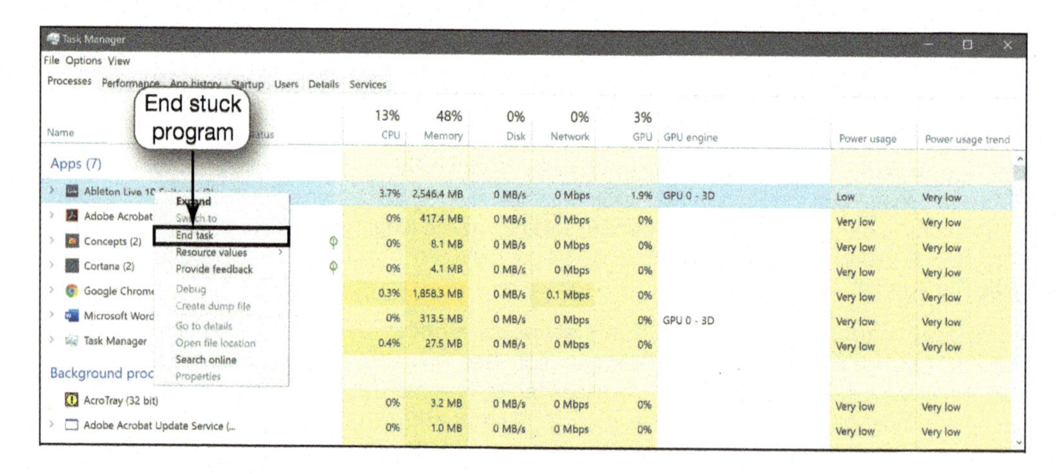

Figure 5.23 You can use Task Manager to close unresponsive programs. *(Courtesy of Microsoft Corp.)*

> *To close an unresponsive program, right-click the taskbar and select **Task Manager**. Right-click the program name and then select **End Task**.*

a special magnetic coating that records the data. Disk defragmentation rearranges fragmented data on the platter so that related file pieces are unified and quicker to access (Figure 5.25). The Defragment and Optimize Drives utility is found in the Windows Administrative Tools folder in the apps list. Before you defragment your hard drive, you should first determine whether the disk needs to be defragmented by selecting Analyze. If your disk is more than 10% fragmented, you should run the disk defragmenter utility.

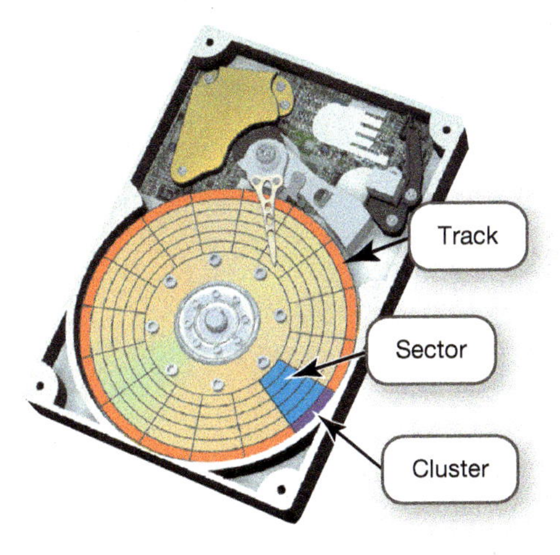

Figure 5.24 On a hard disk platter, data is recorded onto tracks, which are further divided into sectors and clusters.

Note that disk defragmentation doesn't make sense for solid state drives (SSDs) because they do not use platters. But an operation called Trim runs in Windows 10 to make sure the performance of the drive stays optimal even after a number of file writes and deletes. For more about hard disks and defragmenting, check out the Sound Byte, "Hard Disk Anatomy."

Hard Disk Anatomy

In this Sound Byte, you'll watch a series of animations that show various aspects of a mechanical hard drive, including the anatomy of the drive, how a computer reads and writes data to a hard drive, and the fragmenting and defragmenting of a hard drive.

File and System Backup Utilities

How can I protect my data in the event something malfunctions in my system? As noted earlier, when you use the File History utility, you can have Windows automatically create a duplicate of your libraries, desktop, contacts, and website favorites and copy it to another storage device, such as an external hard drive (see Figure 5.26). A backup copy protects your data in the event your hard drive fails or files are accidentally erased. File History also keeps copies of different versions of your files. This means that if you need to go back to the second draft of your history term paper, even though you are now on your fifth draft, File History allows you to recover it. File History needs to be turned on by the user and requires an external hard drive (or your One Drive account) that is always connected to the computer to function. You can choose to back up all folders on the C: drive or select certain ones that you use most often or contain the most sensitive files to back up.

Windows 10 also includes recovery tools that allow you to complete backups of your entire system (system image) that you can later restore in the event of a major hard drive crash.

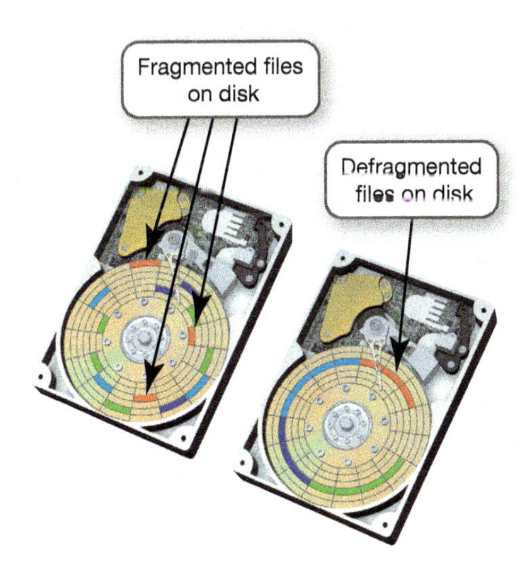

Figure 5.25 Defragmenting the hard drive arranges file fragments so that they are located next to each other. This makes the hard drive run more efficiently.

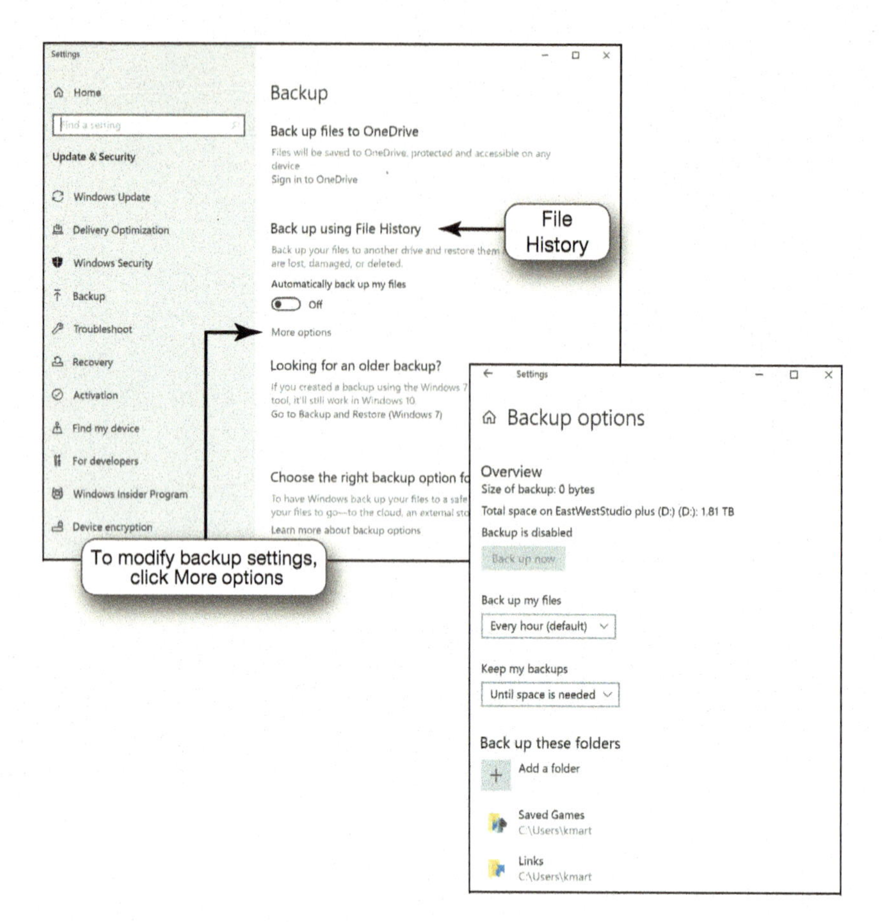

Figure 5.26 File History backs up files to an external hard drive.
> *To turn on File History, choose* ***Settings****, select* ***Update & Security****, and then select* ***Backup****. Under Back Up Using File History, select* ***More Options*** *to customize when backup is done and what is stored. (Courtesy of Microsoft Corp.)*

How can I recover my entire system?

Suppose you've just installed a new software program and your computer freezes. After rebooting the computer, when you try to start the application, the system freezes once again. You uninstall the new program, but your computer continues to freeze after rebooting. What can you do now?

Windows has a utility called System Restore by which to roll your system settings back to a specific date when everything was working properly. A system restore point, which is a snapshot of your entire system's settings, is generated prior to certain events, such as installing or updating software, or automatically once a week if no other restore points were created in that time. You also can create a restore point manually at any time.

If problems occur, if the computer was running just fine before you installed new software or a hardware device, you could restore your computer to the settings that were in effect before the software or hardware installation (see Figure 5.27). System Restore doesn't affect your personal data files (such as Word documents or e-mail), so you won't lose changes made to these files.

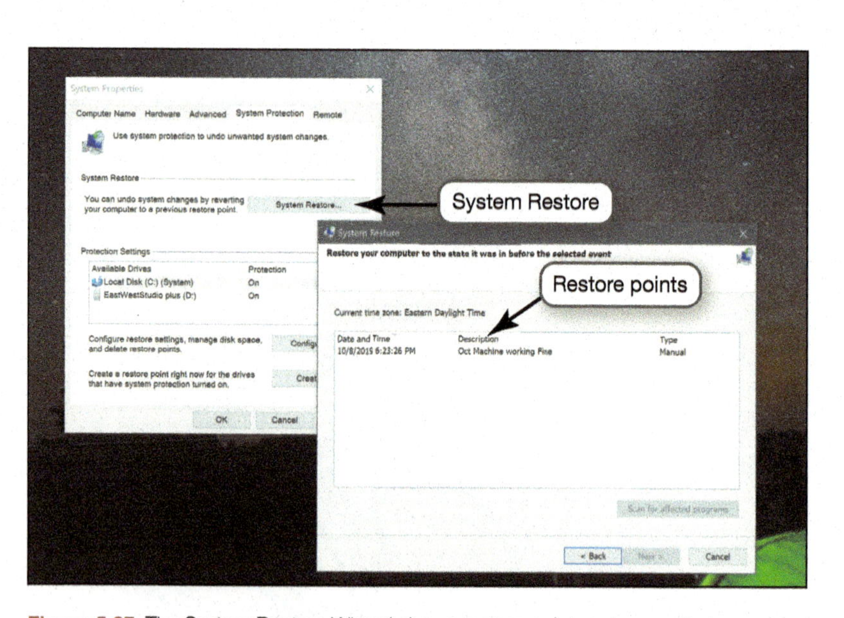

Figure 5.27 The System Restore Wizard shows restore points set manually by a user and automatically by Windows when updates were installed. Setting a restore point is good practice before installing any hardware or software.
> *To access System Restore, search for System restore in the Windows 10 search box and select* ***Create a restore point****. The System Restore Wizard appears, with the restore points shown on the second page of the wizard. (Courtesy of Microsoft Corp.)*

Dig Deeper Windows Tricks

In Windows, there are often many ways to accomplish the same task. Let's dig a bit deeper into the OS to see how you can be most efficient. Once you're more familiar with the design of the OS, you may find some of these keyboard combinations are faster to use than drag and drop or menu systems.

The Windows key (on the bottom row of the keyboard) is the secret to many keyboard shortcuts. You can combine the Windows key with symbols to do very useful things, as shown in Table 5.4.

There are other time-saving tricks as well. Have you tried these?

- **Shake:** Have too many windows open at one time? Grab the title bar of the one you're interested in and give it a quick back and forth shake while holding the mouse button down. The other windows will automatically minimize. Shaking again undoes this change.
- **Dark Mode:** If you prefer, the basic design of Windows can be dark in color. The background of the taskbar, Start menu, and pop-ups will all be set to dark colors. Select Settings and then choose Personalization, Colors, and Dark Mode. You can specify which mode you want for apps as well.
- **Focus Assist:** When you need notifications to be quieted so you can focus on your work, select Settings and then choose System and then Focus Assist. You can allow only certain apps or contacts to send notifications or silence all but alarms you've set.
- **Peek:** Other tidbits are hidden in Windows. For example, if you move your mouse to the bottom right corner of the taskbar, to the right of the notifications icon, there is a hidden button that toggles between a clear desktop with all windows minimized and your current configuration. Named Peek, it can be turned on and off in Settings.

You can find a full list of all Windows keyboard shortcuts, including those that use the Ctrl, Alt, and Shift keys, at Microsoft's support site.

Table 5.4 Windows 10 Shortcuts

Windows Key plus	
D	Minimizes apps to show **Desktop**
E	Opens File **Explorer**
G	Opens Xbox **Game** bar
.	Opens table of emojis and symbols
S	Opens **Search** screen
X	Opens "secret" Start menu, Quick Links menu
Nothing	Opens **Start** menu
Shift-S	Takes a screenshot with **Snip and Sketch**
Arrow keys	Use in combinations to resize and locate windows to the edges of the screen
Tab	Opens **Task** View
Home	Minimizes all but the currently selected window
Ctrl-D	Opens a new **virtual desktop**
V	Shows the full clipboard, including all of the clips stored (make sure Clipboard History is turned on in Settings)
I	Opens **Settings**
Ctrl and Arrow	Moves to next/previous Virtual Desktop
A	Opens **Action** center

Table 5.5 Windows Ease of Access Tools

Magnifier	Narrator
Creates a separate window that displays a magnified portion of the screen 	Reads what is on the screen Can read the contents of a window, menu options, or text you have typed 

Speech Recognition	On-Screen Keyboard	High Contrast
Allows you to dictate text and control your computer by voice 	Allows you to type with a pointing device 	Provides color schemes that invert screen colors for vision-impaired individuals 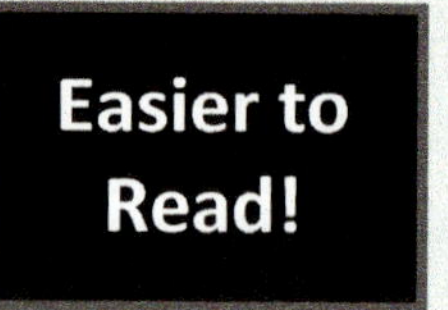

> To locate Ease of Access, click the **Start** button, click Settings, and then select **Ease of Access**. (Imagerymajestic/123RF; Anatoly Maslennikov/123RF; Iqoncept/123RF; Pockgallery/Shutterstock; Courtesy of Microsoft Corp.; Pearson Education)

MacOS includes a backup utility called Time Machine that automatically backs up your files to a specified location. Because Time Machine makes a complete image copy of your system, it can also be used to recover your system in the case of a fatal error. (For more information on backing up your files, see Chapter 9.)

Accessibility Utilities

What utilities are designed for users with special needs? Microsoft Windows includes Ease of Access settings, which provide a centralized location for assistive technology and tools to adjust accessibility settings. The Ease of Access tools, which are accessible from Settings, include tools to help users with disabilities, as shown in Table 5.5. The tools shown in the table are just a sampling of the available tools. Mac OS offers similar features from the Accessibility Options window.

Whether you use Windows, macOS, Linux, or another operating system, a fully featured OS is available to meet your needs. As long as you keep the operating system updated and regularly use the available utilities to fine-tune your system, you should experience little trouble from your OS.

Before moving on to the Chapter Review:

1. Watch Chapter Overview Video 5.2.
2. Then take the Check Your Understanding quiz.

Check Your Understanding // Review & Practice

For a quick review to see what you've learned so far, answer the following questions.

multiple choice

1. Which of the following does Windows use to organize files?
   a. Libraries
   b. Subfolders
   c. Folders
   d. All of the above
2. The Windows app used for locating files and folders is
   a. Disk Manager.
   b. Finder.
   c. Library Explorer.
   d. File Explorer.
3. Which view in File Explorer can you use to sort files by column heading?
   a. Details
   b. Small icons
   c. Tiles
   d. Large icons
4. Which of the following is used to reduce the size of a file or folder?
   a. Defragmenter
   b. Extraction
   c. Compression
   d. Shrinker
5. Which utility is used to exit from an unresponsive program?
   a. System Refresh
   b. Disk Cleanup
   c. Task Manager
   d. File Explorer

(3Dalia/Shutterstock)

chew on this

Windows and macOS both include a number of features and utilities designed for users with special needs. Compare the offerings of each OS. Are there options one OS offers that are not available in the other? Which OS do you think has the best selection? Can you think of any areas that have not been addressed by this assistive technology? Research software from third-party developers to see whether there are other tools that would also be useful. Should these tools be included in the OS? Why or why not?

MyLab IT

Go to **MyLab IT** to take an autograded version of the *Check Your Understanding* review and to find all media resources for the chapter.

For the IT Simulation for this chapter, see MyLab IT.

5 Chapter Review

Summary

Part 1
Understanding System Software

Learning Outcome 5.1 You will be able to explain the types and functions of operating systems and explain the steps in the boot process.

Operating System Fundamentals

Objective 5.1 *Discuss the functions of the operating system.*

- System software is the set of software programs that helps run the computer and coordinates instructions between application software and hardware devices. It consists of the operating system (OS) and utility programs.
- The OS controls how your computer system functions. It manages the computer's hardware, provides a means for application software to work with the CPU, and is responsible for the management, scheduling, and coordination of tasks.
- Utility programs perform general housekeeping tasks for the computer, such as system maintenance and file compression.
- Modern operating systems allow for multitasking—performing more than one process at a time.

Objective 5.2 *Explain the most popular operating systems for personal use.*

- Microsoft Windows is the most popular OS. The most recent release is Windows 10; macOS is designed to work on Apple computers, and Linux is an open source OS based on UNIX and designed primarily for use on personal computers.
- Some operating systems allow interaction with touch-screen interfaces. All personal use operating systems incorporate elements to share and store files on the Internet.
- An OS is designed to run on specific CPUs. Application software is OS-dependent.
- Smartphones and tablets have their own specific mobile operating systems, which allow the user to multitask.

Objective 5.3 *Explain the different kinds of operating systems for machines, networks, and business.*

- Real-time operating systems (RTOSs) require no user intervention.
- A multiuser operating system (network operating system) provides access to a computer system by more than one user at a time.
- UNIX is a multiuser, multitasking OS that is used as a network OS, although it can be used on PCs.
- Mainframes and supercomputers are specialty computers that require mainframe operating systems.

What the Operating System Does

Objective 5.4 *Explain how the operating system provides a means for users to interact with the computer.*

- The OS provides a user interface that enables users to interact with the computer.
- Most OSs today use a graphical user interface (GUI). Common features of GUIs include windows, menus, and icons.

Objective 5.5 *Explain how the operating system helps manage hardware such as the processor, memory, storage, and peripheral devices.*

- When the OS allows you to perform more than one task at a time, it is multitasking. To provide for seamless multitasking, the OS controls the timing of the events on which the processor works.
- As the OS coordinates the activities of the processor, it uses RAM as a temporary storage area for instructions and data the processor needs. The OS coordinates the space allocations in RAM to ensure that there is enough space for the waiting instructions and data. If there isn't sufficient space in RAM for

all the data and instructions, then the OS allocates the least necessary files to temporary storage on the hard drive, called *virtual memory*.
- The OS manages storage by providing a file management system that keeps track of the names and locations of files and programs.
- Programs called *device drivers* facilitate communication between devices attached to the computer and the OS.

Objective 5.6 *Explain how the operating system interacts with application software.*

- All software applications need to interact with the CPU. For programs to work with the CPU, they must contain code that the CPU recognizes.
- Rather than having the same blocks of code appear in each application, the OS includes the blocks of code to which software applications refer. These blocks of code are called *application programming interfaces* (APIs).

Starting Your Computer

Objective 5.7 *Discuss the process the operating system uses to start up the computer and how errors in the boot process are handled.*

- When you start your computer, it runs through a special process called the *boot process*.
- The boot process consists of four basic steps: (1) The basic input/output system (BIOS) is activated when the user powers on the CPU. (2) In the POST check, the BIOS verifies that all attached devices are in place. (3) The OS is loaded into RAM. (4) Configuration and customization settings are checked.
- An authentication process occurs at the end of the boot process to ensure that an authorized user is entering the system.
- Sometimes errors occur in the boot process. Try rebooting the computer or resetting the computer if the problem persists.

Part 2
Using System Software

Learning Outcome 5.2 **You will be able to describe how to use system software, including the user interface, file management capabilities, and utility programs.**

The Windows Interface

Objective 5.8 *Describe the main features of the Windows interface.*

- In Windows 10, the Start menu provides access to your computer's apps, tools, and commonly used programs, and the desktop is the main working area.
- You can customize the Start menu by pinning and resizing tiles and organizing tiles into groups.
- Virtual desktops are used to organize open programs into different working areas.

File Management

Objective 5.9 *Summarize how the operating system helps keep your computer organized and manages files and folders.*

- Files and folders are organized in a hierarchical directory structure composed of drives, libraries, folders, subfolders, and files.
- The C: drive represents the hard drive and is where most programs and files are stored.
- File Explorer is the main tool for finding, viewing, and managing the contents of your computer.
- File Explorer helps you manage your files and folders by showing the location and contents of every drive, folder, and file on your computer.
- There are specific rules to follow when naming files.
- The Recycle Bin is the temporary storage location for files deleted from the hard drive.
- File compression reduces the size of a file by temporarily storing components of a file, and then when the file is extracted (uncompressed), the removed components are brought back into the file.

Utility Programs

Objective 5.10 *Outline the tools used to enhance system productivity, back up files, and provide accessibility.*

- Task Manager is used to exit unresponsive programs.
- Disk Cleanup removes unnecessary files from your hard drive. If not deleted periodically, these unnecessary files can slow down your computer.
- Disk defragmentation utilities rearrange fragmented data so that related file pieces are unified. When a disk is fragmented, it can slow down your computer.
- File History automatically creates a duplicate of your hard drive (or parts of your hard drive) and copies it to another storage device, such as an external hard drive. You can use File History to recover deleted or corrupted files.
- System Restore enables you to roll your system settings back to a specific date (restore point) when everything was working properly.
- Windows Ease of Access settings include tools that help adjust computer settings for users with disabilities.

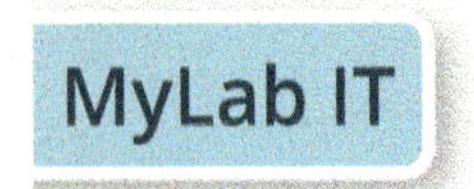

Be sure to check out **MyLab IT** for additional materials to help you review and learn, and don't forget to watch the Chapter Overview Videos.

Key Terms

Chapter Quiz //Assessment

For a quick review to see what you've learned, answer the following questions. Submit the quiz as requested by your instructor. If you are using **MyLab IT**, the quiz is also available there.

multiple choice

1. Which of the following would you NOT see on a Windows 10 Start menu?

a. Task View
b. Power
c. Apps list
d. Tiles

2. When an OS processes tasks in a priority order, it is known as

a. preemptive interrupting.
b. interruptive multitasking.
c. multitasking handling.
d. preemptive multitasking.

3. An example of an open source OS is
   a. Linux.
   b. macOS.
   c. Windows.
   d. DOS.
4. Which of the following is not considered an accessibility utility?
   a. Magnifier
   b. System Restore
   c. Narrator
   d. Speech Recognition
5. Which step happens first during the boot process?
   a. Customization settings checked
   b. POST
   c. Load OS to RAM
   d. CPU activates the BIOS that is stored in ROM
6. Device drivers are
   a. no longer used because of Plug and Play.
   b. often bundled with the operating system.
   c. always installed from a CD.
   d. mapping programs to generate driving instructions.
7. Which of the following is an example of a command-driven interface?
   a. Windows 10
   b. MS-DOS
   c. macOS
   d. iOS
8. A feature in Windows 10 that is used to organize open windows into task-specific groups is called
   a. Program Manager.
   b. Virtual View.
   c. Snap Assist.
   d. Virtual Desktops.
9. When a printer receives a command, it generates a unique signal to the OS, which is called a(n)
   a. interrupt.
   b. event.
   c. slice.
   d. none of the above.
10. C:\ on a Windows PC is like a large filing cabinet and is referred to as the
   a. main directory.
   b. root directory.
   c. library directory.
   d. main path.

true/false

_____ 1. Different versions of Linux are known as distros.
_____ 2. Swapping data and instructions between RAM and the hard drive's swap file is called paging.
_____ 3. Operating systems like Windows and macOS are single-user, single-task OSs.
_____ 4. System restore points can only be created by Windows automatically on a regular schedule.
_____ 5. Once files are placed in the Recycle Bin, they cannot be recovered.

What do you think now?

1. Is your data safe enough if law enforcement agencies can have access to it?
2. Should the government be able to demand that tech companies create algorithms that engineers themselves are opposed to creating?

Team Time

Choosing the Best OS

problem

You're the owner of a technology consulting firm. Your current assignments include advising start-up clients on their technology requirements. The companies include a nonprofit social service organization, a small interior design firm, and a social media advertising agency. One of the critical decisions for each company is the choice of OS.

task

Recommend the appropriate OS for each company.

process

1. Break up into teams that represent the three primary operating systems: Windows, macOS, or Linux.
2. As a team, research the pros and cons of your OS. What features does it have that would benefit each company? What features does it not have that each company would need? Discuss why your OS would be the appropriate (or inappropriate) choice for each company.
3. Develop a presentation that states your position with regard to your OS. Your presentation should have a recommendation and include facts to back it up.
4. As a class, decide which OS would be the best choice for each company.

conclusion

Because the OS is the most critical piece of software in the computer system, the selection should not be taken lightly. The OS that is best for a nonprofit social service organization may not be best for an interior design firm. A social media advertising agency may have different needs altogether. To ensure a good fit, it's important to make sure you consider all aspects of the work environment and the type of work being done.

Ethics Project

Upgrade Your World

In this exercise, you'll research and then role-play a complicated ethical situation. The role you play may or may not match your own personal beliefs, but your research and use of logic will enable you to represent whichever view is assigned.

problem

With the release of Windows 10, Microsoft introduced the Upgrade Your World project, to work with 10 global nonprofit groups and 100 local nonprofits (10 in 10 countries). The organizations were selected because they addressed a key societal issue and had demonstrated a consistent record of creating meaningful change in their community and the world. Each local nonprofit received a $50,000 cash investment plus technology from Microsoft. The program has ended, but Microsoft continues to support nonprofits through its corporate social responsibility efforts.

Which nonprofit organization in your area would you choose as a candidate for support from Microsoft and why?

research areas to consider

- Windows 10 accessibility features
- Microsoft philanthropy
- Tech for Social Impact

process

1. Divide the class into teams.
2. Research the areas cited above and determine a local nonprofit that you think would be deserving of an Upgrade Your World grant.
3. Team members should write a summary that provides background information for their nonprofit suggestion.
4. The team members will present their case to the class or submit a PowerPoint presentation for review by the rest of the class, along with the summary and resolution they developed. The class should determine the pros and cons of awarding the local nonprofit nominee the grant and reach an agreeable conclusion about which nonprofit would be the best choice to suggest to Microsoft.

conclusion

All nonprofits serve need, but some may benefit more than others from a technology grant. Being able to understand and evaluate a decision, while responding in a personally or socially ethical manner, will be an important skill.

Solve This with Excel MyLab IT Grader

Operating Systems: Analyzing Market Share

As a sector analyst for one of the regional banks, you are responsible for reviewing and analyzing data to determine trends. Bankers use your analyses to inform prospective and current clients of possible investment opportunities. Currently, you are looking at the market trends of operating systems and have obtained historical market share data for mobile/tablet operating systems in the United States, global market share data for desktop and table operating systems, and market share data for all operating systems for select countries around the world. You will chart the data to determine trends.

You will use the following skills as you complete this activity:

- Create an Area Chart
- Create a Map Chart
- Create Sparklines
- Add a Shape to a Chart
- Create a 2-D Pie Chart
- Apply Formats and Filters to Charts

Instructions

1. Open *TIA_Ch5_Start.xlsx* and save as **TIA_Ch5_LastFirst.xlsx**.
2. Create a 2-D Area chart, using data in the A3:G7 range.
   Hint: Charts are on the Insert tab in the Chart group. Click Insert Line or Area Chart.
3. Modify the chart as follows:
   a. Position the chart so it fills cells **J2:S19**.
   b. Add a title: **Change in Mobile OS Market Share 2013–2018**. Place "2013–2018" on a separate line.
   Hint: Place insertion point before 2013 and press Shift + Enter to place dates on separate line.
   c. Change the font size of 2013–2018 to **10**.
   d. Filter out the **Others** and **Windows** data so only Android and iOS data appears.
   Hint: Use the Chart Filters icon (the Funnel) on the right of the chart. Clear the Windows and Others check box. Click Apply.
4. Add a **Callout: Line shape** to the chart with the line pointing at the top of the orange area between 2014 and 2015.
   Hint: Callout: Line is in the Callouts section in the Insert Shape group on the Format Chart Tools tab.
5. Add text to the callout: **Android Market Share Takes Off!**
6. Format the callout with **Subtle Effect - Gold, Accent 4 Shape Style.**
7. Create a **2D pie chart,** using the **A15:A21** and **G15:G21** ranges.
   Hint: Select A15:A21 and then press the **Ctrl key** while selecting G15:G21.
8. Modify the chart as follows:
   a. Add a title: **Operating System 2018 Global Market Share.**
   b. Use **Quick Layout 1** to add data % and Series data labels to each data point.
   Hint: Quick Layouts are on the Chart Tools Design tab in the Chart Layouts group.
   c. Move the chart so it is on its own worksheet. Label the worksheet **Global OS.**
   Hint: Use Move Chart on the Chart Tools Design Tab. Do not copy and paste.
9. On the Data worksheet, add a **Column Sparkline** to cell range **H16:H21**, using the data in the **B16:G21** range. Increase the width of column H to **12**.
10. Create a Filled Map Chart from the data in ranges **A26:A39** and **D26:D39**. Place the chart to fill the **I26:S45** range. Add a title: **2018 Android Market Share by Country**.
    a. Click **Chart Elements** and add **Data Labels**.
11. Save the document and submit it, based on your instructor's directions.

Chapter 6

Understanding and Assessing Hardware: Evaluating Your System

 For a chapter overview, watch the **Chapter Overview Videos**.

PART 1

Evaluating Key Subsystems

Learning Outcome 6.1 **You will be able to evaluate your computer system's hardware functioning, including the CPU and memory subsystems.**

PART 2

Evaluating Other Subsystems and Making a Decision

Learning Outcome 6.2 **You will be able to evaluate your computer system's storage subsystem, media subsystem, and reliability and decide whether to purchase a new system or upgrade an existing one.**

MyLab IT All media accompanying this chapter can be found here.

Make This A Location-Aware App on **page 217**

(Wavebreak Media Ltd/123RF; Crstrbrt/123RF; Joppo/Shutterstock; Aeksanderdn/123RF; Leigh Prather/Shutterstock; Ivelin Radkov/Alamy Stock Photo)

What do you think?

As **self-driving cars** become more commonplace, society may face real **ethical dilemmas** about how the **artificial intelligence (AI)** in these vehicles responds to **emergency situations**. Suppose a car comes over a hill and faces a stopped school bus. What should the AI do? Swerve into the woods to the right side to avoid the bus but thereby endanger the occupants of the car? Collide with the bus since the danger to the car occupants is minimal in a rear-end collision? Swerve left into oncoming traffic to avoid the bus, but possibly causing a fatal head-on collision with another vehicle? Who assumes **liability** in case of an accident?

Should cars always be programmed to minimize loss of human life, or should they always afford maximum protection for their occupants? What are the ethical implications of self-driving cars?

?

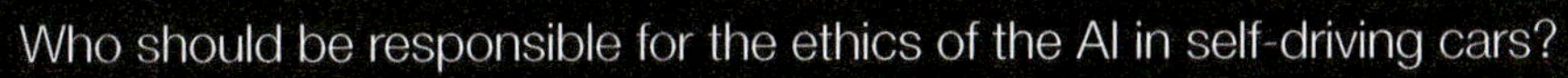

Who should be responsible for the ethics of the AI in self-driving cars?

- *Manufacturers best understand the cars so they should assume responsibility.*
- *Consumers should be able to choose the ethical standards of the AIs in their vehicles.*
- *A government agency should be responsible for setting standards to ensure fairness.*

See the end of the chapter for a follow-up question.

(Rico Ploeg/Alamy Stock Photo)

Part 1

For an overview of this part of the chapter, watch **Chapter Overview Video 6.1.**

Evaluating Key Subsystems

Learning Outcome 6.1 You will be able to evaluate your computer system's hardware functioning, including the CPU and memory subsystems.

It can be tough to know whether your computer is the best match for your needs. New technologies emerge so quickly, and it's hard to determine whether they're expensive extras or tools you need. Do you need Thunderbolt instead of USB? Doesn't it always seem as though your friend's computer is faster than yours? Maybe you could get more out of newer technologies, but should you upgrade the system you have or buy a new machine? In this chapter, you learn how to measure your system's performance and gauge your needs so that you end up with a system you love.

Your Ideal Computing Device

There never seems to be a perfect time to buy a new computer. It seems that if you can just wait a year, computers will be faster and cost less. But is this actually true?

Moore's Law

Objective 6.1 *Describe the changes in CPU performance over the past several decades.*

How quickly does computer performance improve? As it turns out, it is true that if you wait just a while, computers will be faster and cost less. In fact, a rule of thumb often cited in the computer industry, called Moore's Law, describes the pace at which central processing units (CPUs) improve. Named for Gordon Moore, the cofounder of the CPU chip manufacturer Intel, this rule predicts that the number of transistors inside a CPU will increase so fast that CPU processing power will double about every two years. (The number of transistors on a CPU chip helps determine how fast it can process data.)

This rule of thumb has held true for about 60 years. Figure 6.1 shows a way to visualize this kind of exponential growth. If CPU capacity were put into terms of population growth, a group of 2,300 people at the start of CPU development would now be a country of over 1 billion!

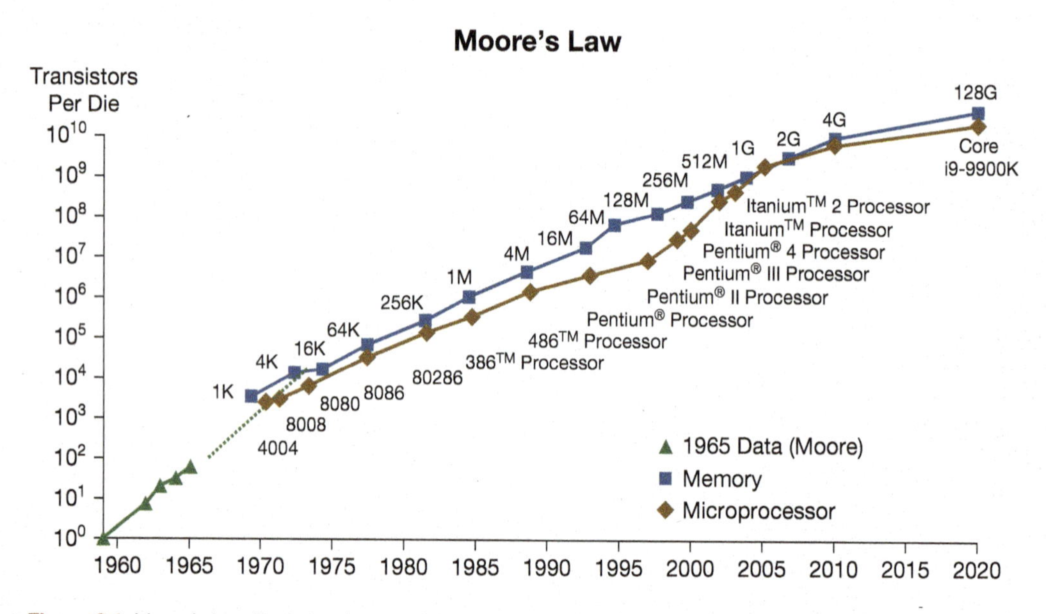

Figure 6.1 Moore's Law illustrates the amazing pace of growth in CPU capabilities.

In addition to the CPU becoming faster, other system components also continue to improve dramatically. For example, the capacity of memory chips such as dynamic random access memory (DRAM)—the most common form of memory found in personal computers—increases about 60% every year. Meanwhile, hard drives have been growing in storage capacity by some 50% each year.

Will Moore's Law always be true? Probably not. Moore himself has predicted that in the future, CPU chips will be manufactured in a different way, thus changing or eliminating the effects of Moore's Law altogether. The scale to which reliable functioning transistors can be produced with silicon has limits (both functional and monetary). Intel is currently producing chips with features just 10 nanometers (nm) wide. Because the company had difficulty rolling out 10 nm chips, it is questionable whether it can go any smaller. Intel and other chip companies are investigating alternative elements for chips such as carbon (in both nanotube and graphene form), indium antimonide (InSb), and indium gallium arsenide (InGaAs). The main advantage with these substances is that they offer higher switching speeds than silicon, at lower power consumption rates. Therefore, the goal may shift toward better-performing transistors rather than sheer quantity.

With technology demands switching to mobile devices and the Internet of Things, the goals of computing chip design have also changed. The drive for conventional CPUs was all about increasing computing capacity. Now chip designers are more concerned about low-power processors (to conserve battery power) and integrating components such as RAM, cellular, Wi-Fi, and sensors (such as accelerometers) into chips. So, although Moore's Law isn't obsolete, the way the industry looks at increases in computing power and functionality of chips is changing.

Selecting a Computing Device

Objective 6.2 *Compare and contrast a variety of computing devices.*

OK, things change fast. How do I know which device is best for me? Consider what kind of user you are and what needs you have. For example, are you a power user who wants a machine for doing video editing and high-end gaming? Or are you a more casual user, mainly using a device for word processing and Internet access? Or are you on the move and need to bring your computer with you everywhere? Figure 6.2 shows a few different types of users—which type (or types) are you?

Now ask yourself, does your current computer match your needs? As we evaluate the pieces of your system in this chapter, it'll become clear whether your current device will need a few upgrades or whether you may need to acquire a new machine.

What are the main types of devices available? As we discussed in Chapter 2, a huge number of choices are on the market (see Figure 6.3):

- Smartphones
- Tablets (like the iPad or Galaxy)
- Ultrabooks (like the MacBook Air)
- 2-in-1s (which can serve as a tablet but also have a full keyboard)
- Laptops (or notebooks)
- Desktops

The main distinction among the available options is based on your need for mobility versus your need for processing power. If you're on the move all the time, don't use keyboard-intensive apps (like word processing), just need to check e-mail and the Internet, and want to have the lightest solution possible, a smartphone may be best for you. But if you use productivity software and need the convenience of a larger screen and a physical keyboard, you might want an ultrabook. At less than three pounds, they're great on weight but don't include as much storage space as conventional laptops. Even lighter are tablets like the iPad, but they have less processing power and, again, may not be able to run all the software you need.

Figure 6.2 What Kind of Technology User Are You?

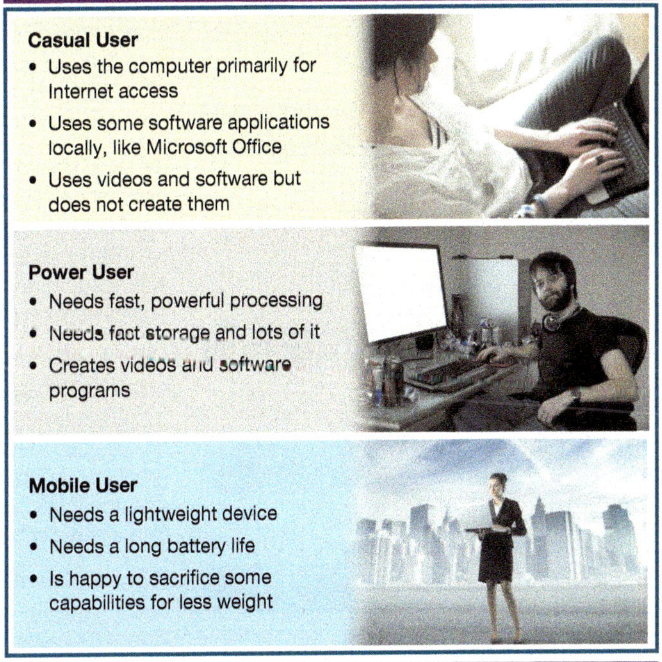

Casual User
- Uses the computer primarily for Internet access
- Uses some software applications locally, like Microsoft Office
- Uses videos and software but does not create them

Power User
- Needs fast, powerful processing
- Needs fast storage and lots of it
- Creates videos and software programs

Mobile User
- Needs a lightweight device
- Needs a long battery life
- Is happy to sacrifice some capabilities for less weight

(Laszlo/Shutterstock, ColorBlind Images/The Image Bank/Getty Images, Ollyy/Shutterstock)

Figure 6.3 Range of Computing Devices

(Dmitry Lobanov/123RF; Pedrosek/123RF; Aleksanderdn/123RF; Stanisic Vladimir/123RF; Olegdudko/123RF; Kaspars Grinvalds/123RF)

Why would anyone consider buying a desktop? Desktop systems are invariably a better value than lighter, more mobile computers. You'll find you get more computing power for your dollar, and you'll have more opportunity to upgrade parts of your system later. In addition, desktops often ship with a 24-inch or larger monitor, whereas portable computers offer screens between 10 and 17 inches.

Figure 6.4 Computer Subsystems

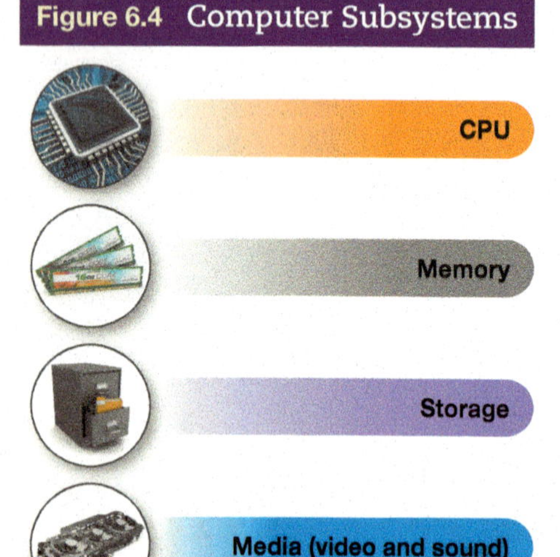

((Mmaxer/Shutterstock; Cigdem/Shutterstock; Maxim Kazmin/123RF; Billion Photos/Shutterstock)

Desktop systems are also more reliable. Because of the vibration that a mobile device experiences and the added exposure to dust, water, and temperature fluctuations that portability brings, mobile devices often have a shorter lifespan than stationary computers. You'll have less worry over theft or loss with a desktop, too. Manufacturers do offer extended warranty plans that cover mobile devices for accidental damage and theft; however, such plans can be costly.

How long should I plan to keep my computing device? You should be able to count on two years, and maybe even four or five years. The answer depends in part on how easy it is to upgrade your system. Take note of the maximum amount of memory you can install in your device. And check whether you can upgrade your device's graphics capabilities down the road. In this chapter, we focus primarily on laptops and desktops, because they provide you with the most options for upgrading a current device without replacing it.

How do I evaluate the performance of my current device? We'll begin by conducting a system evaluation. To do this, we'll look at your computer's subsystems (see Figure 6.4), see what they do, and check how they perform during your typical workday. We'll compare that with what is available on the market, and the path forward for you will become clearer. Even if you're not in the market for a new computer, conducting a system evaluation will help you understand what you might want down the road.

Evaluating the CPU Subsystem

Figure 6.5 Intel CPU chips run many of the laptop and desktop offerings on the market today. *(David Caudery/PC Format Magazine/Future Publishing/Getty Images)*

Let's start by considering your system's processor, or CPU. The CPU is located on the system motherboard and is responsible for processing instructions, performing calculations, and managing the flow of information through your computer. The dominant processors on the market are the Core family from Intel, featuring the i9, i7, and i5 (see Figure 6.5).

How the CPU Works

Objective 6.3 *Describe how a CPU is designed and how it operates.*

How can I find out what CPU my computer has? If you're running Windows, the System About window will show you the type of CPU you have. For example, the computer in Figure 6.6 has an Intel i5 CPU running at 2.6 GHz. AMD is another popular manufacturer of CPUs; you may have one of its processors, such as the Ryzen or the FX series. More detailed information about your CPU, such as its number of cores and amount of cache memory, is not shown on this screen. Let's dive into that.

How does the CPU process data? Any program you run on your computer is a long series of binary code describing a specific set of commands the CPU must perform. These commands may be coming from a user's actions or may be instructions fed from a program while it executes. Each CPU is somewhat different in the exact steps it follows to perform its tasks, but all CPUs must perform a series of similar general steps. Every time the CPU performs a program instruction, it goes through this series of steps:

1. **Fetch.** When any program begins to run, the 1s and 0s that make up the program's binary code must be fetched from their temporary storage location in random access memory (RAM) and moved to the CPU before they can be executed.

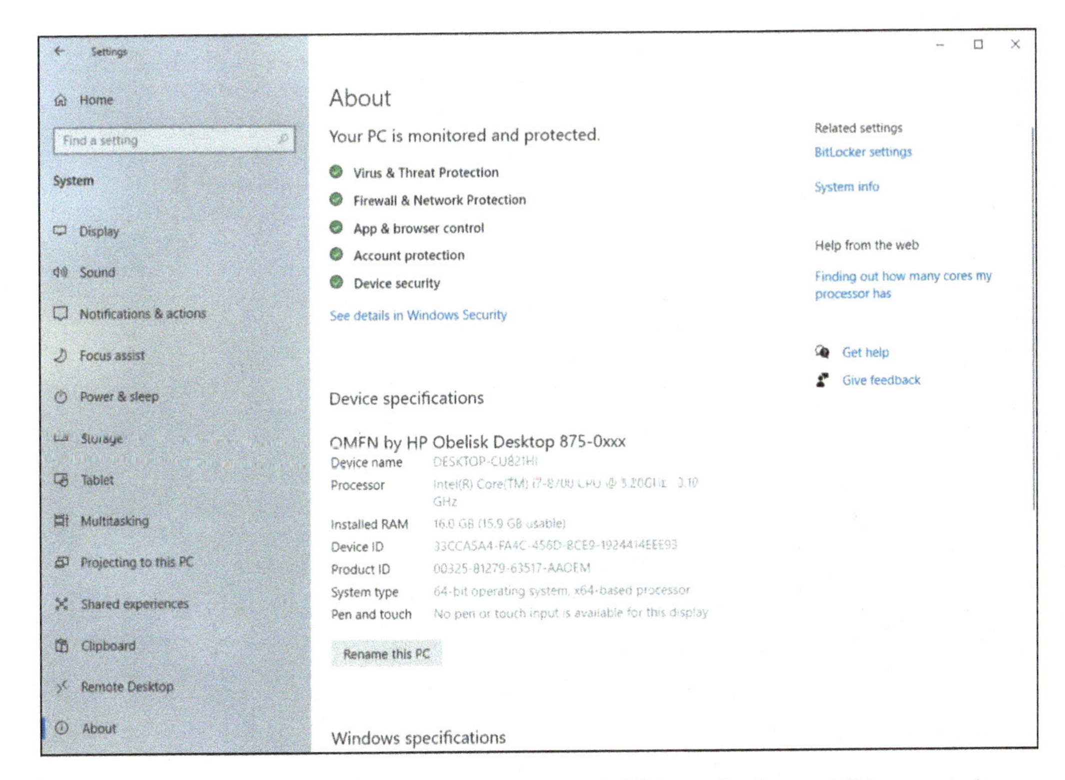

Figure 6.6 The System About window identifies your computer's CPU as well as its speed. This computer has an Intel i5 running at 2.6 GHz. *(Courtesy of Microsoft Corp.)*

> *To view the system window, right-click the* ***Start*** *button. From the menu that appears, choose* ***System****.*

2. **Decode.** Once the program's binary code is in the CPU, it is decoded into commands the CPU understands.
3. **Execute.** Next, the CPU performs the work described in the commands. Specialized hardware on the CPU performs addition, subtraction, multiplication, division, and other mathematical and logical operations.
4. **Store.** The result is stored in one of the registers, special memory storage areas built into the CPU, which are the most expensive, fastest memory in your computer. The CPU is then ready to fetch the next set of bits encoding the next instruction.

This process is called a machine cycle (see Figure 6.7). No matter what program you're running and no matter how many programs you're using at one time, the CPU performs these four steps over and over at incredibly high speeds.

What are the main components in a CPU? The CPU is composed of two units: the arithmetic logic unit (ALU) and the control unit. The ALU is responsible for performing all the arithmetic calculations (addition, subtraction, multiplication, and division). It also makes logic and comparison decisions, such as comparing items to determine whether one is greater than, less than, or equal to another.

The control unit of the CPU manages the switches inside the CPU. The CPU, like any part of the computer system, is designed from a collection of switches. How can the simple on/off switches of the CPU remember the fetch–decode–execute–store sequence of the machine cycle? It is programmed by CPU designers to remember the sequence of processing stages for that CPU and how each switch in the CPU should be set (i.e., on or off) for each stage. With each beat of the system clock, the control unit moves each switch to the correct on or off setting and then performs the work of that stage.

To move from one stage of the machine cycle to the next, the motherboard uses a built-in system clock. This internal clock is a special crystal that acts like a metronome, keeping a steady beat and controlling when the CPU moves to the next stage of processing.

These steady beats, or ticks, of the system clock, known as the clock cycle, set the pace by which the computer moves from process to process. The pace, known as clock speed, is measured in hertz (Hz), which describes how many times something happens per second. Today's system clocks are measured in gigahertz (GHz), or a billion clock ticks per second. Therefore, in a 3-GHz system, there are three billion clock ticks each second.

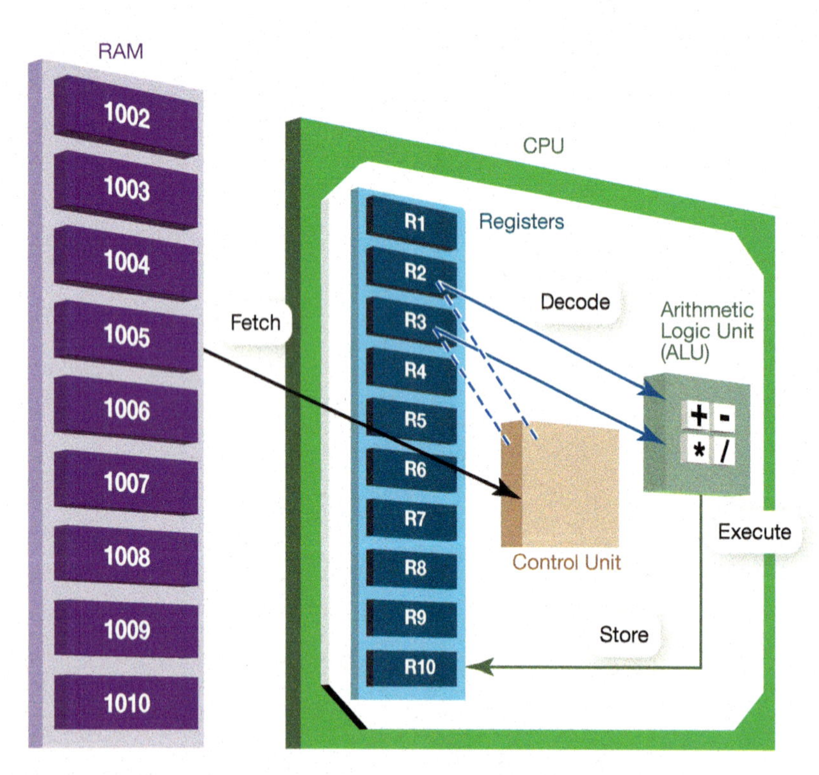

Figure 6.7 The CPU Machine Cycle

What makes one CPU different from another? You pay more for a computer with an Intel i9 than one with an Intel i7 because of its increased processing power. A CPU's processing power is mainly determined by the following:

- Its clock speed
- How many cores it has
- Its amount of cache memory

How does a CPU with a higher clock speed help me? The faster the clock speed, the more quickly the next instruction is processed. CPUs currently have clock speeds of up to 4 GHz (or more).

Some users push their hardware to perform faster, overclocking their processor. Overclocking means that you run the CPU at a faster speed than the manufacturer recommends. It produces more heat, meaning a shorter lifespan for the CPU, and usually voids any warranty, but in gaming systems, you'll see this done quite often.

How does increasing the number of cores in a CPU help me? A core on a CPU contains the parts of the CPU required for processing an instruction. With multiple-core technology, two or more complete processors live on the same chip, enabling the independent execution of two or more sets of instructions at the same time.

If you had clones of yourself sitting next to you working, you could get twice as much (or more) done: That is the idea of multi-core processing. With multi-core processing, applications that are always running behind the scenes, such as virus protection software and your operating system (OS), can have their own dedicated processor, freeing the other processors to run other applications more efficiently. This results in faster processing and smoother multitasking. Chips with quad-core processing capabilities (see Figure 6.8) have four separate parallel processing paths inside them, so they're almost as fast as four separate CPUs. It's not quite four times as fast because the system must do some extra work to decide which processor will work on which part of the problem and recombine the results each CPU produces.

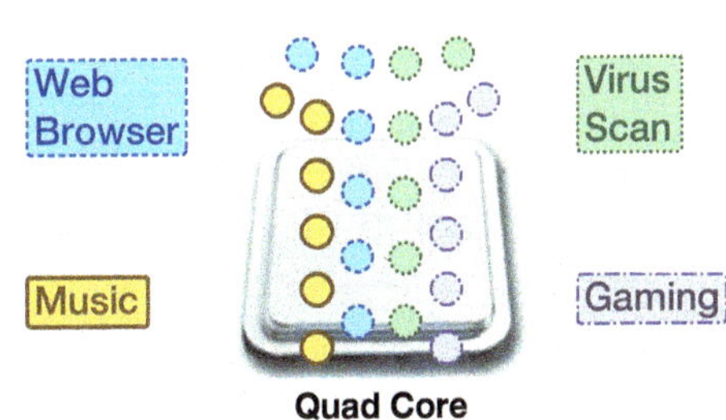

Figure 6.8 Intel Quad Core processors have four cores that can run four programs simultaneously.

Twenty-core processors, like the Intel XEON E5-2698V4, are now available and feature 20 separate processing paths. Multi-processor systems are often used when intensive computational problems need to be solved in such areas as computer simulations, video production, and graphics processing. Extreme gamers also use them in home computers.

Are multi-core CPUs the only way to handle simultaneous processing? Certain types of problems are well suited to a parallel-processing environment, although this approach is not used in personal computing. In parallel processing, there is a large network of computers, with each computer working on a portion of the same problem simultaneously. To be a good candidate for parallel processing, a problem must be able to be divided into a set of tasks that can be run simultaneously. For example, a problem in which millions of faces are being compared with a target image for recognition is easily adapted to a parallel setting. The target face can be compared with many hundreds of faces at the same time. However, if the next step of an algorithm can be started only after the results of the previous step have been computed, parallel processing will present no advantages.

What other factors can affect processing speed? The CPU's cache memory is a form of RAM that gets data to the CPU for processing much faster than bringing the data in from the computer's RAM. There are three levels of cache memory, defined by their proximity to the CPU:

- Level 1 cache is a block of memory built on the CPU chip itself for storage of data or commands that have just been used. That gets the data to the CPU blindingly fast!
- Level 2 cache is located on the CPU chip but is slightly farther away and so takes somewhat longer to access than Level 1 cache. It contains more storage area than Level 1 cache.
- Level 3 cache is also located on the CPU chip itself but is slower to reach and larger in size than Level 2 cache.

Generally, the more expensive the CPU, the more cache memory it will have. As an end user of computer programs, you do nothing special to use cache memory. Unfortunately, because it's built into the CPU chip or motherboard, you can't upgrade cache; it's part of the original design of the CPU. Therefore, as with RAM, when buying a computer, it's important to consider the one with tho most cache memory, everything else being equal.

Besides multiple cores and cache memory, what else can be done to increase the processing power of a CPU? As an instruction is processed, the CPU runs sequentially through the four stages of processing: fetch, decode, execute, and store. Pipelining is a technique that allows the CPU to work on more than one instruction (or stage of processing) at the same time, thereby boosting CPU performance.

For example, without pipelining, it may take four clock cycles to complete one instruction (one clock cycle for each of the four processing stages). However, with a four-stage pipeline, the computer can process four instructions at the same time. The ticks of the system clock (the clock cycle)

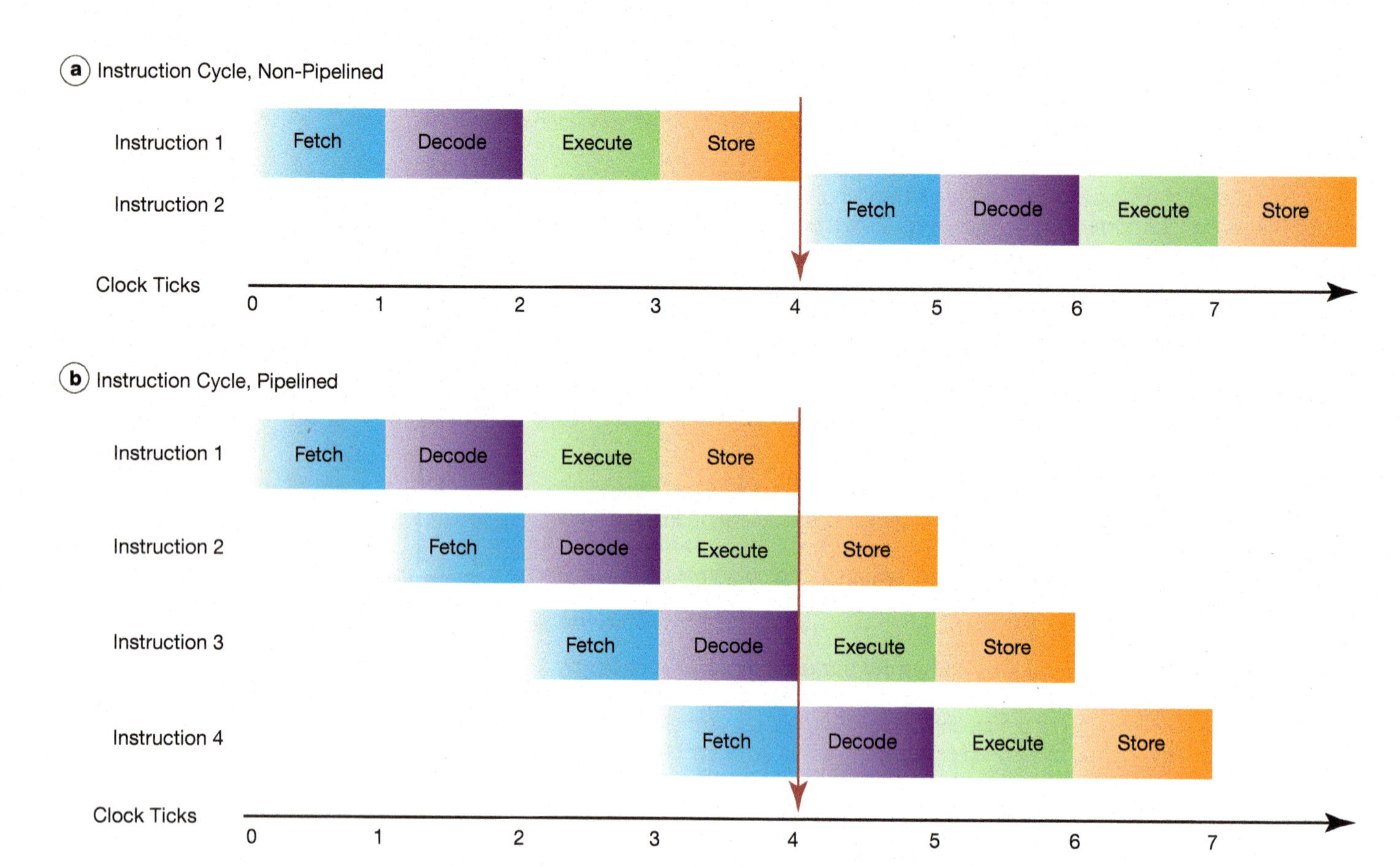

Figure 6.9 Instead of (a) waiting for each instruction to complete, (b) pipelining allows the system to work on more than one set of instructions at a time.

indicate when all instructions move to the next process. By using pipelining, a four-stage processor can potentially run up to four times faster because some instruction is finishing every clock cycle rather than waiting four cycles for each instruction to finish. In Figure 6.9a, a non-pipelined instruction takes four clock cycles to be completed, whereas in Figure 6.9b, the four instructions have been completed in the same time by using pipelining.

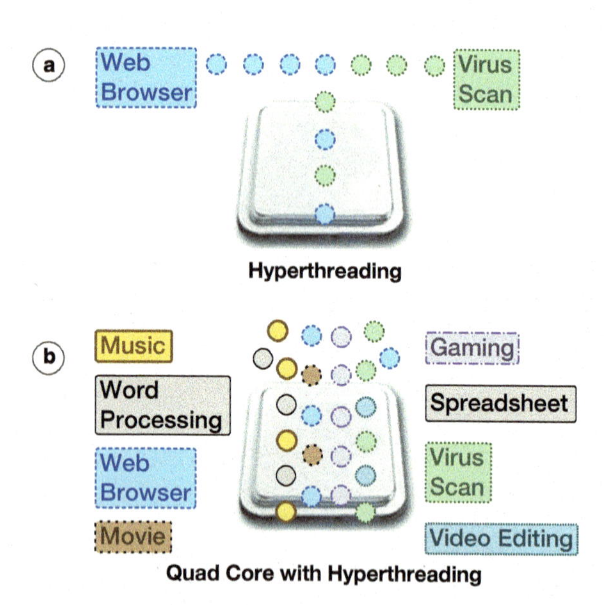

Figure 6.10 (a) Hyperthreading allows work on two processes to happen in one core at the same time, (b) so a four-core, hyperthreaded processor can be working on eight programs at one time.

There is a cost to pipelining a CPU, however. The CPU must be designed so that each stage (fetch, decode, execute, and store) is independent. This means that each stage must be able to run while the other three stages are running. This requires more transistors and a more complicated hardware design.

CPUs began to expand their processing power further when hyperthreading was introduced. Hyperthreading provides quicker processing of information by enabling a new set of instructions in a different program (or thread) to start executing before the previous set has finished. This process takes advantage of certain delays that occur in reading and writing memory to registers with pipelining. As shown in Figure 6.10a, hyperthreading allows two programs to be processed at one time, but they're sharing the computing resources of the chip.

All of the Intel Core processors have multiple cores *and* hyperthreading (see Figure 6.10b). The XEON E5-2698V4 has 20 cores, each one using hyperthreading, so it simulates having 40 processors!

Measuring CPU Performance

Objective 6.4 *Describe tools used to measure and evaluate CPU performance.*

So how do I compare different CPUs? Intel has differentiated its CPUs using generations. The highest generation number indicates the newest (and presumably most powerful) version of that CPU. Picking the best

Bits&Bytes Liquid Cooling

A critical aspect of the design of a computer system is to manage how the heat will be removed. One approach is to water-cool the system (see Figure 6.11). A tube containing liquid is placed in contact with heat-producing parts of the system, such as the CPU. The liquid picks up heat and carries it to a radiator, just as in a car. A fan blows across the fins of the radiator and efficiently disperses the heat. High-performance computers like the Aventum from Digital Storm feature complex water-cooling systems because of the massive amount of heat they generate.

Want to try water cooling in a computer you're building but nervous about water inside your system? Sealed water-cooling solutions are also available so you never have to pour water near the inside of your computer.

Figure 6.11 PC with Liquid Cooling System *(FeelGoodLuck/Shutterstock)*

CPU for the kind of work you do is easier if you research some performance benchmarks. CPU benchmarks are measurements used to compare performance between processors. Benchmarks are generated by running software programs specifically designed to push the limits of CPU performance. Articles are often published comparing CPUs, or complete systems, based on their benchmark performance. Investigate a few, using sites like *cpubenchmark.net*, before you select the chip that's best for you.

How can I tell whether my current CPU is meeting my needs? One way to determine whether your CPU is right for you is to watch how busy it is as you work. You can do this by checking out your CPU usage—the percentage of time your CPU is working.

Your computer's OS has utilities that measure CPU usage. These are incredibly useful, both for considering whether you should upgrade and for investigating whether your computer's performance suddenly seems to drop off for no apparent reason.

On Windows systems, use the Task Manager utility to access this data (see Figure 6.12). The CPU usage graph records your CPU usage for the past minute. (Note that if you have multiple cores and hyperthreading, you'll see only one physical processor listed, but it will show that you have several virtual processors.) Of course, there will be periodic peaks of high CPU usage, but if your CPU usage levels are greater than 90% during most of your work session, a faster CPU will contribute a great deal to your system's performance.

To walk through using the Task Manager, check out the Try This later in this chapter; macOS has a similar utility, named Activity Monitor, in the Utilities folder in the Applications subfolder.

How often should I watch the CPU load? Keep in mind that the workload your CPU experiences depends on how many programs are running at one time. Even though the CPU may meet the specs for each program separately, how you use your machine during a typical day may tax the CPU. If you're having slow response times or decide to measure CPU performance as a check, open the Task Manager and leave it open for a full day. Check in at points when you have many open windows and when you have a lot of networking or disk-usage demand.

So a better CPU means a better-performing system? Your CPU affects only the processing portion of the system performance, not how quickly data can move to or from the CPU. Your system's overall performance depends on many factors, including the amount of RAM installed as well as hard drive speed. Your selection of CPU may not offer significant improvements to your system's performance if there is a bottleneck in processing because of insufficient RAM or hard drive performance, so you need to make sure the system is designed in a balanced way. Figure 6.13 lists factors to consider as you decide which specific CPU is right for you.

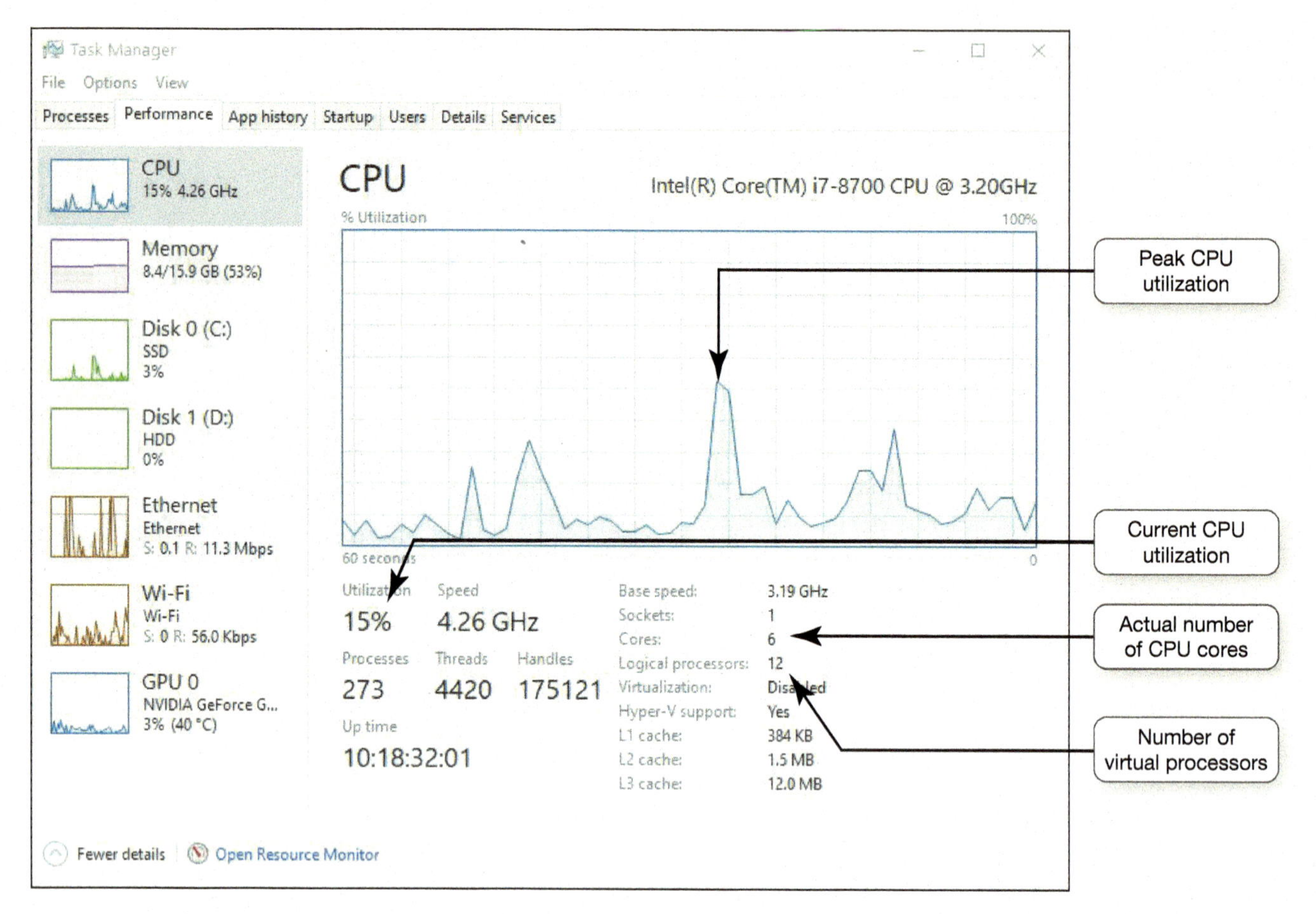

Figure 6.12 The Performance tab of the Windows Task Manager utility shows you how busy your CPU is. *(Courtesy of Microsoft Corp.)*

> *To access the Performance tab, right-click the* ***Start*** *button. From the menu, select* ***Task Manager****, click* ***more details*** *(if necessary), and then click the* ***Performance*** *tab.*

Figure 6.13 Evaluating the CPU

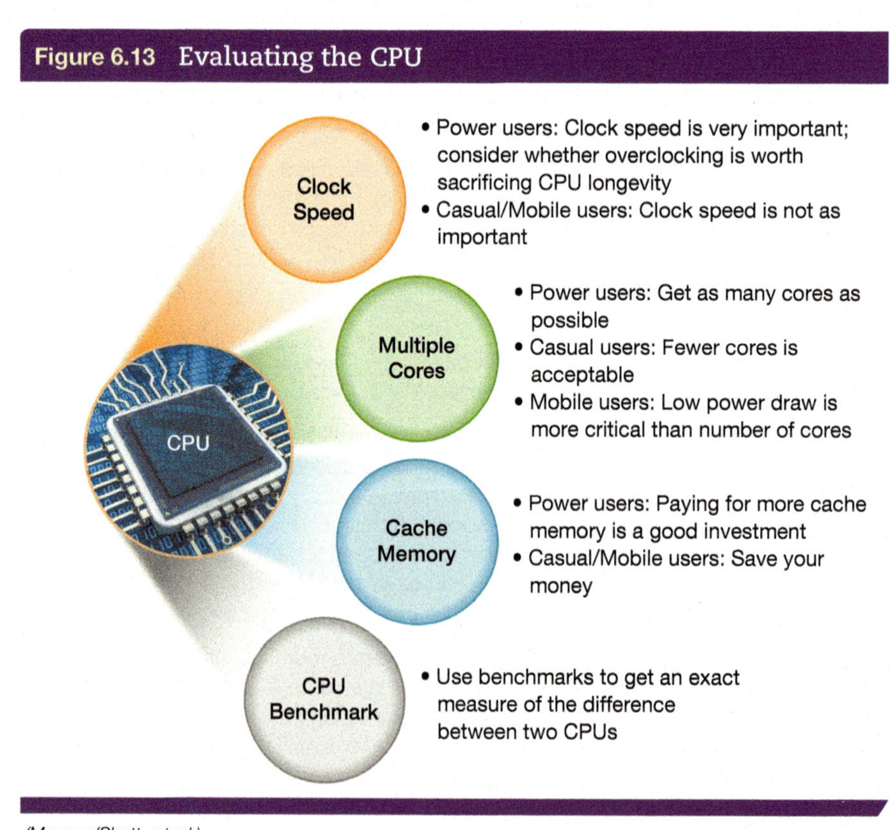

(Mmaxer/Shutterstock)

Evaluating the Memory Subsystem

Let's now take a peek at the memory subsystem of your computer. The memory subsystem can have a terrific impact on your system's processing speed if it is well matched to the power of your CPU. In this section, we look at how to measure that and how to upgrade if you need to.

Random Access Memory

Objective 6.5 *Discuss how RAM is used in a computer system.*

What is RAM? Random access memory (RAM) is your computer's temporary storage space. It really is the computer's short-term memory. When the computer is running, RAM remembers everything that the computer needs to process data, such as data that has been entered and software instructions. But when the power is off, the data stored in RAM disappears. So, RAM is an example of volatile storage. Therefore, systems always include both RAM and nonvolatile storage devices for permanent storage of instructions and data. For example, read-only memory (ROM) is a type of nonvolatile storage that holds the critical startup instructions. Hard drives provide the largest nonvolatile storage capacity in the computer system in laptops and desktops. In mobile devices that lack hard drives, memory chips function as nonvolatile storage.

Why not use a hard drive to store the data and instructions? It's about 1 million times faster for the CPU to retrieve a piece of data from RAM than from a mechanical hard drive. The time it takes the CPU to grab data from RAM is measured in nanoseconds (billionths of seconds), whereas pulling data from a fast mechanical hard drive takes an average of 10 milliseconds (ms), or thousandths of seconds. Even if you are using a solid-state hard drive (SSD), which is about eight times faster than a conventional hard drive, it is still much faster to retrieve data from RAM.

Figure 6.14 shows the various types of memory and storage distributed throughout your system: memory that is part of the CPU (such as CPU registers and cache), RAM, virtual memory, SSDs, and mechanical hard drives. Each of these has its own tradeoff of speed versus price. Because the fastest memory is so much more expensive, systems are designed with much less of it. This principle is influential in the design of a balanced computer system and can have a tremendous impact on system performance.

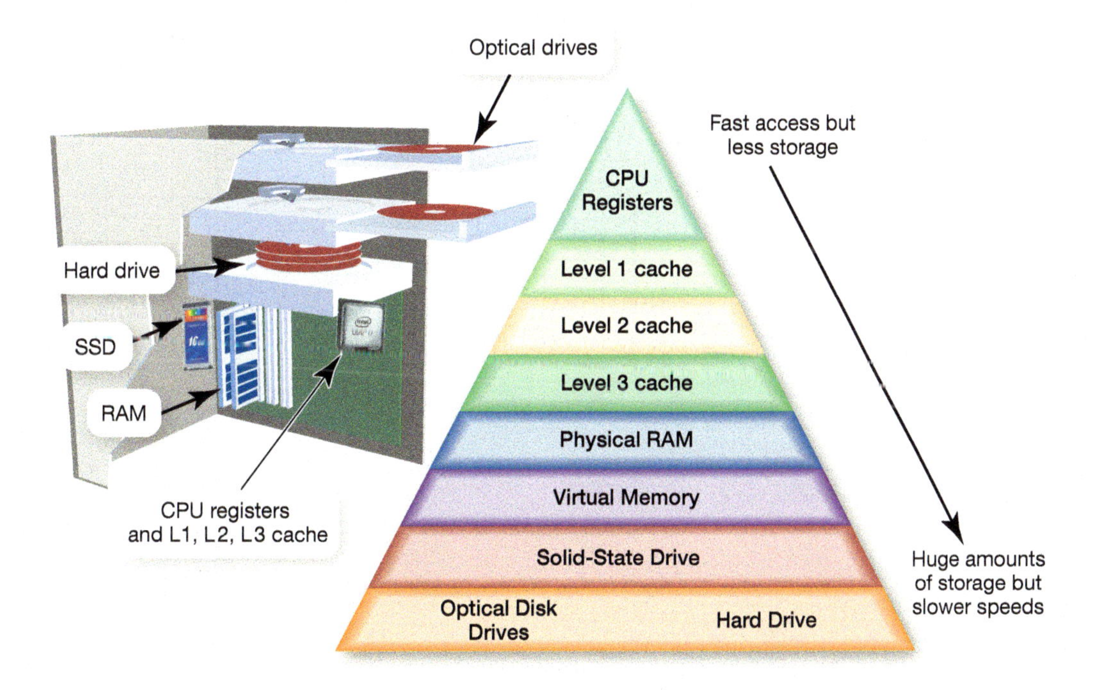

Figure 6.14 A computer system's memory has many levels, ranging from the small amounts in the CPU to the much slower but more plentiful storage of a mechanical hard drive. *(Pearson Education)*

Figure 6.15 A DIMM memory module holds a series of RAM chips and is wrapped with an aluminum heatsink plate to pull away heat. *(Scanrail/123RF)*

Where can I find RAM in my computing device? RAM appears in the device on memory modules (or memory cards), small circuit boards that hold a series of RAM chips and fit into special slots on the motherboard. Most memory modules in today's systems are packaged as a dual inline memory module (DIMM), a small circuit board that holds several memory chips (see Figure 6.15).

How can I tell how much RAM is installed in my computer? The amount of RAM actually sitting on memory modules in your computer is your computer's physical memory. The easiest way to see how much RAM you have is to look in the System window. (On a Mac, choose the Apple menu and then About This Mac.) This is the same window you looked in to determine your system's CPU type and speed and is shown in Figure 6.5. RAM capacity is measured in gigabytes (GB), and most machines sold today have at least 8 GB of RAM.

How can I tell how my RAM is being used? To see exactly how your RAM is being used, open the Resource Monitor and click the Memory tab (see Figure 6.16). The Resource Monitor gives additional details on CPU, disk, network, and memory usage inside your system, and you can use it to see how you're using all the RAM you paid for.

Windows uses a memory-management technique known as SuperFetch. SuperFetch monitors the applications you use the most and preloads them into your system memory so that they'll be ready to be used when you want them. For example, if you have Microsoft Word running, Windows stores

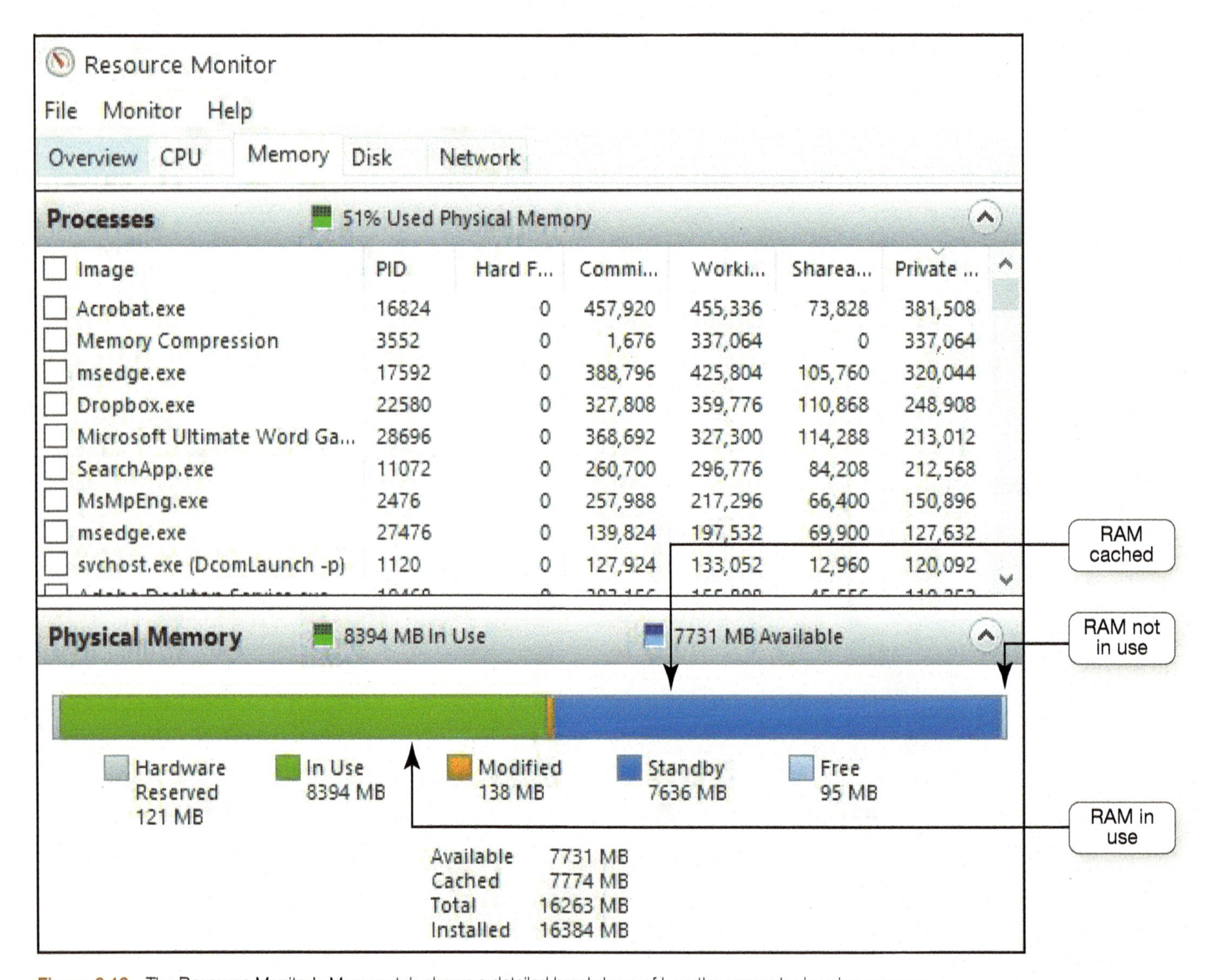

Figure 6.16 The Resource Monitor's Memory tab shows a detailed breakdown of how the computer is using memory.
> *To access the Resource Monitor, type* ***resource monitor*** *in the* ***Start*** *search box and then select it from the list that appears. (Courtesy of Microsoft Corp.)*

Table 6.1 Sample RAM Allocation

Application	RAM Recommended
Windows 10 (high resolution)	2 GB
Microsoft 365	2 GB
Microsoft Edge	1 GB
App for playing videos	2 GB
Adobe Photoshop Elements 2020	4 GB
Total RAM recommended to run all programs simultaneously	**11 GB**

as much of the information related to Word in RAM as it can, which speeds up how fast your application responds. Retrieving information from RAM is much faster than pulling it from the hard drive. You can watch this process at work by using the Resource Monitor. Figure 6.16 shows how the 16 GB of installed RAM is being used:

- 0.12 GB is reserved to run the hardware systems.
- 8.4 GB is running programs.
- 7.6 GB is holding cached data and files ready for quick access.
- 0.095 GB is currently unused.

This is a system that might benefit from additional memory since it has very little unused memory.

How much RAM do I need? At a minimum, your system needs enough RAM to run the OS. Running the 64-bit version of Windows requires a minimum of 2 GB of RAM. However, because you run more applications at one time than just the OS, you'll want to have more RAM than just what's needed for the OS. For example, Table 6.1 shows how much RAM is recommended for the OS, a web browser, and some software.

It's a good idea to have more than the minimum amount of RAM you need now so you can use more programs in the future. Remember, too, that "required" means these are the minimum values recommended by manufacturers; having more RAM often helps programs run more efficiently. High-end systems can come with 64 GB of RAM. The rule of thumb: When buying a new computer, buy as much RAM as you can afford.

Helpdesk MyLab IT

Evaluating Your CPU and RAM

In this Helpdesk, you'll play the role of a helpdesk staffer, fielding questions about what the CPU does and how to evaluate its performance. You'll also field questions about how memory works and how to evaluate how much memory a computer needs.

Adding RAM

Objective 6.6 *Evaluate whether adding RAM to a system is desirable.*

Is there a limit to how much RAM I can add to my computer? The motherboard is designed with a specific number of slots into which the memory cards fit, and each slot has a limit on the amount of RAM it can hold. Some systems ship with the maximum amount of RAM installed, whereas others have room for additional RAM. To determine your specific system limits, check the system manufacturer's website. Also, memory seller sites like Crucial (*crucial.com*) can also help you determine how much RAM your computer model contains, whether you can add more, and the type of memory to add.

In addition, the OS running on your machine imposes its own RAM limit. For example, the maximum amount of RAM for the 32-bit version of Windows 10 Home is 4 GB, whereas the maximum memory limit using the 64-bit version of Windows 10 Pro is 2 TB.

Installing RAM

In this Sound Byte, you learn how to select the appropriate type of memory to purchase, how to order memory online, and how to install it yourself. As you'll discover, the procedure is a simple one and can add great performance benefits to your system.

Is it difficult or expensive to add RAM? Adding RAM is easy (see Figure 6.17). Be sure that you purchase a memory module that's compatible with your computer. Also, be sure to follow the installation instructions that come with the RAM module. Typically, you simply line up the notches and gently push the memory module in place.

RAM is a relatively inexpensive system upgrade. The cost of RAM does fluctuate in the marketplace as much as 400% over time, though, so if you're considering adding RAM, you should watch the prices of memory in advertisements.

Figure 6.17 Adding RAM to a computer is quite simple and relatively inexpensive. On a laptop, you often gain access through a panel on the bottom. *(Editorial Image, LLC/Alamy Stock Photo; Editorial Image, LLC/Alamy Stock Photo)*

Bits&Bytes Telepresence Robots Coming to a Workplace Near You

A *telepresence robot* is a remote-controlled, wheeled device that usually uses a tablet computing device and Internet connectivity to represent a remote worker in the workplace (see Figure 6.18). Telepresence robots provide users with the means to interact by video (including observing) and audio without requiring their physical presence on-site. Although they have been around for a decade, telepresence robots have not made huge inroads into the workplace yet. But this may be changing as telecommuting continues to expand.

Telepresence robots allow virtual employees to feel more physically connected to their human coworkers who are physically in the workplace. The ability to navigate around the workplace and observe what's taking place is an advantage over static teleconference solutions. The ability to move and interact with people directly gives the telepresence robot's user a greater sense of control.

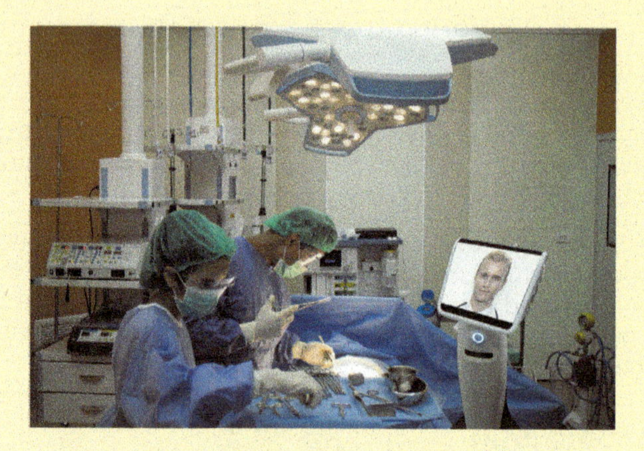

Figure 6.18 Telepresence robots facilitate the dispersion of scarce resources (such as physicians) in remote settings. *(MONOPOLY919/Shutterstock)*

Before moving on to Part 2:

1. Watch Chapter Overview Video 6.1.
2. Then take the Check Your Understanding quiz.

Check Your Understanding // Review & Practice

For a quick review to see what you've learned so far, answer the following questions.

multiple choice

1. Moore's Law describes the pace at which ________ improve.
   a. RAM chips
   b. clock cycles
   c. CPUs
   d. cache modules
2. Which of the following is NOT a stage in the machine cycle?
   a. Fetch
   b. Decode
   c. Decrypt
   d. Store
3. The CPU's ________ memory is a form of RAM that gets data to the CPU for processing much faster than bringing the data in from the computer's RAM.
   a. fetch
   b. ROM
   c. cache
   d. pipeline
4. The limit to how much RAM you can add to your system
   a. depends on the design of the motherboard.
   b. depends on the operating system running on your system.
   c. depends on the amount of memory each memory card slot supports.
   d. all of the above
5. The amount of RAM actually sitting on memory modules in your computer is your computer's ________ memory.
   a. virtual
   b. physical
   c. pipeline
   d. hyperthread

chew on this

(3Dalia/Shutterstock)

Briefly describe the way you would evaluate whether the CPU and memory of your system were meeting your needs. What tools would you need? What operating system programs help you evaluate performance?

MyLab IT

Go to **MyLab IT** to take an autograded version of the *Check Your Understanding* review and to find all media resources for the chapter.

For the IT Simulation for this chapter, see MyLab IT.

Try This

Measure Your System Performance

The Windows Task Manager and the Resource Monitor can help you find a lot of useful information about your computer system. Let's make sure you can use these Windows tools to keep an eye on your system performance. For more step-by-step instructions, watch the Try This video on MyLab IT.

Step 1 Hold the **Windows key** and press **X**. From the pop-up menu, select **Task Manager**. Click **More Details** (if necessary) and then click the **Processes tab**.

Step 2 If you leave this window open while you work, you can pop in and check the history of how your CPU, disk, memory, and network are performing. Let's start by clicking the **Performance tab** and then looking at CPU usage. This computer is only occasionally going over 30%, so the system isn't limited by CPU performance.

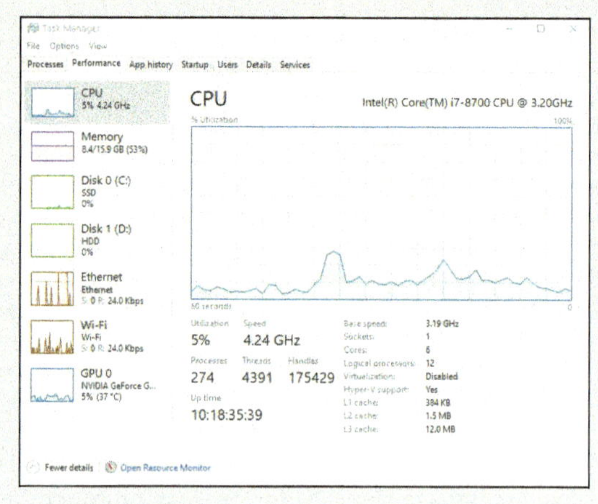

(Courtesy of Microsoft Corp.)

Step 3 Clicking **Memory** in the left panel shows that 8 GB of DDR3 memory is installed in this computer. Notice that memory usage is consistent at about 3.6 GB. Memory is available, so this system isn't limited by memory capacity with its current workload.

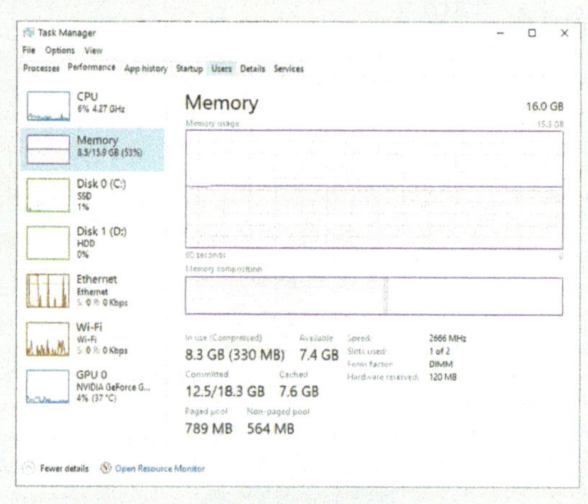

(Courtesy of Microsoft Corp.)

Step 4 Clicking **Disk** in the left panel shows that here we have one disk drive, a 1 TB internal hard drive. The lower graph shows the history of data moving back and forth to the disk, the disk transfer rate. The larger upper graph shows how active the disk is—what percentage of time it is reading and writing. This system was not doing any disk-intensive operations when the screenshot was captured. If the active time is consistently high, upgrading to a faster, larger disk will yield a big performance impact.

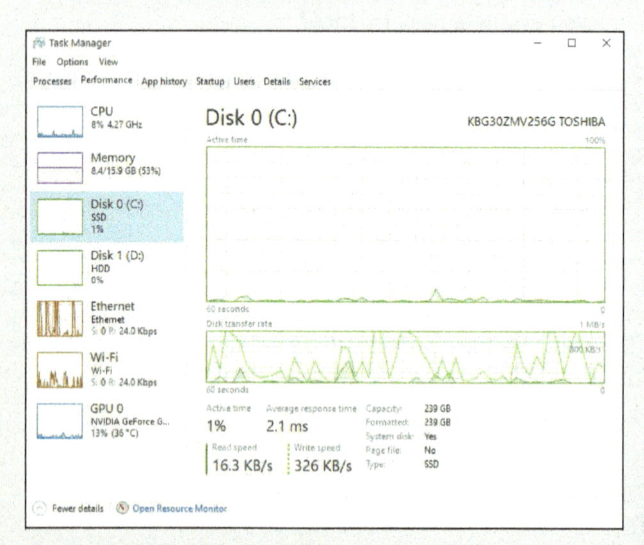

(Courtesy of Microsoft Corp.)

Make This  TOOL: App Inventor 2 or Thunkable

A Location-Aware App

Your smartphone comes equipped with many built-in sensors that can, for example, read accelerations (to tell whether your phone is shaking), location (using GPS satellites), and even atmospheric pressure. App Inventor can work with sensor data and supports a wide set of sensors used in the Lego Mindstorms kits to recognize color or respond to touch and sound.

```
when LocationSensor1 .LocationChanged
  latitude  longitude  altitude
do  set CurrentLatitudeLabel . Text to  get latitude
    set CurrentLongitudeLabel . Text to  get longitude
```

(Copyright MIT, used with permission)

In this exercise, you'll use the LocationSensor component in App Inventor to combine information about your environment into your apps.

The LocationSensor component allows you to work with live GPS data within your mobile app.

```
when ComputeDistButton .Click
do  call GetDistanceTraveled
                 StartLat   LatitudeValue . Text
                 StartLong  LongitudeValue . Text
                 DestLat    LocationSensor1 . Latitude
                 DestLong   LocationSensor1 . Longitude
    set DistTraveledTextBox . Text to  join  get global answer
                                             " miles "
```

(Copyright MIT, used with permission)

To create this app for iOS, go to *Thunkable.com*, a programming platform based on App Inventor.
For the instructions for this exercise, go to MyLab IT.

Part 2

For an overview of this part of the chapter, watch **Chapter Overview Video 6.2.**

Evaluating Other Subsystems and Making a Decision

Learning Outcome 6.2 You will be able to evaluate your computer system's storage subsystem, media subsystem, and reliability and decide whether to purchase a new system or upgrade an existing one.

The audio and video subsystems of your computer affect much of your enjoyment of the machine. Equally important is the storage subsystem, which allows you to save all that content you're generating. Let's evaluate those subsystems and consider what state-of-the-art audio/video would add to your computing experience. Then let's consider how to make sure your system is reliable—nothing interferes with enjoying technology like a misbehaving computer!

Evaluating the Storage Subsystem

Remember, there are two ways data is stored on your computer: temporary storage and permanent storage. RAM is a form of temporary (or volatile) storage. It's critical to have the means to store data and software applications permanently, which we discuss in this section.

Types of Storage Drives

Objective 6.7 *Classify and describe the major types of nonvolatile storage drives.*

Permanent storage options include mechanical hard drives, solid-state hard drives, solid-state hybrid drives, and external hard drives. When you turn off your computer, the data that has been written to these devices will be available the next time the machine is powered on. These devices therefore provide nonvolatile storage.

Today, most desktop system units are designed to support more than one internal hard drive. The Apple Mac Pro has room for four hard drives, and the Thermaltake Level 10 can support six hard drives. Each one simply slides into place when you want to add more storage. To save weight, most laptops have space for only one hard drive.

Mechanical Hard Drives

What makes the hard drive such a popular storage device? With storage capacities exceeding 16 terabytes (TB), a mechanical hard drive has the largest capacity of any storage device. And because it offers the most storage per dollar, the mechanical hard drive is also a more economical device than other options.

How is data stored on a hard drive? A mechanical hard drive is composed of several coated, round, thin plates of metal stacked on a spindle. Each plate is called a platter. When data is saved to a hard drive platter, a pattern of magnetized spots is created on the iron oxide coating of each platter. When the spots are aligned in one direction, they represent a 1; when they're aligned in the other direction, they represent a 0. These 0s and 1s are bits (or binary digits) and are the smallest pieces of data that computers can understand. When data stored on the hard drive platter is retrieved (or read), your computer translates these patterns of magnetized spots into the data you have saved.

How quickly does a hard drive find information? The hard drive's access time, the time it takes a storage device to locate its stored data and make it available for processing, are measured in milliseconds (ms). For large-capacity drives, access times of approximately 12 to 13 milliseconds are typical.

Figure 6.19 The M.2 form factor allows the production of very fast, thin, and light SSD drives that resemble a stick of chewing gum. *(Ifeelstock/Alamy Stock Photo)*

Solid-State Drives

Do mechanical hard drives have the fastest access times? A solid-state drive (SSD) uses electronic memory and has no mechanical motors or moving parts. Having no mechanical motors allows SSDs to offer incredibly fast access times, reaching data in only a 10th of a millisecond (0.1 ms). That's about 100 times faster than mechanical hard drives. SSDs also have a great advantage when booting up, because a mechanical hard drive must wait for motors to bring the platters up to the final rotation speed. The startup time of SSDs is so fast, in fact, that most desktop and laptop systems offer an option to use at least one SSD. This system drive may only be 20 GB, but it holds the operating system and means the wake-up time for the system will be very fast. In addition, SSDs run with no noise, generate very little heat, and require very little power, making them a popular option in laptops.

Installing an SSD Drive

In this Sound Byte, you learn how to install an SSD drive in your computer.

Storage capacities for SSDs range up to around 60 TB, but such a large SSD is very expensive. Typical systems offer an SSD of 256 GB or 512 GB and then a mechanical hard drive, or two, to provide TBs of inexpensive, slower storage space. However, with prices of SSD drives continuing to fall, perhaps soon just having a single SSD drive may be affordable and provide the storage you need.

In the quest for ever thinner and lighter laptops, the typical SSD drive is in the M.2 form factor. First-generation SSD drives were contained in relatively large (2.5- or 3.5-inch) cases to fit the industry-standard hard drive bays in desktops and laptops, but inside these SSD cases was mostly empty space. Now newer motherboards use the M.2 connector (see Figure 6.19) and the drives that connect to it are only about 22 mm wide and range in length from 42 to 80 mm. The drives look like sticks of chewing gum that have sprouted memory modules and a controller chip. This allows the SSD drives to be integrated into the motherboard, which saves space and weight and allows tablet computers to feature SSD drives. In addition, the M.2 drives have even faster data transfer rates than the previous generation of drives.

Solid-State Hybrid Drives Another common storage option is the solid-state hybrid drive (SSHD). An SSHD drive is a combination of both a mechanical hard drive and an SSD into a single device (see Figure 6.20). SSHD drives offer a small amount of SSD storage space, perhaps 8 GB. If it is enough to store the operating system, however, it can have a huge impact on the system boot time. For laptops that will accept only one drive, these drives are a great option if you need large amounts of affordable storage space, because they're currently less expensive than SSD drives.

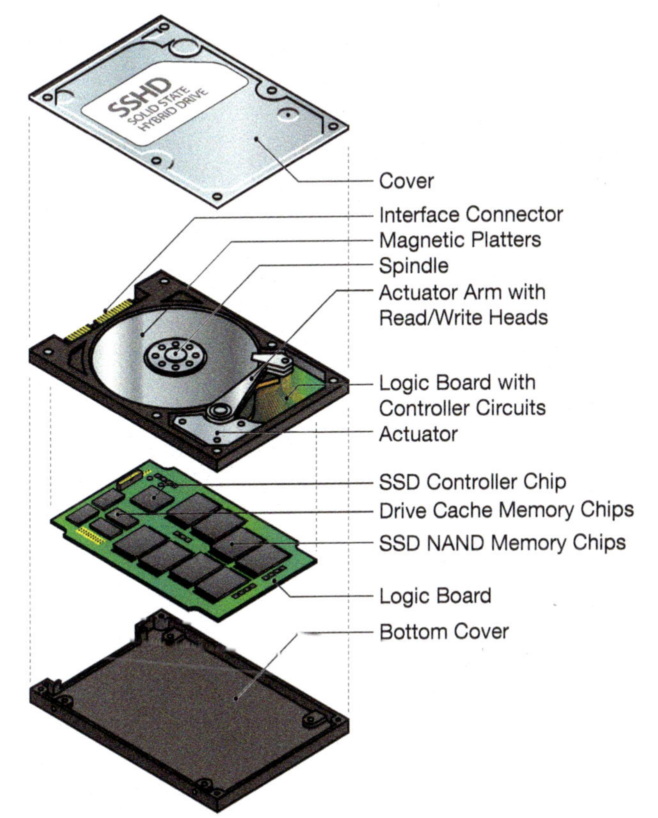

Figure 6.20 The SSHD is a single unit that contains both an SSD electronic drive and a mechanical hard drive. *(Zern Liew/Shutterstock)*

Dig Deeper How Storage Devices Work ... and Fail!

The thin metal platters that make up a mechanical hard drive are covered with a special magnetic coating that enables the data to be recorded onto one or both sides of the platter. Hard-drive manufacturers prepare the disks to hold data through a process called *low-level formatting*. In this process, concentric circles, each called a track, and pie-shaped wedges, each called a sector, are created in the magnetized surface of each platter, setting up a grid-like pattern that identifies file locations on the hard drive. A separate process called *high-level formatting* establishes the catalog that the computer uses to keep track of where each file is located on the hard drive.

Hard drive platters spin at a high rate of speed, some as fast as 15,000 revolutions per minute (rpm). Sitting between the platters are special arms that contain read/write heads (see Figure 6.21).

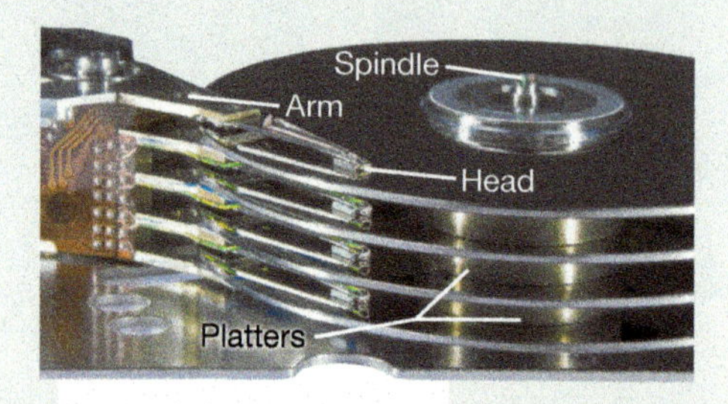

Figure 6.21 A mechanical hard drive is a stack of platters enclosed in a sealed case. Special arms fit between each platter. The read/write heads at the end of each arm read data from and save data to the platters. *(Skaljac/Shutterstock)*

A read/write head moves from the outer edge of the spinning platter to the center as frequently as 50 times per second to retrieve (read) and record (write) the magnetic data to and from the hard drive platter. The average total time it takes for the read/write head to locate the data on the platter and return it to the CPU for processing is called its access time.

Access time is mostly the sum of two factors—seek time and latency:

1. The time it takes for the read/write heads to move over the surface of the disk, moving to the correct track, is called the seek time. (Sometimes people incorrectly refer to this as access time.)
2. Once the read/write head locates the correct track, it may need to wait for the correct sector to spin to the read/write head. This waiting time is called latency (or rotational delay).

The faster the platters spin (or the faster the rpm), the less time you'll have to wait for your data to be accessed. Currently, most hard drives for home systems spin at 7,200 rpm.

The read/write heads don't touch the platters of the hard drive; rather, they float above them on a thin cushion of air at a height of 0.5 micro-inch. Therefore, it's critical to keep your hard drive free from all dust and dirt because even the smallest particle could find its way between the read/write head and the disk platter, causing a head crash—a stoppage of the hard drive that often results in data loss.

SSDs free you from worry about head crashes at all. The memory inside an SSD is constructed with electronic transistors, meaning there are no platters, no motors, and no read/write arms. Instead, a series of cells are constructed in the silicon wafers. If high voltage is applied, electrons move in and you have one state. Reverse the voltage and the electrons flow in another direction, marking the cell as storing a different value. The limiting factor for an SSD's lifespan is how many times data can be written to a cell.

However, both types of drives are subject to possible data loss known as bit rot (or data degradation). Conventional hard drives store data bits on the platters using magnetism. Over time, these bits can degrade and flip their magnetic polarity, which causes a 0 to turn into a 1 (and vice versa). SSD drives store data by trapping an electrical charge (representing either a 0 or 1) on an insulating layer of transistors. This layer can wear out over time and allow the electrons to leak out, causing data loss. Aside from bit rot, both types of drives can lose their data due to physical defects and extreme conditions (like being stored in excessive heat). What should you do about bit rot? Accept that sooner or later, your drive will fail and you will lose data. Replacing drives at least every five years is recommended. But since they can fail any time, having a solid backup strategy (discussed in Chapter 9) is essential.

Storage Needs

Objective 6.8 *Evaluate the amount and type of storage needed for a system.*

How do I know how much storage capacity I have? Typically, hard drive capacity is measured in gigabytes (GB) or terabytes (1 TB = 1,000 GB). Access This PC from File Explorer to display the hard drives, their capacity, and usage information, as shown in Figure 6.22. To get a slightly more detailed view, select a drive and then right-click and choose Properties.

How much storage do I need? You need enough space to store the following:

- The OS
- The software applications you use, such as Microsoft 365, music players, and games
- Your data files
- Your digital music library, photos, videos of television shows and movies, and so on

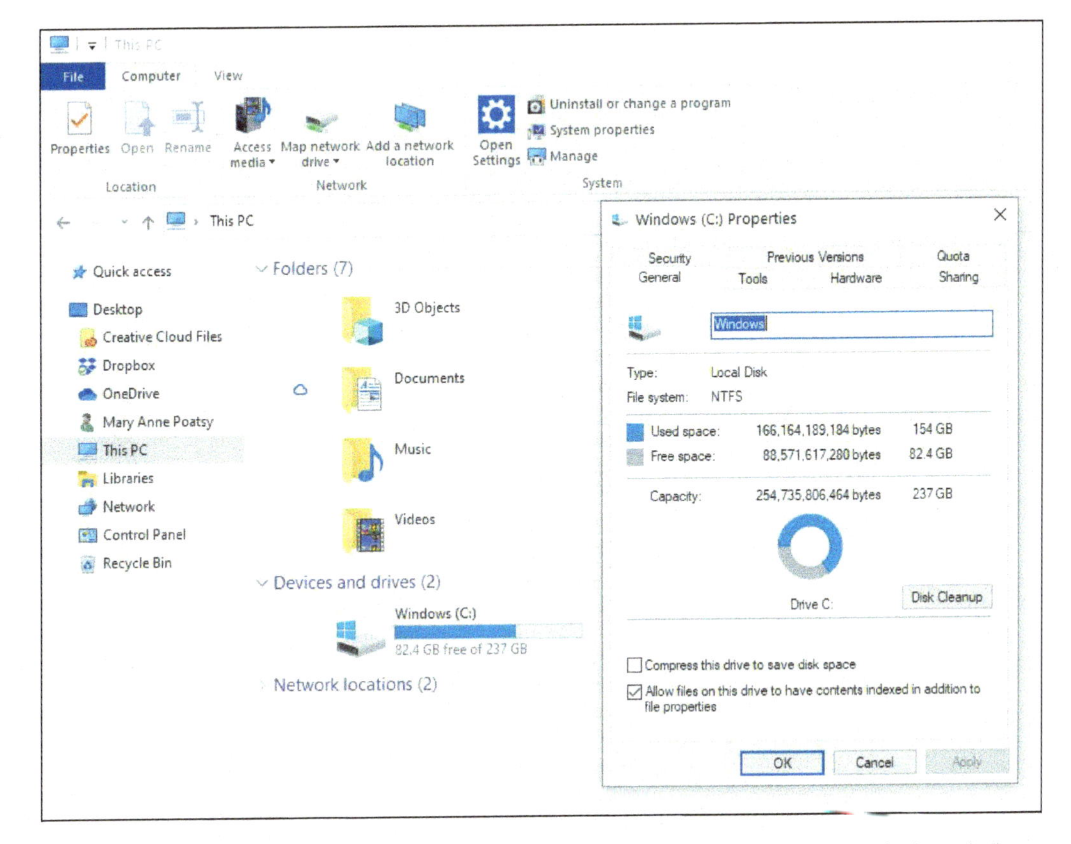

Figure 6.22 In Windows, the free and used capacity of each device in the computer system is shown in the This PC window. The General tab of the Properties dialog box gives you more detailed information.
> *To view storage device details, launch File Explorer and then click* ***This PC****. To view more details, right-click the* ***C drive*** *and select* ***Properties****. (Courtesy of Microsoft Corp.)*

Table 6.2 shows an example of storage calculation. If you plan to have a system backup on the same drive, be sure to budget for that room as well. However, note that if you're going to store your data files only online instead of on your computer, you may not need much hard drive space. For example, if you stream all the movies you watch from Netflix, keep all your data files in Microsoft OneDrive, and use online software like Google Docs to edit, you may need very little hard drive space.

Table 6.2 Sample Hard Drive Space Requirements

Application/Data	Hard Drive Space Required	Heavy Cloud Storage User
Windows 10	20 GB	20 GB
Microsoft 365	3.5 GB	3.5 GB
Adobe Photoshop Elements 2020	4 GB	4 GB
Adobe Premiere Pro CC	10 GB	10 GB
Video library of movies	80 GB (about 40 HD movies)	Streamed through online services
Music library	50 GB (about 7,000 songs)	Stored in cloud (iCloud or Amazon Cloud Drive)
Photographs	5 GB	Stored in iCloud or Dropbox
Total storage in use	**172.5 GB**	**37.5 GB**
Full backup	172.5 GB	Stored in cloud using Carbonite
Total required	**345 GB**	**37.5 GB**

If you still need more local storage than your internal hard drive provides, you can also add an external hard drive to your system. These use a USB port to connect. If you're looking to buy an external hard drive, the USB 3.2 standard supports data transfer rates up to 20 Gb/s. The new USB4 standard supports 40 Gb/s.

What is RAID technology? A redundant array of independent disks (RAID) is a set of strategies for using more than one drive in a system (see Figure 6.23). RAID 0 and RAID 1 are the most popular options for consumer machines. If you purchase two smaller drives, you can combine them by using RAID technology.

If disk performance is very important—for example, when you're doing video editing or sound recording—using two files in RAID 0 could be desirable. RAID 0 is faster because every time data is written to a hard drive, it's spread across the two physical drives (see Figure 6.23a). The write begins on the first drive, and while the system is waiting for that write to be completed, the system jumps ahead and begins to write the next block of data to the second drive. This makes writing information to disk almost twice as fast as using just one hard drive. The downside is that if either of these disks fails, you lose all your data because part of each file is on each drive. So RAID 0 is for those most concerned with performance.

If you're really paranoid about losing data, you should consider having two drives in RAID 1. In a RAID 1 configuration, all the data written to one drive is instantly perfectly mirrored and written to a second drive (see Figure 6.23b). This provides you with a perfect, instant-by-instant backup of all your work. It also means that if you buy two 1-TB drives, you only have room to store 1 TB of data because the second 1-TB drive is being used as the mirror.

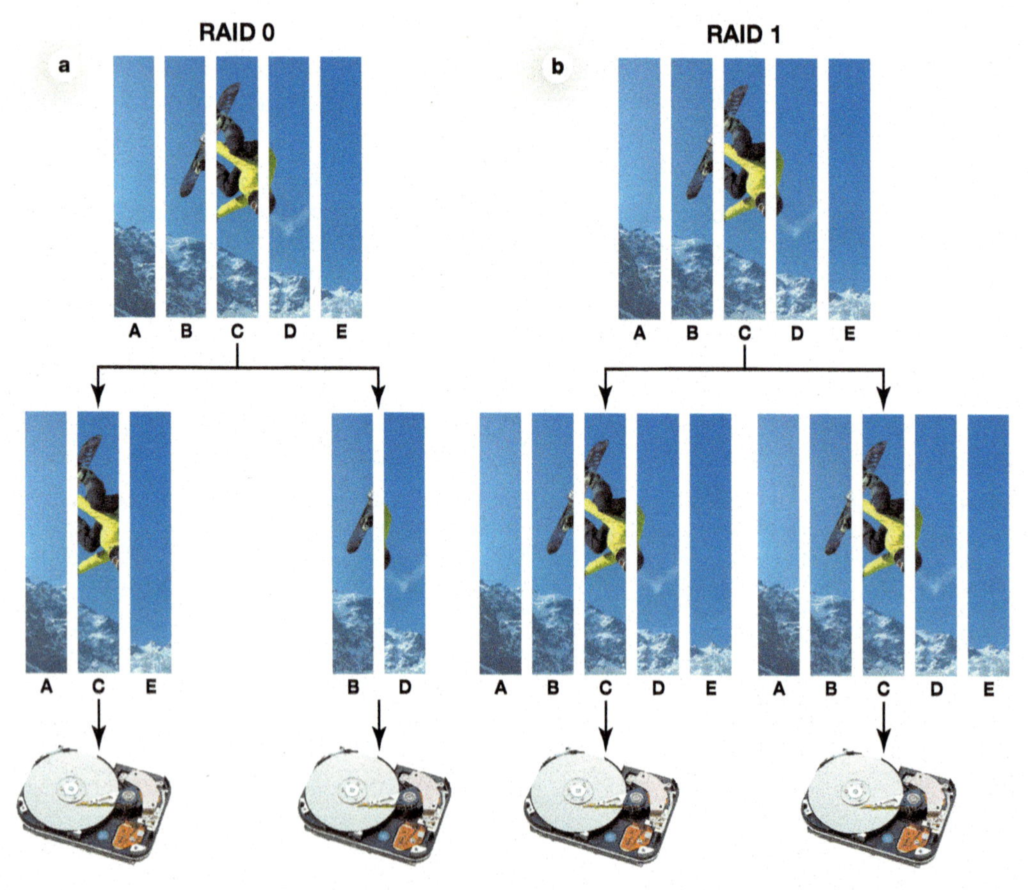

Figure 6.23 (a) A RAID 0 configuration speeds up file read/write time. (b) A RAID 1 configuration gives you an instant backup. *(Sergey Nivens/Shutterstock; Ragnarock/Shutterstock)*

Bits&Bytes Surveillance Drives: Do You Need One?

In standard hard drives in computers, reading and writing data takes place rather evenly. Sometimes your computer writes (stores) data to the drive, and sometimes it pulls (reads) data from the drive. Most computers are switched on a majority of the day but are then shut down at night, providing a rest for the hard drive.

But what if you have computing needs that function 24/7? For example, if your business (or perhaps your home) needs to run a computer to store video surveillance, that device will be operating 24/7 and the hard drive will be subjected to heavier wear and tear. The hard drive will also need to do much more writing (storage) than reading. Surveillance drives are designed to solve this problem. These drives are designed to optimize the writing of data at the expense of reading performance. So although they wouldn't be appropriate for a computer getting average use, they might be just what you need in specialized situations.

RAID 0 and RAID 1 configurations are available on many desktop systems and are even beginning to appear on laptop computers.

So how do my storage devices measure up? Figure 6.24 summarizes the factors you should consider in evaluating your storage subsystem. Don't forget that you can always extend the storage subsystem in your Internet-connected devices by using cloud storage options. If you constantly switch between devices, cloud storage is most likely your best option.

Figure 6.24 Evaluating Storage

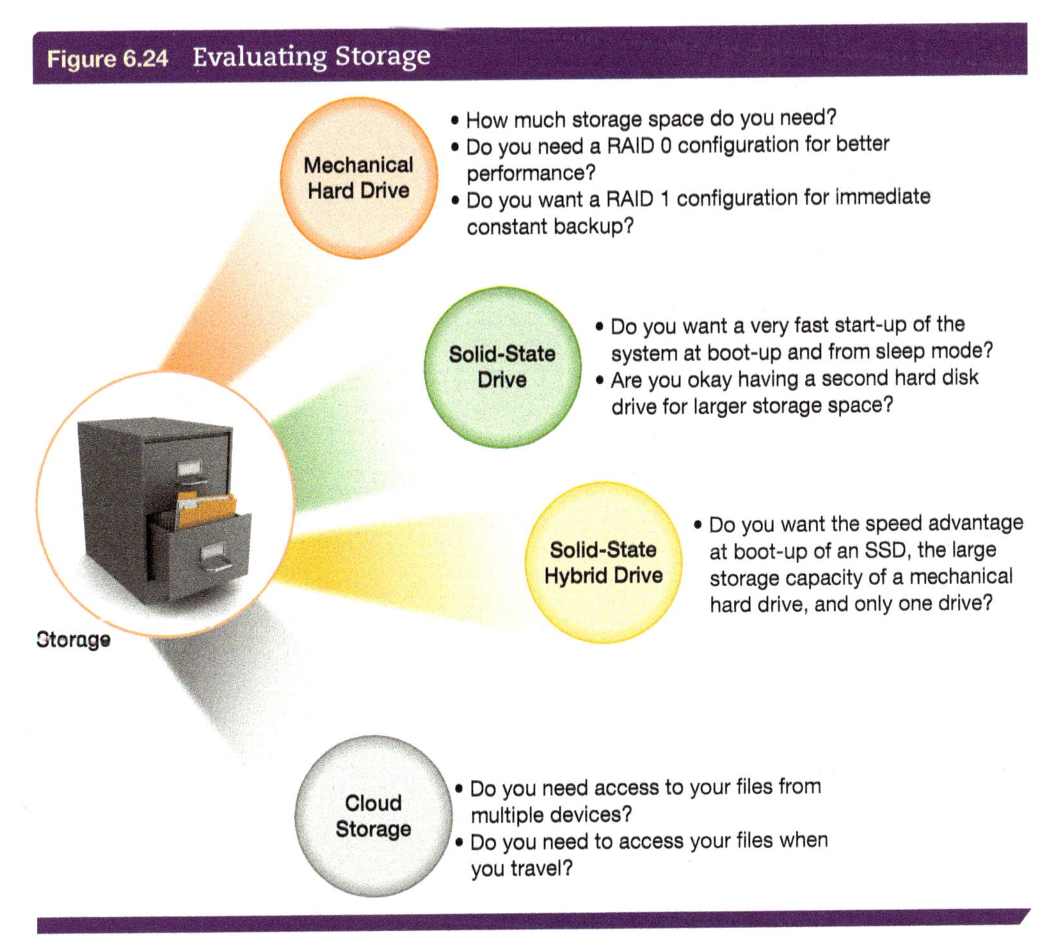

(Maxim Kazmin/123RF)

Evaluating the Media Subsystems

Enjoying interactive media through your computer demands both good video processing hardware and good audio processing hardware. Video quality depends on two components: your video card and your monitor. If you're considering using your computer to display complex graphics, edit high-definition videos, or play graphics-rich games with a lot of fast action, you may want to consider upgrading your video subsystem. Let's start our look at the media subsystems by examining video cards.

Video Cards

Objective 6.9 *Describe the features of video cards.*

What exactly is a video card? A video card (or video adapter) is an expansion card that's installed inside the system unit to translate binary data into the images you view on your monitor. Modern video cards like the ones shown in Figure 6.25 allow you to connect video equipment by using several ports:

- HDMI for high-definition TVs, monitors, or gaming consoles
- DVI for monitors and HDTVs
- DisplayPorts for monitors, HDTVs, or projectors

Figure 6.25 Video cards require their own fan for cooling. They support multiple monitors and multiple styles of ports, like HDMI, DVI, and DisplayPort. *(YamabikaY/Shutterstock)*

How much memory does my video card need? All video systems include their own RAM, called video memory. Several standards of video memory are available, including graphics double data rate 5 (GDDR5) memory and the newer graphics double data rate 6 (GDDR6) memory. Newer GDDR6 video memory chips offer faster data rates (twice as fast) than earlier chips and improvements in energy efficiency.

The amount of video memory on your video card makes a big impact on the resolution the system can support and on how smoothly and quickly it can render video. Most new laptop computers come with video cards equipped with a minimum of 2 GB of video memory. For the serious gamer, 4 GB or more is essential, and cards with 32 GB are available. These high-end video cards allow games to generate smoother animations and more sophisticated shading and texture.

How can I tell how much memory my video card has? You'll find information about your system's video card in Task Manager on a Windows computer (see Figure 6.26). To see this screen,

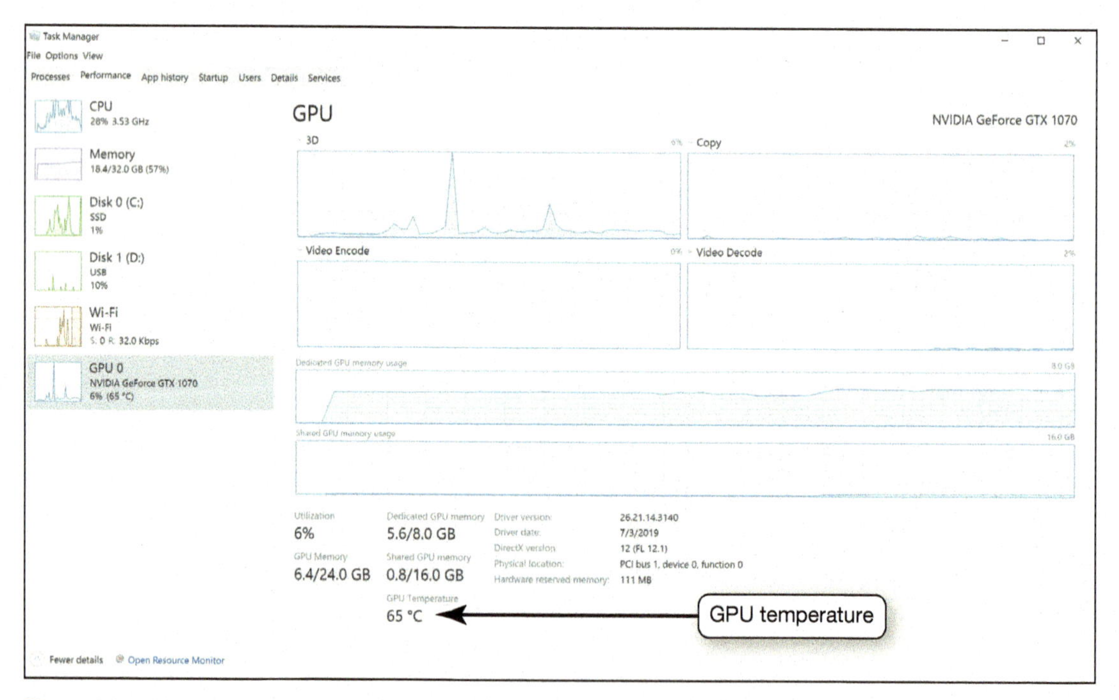

Figure 6.26 Task Manager tells you a wealth of information about the GPU installed in your computer. *(Courtesy of Microsoft Corporation)*

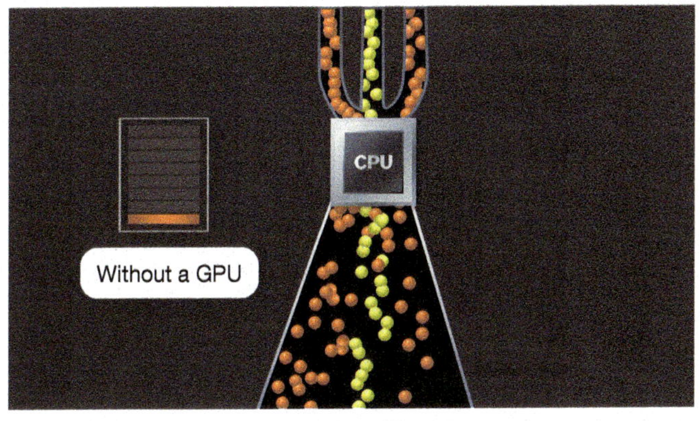

Without a GPU

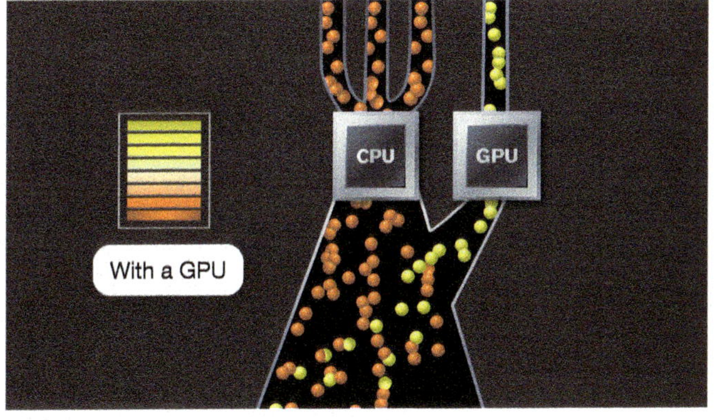

With a GPU

Figure 6.27 The GPU is specialized to handle processing of photos, videos, and video game images. It frees up the CPU to work on other system demands.

launch Task Manager, select the Performance tab, and then select the GPU icon (your computer will only display the GPU icon if your computer has a separate GPU).

How does the CPU handle intensive video calculations? Because displaying graphics demands a lot of computational work from the CPU, video cards come with their own graphics processing unit (GPU). The GPU is a separate processing chip specialized to handle 3D graphics and image and video processing with incredible efficiency and speed. When the CPU is asked to process graphics, it redirects those tasks to the GPU, significantly speeding up graphics processing. Figure 6.27 shows how the CPU can run much more efficiently when a GPU does all the graphics computations.

In addition, special lighting effects can be achieved with a modern GPU. Designers can change the type of light, the texture, and the color of objects based on complex interactions. Some GPU designs incorporate dedicated hardware to allow high-definition movies to be decoded or special physics engines to model water, gravity, and rigid body movements.

Bits&Bytes How Hot Is My GPU?

Serious gaming computers have dedicated GPUs (graphics processing units) to increase graphic performance. But as with other components in your computer, GPUs can generate serious heat. Sustained high heat can degrade the performance of components and shorten their life.

Microsoft has realized how important GPUs are to gamers by displaying statistics about them in Task Manager. In 2020, Windows added a desirable missing statistic to the GPU data in Task Manager: temperature (see Figure 6.26)! Now it is easier for gamers to tell whether they need to boost their cooling systems.

Why do some people use more than one video card in the same system? For users who are primarily doing text-based work, one video card is usually enough. Computer gamers and users of high-end visualization software, however, often take advantage of the ability to install more than one video card. Two or even three video cards can be used in one system. When the system is running at very high video resolutions, such as 5120 × 2880 or higher, multiple video cards working together provide the ultimate in performance.

The two major video chip set manufacturers, NVIDIA and AMD, have each developed its own standards supporting the combining of multiple video cards. For NVIDIA, this standard is named NVLink; for AMD, it is called CrossFireX. If you're buying a new desktop system and might be interested in employing multiple video cards, be sure to check whether the motherboard supports NVLink or CrossFireX.

Can I run a few monitors from one video card? Working with multiple monitors is useful if you often have more than one application running at a time (see Figure 6.28) or even if you just want to expand your gaming experience. Some video cards can support up to six monitors from a single card. The AMD Radeon graphics cards, for example, can merge all six monitors

Figure 6.28 AMD Radeon technology supports six monitors, which can be combined in any way. *(Andriy Popov/123Rf)*

Bits&Bytes Make Your Tablet a Second Monitor

Having two monitors attached to your computer can increase your productivity by allowing you to see more open apps. But before ordering another monitor, consider solutions that use your existing technology.

If you have an iPad or an Android tablet, you can use it as a second screen on your Mac or Windows computer. Head to the app store and install Duet Display. This handy app is an all-software solution to turning your tablet into a second monitor. No physical cables or dongles are needed.

Although not as cable-free as Duet, other Android apps to investigate include spacedesk and Splashtop Wired XDisplay. So make that tablet increase your available display space today!

Figure 6.29 Evaluating the Video Card

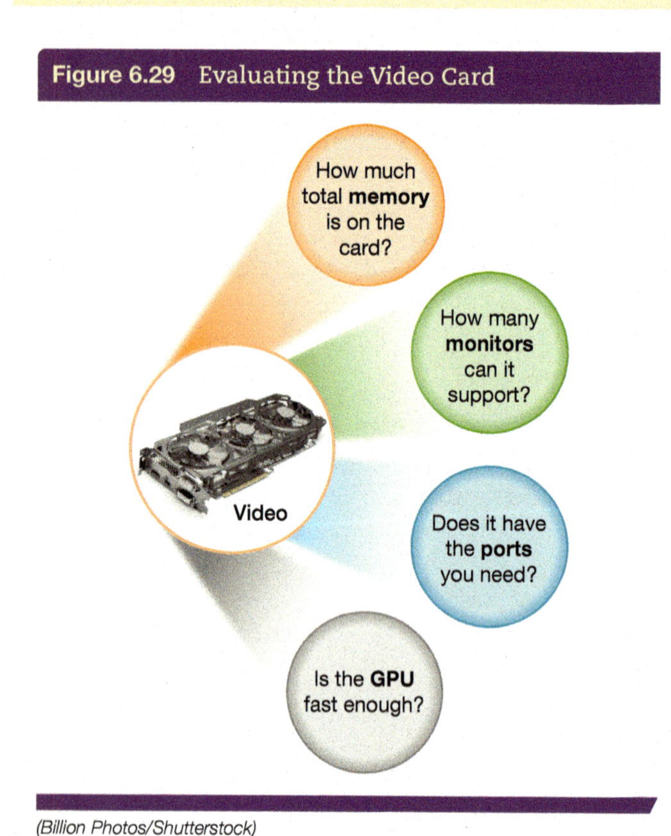

(Billion Photos/Shutterstock)

to work as one screen or combine them into any subset—for example, displaying a movie on two combined screens, Excel on one monitor, Word on another, and a browser spread across the final two.

Can I have a 3D experience from a computer monitor? 3D panels are available for desktop monitors and for some laptops. The 3D wireless vision glasses included with the panels make existing games or 3D movies display in stereoscopic 3D.

How do I know whether I'm putting too much demand on my video card? If your monitor takes a while to refresh when you're editing photos or playing a graphics-rich game, then the video card could be short on memory or the GPU is being taxed beyond its capacity. You can evaluate this precisely, using the software that came with your card or Windows Task Manager. For example, AMD OverDrive software monitors the GPU usage level, the current temperature, and the fan speed.

Review the considerations listed in Figure 6.29 to see whether it might be time to upgrade your video card. On a desktop computer, replacing a video card is fairly simple: Just insert the new video card in the correct expansion slot on the motherboard. For laptop computers, it's usually not practical (or even possible in some cases) to upgrade the video card, because it's integrated directly into the motherboard. Usually your best option is to purchase a new laptop that has a more powerful video card.

Trends in IT — Thunderbolt and USB Ports: Which Ones Do I Need?

It seems as though new, faster ports are always being introduced ...and they are! Here's an overview of the two main data ports currently available to assist you with your selection of a new computing device.

Thunderbolt Ports: Intel used fiber optics—the transmission of digital data through pure glass cable as thin as human hair—to develop Thunderbolt. Thunderbolt supplies much more power to devices than previous ports did. Although initially rolled out on Apple devices, Thunderbolt ports now appear on a wide variety of devices.

Thunderbolt ports have a very slim connector design, allowing laptop designers to make their systems even thinner. Thunderbolt can be used to connect displays and establish Ethernet connections. Thunderbolt 3 supports fast transfer rates of 40 Gb/s. How fast is that? Intel, the company that developed the technology, claims that Thunderbolt can transfer a full-length HD movie in under 30 seconds or copy, in just 10 minutes, a library of music that would take a solid year to play through. Thunderbolt ports also support the transfer of 100 watts of power.

USB Ports: USB 3.2 ports are standard on most devices now, but USB4 ports are the next generation. USB 3.2 and USB4 ports are based on the USB-C connector, but you may still see USB 3.0 ports (colored blue) based on the older USB port design on your devices (especially desktops). The USB-C connector, shown in Figure 6.30, is thin, small, and reversible, so the connector slides into the port in either an up or down orientation.

USB4 ports feature a throughput of 40 Gb/s, which matches Thunderbolt speed and doubles the 20 Gb/s speed of USB 3.2. Intel allowed its Thunderbolt technology to be incorporated in USB4, allowing USB4 to be backward compatible with Thunderbolt 3 as well as with USB 3.2. Both USB 3.2 and USB4 ports support the transfer of 100 watts of power.

The ability to transfer power was a major leap forward for Thunderbolt and USB ports. This allowed computer designers to create computers with a radical port layout: a single port that could power the computer as well as connect peripherals. Because Thunderbolt and USB-C connectors were smaller than previous ports, they quickly became popular on ultrabooks when thinness is critical.

So which port do you need on your next device? It probably doesn't matter that much, because the capabilities of Thunderbolt 3 and USB4 are now comparable. Just make sure that any existing peripherals you have will still work with the backward compatibility of the new ports (which may also require an adapter to fit).

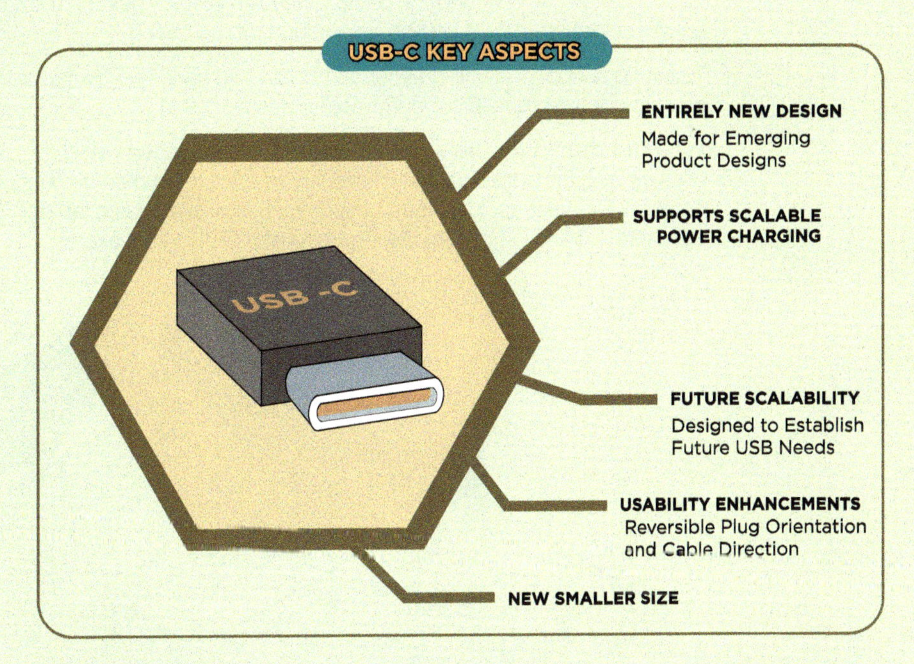

Figure 6.30 USB-C brings a reversible, durable plug capable of charging with 100 watts of power. *(Crystal Eye Studio/Shutterstock)*

Sound Cards

Objective 6.10 *Describe the features of sound cards.*

What does the sound card do? A sound card is an expansion card that enables the computer to drive the speaker system. Sound cards are now routinely integrated with the motherboard on most computers and provide excellent quality sound for most users. However, if you often use your computer to play games, music, and video, you may want to upgrade your speakers or add a higher-quality sound card (which is usually only an option in a desktop computer).

Why do I get such good-quality sound from today's sound cards? Most computers include a 3D sound card. 3D sound technology is better at convincing the human ear that sound is omnidirectional, meaning that you can't tell from which direction the sound is coming. This tends to produce a fuller, richer sound than stereo sound. However, 3D sound is not surround sound.

Helpdesk MyLab IT

Evaluating Computer System Components

In this Helpdesk, you'll play the role of a help desk staffer, fielding questions about the computer's storage, video, and audio devices and how to evaluate whether they match your needs as well as how to improve the reliability of your system.

What is surround sound, then? Surround sound is a type of audio processing that makes the listener experience sound as if it were coming from all directions by using multiple speakers. The current surround-sound standard is from Dolby. Many formats are available, including Dolby Digital EX and Dolby Digital Plus for high-definition audio. Dolby TrueHD is the newest standard. It features high-definition and lossless technology, which means that no information is lost in the compression process.

To create surround sound, another standard, Dolby Digital 7.1, reproduces digital sound from a medium (such as a streaming movie) in eight channels. Seven channels cover the listening field with placement to the left front, right front, left rear, right rear, and center of the audio stage, as well as two extra speakers to the side, as shown in Figure 6.31. The eighth channel holds extremely low-frequency (LFE) sound data and is sent to a subwoofer, which can be placed anywhere in the room.

The name 7.1 surround indicates that there are seven speakers reproducing the full audio spectrum and one speaker handling just lower-frequency bass sounds. There is also a 5.1 surround-sound standard, which has a total of six speakers—one subwoofer, a center speaker, and four speakers for right/left in the front and the back. Dolby Atmos provides the ultimate in cutting edge sound with speakers that project up and bounce sound off the ceiling.

To set up surround sound on your computer, you need two things: a set of surround-sound speakers and, for the greatest surround-sound experience, a sound card that is Dolby Digital–compatible.

What if I don't like the sound from my laptop speakers? The limited size for speakers in a laptop and the added weight of circuitry to drive them means most people use headphones or ear buds for great audio instead of speakers. However, some laptops have built-in higher-quality speakers, like the HP series offering Bang and Olufsen speakers.

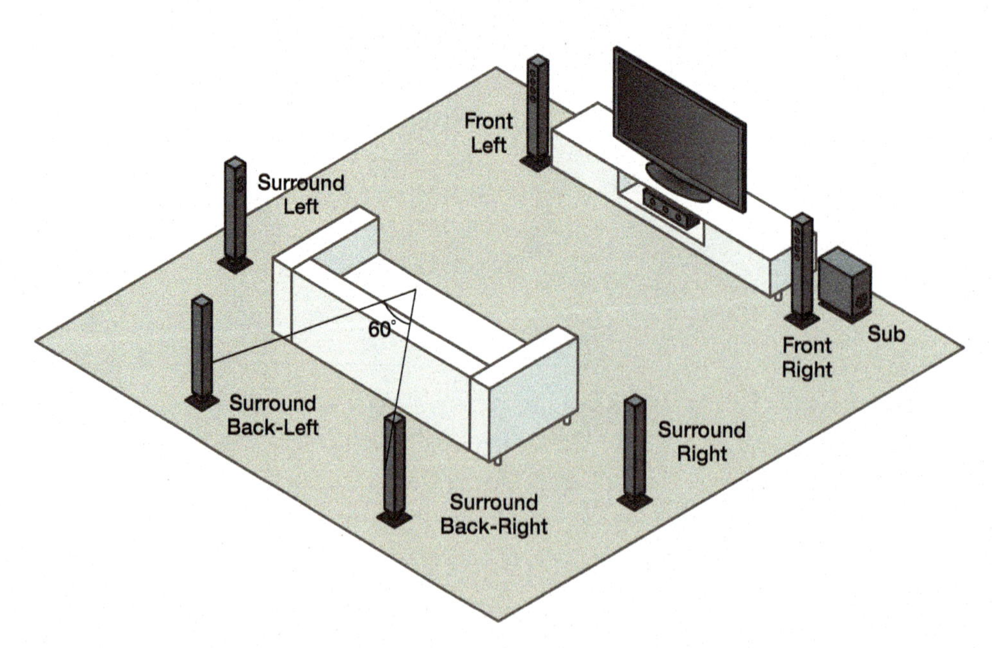

Figure 6.31 Dolby Digital 7.1 surround sound gives you a better quality audio output. *(Zem Liew/Shutterstock)*

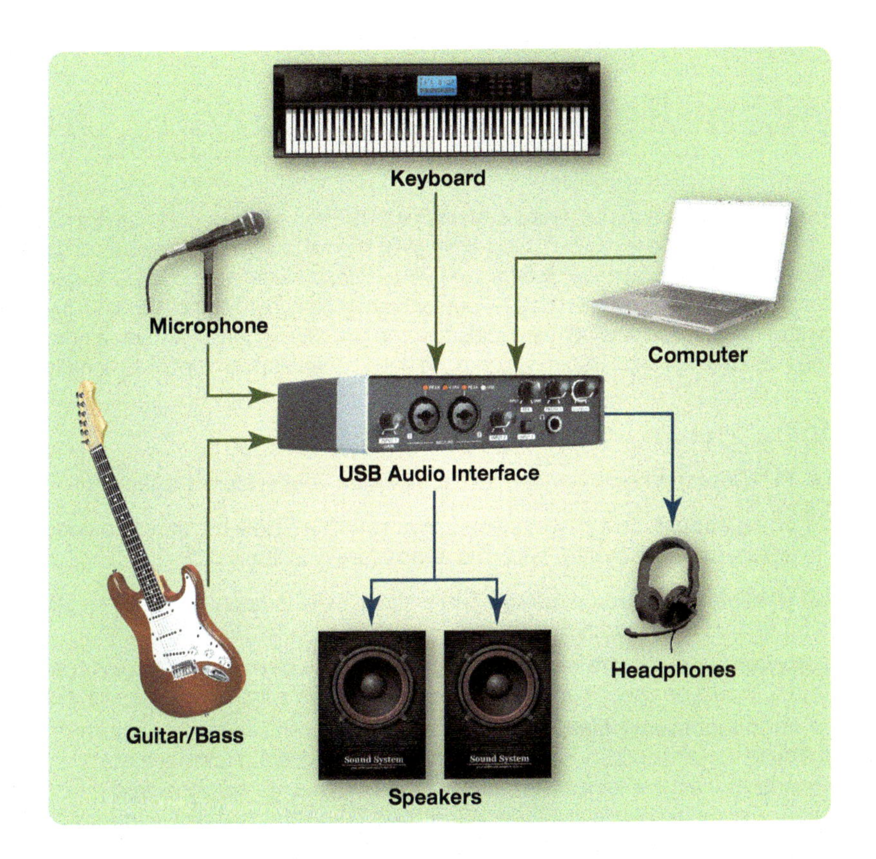

Figure 6.32 Sample Home Recording Studio Setup *(Absentanna/123RF; Sashkin7/123RF; Anna Tonchylova/123RF; Aleksangel/Shutterstock; Mariusz Blach/123RF; Maxx-Studio/Shutterstock; Mike_O/Shutterstock)*

What setup do I need if I want to use my computer for recording my band? You can connect MIDI instruments, high-quality microphones, and recording equipment to your computer through an audio MIDI interface box. MIDI is an electronics standard that allows different kinds of electronic instruments to communicate with each other and with computers. The audio interface box attaches to your computer through a USB port and adds jacks for connecting guitars and microphones. You can edit and mix tracks through many software packages, like Ableton Live or GarageBand. Figure 6.32 shows a simple home recording studio setup.

Figure 6.32 lists the factors to consider when deciding whether your audio subsystem meets your needs.

Figure 6.33 Evaluating the Audio Subsystem

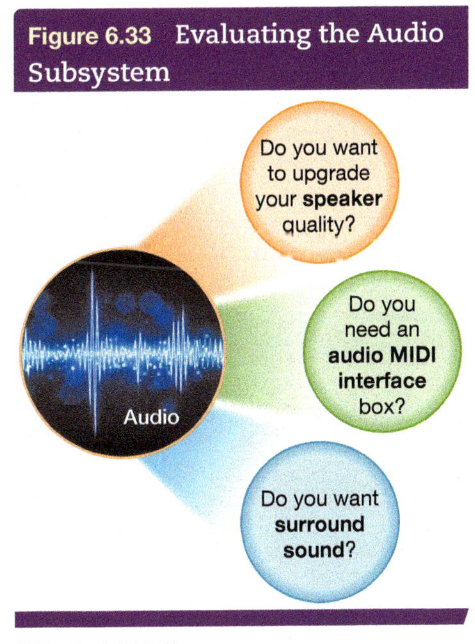

(Nikita Gonin/123RF)

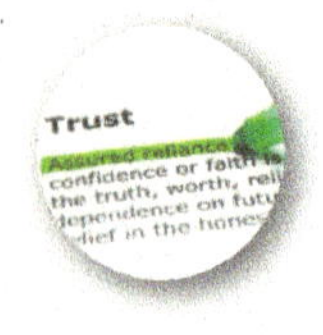

Evaluating System Reliability and Moving On

Many computer users decide to buy a new system because they're experiencing problems such as slow performance, freezes, and crashes. Over time, even normal use can cause your computer to build up excess files and become internally disorganized. This excess, clutter, and disorganization can lead to deteriorating performance or system failure, but fortunately can usually be easily fixed. If you think your system is unreliable, see whether the problem is one you can fix before you buy a new machine. Proper upkeep and maintenance also may postpone an expensive system upgrade or replacement.

Maintaining System Reliability

Objective 6.11 *Describe steps you can take to optimize your system's reliability.*

What can I do to ensure that my system stays reliable? Here are several procedures you can follow to ensure that your system performs reliably (see also Table 6.3):

- **Install a reliable antivirus package.** Make sure it's set to update itself automatically and to run a full system scan frequently.
- **Run spyware and adware removal programs.** These often detect and remove different pests and should be used in addition to your regular antivirus package.
- **Clear out unnecessary files.** Temporary Internet files can accumulate quickly on your hard drive, taking up unnecessary space. Running the Disk Cleanup utility is a quick and easy way to ensure that your temporary Internet files don't take up precious hard drive space. Likewise, you should delete any unnecessary files from your hard drive regularly because they can make your hard drive run more slowly.
- **Run the Disk Defragmenter utility on your hard drive.** When your hard drive becomes fragmented, its storage capacity is negatively affected. When you defragment (defrag) your hard drive, files are reorganized, making the hard drive work more efficiently. But remember that this only makes sense for mechanical drives. There is no need to defrag an SSD drive.
- **Automate the key utilities.** The utilities that need to be run more than once—like Disk Cleanup, Disk Defragmenter, and the antivirus, adware, and spyware programs—can be configured to run automatically at any time interval you want. You can use Windows Task Scheduler to set up a sequence of programs to run one after the other every evening while you sleep, so you can wake up each day to a reliable, secure system.

What can I do when my system crashes? Computer systems are complex. It's not unusual to have your system stop responding occasionally. If rebooting the computer doesn't help, you'll need to begin troubleshooting:

1. **If your system isn't responding, try a System Restore.** Windows automatically creates restore points before any major change to the system takes place, such as when you install a new program or change a device driver. Type Restore Point in the search box, select Create a restore point, and then click the Create button to manually create a restore point at any time. You can then select any restore point and bring your system back to the state it was in at that point. From the System Properties dialog box (accessed through the instructions noted earlier), select System Restore on the System Protection tab to learn more about restoring your system.

Table 6.3 Utilities to Keep Your System Reliable

To Avoid this Problem	Use this Tool	For More Info
Your hard drive is running low on space, making it run slowly	Disk Cleanup utility	Chapter 5
Your system is slowing down; browsers or other programs are behaving strangely	Antivirus software; spyware and adware removal software	Chapter 9
Files are spread across many spots on the hard drive, making the hard drive run slowly	Disk Defragmenter utility	Chapter 5
System not responding	Reset this PC	Chapter 5

For Mac systems, the macOS Time Machine provides automatic backup and enables you to look through and restore (if necessary) files, folders, libraries, or the entire system.

2. **Check that you have enough RAM.** You learned how to do this in the "Evaluating the Memory Subsystem" section earlier in this chapter. Systems with insufficient RAM often crash.
3. **If you see an error code in Windows, visit the Microsoft Knowledge Base** (*support.microsoft.com*). This online resource helps users resolve problems with Microsoft products. For example, it can help you determine what an error code indicates and how you may be able to solve the problem.
4. **Search with Google.** If you don't find a satisfactory answer in the Knowledge Base, try copying the entire error message into Google and searching the larger community for solutions.
5. **As a last resort, consider running Reset this PC.** This Windows utility, shown in Figure 6.34, removes all the changes you've made to the system and brings it back to the state in which it came to you from the factory. It removes all the applications from third-party vendors, but it gives you the option to keep or remove personal files like your music, documents, or videos. Make sure you select the correct option so you don't lose all your personal files!

Can my software affect system reliability? Having the latest version of software makes your system much more reliable. You should upgrade or update your OS, browser software, and application software as often as new patches (or updates) are reported for resolving errors. Sometimes these errors are performance related; sometimes they're potential system security breaches.

If you're having a software problem that can be replicated, use the Steps Recorder to capture the exact steps that led to it. In Windows, type "steps" in the search box and select the Steps Recorder desktop app. Now run the Steps Recorder and go through the exact actions that create the problem you're having. At any particular step, you can click the Add Comment button and add a comment about any part of the screen. The Steps Recorder then produces a documented report, complete with images of your screen and descriptions of each mouse movement you made. You can then e-mail this report to customer support to help technicians resolve the problem.

How do I know whether updates are available for my software? You can configure Windows so that it automatically checks for, downloads, and installs any available updates for itself, the Microsoft Edge browser, and other Microsoft applications such as Microsoft Office. Type "update" in the search box and select Check for updates in System Settings. Click *Advanced options* and customize your update strategy.

So is it time to buy a new computer? Now that you've evaluated your computer system, you need to shift to questions of value. How close does your system come to meeting your needs? Can you even upgrade your existing system? (Tablets and phones usually have very few upgrade

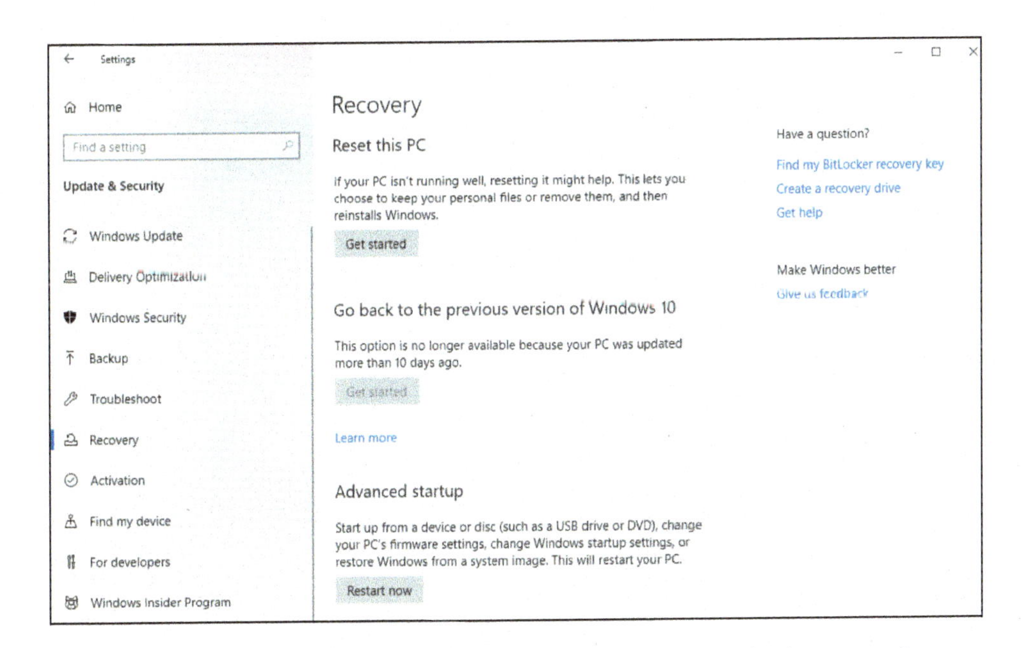

Figure 6.34 The Reset this PC feature in Windows can help when your computer stops responding.
> *To access Recovery, from* ***Settings****, select* ***Update & security*** *and then select* ***Recovery****.* (Courtesy of Microsoft Corp.)

Figure 6.35 Key Items in System Evaluation

CPU

- What is your CPU usage level?

RAM

- Do you have at least 8 GB?

Storage

- Do you need an SSD drive for fast start-up?
- Do you have a fast-access mechanical drive for large storage space?
- Do you need RAID 0 or RAID 1 storage drives for extra-fast performance or mirroring?

Video

- Do you have enough graphics memory?
- Is your GPU powerful enough?
- Do you have HDMI ports?
- How many monitors do you need to run simultaneously?

Audio

- Do you have 7.1 or 5.1 surround sound?

(Oleksandr_Delyk/Shutterstock; Hugh Threlfall/Alamy Stock Photo; Nikkytok/Shutterstock; Fotolia)

options.) How much would it cost to upgrade your current system to match what you'd ideally like your computer to do, not only today but also a few years from now? How much would it cost to purchase a new system that meets these specifications?

To know whether upgrading or buying a new system would have better value for you, price both scenarios. Conduct a thorough system evaluation (see Figure 6.35) to gather the data to help you decide. Purchasing a new system is an important investment of your resources, and you want to make a well-reasoned, well-supported decision.

Getting Rid of Your Old Computer

Objective 6.12 *Discuss how to recycle, donate, or dispose of an older computer.*

What should I do with my old computer? If the result of your system evaluation is that you need a new computer, you're probably thinking of your next new system with excitement. But what can you do with your old machine? You have options. If your old system still works, be sure to consider what benefit you might obtain by having two systems. Would you have a use for the older system?

Also, before you decide to throw it away, consider the environmental impact (see Figure 6.36). Mercury in LCD screens, cadmium in batteries and circuit boards, and flame retardants in plastic housings are all toxic. An alarming, emerging trend is that discarded machines are beginning to create an e-waste crisis.

So how can I recycle my old computer? Instead of throwing your computer away, you may be able to donate it to a nonprofit organization. Here are a few ways to do this:

- Many manufacturers, such as Dell, offer recycling programs and have formed alliances with nonprofit organizations to help distribute your old technology to those who need it.
- Sites like Computers with Causes (*computerswithcauses.org*) organize donations of both working and nonworking computers, printers, and mice.
- You can also take your computer to an authorized computer-recycling center in your area. Earth 911 provides a database you can search to find a local e-cycling center (search.earth911.com).

Figure 6.36 An electronics garbage dump can cause environmental concerns like the leaching of lead, mercury, and other hazardous substances into the ground. *(Dzejdi/Shutterstock)*

Ethics in IT

Bridging the Digital Divide: Hardware for the Masses

Closing schools during a global pandemic quickly brought the issue of the digital divide to the forefront, even in affluent countries like the United States and areas such as Western Europe. Suddenly, students needed to be educated at home through the Internet and videoconferencing. Unfortunately, many students from disadvantaged areas did not possess the devices necessary for electronic home schooling. A mad scramble ensued in many school districts to outfit students with computer hardware.

Given limited school budgets, Chromebooks were purchased in droves, because they are one of the most cost-effective solutions to providing Internet access and videoconferencing capability. However, because the need was unaddressed before the crisis, delays ensued in obtaining enough devices to be distributed to all students. This led to weeks (or months) of substandard remote education while hardware was purchased and distributed.

Unfortunately, two more stumbling blocks quickly presented themselves after the hardware was obtained. Remote education requires Internet access at home to be effective. Many disadvantaged individuals did not have Internet access. School districts quickly brokered deals with Internet providers to enable students to obtain free or discounted Internet access. Many large cable companies (such as Comcast) generously donated Internet access to the disadvantaged free of charge for periods of up to six months.

But the final hurdle was lack of knowledge, both by students and their parents, about how to use the hardware and software necessary for online schooling. School districts had to organize training and support quickly to ensure that online schooling could proceed smoothly.

Naturally, the purchase of equipment and Internet access was unplanned and strained many already thin school district budgets, but what was the alternative? Let students who couldn't afford devices miss out on education? Continue to run live classes and risk student health?

Rarely has there been such widespread evidence of the devastating effects of the digital divide. It has led many to argue that as a society, we must reassess methods for bridging the divide so that all school children can participate in a meaningful learning experience even when in-person instruction is not feasible.

For companies that need to retire large quantities of computers, the risk of creating an environmental hazard is serious. Firms like GigaBiter (*gigabiter.com*) offer a solution. GigaBiter eliminates security and environmental risks associated with electronic destruction by first delaminating the hard drive and then breaking down the computer e-waste into recyclable products. The result of the final step is a sand-like substance that is 100% recyclable.

Can I donate a computer safely, without worrying about my personal data? Before donating or recycling a computer, make sure you carefully remove all data from your hard drive. Built into Windows is an option to help with this. In the search box, type Reset and select Reset this PC. Upon clicking the Get started button, you will have the option to keep or remove your files and then reinstall Windows.

Becoming a victim of identity theft is a serious risk. Credit card numbers, bank information, Social Security numbers, tax records, passwords, and personal identification numbers (PINs) are just some of the types of sensitive information that we casually record to our computers' hard drives. Just deleting files that contain proprietary personal information is not protection enough. Likewise, reformatting or erasing your hard drive does not totally remove data, as was proved by two MIT graduate students. They bought more than 150 used hard drives from various sources. Although some of the hard drives had been reformatted or damaged so that the data was supposedly unrecoverable, the two students were able to retrieve medical records, financial information, pornography, personal e-mails, and more than 5,000 credit card numbers!

The U.S. Department of Defense suggests a seven-layer overwrite for a secure erase. This means that you fill your hard drive seven times over with a random series of 1s and 0s. Fortunately, several programs exist for doing this. For PCs running Windows, look for utility programs like File Shredder or Eraser. Wipe is available for Linux, and ShredIt X can be used for macOS. These programs provide secure hard drive erasures, either of specific files on your hard drive or of the entire hard drive.

What if my hard drive isn't functional? How can I ensure the data won't be recovered? Physical destruction of your hard drive is the best solution. Many computer recycling centers offer hard drive destruction as a service where the drive is tossed into an industrial-strength shredder and is rendered totally irreparable and unusable.

Before moving on to the Chapter Review:

1.  Watch Chapter Overview Video 6.2.
2. Then take the Check Your Understanding quiz.

Check Your Understanding // Review & Practice

For a quick review to see what you've learned so far, answer the following questions.

multiple choice

1. Hard drives are classified as which type of storage?
   a. Volatile
   b. GPU
   c. Nonvolatile
   d. Cache
2. When would you want to consider RAID 1 technology?
   a. When you need the fastest solution for writing data
   b. If you think that SSDs are too expensive
   c. When you need an instant backup of your work
   d. When you want to have only one hard disk drive
3. Which is a type of video port?
   a. HDMI
   b. SSD
   c. GPU
   d. GUI
4. Dolby Digital 7.1 creates
   a. ultra-sharp, high-definition video.
   b. a digital signal from an audio input.
   c. seven-channel surround sound.
   d. eight-channel surround sound.
5. Your Windows computer can be brought to the state it was in when it came to you from the factory with a utility called
   a. Return this PC.
   b. Restore this PC.
   c. Reset this PC.
   d. Reboot this PC.

chew on this

(3Dalia/Shutterstock)

Review the impacts on the environment of your computer during its entire life cycle. How do the production, transportation, and use of the computer affect the increase of greenhouse gas emissions? How does the selection of materials and packaging impact the environment? What restricted substances (like lead, mercury, cadmium, and PVC) are found in your machine? Could substitute materials be used? How would the ultimate green machine be designed?

MyLab IT

Go to **MyLab IT** to take an autograded version of the *Check Your Understanding* review and to find all media resources for the chapter.

For the IT Simulation for this chapter, see MyLab IT.

6 Chapter Review

Summary

Part 1

Evaluating Key Subsystems

Learning Outcome 6.1 **You will be able to evaluate your computer system's hardware functioning, including the CPU and memory subsystems.**

Your Ideal Computing Device

Objective 6.1 *Describe the changes in CPU performance over the past several decades.*

- Moore's Law describes the pace at which CPUs improve by holding more transistors. This rule predicts that the number of transistors inside a CPU will double about every two years.

Objective 6.2 *Compare and contrast a variety of computing devices.*

- A huge number of computing choices are on the market, including smartphones, tablets, ultrabooks, netbooks, 2-in-1s, laptops, and desktops.
- The kind of technology user you are will determine what kind of device you need.

Evaluating the CPU Subsystem

Objective 6.3 *Describe how a CPU is designed and how it operates.*

- The CPU is composed of two units: the control unit and the arithmetic logic unit (ALU). The control unit coordinates the activities of all the other computer components. The ALU is responsible for performing all the arithmetic calculations (addition, subtraction, multiplication, and division). Every time the CPU performs a program instruction, it goes through the same series of steps (a machine cycle): fetch, decode, execute, and store.
- The clock speed of a CPU dictates how many instructions the CPU can process each second.
- A core contains the parts of the CPU required for processing. Modern CPUs have multiple cores.
- Hyperthreading allows two sets of instructions to be run by a single CPU core.
- Pipelining is a technique that allows the CPU to work on more than one instruction (or stage of processing) at the same time, thereby boosting CPU performance.
- The CPU's cache memory is a form of RAM that is part of the CPU chip itself, so retrieving data is much faster than bringing the data in from the computer's RAM.

Objective 6.4 *Describe tools used to measure and evaluate CPU performance.*

- CPU benchmarks are measurements used to compare performance between processors.
- CPU usage is the percentage of time the CPU is busy doing work.
- On Windows systems, the Task Manager utility lets you access this data.

Evaluating the Memory Subsystem

Objective 6.5 *Discuss how RAM is used in a computer system.*

- Random access memory (RAM) is your computer's temporary storage space. RAM is an example of volatile storage. RAM appears in the system on memory modules.
- Physical memory is the amount of RAM installed in the system.
- The Resource Monitor shows how much memory is in use at any time.

Objective 6.6 *Evaluate whether adding RAM to a system is desirable.*

- Adding RAM is simple to do and relatively inexpensive. However, there is a limit to how much RAM can be installed in a device.

Part 2

Evaluating Other Subsystems and Making a Decision

Learning Outcome 6.2 **You will be able to evaluate your computer system's storage subsystem, media subsystem, and reliability and decide whether to purchase a new system or upgrade an existing one.**

Evaluating the Storage Subsystem

Objective 6.7 *Classify and describe the major types of nonvolatile storage drives.*

- Major types of nonvolatile storage include mechanical hard drives, SSDs, and SSHDs.
- Mechanical hard drives are the least expensive and the slowest to access information.
- SSD drives are electronic, so they have no moving parts, produce no heat, and are many times faster than hard drives. However, they are more expensive.
- An SSHD drive is a combination of both a mechanical hard drive and an SSD into a single device.

Objective 6.8 *Evaluate the amount and type of storage needed for a system.*

- Your storage needs will depend on the number and types of programs and data you use. Your local storage needs may be significantly reduced if you use cloud storage for your data files.
- It may be better to have several drives connected, either in RAID 0 for more speed or in RAID 1 for instantaneous backup protection.

Evaluating the Media Subsystems

Objective 6.9 *Describe the features of video cards.*

- A video card translates binary data into images that are displayed on a monitor.
- A video card has specialized video memory that is very fast. Some systems have multiple video cards for even greater performance.
- A video card has a graphics processing unit (GPU), which helps the CPU by handling the graphics workload.
- Certain individual video cards can support multiple monitors.

Objective 6.10 *Describe the features of sound cards.*

- A sound card can support 3D sound as well as surround sound like Dolby 7.1.
- Dolby 7.1 surround sound has one speaker for low-frequency tones and seven additional speakers for a full, immersive experience.
- An audio MIDI interface unit allows you to connect musical instruments, microphones, and headphones to your computer.

Evaluating System Reliability and Moving On

Objective 6.11 *Describe steps you can take to optimize your system's reliability.*

- There are many regular maintenance steps you should take to keep your system reliable. They include using an antivirus program, adware removal software, clearing out unnecessary files, and running a disk defragmenter.

Objective 6.12 *Discuss how to recycle, donate, or dispose of an older computer.*

- A used computer can be recycled through several manufacturers or through nonprofit organizations.
- To recycle or donate a computer safely, you must first remove all applications and personal data. There are options in Windows to help with this.

Be sure to check out **MyLab IT** for additional materials to help you review and learn. And don't forget to watch the Chapter Overview Videos.

Key Terms

3D sound card **228**
access time **218**
arithmetic logic unit (ALU) **206**
audio MIDI interface **229**
bit rot (data degradation) **220**
cache memory **207**
clock cycle **206**
clock speed **206**
control unit **206**
core **207**
CPU benchmarks **209**
CPU usage **209**
CPU usage graph **209**
graphics double data rate 6 (GDDR6) **224**
graphics processing unit (GPU) **225**
hard drive **218**
head crash **220**
hyperthreading **208**
latency (rotational delay) **220**
machine cycle **206**
memory module (memory card) **212**
Moore's Law **202**
nonvolatile storage **211**
overclocking **206**
parallel processing **207**
physical memory **212**
pipelining **207**
platter **218**
random access memory (RAM) **211**
read/write head **220**
redundant array of independent disks (RAID) **222**
registers **206**
sector **220**
seek time **220**
solid-state drive (SSD) **219**
solid-state hybrid drive (SSHD) **219**
sound card **228**
SuperFetch **212**
surround sound **228**
system clock **206**
system evaluation **204**
track **220**
video card (video adapter) **224**
video memory **224**
volatile storage **211**

Chapter Quiz // Assessment

For a quick review to see what you've learned, answer the following questions. Submit the quiz as requested by your instructor. If you are using **MyLab IT**, the quiz is also available there.

multiple choice

1. Moore's Law refers to the
 a. amount of memory on a memory chip.
 b. overall system processing capability.
 c. speed of DRAM.
 d. pace at which CPUs improve.

2. When the CPU performs the work described in the commands, this stage is known as
 a. fetch.
 b. decode.
 c. store.
 d. execute.

3. RAM is
 a. volatile memory.
 b. nonvolatile memory.
 c. neither volatile nor nonvolatile memory.
 d. both volatile and nonvolatile memory.

4. A good way to assess your CPU usage is to
 a. check the Performance tab of the Task Manager.
 b. listen for the sound of a spinning hard disk drive.
 c. check the number reported by the system defrag utility.
 d. feel the temperature of the CPU with your hand.

5. ________ is a set of strategies for using more than one drive in a system.
 a. GPU
 b. RAID
 c. VPU
 d. The machine cycle

6. A GPU is
 a. equivalent to a CPU.
 b. a special processing chip found on video cards.
 c. used as additional storage space when RAM is full.
 d. only found in smartphones.

7. If you want to use your computer for recording your band, you would benefit most from a(n)
 a. HDMI interface.
 b. MIDI interface.
 c. RAID interface.
 d. overclocking interface.

8. Before recycling your Windows computer, you should delete its data by using a program like
 a. File Shredder.
 b. Windows Refresh.
 c. Reclaim.
 d. Windows Reset this PC.

9. Data can be retrieved fastest from
   a. registers.
   b. cache memory.
   c. RAM.
   d. ROM.

10. When you run ________ on your hard drive, files are reorganized, making the hard drive work more efficiently.
   a. Reset this PC.
   b. Disk Eraser
   c. Disk Cleanup
   d. Disk Defragmenter

true/false

_____ 1. Desktop systems are invariably a worse value than lighter, more mobile computers.

_____ 2. SSD drives are preferable to conventional hard drives because they transfer data more quickly.

_____ 3. In Windows, if you experience unreliable behavior, Reset this PC is a "last resort" measure.

_____ 4. There is always a limit to how much RAM you can add to modern computers.

_____ 5. Moore's Law describes the pace at which ROM speed improves.

What do you think now?

Although we don't have self-driving vehicles yet, many vehicles are available with AI-controlled safety features (such as collision-avoidance systems). Recently, a number of accidents have involved cars with intelligent safety features. Studies of these accidents have revealed that in many cases, the car occupants did not understand how the safety systems were designed to function and that they overestimated the protection these systems provide. What do you think is the best way to educate consumers about vehicle safety systems? Who should be responsible for this education?

Team Time

Many Different Devices for Many Different Needs

problem

Even within one discipline, there are needs for a variety of types of computing solutions. Consider the communications department in a large university. Some groups are involved in video production, some groups producing digital music, and some groups creating scripts.

process

1. Split the class into teams. Select one segment of the communications department that your team will represent: video production, digital music, or scripting. The video production team requires its labs to support the recording, editing, and final production and distribution of digital video. The digital music group wants to establish a recording studio (after the model of the Drexel University recording label, Mad Dragon Music Group, at *maddragon-music.com*). The scripting group needs to support a collaborative community of writers and voice-over actors.
2. Analyze the computing needs of that segment, with particular focus on how it needs to outfit its computer labs.
3. Price the systems you would recommend and explain how they will be used. What decisions have you made to guarantee they will still be useful in three years?
4. Write a report that summarizes your findings. Document the resources you used and generate as much enthusiasm as you can for your recommendations.

conclusion

Being able to evaluate a computer system and match it to the current needs of its users is an important skill.

Ethics Project

Benchmarking

In this exercise, you'll research and then role-play a complicated ethical situation. The role you play might not match your own personal beliefs; regardless, your research and use of logic will enable you to represent the view assigned. An arbitrator will watch and comment on both sides of the arguments, and, together, the team will agree on an ethical solution.

problem

We've seen that for complex systems like computers, performance is often determined by using benchmarks, software suites that test a full area of performance. The results of these tests become a major force in marketing and selling the product.

There have been a number of claims of unethical conduct in the area of benchmarking. Companies have been accused of using out-of-date testing software to skew their results. Some companies have manipulated the settings on the machine to raise their score artificially (for example, turning off the display before testing for battery life). Some companies make sure the systems sent out to magazines and other evaluators have better-performing components than you might get off the shelf. Where is the line between gaining a competitive edge and lying when it comes to hardware assessment?

research areas to consider

- BAPCo
- MobileMark
- 2009 Nobel Prize for Physics

process

1. Divide the class into teams. Research the areas cited above from the perspective of an Intel engineer working on a new CPU, an engineer working on a competing CPU, or a benchmark designer.
2. Team members should write a summary that provides documentation for the positions their character takes on the issue of equitable testing of hardware. Team members then should create an outline to use during the role-playing event.
3. Team members should present their case to the class or submit a PowerPoint presentation for review, along with the summary they developed.

conclusion

As technology becomes ever more prevalent and integrated into our lives, more and more ethical dilemmas will present themselves. Being able to understand and evaluate both sides of the argument, while responding in a personally or socially ethical manner, will be an important skill.

Solve This with Excel

MyLab IT Grader

Laptop Alternatives

You need a new laptop that is lightweight but that has enough power to edit the videos you produce for your YouTube channel. You have asked a few of your friends for some suggestions. One friend put together a list of possible computers in an Excel workbook; the other friend created a list as a text file. You will import the text file into the Excel table, sort the data, and then filter the data to display only the computers you are interested in. Using Excel, you will use tables to sort, filter, and display data.

You will use the following skills as you complete this activity:

- Format as Table
- Apply Filters
- Sort Data
- Change Cell Fill Color

Instructions:

1. Open *TIA_Ch6_Start.xlsx*, and save as **TIA_Ch6_ LastFirst.xlsx**.
2. **Format range A1:M41 as a table**, with Blue, **Table Style Medium 9**. Select **Yes** when asked to convert the selection to a table and remove all external connections.
   Hint: To format a range as a table, on the Home tab, in the Styles group, select Format as Table and then select the desired style. Ensure that the *My table has headers* check box is selected.
3. **Sort** the data by multiple levels: **Style (A to Z), Processor Speed (Largest to Smallest)**, and **RAM (Largest to Smallest)**.
   Hint: To Sort data with multiple levels, on the Data tab, in the Sort & Filter group, click Sort and then click Add level. Select the desired column in each sort drop-down list.
4. **Filter** the data to display Intel i7 processors and 12 and 16 GB RAM.
   Hint: To Filter data, ensure that Filter is selected on the Data tab, in the Sort & Filter group. Click the arrow for each column to be filtered and then add or delete check marks for the desired category.
5. **Copy** the header row and four rows that appear after all filters have been applied. Open a new worksheet and paste the copied data in cell A1. Rename the new worksheet **Choices**. Click the **Data worksheet** and press **Esc** to clear the selection.
6. Click the **Choices worksheet**, change the width of **columns C, G, H, I, J, K, L** to **9**, and change the width of **columns D, E, G, and M** to **12**. Select **cells A1:M1** and **Wrap Text**.
   Hint: To change the width of nonadjacent columns, select the first column heading and then hold down Ctrl while selecting the remaining column headings. Click Format in the Cells group and then select Column Width.
7. Select **cells A3:M3** and change the Fill Color to **Yellow**.
8. In cell A8, type **The Toshiba Ultrabook is my choice as it has the most RAM, greatest storage capacity, and the best wireless standard of the four choices.**
9. Save the workbook and submit, based on your instructor's directions.

Chapter

Networking: Connecting Computing Devices

For a chapter overview, watch the **Chapter Overview Videos.**

PART 1

How Networks Function

Learning Outcome 7.1 **You will be able to explain the basics of networking, including the components needed to create a network, and describe the different ways a network can connect to the Internet.**

PART 2

Your Home Network

Learning Outcome 7.2 **You will be able to describe what is necessary to install and configure a home network and how to manage and secure a wireless network.**

MyLab IT All media accompanying this chapter can be found here.

Networked Devices on **page 259**

(Number168/123RF; Anton Shaparenko/Shutterstock; Arcady31/123RF; Leo Blanchette/123RF; Nata-Lia/Shutterstock; McCarony/Shutterstock)

What do you think?

Since **DNA sequencing** has become more affordable, many are spitting into a tube to learn about their **genetic makeup** and receive a glimpse into their **ancestry**. Now, companies are emerging that will enable people to choose from different apps that offer **personalized and customized** services and products based on their DNA results. For example, once you have your DNA results, you can order a customized nutrition or fitness plan or determine whether you're at risk for inherited heart problems. Other apps will use DNA results for less serious products such as determining your wine preferences or creating a scarf or socks that have been printed with your DNA code. There are questions about the **long-term risk** of devaluing DNA tests that are medically necessary to determine inherited diseases, as well as the risk to a person's overall **privacy**. Yet, the lure of owning a product or service that is unique to just you may make some think the risks are worth taking.

Which DNA-generated products would you consider purchasing? (Check all that apply.)

- *An image of what a future baby may look like (requires your partner's DNA)*
- *A wine app that suggests customized varietals*
- *A personalized fitness or diet plan*
- *I would not have my DNA used for any reason*

See the end of the chapter for a follow-up question.

(Ranjith Ravindran/Shutterstock)

Part 1 For an overview of this part of the chapter, watch **Chapter Overview Video 7.1**.

How Networks Function

Learning Outcome 7.1 **You will be able to explain the basics of networking, including the components needed to create a network, and describe the different ways a network can connect to the Internet.**

You access wired and wireless networks all the time—when you use an ATM, print out a document, or use the Internet (the world's largest network). It's important to understand the fundamentals of networking, such as how networks are set up, what devices are necessary to establish a network, and how you can access a network so that you can share, collaborate, and exchange information.

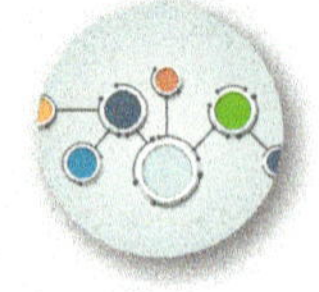

Networking Fundamentals

A typical family engages in many activities that involve sharing and accessing files over and from the Internet and using a variety of Internet-connected devices (see Figure 7.1). What makes all this technology transfer and sharing possible? A computer network!

Understanding Networks

Objective 7.1 *Describe computer networks and their pros and cons.*

What exactly is a network? A computer network is simply two or more computers that are connected by software and hardware so they can communicate with each other. Each device connected to a network is referred to as a node. A node can be a computer, a peripheral such as a printer or a game console, or a network device such as a router (see Figure 7.2).

What are the benefits of networks? There are several benefits to networking computers:

- **Sharing an Internet connection.** Probably the primary reason to set up a network is to share an Internet connection. For example, networks with wireless Internet make connecting smart home devices possible as well as enabling laptops and other portable devices to connect to the Internet wirelessly.

Figure 7.1 With a home network, all family members can connect their computing devices whenever and wherever they want. *(Mdorottya/123RF)*

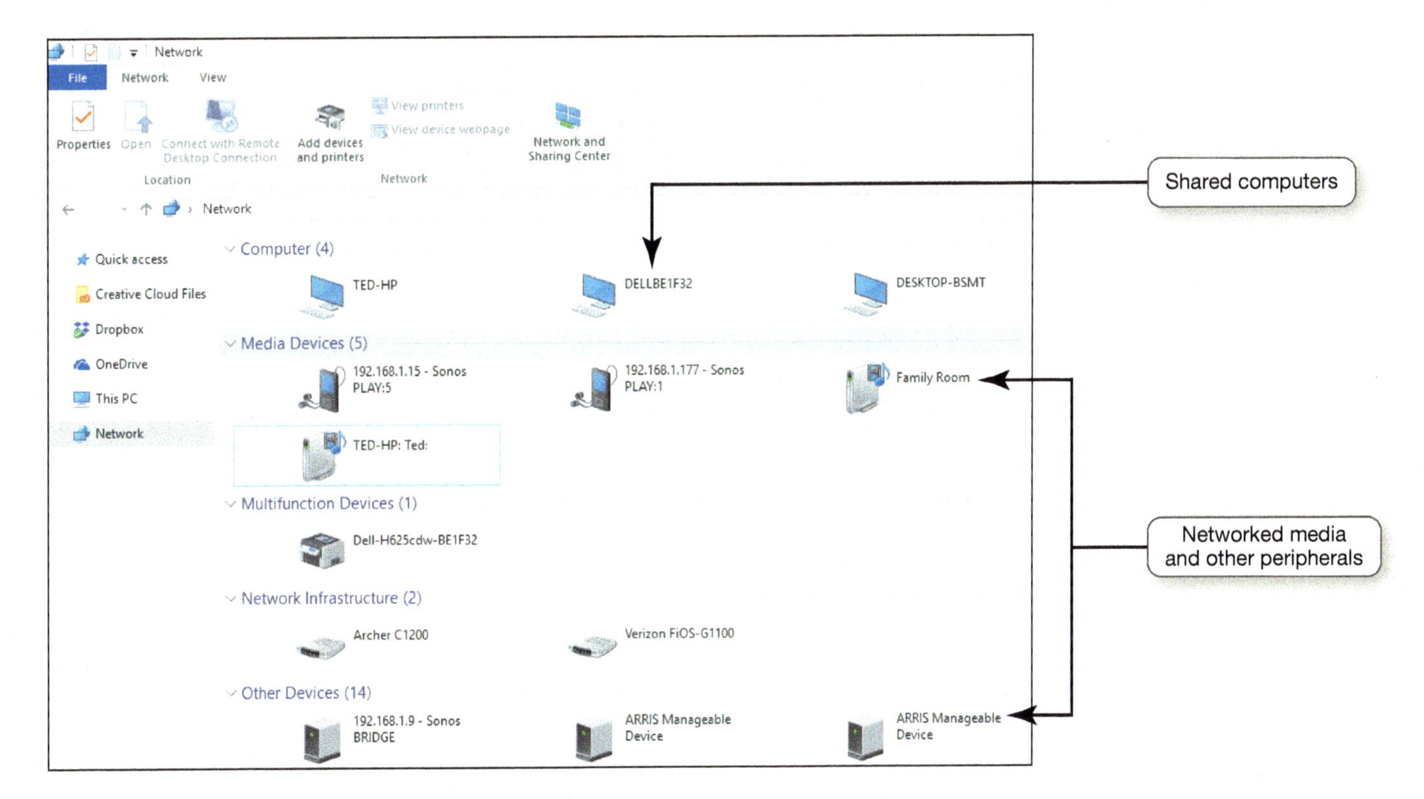

Figure 7.2 File Explorer shows computers, media, and other devices (such as set-top boxes) that are networked for sharing. *(Courtesy of Microsoft Corp.)*

- **Sharing printers and other peripherals.** Networks enable you to share printers and other peripheral devices. For example, to print a document from a laptop that's not connected to a network, you would need to transfer the file to a connected computer using a flash drive or another device that's connected to the printer or carry your laptop to the printer and connect your laptop to it. With a network, you can print directly from your device even if it's not physically connected to the printer.
- **Sharing files.** In many businesses, employees share files between networked computers without having to use portable storage devices such as flash drives to transfer the files. At home, you can set sharing options in Windows or macOS so files can be shared between networked computers. In addition to connecting to home, local, or wide area networks, you can also use services such as OneDrive, Dropbox, and Google Drive to share files that are connected to the Internet (keeping in mind that the Internet is the largest network).
- **Online gaming and home entertainment.** Many popular computer games have modes for multiple players to play together if their computers are networked. Smart TVs, music systems, and other entertainment devices also support wired or wireless networks.
- **Telephone.** For those still with a landline, you can make and receive phone calls through your home network across the Internet with Voice over IP (VoIP) services.
- **Common communications.** Devices running different operating systems can communicate on the same network.

What should be considered before establishing a network? It's hard to imagine living without some form of network these days. The need to establish a network at home, work, or school, or being able to connect in some manner to an existing network, outweighs any difficulties that exist in establishing or maintaining a network, but it's good to understand what difficulties networks might present.

Most basic home networks are simple to set up and, once established, there is very little ongoing maintenance and administration. Some homes that rely on many smart devices or that have multiple users with great demand for Internet access may require additional equipment and perhaps some periodic maintenance, but otherwise they are worry-free.

For large networks that you might find in a business or school, it can be time-consuming and costly to establish the network. In addition, measures must be taken to ensure that the data passing through the network is secure. Last, large networks require ongoing maintenance and administration. Network administration involves tasks such as:

- Installing new computers and devices
- Monitoring the network to ensure that it is performing efficiently
- Updating and installing new software on the network
- Configuring, or setting up, proper security for a network

MyLab IT

Understanding Networking

In this Helpdesk, you'll play the role of a help desk staffer, fielding questions about home networks—their advantages, their main components, and the most common types—as well as about wireless networks and how they're created.

How fast does data move through networks? Data transfer rate (also called bandwidth) is the maximum speed at which data can be transmitted between two nodes on a network. Throughput is the actual speed of data transfer that is achieved. Throughput is always less than or equal to the data transfer rate. Data transfer rate and throughput are usually measured in megabits per second (Mbps) or gigabits per second (Gbps). One of the main factors that determine how fast data moves is the type of network, which we discuss next.

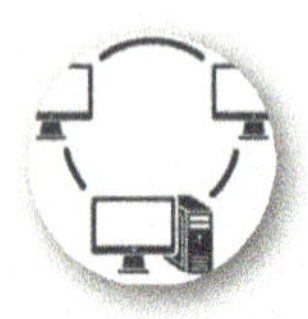

Network Architectures

The network you have in your home differs greatly in terms of its size, structure, and cost from the one on your college campus. This difference is based in part on how the networks are designed or configured. In this section, we look at a variety of network classifications.

Network Designs

Objective 7.2 *Explain the different ways networks are defined.*

What are the different types of networks? Network architectures, or network designs, can be classified by the following:

- The distance between nodes
- The way in which the network is managed (or administered)
- The set of rules (or protocol) used to exchange data between network nodes

Distance

How does the distance between nodes define a network? Networks can range from the smallest network of just one person, in one room with multiple connected devices, to the largest network that spans between cities and even the world. The following are common types of networks (see Figure 7.3):

- A personal area network (PAN) is a network used for communication among devices close to one person, such as smartphones and tablets using wireless technologies such as Bluetooth and Wi-Fi.
- A local area network (LAN) is a network in which the nodes are located within a small geographical area. Examples include a network in a computer lab at school or at a fast-food restaurant.
- A home area network (HAN) is a specific type of LAN located in a home. HANs are used to connect all digital devices in the home, such as computers, peripherals, phones, gaming and entertainment devices, and smart devices.
- A metropolitan area network (MAN) is a large network designed to provide access to a specific geographical area, such as an entire city. Many US cities are now deploying MANs to provide Internet access to residents and tourists. Some MANs employ WiMAX wireless technology that extends local Wi-Fi networks across greater distances.
- A wide area network (WAN) spans a large physical distance. The Internet is the largest WAN, covering the globe. A WAN is also a networked collection of LANs. If a school has multiple

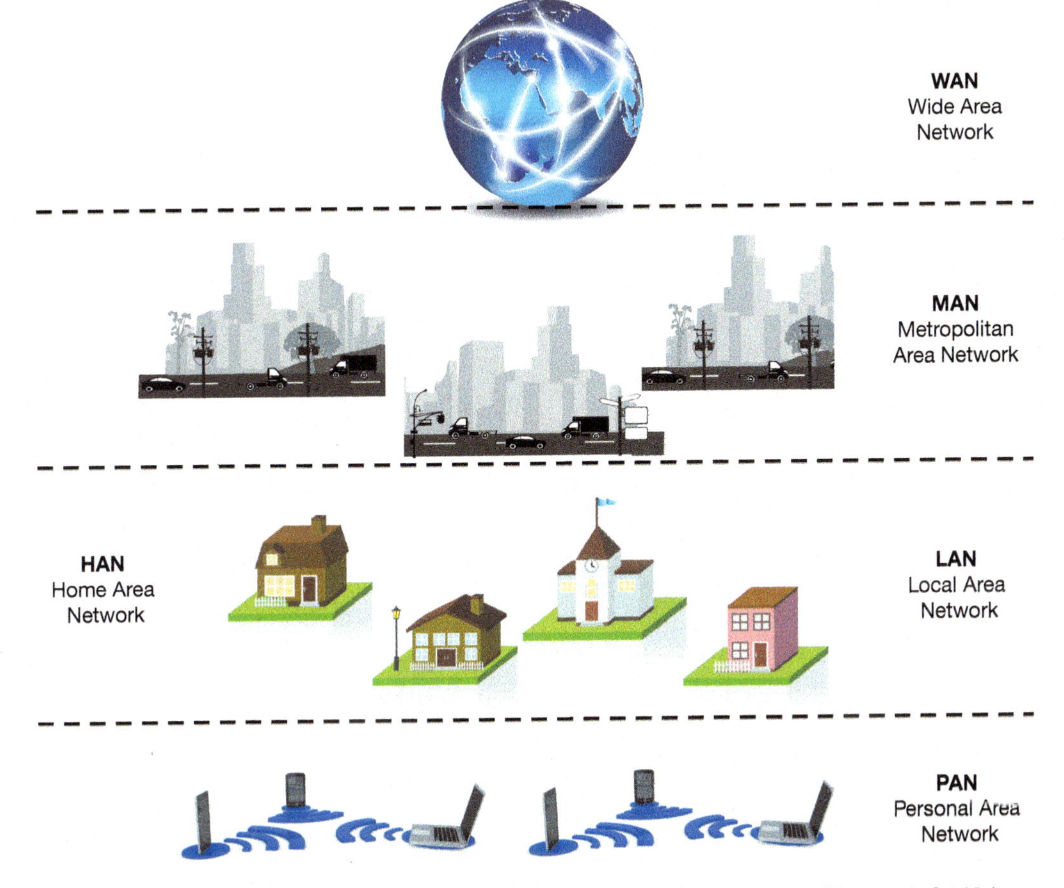

Figure 7.3 Networks can be classified by the distance between their nodes. *(Roy F Wylam/Shutterstock; CataVic/Shutterstock; SiuWing/Shutterstock; Lucadp/Shutterstock)*

campuses located in different towns, each with its own LAN, connecting the LANs of each campus by telecommunications lines allows the users of the LANs to communicate. All the connected LANs would be described as a single WAN.

Levels of Administration

How does the level of administration define a network? A network can be administered, or managed, in two main ways—centrally or locally (see Figure 7.4):

- *Central administration*: In a centrally administered network, tasks performed on one computer can affect the other computers on the network. A client/server network is an example. In a client/server network, a client is a computer on which users accomplish tasks and make requests, whereas the server is the computer that provides information or resources to the client computers as well as central administration for network functions such as printing. Client server networks are often found in medium to large establishments where data security, network expandability, and fast access times are important.
- *Local administration*: In a locally administered network, the configuration and maintenance of the network must be performed on each individual computer attached to the network. A peer-to-peer (P2P) network is an example. In a P2P network, each node connected on the network can communicate directly with every other node on the network. Thus, all nodes on this type of network are peers (equals). When printing, for example, a computer on a P2P network doesn't have to go through the computer that's connected to the printer. Instead, it can communicate directly with the printer. Because they're simple to set up, cost less than client/server networks, and are easier to configure and maintain, P2P networks are the most common type of home network. Very small schools and offices may also use P2P networks.

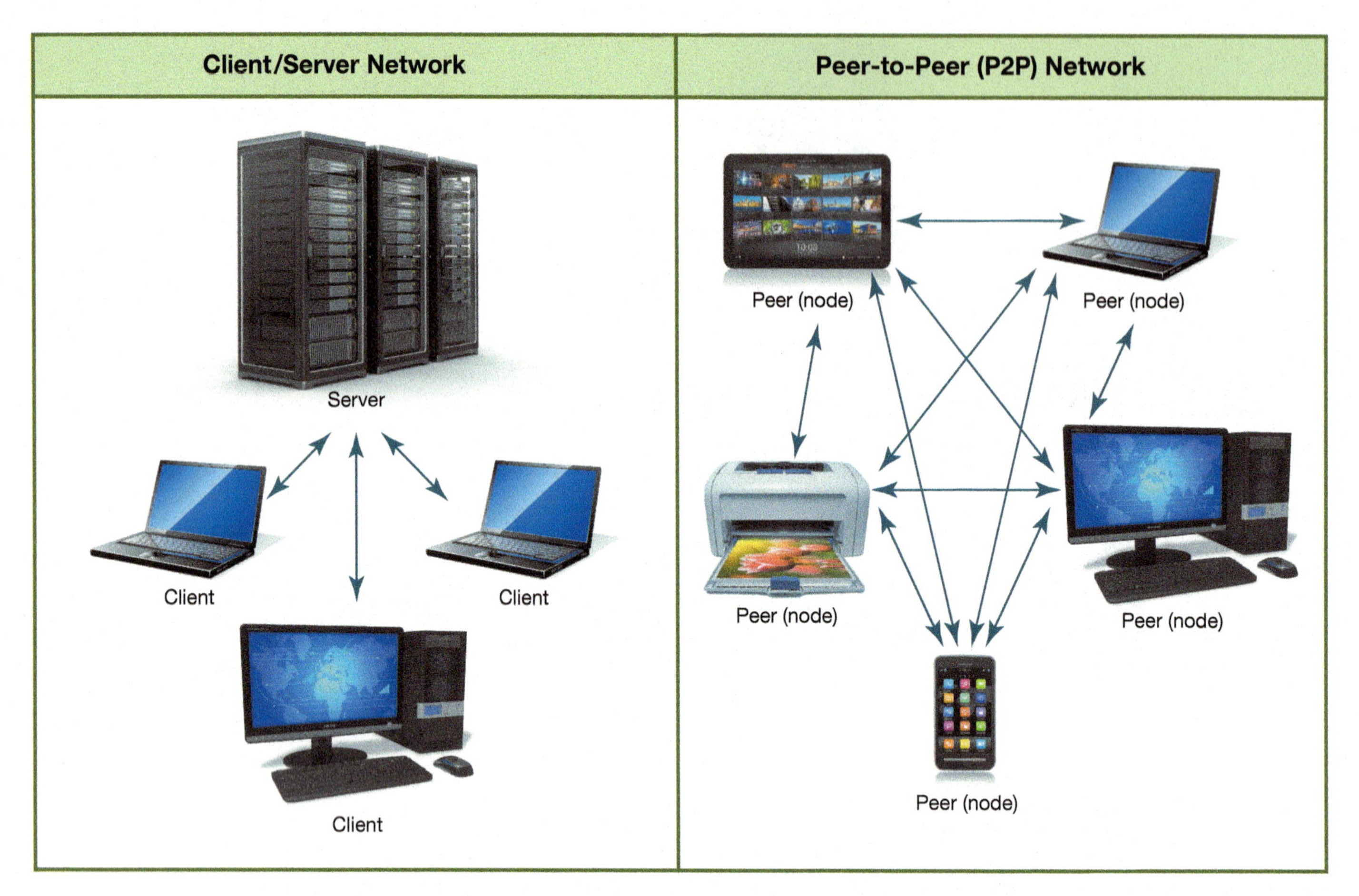

Figure 7.4 Client/server and P2P networks. *(Sashkin/Shutterstock; Tuulijumala/Shutterstock; Scanrail/123RF; Scanrail/123RF; Maksym Dykha/Shutterstock; Scanrail/123RF)*

Ethernet Protocols

What network standard is used in my home network? The vast majority of home and corporate networks are Ethernet networks. An Ethernet network is so named because it uses the Ethernet protocol as the means (or standard) by which the nodes on the network communicate.

The Ethernet protocol was developed by the Institute of Electrical and Electronics Engineers (IEEE), which develops many standard specifications for electronic data transmission that are adopted throughout the world. Establishing standards for networking is important so that devices from different manufacturers will work well together. There are different standards for wired and wireless Ethernet networks.

What is the standard for wired Ethernet networks? The standard for wired Ethernet networks is IEEE 802.3, also known as gigabit Ethernet (GbE). This standard offers a data transfer rate of up to 1 Gbps. For even faster data transfer speeds, 10, 40, and 100 GbE is available, providing maximum data transfer rates of 10, 40, and 100 Gbps, respectively. The 10 and certainly the 40 GbE are intended for businesses. It is possible to have a 10 GbE home network, but the costs may exceed the need. 100 GbE is used for the major transmission lines of the Internet known as the Internet backbone.

What is the standard for wireless Ethernet networks? Wireless networks (referred to as Wi-Fi) are based on the IEEE 802.11 standard (see Table 7.1). In the past, Wi-Fi versions were referred to by one or two letters, such as 802.11n or 802.11ac. The Wi-Fi Alliance, the group that manages Wi-Fi implementation, has simplified Wi-Fi naming conventions. So, 802.11ac is called Wi-Fi 5—as in the fifth generation of Wi-Fi. The previous Wi-Fi version, 802.11n, is called Wi-Fi 4. Wi-Fi 6, also known as 801.11ax, is the latest standard.

How is Wi-Fi 6 different from previous versions? The newer wireless Ethernet standard, Wi-Fi 6, offers faster data transfer speeds overall, but especially in congested Wi-Fi spaces where there are a lot of connected devices. Wi-Fi 6 routers communicate with devices faster and

Dig Deeper P2P File Sharing

Think of how you share files between family members on a home peer-to-peer network. Peer-to-peer (P2P) sharing (see Figure 7.5) uses a similar process to distribute and share files (usually media files such as books, movies, music, and games) between computers all over the world. All nodes on the network are individual computers. No servers are required. However, unlike home P2P networks, which can use the installed operating system, peer-to-peer file sharing requires a special P2P software program, such as Gnutella. This software is used to search for the desired file on other connected computers on the network. Once the desired file is found, it is downloaded to your computer.

Unfortunately, some participants purposely infect files with viruses and spyware, so it's important to exercise caution if you are using such a P2P file sharing system. To combat some of the security risks, some P2P file sharing systems incorporate a centralized server that acts as the intermediary to authenticate users, using passwords or digital signatures, and facilitate the transfer of files. Some P2P systems, known as friend-to-friend (F2F) systems, can be used to allow connections only between users who know each other. Retroshare is a popular F2F system that is open source and works on most platforms.

BitTorrent is another type of P2P file sharing system that uses a central server. However, it is distinguished from other P2P files sharing systems in its download/distribution process. Unlike other centralized systems, BitTorrent does not download an entire file from a single location. Instead, portions of the desired file are gathered from a swarm of connected computers and are downloaded at the same time, thus greatly improving download speeds.

Some people blame P2P file sharing networks for the growth of piracy on the Internet. Because more P2P file sharing sites have made it easier to download copyrighted music and video illegally, sales have dropped significantly. Although the use of BitTorrent and other P2P file sharing systems is legal, it is the users' responsibility to ensure that they are exchanging files that are not copyrighted. Established P2P systems can track downloads back to the user through the computer's IP address, so improper use is easily traced.

P2P networks defend their legality in that they don't run a central server but only facilitate connections between users. Therefore, they have no control over what the users choose to trade. There are legitimate uses of these new avenues of distribution. For example, *BitTorrent (BT) Bundles* are packages of free audio, video, and print content provided for download by musicians through BitTorrent. You can unlock additional content by paying or supplying your e-mail address. Tracks that are easy to remix are provided so fans can create and share their own extensions to the work.

The heavy amount of traffic on P2P sites also played into the debate on Internet neutrality. Internet service providers like Comcast were choosing to throttle, or limit, the speed of data transfer for P2P file exchanges. Users complained this was an illegal use of a public resource and that all data should be treated equally by such Internet service providers.

Figure 7.5 Peer-to-peer (P2P) sharing systems enable users to share files, usually media files such as books, movies, music, and games, between computers all over the world. *(Jozsef Bagota/Shutterstock)*

Table 7.1 Comparison of Wi-Fi Standards

IEEE Standard	802.11a	802.11b	802.11g	802.11n	802.11ac	802.11ax
New Naming Convention	Wi-Fi 1	Wi-Fi 2	Wi-Fi 3	Wi-Fi 4	Wi-Fi 5	Wi-Fi 6
Frequency	5 GHz	2.4 GHz	2.4 GHz	2.4 GHz &5 GHz	2.4 GHz & 5 GHz	2.4 GHz & 5 GHz
Maximum Data Rate	54 Mbps	11 Mbps	54 Mbps	600 Mbps	1.3 Gbps	10–12 Gbps

more efficiently by sending more information with each signal. However, to realize the increase in speed and efficiency, you need both a Wi-Fi 6 router and Wi-Fi 6 devices. Since Wi-Fi 4, Wi-Fi standards are often referred to as *dual band*, because they operate at either a 2.4 GHz or a 5 GHz frequency.

Are there new and emerging wireless standards? WiGig is a wireless option that delivers speeds up to 7 gigabits per second (Gbps) at 60 GHz frequencies (compared to the 2.4 GHz and 5 GHz frequencies that versions of Wi-Fi use). The 60 GHz frequency is less congested so it can transfer more data at once—but only for shorter ranges. Because WiGig works for short distances, it won't replace Wi-Fi. Wi-Fi 6 and WiGig are best when put to work in tandem: WiGig providing very fast transmission speeds for a room-sized area—perfect for delivering streaming media or quick data transfers between devices—and Wi-Fi 6 for all other wireless transmissions.

Another standard, HaLow or low-power Wi-Fi, helps accommodate the continued dependence on wireless devices. Unlike Wi-Fi, HaLow can connect many devices without the signal diminishing. HaLow has twice the range of current Wi-Fi and is also more robust to penetrate walls or other barriers easily, which is perfect for connecting smaller Internet of Things (IoT) devices throughout a home.

Will devices using older Wi-Fi standards still work on a newer network? Devices using older standards will still work with newer network devices. The ability of current devices to use earlier standards in addition to the current standard is known as backward compatibility. It's important to note that the speed of a network connection is determined by the slowest speed of any network device, so although an older Wi-Fi device might work, it will operate with slower data transfer rates and may run into some frequency interference. If you haven't replaced the router in your home network and are noticing slower speeds, you might need to upgrade the router to a newer standard. If the router only supports Wi-Fi 4, for example, all your devices will run at that speed—even if they are capable of working with Wi-Fi 6.

Bits&Bytes **Watch Out for Wi-Fi 6E**

If you've tried to connect your device to Wi-Fi and are presented with a long list of nearby Wi-Fi networks, for example in an airport or crowded venue, you'll most likely experience poor connectivity or, worse yet, not be able to connect at all, because all those wireless networks are fighting for space over the same frequency. To resolve this problem, a new 6 GHz frequency has been approved for wireless devices. Think of the 6 GHz frequency as a highway with four times the number of lanes of either the 2.4 or 5 GHz frequencies. Therefore, 6 GHz will allow far more devices to connect without interfering with each other.

As great as all of this sounds, you won't be seeing Wi-Fi 6E–compatible network devices anytime soon. There are still some regulatory hurdles to tackle, and products that support Wi-Fi 6E will still need to be developed and produced.

Network Components

To function, all networks must include a means to connect the nodes on the network (transmission media). These are special hardware devices that allow the nodes to communicate with each other and to send data, and software that allows the network to run (see Figure 7.6).

Transmission Media

Objective 7.3 *Describe the types of transmission media used in networks.*

How do nodes connect to each other? All network nodes are connected to each other and to the network by transmission media. Transmission media establish a communications channel between the nodes on a network. They can be either wired or wireless. The media used depend on the requirements of a network and its users.

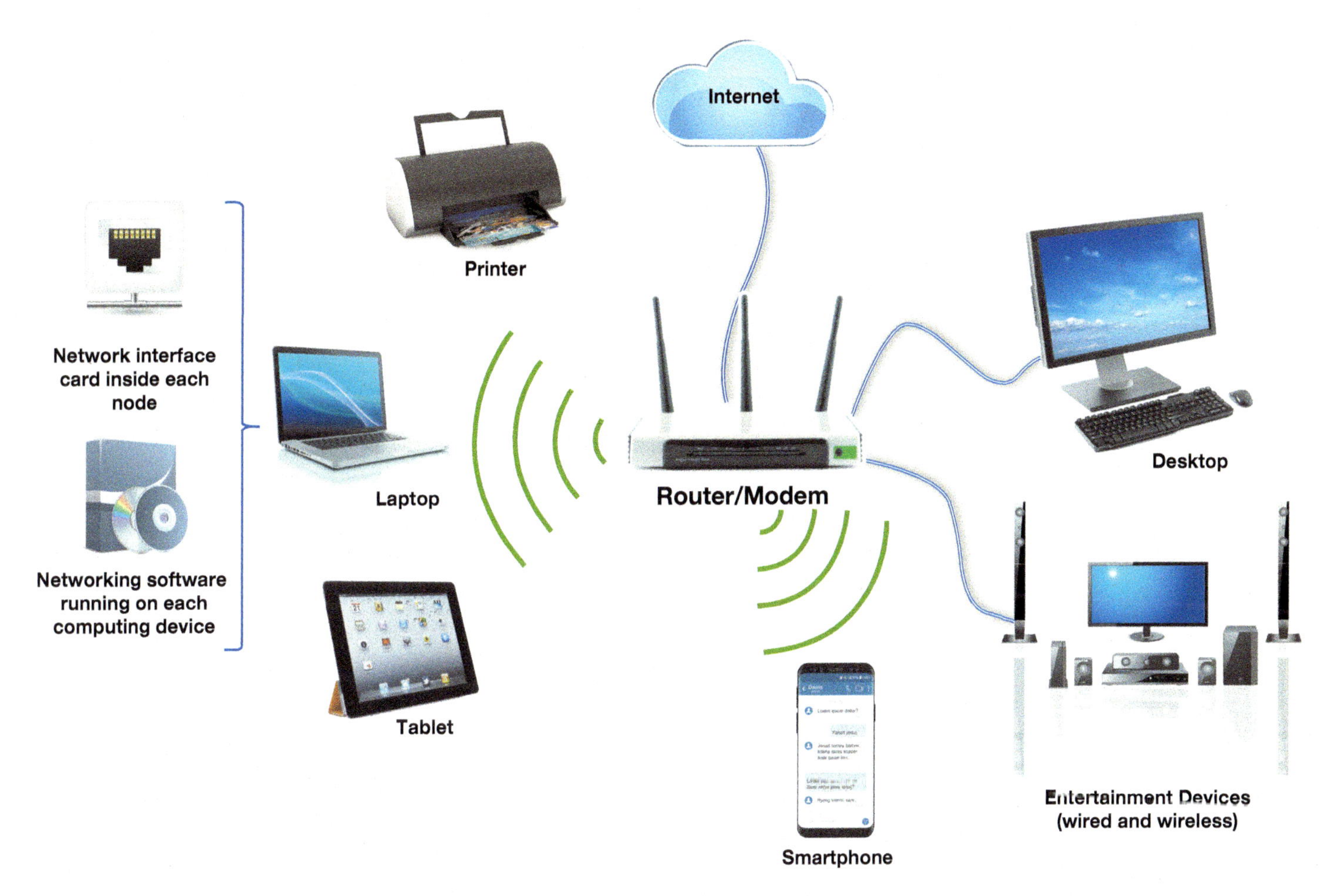

Figure 7.6 Network components. *(RealVector/Shutterstock, Alexander Kharchenko/123RF, Ifong/Shutterstock, Adrian Lyon/Alamy Stock Photo, Norman Chan/Shutterstock MSSA/Shutterstock; Dmitry Rukhlenko/Shutterstock; Stas11/Shutterstock)*

What wired transmission media is used on a network? Wired networks use various types of cable (wire) to connect nodes. As shown in Figure 7.7, the type of network and the distance between nodes determine the type of cable used:

- Unshielded twisted-pair (UTP) cable is composed of four pairs of wires twisted around each other to reduce electrical interference. UTP is slightly different from twisted-pair cable, which is what is used for telephone cable. Twisted-pair cable is made up of copper wires that are twisted around each other and surrounded by a plastic jacket.
- Coaxial cable consists of a single copper wire surrounded by layers of plastic. If you have cable TV, the cable running into your TV or cable box is most likely coaxial cable.
- Fiber-optic cable is made up of plastic or glass fibers that transmit data at extremely fast speeds.

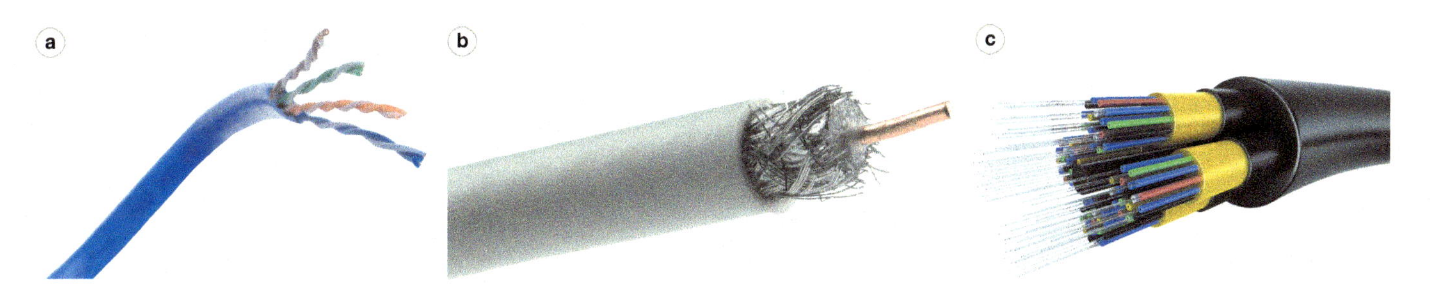

Figure 7.7 Wired transmission media. (a) Unshielded twisted-pair cable. (b) Coaxial cable. (c) Fiber-optic cable. *(Deepspacedave/Shutterstock, Zwola Fasola/Shutterstock, Zentilia/Shutterstock)*

Table 7.2 UTP Cable Types

Cat 5e	Cat 5e cable is the cheapest and is sufficient for many home networking tasks. It is designed for 100 Mbps–wired Ethernet networks that were popular before gigabit Ethernet networks became the standard.
Cat 6	Cat 6 cable is designed to achieve data transfer rates that support a 1 GbE network. Although using Cat 5e cable is sufficient, using Cat 6 cable is probably the better choice for home networking cable, though it's slightly more expensive and more difficult to work with than Cat 5e cable.
Cat 6a	Cat 6a cable is designed for ultrafast Ethernet networks that run at speeds as fast as 10 Gbps and is the most expensive option. Installing a 10 GbE network in the home may be worth the added expense to accommodate the many Internet-dependent devices as well as the current and anticipated use of today's home applications such as gaming and streaming media.
Cat 7	Cat 7 cable supports 10 GbE networks, with the difference of offering greater throughput than Cat 6a.

Is there just one kind of UTP cable? Several types of UTP cable are commonly found in wired Ethernet networks (see Table 7.2). Cat 5e and Cat 6 cable are more common in home networks, whereas Cat 6a and Cat 7 are designed for bigger networks that require more speed. However, with the proliferation of IoT devices in the home, it may not be uncommon to see Cat 6a and Cat 7 cables used for home networks.

What type of transmission media is most common in home networks? Homes today generally have a combination of wired and wireless capabilities. A home network will include a wired Ethernet connection between the main network components such as the router, switches, and wireless access points (described in the next section). In addition, a wired connection may connect a printer or desktop computer (especially if it's used for gaming). For wired connections, the most popular transmission media option is UTP cable, which you can buy in varying lengths with Ethernet connectors (called RJ-45 connectors) already attached. Ethernet connectors resemble standard phone connectors (called RJ-11 connectors) but are slightly larger and have contacts for eight wires (four pairs) instead of four wires.

Wireless (or Wi-Fi) networks do not require physical cables (as the name implies) and uses radio waves to connect computing devices to other devices and to the Internet. A wireless connection will be used to connect the host of portable devices such as laptops and smartphones to the Internet. Wireless connections are also used for stationary devices such as printers, TVs, and some smart home devices such as thermostats and security systems.

Why might a portable device use a wired connection? When you want to achieve the highest possible throughput on your portable device, you may want to use a wired connection, if one is available. Wireless signals have slower throughput than wired connections for the following reasons:

- Wireless bandwidth is shared among devices.
- Wireless signals are more susceptible to interference from magnetic and electrical sources.
- Other wireless networks (such as your neighbor's network) can interfere with the signals on your network.
- Certain building materials (such as concrete and cinderblock) and metal (such as a refrigerator) can decrease throughput. Throughput also varies, depending on the distance between your networking equipment.
- Wireless networks usually use specially coded signals to protect their data, whereas wired connections don't protect their signals. This process of coding signals can slightly decrease throughput, although once coded, data travels at usual speeds.

Basic Network Hardware

Objective 7.4 *Describe the basic hardware devices necessary for networks.*

What equipment do I need to connect to the Internet? There are a few basic pieces of hardware that you need to establish a home network: a modem, router, and a switch (refer back to Figure 7.6). Each piece of equipment plays a specific role in the network. A modem connects your network to the Internet and brings the Internet signal to your home. The specific type of modem will depend on the type of broadband service you have. Your Internet service provider will rent the appropriate modem to you or specify what type of modem to buy to work properly with the Internet service provider's technology. Although your Internet service provider will usually provide you with a device that supports the most recent wireless protocol, if you haven't changed your Internet service in a while, it may be time for a router/modem upgrade to ensure you're receiving the fastest connection.

What devices are needed to share the Internet throughout a home network? To share an Internet signal with your home network and with all the devices in your home, you need network navigation devices: a router and a switch. A router transfers packets of data between two or more networks such as your home network and the Internet.

To distribute the Internet signal to the devices in your home, you need a switch. A switch acts like a traffic signal on a network. It receives data packets from each computer on the network by wired or wireless signals and then sends them to their intended nodes on the same network. During the transmission process, data packets can suffer collisions; subsequently, the data in them is damaged or lost and the network doesn't function efficiently. The switch keeps track of the data packets and, in conjunction with network adapters, helps the data packets find their destinations without running into each other. The switch also keeps track of all the nodes on the network and sends the data packets directly to the node for which they're intended. This keeps the network running efficiently. Most home networks have one device that combines a modem and a router with integrated switches.

Installing a Home Computer Network

Installing a network is relatively easy if you watch someone else do it. In this Sound Byte, you'll learn how to install the hardware and configure Windows for a wired or wireless home network.

What hardware is needed for different nodes on a network to communicate? For all nodes on a network to communicate with each other and access the network, each node needs a network adapter. All desktop and laptop computers as well as smartphones, tablets, and many peripherals sold today have network adapters for both wired and wireless connections built in.

Network Software

Objective 7.5 *Describe the type of software necessary for networks.*

What software is needed to run a home network? Most home networks are P2P networks. One of the advantages of P2P networks is that no specialized network operating system software is required. Instead, any current operating system, such as Windows 10, macOS, or Linux, can perform the tasks and duties of a P2P network. These operating systems already have the necessary capabilities to share files and services between devices.

Do client/server networks need special software? Special network operating system software is needed in a client/server network. Because client/server networks communicate through a centralized server instead of communicating directly with each other, that capability requires more complex software than the standard operating system software running in a P2P network. Therefore, network operating system (NOS) software is installed on the servers in client/server networks. This software handles requests for information, Internet access, and the use of peripherals for the rest of the network nodes. Examples of NOS software include Windows Server and SUSE Linux Enterprise Server.

Connecting to the Internet

One of the main reasons for setting up a network is to share an Internet connection. You've read about the required transmission media, hardware, and software required in a network, but how do you decide what kind of Internet connection you bring into your network? Broadband is the preferred way to access the Internet, but in some situations, other connections may be necessary.

Broadband Internet Connections

Objective 7.6 *Summarize the broadband options available to access the Internet.*

What exactly is broadband Internet access? Broadband, often referred to as high-speed Internet access, refers to a type of connection that offers a means to connect to the Internet with

Table 7.3 Comparing Common Wired Broadband Internet Connection Options

Broadband Type	Transmission Medium	Speed Considerations	Average and Maximum Download Speeds
Fiber-optic	Strands of optically pure glass or plastic	Transmits data by light signals, which do not degrade over long distances	Average speed of 250 Mbps, with maximum of 1,000 Mbps
Cable	Coaxial cable, similar to cable TV wire	Cable connections are shared, so speed can drop during high-usage periods	Average speed of 10 Mbps, with maximum of 500 Mbps
DSL (digital subscriber line)	Copper wire phone line	Speed drops as distance from the main signal source increases	Average speed of 5 Mbps, with maximum of 35 Mbps
Satellite	Wireless signals from orbiting satellites	Speed depends on clear line of sight between receiving satellite dish and orbiting satellite; weather can also disrupt or affect service	Average speed of 500 Kbps, with maximum of 100 Mbps

fast throughput. Broadband has data transmission rates that range from 25 to 500 Mbps. Some businesses and large organizations have a dedicated connection to the Internet, but most homeowners and small businesses purchase Internet access from Internet service providers (ISPs). ISPs may be specialized providers, such as Juno, or companies like Comcast that provide additional services, such as phone and cable TV.

What types of broadband are available? As shown in Table 7.3, the standard wired broadband technologies in most areas are fiber-optic, cable, digital subscriber line (DSL), and satellite service.

How does fiber-optic service work? Fiber-optic service uses fiber-optic cable to transmit data by light signals over long distances. Because light travels so quickly, this technology transmits an enormous amount of data at superfast speeds. When the data reaches your house, it's converted to electrical pulses that transmit digital signals your computer can read. Note that fiber-optic cable is not usually run inside the home. On a fiber-optic network, twisted-pair or coaxial cable is still used inside the home to transport the network signals.

How does cable Internet work? Cable Internet is a broadband service that transmits data over the coaxial cables that also transmit cable television signals; however, cable TV and cable Internet are separate services. Cable TV is a one-way service in which the cable company feeds programming signals to your television. To bring two-way Internet connections to homes, cable companies had to upgrade their networks with two-way data-transmission capabilities.

Figure 7.8 Internet data is transmitted between your personal satellite dish and the satellite company's receiving satellite dish by a satellite that sits in geosynchronous orbit thousands of miles above Earth. *(Rendeep Kumar/123RF)*

How does DSL work? DSL (digital subscriber line) uses twisted-pair cable, the same as that used for regular telephones, to connect your computer to the Internet. The bandwidth of the wires is split into three sections, like a three-lane highway. One lane is used to carry voice data. DSL uses the remaining two lanes to send and receive data separately at much higher frequencies than voice data.

How does satellite Internet work? To take advantage of satellite Internet, you need a satellite dish placed outside your home and connected to your router/modem with coaxial cable. Data from your computer is transmitted between your personal satellite dish and the satellite company's receiving satellite dish by a satellite that sits in geosynchronous orbit thousands of miles above Earth (see Figure 7.8).

How do I choose which broadband connection option is best for me? Check with your local Internet service provider to determine which broadband options are available and what the transfer rates are in your area. Depending on where you live, you might not have a choice of the broadband connection available. Price may be a consideration, as well as determining the right combination of Internet speeds, TV packages, and phone services.

Bits&Bytes Who's Not on Broadband?

Although about 90 percent of Internet users in the United States use high-speed Internet connections such as DSL, cable, or fiber-optic, there are still those who use dial-up because of lack of access to high-speed Internet or because high-speed Internet is too expensive. In addition, the use of home broadband has declined over the past few years. Nearly 20 percent only use their cellular plans and public Wi-Fi to connect to the Internet, forgoing the need to obtain broadband Internet at home.

Wireless Internet Access

Objective 7.7 *Summarize how to access the Internet wirelessly.*

How can I access the Internet wirelessly at home? As mentioned earlier, to access the Internet wirelessly at home, you need to establish Wi-Fi on your home network by using a router that features wireless capabilities. Virtually all laptops, smartphones, game systems, and personal media players sold today are Wi-Fi-enabled and come with wireless capability built in.

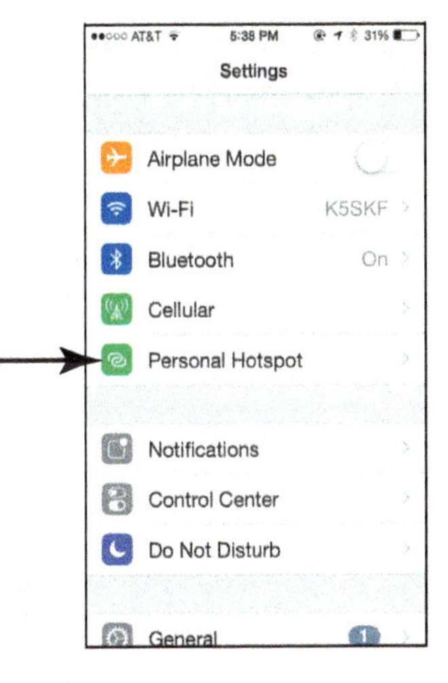

Figure 7.9 You can turn your smartphone into a mobile hotspot. *(Courtesy of Google Inc.)*

How can I access Wi-Fi when I'm away from home? When you're away from home, if you don't want to tether to your cellular plan, you may want to find a Wi-Fi hotspot. Many public places, such as libraries, hotels, and fast-food shops, offer free Wi-Fi access. Websites such as Wi-Fi-FreeSpot or apps such as WiFiGet help you locate a free hotspot wherever you're planning to go. Your ISP may also provide free hotspot service.

When you're not within range of a Wi-Fi hotspot but still need to access the Internet, you may want to consider mobile broadband. Mobile broadband connects you to the Internet through a cellular network to get Internet access. You can purchase mobile broadband data plans for devices such as smartphones, smartwatches, and tablets.

For devices for which a separate plan is not possible, such as laptops, you can still connect to the Internet by using a mobile hotspot. Mobile hotspots enable you to tether, or connect, more than one device to the Internet but require access to a data plan. Although you can buy a separate mobile hotspot device, most smartphones have built-in functionality, enabling you to turn your smartphone into a mobile hotspot (see Figure 7.9).

How do I get mobile broadband? Just as you have an ISP for Internet access for your desktop or laptop computer, you must have a wireless Internet service provider (or wireless ISP) such as T-Mobile, Verizon, or AT&T to connect your smartphone, tablet, or smartwatch to the Internet.

Mobile broadband technologies are categorized by generations. 4G (fourth generation) is the most current service standard; 5G (or fifth generation) wireless data is being rolled out as the latest mobile data service and as a home Internet service (instead of fiber-optic, cable, or DSL). All major cell phone manufacturers have 5G capable devices, and broadband providers are competing heavily to be the first and best at providing 5G service to a broad spectrum of users.

What makes 5G so special? It is expected that 5G will deliver improved wireless service over the current 4G technologies, transmitting data up to 1 Gbps, which, for example, would allow you to download a full-length, high-definition movie in seconds. As a home Internet service, 5G will have incredibly fast speeds to push more data more quickly, making video streaming seamless. Online games will be played in real time. It will also be able to connect more devices at once and without a reduction in speeds as currently experienced with Wi-Fi. In addition, 5G saves up to 90 percent on energy consumption over current standards.

How am I charged for mobile broadband? Mobile broadband providers measure your Internet usage according to how much data you download and upload. A mobile Internet connectivity plan is known as a data plan. You pay a monthly price and are allowed data transfers up to some fixed limit per month. If you exceed your monthly data limit, additional data access is provided at an extra cost. Some plans offer unlimited data access, but some unlimited plans slow down your service after you've exceeded a specified usage limit.

Bits&Bytes Net Neutrality

A few years ago, small web-based businesses had the same access to the Internet as an on-demand movie from Verizon or an online sale at Target, because Internet access was not differentiated by the type of user, the content being uploaded, or the mode of communication. This common access is referred to as *net neutrality*. In 2018, the U.S. Federal Communications Commission (FCC) voted to repeal net neutrality and put in place a tiered structure for Internet access: a premium tier with priority access to a faster Internet and a lower tier with less priority and slower speeds. Access to the faster tier would be costlier than access to the lower tier. In this new structure, big Internet service providers like Comcast and Spectrum could charge a fee for companies to access faster access speeds. So large Internet content companies, like Netflix or Facebook, would presumably pay the fee to avoid being given slower access speeds for their customers. Smaller users who might not be able to afford the better access would have less priority and slower access, thus putting them at a disadvantage. This decision remains controversial and was challenged in 2019. Although the challenge did not overturn the original ruling, it enabled states to set their own Internet legislation. Currently, 27 states have adopted net neutrality legislation.

Before moving on to Part 2:

1. Watch Chapter Overview Video 7.1.
2. Then take the Check Your Understanding quiz.

Check Your Understanding // Review & Practice

For a quick review to see what you've learned so far, answer the following questions.

multiple choice

1. Which of the following transmission media is used most often for wired connections within a home?
   a. Twisted-pair cable
   b. Coaxial cable
   c. Fiber-optic cable
   d. Unshielded twisted-pair cable
2. The type of network used for communication between a smartwatch and smartphone, using Bluetooth, is a
   a. PAN.
   b. MAN.
   c. HAN.
   d. WAN.
3. Which of the following manages how data packets are transmitted and delivered in a network?
   a. Switch
   b. Modem
   c. Home network server
   d. Router
4. Which type of Internet service uses wires similar to telephones?
   a. DSL
   b. Fiber-optic
   c. Cable
   d. Satellite
5. Which of the following needs a cellular plan to connect to the Internet wirelessly?
   a. Wi-Fi
   b. Satellite
   c. DSL
   d. Mobile broadband

chew on this

(3Dalia/Shutterstock)

Today, more than just computers are connected to the Internet. The Internet of Things (IoT) relates to the many smart objects that depend on a wireless connection to the Internet to receive and transmit data. In health care, IoT or smart devices are popping up to help keep patients safe through better monitoring and to help doctors deliver care. However, challenges such as data security and data management can make the costs outweigh the benefits. Think about some health care–related IoT devices and discuss the benefits and related concerns of those devices.

MyLab IT

Go to **MyLab IT** to take an autograded version of the *Check Your Understanding* review and find all media resources for the chapter.

For the IT simulation for this chapter, see MyLab IT.

Try This

Testing Your Internet Connection Speed

Your ISP may have promised you certain download and upload data speeds. How can you tell whether you're getting what was promised? Numerous sites on the Internet, such as SpeedOf.Me and speedtest.net, test the speed of your Internet connection. In this Try This, we will test your Internet connection speed, using SpeedOf.Me. For more step-by-step instructions, watch the Try This video on MyLab IT.

Step 1 Type *SpeedOf.Me* in any browser. SpeedOf.Me is an HTML5 Internet speed test, so it will work on PCs, Apple, and Android devices. Click Start Test to begin.

Step 2 SpeedOf.Me tests upload and download speeds, using sample files of varying sizes, starting with a 128 KB sample until it reaches a sample size that takes more than eight seconds to upload or download (with the largest possible sample size being 128 MB). The results are based on the last sample file. As the test runs, graphics illustrate the process, with the final results appearing when the test is finished.

Step 3 A history chart appears at the bottom of the screen, showing the results of any previous tests you have run on that device. If you want to share your results, you can use the Share button to post them to a social media site or by e-mail.

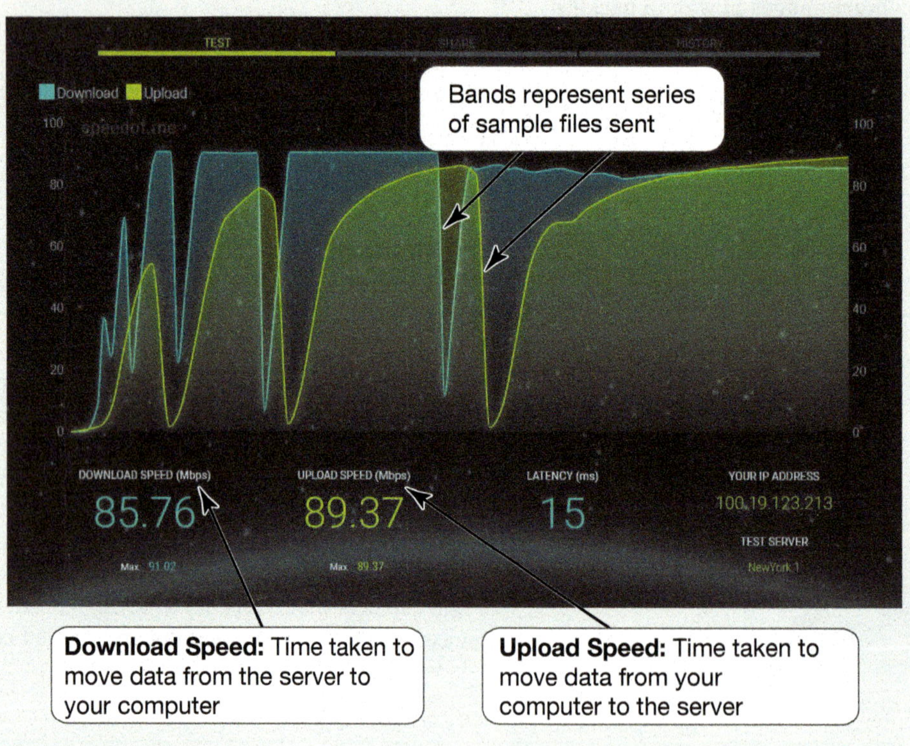

(Courtesy of SpeedOf.Me)

TOOL: Ping and Telnet

Networked Devices

One of the nice features of App Inventor is that, using Wi-Fi, you can instantly see on your target device the changes you make while programming on your PC. But what is going on behind the scenes to allow this connection?

In this exercise, you explore how the AI Companion software works with the App Inventor program to connect your systems. You'll see how networking utilities like Ping and Telnet are used to investigate how network firewalls are set up.

The App Inventor Companion app and program communicate across Wi-Fi to make your programming easier.

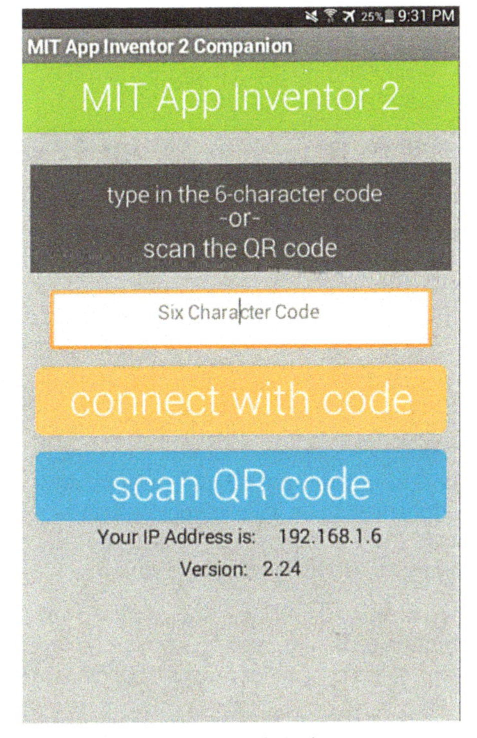

(Copyright MIT, used with permission.)

Command Prompt

```
Microsoft Windows [Version 10.0.10240]
(c) 2015 Microsoft Corporation. All rights reserved.

C:\Users\fly_n>ping 192.168.1.11

Pinging 192.168.1.11 with 32 bytes of data:
Request timed out.
Request timed out.
Reply from 192.168.1.4: Destination host unreachable.
Reply from 192.168.1.4: Destination host unreachable.

Ping statistics for 192.168.1.11:
    Packets: Sent = 4, Received = 2, Lost = 2 (50% loss),

C:\Users\fly_n>_
```

(Courtesy of Microsoft Corp.)

For the instructions for this exercise, go to MyLab IT.

Part 2

 For an overview of this part of the chapter, watch **Chapter Overview Video 7.2**.

Your Home Network

Learning Outcome 7.2 **You will be able to describe what is necessary to install and configure a home network and how to manage and secure a wireless network.**

You know what a network is and the advantages of having one. In this section, we look at installing or updating a home network and keeping it safe.

Installing and Configuring Home Networks

Now that you understand the basic components of a network and your Internet connection options, you're ready to install a network in your home. If you already have a network in your home, it's useful to examine your network devices, settings, and configuration to make sure they're still meeting your needs.

Only a few years ago, most home networks included just a few computers and a printer. However, a network that can manage those devices is often very different from one that can support the smartphones, gaming consoles, tablets, smart TVs, and IoT devices many homes now have connected to their networks. If you're using any of these additional devices and you haven't updated your network equipment or setup in a while, it may be time to do so.

Planning Your Home Network

Objective 7.8 *Explain what should be considered before creating a home network.*

Where do I start? One of the first things you should do when you set up a new network (or evaluate your existing network) is evaluate the number of devices your network will support. To do so, first list all the devices that will connect to the network. Consider not only the obvious devices, such as computers, laptops, tablets, and printers, but also smartphones, smart TVs, IoT devices, wireless stereo equipment, and any other device that connects to the Internet wirelessly. You might be surprised at how many devices you list. Then include any devices you think you may add.

Next, consider how many users will access the network and their usage. Will multiple users stream video at the same time or only one or two users with moderate usage? Are there many devices requiring wireless access all the time? How many devices need wired access for optimal speed?

The results of this type of evaluation will help you choose the right Internet access plan. In addition, it may help you consider issues such as whether you have to update devices or components to keep up with the improvements in Internet access, standards, and devices.

Which wireless Ethernet standard should my network be using? It may be difficult to know exactly which standard every device is running on, but to achieve the greatest efficiency, it's best for all your network nodes—computers, network adapters, routers, and so on—to use the latest wireless Ethernet standard. For example, if you have the fastest Wi-Fi 6 network adapter in your device but the router is a slower Wi-Fi 5, then data will be sent at the speeds supported by the lower standard. If you haven't updated your router in a while, you may want to consider getting a Wi-Fi 5 or Wi-Fi 6 router (see Figure 7.10) to get the fastest connection speeds. For networks with more than 15 connected devices, you should have at least a Wi-Fi 5 router. Most new routers also support WiGig for speedy short-range connections between devices, and Wi-Fi is necessary for longer-range wireless connections.

Figure 7.10 Wi-Fi 6 wireless routers offer the fastest connection speeds and a greater wireless signal range. *(AlexLMX/Shutterstock)*

How can I tell which wireless standard my router supports? Many routers indicate the wireless standard on the device. If yours doesn't, you can search for more information about your router by entering the model number into a search engine. If your router is provided by your ISP and it's an older standard, you should consider asking your ISP to provide you with a new router.

How can I tell which network adapters are installed in my computer? Most PCs, laptops, and tablets contain embedded network adapters, so it's hard to determine visually what kind of network adapter is installed. If you purchased your device a while ago, you might want to ensure

Bits&Bytes Power Your Devices—Wirelessly

Think of when you've been unable to use your mobile phone, smartwatch, Bluetooth speaker, or other device because it's out of power and you don't have access to a charger or the time to wait while the device charges. Wouldn't it be great always to have access to charging power without needing to plug in your device or wait while it charges?

Cota—short for charging over the air—is an emerging technology that brings wireless power to devices without cables, batteries, or charging pads. Cota technology sends power to mobile devices the same way a wireless router sends data, by flowing through the 2.45-GHz spectrum band. A Cota-enabled device has an embedded Cota receiver (a tiny silicon chip). That receiver sends a low-power beacon signal to a Cota transmitter, indicating what kind of charging is needed. When the transmitter gets the signal, it sends power back to the receiver. The Cota beacon can bounce off walls and objects, so power is delivered even while the device is in motion. Cota wireless power technology is already being used with some hearing aids and headphones. The technology isn't quite ready for smartphones and laptops, but those days are getting closer.

that the adapter meets the latest wireless standard or, if you have problems accessing the network, you may want to determine whether the network adapter is working properly or you need a driver update. To see which network adapters are installed in your Windows computer or to determine whether there's a problem with the adapter, use the Device Manager utility (see Figure 7.11). Generally, you should be able to tell which wireless standard the adapter supports. Otherwise, you can search the Internet for information on your specific adapter to determine its capability. If there is a problem with the network adapter, Device Manager displays a warning icon and allows you to update the driver or disable or uninstall the device.

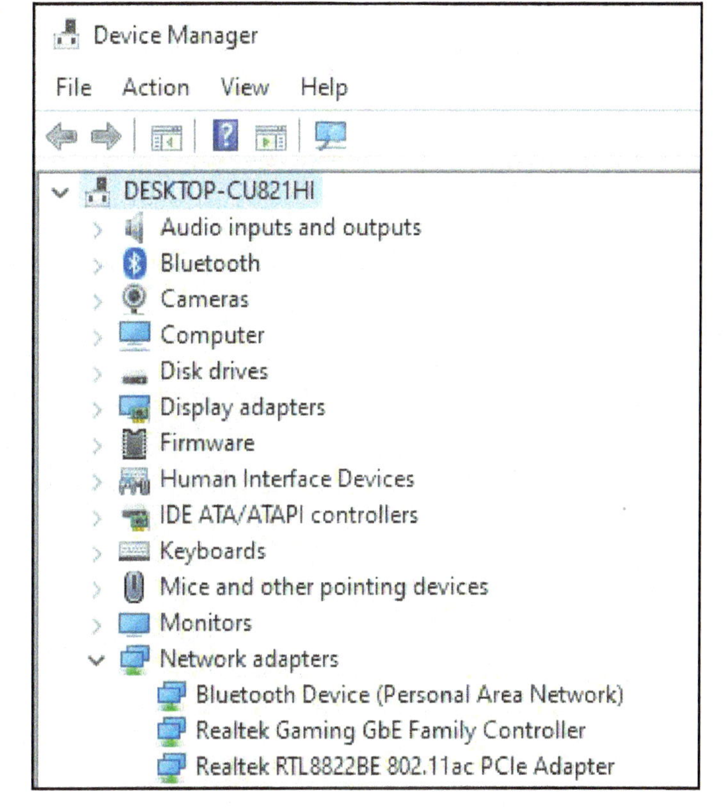

Figure 7.11 The Windows Device Manager shows the wireless and wired network adapters installed on a computer.

> *To access Device Manager, right-click the Windows* ***Start*** *button on the taskbar and select* ***Device Manager****.* (Courtesy of Microsoft Corp.)

Connecting Devices to a Network

Objective 7.9 *Describe how to set up a home network.*

How should a basic home network be set up? Because most newer modems have integrated routers and switches, and only a few wired connections may be needed, setting up a home network is fairly straightforward. As shown in Figure 7.12, the modem/router is the central device of a home network. The router is best placed in a centralized location to increase the wireless signal throughout the home. To reduce interference, the router should be placed away from large metal objects, appliances, and obstructing walls (such as those made of brick). An Ethernet cable connects directly to the modem/router, using the WAN or Internet port. Then all computer devices—such as laptops, tablets, smartphones, HDTVs, gaming consoles, and printers—are connected to the modem/router wirelessly or by wired connections to the LAN ports.

How do I set up a Windows home network for the first time? The process of setting up a network is virtually automated, especially if you're using the same version of Windows on all your computers. The examples in this section assume all computers are running Windows 10.

Before configuring the computers to the network, do the following:

1. Establish a broadband Internet account with a local ISP and either obtain a modem/router from the ISP or purchase a compatible modem/router.
2. Make sure the modem/router is connected to the Internet. If you have separate devices, make sure that your modem is connected to the Internet and that the router is connected to the modem.
3. For any wired connections, plug all devices using Ethernet cables into the router or switch.

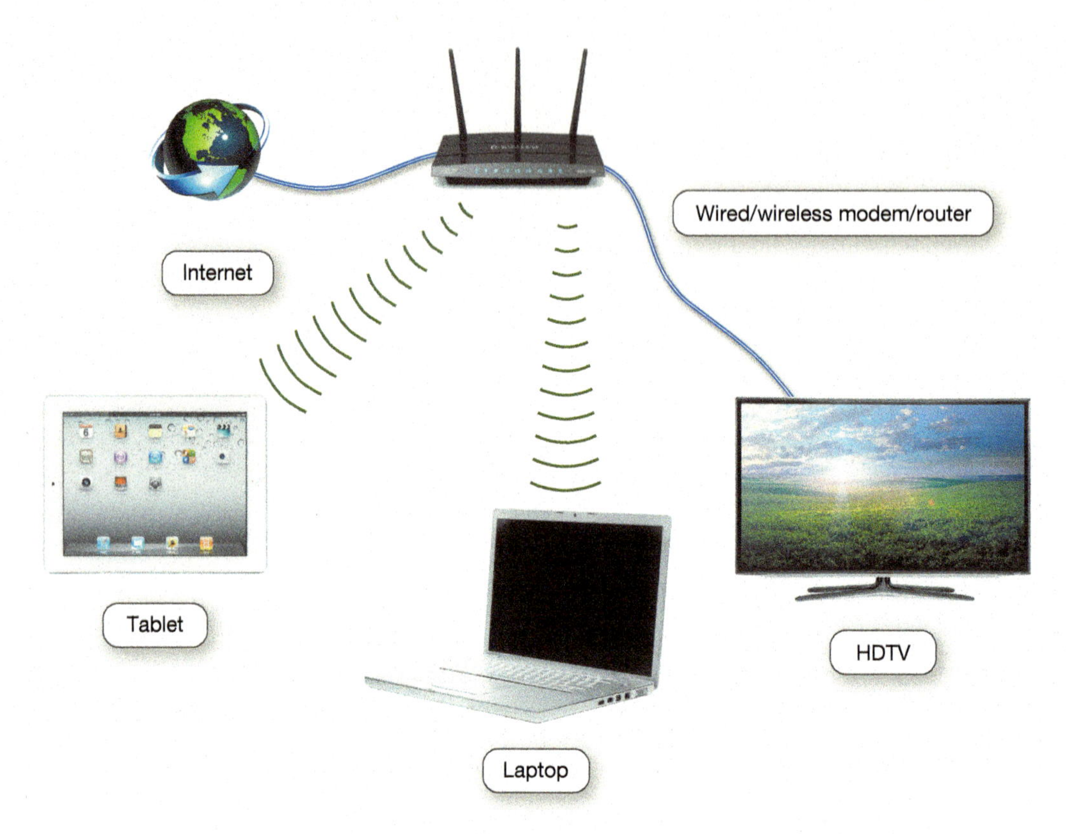

Figure 7.12 A small network with a wired/wireless modem/router attached. *(Scanrail/123RF, Beboy/Shutterstock, Oleksiy Maksymenko/Alamy Stock Photo, Sergey Peterman/Shutterstock, Ruslan Ivantsov/Shutterstock)*

4. Turn on your modem/router to establish the connection to the Internet (or if individual devices, the modem and then the router, allowing each device about one minute each to power up). Then you can power on all other computers and peripherals (printers, scanners, and so on).

If your computer has a wired connection to the network, you should automatically be connected. You should give the network the same secured name you give to your router. (Read the "Troubleshooting Wireless Network Problems" section in this chapter.)

5. To connect wireless devices, ensure that your Wi-Fi is turned on by selecting Wi-Fi in the Network & Internet group in Settings. Then, from the Notification area on the taskbar, click Internet access. A panel opens, displaying the network options. Enter your security passphrase to connect to your wireless network initially. For unsecured networks, you don't need a password. Selecting the Remember this network check box makes any network a preferred network, enabling the computer to connect to the network automatically when that network is available.

Is there a limit to the number of wired devices I can have in my network? Most home routers have an integrated switch with three or four LAN ports on the back to support wired connections. This is usually sufficient for the general house that has at least one computer, maybe a printer, and perhaps some entertainment devices that are connected directly to the router. If you need more LAN ports for wired connections, you can buy a stand-alone switch and plug it into one of the ports on your router (see Figure 7.13). There are many different-sized switches, so purchase one that meets your current needs that also holds a few more for future use.

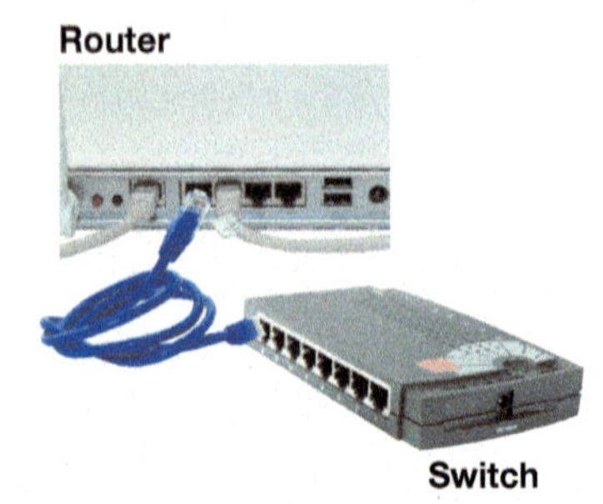

Figure 7.13 You can add ports to your network by connecting a switch to your router. *(Alarich/Shutterstock; Lexan/Shutterstock)*

How many wireless devices can connect to a router in a home network? Most home wireless routers can support up to 253 wireless connections at the same time, although most home networks have far fewer. Regardless of how many devices your home network has, keep in mind that they all share bandwidth when they're connected to a router. Therefore, the more devices connected to a single router that is actively transmitting data, the smaller the portion of the router's bandwidth each device receives. So, for example, when eight wireless devices are all connected to the network, each device has less bandwidth than it would if only six devices were connected to the network.

Bits&Bytes — Mesh Networks: An Emerging Alternative

Mesh networking is the capability to use small radio transmitters instead of wireless routers as network nodes. What is so great about mesh networks is that only one node needs to connect physically to a network connection, and then all other nodes can share wirelessly with each other. This connection-sharing between nodes can extend almost endlessly, because one wired node can share its Internet connection wirelessly with nearby nodes, and then those nodes can share their connection wirelessly with nodes closest to them—thus creating a mesh of connectivity. The truly wireless aspect of mesh networks poses several advantages over the wireless networks that are in place today: They're easier to install because fewer wires need to be run; they can accommodate more nodes; and they enable wireless networks to be created in outdoor and unstructured venues.

In developing nations as well as in some areas in the United States, mesh networks have helped provide access to the Internet in a more timely and inexpensive manner than waiting for the physical connections to be established. Mesh networks can also help promote cellular communications during times of disasters when traditional communications are halted. And if you have an Android mobile phone, you can participate in the Serval Mesh, which allows users to send and receive information without depending on established cellular networks.

Are wireless routers for Windows and macOS networks different? All routers that support Wi-Fi 5 and the newer Wi-Fi 6 should work with computers running the more recent versions of Windows or macOS. However, Apple has routers that are optimized for working with Apple computers. So, if you're connecting Apple computers to a larger home network, you may want to use the Apple AirPort Extreme router, which is Wi-Fi 5–compatible. For smaller home networks using Apple devices, consider using the AirPort Express, which is Wi-Fi 4 compatible. Windows devices can also connect to either router, so it's a great choice for households with both Apples and PCs.

How do I know what's connected to my router? To determine which devices are connected to your router, you need to log on to an account associated with your router's IP address. You can find your router's IP address on the router manufacturer's website. Once you know it, type it into a web browser. You may need to enter a user name and password, but eventually you'll get to a configuration page that lists what wired and wireless devices are in your network.

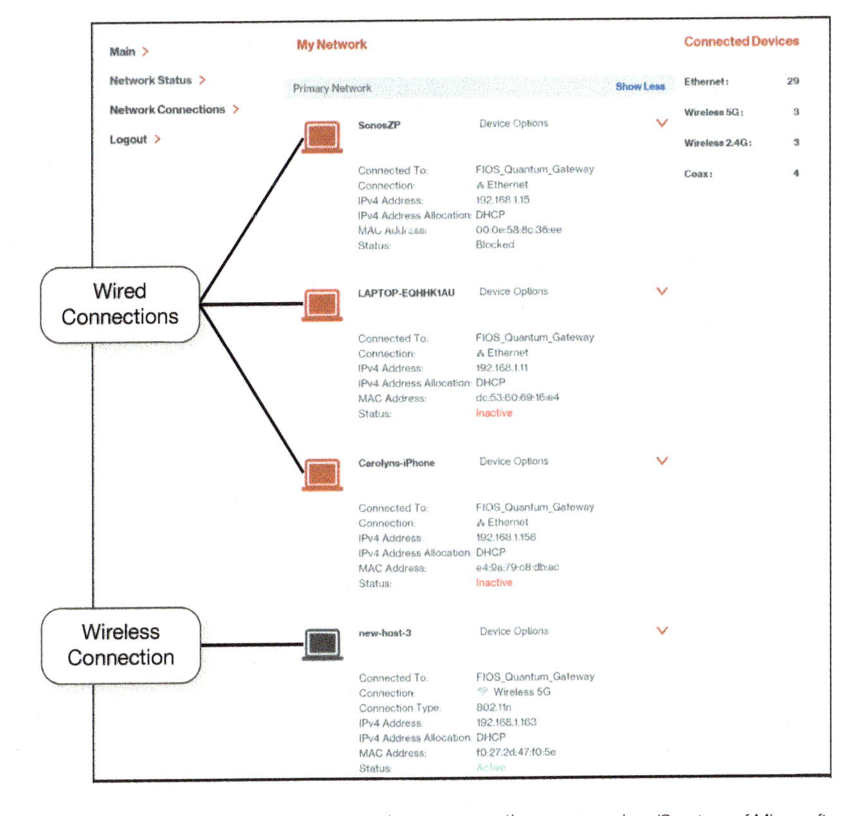

Figure 7.14 Wired and wireless connections can use the same router. *(Courtesy of Microsoft Corp.)*

Figure 7.14 shows a router network listing with wired and wireless devices connected in a home network. You'll notice that each device also has an IP address. You can think of your network as an apartment building. The router's IP address is the building's street address, whereas the IP addresses of the individual devices connected to the router are the apartment numbers. Each device needs an IP address so the router knows to which device to send information.

Other Devices for Your Home Network

What can I attach to my network to facilitate file sharing and backup of data? Network-attached storage (NAS) devices are specialized devices designed to store and manage all network data. Although data can always be stored on individual hard drives in computers on a network, NAS devices provide for centralized data storage and access.

Figure 7.16 Home network servers are often configured with software such as Windows Server. *(Courtesy of Microsoft Corp.)*

Figure 7.15 A network-attached storage device provides centralized data storage and access. *(Leslie Wilk/ Alamy Stock Photo)*

Popular for years on business networks, NAS devices are available for home networks (see Figure 7.15). You can think of them as specialized external hard drives. NAS devices connect directly to the network by a wired Ethernet connection to a router or switch. Specialized software is installed on all computers on the network to ensure that data saved to an individual computer is also stored on the NAS device as a backup.

For Apple computers, the AirPort Time Capsule serves as a NAS device as well as a wireless router. The AirPort Time Capsule works in conjunction with the Time Machine backup feature of macOS.

What other storage devices could I use on a home network? As its name implies, network-attached storage devices are meant specifically for providing a central location for file storage, sharing, and backup. Similar to a NAS, a home network server provides centralized file storage and sharing as well as more sophisticated functionality. A home network server uses a server operating system such as Windows Server or macOS Server (see Figure 7.16). Think of a home network server as another computer on the network, but one that just holds centralized files and applications and is not capable of user interaction like a PC. As such, a home network server can do the following:

- Authenticate and authorize all users and computers on the network
- Provide file storage, sharing, and collaboration tools
- Share applications between connected computers

Note that even though these devices are servers, they don't convert a home P2P network into a client/server network, because these servers don't perform all the functions performed on client/server networks.

Why should I connect digital entertainment devices to my network? When you connect entertainment devices to your network, you can access and share digital content between devices. Connecting these devices to your network also connects them to the Internet so you can access online entertainment. You can also use network-connected devices to play multiplayer games with players all over the world. Smart TVs and other smart devices are continually adding apps and services so that you can play games, view on-demand and online videos, listen to Internet radio, and access social networking sites (see Figure 7.17).

Figure 7.17 Smart TVs have their own apps and enable you to access the web directly. *(Scanrail/123RF)*

Trends in IT How Smart Is Your Home?

The concept of a smart home—where devices and appliances are automated or controlled by apps installed on your smartphone or tablet (see Figure 7.18)—is in full play today, thanks to the power and availability of strong wireless networks and Bluetooth.

At the heart of many smart homes are hubs that are used to control a variety of devices from different manufacturers. A hub receives different wireless signals from smart devices and then translates those into one Wi-Fi signal that your router can understand. In addition, the hub consolidates the controls required by each device and provides a single app for you to interact with, thus simplifying your experience. An alternative to a hub is a home automation platform, such as those offered by Google and Amazon.

With smart home automation, you can dictate how and when a device should engage or respond. For some devices and appliances, just set a schedule, and the rest is automated. Other devices and appliances work using some type of sensor and react to changes in their surrounding environment, such as motion, light, or temperature. For example, some smart home thermostats learn your habits (such as waking and sleeping patterns) and heating and cooling preferences over time and create automatic settings. Some smart home draperies and blinds open and close in reaction to the timing of sunrise and sunset using geolocation settings.

Smart home automation can also alert you to unpredictable events, such as water leaks and unexpected visitors to your home, but smart home technology isn't only about security and efficiency. It's also used to help control your entertainment devices as well.

You can create your own automation by using the IFTTT (If This Then That) app (*IFTTT.com*) or the open-source application openHAB and an IoT device (such as a WeMo motion detector, Nest thermostat, or Philips light bulb).

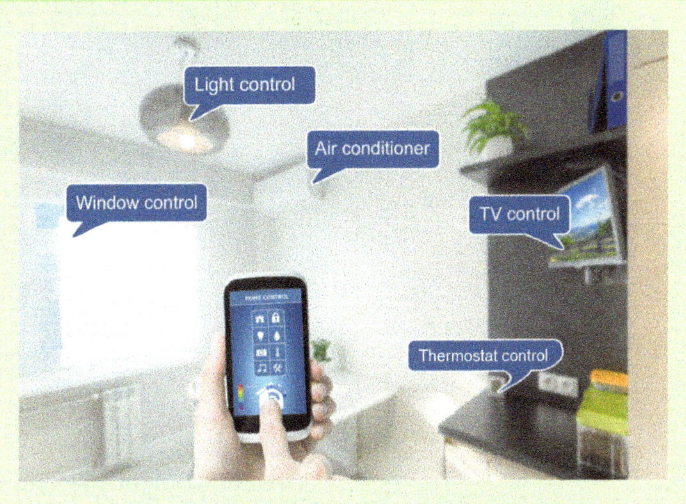

Figure 7.18 You can remotely monitor your home through a dashboard on your mobile device with smart home technology. *(Scyther5/123RF)*

Managing and Securing Wireless Networks

All networks require some maintenance and management, but wireless networks require additional installation, maintenance, and security measures. In this section, we look at how to troubleshoot typical wireless network problems and how to protect wireless networks from common security threats.

Troubleshooting Wireless Network Problems

Objective 7.10 *Describe the potential problems with wireless networks and the means to avoid them.*

What types of problems can I run into when installing wireless networks? The maximum range of current wireless devices is about 350 feet, but as you go farther away from your router, the throughput you achieve decreases. Obstacles between wireless nodes also decrease throughput. Walls, floors, and large metal objects (such as refrigerators) are the most common sources of interference with wireless signals.

What if a node on the network seems slow? Repositioning the node within the same room (sometimes even just a few inches from the original position) can affect communication between nodes. If this doesn't work, move the device closer to the router or to another room in your house. You might also try repositioning the antennas on your router. If these solutions don't work, consider adding additional equipment to your network.

Bedroom

Den

Back Porch

Wireless router

Wireless range extender

Computer A with wireless network adapter

Computer B with wireless range extender

Laptop C with wireless network adapter

Figure 7.19 Because a wireless range extender is installed in the den, Laptop C on the back porch can now connect to the wireless network.

What equipment will help improve a wireless signal throughout the network? There are two types of devices that you can use to amplify or extend the wireless signal to parts of your home that are experiencing poor connectivity: access points and extenders. An access point connects to the router with an Ethernet cable and creates a wireless signal at the location of the access point. Therefore, an access point should be placed in the location where the original Wi-Fi signal becomes weaker. However, because an access point requires a wired connection with the router, given the distance between access point and router, using an access point to improve connectivity may not always be possible.

Alternatively, consider using a wireless range extender, which repeats or amplifies the wireless signal from the router. Because Wi-Fi signals have limited ranges, wireless range extenders work by receiving the Wi-Fi signal near the end of the range and transmitting it a second time to extend the range of the network. For example, as shown in Figure 7.19, Laptop C on the back porch can't connect to the wireless network, even though Computer B in the den can. By placing a wireless range extender in the den, where there is still good connectivity to the wireless network, the wireless signal is amplified and beamed farther out to the back porch. This allows Laptop C to make a good connection to the network. The downside of using an extender over an access point is that it shares the wireless connection and, thus, potentially slows the entire network.

Bits&Bytes Analyzing Network Problems

If you have areas in your home where you have little or no Wi-Fi coverage, try using a Wi-Fi analyzer to gather information before embarking on ways to improve your coverage. Free apps such as WiFi Analyzer (Android) or Wi-Fi Inspector (Windows) provide signal strength and other details that might be helpful in remedying the problem.

Securing Wireless Networks

Objective 7.11 *Describe how to secure wireless home networks.*

Why is a wireless network more vulnerable than a wired network? Packets of information on a wireless network are broadcast through the airwaves. Savvy hackers can intercept and decode information from your transmissions that may allow them to bypass standard protections, such as a firewall, that you have set up on your network. All computers that connect to the Internet need to be secured from intruders. This is usually accomplished by using a firewall, which is a hardware or software solution that helps shield your network from prying eyes. (We discuss firewalls at length in Chapter 9.)

Wireless networks present special vulnerabilities; therefore, you should take additional steps to keep your wireless network safe. Wireless networks with strong signals have wide ranges that may extend outside your house. This makes it possible for someone to access your network without your knowledge. Hackers are known to take advantage of these vulnerabilities, but in areas where residences are close together, wireless signals can reach a neighbor's residence, enticing someone to borrow your wireless signal rather than getting one of their own. Piggybacking is connecting to a wireless network without the permission of the owner. This practice is illegal in many jurisdictions.

Why should I be worried about someone logging on to my wireless network without my permission? If your neighbor is using your network connection, his or her usage could be slowing down your connection speed. Some neighbors might even be computer savvy enough to penetrate your unprotected wireless network and steal personal information, just as any hacker could. Any cyberattack or illegal behavior a hacker initiates from your wireless network could get you in trouble with the authorities. Therefore, to prevent unwanted users gaining access to your wireless network, it's important to take measures to secure it.

How can I secure my wireless network? To secure a wireless network, take the additional precautions described as follows:

1. **Use encryption and security protocols.** Most routers ship with security protocols such as Wired Equivalent Privacy (WEP), Wi-Fi Protected Access (WPA), or Wi-Fi Protected Access version 2 (WPA2). Each uses encryption (a method of translating your data into code) to protect data in your wireless transmissions. WPA3 is the most secure, but WPA2 may be compatible with more client devices. WEP and WPA still offer protection but offer the least protection.
2. **Change your network name (SSID).** Each wireless network has its own name to identify it, known as the service set identifier (SSID). Unless you change this name when you set up your router, the router uses a default network name that all routers from that manufacturer use (such as "Wireless" or "Netgear"). Hackers know the default names and access codes for routers. If you haven't changed the SSID, it's advertising the fact that you probably haven't changed any of the other default settings for your router either.
3. **Disable SSID broadcast.** Most routers are set up to broadcast their SSIDs so that other wireless devices can find them. Disable SSID broadcasting, if your router supports doing so. This makes it more difficult for a hacker to detect your network and nearly impossible for a neighbor to connect to your network inadvertently. Keep in mind that you will have to re-enable this setting when adding new components to the network.
4. **Change the default password on your router.** Routers have default user names and passwords. Hackers can use these to access your router and break into your network. Change the password on your router to something hard to guess. Use at least 12 characters that are a combination of letters, symbols, and numbers.
5. **Use a passphrase.** When you attempt to connect a node to a security-enabled network for the first time, you're required to enter the security key. The security key or passphrase (see Figure 7.20) is the code that computers on your network need to decrypt (decode) data transmissions. Without this key, it's extremely difficult, if not impossible, to decrypt the data transmissions from your network. Usually the manufacturer assigns a passphrase to the router. You can find the passphrase printed on a sticker on the side of your router. You can keep that passphrase or create a passphrase that is more meaningful and easier for you to remember. Keep in mind, however, that you should take similar precautions when creating a passphrase as you do when creating any password. Using upper- and lowercase letters, numbers, and special characters helps create a passphrase that is hard to guess by intruders.
6. **Limit your signal range.** Many routers allow you to adjust the transmitting power to low, medium, or high. Cutting down the power to low or medium could prevent your signal from reaching too far away from your home, making it tougher for interlopers to poach your signal.
7. **Apply firmware upgrades.** Your router has read-only memory that has software written to it. This software is known as firmware. As bugs are found in the firmware (which hackers might exploit), manufacturers issue patches, just as the makers of operating system software do. Periodically check the manufacturer's website and apply any necessary upgrades to your firmware.
8. **Disable remote access.** Some routers have a setting to allow remote access. This makes it easier for the manufacturer to offer technical support. However, hackers can use this to get into your home network. You can always turn on this feature when needed.

If you follow these steps, you'll greatly improve the security of your wireless network. In Chapter 9, we continue to explore many other ways to keep your computer safe from malicious individuals on the Internet and ensure that your digital information is secure.

Securing Wireless Networks

In this Sound Byte, you learn some simple steps to secure your wireless network against intruders.

Managing and Securing Your Wireless Network

In this Helpdesk, you'll play the role of a helpdesk staffer, fielding questions about ways to manage and secure wireless networks.

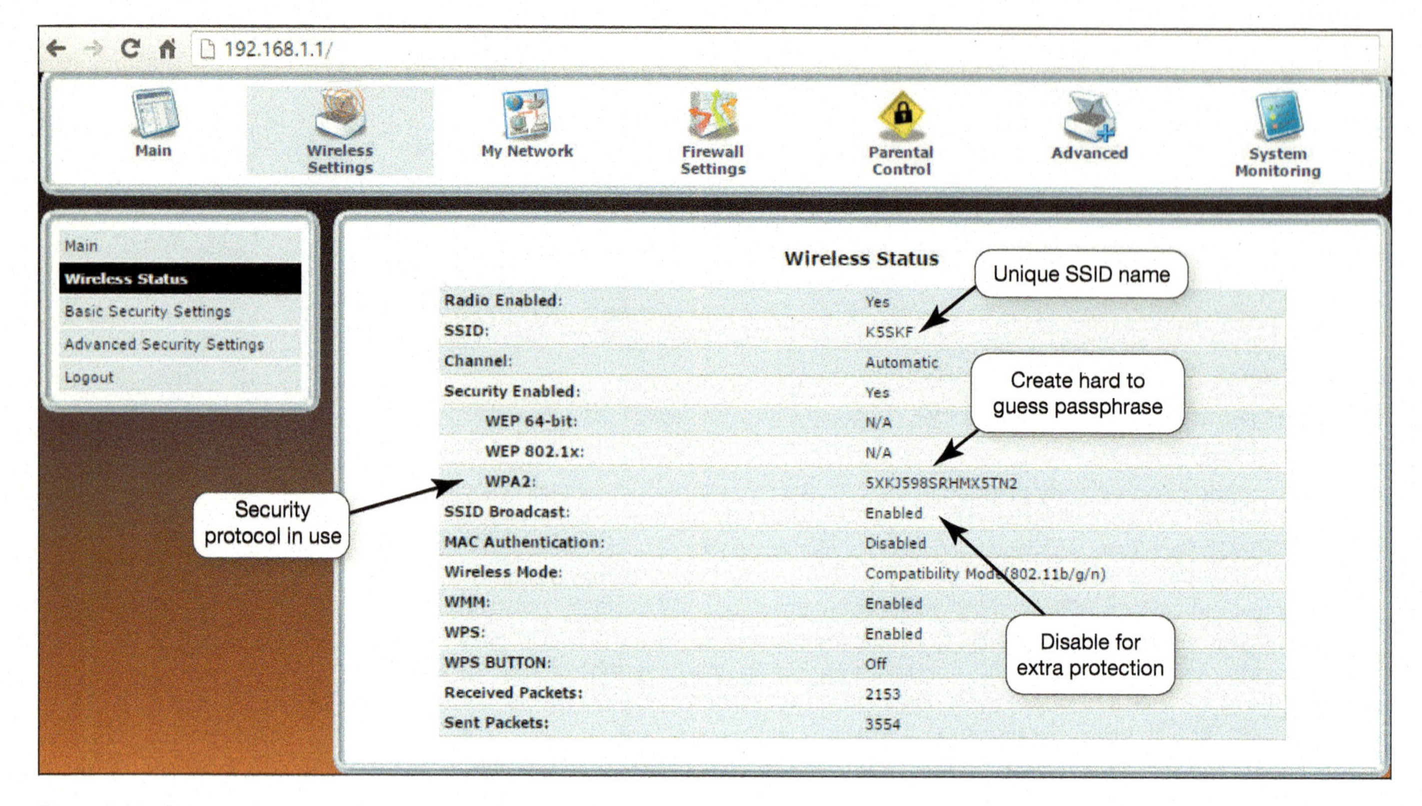

Figure 7.20 By accessing your router, you can configure the security protocols available on your router and change the SSID. (Courtesy of Microsoft Corp.)

Before moving on to the Chapter Review:

1. Watch Chapter Overview Video 7.2.
2. Then take the Check Your Understanding quiz.

Check Your Understanding // Review & Practice

For a quick review to see what you've learned so far, answer the following questions.

multiple choice

1. Which device generally includes a built-in switch?
    a. Modem/router
    b. Wired-connected PC
    c. Network range extender
    d. Network interface card
2. When planning a network, which devices should be considered?
    a. Only the computers that will have a wired connection
    b. Just the computers and printers that will have a wired connection
    c. Just PCs connected with a wired connection and any laptop or tablet connected wirelessly
    d. All devices that will connect with a wired or wireless connection
3. Which of the following might cause a potential problem on a wireless network?
    a. Devices too close to the router
    b. Interference from walls, floors, and large objects
    c. A router using the Wi-Fi 5 standard
    d. A unique password created for the network
4. Which of the following would help improve a wireless signal throughout a network?
    a. Network Internet card
    b. Access point
    c. Network switch
    d. Wireless router
5. In setting up a wireless network, which precaution is used to change the security protocol on the router?
    a. Disabling the passphrase
    b. Changing the default network name
    c. Changing the default network password
    d. Using Wi-Fi Protected Access version 2

chew on this

(3Dalia/Shutterstock)

Networks facilitate the sharing of files between connected computers. Online storage systems such as Dropbox, Google Drive, and OneDrive take advantage of using the Internet to share files. Alternatively, people still use flash drives to transport files physically from one location to another. Discuss advantages and disadvantages of the various options of sharing and transporting files between devices.

Go to **MyLab IT** to take an autograded version of the *Check Your Understanding* review and find all media resources for the chapter.

For the IT Simulation for this chapter, see MyLab IT.

7 Chapter Review

Summary

Part 1
How Networks Function

Learning Outcome 7.1 **You will be able to explain the basics of networking, including the components needed to create a network, and describe the different ways a network can connect to the Internet.**

Networking Fundamentals

Objective 7.1 *Describe computer networks and their pros and cons.*

- A computer network is simply two or more computers that are connected, using software and hardware so they can communicate.
- Advantages of networks include allowing users to share an Internet connection, share printers and other peripheral devices, share files, and communicate with computers regardless of their operating system.
- Disadvantages for larger networks are that they require administration and may require costly equipment.
- Data transfer rate (bandwidth) is the maximum speed at which data can be transferred between network nodes and is always faster than the actual speed of data transfer, known as throughput.

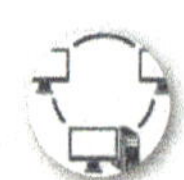

Network Architectures

Objective 7.2 *Explain the different ways networks are defined.*

- Networks can be defined by the distance between nodes:
  - A personal area network (PAN) is used for communication among personal mobile devices using Bluetooth or Wi-Fi wireless technologies.
  - A local area network (LAN) connects nodes that are located in a small geographical area.
  - A home area network (HAN) is a specific type of LAN located in a home.
  - A metropolitan area network (MAN) is a large network in a specific geographical area.
  - A wide area network (WAN) spans a large physical distance.
- Networks can also be defined by the level of administration:
  - **Central:** A client/server network contains two types of computers: a client computer on which users perform specific tasks and a server computer that provides resources to the clients and central control for the network. Most networks that have 10 or more nodes are client/server networks.
  - **Local:** Peer-to-peer (P2P) networks enable each node connected to the network to communicate directly with every other node. Most home networks are P2P networks.
- Ethernet networks are the most common networks used in home networking. Most Ethernet networks use a combination of wired and wireless connections, depending on the data throughput required. Wired connections usually achieve higher throughput than wireless connections. Wired Ethernet home networks use the gigabit Ethernet standard.
- Wireless Ethernet networks are identified by a protocol standard: 802.11a/b/g/n/ac. 802.11ac is the newest standard. WiGig (802.11ad) is a wireless link between devices. WiGig is like Bluetooth, but faster; 802.11af (Super Wi-Fi or White-Fi) and 802.11ah (HaLow or low-power Wi-Fi) help accommodate the continued dependence on wireless devices.

Network Components

Objective 7.3 *Describe the types of transmission media used in networks.*

- Wired networks use various types of cable to connect nodes, including unshielded twisted-pair cable, coaxial cable, and fiber-optic cable. The type of network and the distance between the nodes determine the type of cable used.
- There are several types of UTP cable. Cat 5e is the least expensive and probably most common for

standard home networks, but given the increasing number of devices connected to a home network, Cat 6 and Cat 6a cable may be used, though they will be more costly.
- Wireless networks use radio waves.

Objective 7.4 *Describe the basic hardware devices necessary for networks.*

- A modem is required to connect the home network to the Internet.
- Network navigation devices, such as routers and switches, are necessary for computers to communicate in a network.
- All connected devices must have a network adapter. Network adapters may be used for both wired and wireless connections. Most devices have an integrated network adapter.

Objective 7.5 *Describe the type of software necessary for networks.*

- Home networks need operating system software that supports peer-to-peer networking. All current operating systems, such as Windows 10, macOS, and Linux, support P2P networking.
- Servers on client/server networks require network operating systems (NOS).

Connecting to the Internet

Objective 7.6 *Summarize the broadband options available to access the Internet.*

- Broadband connections include the following types:
  - Fiber-optic cable uses glass or plastic strands to transmit data by light signals.
  - Cable transmits data over coaxial cable, which is also used for cable television.
  - DSL uses twisted-pair wire, similar to that used for telephones.
  - Satellite is a connection option for those who do not have access to faster broadband technologies. Data is transmitted between a satellite dish and a satellite that is in a geosynchronous orbit.

Objective 7.7 *Summarize how to access the Internet wirelessly.*

- Wi-Fi allows users to connect to the Internet wirelessly but is not as fast as a wired connection.
- Mobile broadband is a service delivered by cell phone networks.

Part 2
Your Home Network

Learning Outcome 7.2 **You will be able to describe what is necessary to install and configure a home network and how to manage and secure a wireless network.**

Installing and Configuring Home Networks

Objective 7.8 *Explain what should be considered before creating a home network.*

- Most home network routers should support both wireless and wired access to the Internet.
- For a home network to run efficiently, all connected devices and routers should use the same Ethernet standard, ideally the most current Ethernet standard.
- The Device Manager utility in Windows lists all adapters installed on your computer.

Objective 7.9 *Describe how to set up a home network.*

- The latest versions of Windows make it easy to set up wired and wireless networks.
- All devices are connected to your router, either wirelessly or with a wired connection. Wired connections deliver better throughput than wireless.
- To add additional ports to your network, you can connect a switch to your router.
- Network-attached storage (NAS) devices enable you to store and share data files such as movies and music, as well as provide a central place for file backups.
- Home network servers can be used instead of a NAS device if your needs require more sophisticated functionality.

Managing and Securing Wireless Networks

Objective 7.10 *Describe the potential problems with wireless networks and the means to avoid them.*

- Wireless connections have less throughput; therefore, you may need to consider a wired connection for certain devices.
- Excessive distance from the router, as well as walls, floors, and large metal objects between a device and the router, can interfere with wireless connectivity.
- Access points or wireless range extenders can amplify signals to improve connectivity in areas of poor signal strength.

Objective 7.11 *Describe how to secure wireless home networks.*

- Wireless networks are more susceptible to hacking than wired networks are, because the signals of most wireless networks extend beyond the walls of your home.

- To prevent unwanted intrusions into your network, you should change the default password on your router to make it tougher for hackers to gain access, use a hard-to-guess SSID (network name), disable SSID broadcasting to make it harder for outsiders to detect your network, enable security protocols such as WPA2, WPA or WEP, create a network passphrase, implement media access control, limit your signal range, and apply firmware upgrades.

Be sure to check out **MyLab IT** for additional materials to help you review and learn. And don't forget to watch the Chapter Overview Videos.

Key Terms

4G **255**
5G **255**
access point **266**
backward compatibility **250**
broadband **253**
cable Internet **254**
Cat 5e cable **252**
Cat 6 cable **252**
Cat 6a cable **252**
Cat 7 cable **252**
client/server network **247**
coaxial cable **251**
data plan **255**
data transfer rate (bandwidth) **246**
DSL (digital subscriber line) **254**
Ethernet network **248**
fiber-optic cable **251**
fiber-optic service **254**
firmware **267**
gigabit Ethernet (GbE) **248**
home area network (HAN) **246**
home network server **264**
Internet service provider (ISP) **254**
local area network (LAN) **246**
metropolitan area network (MAN) **246**
mobile broadband **255**
mobile hotspot **255**
modem **252**
network **244**
network adapter **253**
network administration **246**
network architecture **246**
network-attached storage (NAS) device **263**
network navigation device **253**
network operating system (NOS) **253**
node **244**
peer-to-peer (P2P) network **247**
peer-to-peer (P2P) sharing **249**
personal area network (PAN) **246**
piggybacking **267**
router **253**
satellite Internet **254**
service set identifier (SSID) **267**
smart home **265**
switch **253**
throughput **246**
transmission media **250**
twisted-pair cable **251**
unshielded twisted-pair (UTP) cable **251**
wide area network (WAN) **246**
Wi-Fi **248**
wireless Internet service provider (wireless ISP) **255**
wireless range extender **266**

Chapter Quiz // Assessment

For a quick review to see what you've learned, answer the following questions. Submit the quiz as requested by your instructor. If you're using MyLab IT, the quiz is also available there.

multiple choice

1. Which of the following would NOT be a benefit of networking computers?
  a. Sharing an Internet connection
  b. Sharing software licenses
  c. Sharing printers and other peripherals
  d. Sharing files

2. Bethany is in her home, watching the video she took while on vacation, while her brother is playing FIFA Soccer on his Xbox and her dad is checking stock quotes on his iPad. What kind of network does this family most likely have?
  a. Client/server network
  b. P2P network
  c. Both a and b
  d. Neither a nor b

3. What two devices are often combined into one device to connect the network to the Internet and to share the connection between devices on the network?
   a. Router hub and switch
   b. Switch and hub
   c. Router and switch
   d. Modem and router
4. Which required network device is generally integrated in computers and laptops?
   a. Router
   b. Modem
   c. Switch
   d. Network Adapter
5. Which of the following should you NOT do to secure a wireless network?
   a. Implement media access control.
   b. Disable SSID broadcasting.
   c. Create a security encryption passphrase.
   d. Keep the default network name and router password.
6. Which type of broadband Internet offers connections nearing the speed of light?
   a. Cable
   b. Satellite
   c. DSL
   d. Fiber-optic
7. Which of the following would NOT cause interference or poor connectivity between nodes on a wireless network?
   a. Concrete walls
   b. A wireless extender
   c. Some appliances
   d. Nodes that are too far apart
8. Which of the following would most likely use a wired network connection?
   a. Laptop
   b. Smartphone
   c. Desktop computer
   d. Tablet
9. The type of network that would be necessary to connect the networks of two college campuses would be a:
   a. LAN
   b. MAN
   c. WAN
   d. PAN
10. Which of the following describes satellite broadband?
   a. It uses strands of pure glass or plastic.
   b. It requires a dish installed outside the building.
   c. It connects with coaxial cable.
   d. Speed drops as the distance from the main signal source increases.

true/false

_____ 1. You can attach more than four wired devices to a network by using a switch.

_____ 2. P2P networks require specialized network operating system software.

_____ 3. Actual data throughput is usually higher than the stated bandwidth.

_____ 4. A network-attached storage device is like a network-connected hard drive.

_____ 5. Cat 5e and Cat 6 are types of twisted-pair cable.

What do you think now?

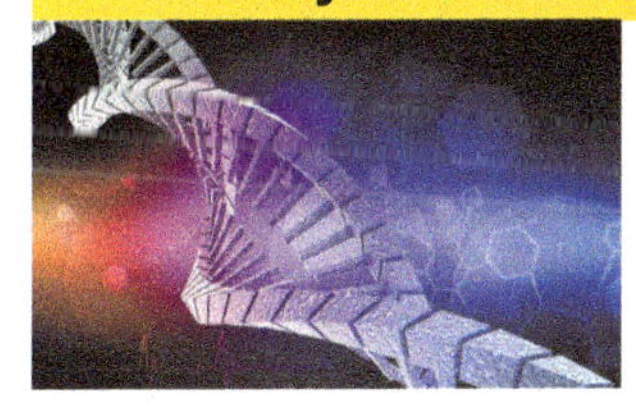

Your genetic data is uniquely yours, so before spitting into a tube with the hope of getting a personalized wine list, consider the risks involved. Research the privacy risks of sharing your DNA with testing companies and explain why you would be likely or unlikely to submit your DNA for testing.

Team Time

Providing Wireless Internet Access to the Needy

problem

Not everyone can afford wireless Internet access, and this can put families, especially ones with school-age children, at a disadvantage.

task

You're volunteering for a charity that wants to install wireless networks in the homes of needy families. The project is funded by charitable grants, with the objective of providing basic broadband Internet access and networking capabilities at no cost to the recipients.

process

Break the class into three teams. Each team will be responsible for investigating one of the following issues.

1. **Internet service providers:** Research broadband Internet service providers that serve the town where your school is. Compare maximum upload and download speeds as well as costs. Be sure to consider the cost of the modems—whether they're purchased or rented. Select what your group considers the best deal.
2. **Networking equipment and network-ready peripherals:** Each home needs to be provided with an 802.11ac-capable wireless router (or modem/router) and a network-ready all-in-one printer. Research three options for these devices, considering price as well as functionality.
3. **Security:** Work in conjunction with the group researching routers to determine the best router to purchase, because it needs to support a security protocol such as WPA2. Consider other types of protection needed, such as antivirus software and firewalls.

Present your findings to your class and come to a consensus about the solution you would propose for the charity.

conclusion

Providing technology to underserved populations on a cost-effective basis will go a long way toward closing the digital divide.

Ethics Project

Firing Employees for Expressing Views on Social Media Sites

In this exercise, you'll research and then role-play a complicated ethical situation. The role you play may or may not match your own personal beliefs; regardless, your research and use of logic will enable you to represent the view assigned. An arbitrator will watch and comment on both sides of the arguments, and together the team will agree on an ethical solution.

problem

Employers are often intolerant of employees who express negative opinions or expose inside information about their employers on social media sites. Given that most jurisdictions in the United States use the doctrine of employment at will (employees can be fired at any time for any reason), many employers are quick to discipline or terminate employees who express opinions with which the company disagrees. When such cases come to court, the courts often find in favor of the employers.

research areas to consider

- Ellen Simonetti and Delta Airlines
- Fired for blogging about work
- Free speech
- Joyce Park or Michael Tunison

process

1. Divide the class into teams. Research the listed areas and devise a scenario in which someone has complained about an employee blogging about a sensitive company issue, such as cleanliness at a food manufacturing facility.
2. Team members should assume the role of an employee or a human resources manager that represents each side of this scenario. Another team member can be an arbitrator who would present an objective point of view for each character. To set the stage for the role-playing event, each team member should write a summary that provides background information for their role and that details their character's behaviors.
3. Team members should present their case to the class or submit a PowerPoint presentation for review, along with the summary and resolution they developed.

conclusion

As technology becomes ever more prevalent and integrated into our lives, more and more ethical dilemmas will present themselves. Being able to understand and evaluate both sides of the argument, while responding in a personally or socially ethical manner, will be an important skill.

Solve This with Word

MyLab IT Grader

Home Networking Guide

You write a newsletter, **The Everyday Technologist**, which specializes in technology tips and tricks for the casual everyday technology user. Your current newsletter is a guide to home networks. The basic document has been created, and now using Microsoft Word you need to add formatting and images to make the guide more visually appealing.

You will use the following skills as you complete this activity:

- Apply Text Effects
- Insert Word Art
- Insert Columns
- Insert Section Breaks
- Insert Column Break
- Insert Images
- Apply Text Wrapping
- Use Custom Margins

Instructions

1. Open *TIA_Ch7_Start.docx*, and save as *TIA_Ch7_ LastFirst.docx*.
2. Select the title Home Networking Guide and convert the text to a **WordArt Object**, using **Fill: Gold, Accent color 4**; **Soft Bevel**.
3. With the WordArt object still selected, click **Layout Options** and select **Top and Bottom**. Click **See more** ... at the bottom of the Layout Options window and, in the Layout dialog box, in the Horizontal section, select **Alignment** and then choose **Centered** from the drop-down box.
4. Apply **Fill: Blue**, **Accent Color 2**, **Shadow Text Effect** to the three headings: Hardware, Wireless vs. Wired, and Powerline Adapters. Increase the font size of each heading to 18.
5. Create **custom margins** so all margins are **.7"**.
6. Position the cursor before the Hardware heading. Insert a **Section Break (Continuous)** and then apply **Two Column** formatting.
7. Place the cursor at the end of the second bullet point paragraph in the Hardware section. Insert the *TIA_Ch7_Modem Router.jpg* image. Change Layout Option of the image to **Square**. Resize the image so the height is **.9"** and the width is **1.2"** and then move the image so that the two top lines of the Router paragraph appear above the image, and the left side of the image is flush with the left margin.
8. Place the cursor at the beginning of the Powerline Adapters paragraph. Insert the *TIA_Ch7_Powerline.jpg* image. Change Layout Option to **Tight**. Change the height of the image to **1"** and then move the image so the right side is flush with the right margin of the document and centered vertically in the Powerline Adapter paragraph.
9. Place the cursor at the beginning of Current Wireless (Wi-Fi) Standards. Insert a **Column Break**. Format the heading with **Bold** and **Italics**.
10. Save and submit for grading.

Chapter

Managing Your Digital Lifestyle: Challenges and Ethics

 For a chapter overview, watch the Chapter Overview Videos.

PART 1

The Impact of Digital Information

Learning Outcome 8.1 **You will be able to describe the nature of digital signals; how digital technology is used to produce and distribute digital texts, music, and video; and the challenges in managing a digital lifestyle.**

PART 2

Ethical Issues of Living in the Digital Age

Learning Outcome 8.2 **You will be able to describe how to respect digital property and use it in ways that maintain your digital reputation.**

All media accompanying this chapter can be found here.

A Video-Playing App on **page 300**

(Valeriy Efremov/123RF, Tetra Image/Getty Images; Kevin Nixon/Future Music Magazine/Getty Images; Ollyy/Shutterstock; Sebastien Decoret/123RF; Jirsak/Shutterstock; John Williams RUS/Shutterstock)

What do you think?

Autonomous (self-driving) vehicles are fast becoming a reality. Proponents of these vehicles tout the increased levels of safety and **decline of accidents**, because AIs working together with one another will result in lower speeds for the same traffic volume and **better decision making** than humans can achieve. But it may take a while to achieve this utopia.

Theoretically, having an AI make the same decisions based on the same set of circumstances should result in fewer accidents than would relying on individual humans to make the best decisions in stressful situations. However, the AI is programmed by humans and therefore is subject to errors. Also, automobiles operate in a very **complex environment** where many things could happen at any time. Accidents (including fatalities) have occurred while testing the prototypes of autonomous vehicles. Is it possible to program an AI to handle every possible situation it might encounter?

Once autonomous vehicles are deemed sufficiently developed to be permitted on public roads, they won't suddenly replace all human-driven vehicles. Developers are concerned that human drivers may bully autonomous vehicles by driving aggressively (such as cutting them off in traffic) if they believe the autonomous vehicles' AI will always err on the side of caution. Integrating autonomous vehicles onto our streets might be tougher than we first thought.

?

Which of the following statements do you agree with?

- *I'd never feel comfortable using a driverless car.*
- *I might feel comfortable putting my car in autonomous mode if I knew I could take control of it.*
- *Autonomous vehicles are safer than human drivers so I'd feel fine using a driverless car.*

See the end of the chapter for a follow-up question.

(Oleksandr Omelchenko/123RF; Who is Danny/Shutterstock)

Part 1 For an overview of this part of the chapter, watch **Chapter Overview Video 8.1.**

The Impact of Digital Information

Learning Outcome 8.1 You will be able to describe the nature of digital signals; how digital technology is used to produce and distribute digital texts, music, and video; and the challenges in managing a digital lifestyle.

Do you realize how many revolutionary changes are happening in all kinds of media as we continue to use digital formats? The entertainment industry has become an all-digital field. The publishing, music, photography, and film industries have all seen radical changes in how content is created and distributed to an audience. In this section, we examine what digital really is and explore how digital technology has changed the production and distribution of media.

Digital Basics

How did we end up in a world in which we're all connected to our devices and able to communicate with each other whenever and almost wherever we want? Part of the story relates to the concept of *digital convergence*, as we explore in this section.

Digital Convergence and the Internet of Things

Objective 8.1 *Describe how digital convergence and the Internet of Things have evolved.*

What is digital convergence? Digital convergence refers to our ability to use a single device to meet all of our media, Internet, entertainment, and phone needs. For example, you see digital convergence in the evolution of smartphones, with which you do just about anything a computer can do. You can also see the push to digital convergence in the migration of digital devices into environments like the cabins of cars (see Figure 8.1a). In fact, devices are beginning to converge so much that an organization—the Digital Living Network Alliance (*dlna.org*)—exists to standardize them.

Even some refrigerators, like the one shown in Figure 8.1b, include LCD touch-screen displays and network adapters so that they can display recipes from websites as well as place a call to the service center and schedule their own repair visit for you. These devices are all part of the so-called *Internet of Things* (IoT).

(a)

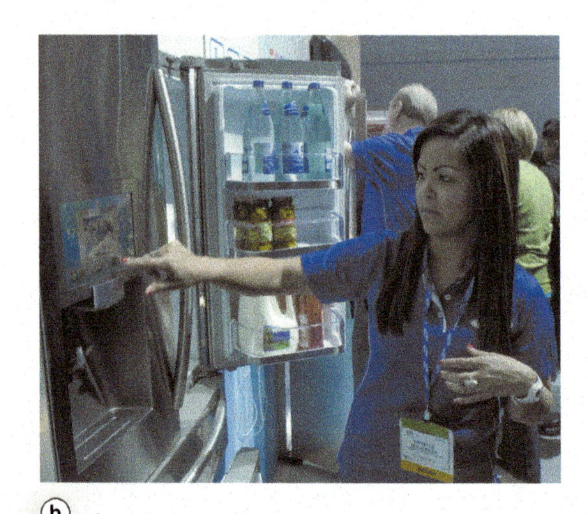

(b)

Figure 8.1 (a) Auto electronics have now converged with tablet technology. Today's vehicles feature touch-screen displays that controls many of the car's electronics systems, providing Internet access, mobile communications, and navigation features, including sensors that detect other vehicles. (b) Some refrigerators are now equipped with touch screens that connect to the Internet. *(Iain Masterton/Alamy Stock Photo, Steve Marcus/Reuters/Alamy Stock Photo)*

Figure 8.2 The Internet of Things

The Things You Can Do

Home entertainment. The DVR is just for starters.

Security cameras. Watch your house with cameras in each room.

Appliances. Trigger or time any function, or set it to alert you.

Security locks. Let people into your home remotely.

Lighting. Turn lights on or off remotely, for hands-free safety.

Smoke detector. Get a fire alert on your phone.

Thermostat. Heat choices when you want, for energy savings.

Irrigation. Program lawn sprinklers for best times of the day.

2000

2004

2008

2012

2016

2020

Each "thing" on the Internet has its own unique address.

It's called an IP (Internet Protocol) address. As more things people own will have IP addresses — not just computers and smart phones, but the above kinds of devices — there'll be more IP addresses assigned.

This graph shows the number of things with an IP address, for each person on the planet (not just online).

That's how many IP addresses are used. So how many IP addresses *can there be?*

Since the beginning of the internet, they've been using a method called IPv4: four clumps of digits, broken up with periods. That method let us have a total of

4,294,967,296

IP addresses, or about 1 for every 2 people on the planet (not just online) in 2016.

But soon, it was clear there were way too few combinations. So on June 16, 2012, they established IPv6, which is six clumps of digits.

With that new standard, the potential number of IP addresses is now

340,282,366,920,938,463,463,
374,607,431,768,211,456

...or, **over 45 octillion** (45×10^{27}) for each person on the planet (not just online) in 2016.

← We are only using enough for seven per person in 2020. Looks like we won't ever run out.

The Internet of Things (IoT) is defined as the interconnection of uniquely identifiable embedded computing devices that transfer data over a network without requiring human-to-human or human-to-computer interaction (see Figure 8.2). "Things" can be anything—machines, appliances, buildings, vehicles, even animals, people, plants, and soil. In addition to all the smart TVs, refrigerators, and thermostats we hear so much about, a thing can be a heart monitor implant that can warn of an oncoming problem or a device in an automobile that monitors driving behaviors so insurance companies can better assess risk. Ultimately, the Internet of Things is about connecting companies, people, and technology in real time through an extension of the Internet into the physical world. By the year 2025, there will be an estimated 75 billion connected devices.

As our appliances, cars, and homes become designed to communicate over a common network, how we manage our media, control our living spaces, and communicate with others will continue to change. Let's start by looking at the foundation for all our computing devices: digital signals.

Digital versus Analog

Objective 8.2 *Explain the differences between digital and analog signals.*

What does it mean to be digital? Any kind of information can be digitized—measured and converted to a stream of numerical values. Consider sound. It's carried to your ears by sound waves, which are actually patterns of pressure changes in the air. Images are our interpretation of the changing intensity of light waves around us. These sound and light waves are called analog waves or continuous waves. They illustrate the loudness of a sound or the brightness of the colors in an image at a given time. They're continuous signals because you would never have to lift your pencil off the page to draw them; they're just long, continuous lines.

First-generation recording devices such as vinyl records and analog television broadcasts were designed to reproduce these sound and light waves. A needle in the groove of a vinyl record vibrates in the same pattern as the original sound wave. Analog television signals are actually waves that tell an analog TV how to display the same color and brightness as are seen in the production studio.

However, it's difficult to describe a wave, even mathematically. The simplest sounds, such as that of middle C on a piano, have the simplest shapes, like the one shown in Figure 8.3a. However, something like the word *hello* generates a highly complex pattern, like the one shown in Figure 8.3b.

What advantages do digital formats have over analog ones? Digital formats describe signals as long strings of numbers. This digital representation gives us a simple way to describe sound and light waves exactly so that sounds and images can be reproduced perfectly any time they're wanted. In addition, we already have easy ways to distribute digital information, such as streaming movies or sending photos in a text message. Digital information can be reproduced exactly and distributed easily. These reasons give digital advantages over an analog format.

When the market for communication devices for entertainment media—like photos, music, and video—switched to a digital standard, we began to have products with new capabilities. Small devices can now hold huge collections of a variety of information. Users now create their own digital assets instead of just being consumers of digital media. We interact with our information any time we like in ways that, prior to the conversion to digital media, had been too expensive or too difficult to learn. The implications of the shift to digital media are continually evolving.

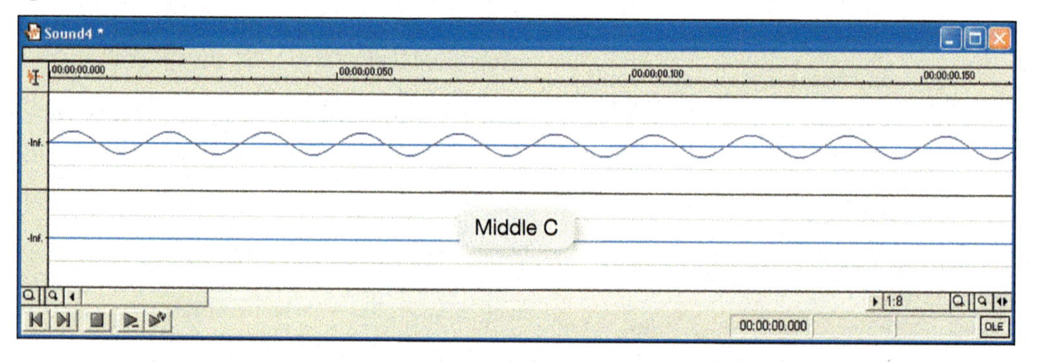

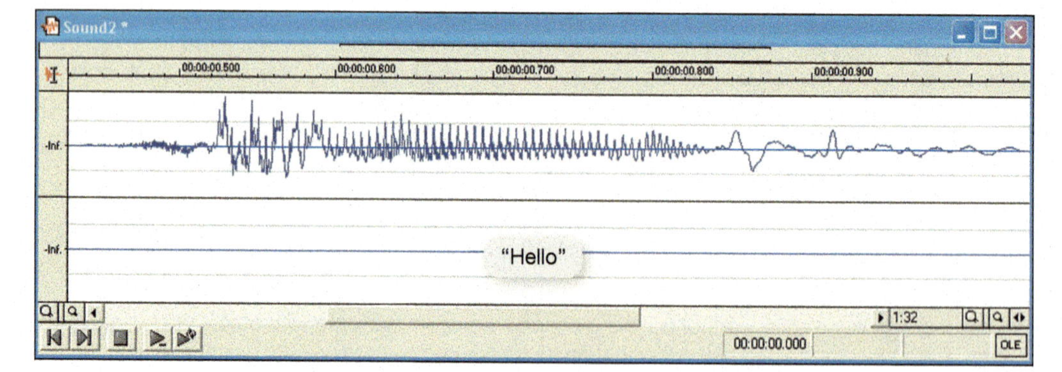

Figure 8.3 (a) An analog wave showing the simple, pure sound of a piano playing middle C. (b) The complex wave produced when a person says "Hello." *(Courtesy of Microsoft Corp.)*

Table 8.1 Analog versus Digital Entertainment

	Analog	Digital
Publishing	Books, magazines	E-books, e-zines
Music	Vinyl record albums and cassette tapes	MP3 files, and streaming music
Photography	35-mm single-lens reflex (SLR) cameras Photos stored on film	Digital SLR cameras, smartphone multi-lens cameras Photos stored as digital files
Video	8-mm, VHS, and Hi8 camcorders Film stored on tapes	8K camcorders and video cameras on phones Film stored as digital files; distributed via streaming services
Radio	AM/FM radio	HD Radio, SiriusXM satellite radio, streaming music stations
Television	Analog TV broadcast	High-definition digital television (HDTV), 4K Ultra HD (UHD) and 8K UHD television

So how has digital impacted entertainment? Today, all forms of entertainment have migrated to the digital domain (see Table 8.1). Phone systems and TV signals are now digital streams of data. MP3 files encode digital forms of music, and cameras and video equipment are all digital. In Hollywood, feature films are shot with digital equipment. Movie theaters receive a hard drive storing a copy of a new release to show using digital projection equipment. Satellite radio systems such as SiriusXM are broadcast in digital formats.

How is analog converted to digital? The answer is provided by a process called *analog-to-digital conversion*. In analog-to-digital conversion, the incoming analog signal is measured many times each second. The strength of the signal at each measurement is recorded as a simple number. The series of numbers produced by the process gives us the digital form of the wave. Figure 8.4 shows analog and digital versions of the same wave. In Figure 8.4a, you see the original, continuous analog wave. You could draw that wave without lifting your pencil. In Figure 8.4b, the wave has been digitized and is no longer a single line; instead, it's represented as a series of points or numbers.

In the next sections, we discuss different types of digital media you interact with or create.

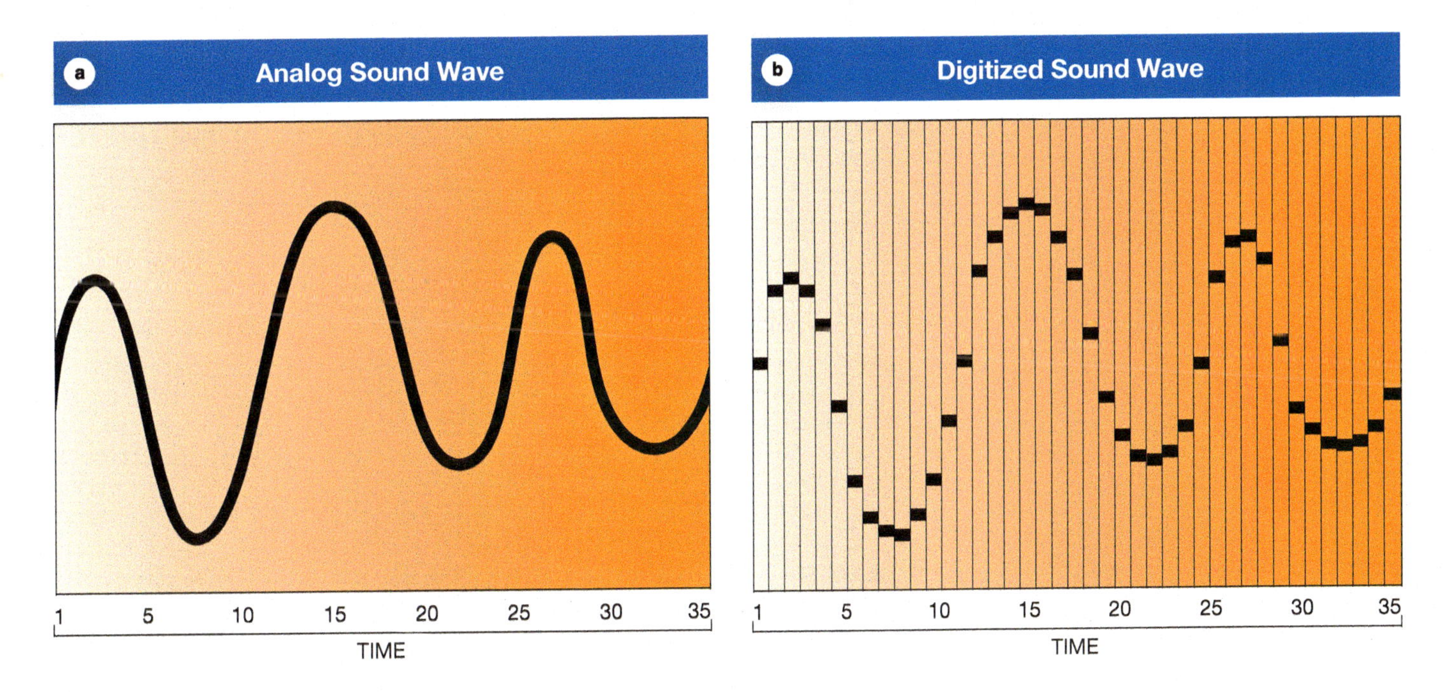

Figure 8.4 (a) A simple analog wave. (b) A digitized version of the same wave.

Digital Publishing

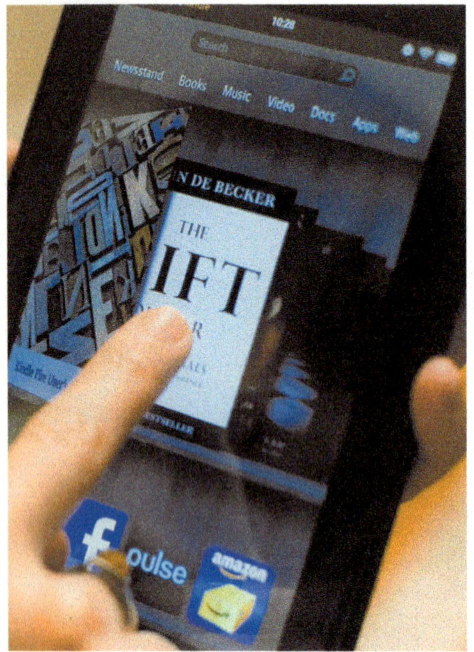

Figure 8.5 E-readers have popularized the digital e-book. *(Kristoffer Tripplaar/Alamy Stock Photo)*

The publishing industry is migrating to digital materials. How does this have an impact on books?

E-readers

Objective 8.3 *Describe the different types of e-readers.*

Are printed books dead? Electronic text (e-text) is textual information captured digitally so that it can be stored, manipulated, and transmitted by electronic devices. With the increasing usage of e-text, the market for printed materials is changing. In fact, Amazon now sells more Kindle e-books than printed books each year. Mainstream authors who traditionally relied on print publishing, such as Stieg Larsson and James Patterson, have each sold over 1 million e-books. John Locke, a self-published author, has also sold over 1 million e-books, giving hope to fledging writers everywhere. But the conversion to digital books is not a 100% shift. Globally, e-books are now accounting for approximately 20% of total book revenue.

What electronic devices are trying to replace books? E-readers (see Figure 8.5) are devices that display e-text and have supporting tools, like note-taking, bookmarks and integrated dictionaries. They're selling at a brisk pace with a range of offerings on the market, including versions of the Amazon Kindle.

Tablets, like the Amazon Fire, are also helping to popularize the e-book and electronic versions of magazines and newspapers. Although traditional print publishers' sales are falling (including their e-book sales), independently published e-book sales are rising at a fast pace.

What features make digital text and e-readers popular? One big allure of digital publishing is distribution. Even a 1,000-page book can be delivered to your e-reader in under a minute. An array of titles is available—millions of books in the Amazon Kindle store alone. In addition, millions of texts without copyright are available for free.

The basic features of e-readers offer many advantages over paper books:

- Integrated dictionaries pull up a definition when you highlight a word. The Kindle, for example, comes with nine free foreign language dictionaries—a help in reading foreign works.
- Note-taking and highlighting are supported, and you can search the text for your own notes or for specific terms. You can also share the notes you make with others.
- URL links or links to a glossary are often live.
- Bookmarks are pushed through cloud technology so you can read on one device and pick up with your most current bookmark on another.
- E-books and audio books of the same title can be linked, so you can switch from reading a book to listening to the book without losing your place.
- For comics and graphic novels, you can magnify panels to read them fully.

How is digital text displayed? Two popular technologies are used for representing digital text:

- **Electronic ink:** Electronic ink (E Ink) is a sharp grayscale representation of text. The page is composed of millions of microcapsules with white and black particles in a clear fluid. Electronic signals can make each spot appear either white or black. E Ink devices reflect the light that shines on the page, like paper. E Ink gives great contrast and is much easier to read in direct sunlight because there's no glare. High-end e-readers also offer a built-in front light so you can read in any light. Examples of devices using E Ink include the Amazon Kindle Paperwhite (see Figure 8.6a).
- **Backlit monitors:** Another option is the high-resolution backlit monitors seen in tablets like the Amazon Fire (see Figure 8.6b). These screens illuminate themselves instead of depending on room lighting. They display color materials, like magazines, with clarity. The glass does reflect glare, though, which makes them hard to use in direct sunlight. Some people experience more

(a)

(b)

Figure 8.6 The two main technologies for e-text display are (a) E Ink grayscale displays, as on the Amazon Kindle, and (b) high-resolution, backlit color screens, as on the Amazon Fire. *(David McNew/ Getty Images News/Getty Images, Oleksiy Maksymenko Photography/Alamy Stock Photo)*

eye fatigue when reading from a backlit device than when using E Ink. Also note that E Ink readers have a battery life of a month or two on a charge, whereas high-resolution color readers hold a charge for 8 to 10 hours.

What kinds of file formats are used in electronic publishing? Digital formats for publishing vary. Amazon uses a proprietary format (.azw extension), so books purchased for a Kindle aren't transportable to a device that isn't running Kindle software. An open format, ePub, is also supported by some e-readers, and ePub reader extensions (plug-ins) exist for all major browsers, as well as for stand-alone ePub reader apps.

Do I need a dedicated device just for reading e-texts? Free software download versions of the Kindle and the NOOK are available that run on either Windows-based or Apple computers. You can also download texts that have no copyrights and read them directly on a computer either as a PDF file or by using browser add-ons like Just Read for Google Chrome. In fact, Just Read can display any web page in e-book format for easier reading.

However, there are some compelling reasons to use a dedicated device for reading e-texts. Dedicated reading devices don't suffer from the intrusion of other forms of communication such as e-mail and social media. E-readers have much longer battery life than tablet computing devices. Also, the displays tend to get washed out less by direct sunlight than conventional tablets do, so they are better for reading outside.

Managing Digital Media

In this Helpdesk, you'll play the role of a help desk staffer, fielding questions about e-readers and e-texts.

Using e-Texts

Objective 8.4 *Explain how to purchase, borrow, and publish e-texts.*

Where do I buy e-books? Many vendors are associated with e-reader devices:

- Amazon Kindle devices connect directly to the Amazon Kindle store.
- Barnes & Noble NOOK devices work with the Barnes & Noble e-bookstore.
- Many publishers sell e-books online that can be read on any kind of device.
- Textbooks can be purchased in e-book format directly from the publisher. Another option is a company like VitalSource, which offers subscriptions to digital texts. When the subscription ends, the book disappears from your device.

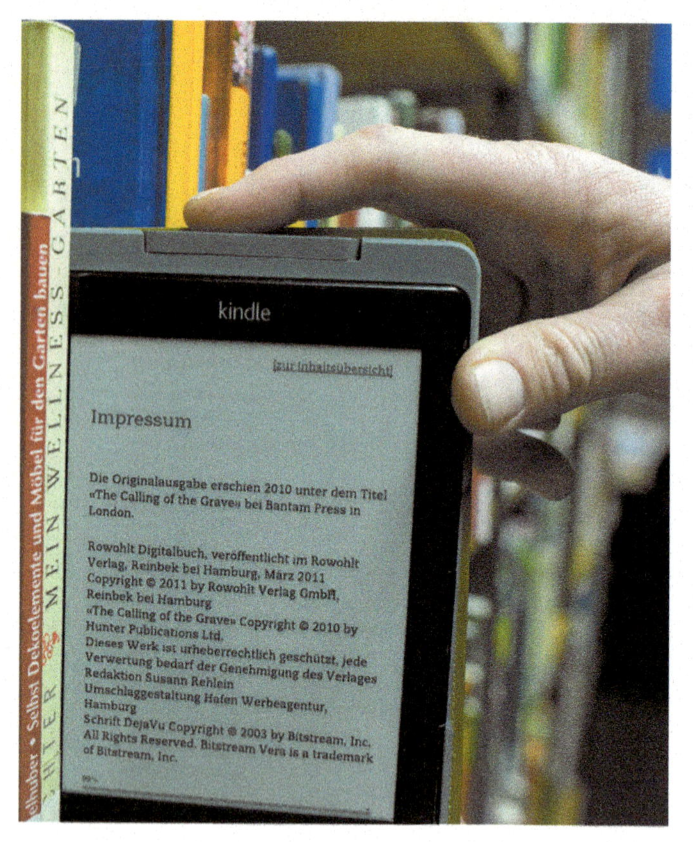

Figure 8.7 E-books and audio books can be borrowed for free at most public libraries. *(Alex Ehlers/DPA/Picture-alliance/Newscom)*

Can I borrow e-books? Libraries offer e-books and audio books (see Figure 8.7). There is never a late fee; the book just times out and disappears from your device when the borrowing period expires. With software like the Libby app (*libbyapp.com*), you can search for the area library that has the book you want. Libraries have a specific number of copies of each e-book title available, so you may be added to a wait list if all the copies are checked out. However, friction exists between publishers and libraries on how to handle lending electronically. It's so convenient that some publishers are refusing to allow their e-books to be distributed through libraries.

Lending of e-books between friends is a popular feature of e-reader systems. Both Barnes & Noble and Amazon support lending books for up to two weeks.

Where can I find free e-books? A great source of free reading is Project Gutenberg (*gutenberg.org*), a collection of over 60,000 free books in ePub, Kindle, and plaintext format. It contains books that are free in the United States because their copyrights have expired. The catalog includes many classic titles, like *Moby Dick* by Herman Melville or science fiction novels by H. G. Wells.

How can I publish my own works? Self-publishing is much easier in the age of digital texts. Many options are available:

- Self-publish into the Amazon Kindle Store in a matter of minutes and earn up to 70% royalty on sales. The Kindle Direct Publishing site (*kdp.amazon.com*) has information on self-publishing by using the Kindle format.
- Use a company like Smashwords (*smashwords.com*). It accepts a Microsoft Word document from you and then makes your book available through many vendors such as the Apple Books store and the Barnes & Noble e-store.
- Use a site like Lulu (*lulu.com*) to learn about social media marketing and other techniques for promoting your book. In addition, it offers services from editors, designers, and marketers and can allow you to offer your e-book for sale as a physical book.

Digital Music

Digital music has upended the recording industry, with new models of distributing music, streaming services, and issues of copyright protection. In this section, we look at the mechanics of creating and distributing digital music.

Creating and Storing Digital Music

Objective 8.5 *Describe how digital music is created and stored.*

How is digital music created? To record digital music, the sound waves created by instruments need to be turned into a string of digital information. Figure 8.8 shows the process of digitally recording a song:

1. Playing music creates analog waves.
2. A microphone feeds the sound waves into a chip called an *analog-to-digital converter (ADC)* inside the recording device.
3. The ADC digitizes the waves into a series of numbers.
4. This series of numbers can be saved in a file and then recorded onto digital media or sent electronically.
5. On the receiving end, a playback device such as a mobile device or a DVD player is fed that same series of numbers. Inside the playback device is a *digital-to-analog converter (DAC)*, a chip that converts the digital numbers to a continuous analog wave.
6. That analog wave tells the receiver how to move the speaker cones to reproduce the original waves, resulting in the same sound as the original.

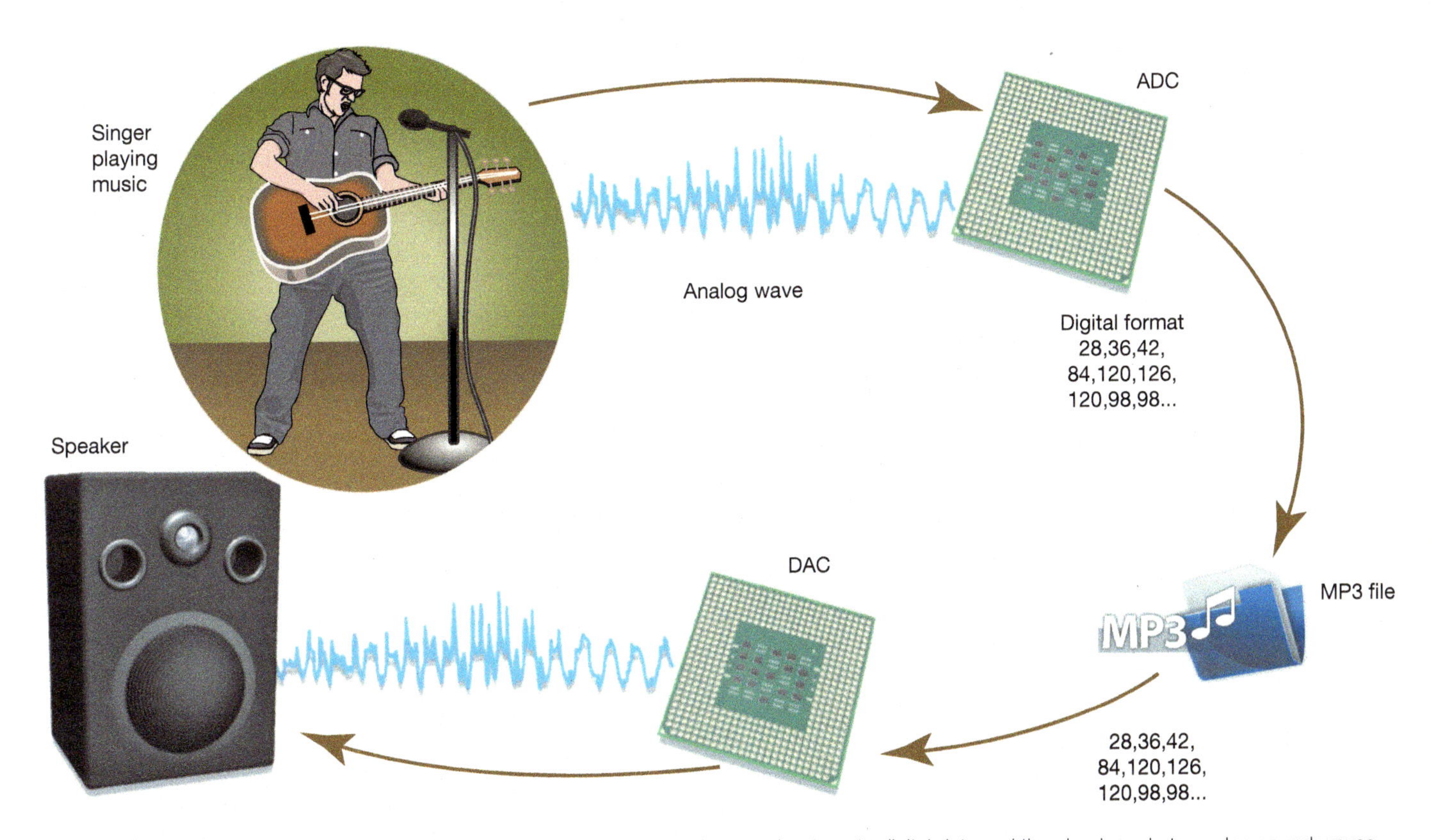

Figure 8.8 During the complete recording process, information changes from analog form to digital data and then back again to analog sound waves.

More precisely, the digital wave will be *close* to exact. How accurate it is, or how close the digitized wave is in shape to the original analog wave, depends on the sampling rate of the ADC. The sampling rate specifies the number of times the analog wave is measured each second. The higher the sampling rate, the more accurately the original wave can be re-created. The improved sound quality that higher sampling can afford also depends on the quality of the output device and speakers, of course. However, higher sampling rates also produce more data and therefore result in bigger files. For example, sound waves on CDs are sampled at a rate of approximately 44,000 times a second. This produces a huge list of numbers for even a single minute of a song.

What file types store digital music? You're no doubt familiar with the MP3 format used to store digital music, but many others exist, such as AAC and WMA. If you buy a song from Apple, for example, you receive an AAC-format file (although these files can be converted to other formats). Many formats—such as DivX, MPEG-4 (which usually has an .mp4 extension), WMV, and Xvid—hold both video and audio information. All file formats compete on sound and video quality and *compression*, which relates to how small the file can be and still provide high-quality playback. Check what kind of files your audio device understands before you store music on it.

How do I know how much digital media my device can hold? The number of songs or hours of video your device can hold depend on how much storage space it has. Some mobile devices also use flash memory or SD cards.

Another factor that determines how much music a player can hold is the quality of the MP3 files. The size of an MP3 file depends on the digital sampling of the song. The same song could be sampled at a rate anywhere between 64 kbps and 320 kbps. The size of the song file will be five times larger if it's sampled at 320 kbps than at 64 kbps. The higher the sampling rate, the better the quality the sound—but the larger the file.

What is the easiest way to access music from multiple devices? You could subscribe to a service that streams the music to you over your Internet connection. Spotify, Amazon Music, and Apple Music are subscription services that let you listen to tracks in their catalogs. The music isn't yours to own, however—if you cancel your subscription, you no longer have access to the music, but because it's streamed to you, it doesn't take up space on your devices or even need to be stored on your devices. Most streaming services offer options to download music so that you can still play the songs on your playlist even if you don't have Internet access.

Figure 8.9 You can wake up to your favorite music through your smart assistant device. *(Shutterstock)*

Distributing Digital Music

Objective 8.6 *Summarize how to listen to and publish digital music.*

What's the best way to listen to digital music? You have many options for listening to music other than with headphones:

- Smart home devices like Google Home and Amazon Echo can access streaming services to which you subscribe to play music (see Figure 8.9).
- Most new cars have fully integrated software systems (like Apple CarPlay) that display and run your playlists, connecting wirelessly to your devices over Bluetooth.
- Systems like Sonos (see Figure 8.10) can mate wirelessly with a mobile device and broadcast sound throughout the house.
- Many audio receivers come with Bluetooth so you can connect a mobile device to them as an audio input source.

If I don't pay for a music download, is it illegal? Although you're required to pay for most music you download or stream, some artists post songs for free. Business models are evolving as artists and recording companies try to meet audience needs while also protecting their intellectual property rights. Several approaches exist. One is to deliver *tethered downloads*, in which you pay for the music and own it but are subject to restrictions on its use. Another approach is to offer *DRM-free* music, which is music without any digital rights management (DRM). DRM is a system of access control that allows only limited use of material that's been legally purchased. It may be that the song can run only on certain devices or a movie can be viewed only a certain number of times. A DRM-free song can be placed on as many devices as you wish.

Will streaming music services eliminate radio stations? Many radio stations have increased listenership by making their stations available through Internet sites and by broadcasting in high-definition quality. And terrestrial radio stations are still free, whereas satellite radio requires a subscription. However, because the Internet allows artists to release new songs to their fans (on sites such as SoundCloud) without relying on radio airtime, many radio stations are competing for listeners with online sources.

What if I want to publicize my band's music? Digital music has made the distribution of recordings simple, using sites like SoundCloud (*soundcloud.com*), CD Baby (*cdbaby.com*), and ReverbNation (*reverbnation.com*). You can quickly create a web page for your band, post songs, and start building a fan base through social media. ReverbNation sends you reports detailing who's listening to your music and what they're saying about it. ReverbNation is also a way to connect with independent recording labels and to find places that want to book your band.

How can I get my band's music on streaming sites like Spotify? Streaming sites usually use artist aggregators like DistroKid and CD Baby rather than dealing with independent artists. Once your music is uploaded to the aggregator, it can be made available to fans on streaming media services at no cost to you. You may even receive small royalty payments from the streaming services when people listen to your songs.

Figure 8.10 Sonos is a multi-room system that streams music wirelessly throughout your home. *(Realimage/Alamy Stock Photo)*

Digital Media

Photography and video have also moved to digital formats, exerting an impact on the movie industry and creating new styles of visual performance.

Digital Photography

Objective 8.7 *Explain how best to create, print, and share digital photos.*

How does digital photography work? Digital cameras don't use film. Instead, they capture images on electronic sensors called charge-coupled device (CCD) arrays and then convert those images to digital data, series of numbers that represent the color and brightness of millions of points in the image. Unlike old-fashioned film cameras, digital cameras can show your images the instant you shoot them. Digital cameras in phones record digital video as well as digital photos. Images and video are usually stored on flash memory in your phone or can be stored in the cloud.

What determines the quality of a digital image? The overall image quality is determined by many factors, including:

- **Lens quality and quantity:** Lens quality on phones has improved significantly and cameras on flagship devices feature three lenses for different scenarios (such as telephoto and ultrawide angle).
- **Image sensor size:** Larger sensors can capture more light, which improves performance.
- **File format and compression used:** Compression algorithms, which reduce the space files take up, can negatively affect the quality of an image.
- **Software algorithms:** Many cell phone cameras' software now use artificial intelligence to adjust for optimal light and color settings. Most software has night or variable-light modes to provide great quality even in low or uneven light settings.
- **Camera resolution:** A digital camera's resolution, or the number of data points it records for each image captured, is measured in megapixels (MP). The word *pixel* is short for *picture element*, which is a single dot in a digital image. High-quality phone cameras are now usually 12 MP or more.

I don't need another device besides my phone to take pictures, do I? The cameras on smartphones have improved significantly in resolution, but for taking high-quality, professional photos, phone cameras can fall short because they often employ smaller image sensors and inferior lenses compared to stand-alone cameras. In addition, many features that serious photographers rely on aren't available in smartphone cameras, such as different types of autofocus and image stabilization algorithms, so if you're a serious amateur photographer or want to sell your work, invest in a digital stand-alone camera.

For instance, professional digital SLR cameras, such as the Canon EOS 5DS, can take photos at resolutions as high as 50.6 MP, which allows for the printing of much larger prints. If you're interested in making only 5" × 7" or 8" × 10" prints, a phone camera with 12 MP is fine. However, low-resolution images become grainy and pixelated when pushed to larger sizes. For example, if you tried to print an 11" × 14" enlargement from a 12 MP image taken with your smartphone camera, you'd start to see individual dots of color instead of a sharp image. The 16- to 24-MP stand-alone cameras on the market have plenty of resolution to guarantee sharp images even with enlargements as big as 11" × 14".

How do I select a digital camera? If you want a stand-alone digital camera, you have a number of camera models from which to choose. Some digital cameras are *fixed-lens cameras*, meaning they come with one lens that you can't change. However, you can switch lenses on digital SLRs, as shown in Figure 8.11.

Figure 8.11 Digital SLR cameras can accommodate a range of separate autofocus lenses. *(Stavklem/Shutterstock)*

In addition to allowing you to switch lenses, the size of the image sensor that actually captures the light is critical. Larger sensors demand larger cameras and lenses. Professional digital SLR cameras use a full-frame sensor, whereas phone cameras use a compact sensor and use a wider-angle lens to capture the full scene, making them less powerful.

Although having such flexibility in selection of lenses and a larger sensor are great advantages, most digital SLR cameras are larger, heavier, and use more battery power than fixed-lens models. Think about how you'll be using your camera and decide which model will serve you best. One resource is Digital Photography Review (*dpreview.com*), a site that compares cameras (including those in phones) and provides feedback from owners.

What file formats are used for digital images? You can choose from several file types with digital cameras to compress the image data into less memory space. When you choose to compress your images, you'll lose some of the detail, but in return you'll be able to fit more images in your storage. The most common file types digital cameras support are raw uncompressed data (RAW) and Joint Photographic Experts Group (JPEG):

- *RAW files* have different formats and extensions, depending on the camera manufacturer. The RAW file records all the original image information, so it's larger than a compressed JPEG file.
- *JPEG files* can be compressed just a bit, keeping most of the details, or compressed a great deal, losing some detail. Most cameras offer a selection of a few JPEG compression levels.

Enhancing Photos with Image-Editing Software

In this Sound Byte, you learn tips on how to use image-editing software best. You learn how to remove red eye from photos and how to incorporate borders, frames, and other enhancements.

Often, cameras also support a very low-resolution option that produces images that aren't useful for printing but are so small that they're easy to e-mail or download quickly. Low-resolution images are often used on mobile-friendly websites because the expectation is that the images will be viewed on smaller screens and need to download quickly to save on data.

How do I share my photos? You probably share photos on social media like Instagram all the time, but other options might be beneficial, especially if you take a lot of photos. Both iOS and Android have photo-sharing features that provide access to photos or video you capture from your devices. Shared Albums is an option within Apple's iCloud storage that allows you to organize photos and videos into albums and share them. Google Photos (*photos.google.com*) is another option. The Google Photos app (available for Android, iOS, Windows, and macOS) automatically uploads photos from your device to the Google Photos website and syncs them to all your other devices. From the website, you can easily share photos. This is especially handy when you upgrade to new devices, because you can maintain access to all the images captured with your old devices.

What are the best options for printing digital photos? If you want to print photos, you have two main options:

1. **Use a photo printer.** Most inkjet printers can print color photos, although they vary in speed, quality, and features. Most high-quality photo printers use six or eight ink cartridges (instead of the standard four) to provide a wider range and depth of colors. Special paper is available specifically for printing photos.
2. **Use a photo-printing service.** Some retail stores, such as CVS, offer digital printing services. The paper and ink used at these locations are higher quality than what's available for home use and produce heavier, glossier prints that won't fade. You can send your photos directly to local merchants for printing, using the merchant's websites that allow uploading from your device or social media account. Online services, such as Flickr (*flickr.com*), store your images and allow you to create hard-copy prints, greeting cards, photo books, calendars, and gifts (like mugs and T-shirts).

Digital Video

Objective 8.8 *Describe how to create, edit, and distribute digital video.*

What devices, sites, and other sources provide digital video content? Digital video surrounds us:

- Television is broadcast in digitally formatted signals.
- The Internet delivers digital video through YouTube, communities like Vimeo (*vimeo.com*), and webcasting sites like Watson Media (*video.ibm.com*).
- Many pay services are available to deliver digital video. These include on-demand streaming from Internet providers such as Apple, Netflix, Hulu, and Amazon.

How do I record my own digital video? Phone cameras record video (as well as still photos), and the latest phones are capable of 8K video resolution. Webcams (integrated into laptops) also work as inexpensive devices for creating digital video. Although editing apps available on your phone are sufficient for causal videos, you may wish to use more sophisticated video-editing software to produce professional-looking video. You can transfer video files from your phone to your computer and, using high-end video-editing software (like Adobe Premiere Pro), edit the video at home. You can then upload the finished video to an appropriate Internet site to share it.

What advantages does using high-end video-editing software offer me? Video-editing software such as Adobe Premiere Pro presents a storyboard or timeline you can use to manipulate your video file, as shown in Figure 8.12. You can review your clips frame by frame or trim them at any point. You can add titles, audio tracks, and animations; order each segment on the timeline; and correct segments for color balance, brightness, or contrast. Examine online tutorial resources such as Izzy Video podcasts (*izzyvideo.com*) and No Film School (*nofilmschool.com*) to learn how to make the most impact with the editing and effects that you apply to your video footage.

What kinds of files will I end up with? Once you're done editing your video, you can save or export it in a variety of formats. Table 8.2 shows some of the popular video file formats in use, along with their file extensions.

Your choice of file format for your video will depend on what you want to do with it. For example, the mp4 file format is a great choice if your file is delivered on the web and you need it to work

Figure 8.12 Adobe Premiere Pro allows you to build a movie from video clips (and still photos) and add soundtracks and special effects. *(Courtesy of Adobe Premiere)*

Table 8.2 Typical File Formats for Digital Video

Format	File Extension	Notes
QuickTime	.qt .mov	You can download the QuickTime player without charge from apple.com/quicktime. The Pro version allows you to build your own QuickTime files.
Moving Picture Experts Group (MPEG)	.mpg .mpeg .mp4	The MPEG-4 video standard was adopted internationally in 2000; it's recognized by most video player software.
Windows Media Video	.wmv	This is a Microsoft file format recognized by Windows Media Player (included with the Windows OS).
Microsoft Video for Windows	.avi	This is a Microsoft file format recognized by Windows Media Player (included with the Windows OS).
WebM	.webm	An open source file format designed to optimize video delivery on the Internet.

well on a variety of devices. The Microsoft AVI format is a good choice if you're sending your file to a wide range of users, because it's the standard video format for Windows and is recognized by most popular free media players used in Windows 10, like VLC Media Player and GOM Player.

Different compression algorithms will have different results on your particular video. Try several to see which one does a better job of compressing your file. A codec (compression/decompression) is a rule, implemented in either software or hardware, that squeezes the same audio and video information into less space. Some information will be lost using compression, and there are several codecs to choose from, each claiming better performance than its competitors. Commonly used codecs include MPEG-4, VP9, and DivX. No one codec is always superior—a codec that works well for a simple interview may not do a good job compressing a live-action scene.

What's the quickest way to get my video out to viewers? Webcasting, or broadcasting your video live to an audience, is a simple option. Use either your phone or a compact camera like a GoPro (see Figure 8.13) and apps like Instagram Live or Periscope to stream live video. Snapchat Stories is also a popular option for distributing casual videos. These apps have features like interactive chat next to the video feed so viewers can interact with you while you're live. On many apps, both the chat and video can be captured and archived for viewers who missed the live broadcast.

Figure 8.13 Compact cameras like the GoPro are small enough to make them part of any activity. *(Cyrus McCrimmon/The Denver Post/Getty Images)*

You can upload to video-sharing sites like YouTube or Vimeo. Of course, it's illegal for you to upload videos you don't own. You also can't post a piece of a copyrighted video publicly. The second part of this chapter presents legal and ethical situations that are important for you to be aware of as a content creator in the digital age.

How is HD different from plain digital? HD stands for high definition. It's a standard of digital television signal that guarantees a specific level of resolution and a specific *aspect ratio*, the ratio of the width of an image to its height. For example, an aspect ratio of 16:9 might mean an image is 16" wide and 9" high. A 1080p HDTV displays 1,920 vertical lines and 1,080 horizontal lines of video on the screen, which is over six times as many pixels as a standard-definition TV. The popular standard, ultra-high-definition television (UHDTV), is also known as 4K or 2160p. A 2160p UHDTV has a resolution of 3840 × 2160 pixels. The aspect ratio used is 16:9 in both formats, which makes the screen wider than it is tall, giving it the same proportions as the rectangular shape of a movie screen (see Figure 8.14). This allows televisions to play movies in widescreen format instead of letterboxing the film with black bars on the top and bottom. Another format is now available: 8K. It features a resolution of 7680 × 4320 pixels and will eventually overtake 4K as the popular standard.

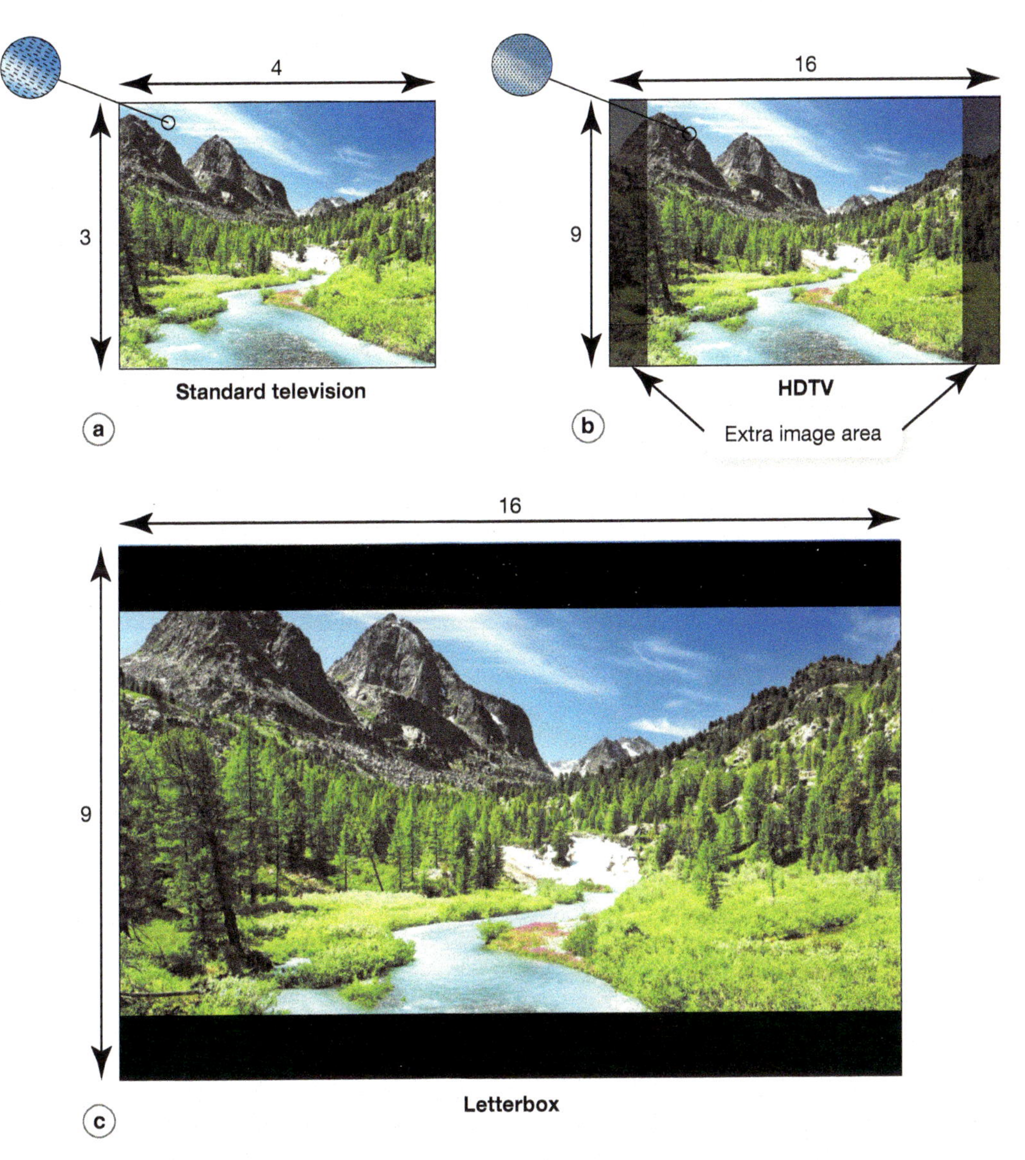

Figure 8.14 (a) Standard-definition TV has a squarer aspect ratio, whereas (b) HDTV and UHDTV match the 16:9 ratio used in the motion picture industry without the need for (c) letterboxing. Top and bottom black bars display due to letterboxing. *(Dmitry Pichugin Shutterstock)*

Bits&Bytes Professional Video-Editing Tools . . . For Free!

Adobe Premiere Pro is the gold standard for professional video-editing software. It is a full-featured product that offers options that beginning video editors probably won't need or won't have time to master. It's also not inexpensive. If you want to experiment with video editing to see whether it's something you want to pursue, you have some free options to choose from.

Shotcut (*shotcut.org*) is a free, open source video editor that runs on multiple operating systems, including Windows, macOS, and Linux. Although its interface is not as slick as commercial software packages, the screen layout is still familiar, with source clips in one window, video previews in another, and a familiar-looking timeline along the bottom of the screen. Shotcut can handle a variety of file formats, and video tutorials are available for learning the application.

DaVinci Resolve (*blackmagicdesign.com*) is a full-featured proprietary video editor that competes with other high-end packages like Premiere Pro and Final Cut Pro. DaVinci Resolve supports 8K resolution and allows you to perform editing, color correction, visual effects, and audio post-production all in the same app. This high-end app is used in post-production for many Hollywood movies and television shows. Although not open source software, the developer offers a free version that is fairly full-featured.

So, try one of these free solutions and see whether you enjoy the world of professional video editing without investing any cash!

Trends in IT Digital Asset Managers Needed!

Some companies routinely generate valuable digital assets. Managing these assets to ensure that they're being properly monetized is now a full-time job. *Digital asset management (DAM)* is an up-and-coming field for media-savvy graduates.

The creative industries were the first to realize they needed DAM. Photographers and graphic designers generate a tremendous amount of content in various formats. Images may need to be saved in different file formats and at different resolutions. The rights granted to third parties for the use of images needs to be tracked. Creation of new digital assets needs to be managed.

However, many other businesses realize the need for DAM as they roll out cohesive brand strategies that need consistency in imagery both on the web and in printed media. Organizing, cataloging, tracking, and sharing image and video files are often done with DAM software, such as Webdam and Libris.

So what skills does it take to become a digital asset manager? Obviously, an understanding of the creation process of digital assets is the first step. Experience managing projects and workflows is also important. The ability to work well with different types of people in the creative, IT, and marketing departments is also a plus, and because many assets are created with Adobe Creative Cloud products, experience with these software packages would make your transition into the job much easier.

If you enjoy working with digital content creators and can organize complex workflows, your next job might just be a lot of DAM fun.

Is there a way to manage all my video media from different sources in one place? Media center software, such as Kodi (kodi.tv), lets you use your computing devices to aggregate your media in a convenient interface. Unlike Netflix or other streaming services, Kodi doesn't include content; rather, it depends on sources of free content or content from services you may already have. Its main advantages are convenience and the ability to run on different devices.

Does anyone need cable TV today? Until recently, most people in the United States bought their television media from cable television providers. But with the proliferation of streaming media companies, cable TV companies are becoming less popular than they once were. Many people are cutting the cord and opting out of cable television altogether.

What does cutting the cord mean? Cutting the cord refers to dropping cable television service and obtaining television programs a different way (Figure 8.15). You can follow some straightforward steps to arrive at your own cord-cutting strategy.

First, assess how you watch TV. Do you mostly watch television at home on a TV set or do you prefer to watch television on your mobile devices? If you like to use mobile devices, you need to assess your data plan with your phone provider.

What size data plan should I buy? Before subscribing to or changing a data plan, you should assess your needs. When you are out of reach of a Wi-Fi network, how often do you:

- download apps, stream music, or play online games?
- watch streaming video?
- download files attached to e-mail or from a website?
- use apps that are communicating with the Internet?

Figure 8.15 Many people save money by cutting the cord and replacing cable television services with Internet streaming services. *(Kristin Chiasson/Shutterstock)*

Begin by estimating how much data you transfer up and down from the Internet each month. To do so, you may be able to use an online estimator supplied by your provider.

The Android OS allows you to see a graph of your data usage and set alarms at specific levels; iOS keeps track of your cellular data usage too. This feature resets automatically each month so you know how much data transfer you have left. Most mobile operating systems keep track of data usage for you.

Be sure you select a data plan that provides adequate service at a good price. There are plans that allow a group of people to share data and pull from a single pool. Some plans transfer unused data over to the next month, whereas others do not. Shop carefully.

What should I do to watch TV at home on a television without cable? You may want to consider buying a digital TV antenna if you like to watch digital content on your TV. These inexpensive antennas attach to the antenna port on your television and are designed to pick up the digital broadcast signals transmitted through the public airwaves. Depending on where you live, you may receive many television channels for free.

If you have smart televisions, you can access television programming over your Internet connection without a cable TV subscription. If you have a TV that isn't smart, one option is to attach a media streaming device (such as a Roku, Amazon Firestick, or Google Chromecast) to your television. These devices, in conjunction with an Internet connection, receive streaming media from the Internet. Not all streaming media is free, but you may be able to save money by not subscribing to a standard cable TV package. Make a list of the channels that you watch regularly. There is no sense in subscribing to channels you rarely (or never) watch.

Finally, decide which streaming services to subscribe to, based on the channels you like to watch. There are many streaming services, such as Netflix, Hulu, Amazon Prime, HBOMax, Disney+, and YouTube, and all offer different channels and programming content. Make sure to check that your chosen options will work for you before canceling your cable TV subscription, and of course, you may still need to get your Internet service from your cable provider.

Managing Your Digital Lifestyle

As technology has become more integrated with our daily lives, we continue to face challenges of seamlessly using technology and protecting our privacy. In this section, we discuss some of the challenges faced in managing an active digital lifestyle.

Digital Challenges and Dilemmas

Objective 8.9 *Discuss the challenges in managing an active digital lifestyle.*

Keeping Data (and Conversations) Private

What are some challenges technology presents, being so integral to our lives? We interact with the IoT—knowingly and sometimes unknowingly—in many aspects of our lives. As we become more connected to the Internet through our vast array of devices, we also become part of the information infrastructure, each of us essentially acting as nodes on the IoT. Although the interconnectivity of the IoT brings with it many benefits, it also poses ethical concerns, including those related to privacy and equal access.

What privacy issues does participation in the IoT raise? As we use our Internet-connected devices, we generate huge amounts of data. Many organizations are acquiring and using that data to help with decision making, resource allocation, and operations. Although we may be used to having Internet advertisers use our search data to send us targeted advertising, how would you feel about companies using information gleaned from an IoT-connected device? For example, say you wear a fitness tracker such as a Fitbit or an Apple Watch. The sensors in these activity monitors are connected wirelessly to smartphones and to the Internet to enable users to track their workout activities. How would you feel if the data collected by your Fitbit or Apple Watch was then shared by your mobile phone carrier with a third-party marketer? How would you feel if you were in a restaurant or grocery store and received text messages providing discounts or coupons for low-calorie offerings from those third-party marketers, simply because your behavior implied that weight loss is one of your goals? Although some might welcome these opportunities, others might consider the offerings an invasion of their privacy.

Also consider the digital assistants (like Alexa) being deployed in IoT devices like the Amazon Echo (see Figure 8.16). Some privacy advocates have voiced concerns about privacy being invaded when the digital assistants are listening and possibly storing conversations. Glitches in the devices have occurred. For example, a woman's conversations were secretly recorded by her device and then sent to a random person in another city. Reports of Amazon Echos randomly laughing have

Figure 8.16 Digital assistants are standing by for our queries, but are they listening when we don't expect them to be? *(Mahathir Mohd Yasin/Shutterstock)*

been attributed to a fault in the listening software, but eerie occurrences like these don't exactly inspire confidence that our privacy is being safeguarded. The revelation by Amazon that Alexa devices record all your conversations with Alexa and store them in the cloud—for the purposes of improving speech recognition—upset many privacy advocates.

How does the IoT raise issues of equal access? The Internet of Things also adds to the ethical dilemma of the *digital divide*—the technological gap between those who can and cannot afford to own these devices. At the very core of the dilemma is the ability for users to access the Internet at all. Significant work has been done to provide developing countries with access to the Internet. But now, the IoT adds new considerations, not only discriminating against certain groups of people who do not have access to the Internet but also against those who cannot afford IoT devices. There have been discussions about providing free nationwide wireless Internet access to bridge this gap, but the creation and delivery of that type of service is complicated and fraught with political and corporate debate, preventing any type of public access anytime soon.

Is anything being done to protect our privacy? Fortunately, states are beginning to realize that consumers need their data to be protected from misuse. California became the first state to pass a sweeping data protection law. Until this point, very few laws have been enacted in the United States that protect consumer data that companies gather. Under this law, consumers are granted the right to be informed about the data that companies are collecting about them, the reasons the data is being collected, and with whom the company is sharing the data. California consumers now gain the right to force companies to delete the information gathered about them as well as the option to prevent sharing and selling their data. The law also restricts collection and sharing information about children under age 16. More than half of the states now have laws covering data protection, but not all of them are as comprehensive as the California law, and companies like Amazon provide the ability for users to change the settings in their smart assistant devices to stop them from recording and saving your conversations. Progress is being made!

Cryptocurrency: Should You Use It?

What is cryptocurrency? In Chapter 3, we introduced cryptocurrency (also called virtual, cyber, or digital currency). Cryptocurrency attempts to eliminate intermediaries or banks to conduct transactions, thus enabling direct anonymous transactions between users. Payments are sent peer to peer from the payee to the recipient without the need to supply personal information; therefore, it is an anonymous way to send money. The currency is not US dollars or euros, but a new currency specifically created for the purpose of making payments. Bitcoin is the most well-known cryptocurrency, but many other types exist, such as Litecoin, Ethereum, Zcash, Dash, Ripple, and Monero (see Figure 8.17).

Figure 8.17 Although cryptocurrencies like Bitcoin and Ethereum are proliferating, they still have some major disadvantages of which users need to be aware. *(Steven Heap/123RF)*

What makes cryptocurrency different from regular currencies? Unlike regular currencies, cryptocurrencies are not under the regulation of governments nor are they issued by central banks of governments. However, cryptocurrencies still function like regular currency because you can use them to pay for goods and services wherever a particular cryptocurrency is accepted. You can also make cryptocurrency payments anonymously. Not having to reveal details of your identity during a transaction makes it virtually impossible for identity theft to occur. This also makes transactions quicker to process.

Dig Deeper: Deep Web versus Dark Web: Are There Places You Shouldn't Go?

Just as in real life, there are places on the Internet you just shouldn't go because they can be dangerous! Let's separate fact from fiction regarding the so-called deep web and dark web.

You are probably in and out of different areas of the deep web every day even though you don't realize it. The deep web is any part of the Internet that you can't find by using a search engine. This would be any areas that are gated off from the open Internet, such as areas protected by a password (such as your bank account) or a paywall (content available on Netflix by subscription). Anything that is not located on the surface of the Internet (easily found by an unrestricted URL) is considered in the deep web.

Obviously, there is nothing scary about accessing your bank account, but you certainly wouldn't want it lying around on the surface of the web where anyone could see it. We need secure areas of the web to protect sensitive information from unauthorized access, hence the deep web. Unfortunately, people often confuse the deep web with the dark web, even incorrectly using these terms interchangeably. Although there is nothing scary about the deep web, the dark web can be extremely frightening. The dark web (see Figure 8.18) is actually part of the deep web; however, it is a very well-concealed part of it. You can't locate the dark web through a search engine and you can't even access it with a regular web browser. Consider it analogous to the back room at a seedy bar where an illegal gambling game is being played. It is definitely not someplace you want to explore.

The dark web consists of anonymously hosted websites that are mostly used for illegal activities or by people who are very concerned about privacy. In most countries it is not illegal to surf the dark web, but many of the activities you can engage in on dark web sites (like buying and selling drugs) are definitely illegal. Criminals prefer the security of the dark web because most sites on it use sophisticated encryption tools like Tor or Freenet. You need special tools to find the dark web and use the sites on it, so you won't stumble onto the dark web accidentally.

One app used to access the dark web is the Tor (The Onion Router) browser, which allows you to locate hidden websites in the **.onion** domain. The name is a play on the structure of an onion that has layers that can be peeled away, reminiscent of the layers of security on the dark web. Estimates are that the Tor browser is used less than 6% of the time to access the dark web. Most users of Tor are using it to surf the surface web anonymously because of its many security features. Unfortunately, these layers of security come at a price: the Tor browser tends to provide much slower response times than conventional browsers.

The interfaces on the dark web remind some users of the early days on the Internet. Sites are simple and unsophisticated. Sites often go down or disappear altogether. It can also be difficult to navigate the dark web. Although you won't have search engines to help you navigate, directories on the dark web such as the Hidden Wiki maintain lists of sites.

So, we need the deep web to protect personal privacy and proprietary assets on the Internet. The dark web we could certainly do without—and it's an excellent place to avoid. Although you can use the Tor browser to protect your privacy when surfing the surface web, other tools such as virtual private networks (VPNs) provide protection without sacrificing response speed.

Figure 8.18 The dark web is mostly used for illegal activities. Fortunately, you won't stumble into the dark web accidentally! *(Leo Lintang/123RF)*

If banks aren't involved, how are cryptocurrency transactions tracked? The time and amount of every finalized cryptocurrency transaction is kept on a massive public online ledger known as a blockchain. The public nature of the blockchain makes it difficult for cryptocurrency transactions to be manipulated by anyone. Unlike traditional banks and financial institutions that tend to keep transactions relatively private, the blockchain ledger is much more transparent to users and forms the permanent record of a particular cryptocurrency's transactions.

Where do you store cryptocurrency? If you want to hold and spend cryptocurrency, you must create a digital wallet, which stores the information needed to complete cryptocurrency transactions. There are two types of wallets: *hot wallets* and *cold wallets*. Hot wallets are digital wallets that are constantly connected to the Internet. Cold wallets are disconnected from the Internet. A hot wallet is analogous to a checking account: you keep small amounts of money in it for making purchases. A cold wallet is more like a savings account, where you keep most of your money safe and secure. Cold wallets are much safer than hot wallets because disconnection from the Internet means hackers can't get to your cryptocurrency.

Why isn't cryptocurrency in more widespread use? Anonymous, fast, and secure—what could be better? As good as it sounds, there are still disadvantages to cryptocurrency:

- **You can't reverse transactions.** Although a benefit for payees, who are guaranteed their payment, this is a drawback for payers.
- **Loss of wallet means loss of currency.** If your digital wallet is inadvertently lost or destroyed, your currency is gone.
- **No insurance.** Unlike deposits in banks, your deposits with cryptocurrency exchanges are not protected. Cryptocurrency exchanges have been hacked, so your money could be lost.
- **Value fluctuates.** Unlike standard currencies whose values tend to remain relatively stable, cryptocurrency values can fluctuate wildly.
- **It's largely unregulated.** Therefore, consumers have very little protection and recourse if something goes amiss.
- **Not widely accepted.** Most merchants do not accept cryptocurrency at this point.

Until some of these disadvantages are mitigated, you should exercise caution when using or investing in any cryptocurrency.

Wearable Technology: Ethical Dilemmas on the Go

What is wearable technology? Wearable technology (or just **wearables**), as its name indicates, is technology incorporated into things you wear. The devices become part of your personal area network (PAN) as they transfer data by Bluetooth to a mobile app or your computer. Popular wearable devices include fitness trackers, such as those from Garmin and Fitbit, and interactive digital fabrics, such as Athos, which have built-in sensors that monitor specific activity as well as your heart rate while you're working out. In addition to health and fitness monitoring, many wearable devices, such as the Apple Watch, let you connect to your e-mail and calendar, among other personal productivity tools (see Figure 8.19). The medical industry is also implementing wearable devices, such as those used to monitor cardiac patients for heart rate and blood pressure. Nearly 25% of American adults use some form of a wearable device, and the trend is increasing at a remarkable rate.

Why are wearables so popular for health monitoring? Many people find that being held accountable, either to a person or a device, encourages them to be more successful in attaining their goals. If you don't have a regular jogging partner, your wearable device can be an encouraging substitute, because it records and even celebrates your progress (for instance, Fitbits set off virtual fireworks when you achieve a daily step goal).

Figure 8.19 Smartwatches are a form of wearable technology that can monitor your activity and connect you to personal productivity tools such as your e-mail and calendar as well as sync to your mobile device. *(DMstudio House/Shutterstock)*

Wearables are also being widely used to monitor people in elder care facilities. People with cognition impairment are prone to getting lost or wandering away. Wearable devices can help staff keep track of residents and even prevent residents from exiting from facilities, because their devices can be programmed to deny them access to certain means of egress (such as elevators and fire exits). In addition, if the wearables are monitoring a resident's vital signs, doctors can be automatically notified of potential medical emergencies such as irregular heart rhythms.

Wearables can even benefit you financially. Some insurance companies are now offering customers the opportunity to use a wearable device to track health progress in exchange for lower insurance premiums. There would seem to be obvious benefits to both the customer and the insurer. If customers are healthier, they probably won't use medical services as often, which results in cost savings for the insurer.

Why are some people concerned about institutions tracking people with wearables? Privacy and security concerns are the most frequently cited. People wonder what the company is doing with the data being collected and whether it is being shared with others without their knowledge or consent. Also, people worry that their data is being adequately secured from hackers. Questions of data ownership are arising. Should you own your personal health data collected by an app, or should the owner of the app (who provided it to you free of charge) be entitled to profit from the personal data collected about you?

Equitable access to the advantages of wearables is a further concern. If a company requires you to use a wearable to receive reduced insurance premiums, this tends to discriminate against disadvantaged individuals who can't afford wearable devices—yet another example of the digital divide that must be closed to ensure fairness.

What can I do to protect myself from misuse of data from my wearables? You should always investigate what is being done with the data being collected by apps connected to your wearable devices. Read the user agreements carefully to ensure that you understand what data is being collected and for what purposes. If the user agreement is unclear, ask questions of the software provider or use a different app.

Bits&Bytes Goodbye Net Neutrality ... Hello Bandwidth Throttling

Net neutrality rules had prevented ISPs from charging more for certain content or providing preferential treatment to certain websites. But with the repeal of net neutrality rules in the United States, there is a concern that ISPs that provide your home Internet service may throttle (intentionally slow down) your bandwidth for certain services (see Figure 8.20).

Even if you paid for unlimited Internet, you don't actually receive unlimited bandwidth. For instance, bandwidth is shared by all cell phone users accessing a single cell tower, so cellular ISPs routinely throttle the amount of bandwidth you receive to your phone to ensure that bandwidth is shared fairly among all users.

Now there is concern that websites that use heavy data streams (like Netflix) may be charged more by the ISPs—and those costs will likely be passed on to consumers, or the ISP may decide to degrade the performance by throttling your connection to Netflix if it decides not to charge Netflix more.

So what can you do? You can urge your congressional representatives to introduce legislation that reinstates net neutrality. Otherwise, you may have to get used to worse performance of certain data streams or higher prices.

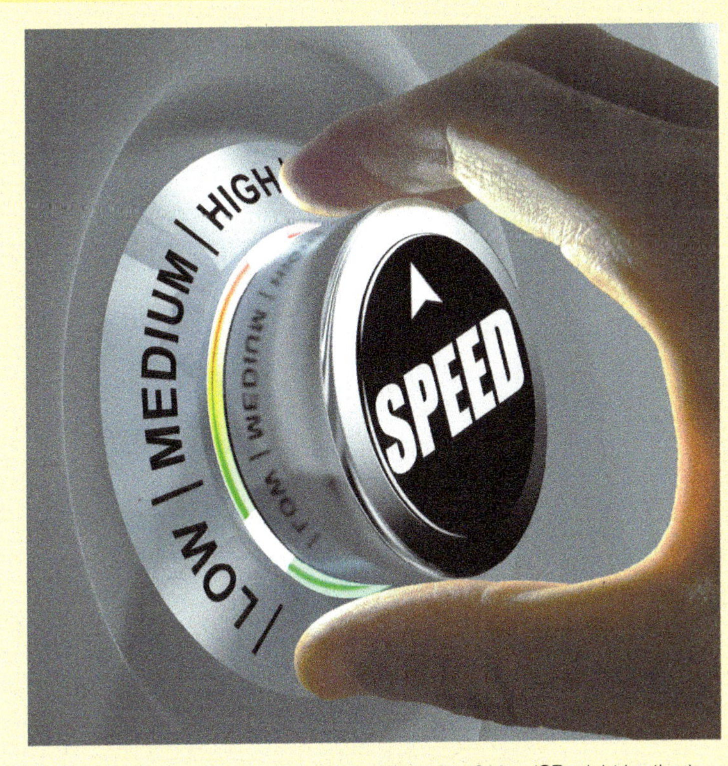

Figure 8.20 Internet speed slower than it should be? Your ISP might be throttling your bandwidth. *(1tjf/123RF)*

Before moving on to Part 2:

1. Watch Chapter Overview Video 8.1.
2. Then take the Check Your Understanding quiz.

Check Your Understanding // Review & Practice

For a quick review to see what you've learned so far, answer the following questions.

multiple choice

1. An analog signal is different from a digital signal because it
   a. is easier to transmit.
   b. has only specific discrete values.
   c. is easier to duplicate.
   d. is continuous.
2. ________ is a sharp grayscale representation of text used in some e-readers.
   a. Electronic ink
   b. Digital ink
   c. Cloud ink
   d. Analog ink
3. When creating a digital sound file, a higher sampling rate will
   a. produce a lower quality file.
   b. produce a higher-quality file.
   c. have no effect; all digital files are the same quality.
   d. create a file that takes up much less space.
4. Digital cameras capture images by using
   a. digital film.
   b. DRM sensors.
   c. LEDs.
   d. charge-coupled device (CCD) arrays.
5. Which of the following is TRUE about cryptocurrency?
   a. Cryptocurrency is regulated by governments just like conventional currency.
   b. Cryptocurrency transactions can be reversed.
   c. Transactions in cryptocurrency are anonymous.
   d. The value of cryptocurrency is relatively stable.

chew on this

(3Dalia/Shutterstock)

What impact will the availability of self-publication have on the writing and recording industries? Will it promote a greater amount of quality content or a flood of amateur work? How could this be managed to ensure that consumers can locate quality works?

Go to **MyLab IT** to take an autograded version of the *Check Your Understanding* review and find all media resources for the chapter.

For the IT simulation for this chapter, see MyLab IT.

Try This

Creating and Publishing a Movie

You've just taken some video and photos at the zoo, and you'd like to organize the media and put it on your YouTube channel. What should you do? In this Try This, we explore Ezvid. For more step-by-step instructions, watch the Chapter 8 Try This video in MyLab IT. Before starting, do the following:

1. Make sure you have video clips and photos available. If you don't have any, you can find free images and video clips at sites like *videezy.com*.
2. Download Ezvid, a free video editor, from *ezvid.com*. Click the **Ezvid for Windows** link and then click the **Download** button. Once installed, type **Ezvid** in the search box to launch.

Step 1 Click the **add pictures or video** button. In the dialog box that appears, browse to where your media files are saved, select the files you wish to import, and then click the **Open** button. The files you've selected will appear on the Media timeline.

Step 2 Click and drag your media to the spot in the timeline where you want it to appear.

Step 3 To create a text slide, click the **add text** button. In the dialog box that appears, type the text you want to appear; change the text color, background color, and font; and then click the **finish** button to add the text slide to your timeline. Click and drag the text slide where you want it to appear.

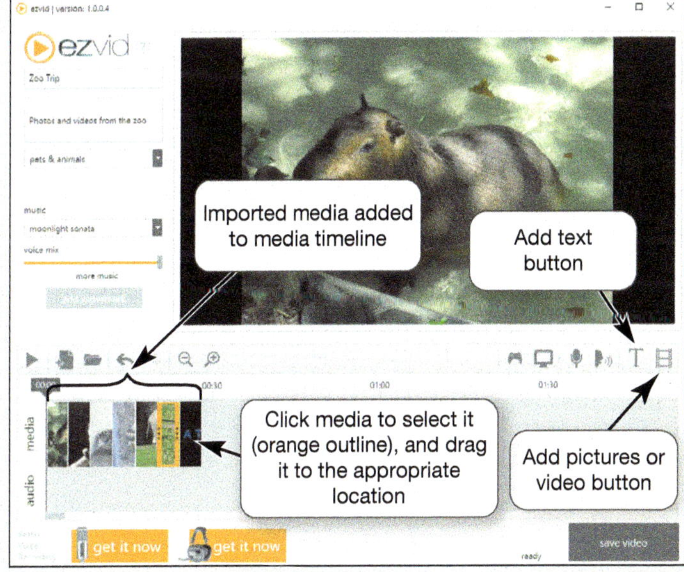

(Courtesy of Ezvid)

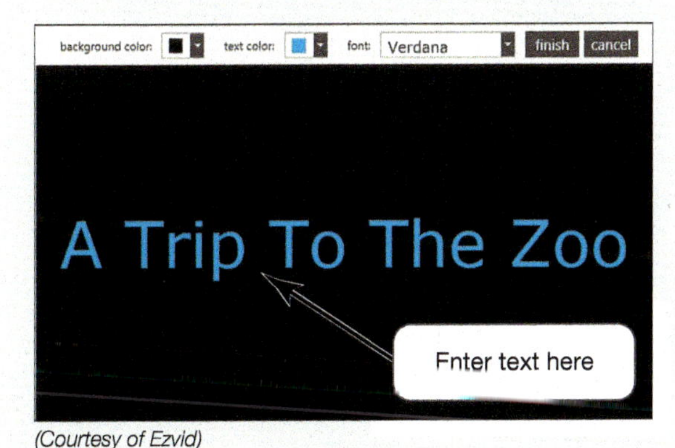

(Courtesy of Ezvid)

Step 4 To adjust the length of time a piece of media displays, click the media and then adjust the width of the clip. Add a video title, description, and category in the appropriate boxes. Select a music track from the ones provided or use your own.

Step 5 To preview your movie, click **Play**. When you're satisfied, click the **save video** button. If you start a new project or close Ezvid, your project will automatically be saved.

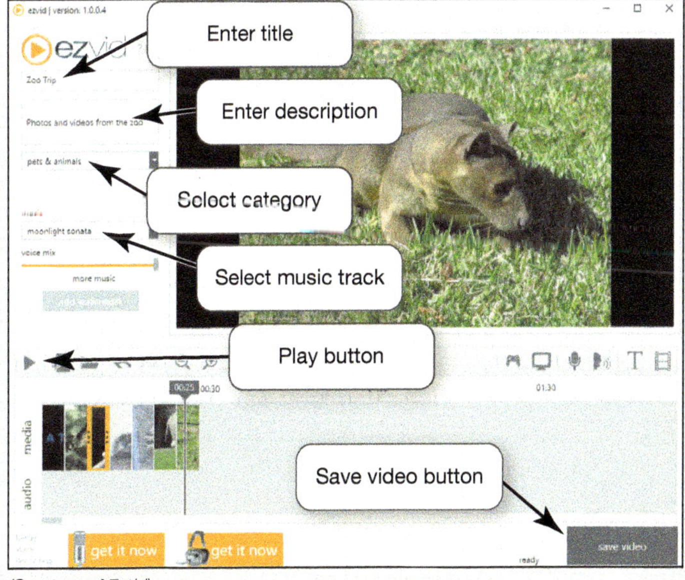

(Courtesy of Ezvid)

Make This  TOOL: App Inventor 2 or Thunkable

A Video-Playing App

Do you want to run video clips inside your app?

In this exercise, you'll continue your mobile app development by using the Video Player component in App Inventor (or Thunkable) to run video clips inside your app. You can use the component to play clips stored on your device, clips that you've recorded with the built-in camera, or clips on the Internet.

The Video Player has controls to pause, play, and fast forward or rewind and lets you manipulate the size of the video screen as well as the volume. So decide how you will make your applications more media-intensive, using the Video Player component in App Inventor!

The Video Player component allows you to display video inside your app.

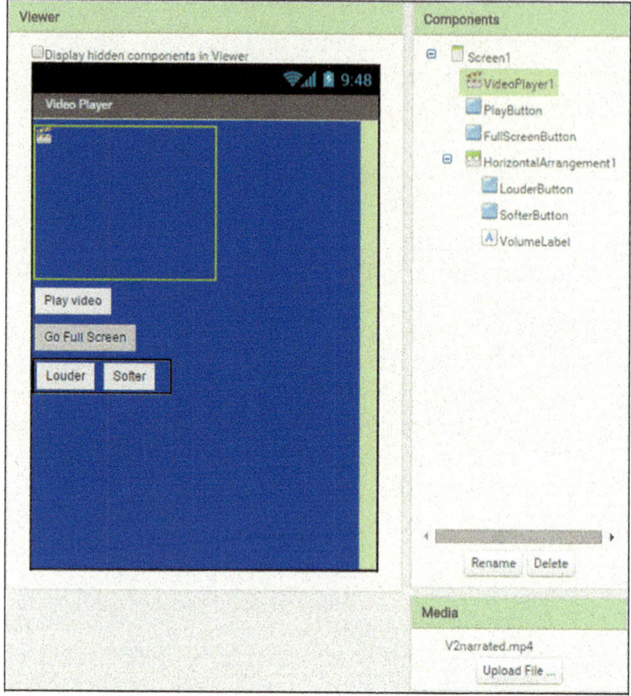

(Copyright MIT, used with permission.)

```
when Camcorder1 .AfterRecording
  clip
do  set VideoPlayer1 . FullScreen to true
    set VideoPlayer1 . Source to get clip
    set VideoPlayer1 . Visible to true
    call Camcorder1 .RecordVideo
```

(Copyright MIT, used with permission.)

To create this app for iOS, go to *Thunkable.com*, a programming platform based on App Inventor.
For the instructions for this exercise, go to MyLab IT.

Part 2

 For an overview of this part of the chapter, watch **Chapter Overview Video 8.2**.

Ethical Issues of Living in the Digital Age

Learning Outcome 8.2 You will be able to describe how to respect digital property and use it in ways that maintain your digital reputation.

In Part 1 of this chapter, you learned about ways to create digital assets. Unfortunately, the proliferation of the digital lifestyle has made it easier to misuse or misappropriate another's digital property. In the following sections, you learn how to protect your digital rights and how to avoid infringing on the rights of others.

Protection of Digital Property

You can easily locate a picture of almost anything by searching Google Images. If you must prepare a research paper on global warming, you can find millions of sources on the subject, including professionally written articles. Does that mean you can freely use all these resources you locate? No, you can't, because they're most likely someone else's property. Just as you wouldn't take someone's bike parked outside your dorm, you can't just take property you find on the Internet.

Property comes in two general types: real and personal. Real property is considered immoveable, such as land or a home, and it's often called *real estate*. Personal property, which is stuff you own, comes in two types: tangible and intangible. We're all familiar with tangible personal property, which is something that has substance like your smartphone or car. Intangible personal property can't be touched—or potentially even seen—yet it still has value. If someone steals your iPad, the action is easily recognizable as theft. However, what if someone tries to steal your ideas? This action moves into the realm of one of the biggest problems in cyberspace: the theft of intangible personal property or intellectual property.

Intellectual Property

Objective 8.10 *Describe the various types of intellectual property.*

What is intellectual property? Intellectual property (IP) is the product of a person's mind and is usually considered an expression of human creativity. Examples of IP are books (or any type of writing), art, music, songs, movies (or any type of video), patents for inventions, formulas (or methods of production), and computer software. IP is considered intangible personal property.

It's important to distinguish intellectual property from the physical medium that carries it. A music CD is not IP. The music that is contained on a CD is the IP, whereas the CD itself is merely a delivery device. Likewise, a poem is IP, but the piece of paper on which it's written is not.

Types of Intellectual Property

How is intellectual property categorized? IP is divided into five broad categories: *copyright, patents, trademarks, service marks*, and *trade dress*. Each category has its own laws of protection.

Copyright protection can be granted to creators of "original works of authorship." In the United States, copyrightable works include:

- Literary works, including computer software
- Musical works, including any accompanying words
- Dramatic works, including any accompanying music
- Pantomimes and choreographic works
- Pictorial, graphic, and sculptural works
- Motion pictures and other audiovisual works
- Sound recordings
- Architectural works

Patents grant inventors the right to stop others from manufacturing, using, or selling (including importing) their inventions for a period of 20 years from the date the patent is filed. Generally, patents aren't renewable but may be extended under certain circumstances.

A trademark is a word, phrase, symbol, or design—or a combination of these—that uniquely identifies and differentiates the goods of one party from those of another. A service mark is essentially the same as a trademark, but it applies to a service as opposed to a product. The Nike swoosh and McDonald's Golden Arches are trademarks, whereas the FedEx logo is a service mark.

Similar to a trademark, a *trade dress* applies to the visual appearance of a product or its packaging. The unique shape of the iconic Coca Cola bottle is an example of trade dress.

Most digital assets you create will be protected by copyright. In this section, we discuss the basic tenets of copyright and how they protect you.

Figure 8.21 Exercising your rights with a video of your band.

Public Performance
- Post the video on YouTube

Reproduction
- Burn DVDs

Distribution to the Public
- Feature exclusive videos on your Patreon site

Derivative Work
- Capture audio tracks from the video soundtrack

Public Display
- Place a still image from the video on a poster

(PopTika/Shutterstock; Denis Dryashkin/Shutterstock; Mega Pixel/Shutterstock; Johavel/Shutterstock; Shutterstock)

Copyright Basics

Objective 8.11 *Explain how copyright is obtained and the rights granted to the owners.*

When does copyright protection begin? Copyright begins when a work is created and fixed into a digital or physical form. For example, if you record a video of your cat, as soon as you save the video to your phone, the video is subject to copyright protection. There is no need for the video to be registered or even published for it to be protected by copyright.

If I own copyright to a work I created, what exactly do I own? Copyright holders own rights that grant them the ability exclusively to do things with the copyrighted work. US law grants these rights to a copyright holder:

- **Reproducing the work.** Applies to copying the entire work or just part of the work. For instance, you might sell prints of one of your paintings on Etsy.
- **Preparing derivative works based on the original work.** This means developing any media based on the original work regardless of what form the original is in. The Flash was originally a character in a DC comic book, but he now appears in movies and television programs.
- **Distributing the work to the public.** This usually involves selling the work. However, the copyright holder could also lend, rent, or give away the work.
- **Public performance of the work.** This applies to any audiovisual work such as plays, movies, and songs. In the case of audio recordings, this also means digital audio transmission.
- **Public display of the work.** This right usually applies to works of art such as paintings, photographs, and sculptures, but this also includes posting images on the Internet.

Figure 8.21 shows ways you could capitalize on a video of your band that you created by exercising your rights granted by copyright.

Can I copyright anything I create? US copyright law does not protect all ideas, but rather the *unique expression* of an idea. You can't copyright common phrases, such as "bad boy" or "good girl"; discoverable facts, such as water boils at 212 degrees Fahrenheit; or old proverbs, such as "To err is human, to forgive divine." However, you can copyright a creative twist on the proverb, such as the one in Figure 8.22.

Figure 8.22 You can copyright this clever twist on an old proverb. *(Quote by Tim Falconer)*

Generic settings and themes of a story also can't be copyrighted. J. R. R. Tolkien wrote *The Lord of the Rings* trilogy, which is about a person who overcomes hardships to save the world from an evil being. In an e-book or video game you create, you're not allowed to use the same characters and plot lines Tolkien used, but you could write about the basic idea of a lone individual overcoming the forces of evil.

How long does copyright last? Current US law grants copyright for the life of the author (creator) plus 70 years for original works. After you die, any copyrights you own are transferred to your heirs. Therefore, if you write a best-selling e-book in 2025 and die in 2095, your heirs can continue to earn money from the book until the year 2165.

Can rights to a copyrighted work be sold? Rights can be sold or granted for free to various individuals or entities in perpetuity or for a limited time. For instance, if your band is going to appear at a bar, you might grant the bar owner the right to play your band video in his bar for a specific period of time leading up to your performance. Or you might grant the right to sell the audio of your performance to a distributor such as DistroKId (*distrokid.com*).

If I work for a company and develop intellectual property, do I own it? The answer to this question depends on the terms of your employment contract. *Works made for hire* or *works of corporate authorship* occur when a company or person pays you to create a work and the company or person owns the copyright when it's completed. Most companies in which IP is developed require employees to sign agreements that any IP developed while working for the company becomes the company's property.

Is buying a copyrighted work the same as buying the copyright? Usually you only buy a physical or digital copy, not the actual copyright. When you buy software, usually what you've purchased is a license to install the software and use it on a specific number of devices. When you purchase music, you've bought the right to listen to the music in a private setting, such as on your phone. When you download a movie or subscribe to Netflix, again you've purchased the right to view the content privately.

However, buying copyrighted works is covered by a rule of law known as the *first sale doctrine*. You have the right to sell, lend, give away, or otherwise dispose of the item you purchase. This principle gives you the right to sell your book to another student at the end of the semester.

Are all works protected by copyright? Works without copyright protection are considered to be in the public domain. Works in the public domain are considered public property; therefore, anyone can modify, distribute, copy, or sell them for profit. Some works that were created never had a copyright holder. Examples are traditional folk songs or stories, the origins of which are unknown and can't be attributed to a specific individual. Copyrighted works can lose their copyright protection after their term of protection expires.

It can be difficult to determine when works enter the public domain. Therefore, you should probably consider a work to be in the public domain only if there is provable, objective evidence that it's in the public domain or it's clearly identified as being in the public domain. The safest assumption is that all intellectual property you find on the Internet is protected by copyright, *unless otherwise stated*.

Protecting Your Work with Copyrights

As we discussed in Part 1 of this chapter, you're potentially creating intellectual property on a regular basis. If your work is valuable and you're using it to earn a living, you should take appropriate steps to protect your copyright.

Isn't copyright protection automatic? Copyright does accrue as soon as you create the intellectual property and set it down in digital or physical form. However, registering copyright provides certain advantages, including the ability to sue someone for violating your copyright.

How do I register works for copyright protection? Instructions and forms for registering work for copyright protection can be found on the US Copyright Office website (*copyright.gov*). Registration can be done online and requires payment of a small fee. Although you can file online, the Library of Congress might still require you to submit a hard copy of your work.

Posting a copyright notice with your work is also a deterrent to copyright infringement, because it establishes that the work is copyrighted. A simple format is to use the word copyright (or the copyright symbol—©), the year the work was first published, the copyright holder's name, and optionally, the location of the copyright holder.

Do I have to pay to register each photo in my portfolio? A copyright registration form can be filed for a collection of works. The stipulation is that the works must be published as a collection. A group of related images, songs, videos, poems, or short stories or a collection of blog postings prior to registration meets this criterion.

What if I don't mind people using my work, but I don't want to constantly answer permission requests? Copyleft—a play on the word *copyright*—is a term for various licensing plans that enable copyright holders to grant certain rights to the work while retaining other rights. Usually, the rights (such as modifying or copying a work) are granted with the stipulation that when users redistribute their work (based on the original work), they agree to be bound by the same

License Conditions

Creators choose a set of conditions they wish to apply to their work.

Attribution (by)

All CC licenses require that others who use your work in any way must give you credit the way you request, but not in a way that suggests you endorse them or their use. If they want to use your work without giving you credit or for endorsement purposes, they must get your permission first.

ShareAlike (sa)

You let others copy, distribute, display, perform, and modify your work, as long as they distribute any modified work on the same terms. If they want to distribute modified works under other terms, they must get your permission first.

NonCommercial (nc)

You let others copy, distribute, display, perform, and (unless you have chosen NoDerivatives) modify and use your work for any purpose other than commercially unless they get your permission first.

NoDerivatives (nd)

You let others copy, distribute, display and perform only original copies of your work. If they want to modify your work, they must get your permission first.

Figure 8.23 The Creative Commons has a variety of licenses, one of which should be right for you. *(Courtesy of Creative Commons)*

terms of the copyleft plan as were used by the original copyright holder. The General Public License (GNU) is a popular copyleft license used for software. For other works, the Creative Commons, a nonprofit organization, has developed a range of licenses that can be used to control rights to works (see Figure 8.23). It provides a simple algorithm to assist you with selecting the proper license for your work (*creativecommons.org/share-your-work*).

What are the advantages and drawbacks of licenses designed by Creative Commons? The advantage to using Creative Commons licenses is that people won't constantly send you permission requests to use your work. The licenses explain how your work can be used. Also, many advocates of copyleft policies feel that creativity is encouraged when people are free to modify other people's work instead of worrying about infringing on copyright.

Opponents of Creative Commons licenses complain that the licenses have affected their livelihoods. If millions of images are on Google Photos with Creative Commons licenses that permit free commercial use, professional photographers might have a tougher time selling their work. Furthermore, Creative Commons licenses are irrevocable. If you make a mistake and select the wrong license for your work, or you later find out a work is valuable and you've already selected a license that allows commercial use, you're out of luck.

Understanding the meaning of copyright, and copyleft, is important so that you respect the rights of others as well as simplify your life in granting permission rights to the works you create.

MyLab IT

Understanding Intellectual Property and Copyright

In this Helpdesk, you'll play the role of a help desk staffer, fielding questions about the basics of intellectual property and how individuals can protect their digital assets with copyright.

Copyright Infringement

Objective 8.12 *Explain copyright infringement, summarize the potential consequences, and describe situations in which you can legally use copyrighted material.*

What happens if you use copyrighted material without the permission of the copyright holder? A violation of the holder's rights is known as copyright infringement. Illegally copying or using software, music, video, and photos tops the list of digital rights violations on the Internet. Here are a few examples of copyright infringement that you may recognize:

- Posting someone else's photo from the Internet on Instagram
- Using a peer-to-peer file-sharing service or torrent to download a movie
- Copying a copyrighted software app and giving it to a coworker

What happens to people who infringe on someone's copyright? Infringing on copyright risks a potentially long and costly legal battle. At best, you might receive a slap on the wrist, but worst-case scenarios involve large fines and prison terms. Penalties vary depending on the type of property being infringed and the country whose laws are in effect.

Is posting a picture on a social media site really copyright infringement? If you don't own the copyright to the picture or the rights to display a copyrighted image, then it's copyright infringement. Although you may not have read a social media site's user agreement, most sites, like Pinterest, have users agree not to post material to which they don't own copyright. Of course, if a website provides a Pinterest link for you to share material, you could argue that it's giving you implicit permission to share its material.

How do courts decide when copyright has been infringed? When judges consider cases of infringement, they generally examine the extent to which there is a substantial similarity between the copyrighted work and the infringing work. Obviously, if you copy an image from the Internet, that is an exact copy and is clearly infringement. What if you wrote a story with a similar plot to an episode of your favorite TV show? The courts would have to determine how similar the characters and plot are to the copyrighted work and whether infringement exists.

Why would anyone risk copyright infringement? Severe monetary penalties exist for copyright infringement, so why take a chance? Here are a few reasons people have indicated they risk copyright infringement:

- **Low likelihood of getting caught.** If you download three songs from a file-sharing website, are you likely to get caught? Probably not, but that doesn't make it ethical. You're still depriving an artist of his or her livelihood. However, some people commit such acts if they think they can escape notice.
- **Everyone is doing it.** If people think "everyone else does it so why shouldn't I?" they might take a chance, especially if the likelihood of getting caught is low.
- **No one would come after an individual.** Many people believe that only large corporations are sued for copyright infringement. But thousands of individuals have been sued by the record and motion picture industries for illegal file sharing. Most of these cases were settled out of court for amounts ranging from $3,000 to $4,000, but court awards have gone higher. The Recording Industry Association of America (RIAA) and movie studios tend to target *supernodes*—people who offer thousands of music or video files on P2P networks for sharing—for their legal action, but they can just as easily target you for downloading a movie or a dozen songs.
- **I'm only downloading one song, picture, and so on—it isn't worth that much.** Whether you're stealing one candy bar from a convenience store or a case of them from a truck, they're both crimes. Because copyright infringement is prohibited by law, copying one song is still stealing.

Is putting a URL that points to a copyrighted website on my social media page copyright infringement? A URL is a direction for finding a specific page on the Internet. It isn't debatable or open to interpretation; therefore, it's considered a fact. Because facts can't be copyrighted, you can list all the textual URLs you want on your website without committing copyright infringement. However, be sure you don't use copyrighted material, such as a logo or character, as a visual link to a website. For instance, using a picture of Mickey Mouse to link to the Disney website may constitute infringement.

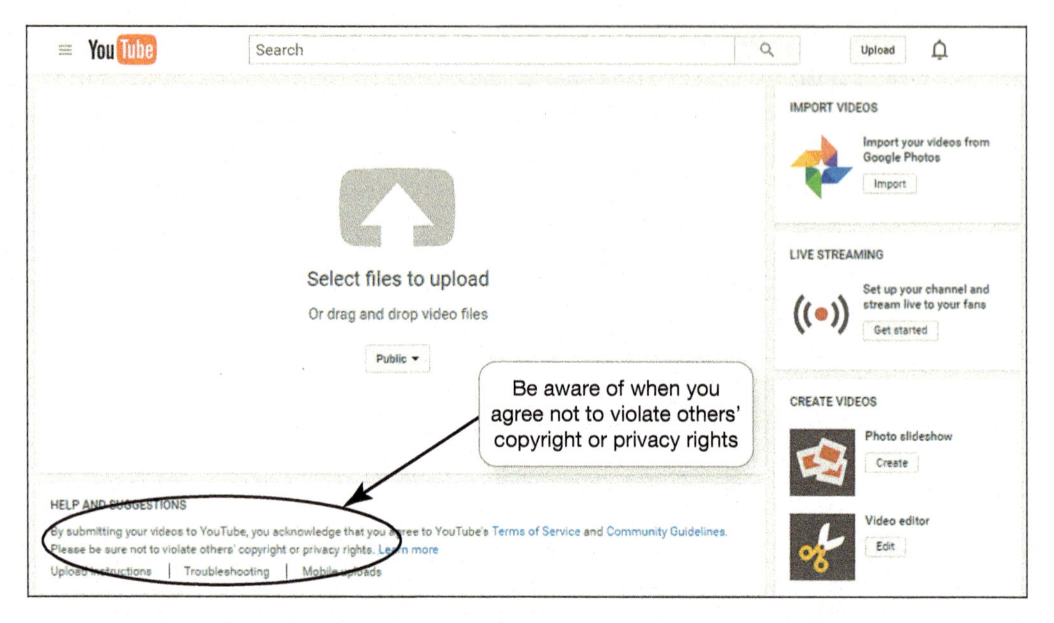

Figure 8.24 Although you might have missed it, when you upload videos to YouTube, you agree to respect copyright and privacy. *(Courtesy of You Tube)*

Music and Video Issues

I uploaded clips from TV shows to YouTube. Could I be liable for copyright infringement? Although many people think that YouTube would be responsible for any infringement, anyone who uploads copyrighted material could be held responsible. Although you might not pay attention when uploading videos, you agree not to infringe on copyright when uploading the files, as shown in Figure 8.24. Therefore, YouTube has potential legal recourse to recover damages from users who ignore the terms of use on its website. If you've uploaded copyrighted material without permission to YouTube or any other file-sharing site, you should remove it immediately.

Why would a band object to uploading a music video to YouTube? Wouldn't it be free publicity for them? Each time someone plays a video that's uploaded illegally to a website, the performer (singer or band), the songwriter, and the music publisher potentially lose money. A song played on a website is considered a live performance. Songwriters and music publishers are entitled to a royalty every time a song is played live, on the radio, on a streaming service (like Spotify), or anywhere on the Internet. Normally, radio stations and streaming services pay royalty fees to performing rights organizations (PROs). The PROs then forward the royalties to the songwriters, music publishers, and other copyright holders. Because you're not paying royalties to a PRO when you view the video on YouTube, someone is losing money.

How significant is the loss of revenue from music and video infringement? A study from Columbia University concluded that 45 percent of US citizens pirate movies on a regular basis. Many popular movies are illegally downloaded tens of millions of times. If everyone pays $5 to watch the movie from a legitimate streaming source, the movie studio could have earned millions of dollars more. Clearly, illegal file swapping is costing people and companies a significant amount of revenue.

Software Issues

How are the rights of software copyright holders usually violated? Illegally using copyrighted software is known as software piracy. If you've ever given a friend a copy of a copyrighted software program or downloaded copyrighted software from a file-sharing site, you've committed infringement.

You need to exercise caution when buying software online to ensure that it's a real, fully licensed copy. Otherwise you might not be able to use the software you bought. Most modern software requires a serial number or product key for installation. Typically, the first time you launch the software after installation, this serial number is checked against a database to ensure that the software hasn't been installed on more computers than the license allows. This is a form of digital rights management (DRM). If you're buying illegally copied software on the Internet, you might not be able to activate it because too many other people have already used the same serial number.

Bits&Bytes Software Piracy: It's More Than Just Downloading and Copying

Copying and downloading software aren't the only acts of software piracy. The Business Software Alliance (*bsa.org*), an organization that has among its goals the prevention of software piracy, provides a few other examples:

- Taking advantage of upgrade offers without possessing a required older version of the software
- Using software designed specifically for academic or noncommercial use for commercial purposes
- Installing a licensed copy of software on more computers than the license permits
- Counterfeiting software (copying it for the purpose of selling it as if it were the real, licensed product)
- Selling computers preloaded with illegal copies of software already installed

How widespread is software piracy? In some countries, such as Indonesia and China, software piracy is rampant. Piracy rates approach 90 percent in some areas of the world. Although less of a problem in the United States on a percentage basis, the United States still leads the world in lost revenue from software piracy.

Photographic Issues

Why is photographic infringement widespread on the Internet? Because it's so easy! Need a picture of the Eiffel Tower for your presentation? Google's image search makes it simple to find a picture. Copy and paste it into your presentation and you're done. Congratulations, you've probably just infringed on someone's copyright!

Who owns the copyright to a photo? When the photo is taken, the photographer owns the copyright, assuming the photographer was taking photos in a public place or had permission from the property owner. If the photographer was working on a work-for-hire basis, usually the copyright belongs to his or her employer.

Do people in the photo have any rights? Being photographed without your consent generally falls under the realm of privacy. The Fourth Amendment to the US Constitution and various other laws and court cases have recognized Americans' rights to certain amounts of privacy. Usually, if a photographer includes you in a photo (see Figure 8.25) that he or she intends to sell for publication, you'll be asked to sign a model release, which usually grants the photographer the right to use an image of the model (or subject of the photo) commercially.

If a picture of you is published without your consent, you might have the basis for a lawsuit. However, be aware that there are times when you agree to allow your photo to be taken and used commercially (usually for publicity), such as when you enter a contest, buy a ticket for an amusement park, or enter an event such as a marathon. So read the conditions on sign-up forms and tickets to ensure that you understand any rights you're surrendering.

Figure 8.25 There is no problem taking a picture of this statue of Abraham Lincoln on display in a public park. But if you want to use the photo for commercial purposes, you need to obtain a model release from the man chatting with Lincoln. *(Alan Evans)*

How do you know whether a photo on the Internet is copyrighted? Usually information is attached to a photo to indicate it's copyrighted, in the public domain, or that all rights are reserved. If no information is attached, the safest course of action is to assume that the photo is protected by copyright and you'd need permission to use the photo. Figure 8.26 shows an example of copyright information for a photo on the Flickr website.

Using Copyrighted Material Legally

Under what circumstances can I use copyrighted material? Many websites that contain copyrighted material also contain lengthy legal documents that delineate the terms of use for the material that you download from the site (see Figure 8.27). It's important to read the terms of use *before* using any copyrighted material on the site. Failure to read the terms of use doesn't absolve you from liability for using copyrighted material without permission.

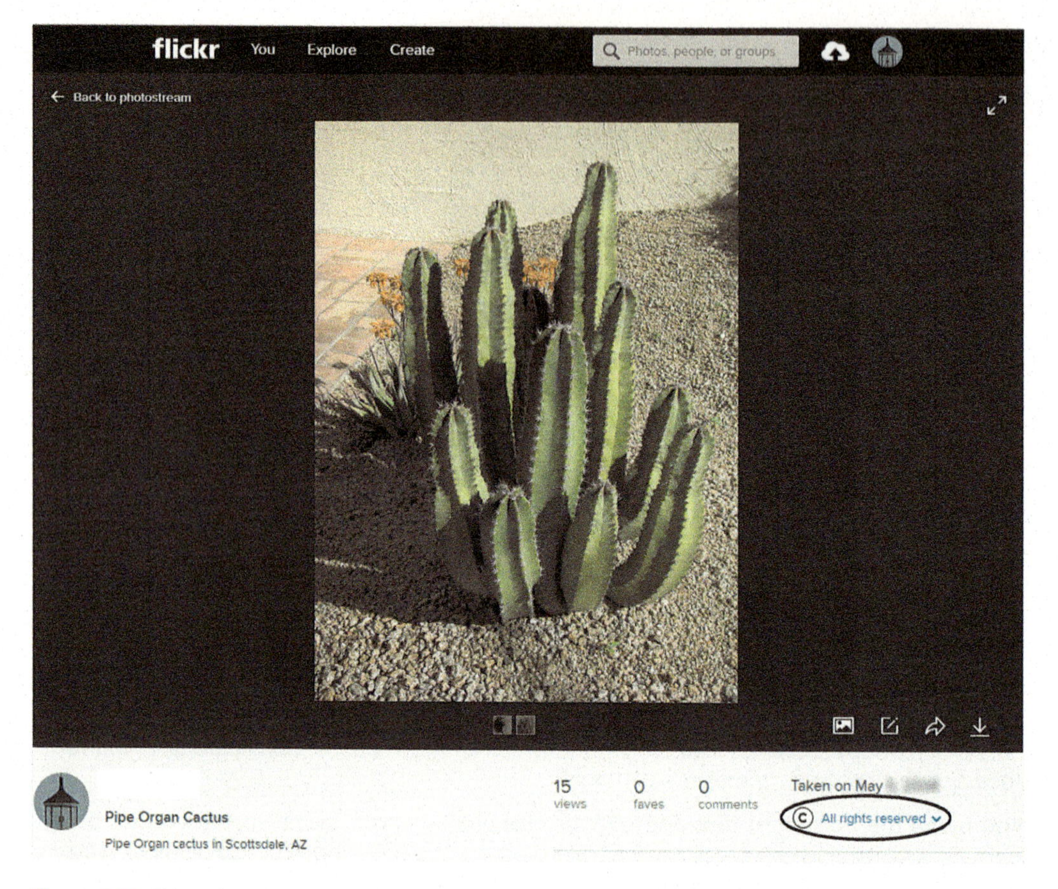

Figure 8.26 Although you can download this photo from Flickr, all rights are reserved so you cannot use it without obtaining permission.
(Courtesy of Flickr)

What if a website does not have terms of use? Ferreting out the terms of use on a site can sometimes be tricky. Look for links that say Terms of Use, Restrictions, Copyright, Rules, Media Room, Usage Guidelines, FAQ, or Contact Us. Sometimes, the usage terms aren't displayed until you attempt to download copyrighted material. If you've done a thorough search and can't find them, contact the organization that maintains the website and ask about the site's terms of use.

Can I use copyrighted material if it isn't permitted in the terms of use or there are no terms of use? Copyright holders can always grant permission to use copyrighted material to an individual or organization. Depending on the material used and the specific nature of the usage, a payment may be required to secure the rights to the copyrighted work. Sometimes, though, simply asking permission is enough to get you the rights to use the work for a specific purpose free of charge.

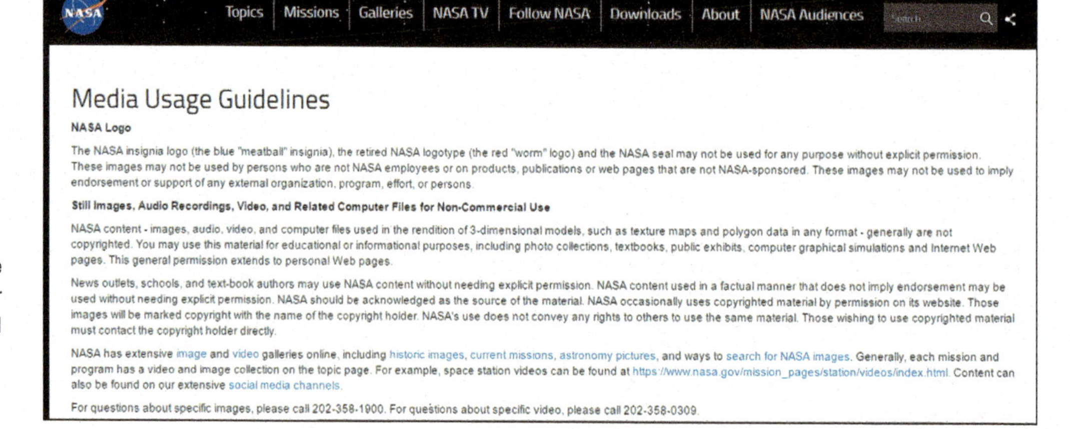

Topics | Missions | Galleries | NASA TV | Follow NASA | Downloads | About | NASA Audiences

Media Usage Guidelines

NASA Logo

The NASA insignia logo (the blue "meatball" insignia), the retired NASA logotype (the red "worm" logo) and the NASA seal may not be used for any purpose without explicit permission. These images may not be used by persons who are not NASA employees or on products, publications or web pages that are not NASA-sponsored. These images may not be used to imply endorsement or support of any external organization, program, effort, or persons.

Still Images, Audio Recordings, Video, and Related Computer Files for Non-Commercial Use

NASA content - images, audio, video, and computer files used in the rendition of 3-dimensional models, such as texture maps and polygon data in any format - generally are not copyrighted. You may use this material for educational or informational purposes, including photo collections, textbooks, public exhibits, computer graphical simulations and Internet Web pages. This general permission extends to personal Web pages.

News outlets, schools, and text-book authors may use NASA content without needing explicit permission. NASA content used in a factual manner that does not imply endorsement may be used without needing explicit permission. NASA should be acknowledged as the source of the material. NASA occasionally uses copyrighted material by permission on its website. Those images will be marked copyright with the name of the copyright holder. NASA's use does not convey any rights to others to use the same material. Those wishing to use copyrighted material must contact the copyright holder directly.

NASA has extensive image and video galleries online, including historic images, current missions, astronomy pictures, and ways to search for NASA images. Generally, each mission and program has a video and image collection on the topic page. For example, space station videos can be found at https://www.nasa.gov/mission_pages/station/videos/index.html. Content can also be found on our extensive social media channels.

For questions about specific images, please call 202-358-1900. For questions about specific video, please call 202-358-0309.

Figure 8.27 The terms-of-use page from nasa.gov clearly spells out your rights regarding photos downloaded from its website. NASA's media usage guidelines are very generous.
(Courtesy of NASA)

Bits&Bytes Your Tax Dollars at Work: Free Media Without Permission!

Need an image or a video but don't want to pay for it? Many (if not all) images posted on US government websites have generous usage guidelines. Need a V-2 rocket photo (see Figure 8.28)? Head to *nasa.gov*. NASA's collection of media is free to use for educational and informational purposes without explicit permission. You can even use NASA media for commercial purposes if it doesn't imply endorsement of your product or company by NASA. Need a photo of Arches National Park? Check out the US Geological Survey site (library.usgs.gov/photo). Any images posted on the site generated by employees of the USGS are considered in the public domain. Picture of a grizzly bear? The US Fish and Wildlife Service can help you out (digitalmedia.fws.gov). Make your tax dollars work for you by saving money on permissions fees for media!

Figure 8.28 US government websites have a wealth of media, like this rocket, that is free to use. *(NASA)*

Whom do I contact for permission? Identifying the appropriate contact from which to obtain permission depends on the nature of the intellectual property. Sometimes it can be difficult to tell who actually owns the copyright, or the particular rights you need, to a piece of media. The creator may not be the copyright holder any longer. He or she may have sold the rights to another party. Figure 8.29 provides suggestions about permission contacts.

What information should I provide in a permission request? Be sure to include these points:

- **Who you are.** Include your contact information (name, address, phone number, and e-mail) and whether you're requesting permission personally or on behalf of an organization.
- **Which work you're requesting permission to use.** Describe the work completely as well as how much of the work you'll be using.
- **Complete details of your usage of the work.** Indicate why you're using the work, whether the work is being used as part of a money-making project, where the work will be used (website, book, magazine, etc.), how frequently the work will be used, and whether you're making modifications to the work.
- **The timing of the request.** Indicate the date you intend to begin using the work and suggest a deadline for responding to the request.

Be sure you receive a written response to your request, authorizing the usage you requested. Never assume that a lack of response indicates tacit approval from the copyright holder. Remember, copyright holders are under no obligation to grant your request or even to respond to it.

Figure 8.29 Permission contacts for various types of media.

Books and Periodicals
- Author or publisher

Website Content
- Webmaster, website owner, or author

Photos, Sculpture, Paintings, or Other Art
- Artist or photographer, publisher (print media), gallery management (on display)

Movies or Video
- Uploader (original video on website), production company, film distributor, screenwriter

(Monticello/Shutterstock; Bannosuke/123RF; Ruslan Gilmanshin/123RF; Axsimen/123RF)

Fair Use

Are there any instances in which I can use copyrighted work without the permission of the copyright holder? Fair use provides for people to use portions of a copyrighted work for specific purposes without receiving prior permission from the copyright holder (see Table 8.3). However, there are no specific rules on exactly what amount of use constitutes fair use, which means each case has to be decided on its own merits.

What types of activities constitute fair use? The following are examples of fair-use activities permitted in the United States:

- Quotation of excerpts in a review or criticism
- Summary of an address or article, with brief quotations, in a news report
- Reproduction by a teacher or student of a small part of a work to illustrate a lesson, commonly known as educational fair use

Table 8.3 Suggested Educational Fair-Use Guidelines

Media	Quantity
Motion media	Up to 10% or 3 minutes, whichever is less
Text material	Up to 10% or 1,000 words, whichever is less
Music, lyrics, and music videos	Up to 10%, but in no event more than 30 seconds, of the music and lyrics from an individual musical work (or in the aggregate of extracts from an individual work)
Illustrations and photographs	One artist: No more than five images Collections of works: No more than 10% or 15 images, whichever is less
Numerical data sets (databases and spreadsheets)	Up to 10% or 2,500 fields or cell entries, whichever is less

Does being a student who uses educational (academic) fair use give me a defense for using any copyrighted material I want? Because there are no fixed guidelines on the quantity of material that can be used and still be considered fair use, guidelines have been developed to assist teachers and students. The Consortium of College and University Media Centers developed suggested guidelines for various types of media, excerpts of which are summarized in Table 8.3.

Again, these are only suggested guidelines and depending on the individual case, you may be able to make a successful argument under the fair use doctrine to use a greater percentage of a given work. Many educational institutions publish their own guidelines, so inquire about your school's guidelines.

Living Ethically in the Digital Era

The rise of digital technology provides new challenges to us all. In this section, we'll explore some of the ethical choices you might be confronted with as you navigate the digital world.

Plagiarism

Objective 8.13 *Explain plagiarism and strategies for avoiding it.*

What exactly is plagiarism? Plagiarism is the act of copying text or ideas from someone else and claiming them as your own. Using ideas from other sources and integrating them into your work is acceptable only if you clearly indicate the content being used (such as through quotation marks) and cite your source. Changing a few words but keeping the essence of someone else's idea is still plagiarism even if you don't copy the text exactly. Although the following examples don't involve copying words or ideas without attribution, they're still examples of plagiarism under the academic definition:

- **Turning in work that someone else did for you.** Copying the Excel file that was due for homework in your computer literacy class from a classmate is still plagiarism, even though the file isn't a text file.
- **Failing to identify a direct quotation with quotation marks.**
- **Copying too much material from other sources.** If a work consists mostly of quotes and ideas from other sources, even though the source has been identified, it's difficult to justify it as original, creative work.

Plagiarism is usually considered an academic offense of dishonesty and isn't punishable under US civil law. Plagiarism is prohibited by almost all academic institutions, and the penalties usually are severe, ranging from receiving a failing grade on the assignment to receiving a failing grade for the course to being dismissed from the institution. Although plagiarism isn't technically copyright infringement, it can easily turn into copyright infringement if too much material is stolen from other sources, such as an entire chapter of a book or an entire research paper. The author may choose to sue you if the market for his or her intellectual property is damaged.

Why is plagiarism such a big issue today when it has been a problem for centuries? Quick access to information on the Internet has made it easier than ever to commit plagiarism. It's possible to copy and paste large quantities of information with just a few mouse clicks.

Another huge problem today is presented by paper mills—websites that sell prewritten or custom-written research papers to students. For custom work, you can even specify what type of grade you would like to get. Some students choose to buy papers to earn them a grade of C if they think that it's less likely to arouse suspicion from their instructor. It's illegal in most states to sell essays that will be turned in by students as their own work, but the mills get around this by putting disclaimers on their websites that the papers should be used only for research purposes and not as the student's work. Fortunately, tools are available to instructors for detecting this type of plagiarism.

What can my instructor do to detect plagiarism? Sometimes, just reading a student's work is a giveaway. If the level of writing improves dramatically from earlier assignments, most instructors become suspicious. Most colleges allow instructors to test students orally on the content of papers they suspect aren't the student's own work. If a student is unfamiliar with the content of the paper and the sources used, charges of academic dishonesty can be brought against the student.

Typing suspicious phrases from a paper into a search engine is an effective way to find uncredited sources. Furthermore, most school libraries subscribe to searchable databases of periodicals that contain the full text of published articles. This aids instructors in ferreting out plagiarism from printed sources.

Educational institutions also often subscribe to specialized electronic tools such as Turnitin (*turnitin.com*). The subscription enables instructors to upload student papers, which are then checked against databases of published journals and periodicals, previously submitted student papers, and websites—both current sites and archived sites that are no longer live. Customized reports, such as the one shown in Figure 8.30, are generated to determine the amount of suspected plagiarism in the paper. Instructors can allow students to upload their papers to check them for inadvertent plagiarism to give them a chance to cite uncredited sources before turning in their final product.

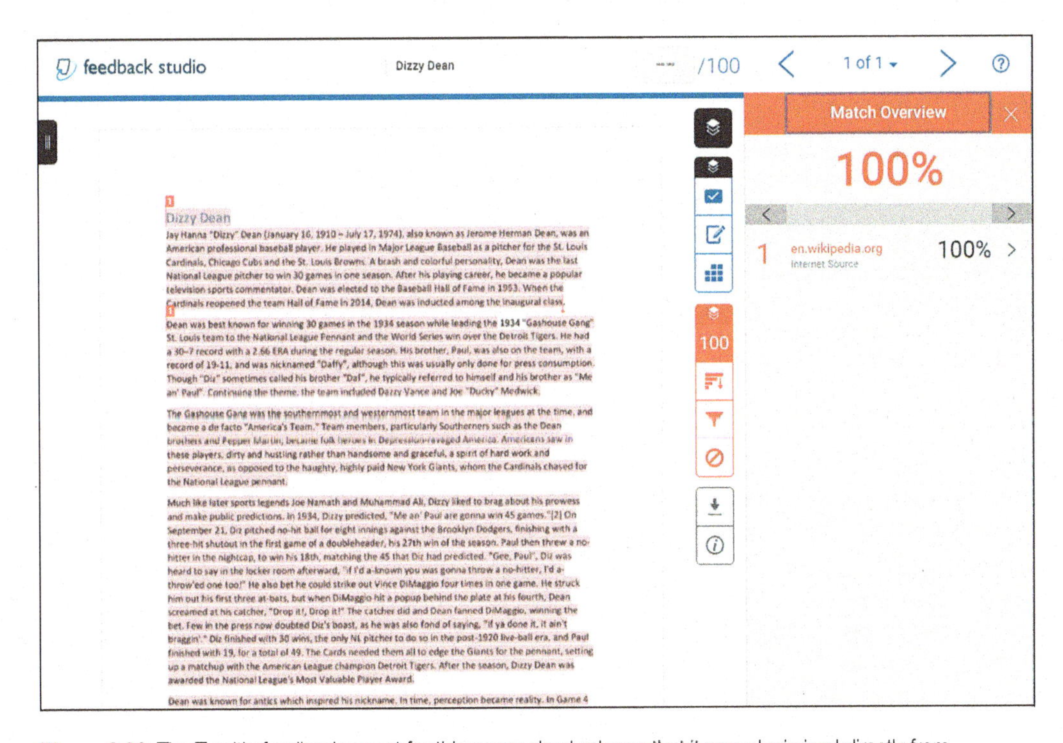

Figure 8.30 The Turnitin feedback report for this paper clearly shows that it was plagiarized directly from Wikipedia. *(Courtesy of Turnitin USA)*

Plagiarism and Intellectual Property

In this Sound Byte, you learn how to document your sources to avoid committing plagiarism. You also learn about the types of licenses available for easily sharing intellectual property.

How can I avoid committing plagiarism? Learn to follow this simple maxim: *When in doubt, cite your source.* If you're taking an exact quote from a work, cite the source. If you're paraphrasing someone else's idea but still retaining the essence of their original, creative idea, cite the source.

How do I cite a printed source properly? Many styles of citations have been developed by organizations such as the American Psychological Association (APA) and the Modern Language Association (MLA). Ask your instructor which style is preferred. Regardless of the style, readers of your work need enough information to find the sources you cite. Microsoft Word has built-in features that make creating citations easy (see the Try This feature in Chapter 4 for more information).

So if plagiarism isn't illegal, why do professors get so upset about it? A summary of the issues follows in Table 8.4.

Table 8.4 Point/Counterpoint: Plagiarism

Issue: Plagiarism	Ethical Question: Is plagiarism wrong?
	• Taking credit for another's work is always unethical. • Plagiarizing published work (implying it's a higher-quality product) puts an unfair burden on other students who have to compete with the plagiarizer. • Creators of intellectual property deserve full credit for their work.
	• There are no laws in the United States that make plagiarism a crime. • Only academics frown on plagiarism. The real world is more concerned with results. • Today's culture fosters an environment of collaborating and sharing. Therefore, how can accessing published information and using it be wrong?

Hoaxes and Digital Manipulation

Objective 8.14 *Describe hoaxes and digital manipulation.*

Everyone seems to post social media stories warning about a variety of issues, such as the police in your state giving out speeding tickets or the plight of sick children with one last wish. Because anyone can put up a web page on any subject, post to social media, or send off a quick e-mail, how can you distinguish fact from fiction? Determining the reliability of information you encounter in cyberspace can be a challenge, but it's important when you use the Internet as a source of scholarly research. Your professor would not be impressed if you gave a presentation on the illusive North American jackalope (see Figure 8.31).

Hoaxes and Other Lies

What is a hoax? A hoax is anything designed to deceive another person either as a practical joke or for financial gain. Hoaxes perpetrated in cyberspace for financial gain are classified as *cyber-crimes*. Hoaxes are most often perpetrated by e-mail and postings on social media.

Many hoaxes become so well known that they're incorporated into society as true events even though they're false. Once this happens, they're known as urban legends. An example of an urban legend is the story about a man who wakes up in a bathtub full of ice water to discover he's had his kidney removed.

What is fake news? Fake news is a type of hoax. Fake news refers to stories that are invented and are intended to spread misinformation rather than fact. Fake news is most often spread

Figure 8.31 Hoax websites often look real and are used in computer fluency classes to teach people the value of fact checking. *(Patti McConville/Alamy Stock Photo)*

through social media, but it has also made its way into traditional news media. The purpose of fake news is to mislead individuals and damage or disparage groups of people, companies, agencies, or governments. Fake news is also propagated for political or financial gain.

How can I tell whether something on social media is a hoax or fake news? Many hoaxes are well written and crafted in such a way that they sound real. Before sharing the post, check it out at sites such as Snopes (*snopes.com*), Hoax-Slayer (*hoax-slayer.com*), or the Museum of Hoaxes (*hoaxes.org*). These sites are searchable, so you can enter a few keywords from the item you suspect may be a hoax and find similar items with an explanation of whether they're true or false. Regarding fake news, there are many fact-checking websites, such as *FactCheck.org* (run by the Annenberg Public Policy Center), that specialize in helping identify fake news stories and helping individuals separate fact from fiction.

How do I know whether something in Wikipedia is true? Wikipedia entries are reviewed by editors, so there's some control over the accuracy of content, but just as with any other source of information, articles in Wikipedia need to be evaluated based on the criteria described in Chapter 3. Many articles on Wikipedia have extensive footnotes and lists of sources, making it possible to investigate the reliability of the source material, so treat Wikipedia as you would any other web page you might want to rely on; evaluate the content thoroughly before relying on it.

Is there a resource for finding quality sites for scholarly research? Several resources on the Internet, many of which are created (or maintained) by librarians, can be useful for locating websites and scholarly publications that are considered reliable, current, and suitable for academic research. Some of the more popular ones are described in Figure 8.32. Using these sites doesn't mean you don't have to check facts, but it does provide you with an excellent starting point for your research.

Figure 8.32 Resources for conducting scholarly research.

Google Scholar (scholar.google.com)

- Specialty search engine that returns results from scholarly sources

- A free site where users share scholarly research and follow the research of others

- Operated by Bielefeld University Library—included information is carefully selected and reviewed

- Searches over 1 billion documents such as books, encyclopedias, newspapers, professional journals, and websites

Because anyone can post anything to the Internet, what's the harm in posting something that is false (see Table 8.5)? Not passing on information that is false is part of being an ethical Internet user. It takes more effort to check potential hoaxes and the accuracy of websites. But think of all the time you'll save your friends and family by not sending out endless streams of bogus information. Certainly your research papers will benefit from using accurate information sources.

Digital Manipulation

What is digital manipulation? Now that we have computer apps that can generate lifelike images for movies and alter photographs, can we believe anything we see? Digital manipulation involves altering media so that the images captured are changed from the way they were originally seen by the human eye.

So manipulation of images started when Photoshop was invented? Hoaxes perpetrated with photographic manipulation have been around since photography was invented. Multiple exposures, compositing negatives, and cutting and pasting multiple photos and then reshooting were common methods of manipulation clever photographers used in the past (see Figure 8.33). It was once common practice for images of family members to be cut and pasted into the same photo to make it seem as though they were together for a family portrait when in fact they never had been.

So when is manipulating a photograph or a video unethical? If changes are made to an image for the purpose of deceiving someone or to alter perceptions of reality, this crosses the boundaries into unethical behavior. For instance, there was a famous image floating around the Internet of George W. Bush reading to elementary school children and the book he was holding was made to appear upside down through digital manipulation. It didn't really happen, but many people believed it did!

Some adjustments that photographers make are purely for aesthetic reasons. *Aesthetics* deals with making things pleasing in appearance. When you use a flash, people in the photo often have red eyes. Eliminating the red eye and changing the brightness and contrast of a photo are things you

Table 8.5 Point/Counterpoint: Posting False Information

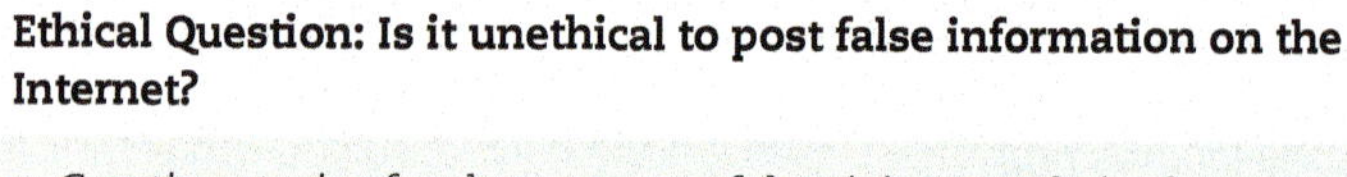

Issue: Posting False Information	Ethical Question: Is it unethical to post false information on the Internet?
 Point	• Creating stories for the purpose of deceiving people is always unethical. • Inaccurate content might be used as the basis for the creation of new content, thereby spreading the contamination even further. • Spreading false information for monetary gain is illegal.
Counterpoint 	• Writers create works of fiction every day. Why is creating hoaxes and rumors any different? • People should be able to judge for themselves and do research to figure out when authors are creating false information for amusement or satire. • Advertisers constantly exaggerate about product attributes. Hoaxes help foster healthy skepticism in the public.

can do right from your camera app. This doesn't alter the reality of the photo; it just improves the appearance. Therefore, this type of digital manipulation isn't considered unethical because it only involves enhancing the photo's aesthetics.

Aren't photos and videos often submitted as evidence in courts? Quite frequently they are and this presents new dilemmas. If a photo or video can be altered, can you rely on it as evidence? Videos are often enhanced to bring out details that are obscured under normal circumstances. Is this fabrication of evidence or just improving aesthetics? Most photos and videos submitted as evidence are examined to determine whether they have been altered. But the techniques for manipulating images are advancing, and soon it may be difficult, if not impossible, to determine whether an image has been altered. Courts are now beginning to wrestle with these issues.

Figure 8.33 Over 100 years before the invention of Photoshop, photo manipulation was alive and well. This purported photo of General Ulysses S. Grant at City Point is a composite of three photographs. *(Library of Congress Prints and Photographs Division [LC-DIG-ppmsca-15886])*

So should you engage in digital manipulation? We explore both sides of the argument in Table 8.6. But as with any other type of information you find on the Internet, be sure to validate the accuracy of photos and videos before relying on them.

Protecting Your Online Reputation

Objective 8.15 *Describe what comprises your online reputation and how to protect it.*

There is already quite a bit of information about you on the Internet. Chances are you've put most of it out there yourself through sites such as Facebook, Instagram, and Twitter. Most people strive to maintain a good reputation in the real world, but in the twenty-first century, your online reputation is just as important as your real-life one.

What is my online reputation? Your online reputation consists of the information available about you in cyberspace that can influence people's opinions of you in the real world. Your online reputation is an extension of your real-life reputation—the view held by the community, the general public, friends, family, and others of your general character. Are you considered an honest person? Do you always speak your mind regardless of the consequences? Are you known for defending the rights of less-fortunate individuals? These are examples of factors that contribute to society's view of the type of individual you are.

Information is constantly added to the Internet about many of us, even by people we don't know. It can be challenging to control the information that contributes to your online reputation. If you go to a party and guests take pictures of you, they may post their pictures to sites and identify (or tag) you in them. Your friends might write about you on their Facebook or Twitter feed. No matter whether this information is true, false, misleading, embarrassing, or disturbing, it's part of your online reputation.

Table 8.6 Point/Counterpoint: Digital Manipulation

Issue: Digital Manipulation	Ethical Question: Is digital manipulation unethical?
 Point	• Images and videos are meant to represent what happened in the real world and therefore should not be subject to alteration. • Photographic and video evidence in court cases would be useless unless we can determine whether they've been altered.
Counterpoint 	• Artists and photographers need to be free to alter the aesthetics of images for artistic purposes. • People should be able to tell when images have been purposely distorted for satirical purposes. Isn't it obvious there is no such thing as an 80-pound housecat?

Why is it so important to protect my online reputation? If you say something or do something you wish you hadn't done in real life, most people will forget about it eventually. However, given the persistence of information on the Internet, pictures, videos, and narratives about you might never disappear. Even if you delete something, it probably still exists somewhere. There are numerous examples of politicians tweeting something they later regret and deleting it from their Twitter feed, only to find many people have captured and saved their tweet already.

Sites such as the Internet Archive (*archive.org*) feature utilities such as the Wayback Machine that show you what a website looked like at a previous point in time. Ever wonder what Yahoo! looked like in the early days? A search on the Wayback Machine shows you that in 1996 it looked like Figure 8.34.

Of course, your reputation varies depending on who is interpreting the cyberspace information about you. Your friends might laugh at that picture of you on Facebook at a recent party, but employers routinely browse through social networking sites to evaluate job applicants as part of their hiring decision process. Just keeping your information private isn't a good solution. Many employers don't interview candidates for whom they find a lack of social information online because they view this as a negative or at least a potential concern. Even after you're hired, if your employer finds things that he or she doesn't like on your sites, you might be fired because of it. According to CareerBuilder, 18 percent of employers surveyed have fired employees due to social media posts.

How can an employer fire me for writing something on my own time? In the majority of states, the employment at-will principle governs the firing of employees. *Employment at-will* means that unless you're covered by an employment contract or a collective bargaining agreement, such as a union contract, employers can fire you at any time for any reason, unless the reason violates a legal statute such as race or age discrimination. Your employer is not required to provide you with

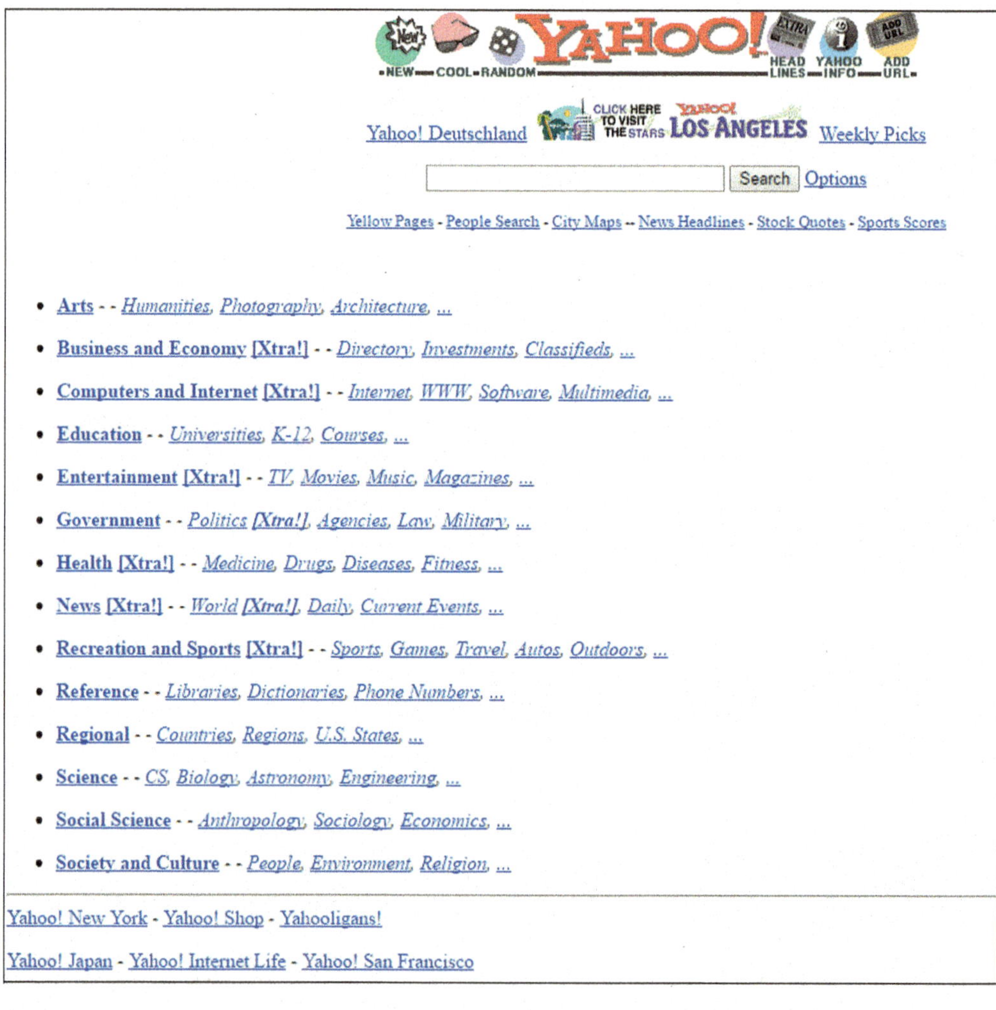

Figure 8.34 A much simpler Yahoo! home page is preserved for posterity in the Internet Archive. Your web pages might live on forever also. *(Courtesy of Yahoo)*

a reason for the firing but can simply tell you not to report for work any longer.

You do have a constitutional right to free speech. However, that doesn't mean your employer will agree with your opinions. Therefore, it's a good idea to familiarize yourself with your employer's social media policies. Many employers now have such policies, and they usually prevent you from saying disparaging things about your employer, work environment, customers, and coworkers on social media sites. Likewise, giving the appearance that your employer supports an activity or political agenda is usually prohibited (see Figure 8.35), so make sure you're aware of the rules at your workplace.

Figure 8.35 You have the right to protest, but your employer's social media policy may prohibit you from posting a picture of yourself wearing a company T-shirt while attending the event. *(Fabio Formaggio/123RF)*

What steps should I take to protect my online reputation? You can do many things if you're proactive:

- **Improve and update your personal profiles.** Make sure your profiles on social media sites are accurate. Update them with new information.
- **Create content on relevant sites.** For instance, if you're an up-and-coming interior designer, post positive comments on relevant sites that assist people with some aspect of design. Make sure your name is associated with things people feel good about reading.
- **Post frequently.** Posting frequently helps minimize negative information, because newer information tends to come up first in search engine results.
- **Be vigilant.** Contact people who've posted negative things about you and see whether you can convince them to modify or delete them.

You need to examine your own reputation periodically. Google yourself and see what information about you is available on the Internet. If you see something that displeases you, ask the poster or the webmaster to remove the offending material. Don't have time to constantly monitor your reputation? Sites such as *Social-Searcher.com* (see Figure 8.36) and YouScan (*youscan.io/en*) provide tools to conduct searches and generate updates regarding information posted about you on social media sites. These tools can also be used to track mentions about specific topics or businesses that interest you.

Keeping your online reputation accurate and not damaging other people's reputations unfairly are integral parts of being an ethical cyber-citizen.

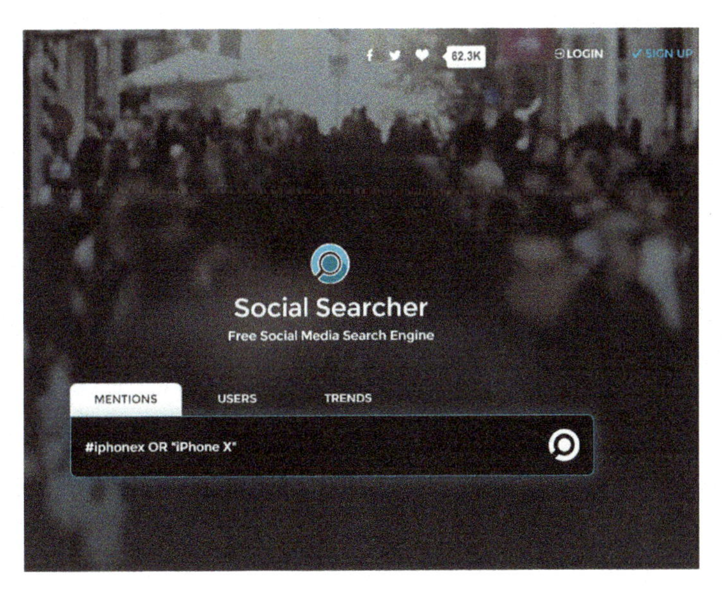

Figure 8.36 Social-Searcher.com is a free social media search engine that helps you track mentions of any person, topic, or business—even yourself! *(© Social Searcher 2011-201963K)*

Bits&Bytes Celebrity Photographic Rights

How do paparazzi get away with photographing celebrities without their consent? By the nature of being a celebrity, courts have decided that these individuals surrender a certain amount of privacy. In the United States, photos that are taken for editorial use in a public place generally enjoy constitutional protection under the right of free speech. Paparazzi are free to photograph celebrities in public places for the purposes of selling their photos to news outlets and news publications. However, paparazzi can't sneak into a celebrity's backyard and photograph them through a window.

There are several exceptions to the free speech doctrine, however. Photos can't have captions that imply something false or libelous about the person in the photo. If they do, then they aren't legally protected free speech. Also, photos of a celebrity can't be used to promote any goods or services without permission of the celebrity, because celebrities have the right of publicity, which grants them control of the commercial use of their name, image, likeness, and other unequivocal aspects of their identity.

Ethics in IT

Acceptable Use Policies: What You Can and Can't Do

With computers and the Internet becoming ubiquitous tools, more and more ethical decisions revolve around how we use technology. Because of the potential for problems related to misuse of technology, most schools and businesses have adopted acceptable use policies—guidelines regarding usage of computer systems—relating to the use of their computing resources. Your school most likely has a policy similar to that shown in Figure 8.37 with which you should familiarize yourself. You can usually find it on your institution's website, either under the Information Technology department section or the Student Policies section. If you can't locate it, check with the help desk personnel at your school; they can likely direct you to the appropriate web page. Among other things, these policies usually cover the following:

- Keeping your account access (logon ID and password) secure from others
- Not running a side business with college/business assets
- Not attempting to gain access to portions of the computer systems you are not authorized to use
- Not illegally copying legally protected materials, such as software
- Not creating, distributing, or displaying threatening, obscene, racist, sexist, or harassing material
- Not installing unauthorized software on the institution's computers
- Not conducting any illegal activities with computing systems

If you have a job, you should also familiarize yourself with your employer's policy, because it may differ from your school's policy. For instance, most schools consider any work products—such as research papers, poems, musical compositions, and so forth—generated with school computing resources to be the property of the students who created them, unless the students were paid to create the works. With many employers, any work products created using company-provided computing resources are deemed to be the property of the company, not the employees who created them.

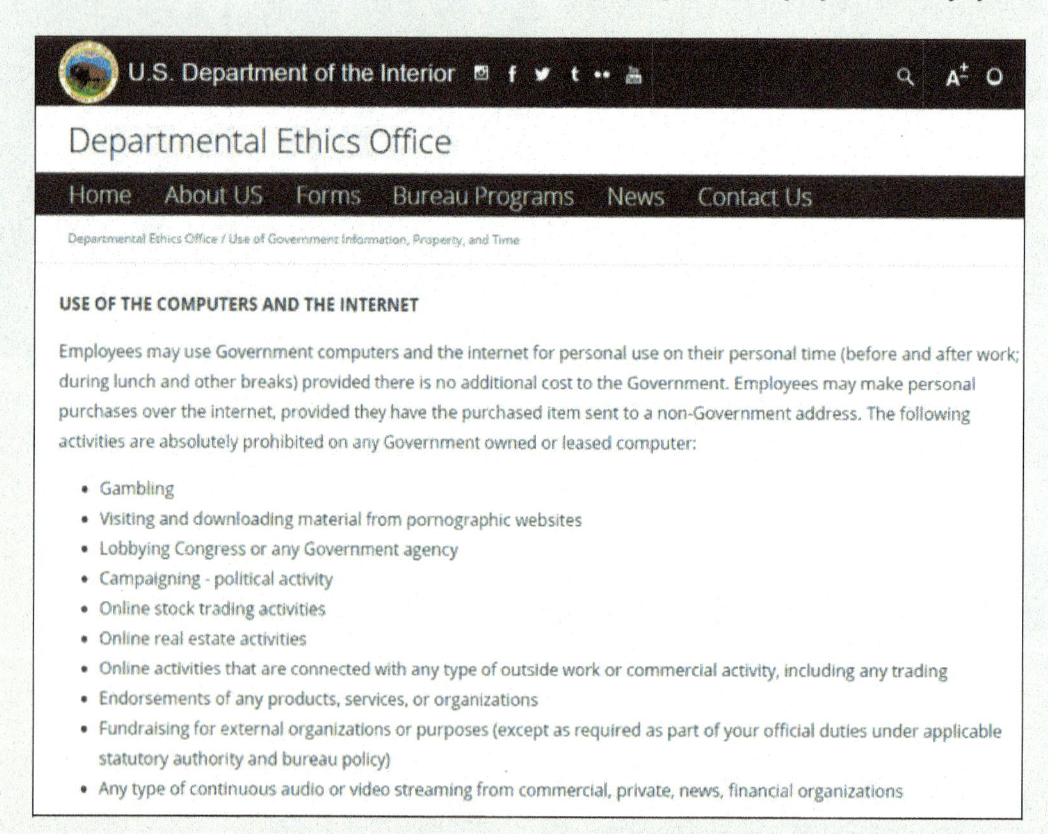
U.S. Department of the Interior

Departmental Ethics Office

Home About US Forms Bureau Programs News Contact Us

Departmental Ethics Office / Use of Government Information, Property, and Time

USE OF THE COMPUTERS AND THE INTERNET

Employees may use Government computers and the internet for personal use on their personal time (before and after work; during lunch and other breaks) provided there is no additional cost to the Government. Employees may make personal purchases over the internet, provided they have the purchased item sent to a non-Government address. The following activities are absolutely prohibited on any Government owned or leased computer:

- Gambling
- Visiting and downloading material from pornographic websites
- Lobbying Congress or any Government agency
- Campaigning - political activity
- Online stock trading activities
- Online real estate activities
- Online activities that are connected with any type of outside work or commercial activity, including any trading
- Endorsements of any products, services, or organizations
- Fundraising for external organizations or purposes (except as required as part of your official duties under applicable statutory authority and bureau policy)
- Any type of continuous audio or video streaming from commercial, private, news, financial organizations

Figure 8.37 A typical acceptable use policy. *(Courtesy of U.S. Department of the Interior)*

Figure 8.38 The Ten Commandments of Computer Ethics.

Thou shalt not use a computer to harm other people

Thou shalt not interfere with other people's computer work

Thou shalt not snoop around in other people's computer files

Thou shalt not use a computer to steal

Thou shalt not use a computer to bear false witness

Thou shalt not copy or use proprietary software for which you have not paid

Thou shalt not use other people's computer resources without authorization or proper compensation

Thou shalt not appropriate other people's intellectual output

Thou shalt think about the social consequences of the program you are writing or the system you are designing

Thou shalt always use a computer in ways that ensure consideration and respect for your fellow humans

If your school or employer doesn't have an acceptable use policy, the Computer Ethics Institute (*computerethicsinstitute.org*) has developed a well-known set of guidelines you can follow. They are known as the Ten Commandments of Computer Ethics (see Figure 8.38).

In the absence of clear policies, following the Ten Commandments of Computer Ethics should help you steer clear of most unethical behavior.

Before moving on to Chapter Review:

1. Watch Chapter Overview Video 8.2.
2. Then take the Check Your Understanding quiz.

Check Your Understanding // Review & Practice

For a quick review to see what you've learned so far, answer the following questions.

multiple choice

1. Which of the following is considered intellectual property?
   a. A song
   b. A sculpture
   c. A movie
   d. All of the above
2. Copyleft
   a. protects you from having others use your work.
   b. is a term for a set of licensing plans.
   c. legalizes unlimited copying of digital music.
   d. all of the above.
3. When is posting a picture on social media considered copyright infringement?
   a. When you don't own the copyright to the photo
   b. When you don't have permission from the copyright holder to post the photo
   c. Neither A nor B
   d. Both A and B
4. Fake news is
   a. illegal under US civil law.
   b. a type of hoax.
   c. not unethical.
   d. not a big problem on social media.
5. The tendency for things like videos and images to remain on the Internet for long periods of time is known as _______ of information.
   a. stagnation
   b. legacy
   c. persistence
   d. inheritance

chew on this

(3Dalia/Shutterstock)

Although copyright accrues without registration, registering your copyright offers additional benefits. Visit *copyright.gov*, prepare a summary of the advantages of registration, and explain how registration would help you pursue an infringement action.

MyLab IT Go to **MyLab IT** to take an autograded version of the *Check Your Understanding* review and find all media resources for the chapter.

For the IT Simulation for this chapter, see MyLab IT.

8 Chapter Review

Summary

Part 1
The Impact of Digital Information

Learning Outcome 8.1 **You will be able to describe the nature of digital signals; how digital technology is used to produce and distribute digital texts, music, and video; and the challenges in managing a digital lifestyle.**

Digital Basics

Objective 8.1 *Describe how digital convergence and the Internet of Things have evolved.*

- Digital convergence has brought us single devices with the capabilities that used to require four or five tools. With more computing power in a mobile processor, a single device can perform a wider range of tasks.
- The Internet of Things (IoT) is embedded computing devices that transfer data over a network without requiring outside interaction. The IoT is integral to the development of smart homes.

Objective 8.2 *Explain the differences between digital and analog signals.*

- Digital media is based on a series of numerical data—that is, number values that were measured from the original analog waveform. As a string of numbers, a digital photo or video file can be easily processed by modern computers.

Digital Publishing

Objective 8.3 *Describe the different types of e-readers.*

- E-readers are devices that store, manipulate, and transmit textual information digitally. Some use electronic ink for a sharp grayscale representation of text. Others use backlit monitors.

Objective 8.4 *Explain how to purchase, borrow, and publish e-texts.*

- There are a variety of formats for e-text files, including .azw and ePub. E-texts can be purchased from many publishers directly or from online stores. Libraries also loan e-texts. You can easily publish your own books, using services such as the Amazon Kindle Store, Smashwords, or Lulu.

Digital Music

Objective 8.5 *Describe how digital music is created and stored.*

- Digital music is created by combining pure digital sounds with samples of analog sounds. It has meant changes for the recording industry, for performers, and for music listeners. It's now inexpensive to carry a music library, to create new songs, and to distribute them worldwide. The sampling rate determines the fidelity of the final recorded digital sound.

Objective 8.6 *Summarize how to listen to and publish digital music.*

- Digital rights management is a system of access control to digital music. Many services, such as SoundCloud, DistroKid, and ReverbNation, are available for openly sharing and streaming your own content.

Digital Media

Objective 8.7 *Explain how best to create, print, and share digital photos.*

- Digital cameras enable you to capture and transfer images to all your digital devices and cloud storage sites instantly. The camera's resolution is important, because it tells the number of data points recorded for each image captured.

Objective 8.8 *Describe how to create, edit, and distribute digital video.*

- By using software for video production, you can create polished videos with titles, transitions, a sound track, and special effects.
- A codec is a compression/decompression rule, implemented in either software or hardware, that squeezes the same audio and video information into less space.
- Vast choices in streaming media allow savvy consumers to cut the cord and save money by forgoing the purchase of television media from cable TV companies.

Managing Your Digital Lifestyle

Objective 8.9 *Discuss the challenges in managing an active digital lifestyle.*

- Although the interconnectivity of the IoT brings with it many benefits, it also poses ethical concerns.
- Although cryptocurrency provides advantages such as anonymity, speed, and security, there are drawbacks, like value fluctuations, lack of insurance, and inability to reverse transactions.
- Wearable technology can provide benefits, such as health tracking, but also raises questions about personal privacy of the data collected.

Part 2

Ethical Issues of Living in the Digital Age

Learning Outcome 8.2 **You will be able to describe how to respect digital property and use it in ways that maintain your digital reputation.**

Protection of Digital Property

Objective 8.10 *Describe the various types of intellectual property.*

- Intellectual property (IP) is a product of a person's mind. Types of IP include material that is protected by copyright, patent, trademark, service mark, and trade dress.

Objective 8.11 *Explain how copyright is obtained and the rights granted to the owners.*

- Copyright begins when a work is created and fixed into a digital or physical form. Copyright can also be registered. Copyright includes the rights to sell, reproduce, and display (or perform) the work publicly.

Objective 8.12 *Explain copyright infringement, summarize the potential consequences, and describe situations in which you can legally use copyrighted material.*

- Using someone's copyrighted work without permission is copyright infringement. Consequences include lawsuits and fines. You can use copyrighted material if you obtain permission from the copyright holder, adhere to terms of use, or qualify under a *fair use* exception.

Living Ethically in the Digital Era

Objective 8.13 *Explain plagiarism and strategies for avoiding it.*

- Plagiarism is the act of copying text or ideas from someone else and claiming them as your own. Citing your sources will help you avoid plagiarism.

Objective 8.14 *Describe hoaxes and digital manipulation.*

- A *hoax* is anything designed to deceive another person. *Digital manipulation* involves altering media so that the images are changed.

Objective 8.15 *Describe what comprises your online reputation and how to protect it.*

- Your online reputation is information available about you online that can influence opinions of you in the real world. Keeping content current, being vigilant, and frequently creating content can help you protect your digital reputation.

Be sure to check out **MyLab IT** for additional materials to help you review and learn. And don't forget to watch the Chapter Overview Videos.

Key Terms

Chapter Quiz // Assessment

For a quick review to see what you've learned, answer the following questions. Submit the quiz as requested by your instructor. If you're using **MyLab IT**, the quiz is also available there.

multiple choice

1. The combination of media, Internet, entertainment, and phone services into a single device illustrates the principle of
- **a.** digital collusion.
- **b.** digital cohesion.
- **c.** digital combination.
- **d.** digital convergence.

2. Which of the following is an *open* electronic publishing format?
- **a.** PDF
- **b.** .azw
- **c.** ePub
- **d.** SMS

3. When analog data is measured and converted to a stream of numerical values, this is referred to as _____ the data.
- **a.** compressing
- **b.** streamlining
- **c.** digitizing
- **d.** upstreaming

4. Increasing the sampling rate when digitizing a file has which effect on the file?
- **a.** It increases the file size.
- **b.** It decreases the file size.
- **c.** It neither increases nor decreases the file size.
- **d.** It depends on the codec used to digitize the file.

5. Which of the following does NOT determine the quality of a digital image?
- **a.** Resolution of the camera
- **b.** Image sensor size
- **c.** File format and compression used
- **d.** The number of pixels in the camera lens

6. Which of the following is a digital video file format?
- **a.** MPEG-4
- **b.** PDF
- **c.** HDMI
- **d.** USB

7. Which of the following is *not* a category of intellectual property?
- **a.** Copyright
- **b.** Trademark
- **c.** Impression
- **d.** Trade dress

8. Using copyrighted works without permission from the copyright holder is possible under the _____ doctrine.
- **a.** shared rights
- **b.** general fairness
- **c.** equitable rights
- **d.** fair use

9. A type of journalism in which stories are invented and intended to spread misinformation rather than fact is known as
- **a.** a public domain tale.
- **b.** a persistence of information story.
- **c.** fake news.
- **d.** an urban legend.

10. It is important to maintain your online reputation
- **a.** because of the persistence of information.
- **b.** because employers often review social media before hiring employees.
- **c.** to ensure that information about you is accurate.
- **d.** all of the above.

true/false

_____ **1.** Digital signals are continuous.
_____ **2.** Plagiarism is illegal in the United States.
_____ **3.** Digital rights management (DRM) is a system of access control that allows unlimited use of the material.
_____ **4.** Copyright lasts for the life of the creator plus 70 years in the United States.
_____ **5.** Copying a movie and giving it to a friend is not fair use.

What do you think now?

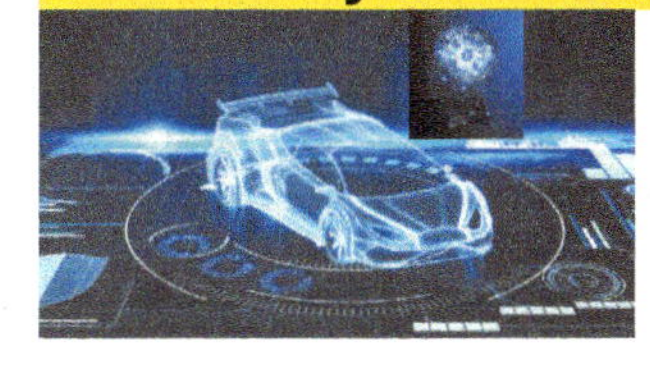

Engineers who design self-driving cars may make decisions about sacrificing speed for controlled acceleration, turns, and braking that result in better fuel economy and less pollution. Which is more important? Travel efficiency or environmental impact? Who should make these decisions—you or the car's designer? Explain your answers.

Team Time

"And One Will Rule Them All"

problem

Digital convergence posits the dream of one device that can do it all, for everyone, but there are so many mobile devices saturating the market that many people are left confused.

task

For each scenario described as follows, the group will select the *minimum* set of devices that would support and enhance the client's life.

process

1. Consider the following three clients:
   - A retired couple who now travels for pleasure a great deal. They want to be involved in their grandchildren's lives and need support for their health, finances, and personal care as they age.
   - A young family with two children, two working parents, and a tight budget.
   - A couple in which each individual is a physician and each adores technology.
2. Make two recommendations for your clients in terms of digital technologies that will enhance their business or their lifestyle. Discuss the advantages and disadvantages of each technology. Consider value, reliability, computing needs, training required, and communication needs as well as expandability for the future.
3. As a group, prepare a final report that considers the recommendations you've made for your clients.

conclusion

With so many digital solutions on the market, recommending options needs to focus on converging to the minimum set of tools that will enhance life without adding complication to it.

Ethics Project

When Everyone Has a Voice

In this exercise, you research and then role-play a complicated ethical situation. The role you play might or might not match your own personal beliefs; in either case, your research and use of logic will enable you to represent the view assigned. An arbitrator will watch and comment on both sides of the arguments, and together the team will agree on an ethical solution.

background

Much of the world's population is now equipped with Internet-connected smartphones. Sensors on these phones could measure for viruses or compute pollution indexes, and the cameras could be used to document a range of human behavior. This could create changes in political movements, art, and culture as everyone's experience is documented and shared.

research areas to consider

- Evgeny Morozov RSA Animate
- Smart Insights mobile marketing statistics
- The Witness Project
- Center for Embedded Networked Sensing

process

1. Divide the class into teams.
2. Research the areas cited and devise a scenario in which mobile access could make an impact politically or environmentally, positively or negatively.
3. To set the stage for the role-playing event, team members should write a summary that provides background information for their character—for example, business owner, politician, reporter, or arbitrator—and that details their character's behaviors. Team members should then create an outline to use during the role-playing event.
4. Team members should arrange a mutually convenient time to meet for the exchange, using a virtual meeting tool or by meeting in person.
5. Team members should present their case to the class or submit a PowerPoint presentation for review by the rest of the class, along with the summary and resolution they developed.

conclusion

As technology becomes ever more prevalent and integrated with our lives, more and more ethical dilemmas will present themselves. Being able to understand and evaluate both sides of the argument, while responding in a personally or socially ethical manner, will be an important skill.

Solve This with PowerPoint

MyLab IT Grader

Intellectual Property and Copyright Basics

You have been asked to make a presentation about intellectual property and copyright basics as they apply to digital images. The presentation is almost done, but you need to make a few adjustments. In addition, you'll need to add a few images from a website and give credit to the source.

You'll use the following PowerPoint skills as you complete this activity:

- Apply Theme
- Create and Modify SmartArt
- Add and Modify Text
- Add New Slide
- Add Zoom Summary Slide
- Insert Images from Web Page
- Apply Design Ideas
- Rearrange Slides
- Apply Slide Transition

1. Open *TIA_Ch8_Start.pptx* and save as **TIA_Ch8_LastFirst.pptx**. Select slide 1 and add your first and last names as the subtitle.
2. Apply the **Berlin Theme** to the entire presentation.
   Hint: Click the Berlin Theme from the Themes group on the Design tab.
3. Select **Slide 2**. Click just before the second sentence, beginning with *IP is considered*, and press **Enter** to create a new bullet point.
4. Create a **New Slide**, using **Title and Content layout** (new slide becomes slide 3). In the Text Pane, type **Examples of Intellectual Property** as the title. Insert a **Basic Block List SmartArt** diagram and enter the following as individual items: **Books, Art, Photographs, Music and Songs, Movies and Videos, Patents, Formulas**, and **Computer Software**.
   Hint: Click the arrow to the left of the SmartArt diagram to open the SmartArt text pane to type text more easily.
5. Change the color of the SmartArt to **Gradient Loop - Accent 2** and change the color of the font to **Black, Background 1.**
6. Select **Slide 4**. Reposition the second bullet so it becomes the first bullet. Create a sub-bullet for the new second bullet and type **This includes photographs and videos posted to the Internet.** Select Copyright Infringement in the third bullet and format with **Bold** and **Italics**.
7. Select **Slide 5**. Enter **Copyright Infringement and Social Media** as the slide title. Increase the font size of the bulleted text to **32**.
8. Select **Slide 6**. Change the title of the slide to **Photographic Copyrights**.
9. Move **Slide 8** so that it becomes Slide 9.
10. Select **Slide 9**. Using any browser, go to **Nasa.gov**. Click **Galleries** from the top menu and then select **Image Galleries**. Click **NASA Images** and then type **rocket** in the search box (uncheck Videos and Audio to search only for images). Right-click any photo of a rocket and select **Copy image**. Click **Slide 9**, click the **Paste arrow**, and select **Picture**. Repeat the procedure with a second rocket image.
11. Select an appealing layout from the **Design Ideas pane**.
    Hint: If the Design Ideas pane does not automatically appear, click Design Ideas from the Designer group on the Design tab.
12. On Slide 9, insert a **Text Box** and type **Images courtesy of Nasa.gov** in the text box. Change the font to **Arial Narrow** and font size to **16**. Place the text box near the photos.
13. Apply the **Wipe** transition and the **From Left** effect option to **all slides**.
14. Click **Slide 1**. Insert a **Summary Zoom** slide, selecting **Slides 2** and **4**. Change the slide layout to **Blank**.
    Hint: Summary Zoom is on the Insert Tab in the Links group.
15. Run through the presentation in Slide Show mode.
16. Save the file and submit for grading.

Chapter

Securing Your System: Protecting Your Digital Data and Devices

 For a chapter overview, watch the **Chapter Overview Videos.**

PART 1

Threats to Your Digital Assets

Learning Outcome 9.1 **You will be able to describe hackers, viruses, and other online annoyances and the threats they pose to your digital security.**

Identity Theft and Hackers 328

Objective 9.1 *Describe how identity theft is committed and the types of scams identity thieves perpetrate.*

Objective 9.2 *Describe the different types of hackers and the tools they use.*

Computer Viruses 333

Objective 9.3 *Explain what a computer virus is, why it is a threat to your security, how a computing device catches a virus, and the symptoms it may display.*

Objective 9.4 *List the different categories of computer viruses, and describe their behaviors.*

Sound Byte: Protecting Your Computer

Online Annoyances and Social Engineering 336

Objective 9.5 *Explain what malware, spam, and cookies are and how they impact your security.*

Objective 9.6 *Describe social engineering techniques, and explain strategies to avoid falling prey to them.*

Helpdesk: Threats to Your Digital Life

PART 2

Protecting Your Digital Property

Learning Outcome 9.2 **Describe various ways to protect your digital property and data from theft and corruption.**

Restricting Access to Your Digital Assets 347

Objective 9.7 *Explain what a firewall is and how a firewall protects your computer from hackers.*

Objective 9.8 *Explain how to protect your computer from virus infection.*

Objective 9.9 *Describe how passwords and biometric characteristics can be used for user authentication.*

Objective 9.10 *Describe ways to surf the web anonymously.*

Helpdesk: Understanding Firewalls

Keeping Your Data Safe 356

Objective 9.11 *Describe the types of information you should never share online.*

Objective 9.12 *List the various types of backups you can perform on your computing devices, and explain the various places you can store backup files.*

Sound Byte: Managing Computer Security with Windows Tools

Protecting Your Physical Computing Assets 360

Objective 9.13 *Explain the negative effects environment and power surges can have on computing devices.*

Objective 9.14 *Describe the major concerns when a device is stolen and strategies for solving the problems.*

MyLab IT All media accompanying this chapter can be found here.

A Password Generator on **page 346**

(Robert Mizerek/123RF; wk1003mike/Shutterstock; Carlos Amarillo/Shutterstock; Solarseven/123RF; Ymgerman/123RF; Graja/Shutterstock)

What do you think?

In the Information Age, **data** is valuable. Your private information—your Social Security number, credit card information, address, and passwords—has great value to companies trying to market you things or to criminals trying to exploit your information for gain. The number of ways that your data could be **exposed** is growing all the time. Even one hack of a major company could be enough to expose your data. The largest attack ever affected all three billion of Yahoo's user accounts. As society struggles to create laws that keep up with the many ways your privacy can be compromised, you need to stay informed and take steps to **protect your information** yourself.

Which of the following has happened to you? (Choose all that apply.)

- *I got a letter notifying me that a company with my credit card info had been hacked.*
- *I logged on to a website from a link sent to me in an e-mail.*
- *I worried about opening an e-mail because it might infect my computer.*
- *I worried my social media settings let too many people see my information.*

See the end of the chapter for a follow-up question.

(Maxim Filipchuk/Alamy Stock Photo)

Part 1

For an overview of this part of the chapter, watch **Chapter Overview Video 9.1**.

Threats to Your Digital Assets

Learning Outcome 9.1 **You will be able to describe hackers, viruses, and other online annoyances and the threats they pose to your digital security.**

The media is full of stories about malicious computer programs damaging computers, criminals stealing people's identities online, and attacks on corporate websites bringing major corporations to a standstill. These are examples of cybercrime—any criminal action perpetrated primarily through the use of a computer. Cybercriminals are individuals who use computers, networks, and the Internet to perpetrate crime. In this part of the chapter, we discuss the most serious types of cybercrime you need to worry about as well as some online annoyances to avoid.

Identity Theft and Hackers

Every year, the Internet Crime Complaint Center (IC3)—a partnership between the FBI and the National White Collar Crime Center (NW3C)—receives hundreds of thousands of complaints related to Internet crime. *Government impersonation scams* involve people pretending to represent official organizations—such as the FBI, the IRS, or Homeland Security—to defraud. *Non-auction/non-delivery scams* involve running auctions (or sales) of merchandise that does not really exist, wherein the perpetrators just collect funds and disappear without delivering the promised goods. *Advance fee fraud* involves convincing individuals to send money as a good-faith gesture to enable them to receive larger payments in return. The scammers then disappear with the advance fees. *Identity theft* involves the stealing of someone's personal information for financial gain. Other complaints received involve equally serious matters such as computer intrusions (hacking), extortion, and blackmail. Figure 9.1 shows that all kinds of industries are having trouble staying current with their security measures. In this section, we look at both identity theft and hacking in more detail.

Figure 9.1 How Prepared Is Industry for Cybercrimes?

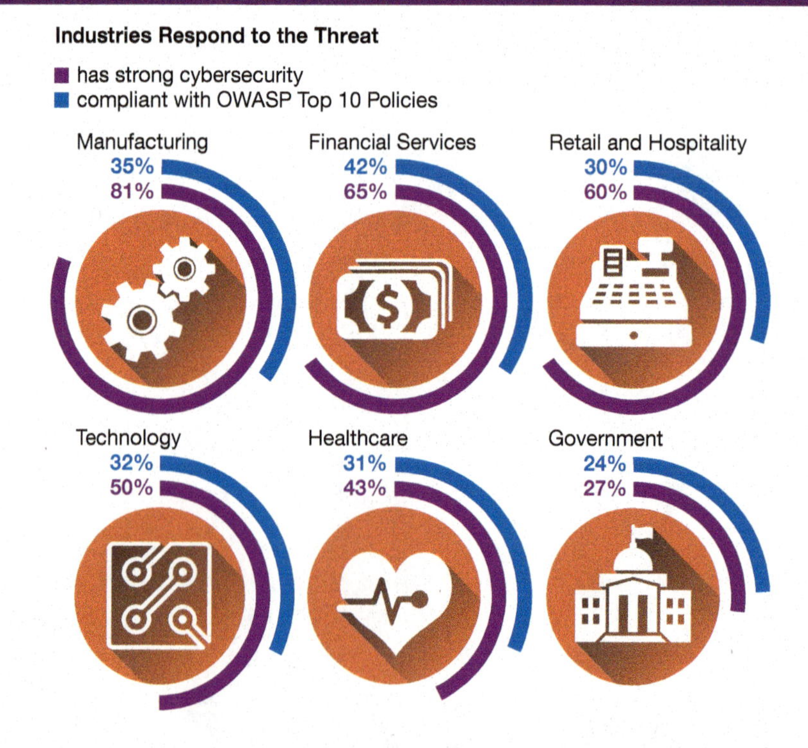

(Robert Mizerek/123RF)

Identity Theft

Objective 9.1 *Describe how identity theft is committed and the types of scams identity thieves perpetrate.*

What is identity theft? Identity theft occurs when a thief steals personal information and poses as you in financial or legal transactions. They may use your name, address, Social Security number, birth date, bank account number, or credit card information in pretending to be you. Many victims of identity theft spend months, or even years, trying to repair their credit and eliminate fraudulent debts.

What types of scams do identity thieves perpetrate? The nefarious acts cover a wide range:

- Counterfeiting your existing credit and debit cards
- Requesting changes of address on your bank and credit card statements, which makes detecting fraudulent charges take longer
- Opening new credit cards and bank accounts in your name and then writing bad checks and not paying off the credit card balances (ruining your credit rating in the process)
- Obtaining medical services under your name, potentially causing you later to lose coverage if the thief's treatment exceeds the limits of your policy's covered services
- Buying a home with a mortgage in your name, then reselling the house and absconding with the money (leaving you with the debt)

Many people believe that the only way your identity can be stolen is by using a computer, but you need to protect your information away from the computer as well. Remember to:

- Protect your debit and credit cards. Most now have apps that allow you to turn off the card as soon as you realize it's lost.
- Protect bank statements and credit card bills, shredding before you dispose of them.
- Never reveal sensitive information over the phone, even if the caller ID indicates a reputable source. A popular scam is to pretend to be the FBI or a Social Security representative in order to collect your private information.
- Be wary of those promising to help you investigate or repair your credit score and asking for personal information.

Although foolproof protection methods don't exist, there are precautions that will help you minimize your risk.

Hacking

Objective 9.2 *Describe the different types of hackers and the tools they use.*

What defines a hacker? A hacker is most commonly defined as anyone who unlawfully breaks into a computer system—either an individual computer or a network (see Figure 9.2).

Are there different kinds of hackers? Not all hackers have criminal intent. There are several categories of hackers:

- White hat hackers (or ethical hackers) break in to systems for unmalicious reasons, such as to test system security vulnerabilities or to expose undisclosed weaknesses. They believe in making security vulnerabilities known either to the company that owns the system or software or to the general public, often to embarrass a company into fixing a problem.
- Black-hat hackers break into systems to destroy information or for illegal gain. The terms *white hat* and *black hat* are references to old Western movies in which the heroes wore white hats and the outlaws wore black hats.
- Gray-hat hackers are a bit of a cross between black and white—they often illegally break into systems merely to flaunt their expertise to the administrator of the system they penetrated or to attempt to sell their services in repairing security breaches.

The laws in the United States consider *any* unauthorized access to computer systems a crime.

Figure 9.2 Hackers use a number of techniques to gain access to private information and systems. *(Macrovector/Shutterstock)*

Bits&Bytes Next Generation White Hat Hackers

CyFi is an 18-year-old high school student in Silicon Valley, but already she is one of the best-known white hat hackers. These ethical hackers use their cybersecurity skills to find bugs and security weaknesses in mobile apps and software and make them safer. CyFi says she feels it is the responsibility of her generation to make the Internet a safer space and runs a program to train younger students about ethical hacking.

Jack Cable, an 18-year-old Stanford University student, and Sam Curry, an 18-year-old in Nebraska, each make six-figure salaries, while still in school, by running security businesses on the side. These white hat hackers work along with companies, scoring big bounties, trying to hack the companies' programs to do things that were not intended. Are you interested in being a white hat hacker?

Could a hacker steal my debit card or bank account number? Hackers often try to break into computers or websites that contain credit card information. If you perform financial transactions online, such as banking or buying goods and services, you probably do so using a credit or debit card. Credit card and bank account information can thus reside on your hard drive or an online business's hard drive and may be detectable by a hacker.

Aside from your home computer, you have personal data stored on various websites. For example, many sites require you to provide a login ID and password to gain access. Even if this data isn't stored on your computer, a hacker may be able to capture it when you're online by using a *packet analyzer (sniffer)* or a *keylogger* (a program that captures all keystrokes made on a computer).

What's a packet analyzer? Data travels through the Internet in small pieces called packets. The packets are identified with an Internet protocol (IP) address, in part to help identify the computer to which they are being sent. Once the packets reach their destination, they're reassembled into cohesive messages. A packet analyzer (sniffer) is a program deployed by hackers that examines each packet and can read its contents. A packet analyzer can grab all packets coming across a particular network—not just those addressed to a particular computer. For example, a hacker might sit in a coffee shop and run a packet sniffer to capture credit card numbers from patrons using the coffee shop's free wireless network.

Wireless networks (see Chapter 7) can be set to use encryption to protect against this. Encrypting data packets protects from this risk because the data is transformed with a complex algorithm before it is sent. The data is then decoded at its destination. A sniffer could only steal the encoded data.

What do hackers do with your information? Once hackers have your debit/credit card information, they can use it to purchase items illegally or can sell the number to someone who will. If hackers steal the login ID and password to an account where you have your bankcard information stored, they can also use your account to buy items and have them shipped to him- or herself instead of to you. If hackers can gather enough information in conjunction with your credit card information, they may be able to commit identity theft.

Although this sounds scary, you can easily protect yourself from packet sniffing by installing a firewall and using a personal VPN to secure all the data you exchange over the Internet. (We discuss these topics later in this chapter.)

Trojan Horses and Rootkits

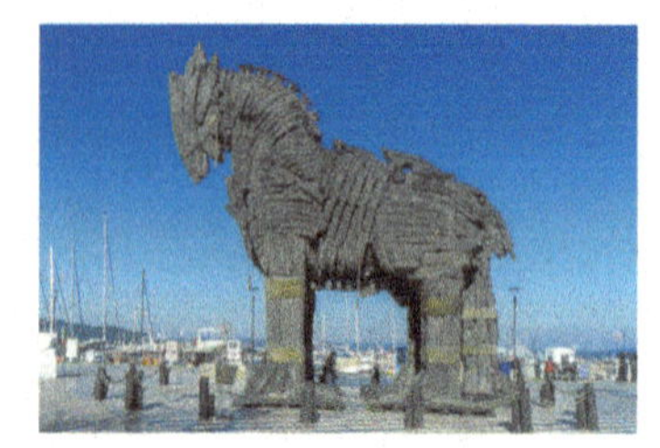

Figure 9.3 The term *Trojan horse* derives from Greek mythology and refers to the wooden horse that the Greeks used to sneak into the city of Troy and conquer it. Computer programs that contain a hidden, and usually dreadful, surprise are referred to as Trojan horses. *(AlpKaya/Shutterstock)*

What other problems can hackers cause if they break into my computer? Hackers often use individuals' computers as a staging area for mischief. To commit widespread computer attacks, for example, hackers need to control many computers at the same time. To this end, hackers often use Trojan horses to install other programs on computers. A Trojan horse is a program that appears to be something useful or desirable, like a game or a screen saver, but while it runs it does something malicious in the background without your knowledge (see Figure 9.3).

What damage can Trojan horses do? Often, the malicious activity perpetrated by a Trojan horse program is the installation of a backdoor program or a rootkit. Backdoor programs and rootkits are programs that allow hackers to gain access to your computer and take almost complete control of it without your knowledge. Using a backdoor program, hackers can access and delete all the files on your computer, send e-mail, run programs, and do just about anything else you can do with your computer. A rootkit is a program that gives an outsider remote control over a computer. A computer that a hacker controls in this manner is referred to as a zombie. Zombies are often used to launch denial-of-service attacks on other computers.

Denial-of-Service Attacks

What are denial-of-service attacks? In a denial-of-service (DoS) attack, legitimate users are denied access to a computer system because hackers are repeatedly making requests of that computer system through a computer they have taken over as a zombie. A computer system can handle only a certain number of requests for information at one time. When it's flooded with requests in a DoS attack, it shuts down and refuses to answer any requests, even if the requests are from a legitimate user. Thus, the computer is so busy responding to the bogus requests for information that authorized users can't gain access.

Could a DoS attack be traced back to the computer that launched it? Yes, launching a DoS attack on a computer system from a single computer is easy to trace. Therefore, most savvy hackers use a distributed denial-of-service (DDoS) attack, which launches DoS attacks from more than one zombie (sometimes thousands of zombies) at the same time.

Figure 9.4 illustrates how a DDoS attack works. A hacker creates many zombies and coordinates them so that they begin sending bogus requests to the same computer at the same time. Administrators of the victim computer often have a great deal of difficulty stopping the attack because it comes from so many computers. Often, the attacks are coordinated automatically by botnets. A botnet is a large group of devices that have been infected by software programs (called *robots* or *bots*) that runs autonomously. Some botnets have been known to span millions of computers.

Because many commercial websites receive revenue from users, either directly (such as by subscriptions to online games) or indirectly (such as when web surfers click advertisements), DDoS attacks can be financially distressing for the owners of the affected websites.

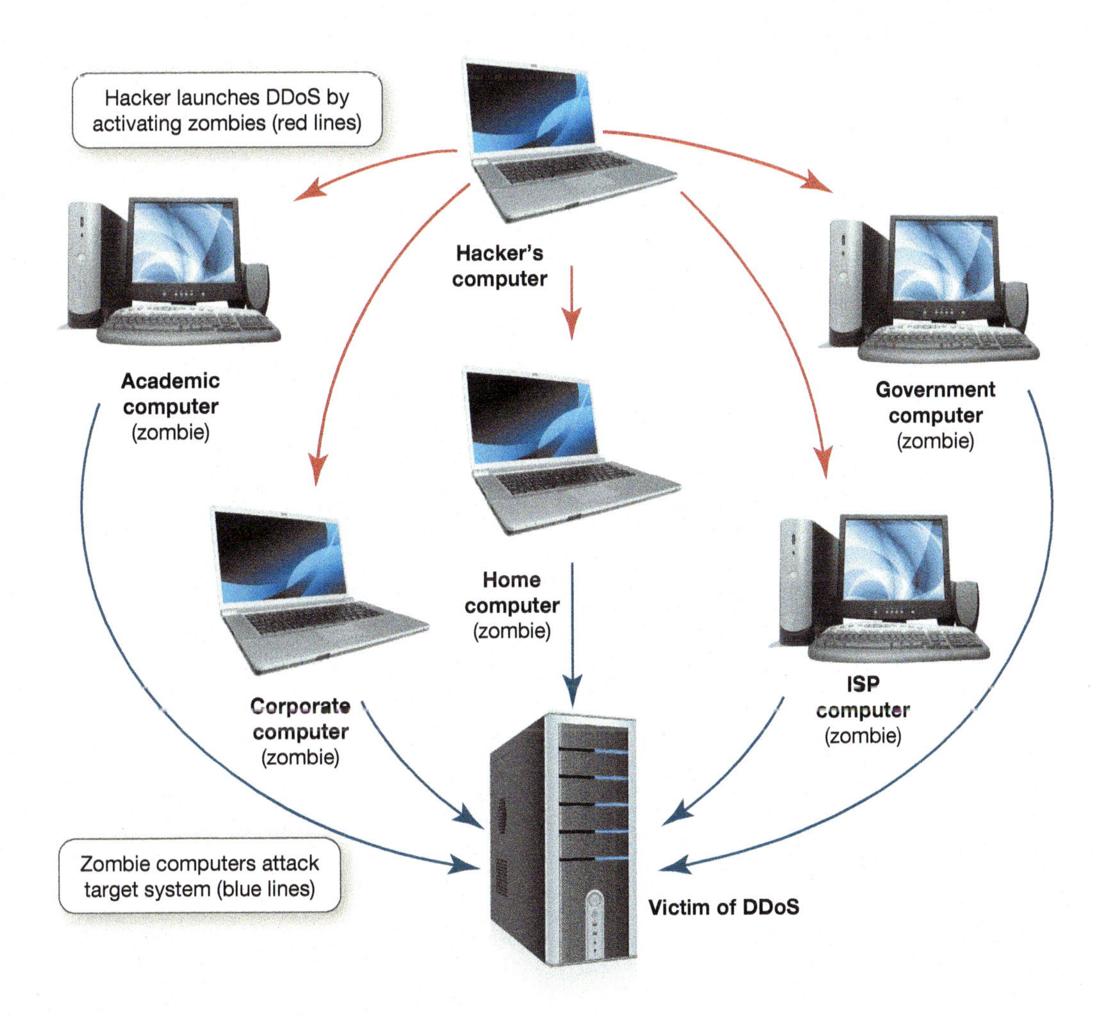

Figure 9.4 Zombie computers are used to facilitate a DDoS attack. *(Vovan/Shutterstock; Nico_blue/iStock/Getty Images; Juffin/DigitalVision Vectors/Getty images)*

How Hackers Gain Computer Access

How does a hacker gain access to a computer? Hackers can gain access to computers directly or indirectly. Direct access involves sitting down at a computer and installing hacking software. It's unlikely that such an attack would occur in your home, but it's always a wise precaution to set up your computer so that it requires a password for a user to gain access.

Indirect access involves subtler methods. Many professional hackers use exploit kits. Exploit kits are software programs that run on servers and search for vulnerabilities of computers that visit the server. Exploit kits look for security holes in browsers and operating systems that haven't yet been patched by the users. When they detect vulnerability, they can deliver spyware, bots, backdoor programs, or other malicious software to your computer. Fortunately, most exploit kits take advantage of known vulnerabilities, so if your antivirus software and operating system are up to date, you should be secure.

Hackers also can access a computer indirectly through its Internet connection. Many people forget that their Internet connection is a two-way street. Not only can you access the Internet, but also people on the Internet can access your computer. Think of your computer as a house. Common sense tells you to lock your home's doors and windows to deter theft when you aren't there. Hooking your computer up to the Internet without protection is like leaving the front door to your house wide open. Your computer obviously doesn't have doors and windows, but it does have logical ports.

What are logical ports? You already know what physical ports are: You use them to attach peripherals, as when you plug your USB flash drive into a USB port. Logical ports are virtual—that is, not physical—communications paths. Unlike physical ports, you can't see or touch a logical port; it's part of a computer's internal organization. Logical ports allow a computer to organize requests for information, so all information arriving from another computer or network that's related to e-mail will be sent to the logical port associated with e-mail.

Logical ports are numbered and assigned to specific services. For instance, logical port 80 is designated for hypertext transfer protocol (HTTP), the main communications protocol for the web. Thus, all requests for information from your browser to the web flow through logical port 80. Open logical ports, like open windows in a home, invite intruders, as illustrated in Figure 9.5. Unless you take precautions to restrict access to your logical ports, other people on the Internet may be able to access your computer through them. Fortunately, you can thwart most hacking problems by installing a firewall.

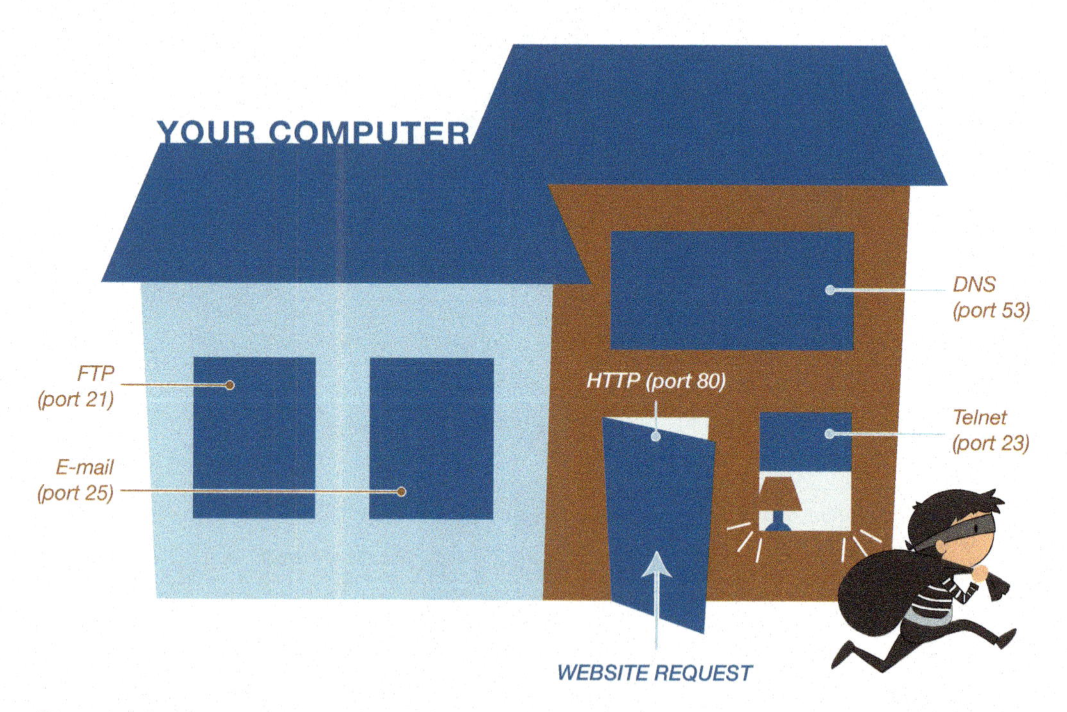

Figure 9.5 Open logical ports are an invitation to hackers.

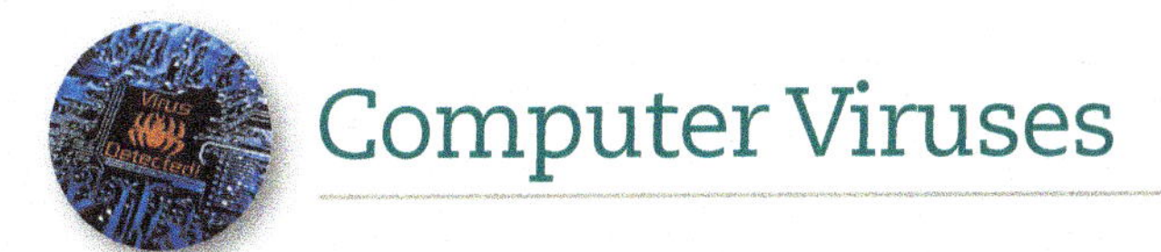

Computer Viruses

Creating and disseminating computer viruses is one of the most widespread types of cybercrimes. Some viruses cause only minor annoyances, whereas others cause destruction or theft of data.

Virus Basics

Objective 9.3 *Explain what a computer virus is, why it is a threat to your security, how a computing device catches a virus, and the symptoms it may display.*

What is a computer virus? A computer virus is a computer program that attaches itself to another computer program (known as the *host* program) and attempts to spread to other computers when files are exchanged. That behavior is very similar to how biological viruses spread, giving this type of threat its name.

What does a computer virus do? A computer virus's main purpose is to replicate itself and copy its code into as many other host files as possible. This gives the virus a greater chance of being copied to another computer system so that it can spread its infection. However, computer viruses require human interaction to spread. Although there might be a virus in a file on your computer, a virus normally can't infect your computer until the infected file is opened or executed.

Why are viruses such a threat to my security? Although virus replication can slow down networks, it's not usually the main threat. The majority of viruses have secondary objectives or side effects, ranging from displaying annoying messages on the computer screen to destroying files or the contents of entire hard drives.

Computer viruses are engineered to evade detection. Viruses normally attempt to hide within the code of a host program to avoid detection, and they are not just limited to computers. Smartphones, tablet computers, and other devices can be infected with viruses. Viruses exist that trick users into downloading an infected file to their phones and then steal their online banking information.

How does my computer catch a virus? If your computer is exposed to a file infected with a virus, the virus will try to copy itself and infect a file on your computer.

Downloading infected audio and video files from peer-to-peer file-sharing sites is a major source of virus infections. Shared flash drives are also a common source of virus infection, as is e-mail. Just opening an e-mail message usually won't infect your computer with a virus, although some new viruses are launched when viewed in the preview pane of your e-mail software. Downloading and executing a file that's attached to the e-mail are common ways that your computer becomes infected. Thus, be extremely wary of e-mail attachments, especially if you don't know the sender. Figure 9.6 illustrates the steps by which computer viruses are often passed from one computer to the next:

1. An individual writes a virus program, attaches it to a music file, and posts the file to a file-sharing site.
2. Unsuspecting, Tony downloads the "music file" and infects his computer when he listens to the song.
3. Tony sends his friend Amit an e-mail with the infected music file and contaminates Amit's tablet.
4. Amit syncs his phone with his tablet and infects his phone when he plays the music file.
5. Amit e-mails the file from his phone to Lily, one of his colleagues at work. Everyone who copies files from Lily's infected work computer, or whose computer is networked to Lily's computer, risks spreading the virus.

I have an Apple computer, so I don't need to worry about viruses, do I? This is a popular misconception! Everyone, even Apple users, needs to worry about viruses. As the macOS operating system has gained market share, the number of virus attacks against it have risen. (For more information, see Chapter 5.)

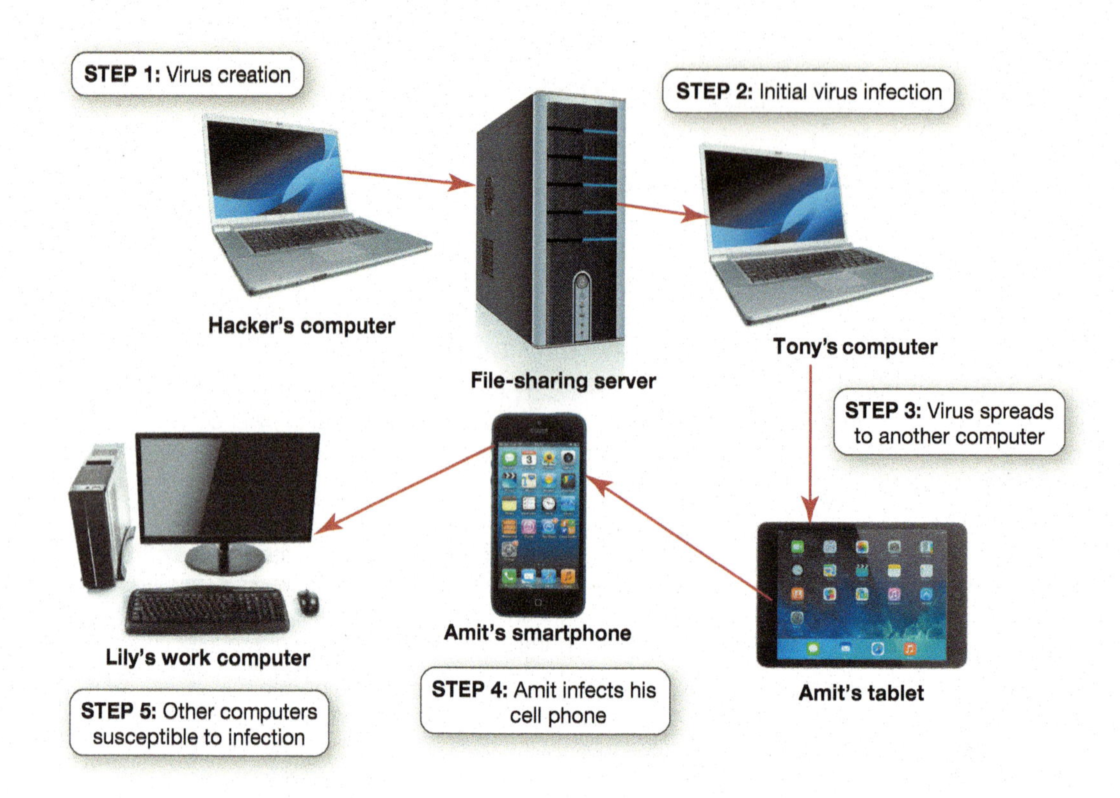

Figure 9.6 Computer viruses are passed from one unsuspecting user to the next. *(Stanca Sanda/Alamy Stock Photo; Ian Dagnall/Alamy Stock Photo; Denis Rozhnovsky/Alamy Stock Photo; Vovan/Shutterstock; Juffin/DigitalVision Vectors/Getty images)*

How can I tell whether my computer is infected with a virus? Sometimes it can be difficult to tell definitively whether your computer is infected with a virus. However, if your computer displays any of the following symptoms, it may be infected with a virus:

- Existing program icons or files suddenly disappear. Viruses often delete specific file types or programs.
- You start your browser and it takes you to a home page you didn't set or it has new toolbars.
- Odd messages, pop-ups, or images are displayed on the screen or strange music or sounds play.
- Data files become corrupt. (However, note that files can become corrupt for reasons other than a virus infection.)
- Programs stop working properly, which could be caused by either a corrupted file or a virus.
- Your system shuts down unexpectedly, slows down, or takes a long time to boot up.

MyLab IT

Protecting Your Computer

In this Sound Byte, you learn how to use a variety of tools to protect your computer, including antivirus software and Windows utilities.

Can I get a virus on my smartphone? Viruses can indeed infect smartphones. So many people now access their bank accounts using a mobile app that smartphones are a likely realm of attack by cybercriminals. Although viruses plaguing smartphones have not yet reached the volume of viruses attacking PC operating systems, with the proliferation of mobile devices, virus attacks are expected to increase.

Trend Micro and McAfee are among the leading companies currently providing antivirus software for mobile devices. Products are designed for specific operating systems; for example, Trend Micro Mobile Security has versions for Android phones and tablets and iOS devices. Often, businesses will have their information technology department install and configure an antivirus solution like this for all the phones used in the organization.

If no antivirus program is available for your phone's OS, the best precautions are commonsense ones. Check the phone manufacturer's website frequently to see whether your smartphone needs any software upgrades that could patch security holes. In addition, remember that you shouldn't download ring tones, games, or other software from unfamiliar websites.

Types of Viruses

Objective 9.4 *List the different categories of computer viruses, and describe their behaviors.*

What types of viruses exist? Although thousands of computer viruses and variants exist, they can be grouped into broad categories based on their behavior and method of transmission.

Boot-Sector Viruses

What are boot-sector viruses? A boot-sector virus replicates itself onto a hard drive's master boot record. The master boot record is a program that executes whenever a computer boots up, ensuring that the virus will be loaded into memory immediately, even before some virus protection programs can load. Boot-sector viruses are often transmitted by an infected flash drive left in a USB port. When the computer boots up with the flash drive connected, the computer tries to launch a master boot record from the flash drive, which is usually the trigger for the virus to infect the hard drive.

Logic Bombs and Time Bombs

What are logic bombs and time bombs? A logic bomb is a virus that is triggered when certain logical conditions are met, such as opening a file or starting a program a certain number of times. A time bomb is a virus that is triggered by the passage of time or on a certain date. For example, the Michelangelo virus was a famous time bomb that was set to trigger every year on March 6, Michelangelo's birthday. The effects of logic bombs and time bombs range from the display of annoying messages on the screen to the reformatting of the hard drive, which causes complete data loss.

Worms

What is a worm? Although often called a virus, a worm is subtly different. Viruses require human interaction to spread, whereas worms take advantage of file transport methods, such as e-mail or network connections, to spread on their own. A virus infects a host file and waits until that file is executed to replicate and infect a computer system. A worm, however, works independently of host file execution and is much more active in spreading itself. Some worms even attack peripheral devices such as routers. Worms can generate a lot of data traffic when trying to spread, which can slow down the Internet.

Script and Macro Viruses

What are script and macro viruses? Some viruses are hidden on websites in the form of scripts. A script is a series of commands—actually, a mini-program—that is executed without your knowledge. Scripts are often used to perform useful, legitimate functions on websites, such as collecting name and address information from customers. However, some scripts are malicious. For example, you might click a link to display a video on a website, which causes a script to run that infects your computer with a virus.

A macro virus is a virus that attaches itself to a document that uses macros. A *macro* is a short series of commands that usually automates repetitive tasks. However, macro languages are now so sophisticated that viruses can be written with them. The Melissa virus became the first major macro virus to cause problems worldwide.

E-Mail Viruses

What is an e-mail virus? In addition to being a macro virus, the Melissa virus was the first practical example of an e-mail virus. E-mail viruses use the address book in the victim's e-mail system to distribute the virus. In the case of the Melissa virus, anyone opening an infected document triggered the virus, which infected other documents on the victim's computer. Once triggered, the Melissa virus sent itself to the first 50 people in the e-mail address book on the infected computer.

Encryption Viruses

What are encryption viruses? When encryption viruses (also called *ransomware*) infect your computer, they run a program that searches for common types of data files, such as Microsoft Word files, and compresses them using a complex encryption key that renders your files unusable. You then receive a message that asks you to send payment to an account if you want to receive the program to decrypt your files. Often, payment is demanded in cyber-currency which makes it difficult for law enforcement officials to trace the payments to an account and catch the perpetrators. Table 9.1 summarizes the major categories of viruses.

Table 9.1 Major Categories of Viruses

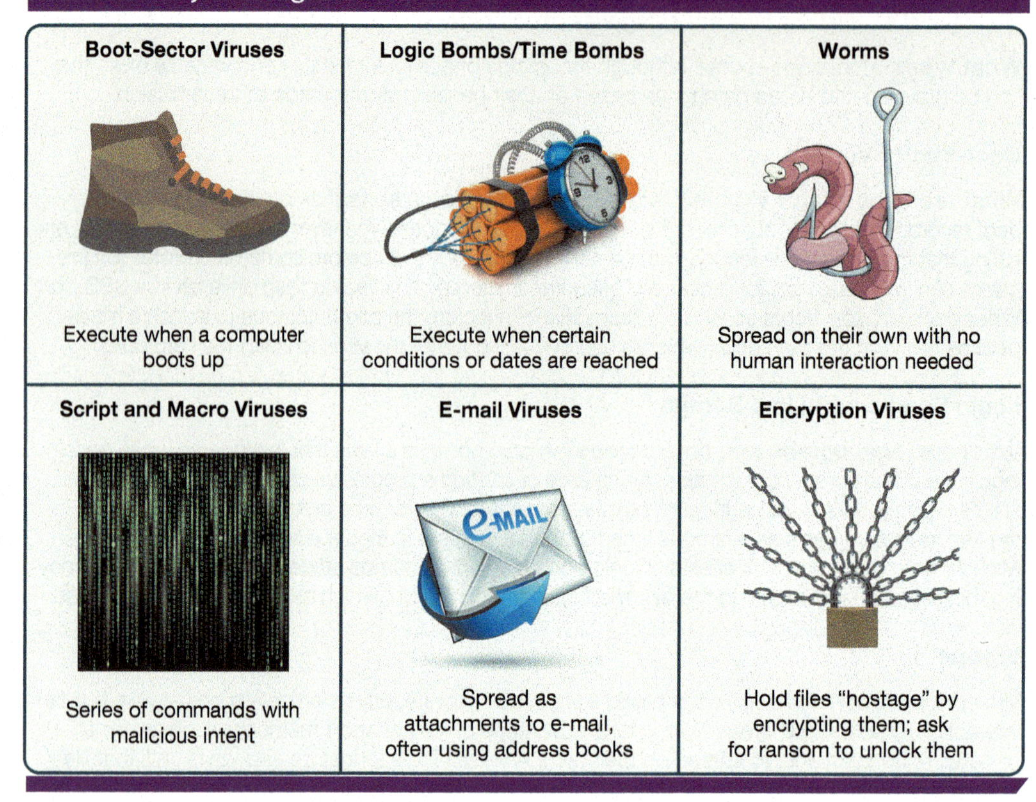

Boot-Sector Viruses	Logic Bombs/Time Bombs	Worms
Execute when a computer boots up	Execute when certain conditions or dates are reached	Spread on their own with no human interaction needed
Script and Macro Viruses	**E-mail Viruses**	**Encryption Viruses**
Series of commands with malicious intent	Spread as attachments to e-mail, often using address books	Hold files "hostage" by encrypting them; ask for ransom to unlock them

(Glinskaja Olga/Shutterstock; Oleksandr_Delyk/Shutterstock; David Martyn Hughes/123RF; Neyro2008/123RF; Bannosuke/Shutterstock; Lukas Gojda/Shutterstock)

Additional Virus Classifications

How else are viruses classified? Viruses can also be classified by the methods they take to avoid detection by antivirus software:

- A polymorphic virus changes its own code or periodically rewrites itself to avoid detection. Most polymorphic viruses infect a particular type of file such as .exe files, for example.
- A multi-partite virus is designed to infect multiple file types in an effort to fool the antivirus software that is looking for it.
- Stealth viruses temporarily erase their code from the files where they reside and then hide in the active memory of the computer. This helps them avoid detection if only the hard drive is being searched for viruses. Fortunately, current antivirus software scans memory as well as the hard drive.

Online Annoyances and Social Engineering

Surfing the web, using social networks, and sending and receiving e-mail have become common parts of most of our lives. Unfortunately, the web has become fertile ground for people who want to advertise their products, track our browsing behaviors, or even con people into revealing personal information. In this section, we look at ways in which you can manage, if not avoid, these and other online headaches.

Online Annoyances

Objective 9.5 *Explain what malware, spam, and cookies are and how they impact your security.*

Malware

What is malware? Malware is software that has a malicious intent (hence the prefix *mal*). There are three primary forms of malware: adware, spyware, and viruses. Adware and spyware are not physically destructive like viruses and worms, which can destroy data. Known collectively as *grayware*, most malware consists of intrusive, annoying, or objectionable online programs that are downloaded to your computer when you install or use other online content such as a free program, game, or utility.

What is adware? Adware is software that displays sponsored advertisements in a section of your browser window or as a pop-up box. It's considered a legitimate, though sometimes annoying, means of generating revenue for developers who don't charge for their software or information. Fortunately, because web browsers such as Safari, Chrome, and Edge have built-in pop-up blockers, the occurrence of annoying pop-ups has been greatly reduced.

Some pop-ups, however, are legitimate and increase the functionality of the originating site. For example, your account balance may pop up on your bank's website. To control which sites to allow pop-ups on, you can access the pop-up blocker settings in your browser (see Figure 9.7) and add websites for which you allow pop-ups. Whenever a pop-up is blocked, the browser displays an information bar or plays a sound to alert you. If you feel the pop-up is legitimate, you can choose to accept it.

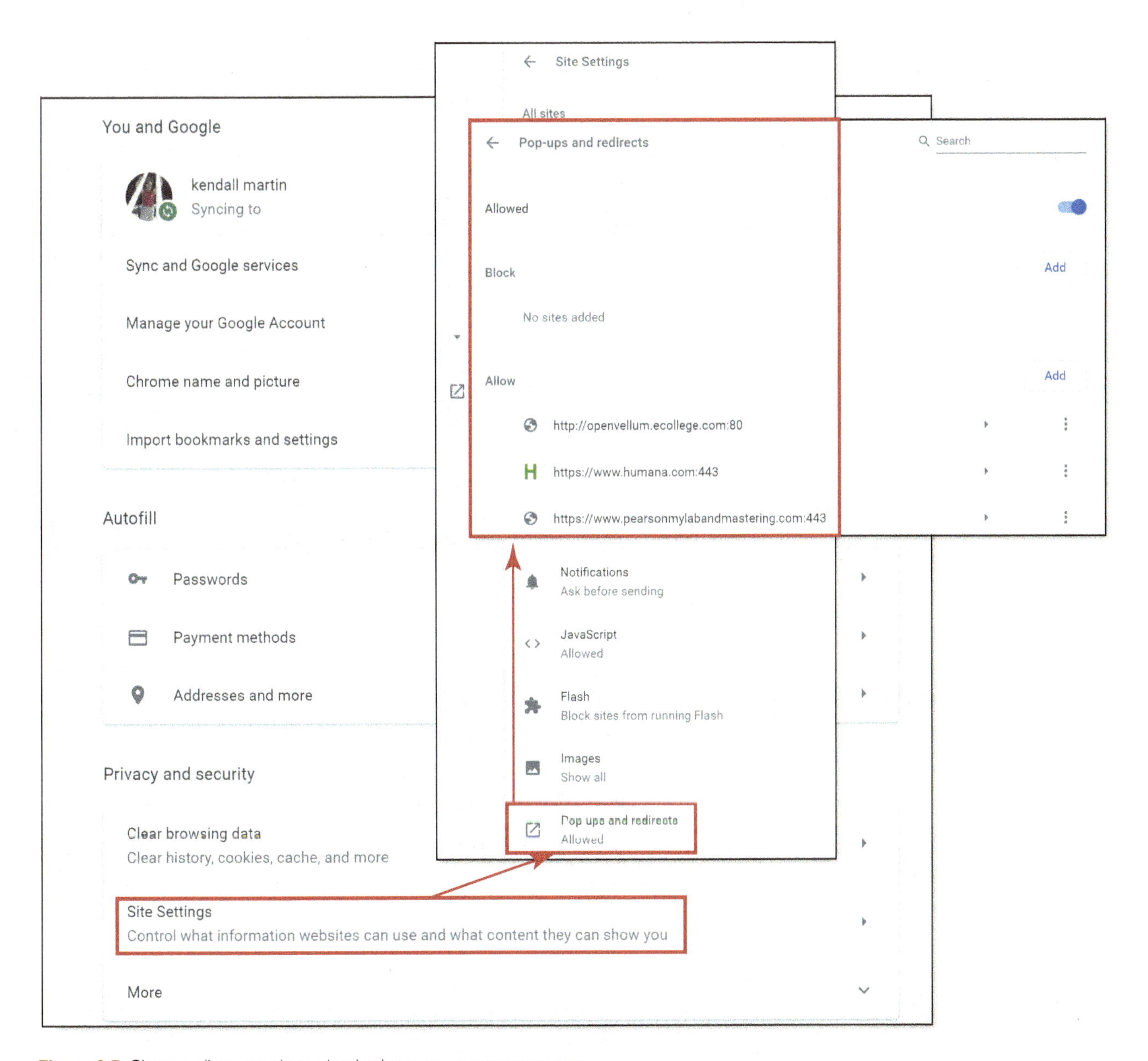

Figure 9.7 Chrome allows you to customize how you manage pop-ups.

> *To display the pop-up management in Chrome, click* ***Settings****, then from the* ***Privacy and security*** *group select* ***Site settings*** *and customize the choices in* ***Popups and redirects****. (Courtesy Google Inc.)*

What is spyware? Spyware is an unwanted piggyback program that usually downloads with other software you install from the Internet and that runs in the background of your system. Without your knowledge, spyware transmits information about you, such as your Internet-surfing habits, to the owner of the program so that the information can be used for marketing purposes. Many spyware programs use tracking cookies (small text files stored on your computer) to collect information. One type of spyware program known as a keystroke logger (keylogger) monitors keystrokes with the intent of stealing passwords, login IDs, or credit card information.

Can I prevent spyware from spying on me? Antispyware software detects unwanted programs and allows you to delete the offending software easily. Most Internet security suites now include antispyware software. Because so many variants of spyware exist, your Internet security software may not detect all types that attempt to install themselves on your computer. Therefore, it's a good idea to install one or two additional stand-alone antispyware programs on your computer.

Because new spyware is created all the time, you should update and run your antispyware software regularly. Windows comes with a program called Windows Defender, which scans your system for spyware and other potentially unwanted software. Malwarebytes, Adaware, and Spybot—Search & Destroy are other antispyware programs that are easy to install and update. Figure 9.8 shows an example of Windows Defender in action.

Spam

How can I best avoid spam? Companies that send out spam—unwanted or junk e-mail—find your e-mail address either from a list they purchase or with software that looks for e-mail addresses on the Internet. Unsolicited instant messages are also a form of spam, called *spim*. If you've used your e-mail address to purchase anything online, open an online account, or have a social media account, your e-mail address may appear on one of the lists that spammers get.

One way to avoid spam in your primary account is to create a secondary e-mail address that you use only when you fill out forms or buy stuff on the web. If your secondary e-mail account is saturated with spam, you can abandon that account with little inconvenience. It's much less convenient to abandon your primary e-mail address.

Another way to avoid spam is to filter it. A spam filter is an option in your e-mail account that places known or suspected spam messages into a special folder (called Spam or Junk Mail). Most e-mail services, such as Gmail and Outlook, offer spam filters (see Figure 9.9).

How do spam filters work? Spam filters and filtering software can catch as much as 95% of spam by checking incoming e-mail subject headers and senders' addresses against databases of

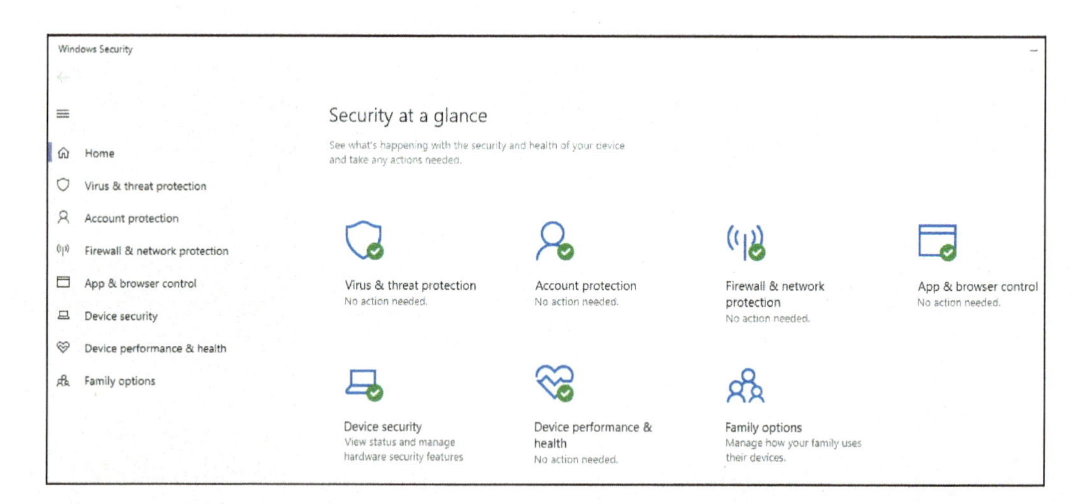

Figure 9.8 Windows Defender Security Center will help you establish secure computing practices. *(Courtesy Google Inc.)*

> *To open the Windows Defender Security Center, go to the Windows search box and type* ***Windows Security***. *Open the Windows Security app.*

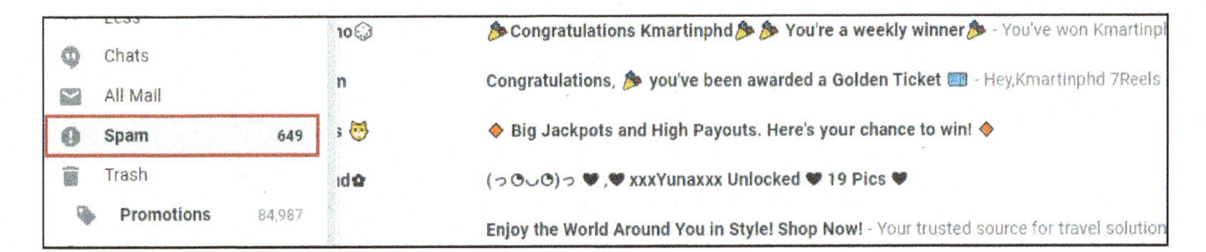

Figure 9.9 In Gmail, messages identified as spam are directed to a folder called Spam for review and deletion. *(Courtesy Google Inc.)*

known spam. Spam filters also check your e-mail for frequently used spam patterns and keywords, such as "for free" and "over 21." Spam filters aren't perfect, and you should check the spam folder before deleting its contents because legitimate e-mail might end up there by mistake. Most programs let you reclassify e-mails that have been misidentified as spam.

How else can I prevent spam? Before registering on a website, read its privacy policy to see how it uses your e-mail address. Don't give the site permission to pass on your e-mail address to third parties. Another strategy is to make sure you don't reply to spam to remove yourself from the spam list. By replying, you're confirming that your e-mail address is active. Instead of stopping spam, you may receive more.

Cookies

What are cookies? Cookies are small text files that some websites automatically store on your hard drive when you visit them. When you log on to a website that uses cookies, a cookie file assigns an ID number to your computer. The unique ID is intended to make your return visit to a website more efficient and better geared to your interests. The next time you log on to that site, the site marks your visit and keeps track of it in its database.

What do websites do with cookie information? Cookies can provide websites with information about your browsing habits, such as the ads you've opened, the products you've looked at, and the time and duration of your visits. Companies use this information to determine the traffic flowing through their website and the effectiveness of their marketing strategy and placement on websites. By tracking such information, cookies enable companies to identify different users' preferences.

Can companies get my personal information when I visit their sites? Cookies do not go through your hard drive in search of personal information such as passwords or financial data. The only personal information a cookie obtains is the information you supply when you fill out forms online.

Do privacy risks exist with cookies? Some sites sell the personal information their cookies collect to web advertisers who are building huge databases of consumer preferences and habits, collecting personal and business information such as phone numbers, credit reports, and the like. The main concern is that advertisers will use this information indiscriminately, thus invading your privacy. You may feel your privacy is being violated by cookies that monitor where you go on a website.

Should I delete cookies from my hard drive? Cookies pose no security threat because it's virtually impossible to hide a virus or malicious software program in a cookie. Because they take up little room on your hard drive and offer you small conveniences on return visits to websites, there is no great reason to delete them. Deleting your cookie files could actually cause you the inconvenience of reentering data you've already entered in website forms. However, if you're uncomfortable with the accessibility of your personal information, you can periodically delete cookies or configure your browser to block certain types of cookies (as shown in Figure 9.10).

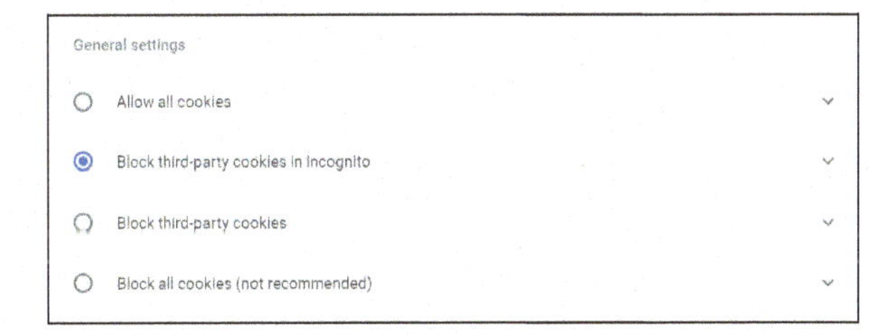

Figure 9.10 Tools are available in your browser (Chrome is shown here) to distinguish between cookies you want to keep and cookies you don't want on your system. (Courtesy Google Inc.)

>From Chrome ***Settings****, look at the* ***Privacy and security*** *block. Select* ***Cookies and other site data*** *to adjust your cookie-handling preferences.*

Social Engineering

Objective 9.6 *Describe social engineering techniques, and explain strategies to avoid falling prey to them.*

What is social engineering? Social engineering is any technique that uses social skills to generate human interaction that entices individuals to reveal sensitive information. Social engineering often doesn't involve the use of a computer or face-to-face interaction. For example, telephone scams are a common form of social engineering because it's often easier to manipulate someone when you don't have to look at them.

How does social engineering work? Most social engineering schemes use a pretext to lure their victims. Pretexting involves creating a scenario that sounds legitimate enough that someone will trust you. For example, you might receive a phone call during which the caller says he is from your bank and that someone tried to use your account without authorization. Or, someone may pose as another person on social media, such as a soldier in need of money to help her family at home, but the images, documents, and details she provides are all faked. The person then tells you he or she needs to confirm a few personal details such as your birth date, Social Security number,

Bits&Bytes I Received a Data Breach Letter . . . Now What?

Data breaches are becoming more common, and companies that are the subject of data breaches now routinely notify customers—usually by physical letter, but sometimes by e-mail. Take steps to secure your credit rating before your credit is compromised.

The safest step is to contact the credit bureaus (Equifax, Experian, and TransUnion) and have a credit freeze put in place. Anyone applying for a new credit card or a mortgage in your name will be denied. If you decide to add a credit card, you can always call again and unfreeze the accounts. You can also place an alert on your information, to show that you think you are at risk for identity theft because of the breach. If someone applies for credit in your name, they will be asked for additional information.

Remember to review your credit reports regularly. You are entitled to one free credit report per year from each of the three big agencies. Go to annualcreditreport.com and request a report from one of the agencies. Then you can repeat the process every four months from a different agency. Be on the lookout for any suspicious activity.

Most Internet security packages can detect and prevent pharming attacks. The major Internet security packages—for example, McAfee and Norton (see Figure 9.11)—also offer phishing-protection tools. When the Norton toolbar appears in your browser, you're constantly informed about the legitimacy of the site you are visiting.

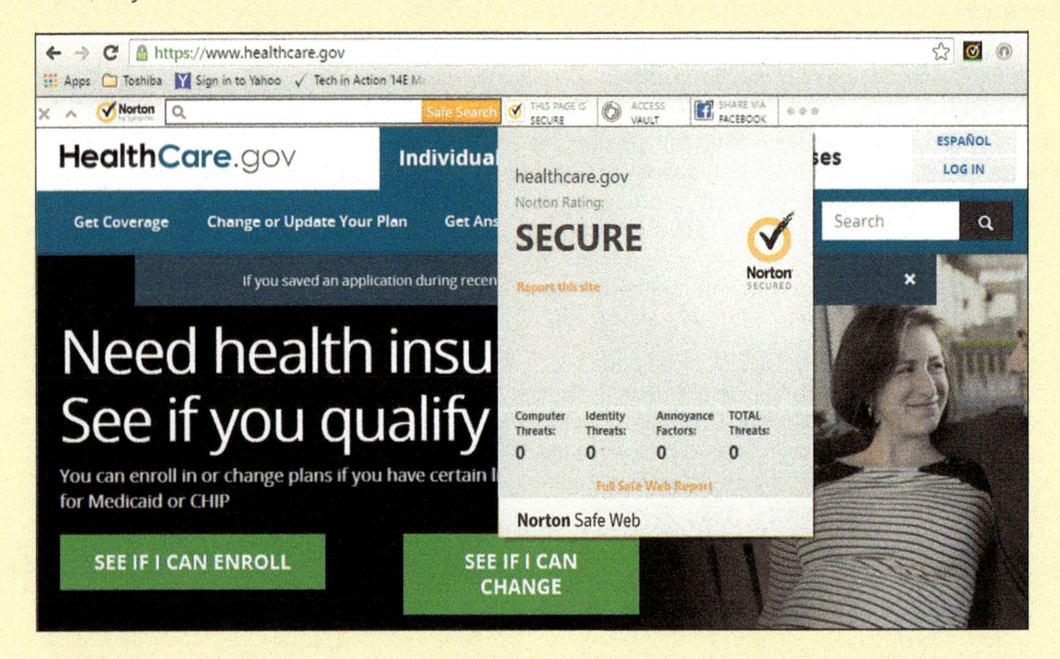

Figure 9.11 Not sure whether you're on the Healthcare.gov website or a cleverly disguised phishing site? Antivirus software can reassure you all is well. *(U.S. Centers for Medicare & Medicaid Services)*

Ethics in IT

Working from Home ... and Being Watched from Home

As many workers in all kinds of industries were sent home to begin telecommuting during the COVID-19 pandemic, employers turned to monitoring software. The rise in telecommuting has grown a large industry of products that track in detail how you conduct yourself, and you should be aware of exactly what your employer knows.

The most frequently cited reason for employee monitoring is to measure productivity. Cyberloafing (or cyberslacking) means using your computer for non-work activities while you're being paid to do your job. Examples of cyberloafing activities include popping over to play some Minecraft or browsing Instagram while on the job. Some employees even do multiple non-work tasks at the same time, which is known as *multi-shirking*. Estimates of business productivity losses due to cyberloafing top $50 billion annually.

Do you have a right to privacy in the workplace? Laws such as the 1986 Electronic Communications Privacy Act (ECPA), which prohibits unauthorized monitoring of electronic communications, have been interpreted by the courts in favor of employers. The bottom line is that employers who pay for equipment and software have the *legal* right to monitor their usage (see Figure 9.12). But now with so many employees at home, employers are often requiring authorization of monitoring software as a condition of employment.

By using software designed for employee monitoring, your boss can receive daily reports on your productivity. It may include GPS data, tracking how long you were out of the house for lunch, and detailed information on how you used your computer. How long were you working on a specific Word document? How many keystrokes per minute did you average during the day? Many monitoring reports take snapshots of your screen every ten minutes to monitor what websites you're visiting.

Figure 9.12 It is legal for employers to monitor your computer usage. *(Beatriz J Gascon/Shutterstock)*

So, is it *ethical* for employers to monitor their employees? Certainly, it seems fair for employers to ensure that they're getting a fair day's work from their employees, just as employees have an obligation to provide a fair effort for a fair wage. But does constant monitoring build a level of trust and unity between employer and employee? Are all employees willing to have their actions recorded so closely to identify a few who are not producing? Should an employer care whether you relax at moments during your working hours if your work is satisfactory?

The bottom line? Carefully read any agreement that allows your employer to monitor you in the workplace, and in your home if that is your workplace. Ask to see the kind of report that is kept on your activities and know how it will be used in evaluating you as an employee.

bank account number, and whatever other information he or she can get out of you. The information obtained can then be used to empty your bank account or commit some other form of fraud. The most common form of pretexting in cyberspace is phishing.

Phishing and Pharming

How are phishing schemes conducted? Phishing (pronounced "fishing") lures Internet users to reveal personal information such as credit card numbers, Social Security numbers, or passwords that could lead to identity theft. The scammers send e-mail messages that look like they're from a legitimate business such as a bank. The e-mail usually states that the recipient needs to update or confirm his or her account information. When the recipient clicks the provided link, they go to a website. The site looks like a legitimate site but is really a fraudulent copy that the scammer has created. Once the e-mail recipient enters his or her personal information, the scammers capture it and can begin using it.

Is pharming a type of phishing scam? Pharming is much more insidious than phishing. Phishing requires a positive action by the person being scammed, such as going to a website mentioned in an e-mail and typing in personal information. Pharming occurs when malicious code is planted on your computer, either by viruses or by your visits to malicious websites, which then alters your browser's ability to find web addresses. Users are directed to bogus websites even when they enter the correct address of the real website. You end up at a fake website that looks legitimate but is expressly set up for the purpose of gathering information.

How can I avoid being caught by phishing and pharming scams? Follow these guidelines to avoid falling prey to such schemes:

- Never reply directly to any e-mail asking you for personal information.
- Don't click a link in an e-mail to go to a website. Instead, type the website address in the browser.
- Check with the company asking for the information and only give the information if you're certain it's needed.
- Never give personal information over the Internet unless you know the site is secure. Look for the closed padlock, https, or a certification seal such as Norton Secured to help reassure you that the site is secure.
- Use phishing filters. The latest versions of Safari, Chrome, and Edge have phishing filters built in, so each time you access a website, the phishing filter checks for the site's legitimacy and warns you of possible web forgeries.
- Use Internet security software on your computer that's constantly being updated.

Helpdesk MyLab IT

Threats to Your Digital Life

In this Helpdesk, you'll play the role of a help desk staffer, fielding calls about identity theft, computer viruses, and malware.

Ransomware and Scareware

What is ransomware? Ransomware is a type of malware that attacks your system by encrypting critical files so the system is no longer operational. You are given directions on how to make a payment so that your files can be restored. Criminals often target services that cannot afford to be offline, such as utilities, hospitals, and even newspapers (see Figure 9.13). Payment is typically demanded in untraceable currency like bitcoin. During the COVID-19 pandemic, hospitals in Germany and Colorado were hit with ransomware attacks.

Ransomware attacks have become more prevalent as Ransomware as a Service (RaaS) sites have appeared. Modeled after SaaS (Software as a Service) delivery, RaaS allows less technically skilled criminals still to conduct ransomware attacks by using the code of a master developer. The developer collects a fee for all of the extortion funds collected by users of the ransomware software.

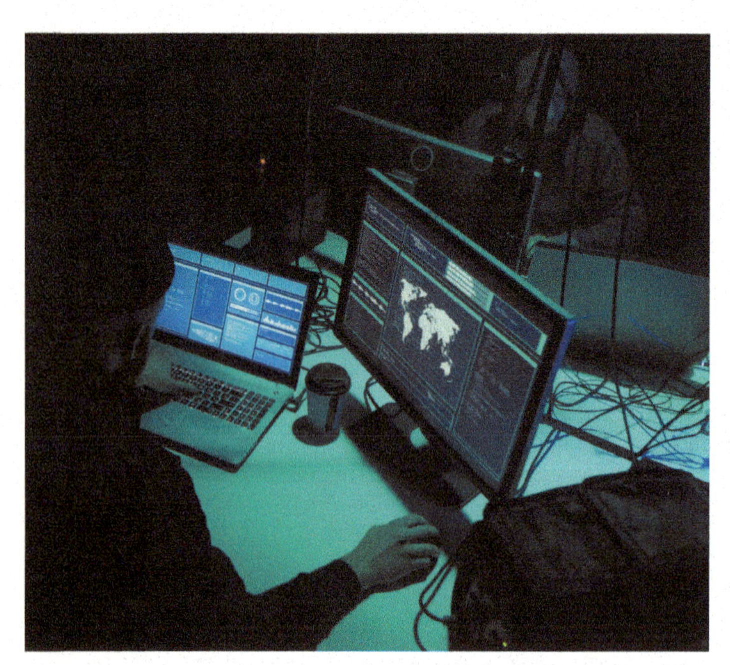

Figure 9.13 Ransomware attacks target businesses that have little ability to tolerate IT system downtime, such as hospitals, water authorities, and newspapers. *(Maksim Shmeljov/Shutterstock)*

What is scareware? Scareware is a type of malware that downloads onto your computer and tries to convince you that your computer is infected with a virus or a type of ransomware. Pop-ups, banners, or other annoying types of messages will flash on your screen, saying frightening things like, "Your computer is infected with a virus . . . immediate removal is required." You're then directed to a website where you can buy fake removal or antivirus tools that provide little or no value.

Scareware is a social engineering technique because it uses people's fear of computer viruses to convince them to part with their money. It is often designed to be extremely difficult to remove from your computer and to interfere with the operation of legitimate security software. Be knowledgeable about any files you download; scareware is often downloaded onto your computer from infected websites or Trojan horse files.

How do I protect myself against scareware? Most Internet security suites and antivirus and antimalware software packages now detect and prevent the installation of scareware. But make sure you never click website banners or pop-up boxes that say "Your computer might be infected, click here to scan your files," because these are often the starting points for installing malicious scareware files on your computer.

Trends in IT Spear Phishing: The Bane of Data Breaches

Most people have vast amounts of personal data residing in the databases of the various companies with which they conduct business. Amazon.com has your credit card and address information. Your bank has your Social Security number, birth date, and financial records. Your local supermarket probably has your e-mail address from when you joined its loyalty club to receive grocery discounts. All this data in various places puts you at risk when companies responsible for keeping your data confidential suffer a data breach.

A data breach occurs when sensitive or confidential information is copied, transmitted, or viewed by an individual who isn't authorized to handle the data. Data breaches can be intentional or unintentional. Intentional data breaches occur when hackers break into digital systems to steal sensitive data. Unintentional data breaches occur when companies controlling data inadvertently allow it to be seen by unauthorized parties, usually due to some breakdown in security procedures or precautions.

Unfortunately, data breaches are quite common, and another one always seems to be in the news. Thousands of data breaches a year are reported by US companies. Hundreds of millions of records are exposed in a typical year. These breaches pose serious risks to the individuals whose data has been compromised, even if financial data is not involved. The data thieves now have the basis with which to launch targeted social engineering attacks even if they just have contact information, such as e-mail addresses.

With regular phishing techniques, cybercriminals just send out e-mails to a wide list of e-mail addresses, whether they have a relationship with the company or not. For example, a criminal might send out a general phishing e-mail claiming that a person's Citibank checking account had been breached. People receiving this e-mail who don't have any accounts at Citibank should immediately realize this is a phishing attack and ignore the instructions to divulge sensitive data.

However, when cybercriminals obtain data on individuals that includes information about which companies those individuals have a relationship with, they can engage in much more targeted attacks known as spear phishing (see Figure 9.14). Spear phishing e-mails are sent to people known to be customers of a company and have a much greater chance of successfully inducing individuals to reveal sensitive data. If cybercriminals obtain a list of e-mail addresses of customers from Barclays Bank, for example, they can ensure that the spear phishing e-mails purport to come from Barclays and will include the customer's full name. This type of attack is much more likely to succeed in fooling people than just random e-mails sent out to thousands of people who might not have a relationship with the company mentioned in the phishing e-mail.

Figure 9.14 Spear phishing is a targeted type of social engineering. *(Vectorpouch/Shutterstock)*

So how can you protect yourself after a data breach? You need to be extra suspicious of any e-mail correspondence from companies involved in the data breach. Companies should never contact you by e-mail or phone asking you to reveal sensitive information or reactivate your online account by entering confidential information.

If you do receive e-mails or phone calls saying they have problems with your accounts, your best course of action is to delete the e-mail or hang up the phone. Then contact the company that supposedly has the problem by a phone number that you look up yourself in legitimate correspondence from the company (say, the toll-free number on your credit card statement). The representatives from your company can quickly tell whether a real problem exists or whether you were about to be the victim of a scam.

Before moving on to Part 2:

1. Watch Chapter Overview Video 9.1.
2. Then take the Check Your Understanding quiz.

Check Your Understanding // Review & Practice

For a quick review to see what you've learned so far, answer the following questions.

multiple choice

1. When a hacker steals personal information with the intent of impersonating another individual to commit fraud, it is known as
   a. impersonation theft.
   b. identity theft.
   c. scareware theft.
   d. malware theft.
2. An attack that renders a computer unable to respond to legitimate users because it is being bombarded with data requests is known as a ________ attack.
   a. stealth
   b. backdoor
   c. denial-of-service
   d. scareware
3. You are at risk for identity theft when you
   a. give your social security number to someone over the phone.
   b. lose a wallet.
   c. are careless about disposing of bank statements.
   d. all the above.
4. A series of commands that are executed without your knowledge is a typical attribute of a(n) ________ virus.
   a. script
   b. boot-sector
   c. time bomb
   d. encryption
5. Software that pretends your computer is infected with a virus to entice you into spending money on a solution is known as
   a. trackingware.
   b. spyware.
   c. adware.
   d. scareware.

chew on this

You most likely have provided personal information to many websites and companies. What information do you wish you had never disclosed? What types of information have companies asked you to provide that you believe was unnecessary? List specific companies and examples of the extraneous information.

(3Dalia/Shutterstock)

MyLab IT

Go to **MyLab IT** to take an autograded version of the *Check Your Understanding* review and to find all media resources for the chapter.

For the IT Simulation for this chapter, see MyLab IT.

Try This

Testing Your Network Security

A properly installed firewall should keep you relatively safe from hackers, but how do you know whether your computer system is safe? Many websites help test your security. In this exercise, we use tools provided by Gibson Research Corporation (*grc.com*) to test the strength of your protection.

Open a web browser and navigate to *grc.com*. From the pull-down Services menu at the top of the screen, select **ShieldsUP!** From the Services drop-down menu on the top left corner, select ShielsUP! and then click the **Proceed button**.

Click the **Common Ports link** to begin the security test. It will take a few moments for the results to appear, depending on the speed of your Internet connection.

You'll now see a report showing your computer's status on the TruStealth analysis test. If you passed, it means that your computer is very safe from attack. However, don't panic if your results say **Failed for TruStealth Analysis**. Consider these common issues:

- A common failure on the TruStealth test is ping reply. This means GRC attempted to contact your computer, using a method called ping, and it replied to the GRC computer. This is not a serious problem. It just means it's a little easier for a hacker to tell that your computer network exists, but that doesn't mean that a hacker can break in. Often, your ISP will insist that your home modem be configured to reply to ping requests to help the company run remote diagnostics if there is ever a problem with your modem. As long as your ports are closed (or stealthed), your system should still be secure.
- As long as all ports are either in stealth mode or closed, you're well protected. For the TruStealth analysis, though, if ports are only closed (not stealthed), the report will result in a failure. Only if some ports report as open do you have to worry that your computer is exposed to hackers. Consult your firewall documentation to resolve the problem.

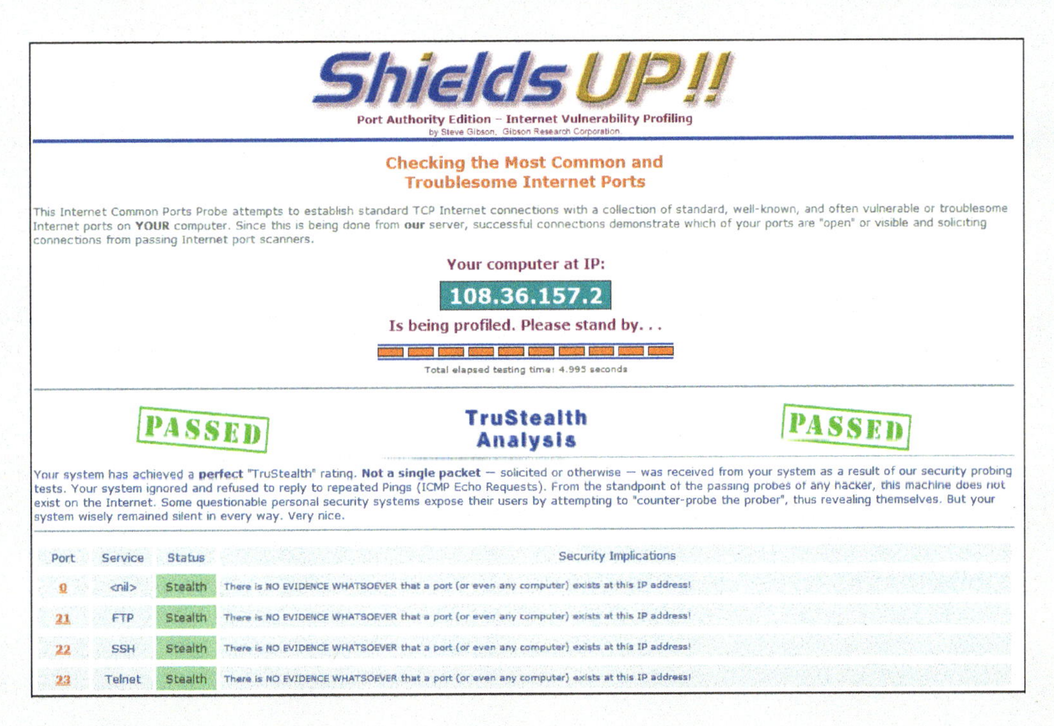

(Gibson Research Corp.)

Make This — App Inventor 2 or Thunkable

A Password Generator

Want to make sure your passwords are randomly selected so they are more secure, but afraid you'll forget what you chose?

In this exercise, you continue your mobile app development by using the TinyDB component in App Inventor to store information on your phone so it can be recalled by your application the next time it runs. This is called *data persistence*, and as you'll see, it is very easy to do!

The TinyDB component allows you to store data on your device to use the next time the app opens.

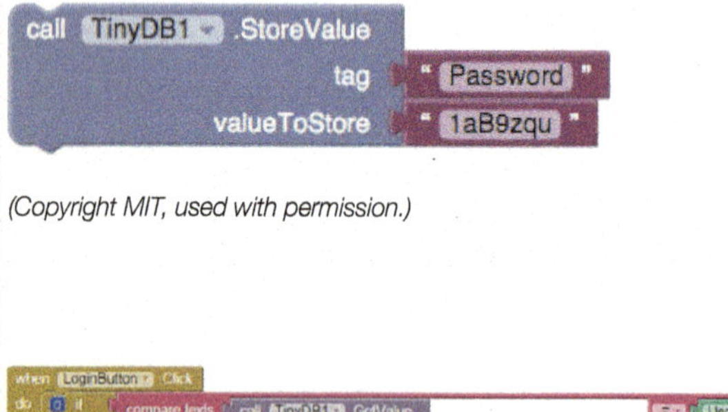

(Copyright MIT, used with permission.)

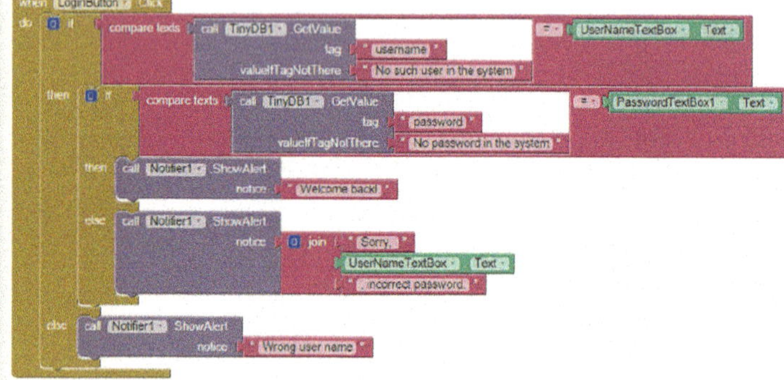

(Copyright MIT, used with permission.)

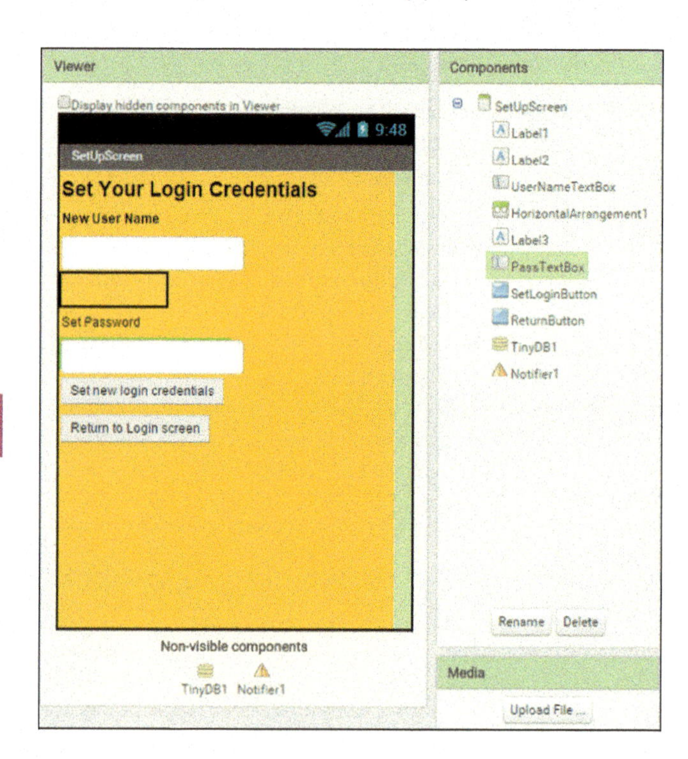

(Copyright MIT, used with permission.)

To create this app for iOS, go to *Thunkable.com*, a programming platform based on App Inventor.

For the instructions for this exercise, go to MyLab IT.

Part 2 For an overview of this part of the chapter, watch **Chapter Overview Video 9.2.**

Protecting Your Digital Property

Learning Outcome 9.2 Describe various ways to protect your digital property and data from theft and corruption.

Often, we can be our own worst enemies when using computing devices. If you're not careful, you might be taken in by thieves or scam artists who want to steal your digital and physical assets. As we discuss in this section, protecting yourself is key, and there are many relatively easy measures you can take to increase your level of security.

Restricting Access to Your Digital Assets

Keeping hackers and viruses at bay is often a matter of keeping them out. You can achieve this by:

- Preventing hackers from accessing your computer (usually through your Internet connection)
- Using techniques to prevent virus infections from reaching your computer
- Protecting your digital information in such a way that it can't be accessed (by using passwords, for example)
- Hiding your activities from prying eyes

In this section, we explore strategies for protecting access to your digital assets and keeping your Internet-surfing activities private.

Firewalls

Objective 9.7 *Explain what a firewall is and how a firewall protects your computer from hackers.*

What is a firewall? A firewall is a software program or hardware device designed to protect computers from hackers. It's named after a housing construction feature that slows the spread of fires from house to house. A firewall designed specifically for home networks is called a personal firewall. By using a personal firewall, you can close open logical ports and make your computer invisible to other computers on the Internet.

Which is better, a software firewall or a hardware firewall? One type isn't better than the other. Both hardware and software firewalls will protect you from hackers. Although installing either a software or a hardware firewall on your home network is probably sufficient, you should consider installing both for maximum protection. This will provide you with additional safety.

Types of Firewalls

What software firewalls are there? Both Windows and macOS include reliable firewalls. The Windows Action Center is a good source of information about the security status of your computer. The status of your Windows Firewall is shown in the Windows Firewall screen (see Figure 9.15). Security packages such as Norton Security Suite and Trend Micro Internet Security also include firewall software. Although the firewalls that come with Windows and macOS will protect your computer, firewalls included in security suites often come with additional features such as monitoring systems that alert you if your computer is under attack.

If you're using a security suite that includes a firewall, the suite should disable the firewall that came with your OS. Two software firewalls running at the same time can conflict with each other and cause your computer to slow down or freeze up.

What are hardware firewalls? You can also buy and configure hardware firewall devices, although most routers sold for home networks include firewall protection. Just like software firewalls, the setup for hardware firewalls is designed for novices, and the default configuration on most routers keeps unused logical ports closed. Documentation accompanying routers can assist more experienced users in adjusting the settings to allow access to specific ports if needed.

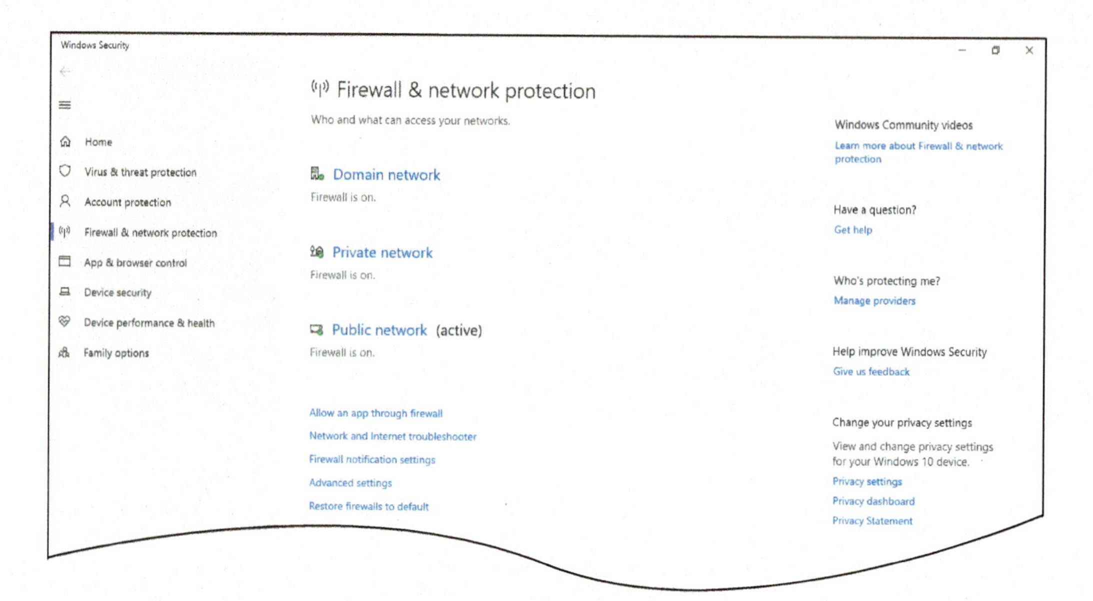

Figure 9.15 The Windows Firewall screen provides the status of your firewall. *(Courtesy of Microsoft Corp.)*

> *To view the Windows Firewall screen, open the* ***Windows Security*** *app then select* ***Firewall and network protection****.*

How Firewalls Work

How do firewalls protect you from hackers? Firewalls are designed to restrict access to a network and its computers. Firewalls protect you in two major ways: (1) by blocking access to logical ports and (2) by keeping your computer's network address secure.

How do firewalls block access to your logical ports? Recall that logical ports are virtual communications paths that allow a computer to organize requests for information from other networks or computers. See Table 9.2 for some logical port numbers and the services they are associated with.

Certain logical ports are very popular in hacker attacks. For example, logical port number 1337 is often used because it looks close to the word "leet," which is hacker talk for "elite," so hackers think it's fun to go after that port.

To block access to logical ports, firewalls examine data packets that your computer sends and receives. Data packets contain the address of the sending and receiving computers and the logical port that the packet will use. Firewalls can be configured so that they filter out packets sent to specific logical ports in a process known as packet filtering.

Firewalls can also be configured to refuse all requests from the Internet that are asking for access to specific ports. That process is referred to as logical port blocking. By using filtering and blocking, firewalls help keep hackers from accessing your computer (see Figure 9.16).

Table 9.2 Common Logical Ports

Port Number	Protocol Using the Port
21	FTP (File Transfer Protocol) control
23	Telnet (unencrypted text communications)
25	SMTP (Simple Mail Transfer Protocol)
53	DNS (domain name system)
80	HTTP (Hypertext Transfer Protocol)
443	HTTPS (HTTP with Transport Layer Security [TLS] encryption)

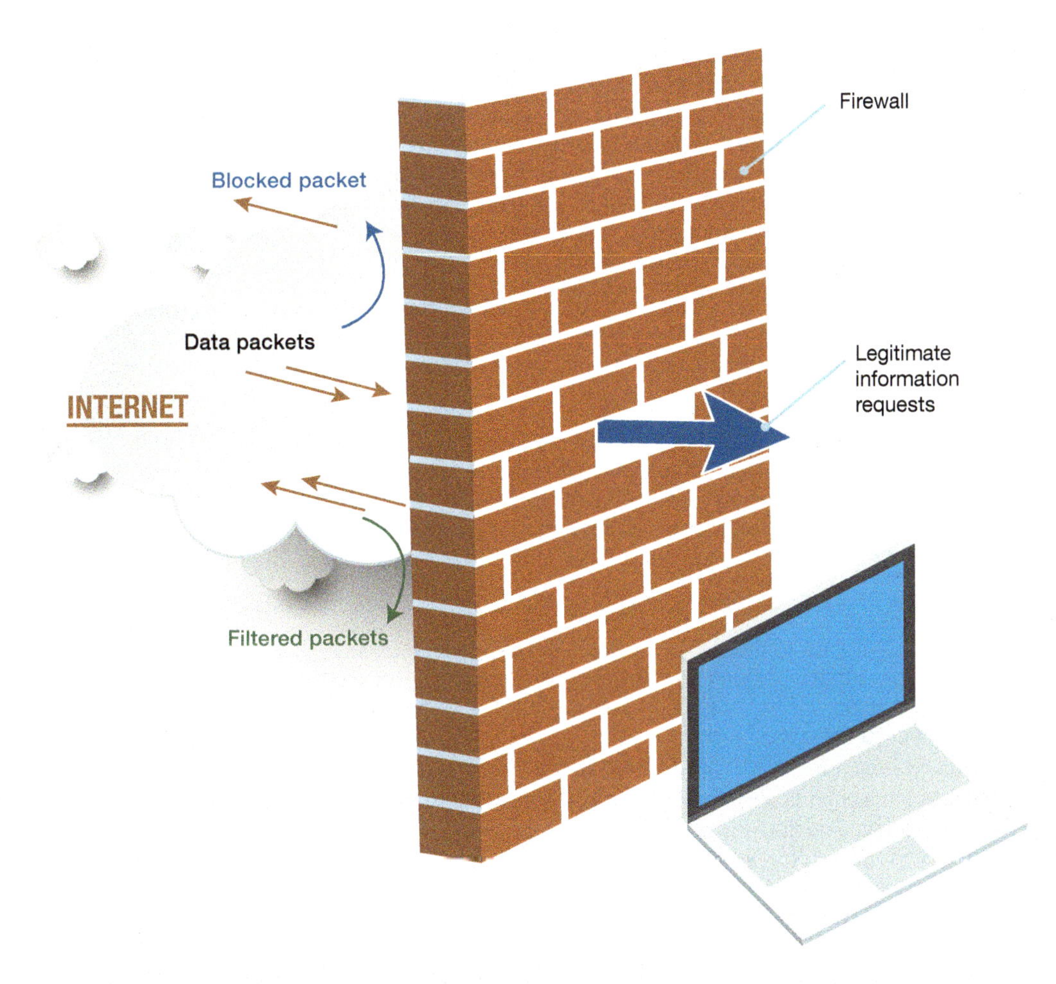

Figure 9.16 Firewalls use filtering and blocking to keep out unwanted data on specific logical ports. That keeps your network safe.

How do firewalls keep your network address secure? Every computer connected to the Internet has a unique address called an *Internet Protocol (IP) address*. Data is routed to the correct computer on the Internet based on the IP address. This is similar to how a letter finds its way to your mailbox. You have a unique postal address for your home. If a hacker finds out the IP address of your computer, they can locate it on the Internet and try to break into it.

Your IP address for your home network is assigned to your router by your Internet service provider (ISP). Then each device on your home network has its own IP address. Firewalls use a process called network address translation (NAT) to assign internal IP addresses on a network. The internal IP addresses are used only on the internal network and therefore can't be detected by hackers. For hackers to access your computer, they must know your computer's internal IP address. With a NAT-capable router/firewall installed on your network, hackers are unable to access the internal IP address assigned to your computer, so your computer is safe. Most routers sold for home use are configured as firewalls and feature NAT.

Preventing Virus Infections

Objective 9.8 *Explain how to protect your computer from virus infection.*

What is the best way to protect my devices from viruses? There are two main ways to protect your computer from viruses: by installing antivirus software and by keeping your software up to date.

Antivirus Software

What antivirus software do I need? Antivirus software is specifically designed to detect viruses and protect your computer and files from harm. Symantec, Trend Micro, and Avast are among the companies that offer highly rated antivirus software packages. Antivirus protection is also included in comprehensive Internet security packages such as Norton Security Suite or Trend Micro Internet Security. These software packages also help protect you from threats other than computer viruses.

Understanding Firewalls

In this Helpdesk, you'll play the role of a help desk staffer, fielding questions about how hackers can attack networks and what harm they can cause as well as what a firewall does to keep a computer safe from hackers.

My new computer came with antivirus software installed, so shouldn't I already be protected? Computers running Windows come with its built-in antivirus software, Windows Defender. This offers built-in virus protection but does not get ratings as strong as many third-party products. Your new system will probably also come with a trial version of a particular company's antivirus program. That software is free only for a limited time, usually 90 or 180 days. After that, you have to pay to upgrade to the full version of the product if you decide you like it. Before you consider paying for the upgrade though, check with your Internet service provider. Many include free antivirus software with your monthly service plan.

How often do I need to run antivirus software? Although antivirus software is designed to detect suspicious activity on your computer at all times, you should run an active virus scan on your entire system at least once a week. By doing so, all files on your computer will be checked for undetected viruses.

Antivirus programs run scans in the background when your CPU is not being heavily used. However, you can also configure the software to run scans at times when you aren't using your system—for example, when you're asleep (see Figure 9.17). (However, it's important to note that your computer must be powered on and not in sleep mode for these virus scans to take place.) Alternatively, if you suspect a problem, you can launch a scan manually and have it run immediately.

How does antivirus software work? The main functions of antivirus software are as follows:

- *Detection.* Antivirus software looks for virus signatures in files. A virus signature is a portion of the virus code that's unique to a particular computer virus. Antivirus software scans files for these signatures when they're opened or executed and identifies infected files and the type of virus infecting them.
- *Stopping virus execution.* If the antivirus software detects a virus signature or suspicious activity, such as the launch of an unknown macro, it stops the execution of the file and virus and notifies you that it has detected a virus. It also places the virus in a secure area on your hard drive so that it won't spread to other files; this procedure is known as quarantining. Usually, the antivirus software then gives you the choice of deleting or repairing the infected file. Unfortunately, antivirus programs can't always fix infected files to make them usable again. You should keep backup copies of critical files so that you can restore them in case a virus damages them irreparably.
- *Prevention of future infection.* Most antivirus software will also attempt to prevent infection by inoculating key files on your computer. In inoculation, the antivirus software records key

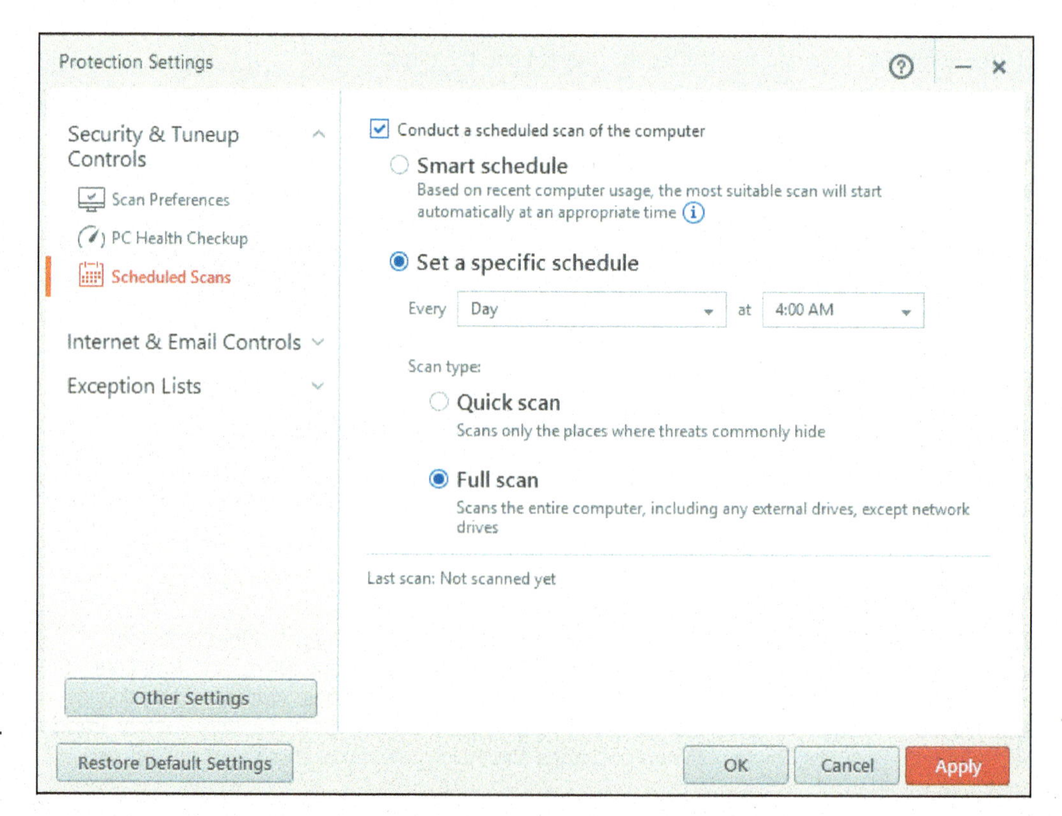

Figure 9.17 You can set up virus scans to run automatically, based on your usage or on a specific schedule. *(Reprinted courtesy of Trend Micro Incorporated)*

attributes about your computer files, such as file size and date created, and keeps these statistics in a safe place on your hard drive. When scanning for viruses, the antivirus software compares the attributes of the files with the attributes it previously recorded to help detect attempts by virus programs to modify your files.

Does antivirus software stop all viruses? Antivirus software catches known viruses effectively; however, new viruses are written all the time. To combat unknown viruses, modern antivirus programs search for suspicious virus-like activities as well as virus signatures. To minimize your risk, you should keep your antivirus software up to date.

How do I make sure my antivirus software is up to date? Most antivirus programs have an automatic update feature that downloads updates for virus signature files every time you go online. Also, the antivirus software usually shows the status of your update subscription so that you can see how much time you have remaining until you need to buy another version of your software. Many Internet security packages offer bonus features such as cloud storage scanners and password managers (see Figure 9.18) to provide you with extra protection.

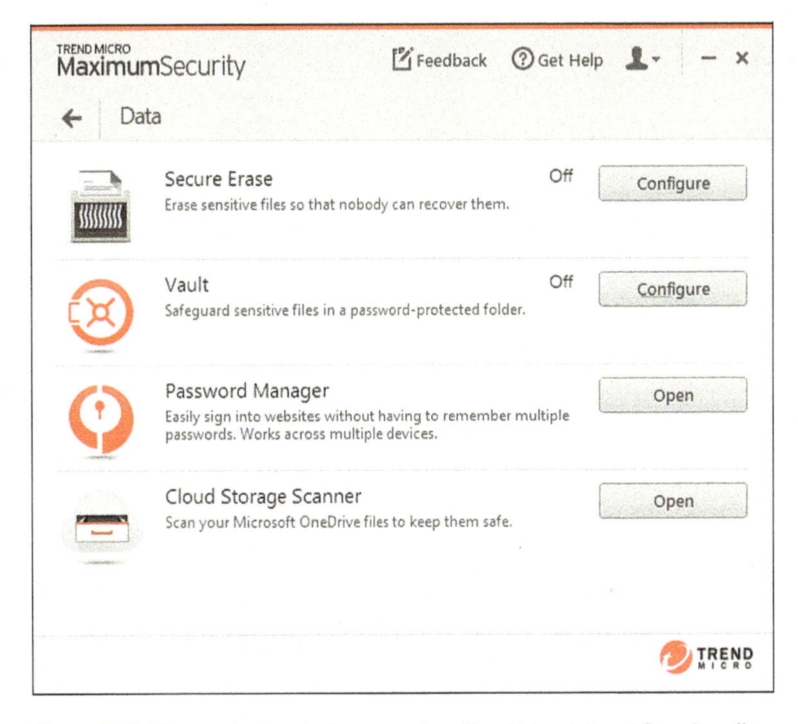

Figure 9.18 Many protection packages, such as Trend Micro Internet Security, offer other types of security features besides basic malware protection. *(Courtesy of Trend Micro Incorporated)*

What should I do if I think my computer is infected with a virus? If you see suspicious behavior from your system, run a complete virus scan. If viruses are detected, you may want to research the situation further to determine whether your antivirus software will eradicate them completely or whether you'll need to take additional manual steps to eliminate the viruses. Most antivirus company websites contain archives of information on viruses and provide step-by-step solutions for removing them if they're particularly stubborn.

How do I protect my phone from viruses? Because smartphones and other mobile devices run operating systems and contain files, they are susceptible to infection by viruses. Cybercriminals hide viruses in legitimate-looking apps for download to mobile devices. Most antivirus software companies now offer antivirus software specifically designed for mobile devices. The Google Play store offers effective free products to protect your Android devices such as 360 Security and Avast Mobile Security.

Software Updates

Why does updating my operating system (OS) software help protect me from viruses? Many viruses exploit weaknesses in operating systems. Malicious websites can be set up to attack your computer by downloading harmful software onto your computer. According to research conducted by Google, this type of attack, known as a drive-by download, affects almost one in 1,000 web pages. To combat these threats, make sure your OS is up to date and contains the latest security patches.

Do OS updates only happen automatically? With Windows 10, updates are downloaded automatically whenever Microsoft provides them. Windows keeps track of the hours you're most busy on your system and tries to schedule your restart for when you're not usually active. You do still have some options, however. You can choose to pick your own restart time manually (see Figure 9.19). You also can set a pause of up to seven days before the update is installed.

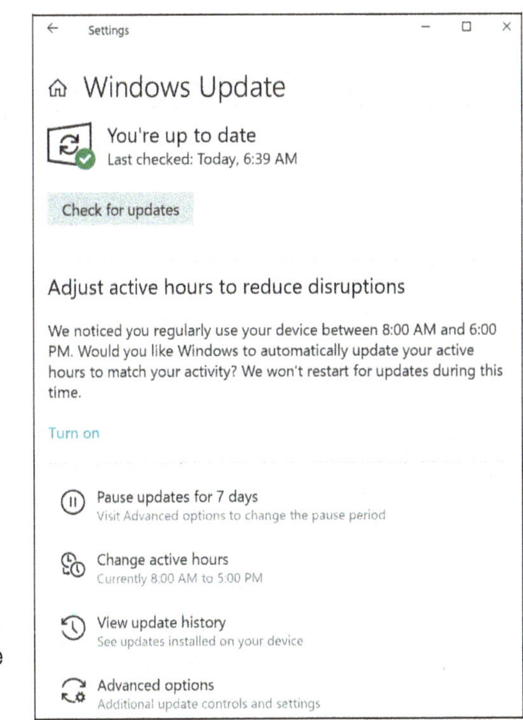

Figure 9.19 The Windows Update screen makes it easy for users to stay abreast of software updates and manage restarts. (Courtesy of Microsoft Corp.)

> *To open the Windows Update screen, open* ***Settings*** *and then select* ***Update & Security****. (Windows 10, Microsoft Corporation)*

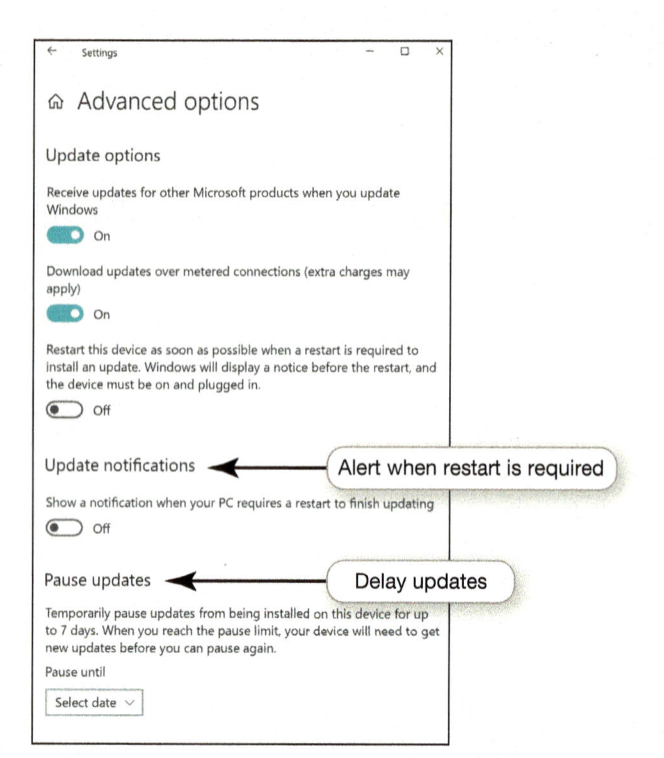

Figure 9.20 The Windows Update Advanced Options screen provides a few more user-controlled update options.

> *To open the Windows Update Advanced options screen, click the* ***Advanced options*** *link of the Windows Update screen (Figure 9.19). (Courtesy of Microsoft Corp.)*

The Advanced Options screen for Windows Update provides these and a few other options (see Figure 9.20). The most notable is the ability to receive updates for other Microsoft products like Microsoft Office.

Authentication: Passwords and Biometrics

Objective 9.9 *Describe how passwords and biometric characteristics can be used for user authentication.*

How do I restrict access to my device? Windows, macOS, and most other operating systems have built-in password protection for files as well as for the entire device. After a certain period of idle time, the device is automatically locked. This provides excellent protection from casual snooping if you need to walk away for a period of time. It's a good idea to use login security on mobile devices because this provides protection of your data if your device is lost or stolen.

How can I use passwords to protect my computer? You no doubt have many passwords you need to remember to access your digital life. However, creating strong passwords—ones that are difficult for hackers to guess—is an essential piece of security that people sometimes overlook. Password-cracking programs have become very sophisticated. In fact, some commonly available programs can test more than one million password combinations per second! Creating a secure password is therefore more important than ever.

When you are creating a password, be mindful that many sites or organizations have rules you must obey. Typically, the password must be at least eight characters and use uppercase, lowercase, numeric, and symbol characters. Longer passwords are more difficult to break, so using a passphrase is a good idea—a sentence that you can easily remember and type quickly.

Websites that need to be very secure, such as those for financial institutions, usually have strong defenses to prevent hackers from cracking passwords, but sites that need less security, such as

Bits&Bytes CAPTCHA: Keeping Websites Safe from Bots

Automated programs called *bots* (or web robots) are used to make tasks easier on the Internet. For example, search engines use bots to search and index web pages. Unfortunately, bots can also be used for malicious or illegal purposes because they can perform some computing tasks faster than humans. For example, bots can be used on ticket-ordering sites to buy large blocks of high-demand concert tickets. They are also often used to post spam in the comment sections of blogs. Fortunately, website owners can deploy CAPTCHA software to prevent such bot activities.

Completely Automated Public Turing test to tell Computers and Humans Apart (CAPTCHA) programs are used to generate distorted text and require it to be typed into a box. Google's reCAPTCHA used to ask you to click a check box (see Figure 9.21a) to demonstrate you're human and pass the test. Google has moved to a more sophisticated technique like monitoring how you move the mouse or what your browsing habits are to determine whether you're a human or a bot. Because some tests, like asking you to type in some distorted text (see Figure 9.21b), are impossible for the vision-impaired, audio CAPTCHAs are also available. If you want to try integrating a CAPTCHA program into your website (to protect your e-mail address), go to google.com/recaptcha, which offers free CAPTCHA tools to help you protect your data.

Figure 9.21 (a) CAPTCHA programs like reCAPTCHA monitor a user's website behavior and try to determine whether a user is human. (b) If the app isn't sure whether you're a bot, you may be issued a challenge. *(Metrue/Shutterstock)*

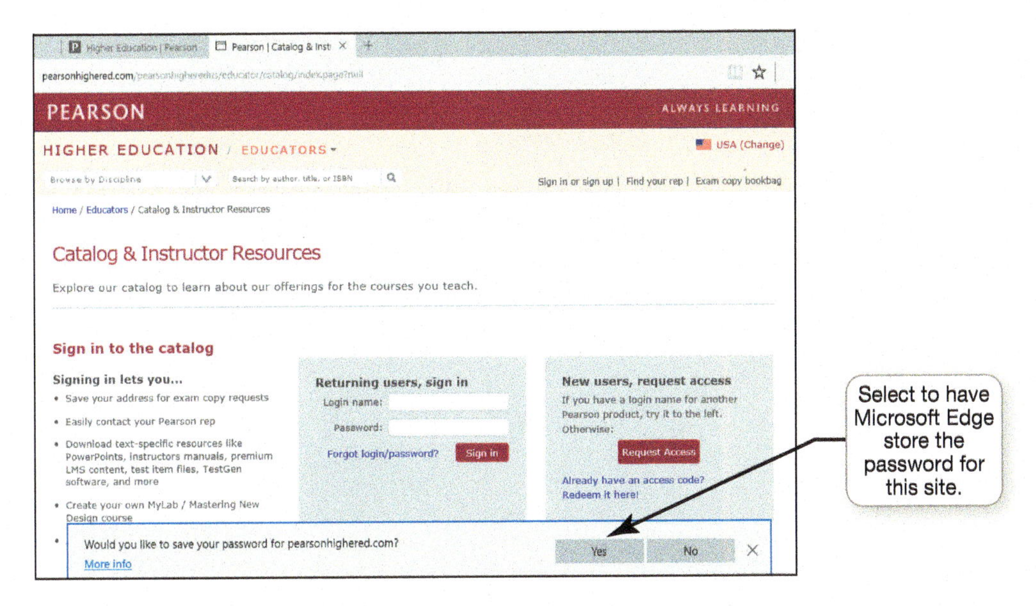

Figure 9.22 The password dialog box in Microsoft Edge appears whenever you type in a password for a website for which Microsoft Edge hasn't yet stored the password. *(Courtesy of Pearson Higher Education website)*

Bits&Bytes Chrome Safety Check

The latest version of Google Chrome adds a feature called Safety Check. Found on the Settings page of Chrome, it can help keep you safe from data breaches and more.

Run the Safety Check and it will make sure your version of Chrome is the latest and most up to date. It will check your stored passwords and prompt you to change passwords that were involved in known data breaches. Turn on Safe Browsing and Google warns you before heading to sites that are known to be dangerous. Run Safety Check often to make sure you are doing all you can to stay protected while on the Internet.

casual gaming or social networking sites, might have less protection. Hackers attack poorly defended sites for passwords because many people use the same password for every site they use. Don't let a hacker get your password from a poorly secured gaming site, and then be able to access your bank account with the same password!

How do I manage so many passwords? Products like 1password (1password.com) are available to both generate strong passwords and store them for you. There is only one key to open 1password and then it fills in all your passwords automatically.

If you don't want to purchase an additional product, web browsers provide password-management tools. When you go to a website that requires a login, the browser will display a dialog box asking whether you want the browser to remember the password for the site (see Figure 9.22). The next time you return to the site, it will automatically fill in the information for you. Consider combining this with two-factor authentication of access to your browser account to be most secure.

What other ways are there to log on securely to your device? There are a number of security options beyond typing a password. Windows allows you to use gestures on touchscreen devices. You select a picture and then draw three gestures on it—either straight lines, circles, or taps to access your device).

Figure 9.23 Modern mobile devices include iris scanning as an option for securing your device. *(Titima Ongkantong/Shutterstock)*

A biometric authentication device reads a unique personal characteristic such as a fingerprint or the iris pattern in your eye and converts its pattern to a digital code (see Figure 9.23). Because fingerprints and iris patterns are unique, these devices provide a high level of security. They also eliminate the

Figure 9.24 Windows Hello is "Looking for you..." because it uses facial recognition to log you into your laptop. *(JL IMAGES/Shutterstock)*
> *From the* ***Start*** *menu, open the* ***Settings*** *app. Go to* ***Accounts*** *and select* ***Sign-in options****. Select* ***Windows Hello Face****. (Windows 10, Microsoft Corporation)*

human error that can occur in password protection. You might forget your password, but you won't forget to bring your fingerprints when you're working on your computer!

Biometric verification is available on most devices now. Many smartphones include fingerprint readers and facial recognition systems. Windows 10 offers facial recognition for login, called Hello, on devices (see Figure 9.24) that have integrated cameras. Table 9.3 summarizes security measures in the marketplace.

Anonymous Web Surfing: Hiding from Prying Eyes

Objective 9.10 *Describe ways to surf the web anonymously.*

Should I be concerned about surfing the Internet on public computers? If you use shared computers in public places, such as libraries or coffee shops, you never know what nefarious tools have been installed by hackers. When you browse the Internet, traces of your activity are left behind on that computer, often as temporary files. A wily hacker can glean sensitive information long after you've finished your surfing session. In addition, many employers routinely review the Internet browsing history of employees to ensure that workers are spending their time on the Internet productively.

Table 9.3 Security Measures

Style	Devices	Advantages	Disadvantages
Password	Any	Can be long personal phrase	Takes time to type; must be updated often
Fingerprint	Smartphones, tablets	Unique; cannot be forgotten	Could possibly be lifted and used by another person
Gesture	Windows 10 devices, many mobile devices	Quick; hard to forget	Can be compromised if someone watches you log on
Facial recognition	iOS and Android devices; Windows	Quick; no need to remember anything	Some algorithms have trouble matching faces of people of certain ethnic groups
Iris scan	Android	Quick; unique; difficult to steal	Can be difficult if user wears eyeglasses

Figure 9.25 Modern browsers include privacy tools to support anonymous web surfing. *(Courtesy of Google Inc.)*

What tools can I use to keep my browsing activities private when surfing the Internet? The current versions of Mozilla Firefox, Microsoft Edge, and Google Chrome include privacy tools (called Private Browsing, InPrivate, and Incognito, respectively) that help you surf the web anonymously (see Figure 9.25). When you choose to surf anonymously, all three browsers open special versions of their browser windows that are enhanced for privacy. When surfing in these windows, records of websites you visit and files you download don't appear in the web browser's history files. Furthermore, any temporary files generated in that browsing session are deleted when you exit the special window.

Are there any other tools I could use to protect my privacy? If you find a need to work on public workstations, you may want to travel with a portable flash drive. Store sensitive Internet files, such as cookies, passwords, Internet history, and browser caches, on your personal device, not on the computer you're using. Look for USB flash drives that use encryption and password protection so your information is secure even if the drive is lost. There are also extensions for Chrome that allow you to clear the cache, your history, and any URLs you've typed with one click from the machine you're on. All traces of your Internet activity will be removed.

Is there anything else I can do to keep my data safe on shared computers? Another free practical solution is to take the Linux OS with you on a flash drive and avoid using the public or work computer's OS. The interfaces of many Linux distros look almost exactly like Windows and are easy to use. There are several advantages to using a Linux-based OS on a public or work computer:

- When you run software from your own storage medium, such as a flash drive, you avoid reading and writing to the hard disk of the computer. This significantly enhances your privacy because you don't leave traces of your activity behind.
- Your risk of picking up viruses and other malicious programs is significantly reduced because booting a computer from a flash drive completely eliminates any interaction with the computer's OS. This, in turn, significantly reduces the chance that your flash drive will become infected by any malware running on the computer.
- Virus and hacking attacks against Linux are far less likely than attacks against Windows. Because Windows has about 90 percent of the OS market, people who write malware tend to target Windows systems.

Pendrivelinux (*pendrivelinux.com*) is an excellent resource that offers many versions of Linux for download and includes step-by-step instructions on how to install them on your flash drive. If you're a Mac user, the Elementary OS Luna distro of Linux provides a close approximation of macOS, so you can feel right at home.

Bits&Bytes Multi-Factor Authentication: Don't Rely Solely on Passwords!

Computer system security depends on authentication—proving users are who they say they are. There are three independent authentication factors:

- Knowledge: something the user knows (password, PIN)
- Possession: something the user has (ATM card, mobile phone)
- Inherence: something only the user is (biometric characteristics, such as fingerprints or iris pattern)

Multifactor authentication requires two of these three above factors to be demonstrated before authorization is granted (see Figure 9.26). Commonly, a multifactor login would require you to know the password and then also the code sent to your mobile phone.

After you enter your password, the company will text a code to your phone (something you have) to serve as the second authentication step. Even this can be compromised when hackers exploit SIM jacking. If they know enough about you, a hacker can convince the phone provider to assign your phone number to a new SIM card, one the hacker owns. You may want to consider purchasing a physical key like the YubiKey as your second factor. You will then need to know your password and be required to insert your special USB key into your computer to log on to your accounts.

Multifactor authentication is much safer than single-factor authentication, so use it when it is offered to you to enhance your account security.

Figure 9.26 Multifactor Authentication

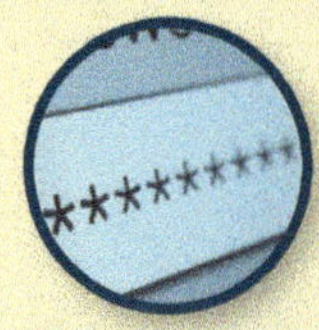

Something you KNOW:
Something the user knows
(password, PIN)

Something you HAVE:
Something the user has
(ATM card, mobile phone, Yubi key)

Something you ARE:
Something only the user is
(fingerprints, iris patterns)

Strong Authentication:
Two of the three factors

(Lolostock/Shutterstock; Jmiks/Shutterstock; Jamie Cross/123RF; Art Pencil Studio/Shutterstock)

How can I protect sensitive data transmissions if I have to use a public wireless network? Virtual private networks (VPNs) are secure networks that are established using the public Internet infrastructure. VPNs use specialized software, servers, and data transmission protocols to send information on the public Internet in such a manner that the data is as secure as sending it on a private network. VPNs were only used by businesses, but with public concerns about information security on the rise, many VPN software providers (such as Private Internet Access and NordVPN) are marketing affordable solutions to individuals. If you routinely transmit sensitive information, you should consider a personal VPN solution.

Make sure to use some of these methods to keep your activities from prying eyes and restrict access to your digital information.

Keeping Your Data Safe

People are often too trusting or just plain careless when it comes to protecting private information about themselves or their digital data. In this section, we discuss ways to keep your data safe from damage, either accidental or intentional.

Protecting Your Personal Information

Objective 9.11 *Describe the types of information you should never share online.*

If a complete stranger walked up to you on the street and asked you for your address and phone number, would you give it to him or her? Of course you wouldn't! But many people are much less

Information Identity Thieves Crave

- Social Security Number
- Full Date of Birth
- Phone Number
- Street Address

Never make this information visible on websites!

Other Sensitive Information

- Full Legal Name
- E-mail Address
- Zip Code
- Gender
- School or Workplace

Only reveal this information to people you know—don't make it visible to everyone!

Figure 9.27 Internet Information Sharing Precautions *(Pancaketom/123RF; Fredex8/123RF)*

careful when it comes to sharing sensitive information online. And often they inadvertently share information that they really only intended to share with their friends. With cybercrimes like identity theft rampant, you need to take steps to protect your personal information.

What information should I never share on websites? A good rule of thumb is to reveal as little information as possible, especially if the information would be available to everyone. Figure 9.27 gives you some good guidelines.

Your Social Security number, phone number, date of birth, and street address are four key pieces of information that identity thieves use to steal an identity. This information should never be shared in a public area on any website.

How can I tell who can see my information on a social network? Social networking sites like Facebook and Instagram make privacy settings available in your profile settings. If you've never changed your default privacy settings, you're probably sharing information more widely than you should.

How can I protect my information? To begin, consider changing your privacy settings in your profile from some of the default options. In general, it's a bad idea to make personal information available to the public. It's a good idea to consider using two-factor authorization as well, to reduce the chance of having your account hijacked.

Sound Byte MyLab IT

Managing Computer Security with Windows Tools

In this Sound Byte, you learn how to monitor and control your computer security, using features built into the Windows OS.

Backing Up Your Data

Objective 9.12 *List the various types of backups you can perform on your computing devices, and explain the various places you can store backup files.*

How might I damage my data? A hacker can gain access to your computer and steal your data, but a more likely scenario is that you'll lose your data unintentionally. You may accidentally delete files or drop your laptop on the ground, damaging the hard drive. You should have a strategy for backing up your files (see Table 9.4). Backups are copies of files that you can use to replace the originals if they're lost or damaged.

What types of files do I need to back up? Two types of files need backups:

1. Data files include files you've created or purchased, such as research papers, spreadsheets, music and photo files, contact lists, address books, e-mail archives, and your Favorites list from your browser.
2. Program files include files used to install software. Most manufacturers allow you to re-download the installation files if you need to reinstall the program, but some don't or charge you an extra fee for that service. Making sure you have your own backup of your system protects you in either case.

Table 9.4 An Effective Backup Strategy

Files to Back Up
• **Program files:** Installation files for productivity software (i.e., Microsoft Office) • **Data files:** Files you create (term papers, spreadsheets, etc.)

Types of Backups
• **Full backup:** All appliation and data files • **Image (system):** Snapshot of your entire computer, including system software

Where to Store Backup Files
• Online (in the cloud) • External hard drives • Network-attached storage devices or home servers

Table 9.5 A Comparison of Typical Data Backup Locations

Backup Location	Pros	Cons
Online (in the cloud) 	• Files stored at a secure, remote location • Files/backups accessible anywhere through a browser	• Most free storage sites don't provide enough space for image backups
External hard drive 	• Inexpensive, one-time cost • Fast backups with USB 3.0 devices connected directly to your computer	• Could be destroyed in one event (fire/flood) with your computer • Can be stolen • Slightly more difficult to back up multiple computers with one device
Network-attached storage device and home server 	• Makes backups much easier for multiple computing devices	• More expensive than a stand-alone external hard drive • Could be destroyed in one event (fire/flood) with your computer • Can be stolen

(Mipan/Shutterstock; Gleb Semenov/123RF; Prapass/Shutterstock)

What types of backups can I perform? There are two important types of backups:

1. A full backup means that you create a copy of all your application and data files. This is followed by a schedule of incremental backups (or partial backups). These involve only backing up files that have changed or have been created since the last backup was performed.
2. An image backup (or system backup) means that all system files are backed up, not just the application and data files. An image backup ensures that you capture a complete snapshot of everything that makes your computer run—the operating system, the applications, and the data. The idea of imaging is to make an exact copy of the setup of your computer so that in the event of a total hard drive failure, you can copy the image to a new hard drive and have your computer configured exactly the way it was before the crash.

Where should I store my backups? You have three main choices for where to back up your files (see Table 9.5).

1. *Online (in the cloud).* To be truly secure, backups should be stored online. Because the information is stored online, it's in a secure, remote location, so data isn't vulnerable to the disasters that could harm data stored in your home. Image backups may not fit within the storage limits offered for free by cloud providers. Look for a service that specifically provides mirror-image backups, which will include a copy of your full operating system. For example, Carbonite (carbonite.com) offers certain plans that include online storage of a full system backup.
2. *External hard drives.* External hard drives, or even large-capacity flash drives, are popular backup options. Although convenient and inexpensive, using external hard drives for backups still presents the dilemma of keeping the hard drive in a safe location. Also, external hard drives can fail, possibly leading to loss of your backed-up data. Therefore, using an external hard drive for backups is best done in conjunction with an online backup strategy for added safety.
3. *Network-attached storage (NAS) devices and home servers.* NAS devices are essentially large hard drives connected to a network of computers instead of to just one computer, and they can be used to back up multiple computers simultaneously. Home servers also act as high-capacity NAS devices for automatically backing up data and sharing files.

How do I create a backup of my data files? Use the Windows 10 File History utility (see Figure 9.28). You may want to connect an external hard drive to your computer or a NAS device (or home server) to your network to store your backup, or you may just want to back up to a USB flash drive, depending on how much data you have to store. Once configured, your data files will be backed up as often as you indicate. You can also restore files that you've backed up from the File History utility. With a few clicks you can retrieve any file back to your system.

How often should I back up my data files? File History can be set to save files you've changed every 10 minutes or every day automatically, whatever you select. File History keeps previous versions of the file on the backup drive so you can revert to an older copy of the file if you need to do so.

How do I create an image backup? Windows also includes the System Image backup utility, which provides a quick and easy way to perform image backups. A full system image backup includes more than just data files. It also has the Windows files and drivers to completely restore your operating system. You can access this utility from the Backup and Restore screen (see Figure 9.29a) by clicking *Create a system image*. Before starting this utility, make sure your external drive or NAS device is connected to your computer or network and is powered on. To create the image backup:

1. Click the *Create a system image* link. Select the location (drive) for your backup files and click *Next* to proceed.
2. On the second screen (see Figure 9.29b), you can select the drives (or partitions of drives) from your computer to be backed up. Notice that all the drives/partitions that are required for Windows to run are preselected for you. Windows will back up all data files and system files on all selected drives/partitions. Click *Next* to proceed.
3. On the third screen, click *Start backup* to start your system image.

After the system image backup runs for the first time, you'll see the results of the last backup and the date of the next scheduled backup on the Backup and Restore (Windows 7) screen (see Figure 9.29a). If the scheduled backup time is not convenient for you, click the *Change settings* link to select an alternative time.

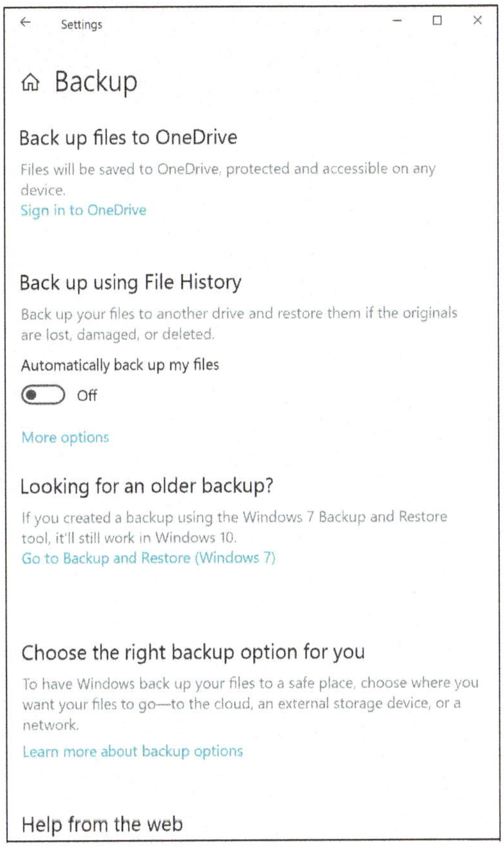

Figure 9.28 You can use the Windows File History utility to back up files and restore files from a previous backup. *(Courtesy of Microsoft Corporation) > From the Windows search bar, search for **Backup Settings**.*

From the Backup and Restore screen, you can also create a system repair disc, which contains files that can be used to boot your computer in case of a serious Windows error.

How often should I create an image backup? Because your program and OS files don't change as often as your data files, you can perform image backups on a less frequent basis. You might consider scheduling an image backup of your entire system on a weekly basis, but you should definitely perform one after installing new software.

What about backing up Apple computers? For macOS users, backups are very easy to configure. The Time Machine feature in macOS detects when an external hard drive is connected to the computer or a NAS device is connected to your network. You're then asked whether you want this to be your backup drive. If you answer yes, all of your files (including OS files) are automatically backed up to the external drive or NAS device.

Should I back up my files that are stored on my school's network? If you're allowed to store files on your school's network, most likely these files are backed up regularly. You should check with your school's network administrators to determine how often they're backed up and how you would request files to be restored from the backup if they're damaged or deleted. Don't rely on these network backups to bail you out if your data files are lost or damaged. It may take days for the network administrators to restore your files. It's better to keep backups of your data files yourself, especially homework and project files, so that you can immediately restore them. Storing your school projects in an organized portfolio with a cloud service like Dropbox or Google Drive is the best way to keep them safe and to be prepared for creating the résumé you'll need in the future.

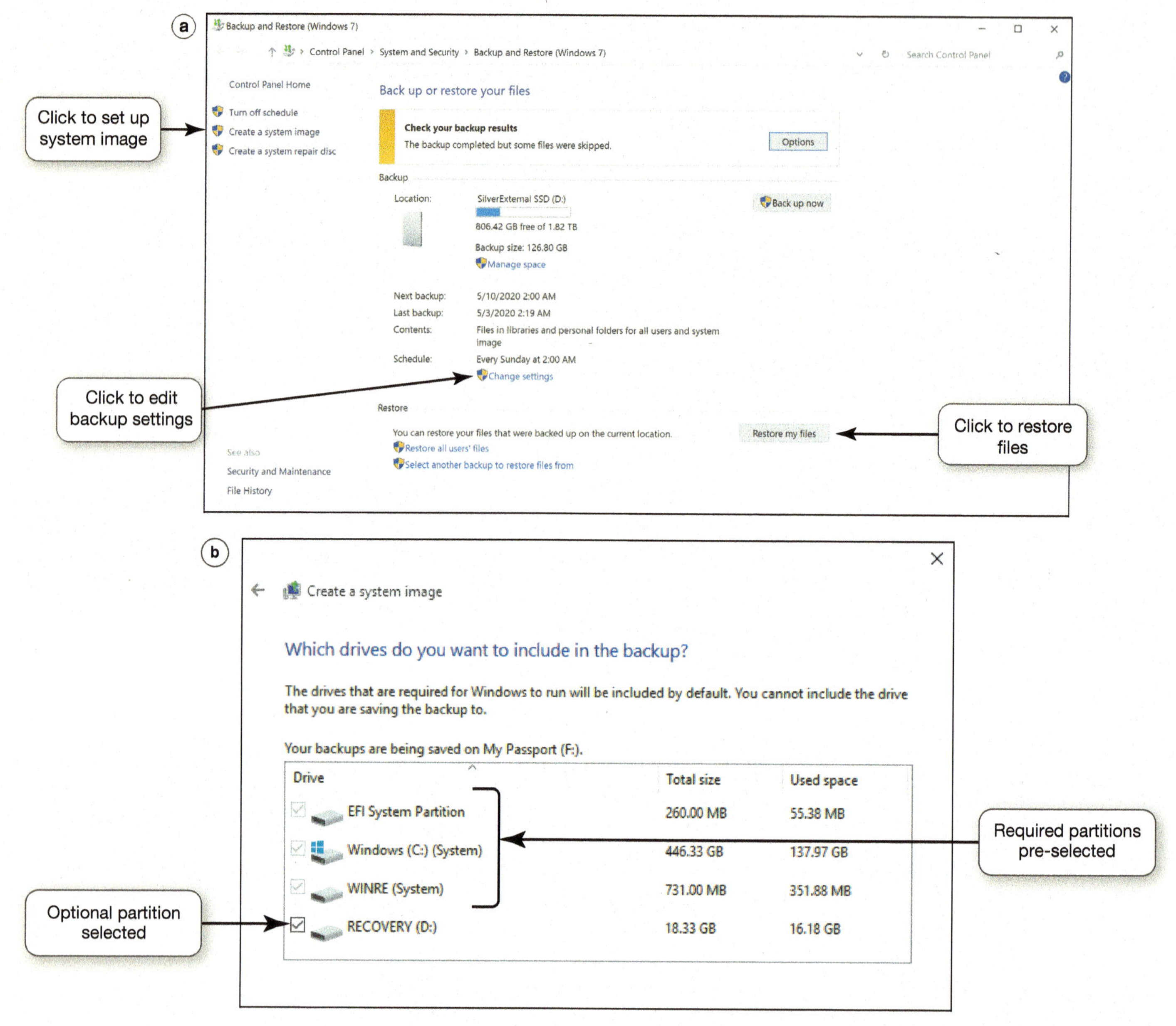

Figure 9.29 (a) The Backup utility allows you to customize many details of the backup regimen. (b) Restore allows you to select the folders or files to be pulled back from a specific date.
> *From the* ***Backup*** *screen (Figure 9.28), click* ***Go to Backup and Restore****. (Courtesy of Microsoft Corp.)*

Protecting Your Physical Computing Assets

It's essential to ensure a safe environment for all your computing devices. This includes protecting them from environmental factors, power surges, power outages, and theft.

Environmental Factors and Power Surges

Objective 9.13 *Explain the negative effects environment and power surges can have on computing devices.*

Why is the environment critical to the operation of my computer equipment? Sudden movements, such as a fall, can damage your computing device's internal components. Make sure that your computer sits on a flat, level surface, and if it's a laptop or a tablet, carry it in a protective case.

Electronic components don't like excessive heat or excessive cold. Don't leave computing devices and phones in a car during especially hot or cold weather because components can be damaged by extreme temperatures. Unfortunately, computers generate a lot of heat, which is why they have fans to cool their internal components. Chill mats that contain cooling fans and sit underneath laptop computers are useful accessories for dissipating heat. Make sure that you place your desktop computer where the fan's intake vents, usually found on the rear of the system unit, are unblocked so air can flow inside.

Naturally, a fan drawing air into a computer also draws in dust and other particles, which can wreak havoc on your system. Therefore, keep the room in which your computer is located as clean as possible. Finally, because food crumbs and liquid can damage keyboards and other computer components, consume food and beverages away from your computer.

How do I protect my devices from a power surge? Power surges occur when electric current is supplied in excess of normal voltage. Old or faulty wiring, downed power lines, malfunctions at electric company substations, and lightning strikes can all cause power surges. A surge protector is a device that protects your computer against power surges.

Figure 9.30 Surge protectors are critical for protecting your electronic devices. *(Ifeelstock/Alamy Stock Photo)*

Note that you should replace your surge protectors every two to three years. Also, after a major surge, the surge protector will no longer function and must be replaced. It's wise to buy a surge protector that includes indicator lights, which illuminate when the surge protector is no longer functioning properly. You may even want a whole-house surge protector to protect all electrical devices in the house.

What is a UPS? UPS stands for uninterruptible power supply. It acts as a battery backup if your power goes out. The UPS automatically switches in and allows you to shut down your devices carefully, without hardware failures. You can purchase surge protection and UPS backup in a single device (see Figure 9.30).

If you live in an area where there are often fluctuations in the actual voltage delivered, a UPS is important to protect the life of your electronic devices. If you are concerned about using your devices even if the power is down for long periods, consider a generator for the home or consider whole-house batteries like the Tesla Powerwall (see Figure 9.31).

Preventing and Handling Theft

Objective 9.14 *Describe the major concerns when a device is stolen and strategies for solving the problems.*

What if my computing device is stolen? The theft of tablets, smartphones, notebook computers, and other portable computing devices is always a risk. The resale value for used electronic equipment is high, and the equipment can be easily sold online. Because they're portable, laptops, tablets, and phones are easy targets for thieves. The main security concerns with mobile devices are:

1. Keeping them from being stolen
2. Keeping data secure in case they are stolen
3. Finding a device if it is stolen

What type of alarm can I install on my mobile device? Motion alarm software is a good, inexpensive theft deterrent. Apps such as Don't Touch My Phone help secure your device, for example alerting you if someone moves your device while you've left it charging at an outlet.

How can I secure the data on my mobile devices? The easiest way to keep the data secure is to make sure you have a login passcode set for your device. If your device supports biometrics like fingerprint detection for login, that's another way to make sure the contents are safe even if it's stolen.

Figure 9.31 A solar home can store energy in a battery like the Tesla Powerwall to use if there is a power outage. *(Widmann/CHL/Alamy Stock Photo)*

Even if somehow the thief gets entry into the device, you can protect the data from prying eyes. Encrypting the data on your mobile device can make it practically impossible for thieves to obtain the data. Encryption involves transforming your data by using an algorithm that can only be unlocked by a secure passcode. Encrypted data can't be read unless it's decrypted, which requires the passcode. If you encrypt your SD card, for example, it cannot be read if it is removed and placed in another device. On both iOS and Android devices, the operating system offers a built-in encryption option.

How can my device help me recover it when it is stolen? Software is available that enables your device to report back its current location. For iOS devices, Apple offers the Find My iPhone service. Enabling this service provides you with numerous tools to assist you in recovering and protecting your mobile devices. Did you forget where you left your iPad? Just sign in with your Apple ID at the iCloud website to see a map showing the location of your iPad. You can send a message to your device, remotely password-lock the device, or wipe all data from the device to protect your privacy completely. For Android devices, the Android Device Manager offers similar features. You can type "Find my phone" into the omnibox (the browser search bar) of Google Chrome, and a map with the location of your Android device will be shown (see Figure 9.32). For Windows 10 devices, a similar Find My Device command is located in the Update and Security section of Settings.

How can I ensure that I've covered all aspects of protecting my digital devices? Table 9.6 provides a guide to ensure that you haven't missed critical aspects of security. If you've addressed all of these issues, you can feel reasonably confident that your data and devices are secure.

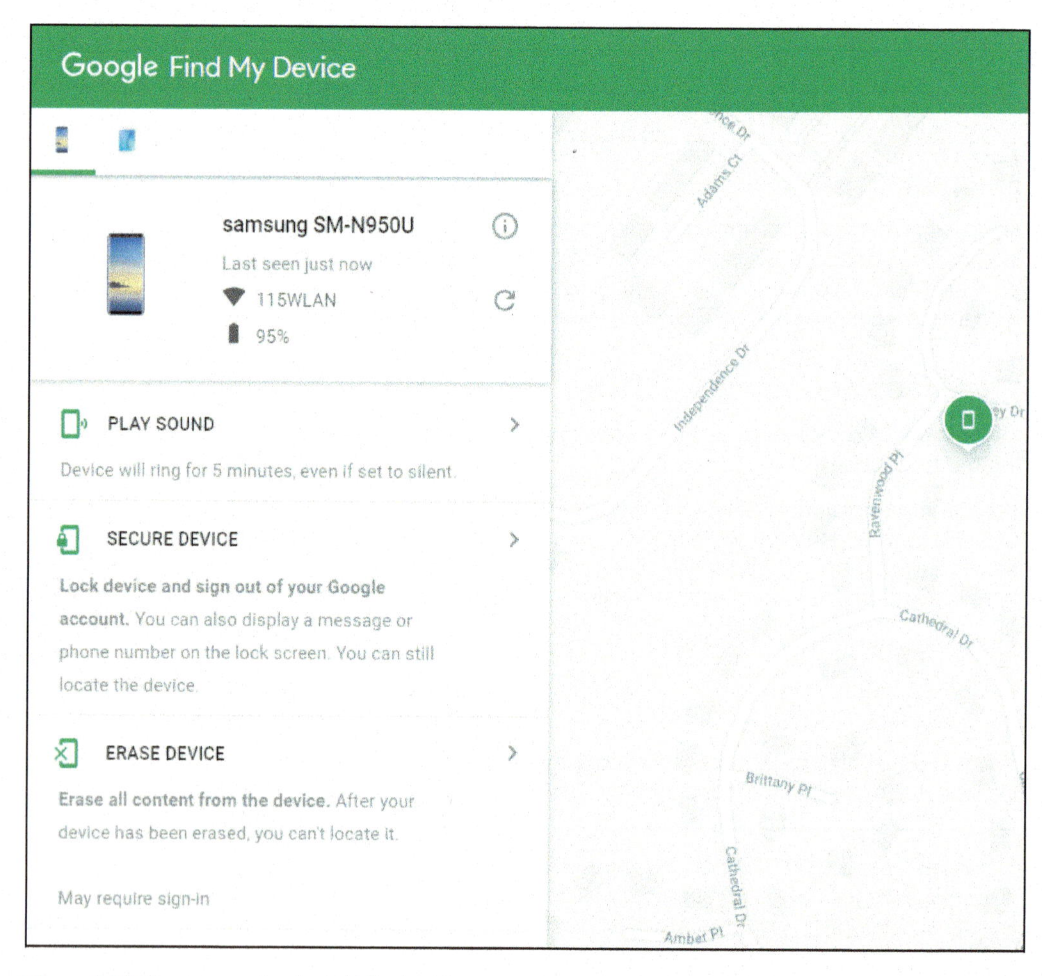

Figure 9.32 You can use the Android Device Manager to locate any Android device and take steps to find and return it to its owner. *(Courtesy of Google Inc.)*

Dig Deeper Digital Forensics: How It Works

On law enforcement TV shows, you often see computer technicians working on suspects' computers to assist detectives in solving crimes. It may look simple, but the science of digital forensics is a complex, step-by-step process that ensures that evidence is collected within the confines of the law.

Forensic means that something is suitable for use in a court of law. There are many branches of forensic science. For example, forensic pathologists provide evidence about the nature and manner of death in court cases involving deceased individuals. Digital forensics involves identifying, extracting, preserving, and documenting any kind of digital evidence. Digital forensics is performed by individuals known as *digital forensic scientists*, who rely primarily on specialized software to collect their evidence.

Phase 1: Obtaining and Securing Computer Devices

The first step in a digital forensics investigation is to seize any equipment that law enforcement officials believe contains pertinent evidence. Police are required to obtain a warrant to search an individual's home or place of business. Warrants must be very specific by spelling out exactly where detectives can search for evidence and exactly what type of evidence they're seeking. If a warrant indicates that the police may search an individual's home for his laptop computer, they can't then confiscate a digital doorknob camera. It is important to specify in the warrant all types of storage devices where potential evidence might be stored, such as external hard drives, flash drives, and servers as well as cloud services where Amazon Alexa recordings or security surveillance cameras may store recordings.

Once permission to collect the computers and devices containing possible evidence has been obtained, law enforcement officials must exercise great care when collecting the equipment. They need to ensure that no unauthorized persons can access or alter the computers or storage devices. The police must make sure the data and equipment are safe; if the equipment is connected to the Internet, the connection must be severed without data loss or damage. It's also important for law enforcement officials to understand that they may not want to power off equipment, because potential evidence contained in RAM may be lost. After the devices are collected and secured, the digital forensic scientists take over the next phase of the investigation.

Phase 2: Cataloging and Analyzing the Data

It's critical to preserve the data exactly as it was found, or attorneys may argue that the computer evidence was subject to tampering or altering. Because just opening a file can alter it, the first task is to make a copy of all computer systems and storage devices collected (see Figure 9.33). The investigators then work from the copies to ensure that the original data always remains preserved exactly as it was when it was collected.

In the era of cloud computing, there is another aspect of data collection: retrieving permissions to access data that smart devices around the scene may have stored. Many IoT (Internet of Things) devices are in an always-on state and record sounds, images, video, or even just access times. Evidence that a specific device was used at a certain time may vindicate, or indict, a potential suspect.

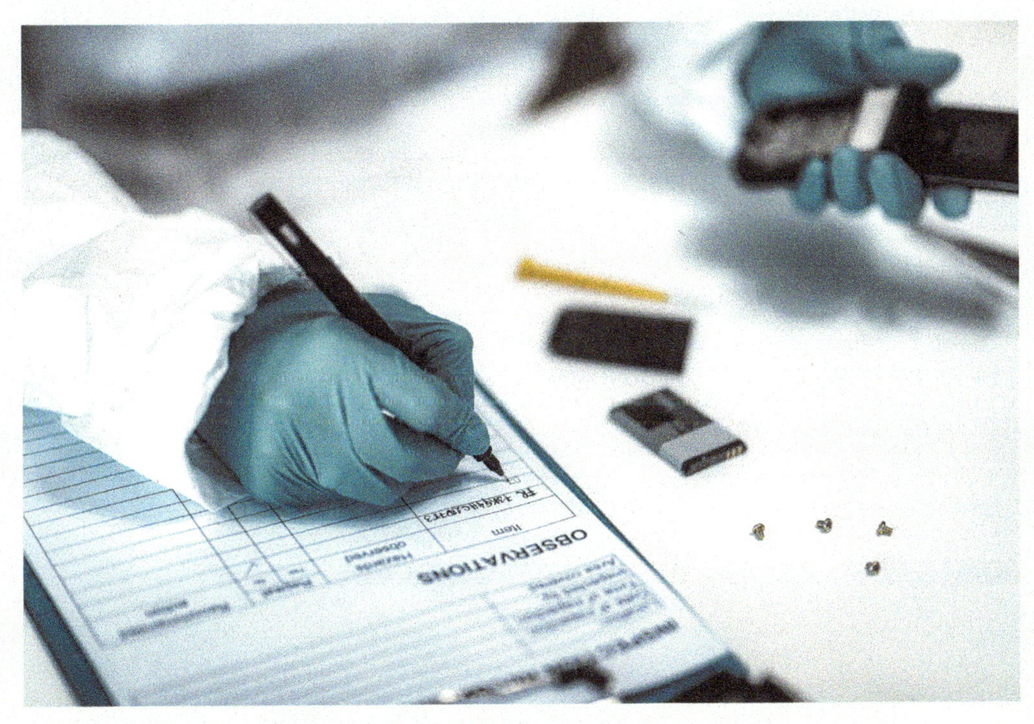

Figure 9.33 Digital forensics experts document evidence of criminal activity from digital devices. *(Microgen/Shutterstock)*

After obtaining a copy to work from, forensics professionals attempt to find every file on the system, including deleted files. Files on a computer aren't actually deleted, even if you empty the Recycle Bin, until the section of the hard disk they're stored on is overwritten with new data. Therefore, using special forensic software tools such as the SANS Investigative Forensic Toolkit (SIFT), forensic scientists catalog all files found on the system or storage medium and recover as much information from deleted files as they can. Software like the Forensic Toolkit (FTK) can readily detect hidden files and perform procedures to crack encrypted files or access-protected files and reveal their contents.

The most important part of the process is documenting every step. Forensic scientists must clearly log every procedure performed because they may be required to provide proof in court that their investigations did not alter or damage information contained on the systems they examined. Detailed reports should list all files found, how the files were laid out on the system, which files were protected or encrypted, and the contents of each file. Finally, digital forensic professionals are often called on to present testimony in court during a trial.

Criminals are becoming more sophisticated and are now employing anti-forensics techniques to foil investigators. Although techniques for hiding or encrypting data are popular, the most insidious anti-forensics techniques are programs designed to erase data if unauthorized persons (i.e., not the criminal) access a computer system or if the system detects forensics software in use. When investigators detect these countermeasures, they must often use creative methods and custom-designed software programs to retrieve and preserve the data.

Digital forensics is an invaluable tool to law enforcement in many criminal investigations, but only if the correct procedures are followed and the appropriate documentation is prepared.

Table 9.6 Computer Security Checklist

Firewall
• Do all your computers and tablets have firewall software installed and activated before connecting to the Internet? • Is your router also able to function as a hardware firewall? • Have you tested your firewall security by using the free software available at grc.com?
Virus and Spyware Protection
• Is antivirus and antispyware software installed on all your devices? • Is the antivirus and antispyware software configured to update itself automatically and regularly? • Is the software set to scan your device on a regular basis (at least weekly) for viruses and spyware?
Software Updates
• Have you configured your operating systems (Windows, macOS) to install new software patches and updates automatically? • Is other software installed on your device, such as Microsoft Office or productivity apps, configured for automatic updates? • Is the web browser you're using the latest version?
Protecting Your Devices
• Are all computing devices protected from electrical surges? • Do your mobile devices have alarms or tracking software installed on them?

Before moving on to the Chapter Review:

1. Watch Chapter Overview Video 9.2.
2. Then take the Check Your Understanding quiz.

Check Your Understanding//Review & Practice

For a quick review to see what you've learned so far, answer the following questions.

multiple choice

1. Firewalls work by closing ____ in your computer.
   a. software gaps
   b. logical ports
   c. logical doors
   d. backdoors
2. Antivirus software looks for ____ to detect viruses in files.
   a. virus signatures
   b. virus artifacts
   c. virus bots
   d. virus VPNs
3. ____ involve using a physical attribute such as a fingerprint for authentication.
   a. Backdoors
   b. Biometrics
   c. Rootkits
   d. Trojan horses
4. A VPN is
   a. a secure network created by using public, open networks.
   b. a temporary file that is created and immediately deleted.
   c. a style of multifactor authentication.
   d. a backup protocol.
5. Once you encrypt your SD card, those files
   a. are illegal.
   b. are compressed and can be stored in less space.
   c. cannot be read if the SD card is put in another phone.
   d. are backed up.

chew on this

To keep your data secure, you have many choices. How will you structure the combination of multifactor authentication, biometrics, and password managers to make sure your devices are safe from prying eyes? How would your answer be different if you ran a business of 20 people?

MyLab IT

Go to **MyLab IT** to take an autograded version of the *Check Your Understanding* review and to find all media resources for the chapter.

For the IT Simulation for this chapter, see MyLab IT.

9 Chapter Review

Summary

Part 1
Threats to Your Digital Assets

Learning Outcome 9.1 **You will be able to describe hackers, viruses, and other online annoyances and the threats they pose to your digital security.**

Identity Theft and Hackers

Objective 9.1 *Describe how identity theft is committed and the types of scams identity thieves perpetrate.*

- Identity theft occurs when a thief steals personal information about you and runs up debts in your name. Identity thieves can obtain information by stealing mail, searching through trash, or tricking people into revealing information over the phone or by e-mail. Identity thieves counterfeit existing credit cards, open new credit card and bank accounts, change the address on financial statements, obtain medical services, and buy homes with a mortgage in the victim's name.

Objective 9.2 *Describe the different types of hackers and the tools they use.*

- White-hat hackers break into systems for unmalicious reasons such as testing security to expose weaknesses. Black-hat hackers break into systems to destroy information or for illegal gain. Gray-hat hackers often break into systems just for the thrill or to demonstrate their prowess.
- Packet analyzers (sniffers) are programs used to intercept and read data packets, and they travel across a network. Trojan horses are programs that appear to be something else but are really a tool for hackers to access your computer. Backdoor programs and rootkits are tools hackers use to gain access to and take total control of a computer system. Denial-of-service attacks overwhelm computer systems with so many requests for data that legitimate users can't access the system.

Computer Viruses

Objective 9.3 *Explain what a computer virus is, why it is a threat to your security, how a computing device catches a virus, and the symptoms it may display.*

- A computer virus is a computer program that attaches itself to another computer program and attempts to spread to other computers when files are exchanged. Computer viruses can display annoying messages, destroy your information, corrupt your files, or gather information about you. Computers catch viruses when exposed to infected files. This can occur from downloading infected files, downloading and running infected e-mail attachments, or sharing flash drives that contain infected files. Symptoms of virus infection include (1) disappearing files or app icons; (2) the browser reset to an unusual home page; (3) odd messages, pop-ups, or images that appear; (4) data files that become corrupt; and (5) programs that stop working properly.

Objective 9.4 *List the different categories of computer viruses, and describe their behaviors.*

- Boot-sector viruses copy themselves onto the master boot record of a computer and execute when the computer is started. Logic bombs and time bombs are viruses triggered by the completion of certain events or by the passage of time. Worms can spread on their own without human intervention, unlike conventional viruses. Macro viruses lurk in documents that use macros (short series of commands that automate repetitive tasks). E-mail viruses access the address book of a victim to spread to the victim's contacts. Encryption viruses render files unusable by compressing them with complex encryption keys. A multi-partite virus is designed to infect multiple file types in an effort to fool the antivirus software that is looking for it. Polymorphic viruses periodically rewrite themselves to avoid detection. Stealth viruses temporarily erase their code and hide in the active memory of the computer.

Online Annoyances and Social Engineering

Objective 9.5 *Explain what malware, spam, and cookies are and how they impact your security.*

- Malware is software that has a malicious intent. Adware is software that displays sponsored advertisements in a section of your browser window or as

a pop-up box. Spyware collects information about you, without your knowledge, and transmits it to the owner of the program.
- Spam is unwanted or junk e-mail. Spim is unsolicited instant messages, which is also a form of spam. Spam filters in e-mail systems forward junk mail to their own folder. Never reply to spam or click unsubscribe links, because this usually just generates more spam.
- Cookies are small text files that some websites automatically store on your hard drive when you visit them. Cookies are usually used to keep track of users and personalize their browsing experience. Cookies do not pose a security threat, although some individuals view them as privacy violations.

Objective 9.6 *Describe social engineering techniques and explain strategies to avoid falling prey to them.*

- Social engineering is any technique that uses social skills to generate human interaction that entices individuals to reveal sensitive information. Phishing lures people into revealing personal information through bogus e-mails that appear to be from legitimate sources and direct people to scammer's websites. Pharming occurs when malicious code is planted on your computer, either by viruses or by the user visiting malicious websites, and then alters your browser's ability to find web addresses. Ransomware is a type of malware that attacks your system by encrypting critical files so the system is no longer operational; criminals then demand payment before restoring the system. Scareware attempts to convince you that your computer is infected with a virus and then tries to sell you a solution. Most Internet security software packages have built-in protection to help avoid these dangers.

Part 2
Protecting Your Digital Property

Learning Outcome 9.2 **Describe various ways to protect your digital property and data from theft and corruption.**

Restricting Access to Your Digital Assets

Objective 9.7 *Explain what a firewall is and how a firewall protects your computer from hackers.*

- A firewall is a software program or hardware device designed to protect computers from hackers. Firewalls block access to your computer's logical ports and help keep your computer's network address secure. Firewalls use packet filtering to identify packets sent to specific logical ports and discard them. Firewalls use network address translation to assign internal IP addresses on networks. The internal IP addresses are much more difficult for hackers to detect.

Objective 9.8 *Explain how to protect your computer from virus infection.*

- Antivirus software is specifically designed to detect viruses and protect your computer and files from harm. It detects viruses by looking for virus signatures: code that specifically identifies a virus. It also stops viruses from executing and quarantines infected files in a secure area on the hard drive. Virus software must be constantly updated to remain effective.

Objective 9.9 *Describe how passwords and biometric characteristics can be used for user authentication.*

- Utilities built into web browsers and Internet security software can be used to manage your passwords.
- A biometric authentication device reads a unique personal characteristic to identify an authorized user. Fingerprint readers, iris scanners, and facial recognition software are common examples of biometric security devices.

Objective 9.10 *Describe ways to surf the web anonymously.*

- Privacy tools built into web browsers help you surf anonymously by not recording your actions in the history files. USB devices containing privacy tools are available that prevent the computer you are using from storing any information about you. Virtual private networks (VPNs) are secure networks that can be used to send information securely across the public Internet.

Keeping Your Data Safe

Objective 9.11 *Describe the types of information you should never share online.*

- Reveal as little information as possible about yourself. Your Social Security number, phone number, date of birth, and street address are four key pieces of information that identity thieves need to steal an identity.

Objective 9.12 *List the various types of backups you can perform on your computing devices, and explain the various places you can store backup files.*

- An incremental backup involves backing up only files that have changed or have been created since the last full backup was performed. An image backup (or system backup) means that all system, application, and data files are backed up, not just the files that changed. You can store backups online (in the cloud), on external hard drives, or on network-attached storage (NAS) devices.

Protecting Your Physical Computing Assets

Objective 9.13 *Explain the negative effects environment and power surges can have on computing devices.*

- Computing devices should be kept in clean environments free from dust and other particulates and should not be exposed to extreme temperatures (either hot or cold). You should protect all electronic devices from power surges by hooking them up through surge protectors, which will protect them from most electrical surges that could damage the devices.

Objective 9.14 *Describe the major concerns when a device is stolen and strategies for solving the problems.*

- The four main security concerns regarding computing devices are (1) keeping them from being stolen, (2) keeping data secure in case they are stolen, (3) finding a device if it is stolen, and (4) remotely recovering and wiping data off a stolen device.
- Software is available for installation on devices that will (1) set off an alarm if the device is moved; (2) help recover the device, if stolen, by reporting the computer's whereabouts when it is connected to the Internet; and (3) allow you to lock or wipe the contents of the device remotely.

Be sure to check out **MyLab IT** for additional materials to help you review and learn. And don't forget to watch the Chapter Overview Videos.

Key Terms

adware **337**
antivirus software **349**
backdoor program **330**
backup **357**
biometric authentication device **353**
black-hat hacker **329**
boot-sector virus **335**
botnet **331**
cookie **339**
cybercrime **328**
cyberloafing (cyberslacking) **341**
data breach **343**
data file **357**
denial-of-service (DoS) attack **331**
digital forensics **363**
distributed denial-of-service (DDoS) attack **331**
drive-by download **351**
e-mail virus **335**
encryption virus **335**
exploit kits **332**
firewall **347**
full backup **358**
gray-hat hacker **329**
hacker **329**
identity theft **329**
image backup (system backup) **358**
incremental backup (partial backup) **358**
inoculation **350**
keystroke logger (keylogger) **338**
logic bomb **335**
logical port **332**
logical port blocking **348**
macro virus **335**
malware **336**
master boot record **335**
multi-factor authentication **356**
multi-partite virus **336**
network address translation (NAT) **349**
packet **330**
packet analyzer (sniffer) **330**
packet filtering **348**
personal firewall **347**
pharming **342**
phishing **341**
polymorphic virus **336**
pretexting **340**
program file **357**
quarantining **350**
ransomware **342**
RaaS (Ransomware as a Service) **342**
rootkit **330**
scareware **342**
script **335**
social engineering **340**
spam **338**
spam filter **338**
spear phishing **343**
spyware **338**
stealth virus **336**
surge protector **361**
time bomb **335**
Trojan horse **330**
virtual private network (VPN) **356**
virus **333**
virus signature **350**
white-hat hacker (ethical hacker) **329**
whole-house surge protector **361**
worm **335**
zombie **330**

Chapter Quiz//Assessment

For a quick review to see what you've learned, answer the following questions. Submit the quiz as requested by your instructor. If you're using MyLab IT, the quiz is also available there.

multiple choice

1. Hackers
   - **a.** illegally break into a computer or network.
   - **b.** use VPNs to gather information.
   - **c.** are security experts in digital forensics.
   - **d.** are another name for a zombie attack.
2. Viruses that load from USB drives left connected to computers when computers are turned on are known as
   - **a.** script viruses.
   - **b.** polymorphic viruses.
   - **c.** encryption viruses.
   - **d.** boot-sector viruses.
3. A packet analyzer is called a sniffer because
   - **a.** it filters out spam.
   - **b.** it starts a denial-of-service attack.
   - **c.** it captures and reads data inside each packet.
   - **d.** it detects high-speed transmissions.
4. Ransomware as a Service has allowed the increase of
   - **a.** sharing flash drives.
   - **b.** using software by subscription.
   - **c.** ransomware attacks.
   - **d.** getting malware protection from the cloud.
5. A is a program that takes complete control of your computer without your knowledge.
   - **a.** botnet
   - **b.** rootkit
   - **c.** VPN
   - **d.** sniffer
6. Multifactor authentication requires you to have a combination of
   - **a.** your password and your PIN.
   - **b.** something you know, something you are, and something you own.
   - **c.** methods of payment.
   - **d.** gestures.
7. Authentication techniques that rely on personal biological traits are called
   - **a.** Trojan horses.
   - **b.** scanners.
   - **c.** botnets.
   - **d.** biometrics.
8. Antivirus software prevents infection by recording key attributes about your files and checking to see whether they change over time in a process called
   - **a.** updating.
   - **b.** quarantine.
   - **c.** signature.
   - **d.** inoculation.
9. A surge protector may look like just another power strip, but it also can
   - **a.** detect viruses.
   - **b.** prevent theft.
   - **c.** save your device if a voltage spike is sent through the electrical system.
   - **d.** help manage heat.
10. Firewalls use a process of to assign IP addresses to the devices internal to the network so hackers will not know what they are.
    - **a.** packet filtering
    - **b.** port blocking
    - **c.** VPN
    - **d.** network address translation (NAT)

true/false

_____ **1.** The logical port of your computer is always port 8000.

_____ **2.** Your Social Security number should never be shared on a website, in an e-mail, or through messaging.

_____ **3.** Sending e-mail to lure people into revealing personal information is a technique known as phishing.

_____ **4.** Encrypting data is not an appropriate measure for mobile devices such as smartphones.

_____ **5.** Virtual private networks make it just as secure to send information across a public network as it is on a secure private network.

What do you think now?

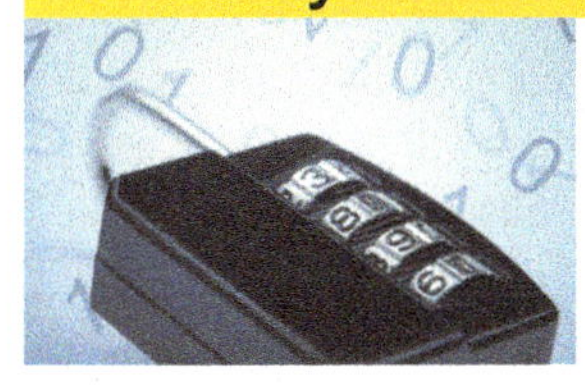

1. After reading this chapter, are you planning to use two-factor authentication? On which sites or products?
2. Check your mobile devices for the types of bio-authentication they provide. Are you using these in place of traditional passcodes?

Team Time

Protecting a Network

problem

Along with easy access to the web comes the danger of theft of digital assets.

task

A school alumnus is in charge of the county government computer department. The network contains computers running Windows 10, Windows 8, Windows 7, and macOS. He asked your instructor for help to ensure that his computers and network are protected from viruses, malware, and hackers. He is hoping that there may be free software available that can protect his employees' computers.

process

1. Break the class into three teams. Each team will be responsible for investigating one of the following issues:
   a. **Antivirus software.** Research alternatives that protect computers from viruses. Find three alternatives and support your recommendations with reviews that evaluate free packages.
   b. **Anti-malware software.** Research three free anti-malware alternatives and determine whether the software can be updated automatically. You may want to recommend that the county purchase software to ensure that a minimum of employee intervention is needed to keep it up to date.
   c. **Firewalls.** Determine whether the firewall software provided with Windows 10, Windows 8, Windows 7, and macOS is reliable. If it is, prepare documentation (for all four OSs) for county employees to determine whether their firewalls are properly configured. If additional firewall software is needed, research free firewall software and locate three options that the county can deploy.
   d. Present your findings to the class and provide your instructor with a report suitable for eventual presentation to the manager of the county office network.

conclusion

With the proliferation of viruses and malware, it is essential to protect computers and networks. Free alternatives might work, but you should research the best protection solution for your situation.

Ethics Project

Content Control: Censorship to Protect Children

In this exercise, you research and then role-play a complicated ethical situation. The role you play might or might not match your own personal beliefs; in either case, your research and use of logic will enable you to represent the view assigned. An arbitrator will watch and comment on both sides of the arguments, and together the team will agree on an ethical solution.

problem

Many parents use web-filtering software (content-control software) to protect their children from objectionable Internet content. The US government requires libraries to use content-control software as a condition to receive federal funds. Some states, such as Virginia, have passed laws requiring libraries to install filtering software even if they did not receive federal funds. Upon installation of the software, it's up to the library administrators to decide what content is restricted. Therefore, content restriction can vary widely from library to library.

research areas to consider

- US Supreme Court case *United States v. American Library Association*
- Content-control software and First Amendment rights
- Violation of children's free speech rights
- Children's Internet Protection Act (CIPA)

process

1. Divide the class into teams. One team will play the role of parents of elementary school students. Another team will represent the library administration, and another will be the users of the local town library.
2. Each team will be presenting its position on how access to the Internet should be managed for the local library, where there are many children, as well as adults, using services. The team position needs to be grounded in more than just opinion. Research so your position is supported by specific data, facts, and examples. Consider both the benefits and challenges of your position and the impact it will have on the other groups' interests.
3. Team members should present their case to the class or submit a PowerPoint presentation for review by the class. Be sure to include a summary statement of your position.

conclusion

As technology becomes ever more prevalent and integrated with our lives, more and more ethical dilemmas will present themselves. Being able to understand and evaluate both sides of the argument, while responding in a personally or socially ethical manner, will be an important skill.

Solve This with Word

MyLab IT Grader

Computer Security

You have been asked to prepare a report on computer security, using Microsoft Word. You have written the majority of the report, but you have to make some final modifications and refinements such as adding the results of some research of good antivirus software, adding a cover page, and generating a table of contents.

You will use the following skills as you complete this activity:

- Create and Modify Tables
- Create and Update Table of Contents
- Insert Tab Stops
- Add Footnote
- Add Watermark
- Use Find and Replace
- Insert Cover Page

Instructions

1. Open *TIA_Ch9_Start.docx* and save as **TIA_Ch9_LastFirst.docx**.
2. Find all instances of *e-mail* and replace them with **e-mail** (do not apply bold formatting).
   Hint: Click Replace in the Editing group on the Home tab. Use e-mail as the Find term, and use e-mail as the Replace term. Click Replace All. You will make four replacements.
3. Find the first instance of malware. Place the cursor after the word *malware* (before the period) and insert the footnote: **Malware is defined as software that is intended to gain access to, or damage or disable, computer systems for the purposes of theft or fraud.** Close the Navigation pane.
   Hint: Use Find in the Editing group on the **Home** tab to locate malware.
   Hint: Click Insert Footnote in the Footnotes group on the References tab.
4. In the *Types of Viruses* section, highlight the six lines of text that outline the categories of computer viruses and variants. Add a **Right Align Tab Stop** at 2".
   Hint: To set the tab stop, display the ruler and select tab stop style from the Select tab box to the left of the ruler; click at the desired position on the ruler.
5. Place the cursor at the end of the paragraph in the *Antivirus Software* section. Press **Enter** and then insert a **3 × 4 Table**. Type **Product Name, Description/Review**, and **Cost** in the top three cells. Adjust the width of the Description/Review column to 3.5" and the width of the Cost column to 1".
   Hint: Click Table in the Tables group on the Insert tab and drag to select the desired grid.
6. Add a row at the top of the table, **Merge Cells**. Type **Antivirus Software Reviews** and **Align Center** the contents. Format the table with **Grid Table 4 - Accent 1** style.
   Hint: Click Insert above in the Rows & Columns group on the **Table Tools Layout** tab.
7. Open a browser and go to www.pcmag.com/reviews/antivirus. Research four antivirus software programs and place the software name, review, and cost of the software in the respective columns in the table.
   Hint: Press Tab at the end of the third row to add an additional line to the table to accommodate a fourth review.
8. Press **Ctrl + Home** and then insert the **Banded Cover Page**. Ensure that *Computer Security* appears as the title, and *your name* appears as the Author. **Delete** the Company and Address placeholders.
9. Insert **Page Numbers** at the bottom of the page, using the **Plain Number 3** format. Ensure that Different First Page is selected. Close Header and Footer.
10. On page 2 of the document, insert a **Page Break** before the report title, *Computer Security*.
11. Place the cursor at the top of the new blank page and insert **Table of Contents**, using the Contents format.
    Hint: Click Table of Contents in the Table of Contents group on the References tab.
12. Press **Ctrl + End**, scroll up and change the heading style of *Firewalls* to **Heading 2**, change the heading style of *Software Firewalls* and *Hardware Firewalls* to **Heading 3**.
13. Update the Table of Contents to reflect the changes in headings; ensure that you update the entire table.
    Hint: Click anywhere in the Table of Contents. Click Update table and then click Update entire table.
14. Add a **Draft 1 Watermark** to the report.
    Hint: Click Watermark in the Page Background group on the Design tab.
15. Save the document and submit it, based on your instructor's directions.

Chapter

10 Behind the Scenes: Software Programming

For a chapter overview, watch the Chapter Overview Videos.

PART 1

Understanding Programming

Learning Outcome 10.1 **You will be able to describe the life cycle of a software project and identify the stages in the program development life cycle.**

Life Cycle of an Information System 374

Objective 10.1 *Describe the importance of programming to both software developers and users.*

Objective 10.2 *Summarize the stages of the system development life cycle.*

Life Cycle of a Program 376

Objective 10.3 *Define programming and list the steps in the program development life cycle.*

Objective 10.4 *Describe how programmers construct a complete problem statement from a description of a task.*

Objective 10.5 *Explain how programmers use flow control and design methodologies when developing algorithms.*

Objective 10.6 *Discuss the categories of programming languages and the roles of the compiler and the integrated development environment in coding.*

Objective 10.7 *Identify the role of debugging in program development.*

Objective 10.8 *Explain the importance of testing and documentation in program development.*

Sound Byte: Using the Arduino Microcontroller

Helpdesk: Understanding Software Programming

PART 2

Programming Languages

Learning Outcome 10.2 **You will understand the factors programmers consider when selecting an appropriate programming language for a specific problem and will be familiar with some modern programming languages.**

Many Programming Languages 396

Objective 10.9 *Discuss the driving factors behind the popularity of various programming languages.*

Objective 10.10 *Summarize the considerations in identifying an appropriate programming language for a specific setting.*

Sound Byte: Programming with the Processing Language

Exploring Programming Languages 399

Objective 10.11 *Compare and contrast modern programming languages.*

Objective 10.12 *State key principles in the development of future programming languages.*

Helpdesk: A Variety of Programming Languages

All media accompanying this chapter can be found here.

(Dizain/Shutterstock; Dondesigns/DigitalVision Vectors/Getty Images; Stuart Kinlough/Alamy Stock Photo; Geozacho/123RF)

What do you think?

Programming has had the aura of a cryptic, mysterious skill only open to those with a deep understanding of logic and mathematics. It demanded learning strange languages that were hard to read and very tricky to master. That is changing as programming matures as a science. **Visual programming** languages remove much of the burden of memorizing weird syntaxes and allow the programmer to drag and drop visual objects to accomplish sophisticated tasks. The use of standardized **APIs** (application programming interfaces) allows programmers to work at a higher level of thinking, with much of the detail work packaged up in a single function call. Web programming **frameworks** make it much faster for a new programmer to produce complicated sites and analyze how they are used.

The need for programming is changing as the **Internet of Things** grows and we have almost constant access to the Internet throughout our day. What impact will this have on you?

Which of the following do you feel will be important for you over the next 10 years?

- *Knowing how to use software but not how to program*
- *Knowing how to program but only to customize software (e.g., macros in Word)*
- *Knowing how to program but only because I'll use it in my career*
- *Knowing how to program because I'll use it in my career and personal life*

See the end of the chapter for a follow-up question.

(Oleksiy Maksymenko/Alamy Stock Photo)

Part 1

For an overview of this part of the chapter, watch **Chapter Overview Video 10.1**.

Understanding Programming

Learning Outcome 10.1 You will be able to describe the life cycle of a software project and identify the stages in the program development life cycle.

Every day, we face a wide array of tasks. Some tasks are complex and require creative thought and a human touch. However, tasks that are repetitive, work with electronic information, and follow a series of clear steps are candidates for automation with computers—automation achieved through programming. In this part of the chapter, we look at some of the basics of programming.

Life Cycle of an Information System

A career in programming offers many advantages—jobs are plentiful, salaries are strong, and telecommuting is often easy to arrange. Even if you're not planning a career in programming, knowing the basics of programming is important to help you use computers productively.

The Importance of Programming

Objective 10.1 *Describe the importance of programming to both software developers and users.*

Figure 10.1 Drones are playing a larger role in transportation, videography, agriculture, and other fields. They require programming to integrate sensor and geographic information. *(Suwin Puengsamrong/Alamy Stock Photo)*

Why would I ever need to create a program? Computer programs already exist for many tasks. However, for users who can't find an existing software product to accomplish a task, programming is mandatory. For example, imagine that a farmer needs a drone to fly a specific pattern to inspect his or her crops, automatically taking photos and sensor readings at specific points to create a weekly report. No software product exists that will accumulate and relay information in just this manner. Therefore, a team of programmers will have to create smart farming software (see Figure 10.1).

If I'm not going to be a programmer, why should I learn about programming? If you plan to use only existing, off-the-shelf software, having a basic knowledge of programming enables you to add features that support your personal needs. For example, most modern software applications enable you to automate features by using custom-built mini-programs called macros. By creating macros, you can execute a complicated sequence of steps with a single command. Understanding how to program macros enables you to add custom commands to, for example, Word or Excel automate frequently performed tasks, providing a huge boost to your productivity. If you plan to create custom applications from scratch, having a detailed knowledge of programming will be critical to the successful completion of your projects.

System Development Life Cycle

Objective 10.2 *Summarize the stages of the system development life cycle.*

What is an information system? Generally speaking, a system is a collection of pieces working together to achieve a common goal. Your body, for example, is a system of muscles, organs, and other groups of cells working together. An information system includes data, people, procedures, hardware, and software that help in planning and decision making. Information systems help run an office and coordinate online-purchasing systems. They are behind database-driven applications that Amazon and Netflix use, but how are these complicated systems developed? Through a particular step-by-step process.

Why do I need a process to develop a system? Because teams of people are required to develop information systems, an organized process is necessary to ensure that development proceeds in an orderly fashion. Software applications also need to be available for multiple operating systems, to work over networked environments, and to be free of errors and well supported. Therefore, a process often referred to as the system development life cycle (SDLC) is used.

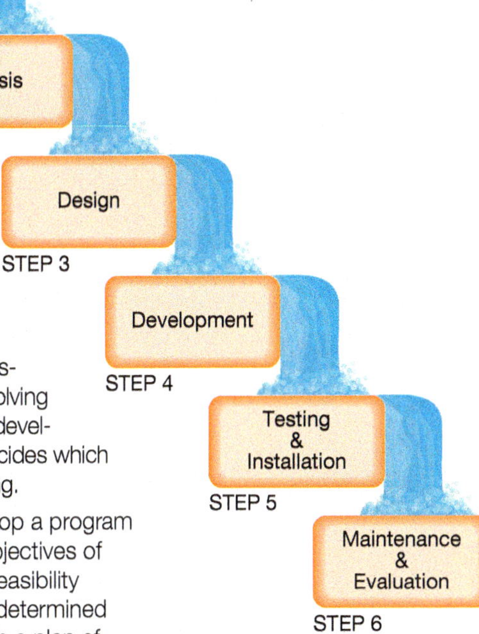

Figure 10.2 Each step of the system development life cycle must be completed before you can progress to the next.

What steps constitute the SDLC? There are six steps in a common SDLC model, as shown in Figure 10.2. This system is sometimes referred to as a waterfall system because each step depends on completion of the previous step before it can be started.

1. *Problem and opportunity identification.* Corporations are always attempting to break into new markets, develop new customers, or launch new products. At other times, systems development is driven by a company's desire to serve its existing customers more efficiently or to respond to problems with a current system. Whether solving an existing problem or exploiting an opportunity, large corporations typically form a development committee to evaluate systems development proposals. The committee decides which projects to take forward based on available resources such as personnel and funding.
2. *Analysis.* In this phase, analysts explore the problem or need in depth and develop a program specification. The program specification is a clear statement of the goals and objectives of the project. The first feasibility assessment is also performed at this stage. The feasibility assessment determines whether the project should go forward. If the project is determined to be feasible, the analysis team defines the user requirements and recommends a plan of action.
3. *Design.* The design phase generates a detailed plan for programmers to follow. The proposed system is documented using flowcharts and data-flow diagrams. Flowcharts are visual diagrams of a process, including the decisions that need to be made along the way. Data-flow diagrams trace all data in an information system from the point at which data enters the system to its final resting place (storage or output). The data-flow diagram in Figure 10.3 shows the flow of concert ticket information.

   The design phase details the software, inputs and outputs, backups and controls, and processing requirements of the problem. Once the system plan is designed, a company evaluates existing software packages to determine whether it needs to develop a new piece of software or can buy something already on the market and adapt it to fit its needs.

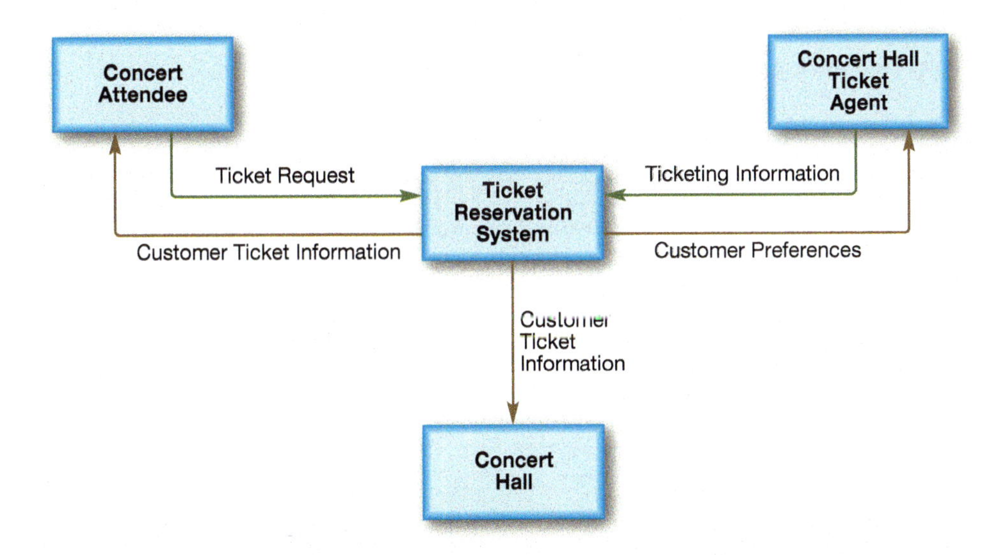

Figure 10.3 Data-flow diagrams illustrate the way data travels in an information system.

Bits&Bytes **The Agile Scrum**

A software methodology named *agile* has appeared as an alternative to the waterfall-style SDLC. Small teams are formed and build a list of top priority items for the project. They select the first few and sprint to finish the coding in a fixed window of time. Daily check-in meetings called the scrum are held to keep communication lines open. The team reviews how anything they have learned might change the project and then returns to the list to select the next round of priority items to code. With this methodology, testing is done in every pass, not just at the end as in the waterfall approach, making it agile.

4. *Development.* It is during this phase that actual programming takes place. This phase is also the first part of the program development life cycle, described in detail later in the chapter. The documentation work is begun in this phase by technical writers.
5. *Testing and installation.* Testing the program ensures that it works properly. It is then installed for official use.
6. *Maintenance and evaluation.* In this phase, program performance is monitored to determine whether the program is still meeting the needs of end users. Errors that weren't detected in the testing phase but that are discovered during use are corrected. Additional enhancements that users request are evaluated so that appropriate program modifications can be made.

Can the SDLC vary from these steps? The waterfall model is an idealized view of software development. Most developers follow some variation of it. For example, a design team may spiral, when a group that's supporting the work of another group will work concurrently with the other group on development. This contrasts with workflows in which the groups work independently, one after the other. Often, there is a backflow up the waterfall because even well-designed projects can require redesign and specification changes midstream.

Some people criticize the waterfall model for taking too long to provide actual working software to the client. This may contribute to scope creep, an ever-changing set of requests from clients for additional features as they wait longer and longer to see a working prototype. Other developmental models are being used in the industry to address these issues (see the Bits & Bytes, "The Agile Scrum").

Life Cycle of a Program

Programming takes place during the fourth step in the SDLC and requires its own step-by-step process.

The Program Development Life Cycle

Objective 10.3 *Define programming and list the steps in the program development life cycle.*

How do programmers tackle a programming project? Programming is the process of translating a task into a series of commands that a computer will use to perform that task. It involves identifying which parts of a task a computer can perform, describing those tasks in a highly specific and complete manner, and, finally, translating this description into the language spoken by the computer's central processing unit (CPU).

Programming often begins with nothing more than a problem or a request, such as, "Can you tell me how many transfer students have applied to our college?" A proposal will then be developed for a system to solve this problem, if one doesn't already exist. Once a project has been deemed feasible and a plan is in place, the work of programming begins.

What are the steps in the programming process? Each programming project follows several stages, from conception to final deployment. This process is sometimes referred to as the program development life cycle (PDLC):

1. *Describing the problem.* First, programmers must develop a complete description of the problem. The problem statement identifies the task to be automated and describes how the software program will behave.

2. *Making a plan.* The problem statement is next translated into a set of specific, sequential steps that describe exactly what the computer program must do to complete the work. The steps are known as an algorithm. At this stage, the algorithm is written in natural, ordinary language (such as English).
3. *Coding.* The algorithm is then translated into programming code, a language that is friendlier to humans than the 1s and 0s that the CPU speaks but is still highly structured. By coding the algorithm, programmers must think in terms of the operations that a CPU can perform.
4. *Debugging.* The code then goes through a process of debugging in which the programmers repair any errors found in the code.
5. *Testing and documentation.* The software is tested by both the programming team and the people who will use the program. The results of the entire project are documented for the users and the development team. Finally, users are trained so that they can use the program efficiently.

Figure 10.4 illustrates the steps of a program development life cycle.

Now that you have an overview of the process involved in developing a program, let's look at each step in more detail.

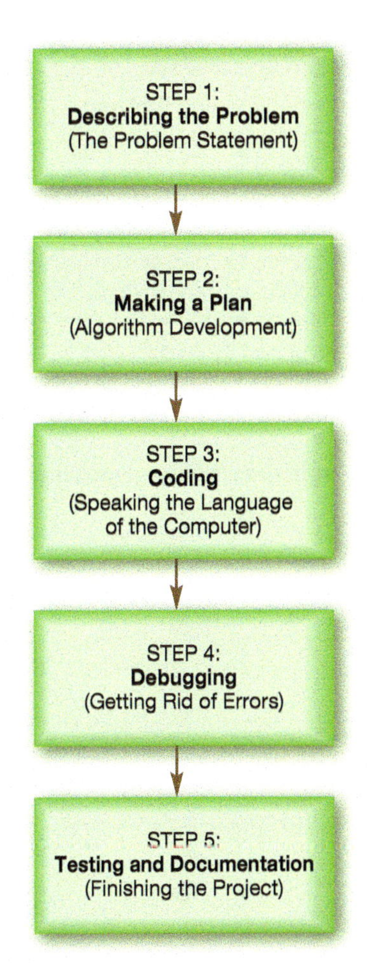

Figure 10.4 The stages that each programming project follows, from conception to final deployment, are collectively referred to as the program development life cycle.

Using the Arduino Microcontroller

In this Sound Byte, you program the Arduino microcontroller to control LED lights and motors.

The Problem Statement

Objective 10.4 *Describe how programmers construct a complete problem statement from a description of a task.*

Why is a problem statement necessary? The problem statement is the starting point of programming work. It's a clear description of what tasks the computer program must accomplish and how the program will execute those tasks and respond to unusual situations. Programmers develop problem statements so they can understand the goals of their programming efforts better.

What kinds of problems can computer programs solve? Tasks that are repetitive, that work with electronic information, and that follow a series of clear steps are good candidates for computerization. This might sound as if computers help us with only the dullest and most simplistic tasks. However, many sophisticated problems can be broken down into a series of easily computerized tasks. For example, pharmaceutical companies design drugs and vaccines by using complex computer programs that model molecules. Using simulation software to perform dry chemistry, chemists can quickly create new drugs and determine whether they will have the desired pharmacological effects. Scientists then select the most promising choices and begin to test those compounds in the wet laboratory.

What kinds of problems can computers not solve? Computers can't yet act with intuition or be spontaneously creative. They can attack highly challenging problems, such as making weather predictions or playing chess, but only in a way that takes advantage of what computers do best—making fast, reliable computations. Computers don't think as humans do. They can only follow instructions and algorithms.

How do programmers create problem statements? The goal in creating a useful problem statement is to have programmers interact with users to describe three things relevant to creating a useful program:

1. Data is the raw input that users have at the start of the job.
2. Information is the result, or output, that the users require at the end of the job.
3. Method, described precisely, is the process of how the program converts the inputs into the correct outputs.

For example, say you want to compute how much money you'll earn working at a parking garage. Your salary is $7.50 per hour for an eight-hour shift, but if you work more than eight hours a day, you get time and a half, which is $11.25 per hour, for the overtime work. To determine how much money you will make in any given day, you could create a simple computer program to do the work for you. In this example, what are the three elements of the problem statement?

1. *Data (input):* The data you have at the beginning of the problem, which is the number of hours you worked and the pay rate.
2. *Information (output):* The information you need to have at the end of the problem, which is your total pay for the day.

Understanding Software Programming

In this Helpdesk, you'll play the role of a helpdesk staffer, fielding questions about the life cycle of a program, the role a problem statement plays in programming, how programmers create algorithms and move from algorithm to code to the 1s and 0s a CPU can understand, and the steps involved in completing the program.

3. *Method (process):* The set of steps that take you from your input to your required output. In this case, the computer program would check whether you worked more than eight hours. This is important because it would determine whether you would be paid overtime. If you didn't work overtime, the output would be $7.50 multiplied by the total number of hours you worked ($60.00 for eight hours). If you did work overtime, the program would calculate your pay as eight hours at $7.50 per hour for the regular part of your shift, plus an additional $11.25 multiplied by the number of overtime hours you worked.

How do programmers handle bad inputs? In the problem statement, programmers also must describe what the program should do if the input data is invalid. This part of the problem statement is referred to as error handling.

The problem statement also includes a testing plan that lists specific input numbers that the programmers would typically expect the user to enter. The plan then lists the precise output values that a perfect program would return for those input values. Later, in the testing process, programmers use the input and output data values from the testing plan to determine whether the program they created works the way it should. We discuss the testing process later in this chapter.

Does the testing plan cover every possible use of the program? The testing plan can't list every input the program could ever encounter. Instead, programmers work with users to identify the categories of inputs that will be encountered, find a typical example of each input category, and specify what kind of output must be generated. In the parking garage pay example, the error-handling process would describe what the program would do if you happened to enter "–8" (or any other nonsense character) for the number of hours you worked. The error handling would specify whether the program would return a negative value, prompt you to reenter the input, or shut down.

We could expect three categories of inputs in the parking garage example. The user might enter:

1. A negative number for hours worked that day
2. A positive number equal to or less than eight
3. A positive number greater than eight

The testing plan would describe how the error would be managed or how the output would be generated for each input category.

Is there a standard format for a problem statement? Most companies have their own format for documenting a problem statement. However, all problem statements include the same basic components: the data that is expected to be provided (inputs), the information that is expected to be produced (outputs), the rules for transforming the input into output (processing), an explanation of how the program will respond if users enter data that doesn't make sense (error handling), and a testing plan. Table 10.1 shows a sample problem statement for our parking garage example.

Table 10.1 Complete Problem Statement for the Parking Garage Example

	Problem Statement		
Program Goal	Compute the total pay for a fixed number of hours worked at a parking garage.		
Input	Number of hours worked (a positive number)		
Output	Total pay earned (a positive number)		
Process	Total pay earned is computed as $7.50 per hour for the first eight hours worked each day. Any hours worked consecutively beyond the first eight are calculated at $11.25 per hour.		
Error Handling	The input (number of hours worked) must be a positive real number. If it is a negative number or another unacceptable character, the program will force the user to reenter the information.		
Testing Plan	**Input**	**Output**	**Notes**
	8	8 * 7.50	Testing positive input
	3	3 * 7.50	Testing positive input
	12	8 * 7.50 + 4 * 11.25	Testing overtime input
	–6	Error message/ask user to reenter value	Handling error

Algorithm Development

Objective 10.5 *Explain how programmers use flow control and design methodologies when developing algorithms.*

How does the process start? Once programmers understand exactly what the program must do and have created the final problem statement, they can begin developing a detailed algorithm—a set of specific, sequential steps that describe in natural language exactly what the computer program must do to complete its task. Let's look at some ways in which programmers design and test algorithms.

Do algorithms appear only in programming? You design and execute algorithms in your daily life. For example, say you're planning your morning. You know you need to (1) get gas for your car, (2) pick up a mocha latte at the café, and (3) stop by the bookstore and buy a textbook before your 9:00 A.M. accounting lecture.

You quickly think over the costs and decide it will take $150 to buy all three. In what order will you accomplish all these tasks? How do you decide? Should you try to minimize the distance you'll travel or the time you'll spend driving? What happens if you forget your debit card?

Figure 10.5 presents an algorithm you could develop to make decisions about how to accomplish these tasks. This algorithm lays out a specific plan that encapsulates all the choices you need to make in the course of completing a particular task and shows the specific sequence in which these tasks will occur. At any point in the morning, you could gather your current data (your inputs)—"I have $20 and my Visa card, but the ATM is out of my way"—and the algorithm would tell you unambiguously what your next step should be.

What are the limitations of algorithms? An algorithm is a series of steps that's completely known: At each point, we know exactly what step to take next. However, not all problems can be described as a fixed sequence of predetermined steps; some involve random and unpredictable events. For example, although the program that computes your parking garage take-home pay each day works flawlessly, programs that predict stock prices are often wrong because many random events (inputs), such as a flood in India or a shipping delay in Texas, can change the outcomes (outputs).

How do programmers represent an algorithm? Programmers have several visual tools at their disposal to help them document the decision points and flow of their algorithms:

- Flowcharts provide a visual representation of the patterns the algorithm comprises. Figure 10.5 presents an example of a flowchart used to depict the flow of an algorithm. Specific symbols indicate program behaviors and decision types. Diamonds indicate that a yes/no decision will be performed, and rectangles indicate an instruction to follow. Figure 10.6 lists additional flowcharting symbols and explains what they indicate. Many software packages make it easy for programmers to create and modify flowcharts. Microsoft Visio is one popular flowcharting program.
- Pseudocode is a text-based approach to documenting an algorithm. In pseudocode, words describe the actions that the algorithm will take. Pseudocode is organized like an outline, with differing levels of indentation to indicate the flow of actions within the program. There is no standard vocabulary for pseudocode. Programmers use a combination of common words in their natural language and the special words that are commands in the programming language they're using.

Flow Control

How do programmers handle complex algorithms? When programmers develop an algorithm, they convert the problem statement into a list of steps the program will take.

Problems that are complex involve choices, so they can't follow a sequential list of steps. Algorithms therefore include decision points, places where the program must choose from a list of actions based on the value of a certain input.

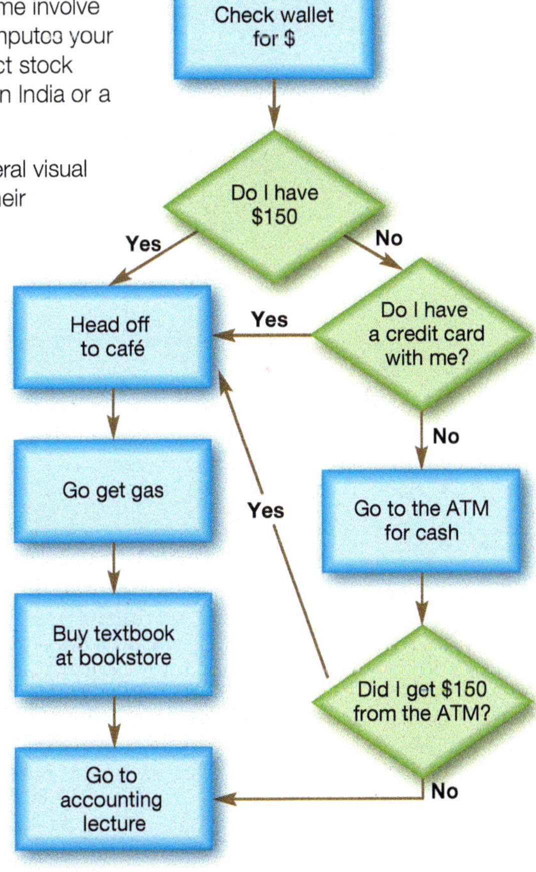

Figure 10.5 An algorithm you might use to plan your morning would include all the decisions you might need to make and would show the specific sequence in which these steps would occur.

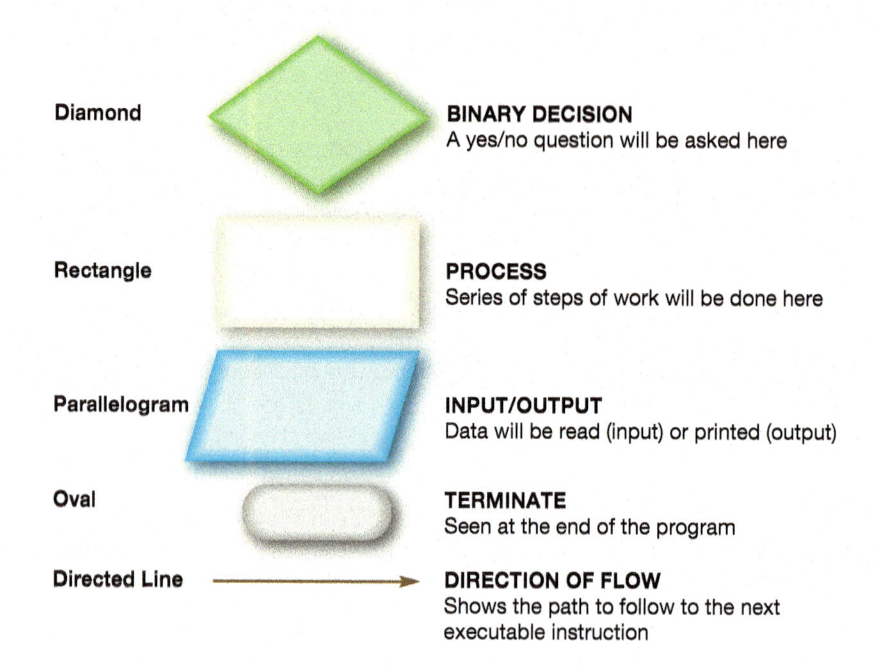

Figure 10.6 Standard Symbols Used in a Flowchart

Figure 10.7 shows the steps of the algorithm in our parking garage example. If the number of hours you worked in a given day is eight or fewer, the program performs one simple calculation: It multiplies the number of hours worked by $7.50. If you worked more than eight hours in a day, the program takes a different path and performs a different calculation.

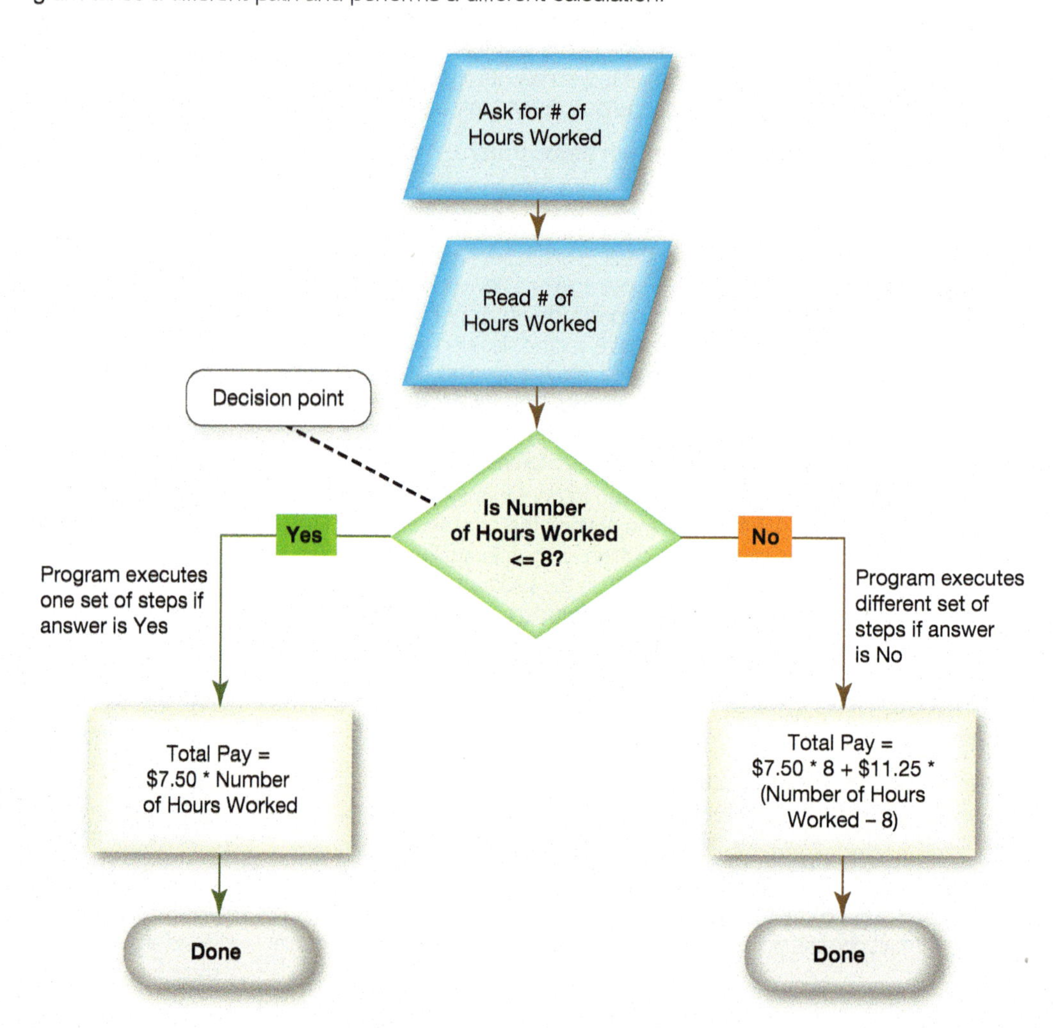

Figure 10.7 Decision points force the program to travel down one branch of the algorithm or another.

What kinds of decision points are there? Two main types of decisions change the flow of an algorithm:

1. *Binary decisions.* One decision point that appears often in algorithms is like a fork in the road. Such decision points are called binary decisions because they can be answered in one of only two ways: yes (true) or no (false). For example, the answer to the question, "Did you work at most eight hours today?" (Is number of hours worked ≤ 8 hours?), shown in Figure 10.7, is a binary decision because the answer can be only yes or no. If the answer is yes, the program follows one sequence of steps; if the answer is no, it follows a different path.
2. *Loops.* A second decision point that often appears in algorithms is a repeating loop. In a loop, a question is asked, and if the answer is yes, a set of actions is performed. Once the set of actions has finished, the question is asked again, creating a loop. As long as the answer to the question is yes, the algorithm continues to loop around and repeat the same set of actions. When the answer to the question is no, the algorithm breaks free of the looping and moves on to the first step that follows the loop.

In our parking garage example, the algorithm would require a loop if you wanted to compute the total pay you earned in a full week of work rather than in just a single day. For each day of the week, you would want to perform the same set of steps. Figure 10.8 shows how the idea of looping would be useful in this part of our parking garage program. On Monday, the program would set the Total Pay value to $0.00. It would then perform the following set of steps:

1. Read the number of hours worked for that day.
2. Check that we are still in week 1. If we are, then:
   - Compute the pay earned that day.
   - Calculate the Total Pay for the week so far.
   - Bump up to the next day.

On Tuesday, the algorithm would loop back, repeating the same sequence of steps it performed on Monday, adding the amount you earned on Tuesday to the Total Pay amount. The algorithm would continue to perform this loop for each day (seven times) until it hits Monday again. At that point, the decision "Are we still in the same week?" would become false. The program would stop, calculate the total pay for the entire week of work, and print the weekly paycheck. As you can see, there are three important features to look for in a loop:

1. A beginning point, or initial value. In our example, the total pay for the week starts at an initial value of $0.00.
2. A set of actions that will be performed. In our example, the algorithm computes the daily pay each time it passes through the loop.
3. A test condition, or a check to see whether the loop is completed. In our example, the algorithm should run the loop seven times, no more and no fewer.

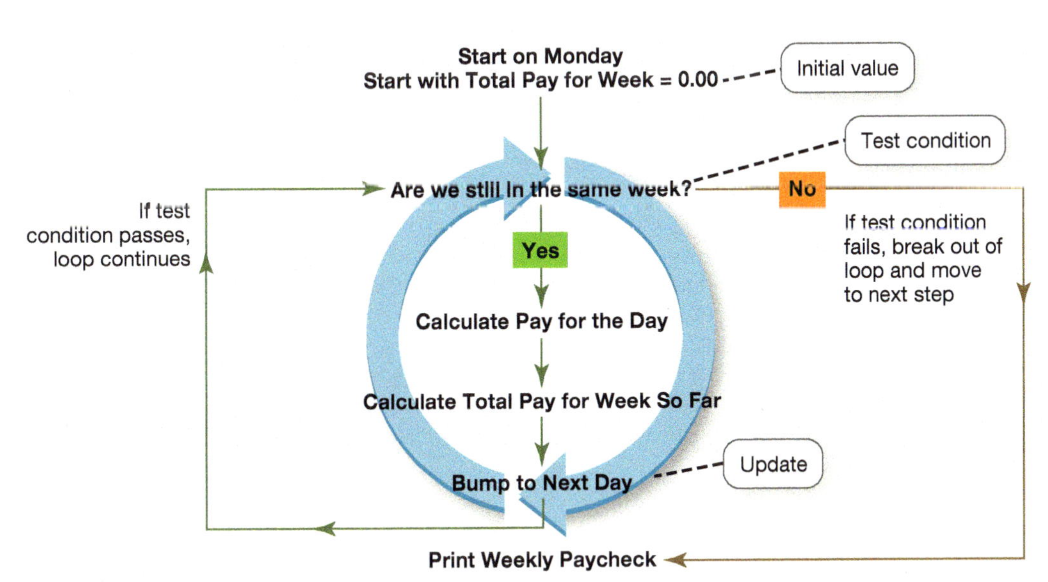

Figure 10.8 We stay in the loop until the test condition is no longer true. We then break free from the loop and move on to the next step in the algorithm, which is outside of the loop.

Bits&Bytes Build a Coding Portfolio

Once you begin to program your own projects and make alterations to others' projects, you will want to start collecting them in a formal portfolio. Your portfolio should document how you're growing as a programmer and should serve as a tool for finding a job or getting into a special course. Many programmers develop their projects, using GitHub (*github.com*) and use GitHub Pages (*pages.github.com*) to create and host an online code portfolio quickly.

Devpost (*devpost.com*) is another site that offers free portfolio management, enabling you to showcase your work and your projects in progress. You can insert screenshots and videos to explain the story behind the code. Devpost is also a great site to find your next hackathon, which you can feature in your Devpost portfolio!

Every higher-level programming language supports both making binary yes/no decisions and handling repeating loops. Control structures are keywords in a programming language that allow the programmer to direct the flow of the program based on a decision.

How do programmers create algorithms for specific tasks? It's difficult for human beings to force their problem-solving skills into the highly structured, detailed algorithms that computing machines require. Therefore, several methodologies have been developed to support programmers, including top-down design and object-oriented analysis.

Design Methodology: Top-Down Design

What is top-down design? Top-down design is a systematic approach in which a problem is broken into a series of high-level tasks. In top-down design, programmers apply the same strategy repeatedly, breaking each task into successively more detailed subtasks. They continue until they have a sequence of steps that are close to the types of commands allowed by the programming language they'll use for coding. Previous coding experience helps programmers know the appropriate level of detail to specify in an algorithm generated by top-down design.

How is top-down design used in programming? Let's consider our parking garage example again. Initially, top-down design would identify three high-level tasks: Get Input, Process Data, and Output Results (see Figure 10.9a).

By applying top-down design to the first operation, Get Input, we'd produce the more detailed sequence of steps shown in the first step of Figure 10.9b: Announce Program, Give Users Instructions, and Read the Input NumberHoursWorkedToday. When we try to refine each of these steps, we find that they are just print-and-read statements and that almost every programming language will support them. Therefore, the Get Input operation has been converted to an algorithm.

Next, we move to the second high-level task, Process Data, and break it into subtasks. In this case, we need to determine whether overtime hours were worked and then compute the pay accordingly. We continue to apply top-down design on all tasks until we can no longer break tasks into subtasks, as shown in Figure 10.9c.

Design Methodology: Object-Oriented Analysis

What is object-oriented analysis? With object-oriented analysis, programmers first identify all the categories of inputs that are part of the problem the program is meant to solve. These categories are called classes. With the object-oriented approach, the majority of design time is spent identifying the classes required to solve the problem, modeling them, and thinking about what relationships they need to be able to have with each other. Constructing the algorithm becomes a process of enabling the objects to interact. For example, the classes in our parking garage example might include a TimeCard class and an Employee class.

Classes are defined by:

- Information (data) within the class.
- Actions (methods) associated with the class.

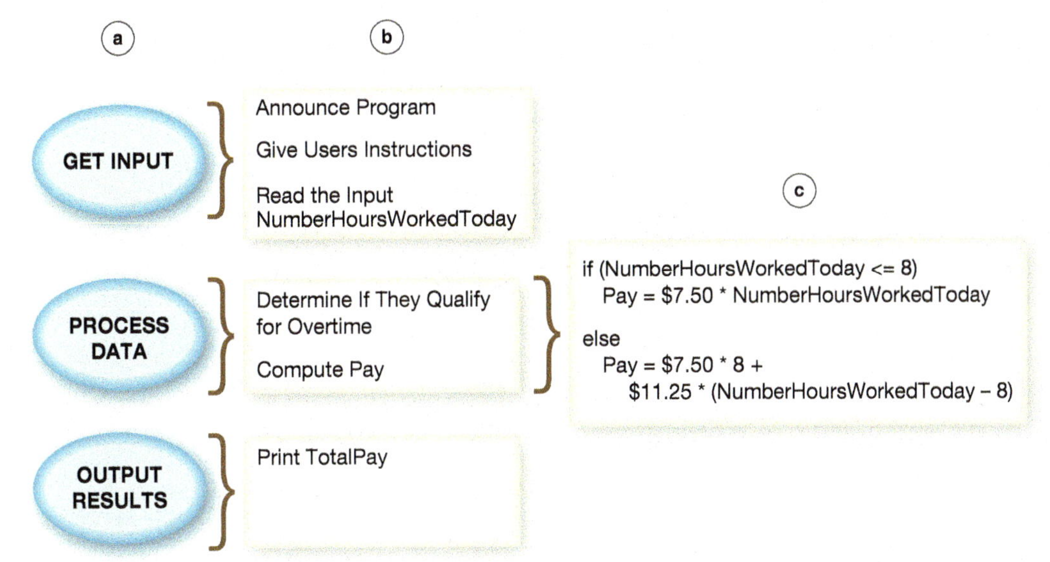

Figure 10.9 (a) A top-down design is applied to the highest level of task in our parking garage example, (b) the tasks are further refined into subtasks, and (c) subtasks are refined into a sequence of instructions—an algorithm.

For example, as shown in Figure 10.10, data for an Employee would include a Name, Address, and Social Security Number, whereas the methods for the Employee would be GoToWork(), LeaveWork(), and CollectPay(). The data of the class describes the class, so classes are often characterized as nouns, whereas methods are often characterized as verbs—the ways that the class acts and communicates with other classes.

Programmers may need to create several examples of a class. Each of these examples is an object. In Figure 10.10, John Doe, Jane Doe, and Bill McGillicutty are each Employee objects (specific examples of the Employee class). Each object from a given class is described by the same pieces of data and has the same methods; for example, John, Jane, and Bill are all Employees and can use the GoToWork(), LeaveWork(), and CollectPay() methods. However, because they all have different pay grades (PayGrade 5, PayGrade 10, and PayGrade 4, respectively) and different Social Security numbers, they are all unique objects.

Why would a developer select the object-oriented approach over top-down design? An important aspect of object-oriented design is that it leads to reusability. Object-oriented analysis forces programmers to think in general terms about their problem, which tends to lead to more general and reusable solutions. Because object-oriented design generates a family of classes for each project, programmers can easily reuse existing classes from other projects, enabling them to produce new code quickly.

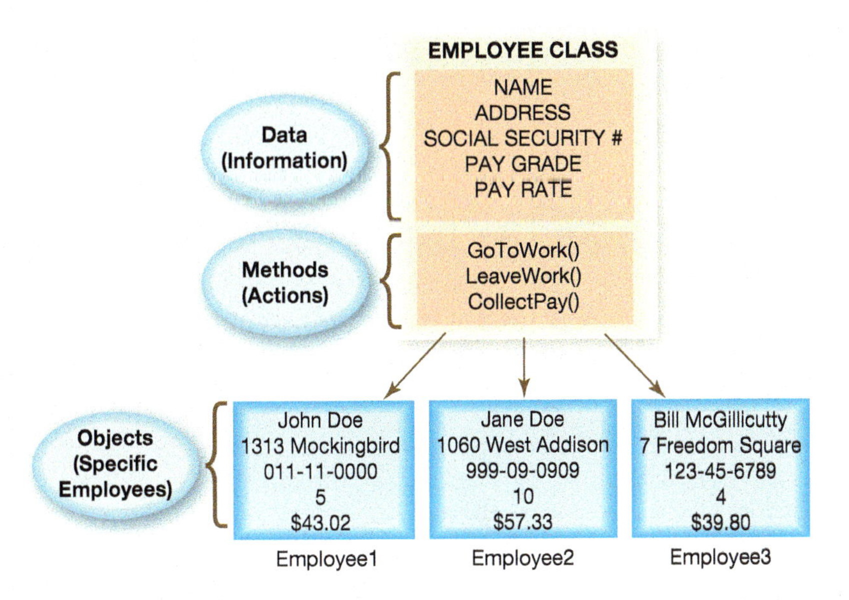

Figure 10.10 The Employee class includes the complete set of information (data) and actions (methods or behaviors) that describe an Employee.

Dig Deeper

The Building Blocks of Programming Languages: Syntax, Keywords, Data Types, and Operators

Programming languages are evolving constantly. New languages emerge every year, and existing languages change dramatically. Therefore, it would be extremely difficult and time-consuming for programmers to learn every programming language. However, all languages have several common elements: rules of syntax, a set of keywords, a group of supported data types, and a set of allowed operators. By learning these four concepts, a programmer will be better equipped to approach any new language.

The transition from a well-designed algorithm to working code requires a clear understanding of the rules (syntax) of the programming language being used. Syntax is an agreed-upon set of rules defining how a language must be structured. The English language has a syntax that defines which symbols are words (for example, *gorilla* is a word but *allirog* is not) and what order words and symbols (such as semicolons and commas) must follow.

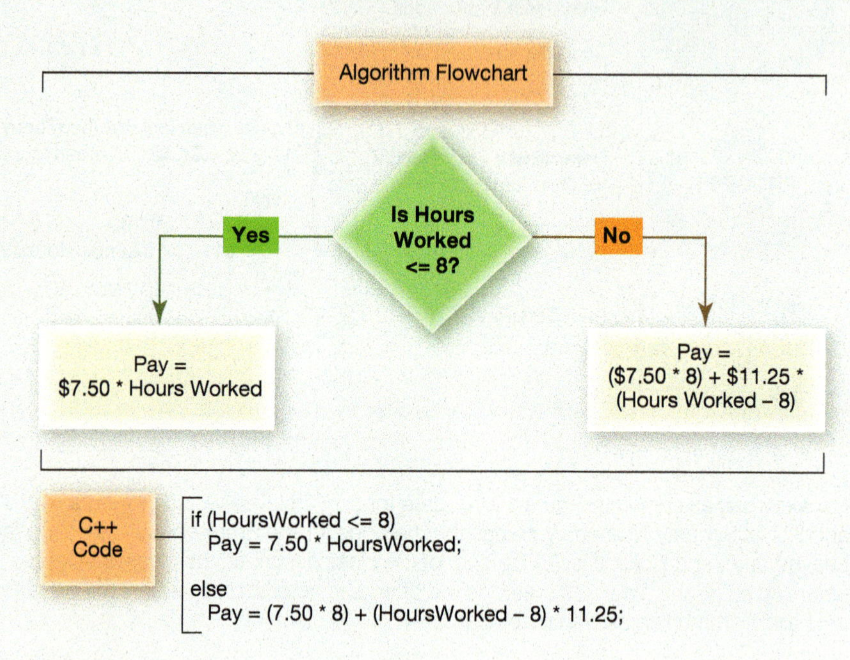

Figure 10.11 The binary decision in the algorithm has been converted into C++ code.

Likewise, all programming languages have a formal syntax that programmers must follow when creating code statements, which are sentences in a code. Syntax errors are violations of the strict, precise set of rules that define the programming language. In a programming language, even misplacing a single comma or using a lowercase letter where a capital letter is required will generate a syntax error and make the program unusable.

Keywords are a set of words that have predefined meanings for a particular language. Keywords translate the flow of the algorithm into the structured code of the programming language. For example, when the algorithm indicates where a binary decision must be made, the programmer translates that binary decision into the appropriate keyword(s) from the language.

To illustrate further, in the C++ programming language, the binary decision asking whether you worked enough hours to qualify for overtime pay would use the keywords if else. At this point in the code, the program can follow one of two paths: If you indicated through your input that you worked fewer than or equal to eight hours, the program would take one path; if not (else), it would follow another. Figure 10.11 shows this binary decision in the algorithm and the corresponding lines of C++ code that use the "if else" keywords.

Loops are likewise translated from algorithm to code by using the appropriate keyword from the language. For example, in the Visual Basic programming language, programmers use the keywords For and Next to implement a loop. After the keyword For, an input or output item is given a starting value, and then the statements, or "sentences," in the body of the loop are executed. When the command Next is run, the program returns to the For statement and increments the value of the input or output item by 1. It then tests that the value is still inside the range given. If it is, the body of the loop is executed again. This continues until the value of the input or output item is outside the range listed. The loop is then ended, and the statement that follows the loop is run.

In the parking garage example, the following lines of Visual Basic code loop to sum up the total pay for the entire week:

```
For Day = 1 to 7
    TotalPay = TotalPay + Pay
Next Day
```

In this statement, the starting value of the input item Day is 1, and the program loops through until Day equals 7. Then, when Day equals 8, we move out of the loop, to the line of code that comes after the loop.

Often, a quick overview of a language's keywords can reveal the unique focus of that language. For example, the C++ language includes the keywords *public*, *private*, and *protected*, which indicate that the language includes a mechanism for controlling security.

Each time programmers want to store data in their program, they must ask the OS for storage space at a RAM location. Data types describe the kind of data being stored at the memory location. Each programming language has its own data types (although there is some degree of overlap). For example, C++ includes data types that represent integers, real numbers, characters, and Boolean (true–false) values. These C++ data types show up in code statements as *int* for integer, *float* for real numbers, *char* for characters, and *bool* for Boolean values.

Because it takes more room to store a real number such as 18,743.23 than it does to store the integer 1, programmers use data types in their code to indicate to the OS how much memory it needs to allocate. Programmers must be familiar with all the data types available in the language so that they can assign the most appropriate one for each input and output value, without wasting memory space.

Operators are the coding symbols that represent the fundamental actions of the language. Each programming language has its own set of operators. Many languages include common algebraic operators such as +, –, *, and / to represent the mathematical operations of addition, subtraction, multiplication, and division, respectively. Some languages, however, introduce new symbols as operators. The language APL (short for A Programmer's Language) was designed to solve multidimensional mathematical problems. APL includes less familiar operators such as rho, sigma, and iota, each representing a complex mathematical operation. Because it contains many specialized operators, APL requires programmers to use a special keyboard (see Figure 10.12).

Figure 10.12 APL requires programmers to use an APL keyboard that includes the many specialized operators in the language. *(Primsky/Shutterstock)*

Programming languages sometimes include other unique operators. The C++ operator >>, for example, is used to tell the computer to read data from the keyboard or from a file. The C++ operator && is used to tell the computer to check whether two statements are both true. "Is your age greater than 20 AND less than 30?" is a question that requires the use of the && operator.

In the following C++ code, several operators are being used. The > operator checks whether the number of hours worked is greater than 0. The && operator checks that the number of hours worked is simultaneously positive AND less than or equal to 8. If that happens, then the = operator sets the output Pay to equal the number of hours multiplied by $7.50:

```
if (Hours > 0 && Hours ≤ 8)
    Pay = Hours * 7.50;
```

Knowing operators such as these, as well as the other common elements described earlier, helps programmers learn new programming languages.

How does a programmer take advantage of reusability? Programmers must study the relationships between objects. Hierarchies of objects can be built quickly in object-oriented languages, using the mechanism of inheritance. Inheritance means that a new class can automatically pick up all the data and methods of an existing class and then can extend and customize those to fit its own specific needs.

The original class is called the base class, and the new, modified class is called the derived class, as illustrated in Figure 10.13. You can compare this with making soup. For example, you have a basic recipe for soup stock (base class: Soup Stock). However, in your family, some people like lentil soup (derived class: Lentil Soup), and others like minestrone (derived class: Minestrone Soup). All soups share the attributes of the basic soup stock. However, instead of creating two entirely new recipes—one for lentil soup and one for minestrone—the two varieties inherit the basic stock (base class) recipe; the recipe is then customized to make lentil and minestrone (derived classes).

Figure 10.13 In object-oriented programming, a single base class—for example, Soup Stock—helps you quickly create many additional derived classes, such as Lentil Soup or Minestrone Soup. *(Valentina Razumova/Shutterstock)*

Coding

Objective 10.6 *Discuss the categories of programming languages and the roles of the compiler and the integrated development environment in coding.*

How is a person's idea translated into CPU instructions? Once programmers create an algorithm, they select the best programming language for the problem and then translate the algorithm into that language. Translating an algorithm into a programming language is the act of coding. Programming languages are somewhat readable by humans but then are translated into patterns of 1s and 0s to be understood by the CPU.

Although programming languages free programmers from having to think in binary language—the 1s and 0s that computers understand—they still force programmers to translate the ideas of the algorithm into a highly precise format. Programming languages are quite limited, allowing programmers to use only a few specific keywords, while demanding a consistent structure.

How exactly do programmers move from algorithm to code? Once programmers have an algorithm, in the form of either a flowchart or a series of pseudocode statements, they identify the key pieces of information the algorithm uses to make decisions:

- What steps are required for the calculation of new information?
- What is the exact sequence of the steps?
- Are there points where decisions have to be made?
- What kinds of decisions are made?
- Are there places where the same steps are repeated several times?

Once programmers identify the required information and the flow of how it will be changed by each step of the algorithm, they can begin converting the algorithm into computer code in a specific programming language.

Categories of Programming Languages

What exactly is a programming language? A programming language is a kind of code for the set of instructions the CPU knows how to perform. Computer programming languages use special words and strict rules so that programmers can control the CPU without having to know all its hardware details.

What kinds of programming languages are there? Programming languages are classified into several major groupings, sometimes referred to as generations. With each generation of language development, programmers have been relieved of more of the burden of keeping track of what the hardware requires. The earliest languages—assembly language and machine languages—required the programmer to know a great deal about how the computer was constructed internally and how it stored data. Programming is becoming easier as languages continue to become more closely matched to how humans think about problems.

How have modern programming languages evolved? There are five major categories of languages:

1. A first-generation language (1GL) is the actual machine language of a CPU, the sequence of bits (1s and 0s) that the CPU understands.
2. A second-generation language (2GL) is also known as an assembly language. Assembly languages allow programmers to write their programs by using a set of short, English-like commands that speak directly to the CPU and that give the programmer direct control of hardware resources.
3. A third-generation language (3GL) uses symbols and commands to help programmers tell the computer what to do. This makes 3GL languages easier for humans to read and remember. Most programming languages today, including BASIC, FORTRAN, COBOL, C/C++, and Java, are considered third generation.
4. Fourth-generation languages (4GLs) include database query languages and report generators. Structured Query Language (SQL) is a database programming language that is an example of a 4GL. The following SQL command would check a huge table of data on the employees and build a new table showing all those employees who worked overtime:

   ```
   SELECT EmployeeName, TotalHours FROM EMPLOYEES WHERE
   "TotalHours" Total More Than 8
   ```

5. Fifth-generation languages (5GLs) are considered the most natural of languages. In a 5GL, a problem is presented as a series of facts or constraints instead of as a specific algorithm. The system of facts can then be queried (asked) questions. Prolog (PROgrammingLOGic) is an example of a 5GL. A Prolog program could be a list of family relationships and rules such as, "Mike is Sally's brother. A brother and a sister have the same mother and father." Once a user has amassed a huge collection of facts and rules, he or she can ask for a list of all Mike's cousins, for example. Prolog would find the answers by repeatedly applying the principles of logic instead of following a systematic algorithm that the programmer provided.

Table 10.2 shows small code samples of each generation of language.

Table 10.2 Sample Code for Different Language Generations

Generation	Example	Sample Code
1GL	Machine	**Bits** describe the commands to the CPU. **1110 0101 1001 1111 0000 1011 1110 0110**
2GL	Assembly	**Words** describe the commands to the CPU. **ADD Register 3, Register 4, Register 5**
3GL	FORTRAN, BASIC, C/C++, Java	**Symbols** describe the commands to the CPU. **TotalPay = Pay + OvertimePay**
4GL	SQL	**More powerful commands** allow complex work to be done in a single sentence. **SELECT isbn, title, price, price*0.06 AS sales_tax FROM books WHERE price>100.00 ORDER BY title;**
5GL	Prolog	Programmers can build applications **without specifying an algorithm**. Find all the people who are Mike's cousins as:?-cousin (Mike, family)

Portability, Variable Declarations, and Commenting

Do programmers have to use a higher-level programming language to solve a problem with a computer? No, experienced programmers sometimes write a program directly in the CPU's assembly language. However, the main advantage of higher-level programming languages—C or Java, for example—is that they allow programmers to think in terms of the problem they're solving rather than having to worry about the internal design and specific instructions available for a given CPU. In addition, higher-level programming languages can easily produce a program that will run on differently configured CPUs. For example, if programmers wrote directly in the assembly language for an Intel i7 CPU, they would have to rewrite the program completely if they wanted it to run on a Sun workstation with the SPARC64 CPU. Thus, higher-level programming languages offer portability—the capability to move a completed solution easily from one type of computer to another.

What happens first when you write a program? All of the inputs a program receives and all of the outputs the program produces need to be stored in the computer's RAM while the program is running. Each input and each output item that the program manipulates, also known as a variable, needs to be announced early in the program so that memory space can be set aside. A variable declaration tells the operating system that the program needs to allocate storage space in RAM. The following line of code is a variable declaration in the Java language:

```
int Day;
```

The one simple line makes a lot of things happen behind the scenes:

- The variable's name is Day. The "int" that precedes the Day variable indicates that this variable will always be an integer (a whole number such as 15 or –4).
- The statement "int Day;" asks for enough RAM storage space to hold an integer. After the RAM space is found, it is reserved. As long as the program is running, these RAM cells will be saved for the Day variable, and no other program can use that memory until the program ends. From that point on, when the program encounters the symbol Day, it will access the memory it reserved as Day and find the integer stored there.

Can programmers leave notes to themselves inside a program? Programmers often insert a comment into program code to explain the purpose of a section of code, to indicate the date they wrote the program, or to include other important information about the code so that fellow programmers can more easily understand and update it.

Comments are written into the code in plain English. The compiler, a program that translates codes into binary 1s and 0s, just ignores comments. Comments are intended to be read only by human programmers. Languages provide a special symbol or keyword to indicate the beginning of a comment. In C++, the symbol// at the beginning of a line indicates that the rest of the line is a comment.

What would completed code for a program look like? Figure 10.14 presents a completed C++ program for our parking garage example. We discuss later how you would actually enter this program on your system. This program, which is written in a top-down style, does not use objects.

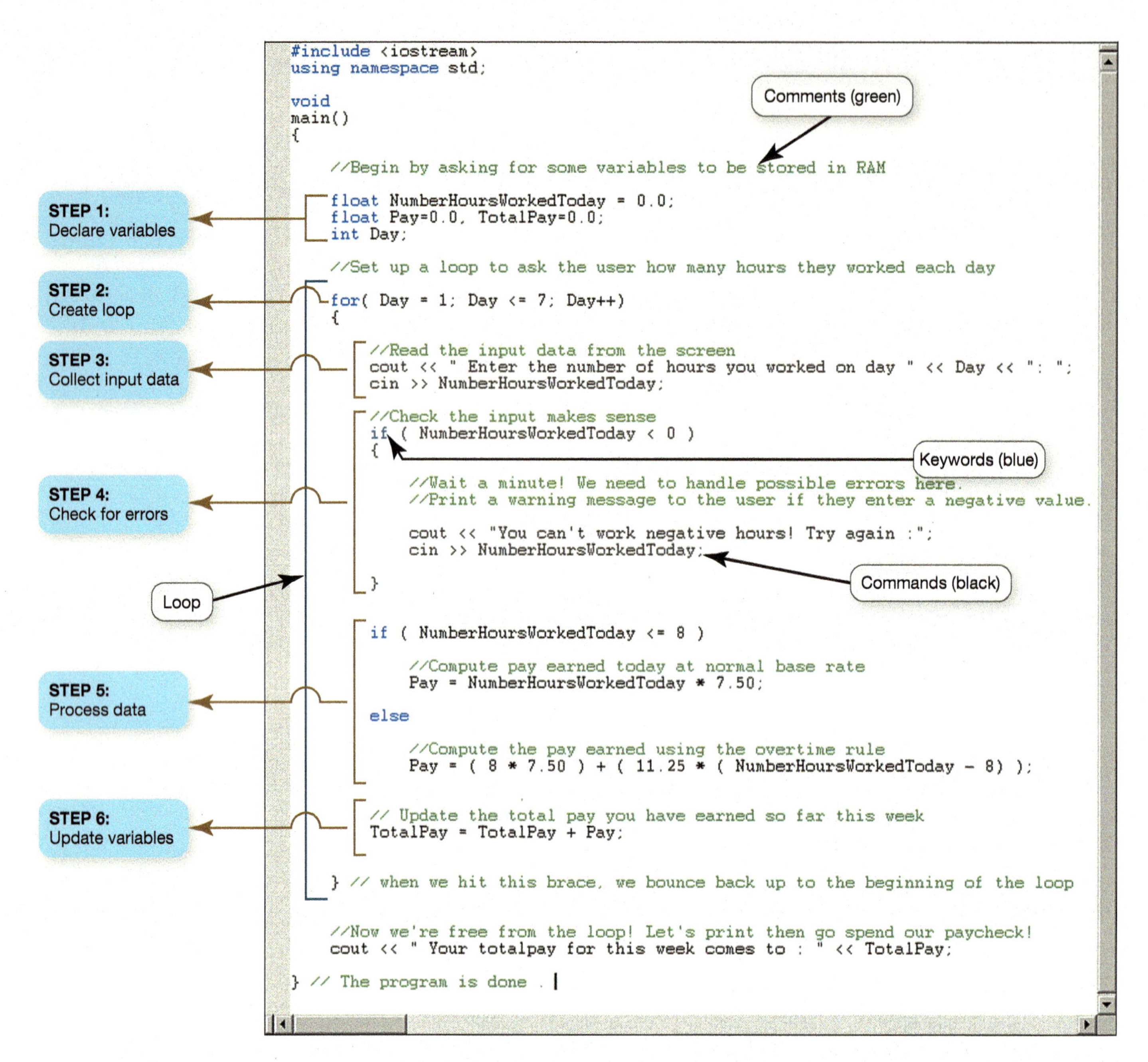

Figure 10.14 A complete C++ program that solves the parking garage pay problem.

Each statement in the program is executed sequentially (that is, in order from the first statement to the last) unless the program encounters a keyword that changes the flow. In the figure:

1. The program begins by declaring the variables needed to store the program's inputs and outputs in RAM.
2. Next, the *for* keyword begins a looping pattern. All the steps between the very beginning bracket ({) and the end one (}) will be repeated seven times to gather the total pay for each day of the week.
3. The next section collects the input data from the user.
4. The program then checks that the user entered a reasonable value. In this case, a positive number for hours worked must be entered.
5. Next, the program processes the data. If the user worked eight or fewer hours, he or she is paid at the rate of $7.50 per hour, whereas hours exceeding eight are paid at $11.25 per hour.
6. The final statement updates the value of the TotalPay variable. The end bracket indicates that the program has reached the end of a loop. The program will repeat the loop to collect and process the information for the next day. When the seventh day of data has been processed, the Day variable will be bumped up to the next value, 8. This causes the program to fail the test (Day ≤7). At that point, the program exits the loop, prints the results, and quits.

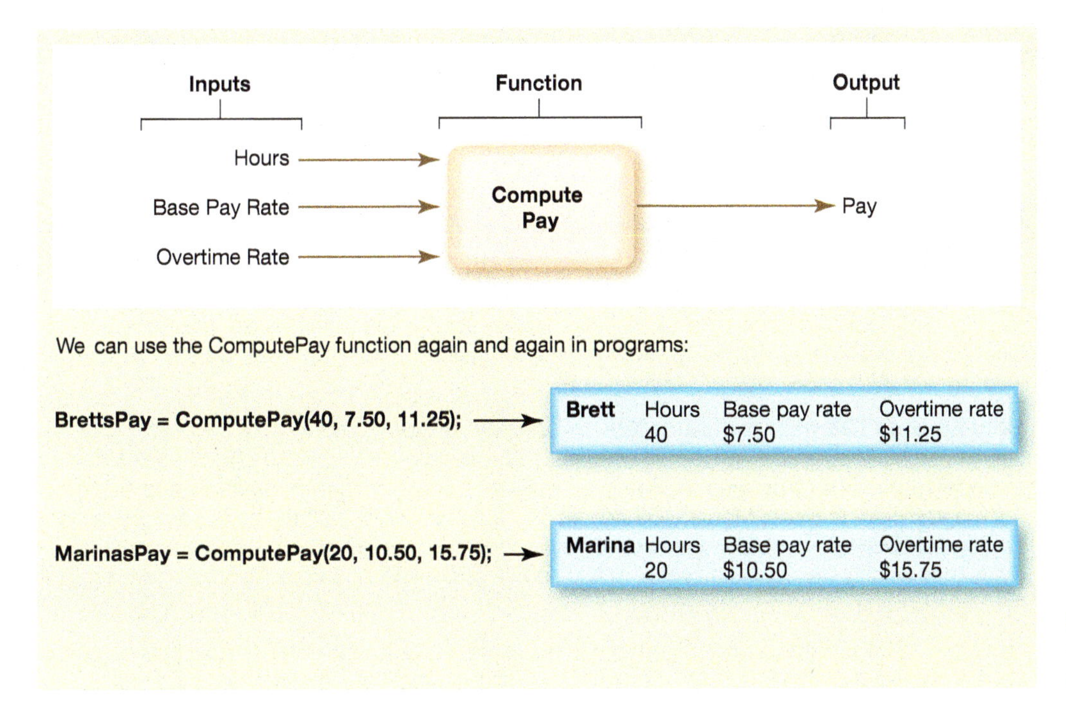

Figure 10.15 A function is a reusable component that can be used in different settings.

Are there ways in which programmers can make their code more useful for the future? One aspect of converting an algorithm into good code is the programmer's ability to design general code that can easily be adapted to new settings. Sections of code that will be used repeatedly, with only slight modification, can be packaged into reusable containers, or components. Depending on the language, these reusable components are referred to as functions, methods, procedures, subroutines, modules, or packages.

In our program, we could create a function that implements the overtime pay rule. As it stands in Figure 10.15, the code works only in situations in which the hourly pay is exactly $7.50 and the bonus pay is exactly $11.25. However, if we rewrote this part of the processing rules as a function, we could have code that would work for any base pay rate and any overtime rate. If the base pay rate or overtime rate changed, the function would use whichever values it was given as input to compute the output pay variable. Such a function, as shown in Figure 10.15, may be reused in many settings without changing any of the code.

The Compiler

How does a programmer move from code in a programming language to the 1s and 0s the CPU can understand? Compilation is the process by which code is converted into machine language—the language the CPU can understand. A compiler is a program that understands both the syntax of the programming language and the exact structure of the CPU and its machine language. It can read the source code, which comprises the instructions programmers have written in the higher-level language and can translate the source code directly into machine language—the binary patterns that will execute commands on the CPU. (You can learn more about the binary number system in Chapters 2 and 4.)

Each programming language has its own compiler. At this stage, the compiler creates a file that can be distributed: the executable program. The executable program, the binary sequence that instructs the CPU to run its code, can't be read by human eyes because it consists of pure binary codes. Executable programs are stored as *.exe or *.com files on Windows systems.

Does every programming language have a compiler? Some programming languages don't have a compiler but use an interpreter instead. An interpreter translates the source code into an intermediate form, line by line. Each line is then executed as it's translated. The compilation process takes longer than the interpretation process because in compilation, all the lines of source code are translated into machine language before any lines are executed. However, the finished compiled program runs faster than an interpreted program because the interpreter is constantly translating and executing as it goes.

If producing the fastest executable program is important, programmers will choose a language that uses a compiler instead of an interpreter. For development environments in which many changes are still being made to the code, interpreters have an advantage because programmers don't have to wait for the entire program to be recompiled each time they make a change. With interpreters, programmers can immediately see the results of their program changes as they're making them.

Integrated Development Environments

What tools make the coding process easier? Modern programming is supported by a collection of tools that make the writing and testing of software easier. An integrated development environment (IDE) is a developmental tool that helps programmers write and test their programs. One IDE can often be configured to support many languages. Figure 10.16 shows the IDE jGRASP working with Java code.

How does an IDE help programmers when they're typing the code? Code editing is the step in which a programmer physically types the code into the computer. An IDE includes an editor, a special tool that helps programmers as they enter the code, highlighting keywords and alerting the programmers to typos. Modern IDE editors also automatically indent the code correctly, align sections of code appropriately, and apply color to code comments to remind programmers that these lines won't be executed as code. In addition, IDEs provide auto-completion of code, suggest solutions to common errors, and more.

How does the IDE help programmers after code editing is finished? Editing is complete when the entire program has been keyed into the editor. At that time, the programmer presses a button in the IDE, and the compilation process begins. The IDE shows how the compilation is progressing, which line is currently being compiled, how many syntax errors have been identified, and how many warnings have been generated. A warning is a suggestion from the compiler that

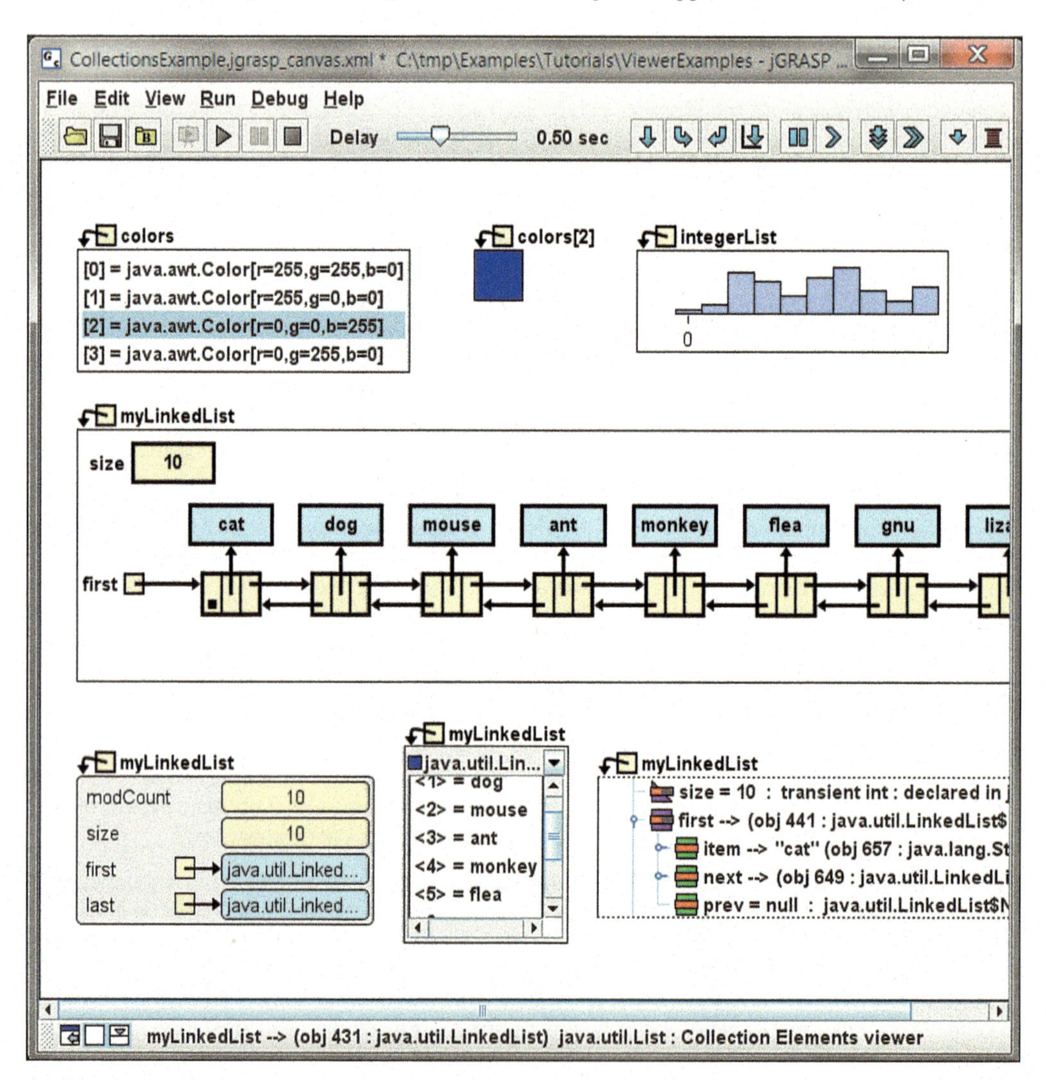

Figure 10.16 The jGRASP IDE is a free tool that helps the programmer visualize the code as it runs to help eliminate logical errors. *(From jGRASP [http://www.jgrasp.org/viewers.html]. Copyright © by Auburn University. Used by permission Auburn University.)*

the code might not work in the way the programmer intended even though there's no formal syntax error on the line.

A syntax error is a violation of the strict, precise set of rules that define the language. Programmers create syntax errors when they misspell keywords (such as typing BEEGIN instead of BEGIN) or use an operator incorrectly (such as typing x =, y + 2 instead of x = y + 2). Once compilation is finished, the IDE presents all the syntax errors in one list. The programmer can then click any item in the list to see a detailed explanation of the type of error. When the programmer double-clicks an item in the list, the editor jumps to the line of code that contains the error, enabling the programmer to repair syntax errors quickly.

Debugging

Objective 10.7 *Identify the role of debugging in program development.*

What is debugging? Once the program has compiled without syntax errors, it has met all the syntax rules of the language. However, this doesn't mean that the program behaves in a logical way or that it appropriately addresses the task the algorithm described. If programmers made errors in the strategy used in the algorithm or in how they translated the algorithm to code, problems will occur. The process of running the program over and over to find and repair errors and to make sure the program behaves in the way it should is termed debugging (see Figure 10.17).

Figure 10.17 Debugging—the process of correcting errors in a program—combines logic with an understanding of the problem to be solved. *(Nullplus/Shutterstock)*

How do programmers know the program has solved the problem? At this point in the process, the testing plan that was documented as part of the problem statement becomes critically important to programmers. The testing plan clearly lists input and output values, showing how the users expect the program to behave in each input situation. It's important for the testing plan to contain enough specific examples to test every part of the program.

In the parking garage problem, we want to make sure the program calculates the correct pay for a day when you worked eight or fewer hours and for a day when you worked more than eight hours. Each of these input values forces the program to make different decisions in its processing path. To be certain the program works as intended, programmers try every possible path.

For example, once we can successfully compile the example code for the parking garage problem, we can begin to use our testing plan. The testing plan indicates that an input of 3 for NumberHoursWorkedToday must produce an output of Pay = 3 * $7.50 = $22.50. In testing, we run the program and make sure that an input value of 3 yields an output value of $22.50. To check that the processing path involving overtime is correct, we input a value of 12 hours. That input must produce Pay = 8 * $7.50 + (12 – 8) * $11.25 = $105.00.

A complete testing plan includes sample inputs that exercise all the error handling required as well as all the processing paths. Therefore, we would also want to check how the program behaves when NumberHoursWorkedToday is entered as –6.

If the testing plan reveals errors, why does the program compile? The compiler can't think through code or decide whether what the programmer wrote is logical. The compiler can only make sure that the specific rules of the language are followed.

For example, if in the parking garage problem we happened to type the if statement as

```
if (NumberHoursWorkedToday > 88)
    //Use the Overtime Pay rule
```

instead of

```
if (NumberHoursWorkedToday > 8)
    //Use the Overtime Pay rule
```

the compiler wouldn't see a problem. It doesn't seem strange to the compiler that you only get overtime after working 88 hours a day. These logical errors in the problem are caught only when the program executes.

Another kind of error caught when the program executes is a runtime error. For example, it's easy for a programmer accidentally to write code that occasionally divides by zero, a big no-no mathematically! That kind of forbidden operation generates a runtime error message.

Are there tools that help programmers find logical errors? Most IDEs include a tool called a debugger that helps programmers dissect a program as it runs. The debugger pauses the program while it's executing and allows the programmer to examine the values of all the variables. The

Bits&Bytes Coding with a Purpose

Hackathons, events at which programmers gather and work collectively for a day or even a full weekend, are beginning to specialize and focus on areas that address social needs. The MIT Hacking Medicine event works on developing projects that can help more people access medical interventions. Hack the Gap is the largest hackathon for women and non-binary programmers. Earth Hack develops coding solutions to environmental problems, and Code for America creates solutions for developing better government services.

Hackathons can help you build your programming skills incredibly quickly. If you don't have a team, most hackathons link you up with other programmers on the spot. Visit Devpost (*devpost.com*) to find the coding events nearest you—and join in the fun!

programmer can then run the program in slow motion, moving it forward just one line at a time to see the exact sequence of steps being executed and the outcome of each calculation. He or she can then isolate the precise place in which a logical error occurs, correct the error, and recompile the program.

Testing and Documentation

Objective 10.8 *Explain the importance of testing and documentation in program development.*

What is the first round of testing for a program? Once debugging has detected all the runtime errors in the code and they have been repaired, it's time for users to test the program. This process is called *internal testing*. In internal testing, a group in the software company uses the program in every way it can imagine—including how the program was intended to be used and in ways only new users might think up. The internal testing group makes sure the program behaves as described in the original testing plan. Any differences in how the program responds are reported back to the programming team, which makes the final revisions and updates to the code.

The next round of testing is external testing. In this testing round, people like the ones who will eventually purchase and use the software must work with it to determine whether it matches their original vision.

What other testing does the code undergo? Before its final commercial release, software is often provided free or at a reduced cost in a beta version to certain test sites or to interested users. By providing users with a beta version of software, programmers can collect information about remaining errors in the code and make a final round of revisions before officially releasing the program. Often, popular software packages are available for free beta download for months before the official public release.

What happens if problems are found after beta testing? The manufacturer will make changes before releasing the product to other manufacturers, for installation on new machines, for example. That point in the release cycle is called release to manufacturers (RTM). After the RTM is issued, the product is in general availability (GA) and can be purchased by the public.

Users often uncover problems in a program even after its commercial release to the public. These problems are addressed with the publication of software updates or service packs. Users can download these software modules to repair errors identified in the program code. Most operating systems now push updates directly to users to make sure key security patches are automatically installed.

After testing, is the project finished? Once testing is complete, but before the product is officially released, the work of documentation is still ahead. At this point, technical writers create internal documentation for the program that describes the development and technical details of the software, how the code works, and how the user interacts with the program. In addition, the technical publishing department produces all the necessary user documentation that will be distributed to the program's users. User training begins once the software is distributed. Software trainers take the software to the user community and teach others how to use it efficiently.

Before moving on to Part 2:

1. Watch Chapter Overview Video 10.1.
2. Then take the Check Your Understanding quiz.

Check Your Understanding // Review & Practice

For a quick review to see what you've learned so far, answer the following questions.

multiple choice

1. Custom-built mini-programs that can execute a complicated series of commands in a tool like Microsoft Word or Excel are called
   a. executables.
   b. compilers.
   c. macros.
   d. debuggers.
2. A flow chart is used for
   a. visually documenting an algorithm.
   b. locating runtime errors in code.
   c. developing the relationships between classes.
   d. compiling source code.
3. Data is different from information, because
   a. data is tested and information is not.
   b. an algorithm can use information but cannot use data.
   c. data is never biased and information is.
   d. data is the raw input and information is the output the users require.
4. In object-oriented analysis, classes are defined by
   a. objects and data.
   b. operators and objects.
   c. data and methods.
   d. behaviors and keywords.
5. Compiling a program turns
   a. top-down design into object-oriented design.
   b. source code into an executable program.
   c. a first-generation language into a fourth-generation language.
   d. none of the above

chew on this

Think of the last time you faced a big shopping trip. What would the key classes be if you were going to design a software program that helped you manage your shopping excursion? What data and methods would belong in each class?

MyLab IT Go to **MyLab IT** to take an autograded version of the *Check Your Understanding* review and find all media resources for the chapter.

For the IT Simulation for this chapter, see MyLab IT.

Try This

Programming with Corona

You've already been doing mobile app development in the Make This activities, using App Inventor. In this Try This, you will use the product Corona to create an application that could be run on either an Android or an iOS device. For more step-by-step instructions, watch the Try This video.

Note: You need to be confident in installing and uninstalling demo software to your system to proceed. Refer to Chapter 4 if you need more support. In addition, be aware that you will be asked to provide an e-mail address to qualify for the free trial version of Corona.

Step 1 First install the Notepad++ application. From **notepad-plus-plus.org**, click **Download** to install the Notepad++ editor. Launch Notepad++ and click **File->New**.

Step 2 Head to *coronalabs.com*. Enter your e-mail address; click the **Download** button and follow the download instructions.

Step 3 Run the Corona simulator.

Step 4 In Notepad++, type the following:

```
myText = display.newText("TECH IN ACTION", 400, 400, native.systemFont, 88)
myText:setTextColor(1, 0.1, 0.1)
```

Step 5 In Notepad++, click **File-> Save As**. In the Save As type drop-down, select **LUA source File (*.lua)** and save this file with the name **main**. The file *main.lua* is now created.

Step 6 Move to the Corona Simulator window and click **File->Open Project**. Navigate to *main.lua* and open it.

You have just written a complete iPhone application! You can also run it on an iPad or any Android phone or tablet. In the Corona Simulation window, click **View->View As** and see how the program output looks on a few different devices.

Step 7 Play with changing the color by altering the values in setTextColor. Try adding the line **myText.rotation = 45.**

Visit *docs.coronalabs.com/api* to see the full set of available built-in methods. Look over the physics library and begin to imagine what you could create!

(Corona Labs Inc)

Make This  TOOL: App Inventor 2 or Thunkable

A Notepad

Drawing and writing on a touch device is convenient and creative.

In this exercise, you continue your mobile app development by using the Canvas component in App Inventor to build a notepad that lets you switch pen colors and draw shapes.
The Canvas component lets you draw on the screen. Combine that with statements like "if" and "for" and you can make more sophisticated behaviors.
Use the Canvas component in App Inventor to combine control statements with drawing commands to create your own images.

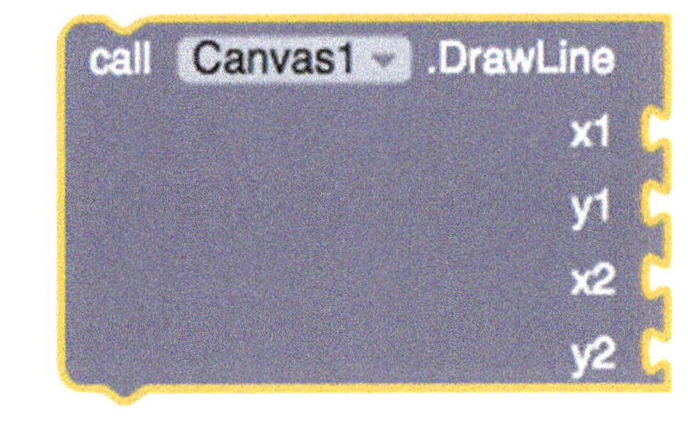

(Copyright MIT, used with permission.)

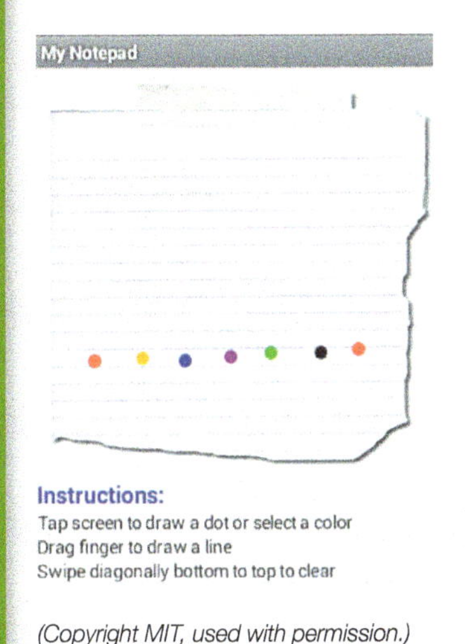

(Copyright MIT, used with permission.)

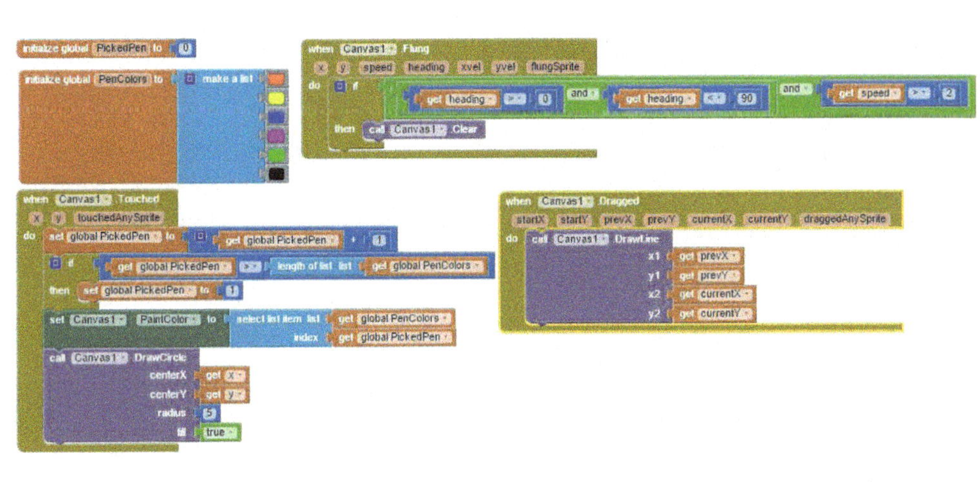

(Copyright MIT, used with permission.)

To create this app for iOS, go to *Thunkable.com*, a programming platform based on App Inventor.

For the instructions for this exercise, go to MyLab IT.

Part 2

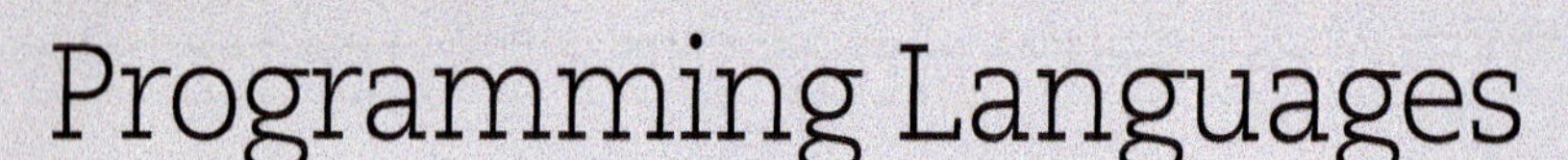

Programming Languages

Learning Outcome 10.2 You will understand the factors programmers consider when selecting an appropriate programming language for a specific problem and will be familiar with some modern programming languages.

Earlier in the chapter, you learned about the five main categories (generations) of programming languages. In this section, we discuss the specific programming languages that are members of these different generations.

Many Programming Languages

In any programming endeavor, programmers want to create a solution that meets several objectives. The software needs to:

- Run quickly
- Be reliable
- Be simple to expand later when the demands on the system change
- Be completed on time
- Be finished for the minimum possible cost

Because it's difficult to balance these conflicting goals, many programming languages have been developed. Although programming languages often share common characteristics, each language has specific traits that allow it to be the best fit for certain types of projects. The ability to understand enough about each language to match it to the appropriate style of problem is an exceptionally powerful skill for programmers.

Need for Diverse Languages

Objective 10.9 *Discuss the driving factors behind the popularity of various programming languages.*

Which languages are popular today? One quick way to determine which languages are popular is to examine job postings for programmers. As of this writing, the languages most in demand include C/C++ and Java. In specific industries, certain languages tend to dominate the work. In the banking and insurance industries, for example, the COBOL programming language is still common, although most other industries rarely use it anymore. Figure 10.18 shows the popularity of different languages in the software industry.

Programming with the Processing Language

In this Sound Byte, you use the Processing language to create interactive art.

How do I know which language to study first? A good introductory programming course will emphasize many skills and techniques that will carry over from one language to another. You should find a course that emphasizes design, algorithm development, debugging techniques, and project management. All of these aspects of programming will help you in any language environment.

Visual languages like App Inventor or Scratch are often used to introduce key programming concepts. Many colleges and universities have opted to have students begin with Java, C++, or Python. Students interested in the arts often use the Processing (*processing.org*) language as a first experience with coding.

How does anyone learn so many languages? Professional programmers can become proficient at new languages because they've become familiar with the basic components, discussed in this chapter's Dig Deeper feature, that are common to all languages: syntax, keywords, operators, and data types.

Figure 10.18 Programming Language Popularity

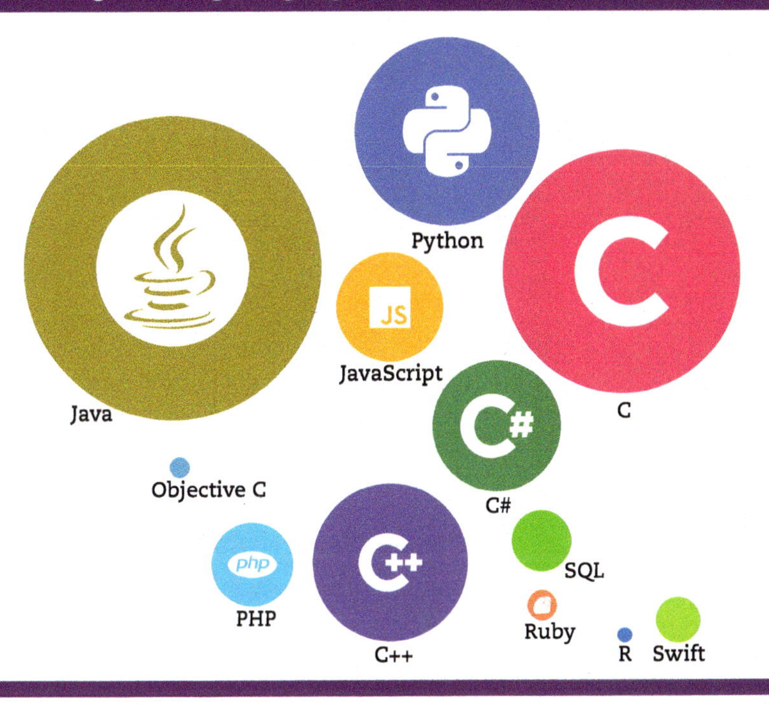

Selecting the Right Language

Objective 10.10 *Summarize the considerations in identifying an appropriate programming language for a specific setting.*

How do programmers know which language to select for a specific project? A programming team considers several factors before selecting the language it will use for a specific project:

- *Space available.* Not all languages produce code that takes up the same amount of space. Therefore, the target language should be well matched to the amount of space available for the

Bits&Bytes Learn Programming with Patch Cords

Not all programming languages demand that you focus on syntax, the semicolons and commas and grammatical details of a language. Visual programming languages are primarily written using snap and click pieces, like the tiles in App Inventor that are used in the Make This exercises in this book.

One very popular visual programming language that is popular for audio and visual processing is Max/MSP from Cycling '74. Without worrying about syntax, a Max programmer can create a visualization that interacts with a piece of electronic music or can design a multi-channel synthesizer. Patch cords are dragged from an outlet of one processing block to the inlet of another, sending data, audio, and video signals. The example in Figure 10.19 applies the input from a laptop camera to a kaleidoscope effect and displays the resulting video.

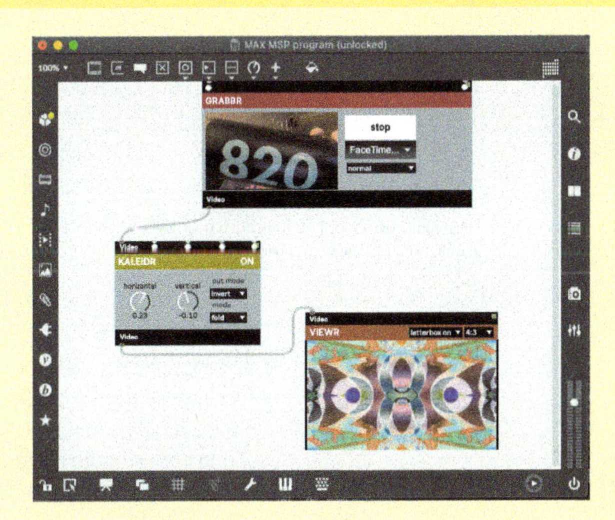

Figure 10.19 Max/MSP is a visual programming language for audio and video processing.

Ethics in IT

When Software Runs Awry

In the field of software programming, mistakes can be costly. Consider these examples:

- In 1996, Europe's unmanned rocket, the Ariane 5, had to be destroyed seconds after launch, along with its cargo of four scientific satellites. Cost: $500 million. The problem occurred when the guidance computer tried to convert rocket velocity from a 64-bit to a 16-bit value. Programmers had not properly coded for this, and an overflow error resulted. When the guidance system shut down, control passed to the backup unit, which also failed because it was running the same algorithm.
- In 2020, London's Heathrow Airport had to ground or delay hundreds of flights because the software that runs the departure boards and check-in systems malfunctioned. The year before saw a British Airways software failure that stopped all online check-ins, leading to enormous check-in lines and delays.
- In 2012, the Knight Capital Group had a bug in the software it used to trade stocks. The error cost the firm $440 million in 30 minutes of trading when the software began to send out orders all at once that were meant to be placed across several days.

Could better software engineering practices have prevented these failures? Many factors contributed to this set of accidents, including:

- Simple programming errors
- Inadequate safety engineering
- Poor human–computer interaction design
- Inadequate or inappropriate testing
- Too little focus on safety by the manufacturing organization
- Inadequate reporting structure at the company level

Who should be held responsible for defective software? Is it the corporate management, who did not institute a defined software process? Is it the production managers, who forced tight schedules that demanded risky software engineering practices? What about the software engineers who wrote the defective code? What about the users of the software? Can they be held responsible for accidents? What if they made changes to the system?

Organizations of engineers and software designers have tried to define ethical standards that will minimize the risk of negative impact from their work. The very first article of the code of ethics of the Institute of Electrical and Electronics Engineers (IEEE) alerts engineers to their responsibility for making decisions that protect the health and safety of others. The Association for Computing Machinery (ACM) and the IEEE have established eight principles for ethical software engineering practices:[1]

1. *Public:* Software engineers shall act consistently with the public interest.
2. *Client and Employer:* Software engineers shall act in a manner that is in the best interests of their client and employer consistent with the public interest.
3. *Product:* Software engineers shall ensure that their products and related modifications meet the highest professional standards possible.
4. *Judgment:* Software engineers shall maintain integrity and independence in their professional judgment.
5. *Management:* Software engineering managers and leaders shall subscribe to and promote an ethical approach to the management of software development and maintenance.
6. *Profession:* Software engineers shall advance the integrity and reputation of the profession consistent with the public interest.
7. *Colleagues:* Software engineers shall be fair to and supportive of their colleagues.
8. *Self:* Software engineers shall participate in lifelong learning regarding the practice of their profession and shall promote an ethical approach to the practice of the profession.

Can you think of other steps the software industry could take to make sure its work contributes positively to our society?

[1] "Software Engineering Code of Ethics and Professional Practice," by IEEE-CS/ACM Joint Task Force on Software Engineering Ethics and Professional Practices. Copyright © 2011 by the Institute of Electrical and Electronics Engineers. Reprinted by permission.

final program. For example, if the program will be embedded in a chip for use in a cell phone, it's important for the language to create space-efficient programs.

- *Speed required.* Although poorly written code executes inefficiently in any language, some languages can execute more quickly than others. Some projects require a focus on speed rather than on size. These projects require a language that produces code that executes in the fastest possible time.
- *Organizational resources available.* Another consideration is the resources available in a manager's group or organization. Selecting a language that's easy to use and that will be easy to maintain if there's a turnover in programmers is an important consideration.

- *Type of target application.* Certain languages are customized to support a specific environment (UNIX or Windows, for instance). Knowing which languages are most commonly used for which environments can be helpful. We discuss this more in the next section.

Can I just point and click to create an application? Many languages have a development environment that features a drag-and-drop–style interface for designing the visual layout of an application. Programmers use a mouse to lay out the screen and position scroll bars and buttons. The code needed to explain this to the computer is then written automatically. Visual programming languages, like Scratch and App Inventor, go even further. They use graphical blocks to represent control elements and variables. Programming consists of clicking these blocks together to define program behavior.

Exploring Programming Languages

How are modern programming languages the same and how are they different? What will future programming languages look like? Let's find out.

Tour of Modern Languages

Objective 10.11 *Compare and contrast modern programming languages.*

All programming languages have many features in common. Let's take a tour of many of the popular languages and learn what makes each one special for particular situations.

Visual Basic .NET

Why do programmers choose Visual Basic? Programmers often like to build a prototype, or small model, of their program at the beginning of a large project. Prototyping is a form of rapid application development (RAD), an alternative to the waterfall approach of systems development that was described at the beginning of this chapter. Instead of developing detailed system do++Ωcuments before they produce the system, developers create a prototype first, and then generate system documents as they use and remodel the product.

Visual Basic (VB) was developed as a language that could be learned quickly and used to build a wide range of Windows applications. The interface is easy for a new programmer to learn and use. It began as the BASIC (Beginner's All-purpose Symbolic Instruction Code) language and grew to add easy-to-manipulate visual programming features, hence Visual Basic. Microsoft no longer supports Visual Basic, but you will find a related version of the language, Visual Basic for Applications (VBA), used to construct custom solutions for the Microsoft Office suite.

What is VB.NET? VB.NET is a more modern, sophisticated, and full-featured object-oriented language. It has its roots in Visual Basic but has significant changes. VB.NET employs the Microsoft .NET (pronounced "dot net") Framework, a software development environment designed to let websites talk to each other easily. The .NET Framework includes web services, programs that a website uses to make information available to other websites. Web services provide a standard way for software to interact. The rules are documented in an application programming interface (API). For example, a web application could use the Google web service to search for information or to check the spelling of a word. The Google web service returns the requested information to your program in a standard format. Figure 10.20 shows just a small list of the many web services available to programmers.

Figure 10.20 Web Services

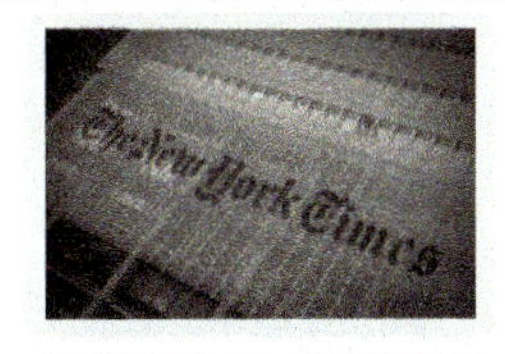

***The New York Times* API** offers access to articles, book lists, and movie reviews.

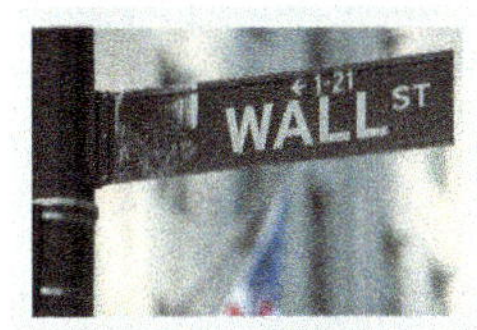

Yahoo Finance API provides stock quotes, historic quotes, exchange rates, and company reports.

Wunderground API provides data on tides, forecasts, hurricanes, and weather alerts.

Chicago Open City gives access to plow locations, crime statistics, public school data, and more. Most cities offer a similar API.

(Dipak Pankhania/Alamy Stock Photo; Mbbirdy/E+/Getty Images; Solarseven/123RF; Gary Hebding Jr./Alamy Stock Photo)

C and C++

What languages do programmers use if the problem requires a lot of number crunching? A Windows application that demands raw processing power to execute difficult, repetitive numerical calculations is most often a candidate for C/C++. For example, applications that simulate human cells and drug interactions have to solve elaborate mathematical equations many thousands of times each second and are, therefore, excellent candidates for programming using C/C++. Several companies sell C/C++ design tools equipped with an environment that makes Windows programming easier with a visual interface.

Why was the C language developed? The predecessor of C++, C, was originally developed for system programmers. It was defined by Brian Kernighan and Dennis Ritchie of AT&T Bell Laboratories in 1978 as a language that would make accessing the operating system easier. It provides higher-level programming language features (such as if statements and for loops) but still allows programmers to manipulate the system memory and CPU registers directly. This mix of high- and low-level access makes C highly attractive to power programmers. Most modern operating systems (Windows, macOS, and Linux) were written in C.

The C++ language takes C to an object-oriented level. Bjarne Stroustrup, the developer of C++, used all the same symbols and keywords as C, but he extended the language with additional keywords, better security, and more support for the reuse of existing code through object-oriented design.

Are C and C++ natural choices for when I'm looking to learn my first language? Neither C nor C++ was intended as a teaching language. The notation and compactness of these languages make them relatively difficult to master. They're in demand in industry, however, because C/C++ can produce fast-running code that uses a small amount of memory. Programmers often choose to learn C/C++ because their basic components (operators, data types, and keywords) are common to many other languages.

Java and C#

What language do programmers use for applications that need to collect information from networked computers? Java would be a good choice for these types of applications. James Gosling of Sun Microsystems introduced Java in the early 1990s. It quickly became popular because its object-oriented model enables Java programmers to benefit from its large set of existing classes. For example, a Java programmer could begin to use the existing network connection class with little attention to the details of how that code itself was implemented. Classes exist for many graphical objects, such as windows and scroll bars, and for network objects such as connections to remote machines. Observing Java's success, Microsoft released a language named C# (pronounced "see sharp") that competes with Java.

Can a Java application work on any type of computer? An attractive feature of Java is that it's architecture-neutral. This means that Java code needs to be compiled only once, after which it can run on many CPUs (see Figure 10.21). The Java program doesn't care which CPU, OS, or user interface is running on the machine on which it lands. This is possible because the target computer runs a Java virtual machine (VM), software that can explain to the Java program how to function on any specific system. A Java VM installed with Microsoft Edge, for example, allows Edge to execute any Java applet (a small Java-based program) it encounters on the Internet. Although Java code

Bits&Bytes **Repl.It**

Do you want to try the Java programming language but hesitate when it comes to installing the compiler and an IDE just to try it? Or maybe you wish you could play with a few examples in Python or Nodejs, but you do not want to commit to making changes to your computer system to do it. Explore Repl.It (*repl.it*), a free in-browser IDE that lets you try programming in any of over 50 languages. Simple examples are provided for each language. You can easily invite friends to join you and explore programming together.

As you become more sophisticated in your programming, Repl.it can grow with you. It supports external APIs and source code control, using services like GitHub. Try out a new language, with no commitment!

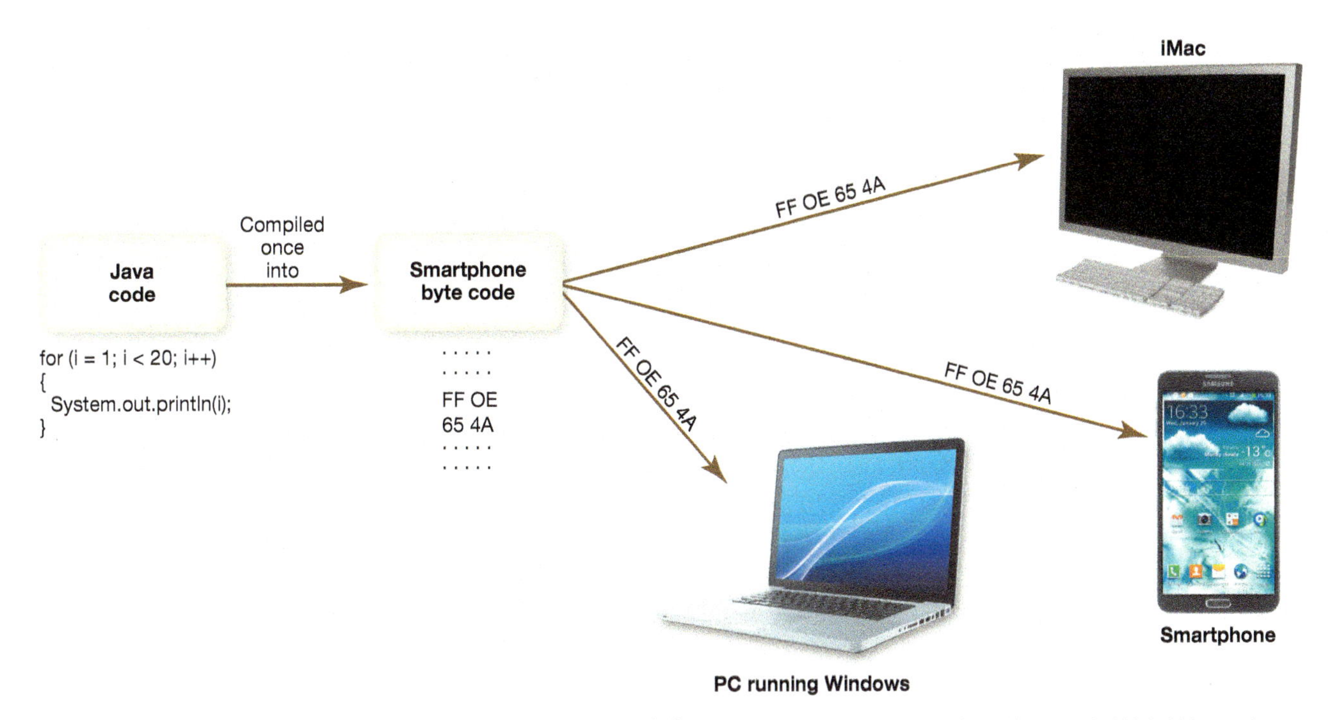

Figure 10.21 Java programs can be compiled once and run on many platforms. *(D. Hurst/Alamy Stock Photo; Ifong/Shutterstock; Oleksiy Maksymenko Photography/Alamy Stock Photo)*

doesn't perform as fast as C++, the advantage of needing to compile only once before a program can be distributed to any system is extremely important.

Objective C and Swift

What's the most popular language for writing macOS applications? Objective C is the language most often used to program applications to run under macOS (see Figure 10.22). It's an object-oriented style of language, a superset of the C language, so it includes all the keywords of C and then adds more keywords and features. It's often used together with a library called Cocoa. The Cocoa library, or framework, enables users to program for the macOS graphical user interface. The Cocoa Touch extension introduces a framework of methods that support gesture recognition for touch devices like the iPhone or iPad.

Swift is a newer programming language Apple introduced for developing for iOS and macOS. Easier to read and maintain than Objective C, Swift uses a playground area where you can quickly write and test ideas before moving the code into the main project.

Is there a favorite IDE for Objective C? Many IDE tools support Objective C, but macOS ships with a tool named Xcode that's often used to develop Objective C applications for macOS. Xcode is available free from the online Mac App Store.

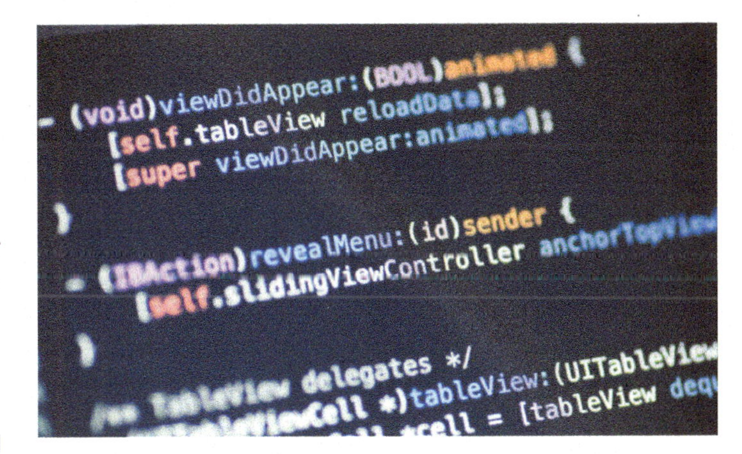

Figure 10.22 Objective C is the programming language that is most often used to program applications to run under macOS. *(Photovibes/Shutterstock)*

HTML

What's the most basic formatting language for web applications? A document that will be presented on the web must be written using special symbols called *tags*. Tags control how a web browser will display the text, images, and other content tagged in Hypertext Markup Language (HTML).

Although knowledge of HTML is required to program for the web, HTML itself isn't a programming language. It is just a series of tags that modify the display of text. HTML was the original standard defining these tags. Many good HTML tutorials are available on the web at sites such as Learn the Net (*learnthenet.com*) and the W3 Schools (*w3schools.com*). These sites include lists of the major tags that can be used to create HTML documents.

Are there tools that help programmers write in HTML? Several programs are available to assist in the generation of HTML. Adobe Dreamweaver presents web page designers with an interface that's similar to a word processing program. Web designers can quickly insert text, images, and hyperlinks, as shown in Figure 10.23. The program automatically inserts the corresponding HTML tags. For simple, static web pages, no programming is required. Online services like Squarespace and Wix enable you to design and maintain a website with powerful features with no direct programming.

Will HTML continue evolving? Yes, a working committee looks constantly at repairing errors and improving the HTML standard. HTML5 was a significant advancement. For example, new HTML5 tags <audio> and <video> allow pages to integrate media easily and enable designers more easily support manipulating those elements. HTML5 eliminates the need to install third-party browser plug-ins and supports features like drag-and-drop elements and document editing.

Scripting Languages

Which programming languages do programmers use to make complex web pages? To make their web pages more visually appealing and interactive, programmers use scripting languages to add more power and flexibility to their HTML code. A scripting language is a simple programming language that's limited to performing a set of specialized tasks. Scripts allow decisions to be made and calculations to be performed. Several popular scripting languages work well with HTML, including JavaScript and PHP (Hypertext Preprocessor).

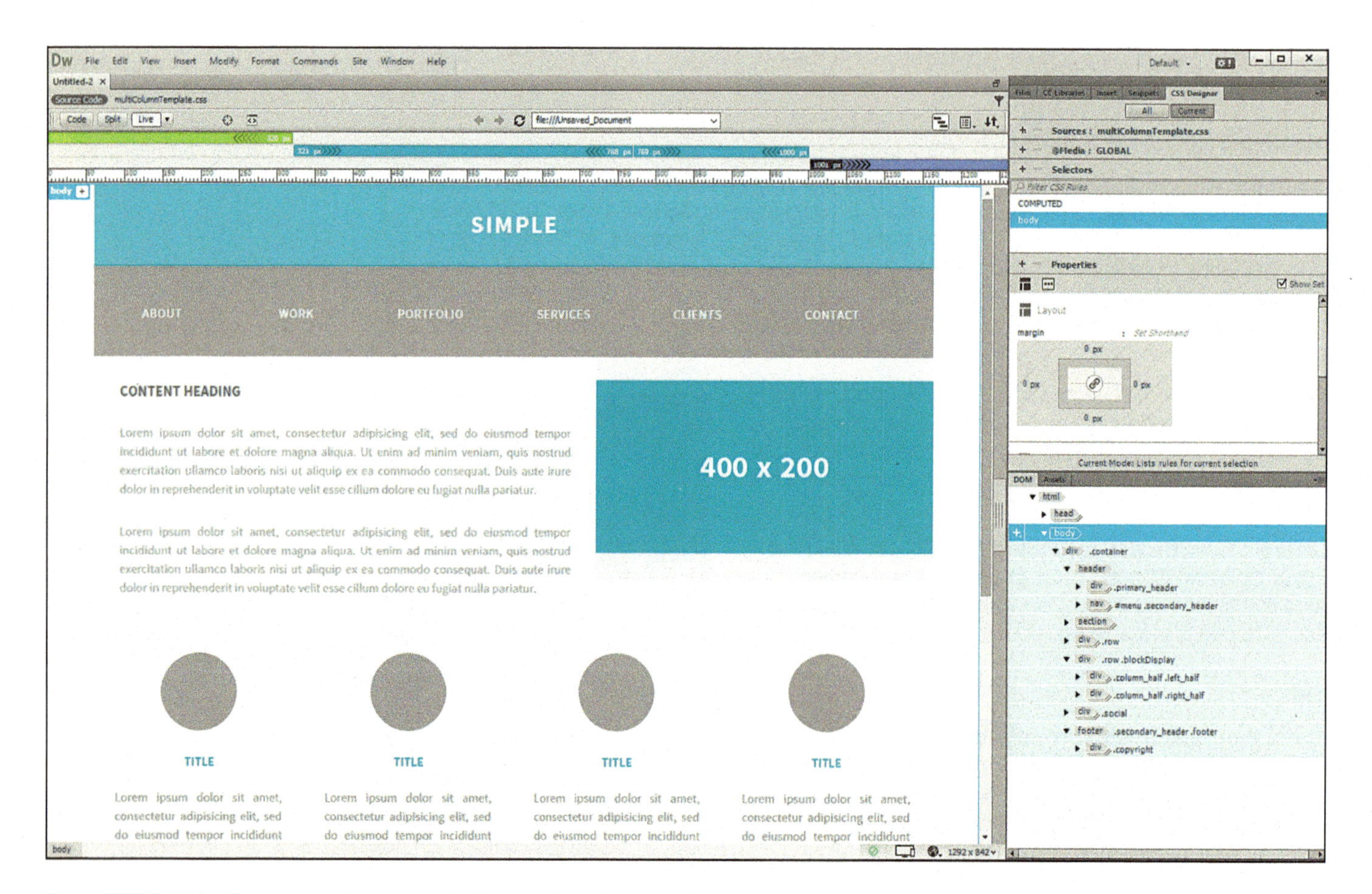

Figure 10.23 Adobe Dreamweaver is a popular tool for creating web pages. *(© 2018 Adobe. All rights reserved. Adobe and Illustrator are either registered trademarks or trademarks of Adobe in the United States and/or other countries.)*

JavaScript is a scripting language that's customized to work with the elements of a web page and is often used to add interactivity to web pages. The syntax, keywords, data types, and operators are similar to those of Java. In addition, JavaScript has a set of classes that represent the objects often used on web pages: buttons, check boxes, and drop-down lists.

The JavaScript button class, for example, describes a button with a name and a type—for example, whether it's a regular button or a Submit or Reset button. The language includes behaviors, such as click(), and can respond to user actions. For example, when a user moves his or her mouse over a button and clicks to select it, the button knows the user is there and jumps in and performs a special action (such as playing a sound).

Are there other scripting languages besides JavaScript? PHP, discussed in the following section, is another scripting language that's commonly used. It's a free, open-source product that runs very efficiently on multiple platforms, including Windows, UNIX, and Linux. Scripting languages like PHP and JavaScript allow dynamic decision making to be incorporated into the web page design. Dynamic decision making means that the page can decide how to display itself based on the choices the reader makes. Programmers who are more familiar with Visual Basic than with Java or C++ might use VBScript, a subset of Visual Basic.

ASP, JSP, and PHP

How are interactive web pages built? To build websites with interactive capabilities, programmers use Active Server Pages (ASP), JavaServer Pages (JSP), or the PHP (PHP: Hypertext Preprocessor) scripting language to adapt the HTML page to the user's selections. The user supplies information that's translated into a request by the main computer at the company that owns the website, often using a database query language such as SQL. Scripting code in ASP, JSP, or PHP controls the automatic writing of the custom HTML page that's returned to the user's computer.

What does additional programming bring to a web page? Modern web applications interact with the user, collecting information and then customizing the content displayed based on the user's feedback. For example, the client/server-type application shown in Figure 10.24 shows the web page of the ABC Bike Company, collecting a customer's bicycle inquiry for red bikes. The program then asks ABC's main server for a list of red bicycles sold by the company. A PHP program running on the server creates a new HTML page and delivers that to the user's browser, telling the customer what red bikes (including details such as model and size) ABC currently sells.

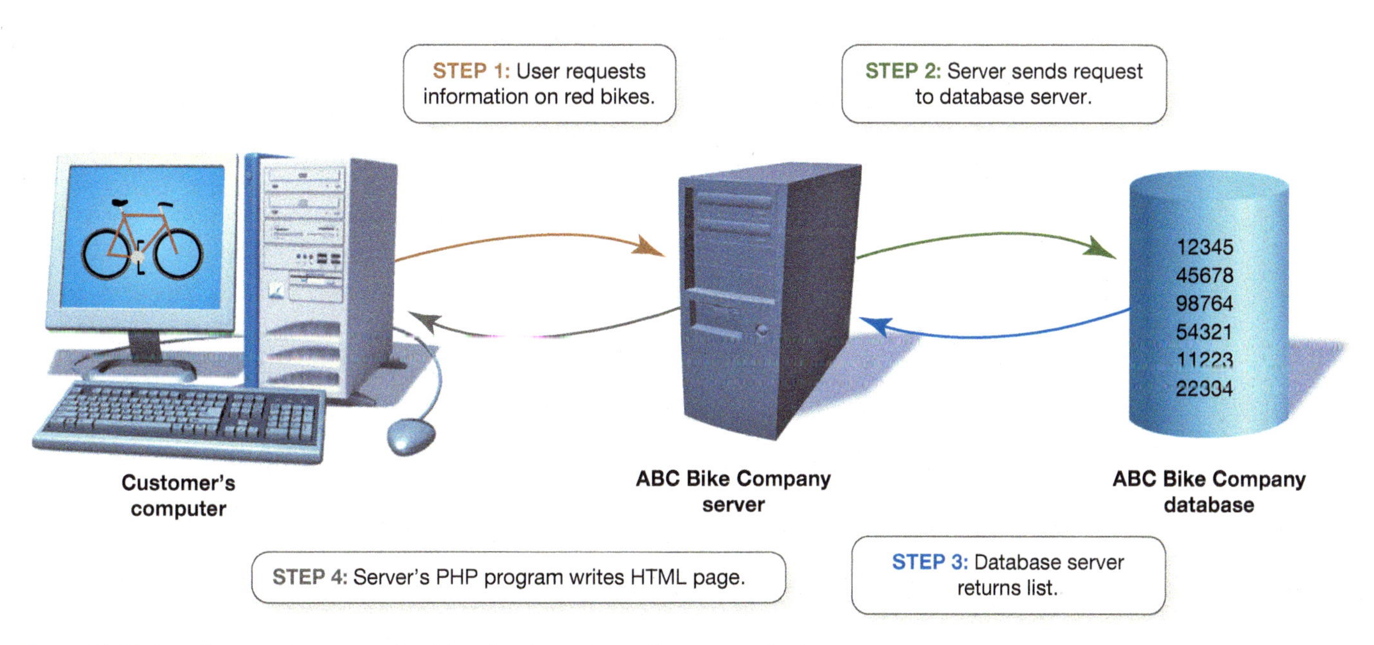

Figure 10.24 An online store is an example of the client/server type of Internet application.

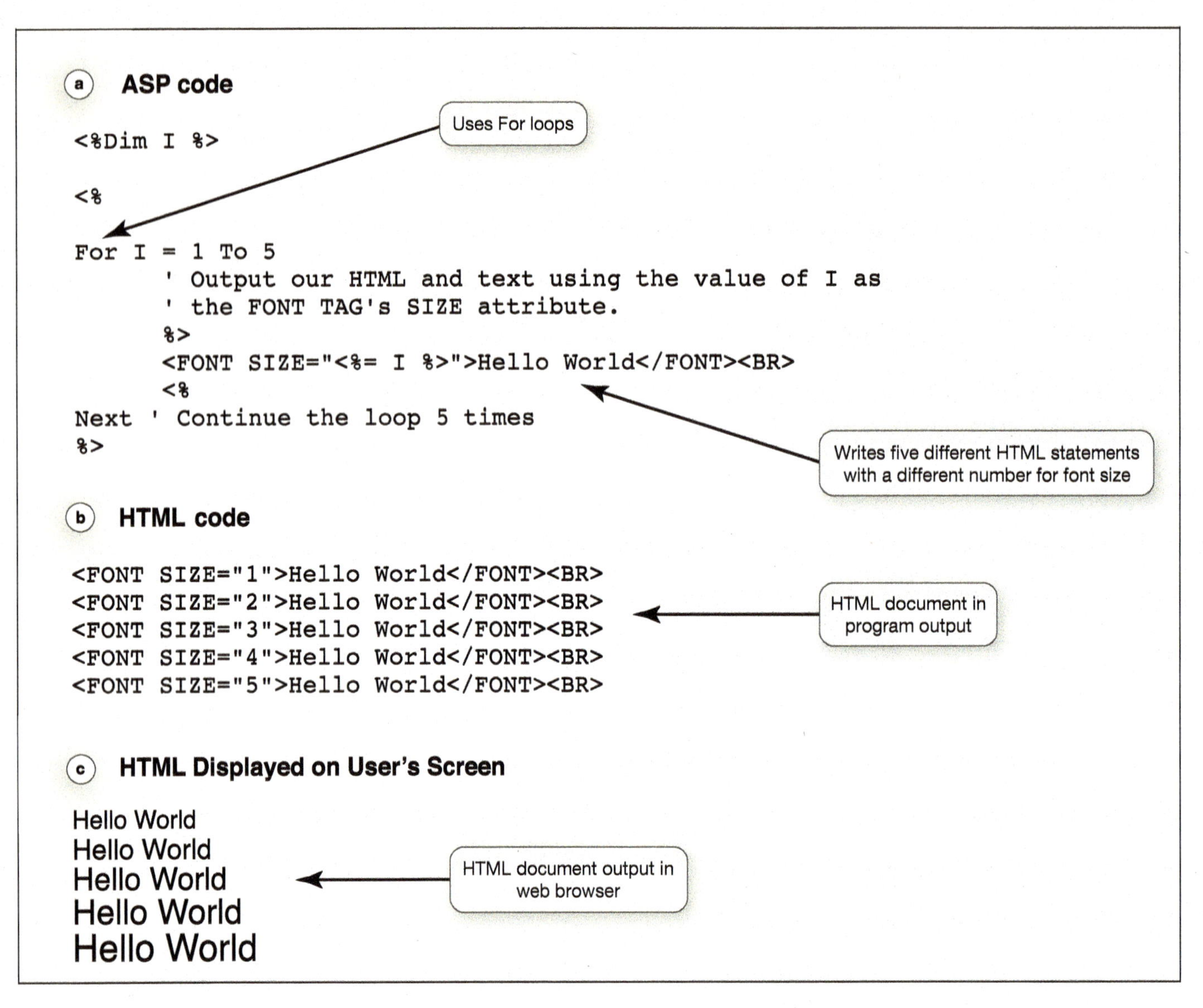

Figure 10.25 An ASP program can (a) write HTML code as its (b) output. (c) This image illustrates how the HTML page would show up in a browser.

Server programs, written in ASP or PHP, can output HTML code. They use what the user has told them (from the list boxes, check boxes, and buttons on the page) to make decisions. Based on those results, the server software decides what HTML to write. A small example of ASP, writing its own HTML code, is shown in Figure 10.25.

AJAX, XML, and JSON

What if a programmer wants to create a web application that smoothly updates and communicates with other computers? Many websites feature elaborate animations that interact with visitors. They may reach out to several other web resources to gather information or request services such as translation. The collection of technologies referred to as AJAX (Asynchronous JavaScript and XML) and the continued evolution of HTML5 allow such web applications to update information on a page without requiring the user to do a page refresh or leave the page. By using existing technologies to do more processing in the browser, users have a more responsive experience.

How do programs gather information from other computers? The markup language called eXtensible Markup Language (XML) enables designers to define their own data-based tags, making it much easier for a program running on a server computer (a web service) to transfer the key information on its page to another site. When websites communicate with humans, HTML works well because the formatting it controls is important. People respond immediately to the visual styling of textual information; its layout, color, size, and font design all help to transfer the message of the page to the reader. When computer programs want to communicate with one another, however, all these qualities just get in the way.

With XML, groups can agree on standard systems of tags that represent important data elements. For example, the XML tags <stock> and </stock> might delimit key stock quote information. Mathematicians have created a standardized set of XML tags named MathML for their work, and biometrics groups continue to refine XML standards to describe and exchange data such as DNA and face scans. Without XML, a website that wanted to look up current stock pricing information at another site would have to retrieve the HTML page, sort through the formatting information, and try to recognize which text on the page identified the data needed. JavaScript Object Notation (JSON) is another standard for exchanging information between a server computer process and a client. The information is delivered as a series of names and their values, making it easy for programs to parse and easy for humans to read.

Table 10.3 shows a table of popular programming languages and technologies with their features and the typical settings in which they're used.

Mobile Application Development

How do programmers build applications for mobile devices? Special languages and supporting tools are speeding the development of applications for mobile devices like smartphones and tablets. Programmers need to be able to take advantage of specific features like GPS capability, compasses, software keyboards, and touch-sensitive screens. In addition, the user interface has to take the smaller screen size of mobile devices into account.

Table 10.3 Popular Programming Languages

Programming Language	Features	Typical Setting
C/C++ and C#	• Can create compact code that executes quickly • Provides high- and low-level access	• Used in industrial applications such as banking and engineering
Java	• Is architecture-neutral • Is object-oriented	• Used to create applets that can be delivered over the web
Objective C	• Has a framework for writing iOS applications	• Used to create applications for macOS and Apple mobile devices
Visual Basic .NET	• Is easy to learn and use • Is object-oriented • Has a drag-and-drop interface	• Used in prototype development • Used to design graphical user interfaces
Web Technologies	**Features**	**Typical Setting**
AJAX	• Uses a combination of existing technologies, like JavaScript, CSS, and XML	• Creates websites that can update without the user refreshing the page
HTML5	• Latest version of HTML	• Introduces tags like <video> and supports drag and drop
VBScript	• Is similar in syntax to Visual Basic • Has classes that represent buttons, drop-down lists, and other web page components	• Creates code that lives on the client machine and adds interaction to web pages
XML	• Enables users to define their own tags	• Facilitates exchange of information from web services
JSON	• Format defined with name/value pairs	• Very common format for exchange of information from web services

What development tools are used for creating mobile apps for Apple's iOS platform? To start a complex project like an iPhone app requires a detailed prototype. Each of the many screens, all the user interface elements, and all the content need to be organized and linked smoothly. Often, programmers begin with a prototype, created quickly with drag-and-drop elements, using products like Proto.io or Interface Builder. Proto.io (*proto.io*) is a browser-based tool that enables you to construct a working simulation of your application. Interface Builder is part of the Apple Xcode development tool; it requires a bit more expertise to use than Proto.io but can also rapidly create a prototype.

When it's time to begin writing the code for an iOS app, programmers turn to Objective C and use the Apple Xcode development toolset. Designers use Xcode to code and debug the behavior of the application and simulate the application in a software version of the target device. After the program is running, its performance can be profiled for speed, memory usage, and other possible problems.

Are there different tools for building apps for Android devices? Yes, the Android software development kit (SDK) is required to build apps targeting Android smartphones and tablets. There are many ways programmers work with the Android SDK, including using well-known IDEs like Eclipse with special plug-ins or using the Android Studio, shown in Figure 10.26. Information on the latest version of the Android SDK, as well as tutorials, guides, and other resources, is available at the Android Developers page (*developer.android.com*).

I'm not a programmer, but can I make a simple app? Absolutely. Tools like Corona and App Inventor produce amazing games and apps quickly (see the Try This on page 394 for more information). Tens of millions of apps constructed using Corona have been downloaded. The apps you have written using App Inventor in the Make This exercises can be uploaded to the Google Play store for download worldwide.

If your goal is to make a mobile app that's very simple or even specific to one occasion, like a wedding, there are web-based products that make that quick and easy. Magmito (*magmito.com*) supports development of a simple app with text and graphics and requires no programming knowledge.

Does an application need to be rewritten for every kind of mobile device? Programming environments like Corona and Magmito support publishing an application to several types of devices. Although these tools can be great time-savers for very simple applications, for programmers using specific features that make that device unique or for those concerned with extracting ultimate performance, custom programming for each environment is still required.

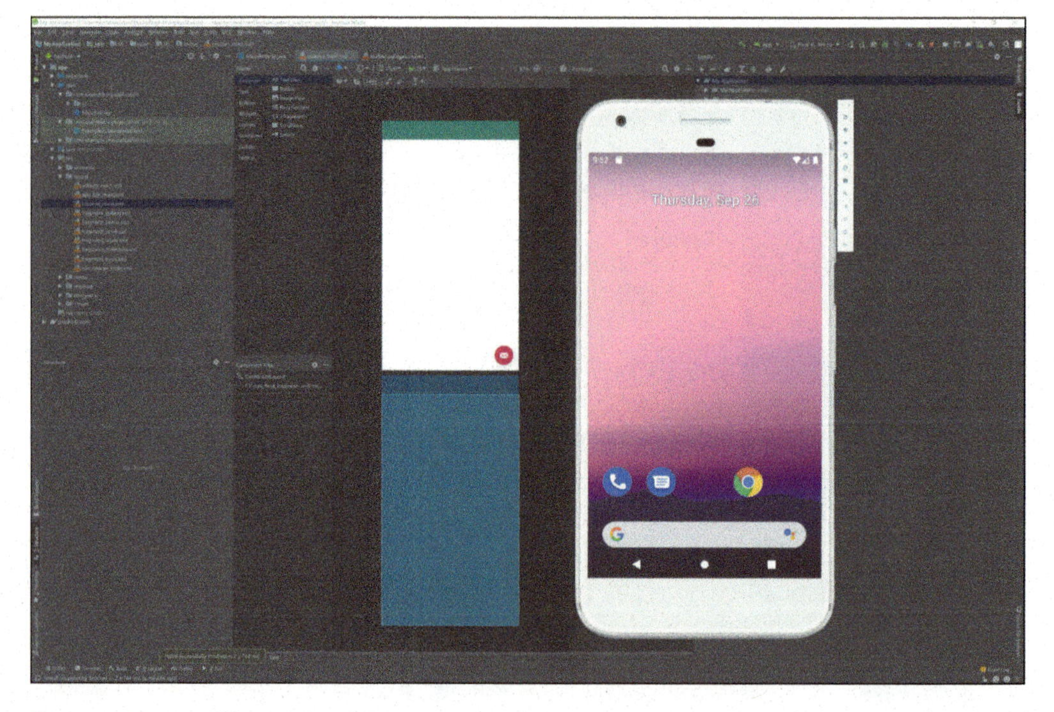

Figure 10.26 Android Studio integrates all the tools required to program Android applications from scratch, targeting tablets, phones, TVs, wearables, and even Android Auto. *(Courtesy of Google Inc.)*

Trends in IT — Emerging Technologies: Unite All Your Video Game Design Tools

Unity is a game-development environment that supports home video game production by combining many tools into one package, giving you what you need to edit, test, and play your game idea quickly. It can generate real-time shadowing effects, for example, and preprogrammed scripts are integrated. After you've polished your production, Unity allows you to publish it for web, mobile, or console platforms. With over five million developers using Unity, and over 700 million people playing Unity-made games, the potential of this tool is amazing (see Figure 10.27).

You'll start with the built-in Unity editor so you can quickly assemble a 3D environment. Unity gives you access to its Asset Store right from the editor window. Art assets are supplied so that you instantly have materials to create your own landscape. Tutorials and sample projects are available, as well as character models and textures. Grab a sound effect or some code to simulate a great explosion effect. If you create assets you'd like to distribute, you can put them up for sale in the Asset Store.

Unity incorporates lighting control and shading and the application of textures to develop rich terrains. Its built-in physics engine lets objects respond to gravity, friction, and collisions. One example is the wheel collider, a module that simulates the traction of real car tires. Audio processing is included, with effects like echo, reverb, and a chorus filter, as are tools to make real-time networking available from your game.

Programming is handled through several flexible scripting languages—JavaScript, C#, and an implementation of Python named Boo. It has a visual interface that lets you drag and drop objects to set variables, or you can choose to sync with an external IDE like Visual Studio. There's an integrated debugger so that you can pause your game, check variable values, and quickly repair errors. When it's time to optimize the performance before your final release, you can use the built-in profiler: It reports statistics on where the processing time is being spent so that you can fine-tune game play.

Figure 10.27 Unity is a free game engine that can build web, mobile, or console games. *(Courtesy of Unity Technologies)*

Unity is available in a free downloadable version (*unity3d.com*) and in pay versions. The Unity 3D Pro version adds features and can also be licensed and used by companies. The free version of Unity can be used to create games that can be sold—no royalty or revenue sharing is required. Modules also let you easily port your code to a number of target platforms: PC, iOS, Android, and the web, as well as Xbox 360, PS4, and Wii U.

You'll also find an active community to support you as you learn. Live online training courses are offered, as are forums for community discussion. There is also a dedicated Internet relay chat (IRC) channel for live chatting with Unity users. If you need help distributing your finished game, you can join Union, a service that takes your game to new markets for you.

If you're interested in video game development, architectural visualizations, or just creating interactive animations, download Unity and realize the power of programming.

Future of Programming Languages

Objective 10.12 *State key principles in the development of future programming languages.*

What will the next great language be? It is never easy to predict which language will become the next great language. Software experts predict that as software projects continue to grow in size, the amount of time needed to compile a completed project will also grow. It's not uncommon for a large project to require 30 minutes or more to compile. Interpreted languages, however, take virtually no compile time because compilation occurs while the code is being edited. As projects get larger, the capability to be compiled instantaneously will become even more essential. Thus, interpreted languages such as Python, Ruby, and Smalltalk could become more important in the coming years.

Will visual programming languages become more popular? Visual programming languages (VPLs) have a history of more than 50 years. The idea of "programming without code"—also the name of an MIT open source project to promote VPLs—is maturing but is not yet a tool for most professional software development environments. The future may bring a push for the elegance of visual programming. VPLs use visual modeling, creating the code by using images to represent each statement, which helps programmers understand complex designs. We have already

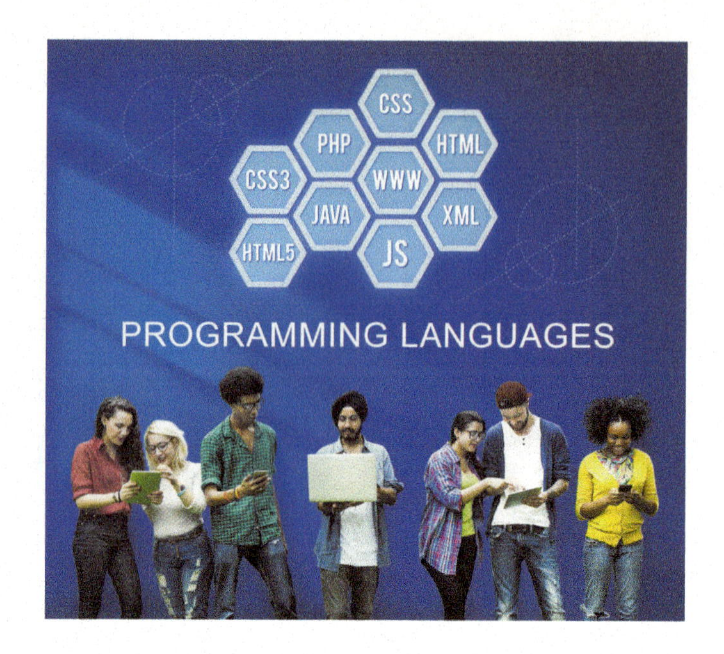

Figure 10.28 Just as there is diversity in human programmers, each programming language has its own personality. *(Rawpixel.com/ Shutterstock)*

seen a move toward visual modeling in audio production in the development of software for adding effects like echo or reverse to music snippets. Programming environments like Max/MSP are visual and have a growing and enthusiastic user base.

Will all languages someday converge into one? Certain characteristics of modern programming languages correspond well with how programmers actually think. These traits support good programming practices and are emerging as common features of most modern programming languages. The object-oriented paradigm is one example. Both VB.NET and COBOL are now object-oriented languages.

There will always be a variety of programming languages, each with its own distinct characteristics. As Figure 10.28 shows, this also mirrors the diversity of the programmer community.

Forcing a language to be so general that it can work for any task also forces it to include components that make it slower to compile, produce larger final executables, and require more memory to run. Having a variety of languages and mapping a problem to the best language create the most efficient software solutions.

> **Helpdesk** MyLab IT
>
> **A Variety of Programming Languages**
>
> In this Helpdesk, you'll play the role of a helpdesk staffer, fielding questions about the variety of modern programming languages: what makes one popular, which language is best for a certain situation, and how they all compare.

So what do I do if I want to learn languages that will be relevant in the future? No particular set of languages is best to learn, and there is no one best sequence in which to learn them. The Association for Computing Machinery (*acm.org*) encourages educators to teach a core set of mathematical and programming skills and concepts, but school and university departments are free to offer a variety of languages.

When you're selecting which programming languages to study, some geographical and industry-related considerations come into play. For example, in an area in which a large number of pharmaceutical companies exist, there may be a demand for the Massachusetts General Hospital Utility Multi-Programming System (MUMPS). This language is often used to build clinical databases, an important task in the pharmaceutical industry. Review the advertisements for programmers in area newspapers and investigate resources such as Glassdoor (*glassdoor.com*) to identify languages in demand in your area.

Regardless of whether you pursue a career in programming, having an understanding of how software is created will help you in many IT careers. Software is the set of instructions that allows us to use our hardware. Programming skills give you the power to understand, create, and customize a computer system.

> **Before moving on to the Chapter Review:**
>
> 1. Watch Chapter Overview Video 10.2.
> 2. Then take the Check Your Understanding quiz.

Check Your Understanding // Review & Practice

For a quick review to see what you've learned so far, answer the following questions.

multiple choice

1. Rapid application development (RAD) encourages developers to
   a. use an integrated debugging environment.
   b. write code in visual programming languages.
   c. create code that executes as fast as possible.
   d. create a prototype first during the development of a system.
2. Selecting the right programming language for a project depends on
   a. who you have working for you and what they know.
   b. the final requirements for performance and space used.
   c. the environment of the finished project.
   d. all of the above
3. Which is NOT an advantage of Java?
   a. Java is architecture-neutral.
   b. Java needs to compile only once prior to distribution.
   c. Java performs faster than C++.
   d. Java supports network communications.
4. Which is TRUE about XML?
   a. XML supports the development of rich multimedia.
   b. XML enables designers to define their own data-based tags.
   c. XML makes it possible to update web pages without refreshing.
   d. XML has classes that represent drop-down lists and other web elements.
5. Patterns of repeating actions are handled in programming languages, using
   a. an object-oriented model.
   b. encapsulation.
   c. binary decision handling.
   d. control structures like loops.

chew on this

Is there one programming language that can outperform all others? How would you select the first programming language to learn? Will the number of programming languages you need to know as a professional programmer keep increasing or settle down to just a few choices?

MyLab IT Go to **MyLab IT** to take an autograded version of the *Check Your Understanding* review and find all media resources for the chapter.

For the IT Simulation for this chapter, see MyLab IT.

10 Chapter Review

Summary

Part 1
Understanding Programming

Learning Outcome 10.1 **You will be able to describe the life cycle of a software project and identify the stages in the program development life cycle.**

Life Cycle of an Information System

Objective 10.1 *Describe the importance of programming to both software developers and users.*

- Programming skills allow you to customize existing software products to accomplish required tasks. With a beginning-level knowledge of programming, you can create macros, customized mini-programs that speed up redundant tasks.

Objective 10.2 *Summarize the stages of the system development life cycle.*

- An information system includes data, people, procedures, hardware, and software. The set of steps followed to ensure that development proceeds in an orderly fashion is the system development life cycle (SDLC). There are six steps in the SDLC waterfall model:

  1. A problem or opportunity is identified.
  2. The problem is analyzed, and a program specification document is created to outline the project objectives.
  3. A detailed plan for programmers to follow is designed using flowcharts and data-flow diagrams.
  4. Programmers use the developed plan to develop the program, and the program is then documented.
  5. The program is tested to ensure that it works and that it's installed properly.
  6. Ongoing maintenance and evaluation ensure a working product.

Life Cycle of a Program

Objective 10.3 *Define programming and list the steps in the program development life cycle.*

- Programming is the process of translating a task into a series of commands that a computer will use to perform that task. The PDLC involves five steps: describing the problem, developing the algorithm, coding, debugging, and testing and documentation.

Objective 10.4 *Describe how programmers construct a complete problem statement from a description of a task.*

- The problem statement is an explicit description of what tasks the computer program must accomplish and how the program will execute these tasks and respond to unusual situations. The problem statement describes the input data that users will have at the start of the job, the output that the program will produce, and the exact processing that converts these inputs to outputs. The problem statement identifies potential errors and plans to address these errors.

Objective 10.5 *Explain how programmers use flow control and design methodologies when developing algorithms.*

- Programmers create an algorithm by converting a problem statement into a list of steps and identifying where decision points occur. Yes/no binary decisions are common, and often the pattern of a repeating action loop is recognized. Algorithms are documented in the form of a flowchart or in pseudocode. Programmers use either top-down or object-oriented analysis to produce the algorithm.

Objective 10.6 *Discuss the categories of programming languages and the roles of the compiler and the integrated development environment in coding.*

- Computer code uses special words and strict rules to enable programmers to control the CPU without having to know all of its hardware details. Programming languages are classified in several groupings, sometimes referred to as *generations*. The first generation is machine language—the binary code of 1s and 0s that the computer understands. Assembly language is the next generation; it uses short, English-like commands that speak to the CPU and give the programmer direct control of hardware resources. Each successive generation has relieved programmers of some of the burden of keeping track of what the hardware requires and more closely matches how humans think about problems.
- Compilation is the process by which code is converted into machine language, the language the CPU

can understand. A compiler understands both the syntax of the programming language and the structure of the CPU and its machine language. It translates the instructions written by programmers in the higher-level language into machine language, the binary patterns that execute commands on the CPU. Each programming language has its own compiler.
- An integrated development environment (IDE) is a developmental tool that helps programmers write and test their programs.

Objective 10.7 *Identify the role of debugging in program development.*

- If programmers make errors in the algorithm or in translating the algorithm to code, problems will occur. Programmers debug the program by running it constantly to find errors and to make sure the program behaves the way it should.

Objective 10.8 *Explain the importance of testing and documentation in program development.*

- Once debugging has detected all the code errors and they have been repaired, users—both within the company and outside the company—test the program in every way they can imagine, both as the program was intended to be used and in ways only new users might think up. Before its commercial release, software is often provided at a reduced cost or at no cost in a beta version to certain test sites or to interested users for a last round of testing.
- Once testing is complete, technical writers create internal documentation for the program and external documentation that will be provided to users of the program. User training, which begins once the software is distributed, teaches the user community how to use the software efficiently.

Part 2
Programming Languages

Learning Outcome 10.2 **You will understand the factors programmers consider when selecting an appropriate programming language for a specific problem and will be familiar with some modern programming languages.**

Many Programming Languages

Objective 10.9 *Discuss the driving factors behind the popularity of various programming languages.*

- There are many languages because of the many trade-offs that must be made among speed of execution, security, ease of programming, interface, and ability to grow and support more features later.

Objective 10.10 *Summarize the considerations in identifying an appropriate programming language for a specific setting.*

- A programming team reviews several considerations before selecting the language to be used. Certain languages are best used for certain problems. The target language should be well matched to the amount of space available for the final program. Some projects require the selection of a language that can produce code that executes in the fastest possible time. Selecting a language with which the programmers are familiar is also helpful.

Exploring Programming Languages

Objective 10.11 *Compare and contrast modern programming languages.*

- VB.NET, C/C++, and Java enable programmers to include control features such as scroll bars and expanding and collapsing menus.
- Objective C is used in applications for mobile devices that use iOS and applications that run under macOS.
- Programmers use HTML tags to structure web pages. HTML5 includes more advanced tags like <audio> and <video>.
- For more complex web development, scripting programs such as JavaScript, PHP, and VBScript are popular.
- AJAX uses a combination of technologies to create websites that can update without the user refreshing the page.
- XML and JSON allow programmers to create their own tags so that web pages can exchange information, not just formatting details.

Objective 10.12 *State key principles in the development of future programming languages.*

- New languages may lean toward being interpreted instead of compiled because of the speed gained in development. Many languages have moved toward the object-oriented paradigm.

Be sure to check out **MyLab IT** for additional materials to help you review and learn. And don't forget to watch the Chapter Overview Videos.

Key Terms

action **381**
Active Server Pages (ASP) **403**
AJAX (Asynchronous JavaScript and XML) **404**
algorithm **377**
application programming interface (API) **399**
architecture-neutral **400**
assembly language **386**
base class **385**
beta version **392**
binary decision **381**
C **400**
C++ **400**
C# **400**
class **382**
code editing **390**
coding **385**
comment **387**
compilation **389**
compiler **389**
control structure **382**
data **377**
data-flow diagram **375**
data type **384**
debugger **391**
debugging **391**
decision point **379**
derived class **385**
documentation **392**
dynamic decision making **403**
editor **390**
error handling **378**
executable program **389**
eXtensible Markup Language (XML) **404**
fifth-generation language (5GL) **386**
first-generation language (1GL) **386**
flowchart **379**
fourth-generation language (4GL) **386**
general availability (GA) **392**
Hypertext Markup Language (HTML) **401**
if else **384**
information **377**
information system **374**
inheritance **385**
initial value **381**
integrated development environment (IDE) **390**
interpreter **389**
Java **400**
Java applet **400**
JavaScript **403**
JavaScript Object Notation (JSON) **405**
JavaServer Pages (JSP) **403**
keyword **384**
logical error **391**
loop **381**
machine language **386**
macro **374**
method **377**
object **383**
Objective C **401**
object-oriented analysis **382**
operator **385**
PHP (PHP: Hypertext Preprocessor) **403**
portability **387**
problem statement **377**
Processing **396**
program development life cycle (PDLC) **376**
program specification **375**
programming **376**
programming language **386**
prototype **399**
pseudocode **379**
rapid application development (RAD) **399**
release to manufacturers (RTM) **392**
reusability **383**
runtime error **391**
scope creep **376**
scripting language **402**
second-generation language (2GL) **386**
service pack **392**
source code **389**
statement **384**
Structured Query Language (SQL) **386**
Swift **401**
syntax **384**
syntax error **384**
system development life cycle (SDLC) **375**
test condition **381**
testing plan **378**
third-generation language (3GL) **386**
top-down design **382**
variable **387**
variable declaration **387**
VB.NET **399**
VBScript **403**
Visual Basic (VB) **399**
visual programming languages **399**
web service **399**

Chapter Quiz // Assessment

For a quick review to see what you've learned, answer the following questions. Submit the quiz as requested by your instructor. If you're using **MyLab IT**, the quiz is also available there.

multiple choice

1. You need to know how to program to
   a. create animations in PowerPoint.
   b. create unique software solutions for specialized problems.
   c. use Microsoft Word to type text and select various fonts.
   d. open a template file for budgeting in Microsoft Excel.
2. What is the first step of the SDLC?
   a. Design
   b. Analysis
   c. Development
   d. Problem identification
3. The steps of the program development life cycle start with
   a. algorithm development.
   b. coding.
   c. describing the problem.
   d. testing.
4. How the program should behave if the input data is invalid is part of
   a. process.
   b. scope creep.
   c. beta testing.
   d. error handling.
5. Inheritance is a useful feature of a language because it promotes
   a. easier compilation.
   b. use of HTML5.
   c. proper control structures.
   d. reuse of existing code.
6. For complex web development, programmers often use
   a. scripting languages.
   b. Visual Basic.
   c. DNA.
   d. Swift.
7. Which of the following are common to all programming languages?
   a. Architecture neutrality
   b. Syntax, keywords, and operators
   c. SDLC and PDLC
   d. JSON and APIs
8. In selecting the most appropriate language, it is useful to know
   a. how quickly the final code must run.
   b. how much space the final code delivered can occupy.
   c. if there is a specific target environment (say, a specific operating system).
   d. all of the above
9. Java is considered architecture-neutral because it
   a. can be recompiled to run on many CPUs.
   b. can run on many generations of Windows.
   c. only needs to be compiled once and can run on many kinds of CPUs.
   d. can run in any browser.
10. The database programming language SQL stands for is
    a. Structured Question Lingo.
    b. Structured Query Language.
    c. Selected Query Launch.
    d. Simple Quick Language.

true/false

_____ **1.** A logical error can be detected and repaired by the compiler.
_____ **2.** A JSON version of software is often released so users can help debug it.
_____ **3.** AJAX allows you to create websites that can refresh without the user taking an action.
_____ **4.** Java is popular because it is the language used to develop iOS mobile applications.
_____ **5.** Many programming languages are moving away from the object-oriented paradigm.

What do you think now?

1. What kind of programming do you imagine needing to know in your professional life?
2. What kind of programming will you use for hobbies, home projects, and managing your personal life?

Team Time

Working Together for Change

problem

Your team has been selected to write a program that tells a vending machine how to make change. The program needs to deliver the smallest number of coins for each transaction.

task

Divide the class into three teams: Algorithm Design, Coding, and Testing. The responsibilities of each team are outlined as follows.

process

1. The Algorithm Design team must develop a document that presents the problem as a top-down design sequence of steps, and another document that uses object-oriented analysis to identify the key objects in the problem. Each object needs to be represented as data and behaviors. Inheritance relationships between objects should be noted as well. You can use flowcharts to document your results.
2. The Coding team needs to decide which programming language would be most appropriate for the project. This program needs to be fast and take up little memory. Use the web to gather information to defend your selection. You may also consider using a product such as Visual Logic (*visuallogic.org*) or App Inventor to develop the prototype of the system.
3. The Testing team must create a testing plan for the program, including a table listing combinations of inputs and correct outputs.
4. As a group, discuss how each team would communicate its results to the other teams. Once one team has completed its work, are the team members finished? How would the tools of a site such as GitHub (*github.com*) help your development team?

conclusion

Any programming project requires teams to produce an accurate and efficient solution to a problem. The interaction of team members is vital to successful programming.

Ethics Project

Software That Kills

In this exercise, you'll research and then role-play a complicated ethical situation. The role you play might not match your personal beliefs, but your research and use of logic will enable you to represent the view assigned. An arbitrator will watch and comment on both sides of the argument, and together the team will agree on an ethical solution.

problem

The Therac-25 was a computerized radiation-therapy machine. In a two-year period, six people were killed or badly harmed due to a flaw in its software. The dose that one patient had received was estimated to be in the 15,000- to 20,000-rad range. This is roughly 100 times more than the expected dose and 15 to 20 times more than a fatal dose. The error-prone software remained on the market for months before the problem was recognized, acknowledged, and solved.

Does this situation indicate criminal conduct? Or is it a situation that needs to be resolved in civil court (for monetary damages)? Is this an example of the price society pays for using complex technology, so no blame should be assigned?

research areas to consider

- Therac-25 case resolution
- Toyota recalls for unanticipated acceleration
- Software errors in avionic software systems
- ACM ethical guidelines

process

1. Divide the class into teams. Research the areas cited above from the perspective of a software developer, the people impacted, or the arbitrator.
2. Team members should write a summary that provides factual support for their character's position regarding the fair and ethical design and use of technology. Then, team members should create an outline to use during the role-playing event.
3. Team members should present their case to the class or submit a PowerPoint presentation for review by the rest of the class, along with the summary and resolution they developed.

conclusion

As technology becomes ever more prevalent and integrated into our lives, ethical dilemmas will present themselves to an increasing extent. Being able to understand and evaluate both sides of the argument, while responding in a personally or socially ethical manner, will be an important skill.

Solve This with Excel

MyLab IT Grader

Time Sheets

You've been asked to automate a time card and record a log of the weekly payments automatically, using Microsoft Excel. You have made great progress on the time card and now need to add data validation, VLOOKUPs to automate Employee data entry, an IF function to calculate the daily payments, and a macro to copy the time card data to a payment log.

You will use the following skills as you complete this activity:

- Customize the Ribbon
- Display the Developer Tab
- Use Absolute Cell Reference
- Create a Data Validation
- Create IF and VLOOKUP Functions
- Fill Date Series
- Record a Macro
- Use Paste Values

Instructions

1. Open *TIA_Ch10_Start.xslx* and save as **TIA_Ch10_LastFirst.xslx**.
2. To display the Developer tab, click **File**, click **Options**, and click **Customize Ribbon**. In the Customize the Ribbon pane, click to select the **Developer check box**. Click **OK**.
3. Click the **Time Card worksheet**; in **cell B3**, type **Ted Hoyt**. In cell B4, create a **VLOOKUP function** to look up the value of cell B3 in the A2:C16 range on the Employee Data worksheet and return the value in the third column of that range. Use False as the Range_Lookup argument.
   Hint: The function will be = VLOOKUP(B3,'Employee Data'!A2:C15,3,FALSE)
4. In **cell B5**, create a VLOOKUP function to look up the value of cell B3 in the A2:C16 range on the Employee Data worksheet and return the value from the second column. Use False as the Range_Lookup argument.
   Hint: The function will be = VLOOKUP(B3,'Employee Data'!A2:C15,2,FALSE)
5. Click **cell C9** on the Time Card worksheet, type **4/11/2021**, and then use the **Fill Handle** to copy the series of dates in **cells C10:C13**.
6. Select **range D9:D13**. Click the **Data tab** and select **Data Validation** in the Data Tools group. In the Data Validation dialog box, on the Settings tab, click the **Allow arrow** and select **Decimal**. Click the **Data arrow** and select **between**; type **0** in the Minimum box and **12** in the Maximum box. Click the Error Alert tab, ensure that a Stop style is selected, and type the title: **Invalid Entry** and an Error message: **Total hours worked cannot exceed 12 hours**.
7. In **cell D9**, type **16** and press **Enter**. The Data Validation Error box appears. Click Retry, type **10**, and then press **Enter**. In **cells D10:D13**, type **8, 12, 8**, and **9**, respectively. Note that Overtime Hours fills in automatically, calculating the hours in excess of eight hours.
8. In **cell F9**, use an **IF function** to test whether the value in D9 is less than or equal to 8. If this condition is true, then multiply D9 times E4 (the regular rate). If the condition is false, then multiply 8 times E4 and add to that E9*E5 (the overtime rate). Make sure you use absolute cell references in the formula where needed. Use the **Fill Handle** to copy the formula to **range F10:F13**.
   Hint: The formula is = IF(D9≤8,D9*E4,(8*E4)+(E9*E5))
9. Click the **Payment Log worksheet**. Click **cell A1**.
10. Click the **Developer** tab and click **Record Macro** in the Code group. Name the macro **CopyTimeCard** and assign a shortcut key of Ctrl + **t**. With the Macro recorder on, complete the following steps: Click the **Time Card worksheet**, select and **copy range A20:F20**, click the **Payment Log worksheet**, and then press **Ctrl** + ↓ to go to the last row with values. Click the **Developer tab**, click **Use Relative References** in the Code group, and then press the ↓ **key** to move to the first blank cell in the worksheet. Click **Use Relative References** in the Code group to deselect it. Click the **Home tab**, click the **Paste arrow**, and select **Values**. Click **cell A1**, select the **Time Card worksheet**, press **Esc**, and then click **cell B3**. Click the **Developer tab** and then click **Stop recording** in the Code group.
    Hint: You may want to practice the macro steps before recording. If you make a mistake, click Macros, select the CopyTimeCard macro, click Delete, and try recording the macro again.
11. Test the macro by typing **Brett Martin** in cell B3. In cells D9:D13, type **8, 10, 8, 8**, and **9**, respectively. Click **cell B3** and then press **Ctrl + t** to run the macro. Click the **Payment Log worksheet** to ensure that the Brett Martin data was successfully transferred.
12. Click the **Developer tab**, click **Macros** in the Code group, select the **CopyTimeCard** macro (including Sub to End Sub lines), and click **Edit. Copy** the macro code, **close** the VBA window, **add** a new worksheet, and **paste** in cell A1 of the new worksheet. Name the worksheet **Macro**.
13. Save the workbook. *Note:* Answer **Yes** when asked to save as a Macro-free workbook. Submit based on your instructor's directions.

Chapter

11 Behind the Scenes: Databases and Information Systems

For a chapter overview, watch the **Chapter Overview Videos**.

PART 1

Database Fundamentals

Learning Outcome 11.1 **You will be able to explain the basics of databases, including the most common types of databases and the functions and components of relational databases in particular.**

The Need for Databases 418

Objective 11.1 *Explain what a database is and why databases are useful.*

Database Types 420

Objective 11.2 *Describe features of flat databases.*

Objective 11.3 *Describe features of relational databases.*

Objective 11.4 *Describe features of object-oriented databases.*

Objective 11.5 *Describe features of multidimensional databases.*

Objective 11.6 *Describe how dynamic, web-created data is managed in a database.*

Helpdesk: Using Databases

Using Databases 425

Objective 11.7 *Describe how relational databases organize and define data.*

Objective 11.8 *Describe how data is inputted and managed in a database.*

Sound Byte: Creating and Querying an Access Database

PART 2

How Businesses Use Databases

Learning Outcome 11.2 **You will be able to explain how businesses use data warehouses, data marts, and data mining to manage data and how business information systems and business intelligence are used to make business decisions.**

Data Warehousing and Storage 440

Objective 11.9 *Explain what data warehouses and data marts are and how they are used.*

Objective 11.10 *Describe data mining and how it works.*

Helpdesk: How Businesses Use Databases

Using Databases to Make Business Decisions 445

Objective 11.11 *Describe the main types of business information systems and how they are used by business managers.*

Sound Byte: Analyzing Data with Microsoft Power BI Suite

MyLab IT All media accompanying this chapter can be found here.

A Family Shopping List on page 439

(Marek Uliasz/Alamy Stock Photo; Alexmit/123RF; Alphaspirit/123RF; Nicescene/Shutterstock; Master_art/Shutterstock)

What do you think?

Think about how much **personal data** is created from all our devices and apps. This data is stored in **large data storage centers** and is used for marketing, product development, and communication purposes. But how is our data protected?

In the United States, no central data protection legislation exists. Rather, **data protection is regulated** by state and federal laws. However, the United Kingdom's General Data Protection Regulation (GDPR) standardizes data protection across the UK and all European Union (EU) countries. The GDPR allows individuals to specify what personal data is stored and how it is used. In addition, privacy statements must be in easily understood language, and customers must **actively opt in** to data collection and can withdraw at any time.

The GDPR applies to every company that has a digital presence in the EU. This includes many US-based companies such as **Facebook, Google, and Uber** as well as many online retailers, health care providers, insurers, and banks. Because of this, the GDPR is in effect setting a **new global standard** for data protection. Meanwhile, in the United States, the Federal Trade Commission is conducting a comprehensive review of consumer privacy that will likely result in changes to current US laws.

How do you feel about your personal data?

- *I am not concerned about my personal data.*
- *Only my Social Security number and credit card data need to be protected.*
- *Data breaches never seem to affect me.*
- *I want to take full control over my personal data.*

See the end of the chapter for a follow-up question.

(Laymanzoom/Shutterstock)

Part 1

For an overview of this part of the chapter, watch **Chapter Overview Video 11.1.**

Database Fundamentals

Learning Outcome 11.1 **You will be able to explain the basics of databases, including the most common types of databases and the functions and components of relational databases in particular.**

You interact with databases every day: when you retrieve a friend's phone number from your contacts, post to Instagram, or add a song to your Spotify playlist. In this part of the chapter, we look at the basics of databases and how databases can be used to provide relevant information quickly and easily.

The Need for Databases

Many actions we frequently perform, such as using an ATM, shopping online, and registering for classes, generate data that needs to be stored, managed, and used by others. Most likely, a system has been created that receives and stores the inputted data and enables that data to be processed and used by others, effectively turning data into information. In this section, we look at why databases are needed as well as their specific advantages.

Database Basics

Objective 11.1 *Explain what a database is and why databases are useful.*

Why do I need to know about databases? A database is a collection of related data that can be stored, sorted, organized, and queried. By creating an organized structure for data, databases make data more meaningful and therefore more useful. Because you interact with databases every day, understanding how databases work and what you can do with databases will help you use the ones you interact with more effectively. For example, categorizing, sorting, and filtering are key attributes of most databases. If a database isn't set up correctly or if you don't know how to use the database, you may not get the type of information you're looking for.

Amazon (see Figure 11.1) is an example of a complex database. When consumers want to view or purchase an item, they can specify the department the item falls into to help narrow the search,

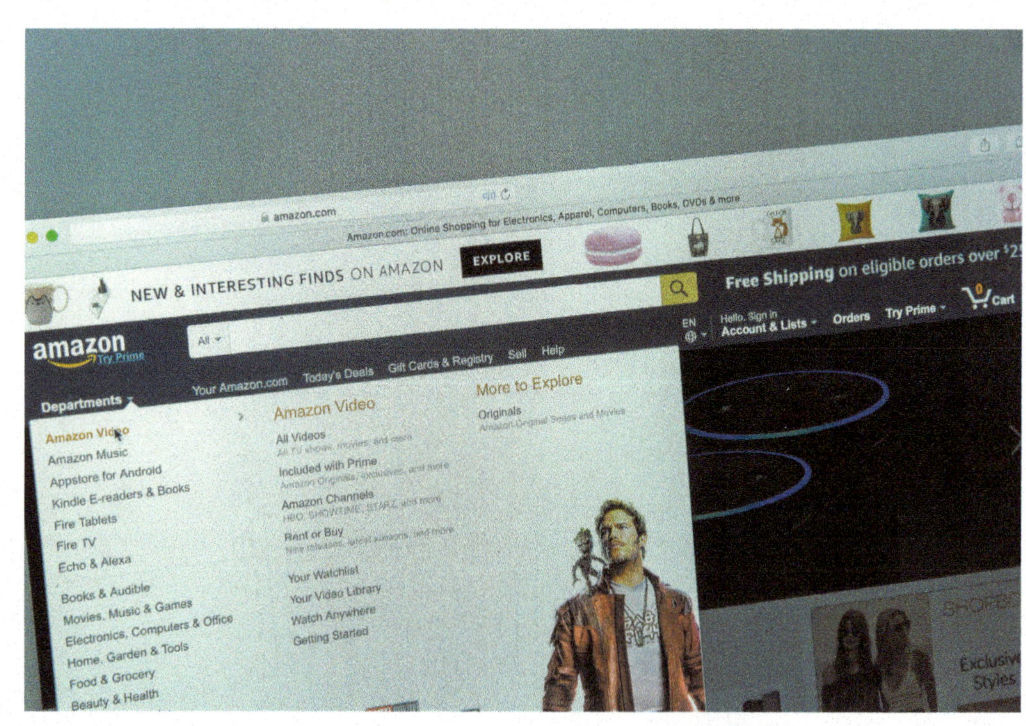

Figure 11.1 Amazon and other electronic commerce websites use databases to enhance customer experience and manage inventory. *(Casimiro PT/Shutterstock)*

such as books, grocery, or electronics, and then specify one or more subcategories. If the database doesn't have the right categories for consumers to choose from, they won't easily be able to locate the item they want. Most online retailers use some form of a sophisticated database to enhance the consumer experience but also to help manage inventory.

What are the main advantages of databases? Databases provide several main advantages:

- They manage large amounts of data efficiently.
- They enable information sharing.
- They promote data integrity.

How can databases manage large amounts of data efficiently? Often, large amounts of data are complex and need to be organized in specific ways to be used most efficiently. A database organizes large amounts of data into related groups. Databases also have means to input data and extract subsets of data and have mechanisms to keep data accurate as it's inputted, updated, and manipulated.

How do databases make information sharing possible? With most types of databases, data is kept in one location such as a central computer or database system, providing data centralization. A centralized source ensures that everyone can access the same data (see Figure 11.2). For example, in a college database, each department that needs to use student information accesses it from the same set of data. Centralized data also increases efficiency because the need to make updates to the

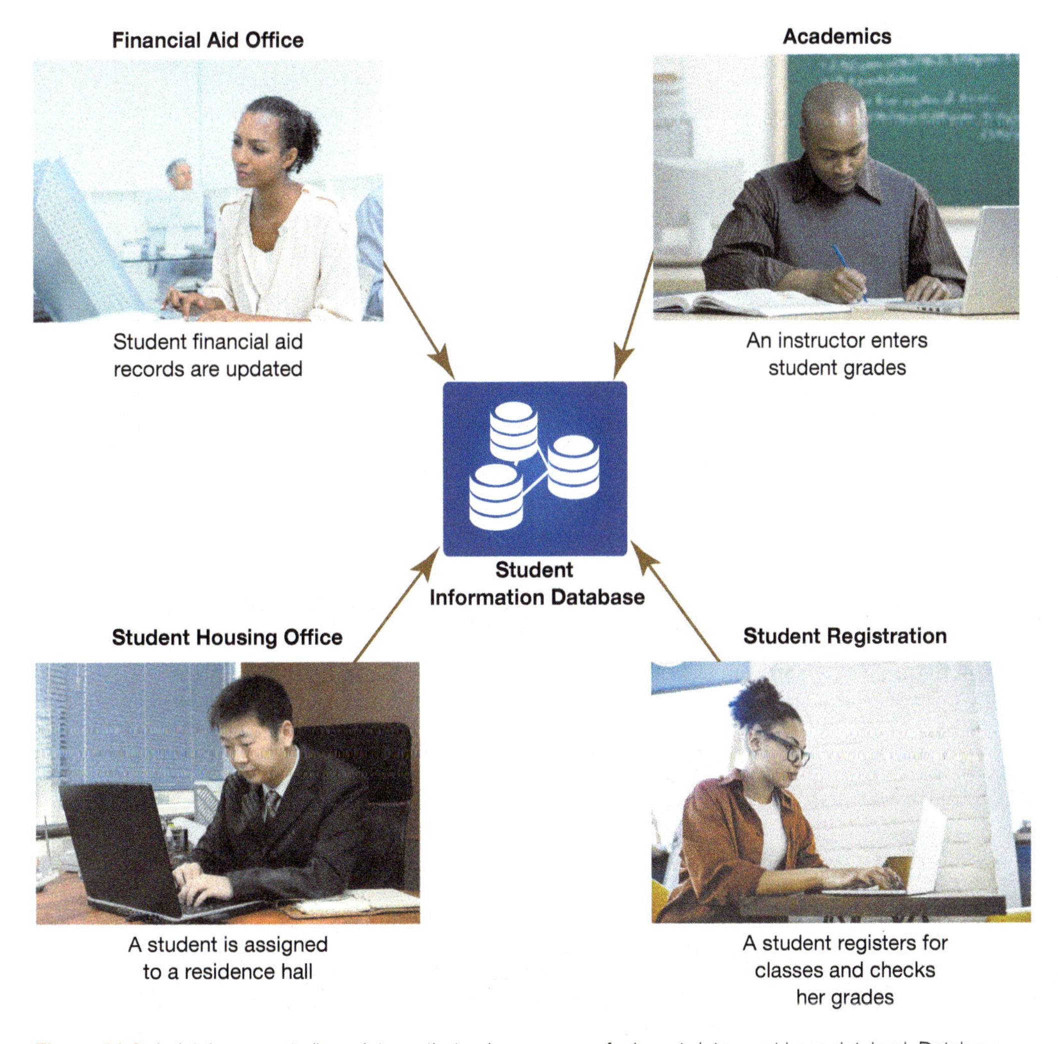

Figure 11.2 A database centralizes data so that only one copy of relevant data must be maintained. Database users therefore access the same up-to-date information. *(Wavebreakmedia/Shutterstock; AVAVA/Shutterstock; KK Tan/Shutterstock; Phipatbig/Shutterstock; Shutterstock)*

same data in multiple files is eliminated. In addition, many databases also provide the ability to control who has access to the data, helping to preserve the security of the data.

How do databases ensure that data is accurate? Data centralization goes a long way toward ensuring data integrity. Data integrity means that the data contained in the database is accurate and reliable. If an address changes in a database with a centralized source of data, the change would be done in one place. With multiple lists, it's quite possible that data will be updated in one list and not in another. Data integrity checks would prevent a duplicate entry of employee ID 57 if one already exists. Similarly, data integrity would prevent an hourly wage of $23 to be applied if the wage range was $10–22, because $23 falls outside the wage range. We discuss these data integrity checks as they apply to Microsoft Access databases later in the chapter.

Are any disadvantages associated with databases? Large and complex databases can be time-consuming and expensive to set up and administer. Great care must be exercised in the design of such databases to ensure that they'll function as intended. Although average individuals can design small databases, for larger databases, it's necessary to have an experienced database administrator (DBA or database designer)—an IT professional who is responsible for designing, constructing, and maintaining databases. DBAs review and manage the database on an ongoing basis to ensure that data is flowing smoothly into and out of the database. DBAs can monitor table and CPU usage to help determine whether database performance is acceptable.

Database Types

Databases range from the very simple to the very complex. The type of database that is used depends on the type of data that needs to be managed. In this section, we look at several types of databases.

Flat Databases

Objective 11.2 *Describe features of flat databases.*

What database is best to manage simple data? Not every situation demands a complex database system. For simple tasks, a flat database, often represented as a list or simple table, is adequate. Most word processing and spreadsheet applications have tools to help you create a flat database. A table you create in Microsoft Word can serve as a flat database, as can a spreadsheet you create in Microsoft Excel.

Figure 11.3 shows a simple Books to Buy list you might create in Excel before beginning college. This flat database works well because it's simple and suited to just one purpose: to provide you with a list of books you need to buy for a semester.

When is a flat database not sufficient for organizing data? A flat database isn't an efficient solution if it contains complex information that needs to be organized or requires more than one person to access it. In addition, flat databases can lead to several problems:

- **Duplication.** Flat databases often lead to duplicate data. For example, as shown in Figure 11.4a, each time Julio Garza registers for a class, his name and address are entered into the

	A	B	C	D
2	**Books To Buy**			
3				
4	**Class**	**Title**	**Author**	**Bought**
5	American Literature 1	Anthology of American Literature, Volume 1	McMichael et al.	
6	American Literature 1	Moby Dick	Melville	Yes
7	American Literature 1	The Last of the Mohicans	Cooper	Yes
8	English Composition 1	The Academic Writer's Handbook	Rosen	
9	English Composition 1	The Write Stuff: Thinking Through Essays	Sims	Yes
10	Intro to Business	Better Business	Solomon, Poatsy, Martin	Yes

Figure 11.3 A basic list created as a spreadsheet in Microsoft Excel or as a table in Microsoft Word are examples of a flat database and are often sufficient to organize simple data. *(Courtesy of Microsoft Corp.)*

Class Registration list. He also provides the same data to the housing and dining manager when he makes his residence hall and meal plan selections, as shown in Figure 11.4b. His information is repeated several times and on two lists. Such unnecessary duplication of data, referred to as data redundancy, can be problematic. Imagine the time wasted by entering the same data multiple times, as well as the time required for the database to search through the duplicated information. Consider, too, the increased likelihood that someone will make a data-entry mistake, thus leading to the next problem: data inconsistency.

- **Inconsistency.** When there are multiple lists or tables with the same information, and not all are updated when data changes or data is entered incorrectly, inconsistent data results. For example, if a student moves, his or her data will need to be updated in all the lists that contain his or her address. It is possible to overlook one or more lists or even one or more entries in the same list. This would lead to a state of data inconsistency, when different versions of the same data appear in the same or different lists. With inconsistent data, it is difficult to determine which data is correct. Notice in Figure 11.4 that Susan Finkel's address on the registration list differs from that on the residence hall and meal plan list, resulting in data inconsistency and with no indication as to which address is correct.

  In addition, correct data can be entered in an inconsistent format. Look at students Julio Garza and Mei Zhang in Figure 11.4a. Both are registered for PSY 101, but different course names appear: *Intro to Psych* and *Intro to Psychology*. If a search were made for Intro to Psych, the data for Intro to Psychology would not be included in that search because the names are not exactly the same. Confusion arises when data is inconsistently entered. Ensuring data consistency is difficult to do with a list.
- **Inappropriate data.** In Figure 11.4b, each student has selected a meal plan; however, the meal plan assigned to Mei Zhang doesn't exist. With a list, there are few checks to make sure that data entered is valid.
- **Incomplete data.** With a flat database, there are few checks to make sure that required data is entered. In Figure 11.4a and Figure 11.4b, Leanne O'Connor has enrolled in the college, but there are no entries for courses or a meal plan. A flat database allows for such incomplete records to exist, thus leading to potential problems later.

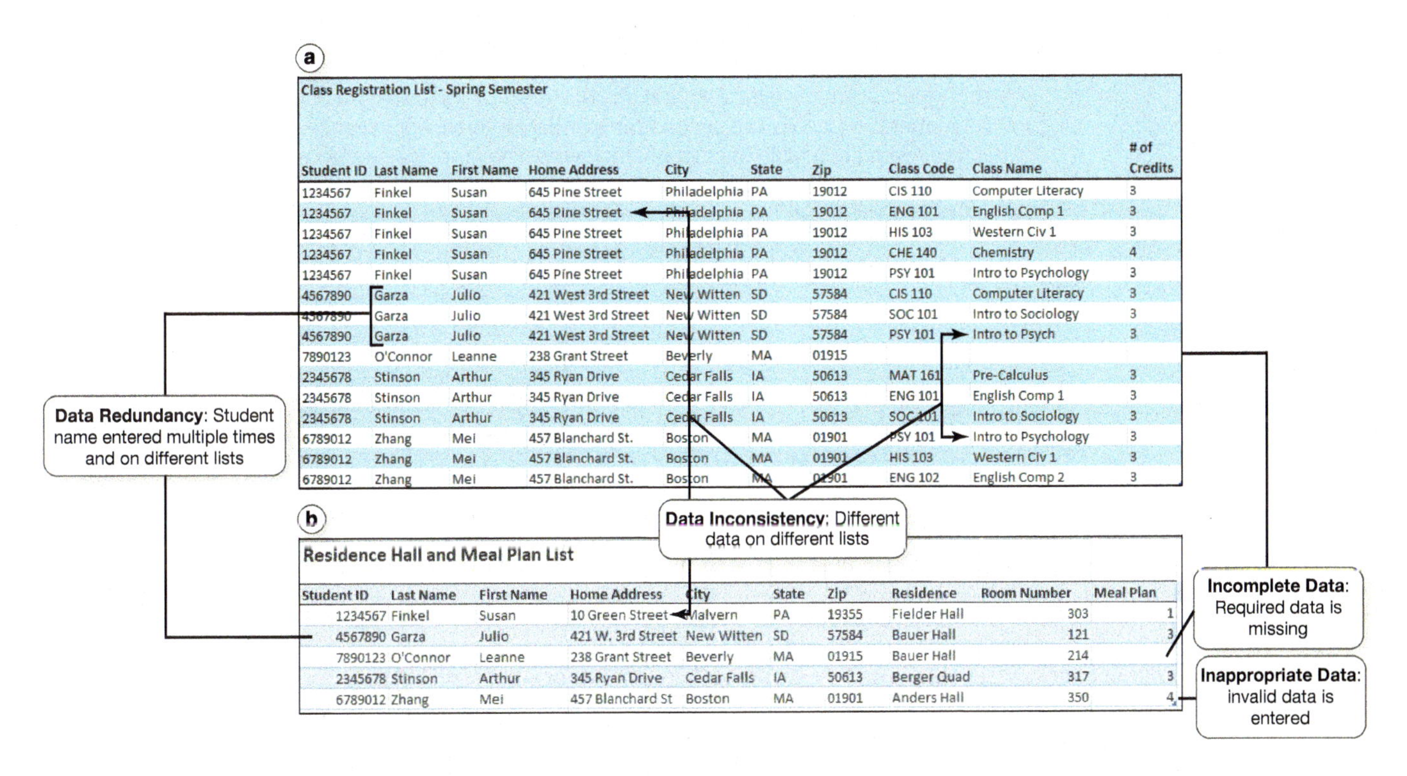

Class Registration List - Spring Semester

Student ID	Last Name	First Name	Home Address	City	State	Zip	Class Code	Class Name	# of Credits
1234567	Finkel	Susan	645 Pine Street	Philadelphia	PA	19012	CIS 110	Computer Literacy	3
1234567	Finkel	Susan	645 Pine Street	Philadelphia	PA	19012	ENG 101	English Comp 1	3
1234567	Finkel	Susan	645 Pine Street	Philadelphia	PA	19012	HIS 103	Western Civ 1	3
1234567	Finkel	Susan	645 Pine Street	Philadelphia	PA	19012	CHE 140	Chemistry	4
1234567	Finkel	Susan	645 Pine Street	Philadelphia	PA	19012	PSY 101	Intro to Psychology	3
4567890	Garza	Julio	421 West 3rd Street	New Witten	SD	57584	CIS 110	Computer Literacy	3
4567890	Garza	Julio	421 West 3rd Street	New Witten	SD	57584	SOC 101	Intro to Sociology	3
4567890	Garza	Julio	421 West 3rd Street	New Witten	SD	57584	PSY 101	Intro to Psych	3
7890123	O'Connor	Leanne	238 Grant Street	Beverly	MA	01915			
2345678	Stinson	Arthur	345 Ryan Drive	Cedar Falls	IA	50613	MAT 161	Pre-Calculus	3
2345678	Stinson	Arthur	345 Ryan Drive	Cedar Falls	IA	50613	ENG 101	English Comp 1	3
2345678	Stinson	Arthur	345 Ryan Drive	Cedar Falls	IA	50613	SOC 101	Intro to Sociology	3
6789012	Zhang	Mei	457 Blanchard St.	Boston	MA	01901	PSY 101	Intro to Psychology	3
6789012	Zhang	Mei	457 Blanchard St.	Boston	MA	01901	HIS 103	Western Civ 1	3
6789012	Zhang	Mei	457 Blanchard St.	Boston	MA	01901	ENG 102	English Comp 2	3

Residence Hall and Meal Plan List

Student ID	Last Name	First Name	Home Address	City	State	Zip	Residence	Room Number	Meal Plan
1234567	Finkel	Susan	10 Green Street	Malvern	PA	19355	Fielder Hall	303	1
4567890	Garza	Julio	421 W. 3rd Street	New Witten	SD	57584	Bauer Hall	121	3
7890123	O'Connor	Leanne	238 Grant Street	Beverly	MA	01915	Bauer Hall	214	
2345678	Stinson	Arthur	345 Ryan Drive	Cedar Falls	IA	50613	Berger Quad	317	3
6789012	Zhang	Mei	457 Blanchard St	Boston	MA	01901	Anders Hall	350	4

Figure 11.4 (a) A class registration list and (b) a list of residence hall assignments are two flat databases a college might create to keep track of student information, but doing so could lead to data problems. *(Courtesy of Microsoft Corp.)*

Relational Databases

Objective 11.3 *Describe features of relational databases.*

What is a relational database? A relational database, such as Microsoft Access, operates by organizing data into various tables based on logical groupings. Tables are related to each other based on shared sets of data. Relational databases are the most commonly used type of database.

When complex data needs to be organized or accessed by others, using a relational database helps to ensure that data is complete and accurate. In our earlier example, all student contact information would be grouped into one table. The housing and dining information also would be grouped into a separate table, but with only the Student ID included and not all the student contact information. This change requires the student contact information to be linked to the housing and dining table so the housing and dining manager can access student contact information when needed. In relational databases, a link between tables that defines how the data is related is referred to as a relationship. A common field between the two tables is used to create the link. In Figure 11.5, StudentID is the common field in the Residence Assignments and Student Information tables, and ResidenceID is the common field in the Residence Halls and Residence Assignments tables.

Are there different types of relationships in relational databases? A relationship in relational databases can take one of three forms:

1. A one-to-many relationship is characterized by a record appearing only once in one table while having the capability of appearing many times in a related table. For example, as shown in Figure 11.5, the "1" next to ResidenceID in the Residence Halls table indicates a residence hall can be listed only once in the Residence Halls table. The "∞" next to ResidenceID in the Residence Assignments table indicates that many records can have the same ResidenceID in the Residence Assignments table. Each residence hall will be listed more than once in the Residence Assignments table to account for every student who will be living there. One-to-many relationships are the most common type of relationship in relational databases.
2. A one-to-one relationship indicates that for each record in a table, there is only one corresponding record in a related table. For example, each student is assigned a room. Because a student can live in only one room at a time, a one-to-one relationship is established between the student and the resident assignment, as shown in Figure 11.5 by the "1" next to the StudentID field in the Residence Assignments table and the Student Information table.
3. A many-to-many relationship is characterized by multiple records in one table being related to multiple records in a second table and vice versa. For instance, a table of students could be related to a table of student employers. The employers could employ many students, and students could work for more than one employer. Generally, to set up a many-to-many relationship in Access, you must create a third, associate table that connects two tables. In our example, the associate table would include the EmployerID and the StudentID fields.

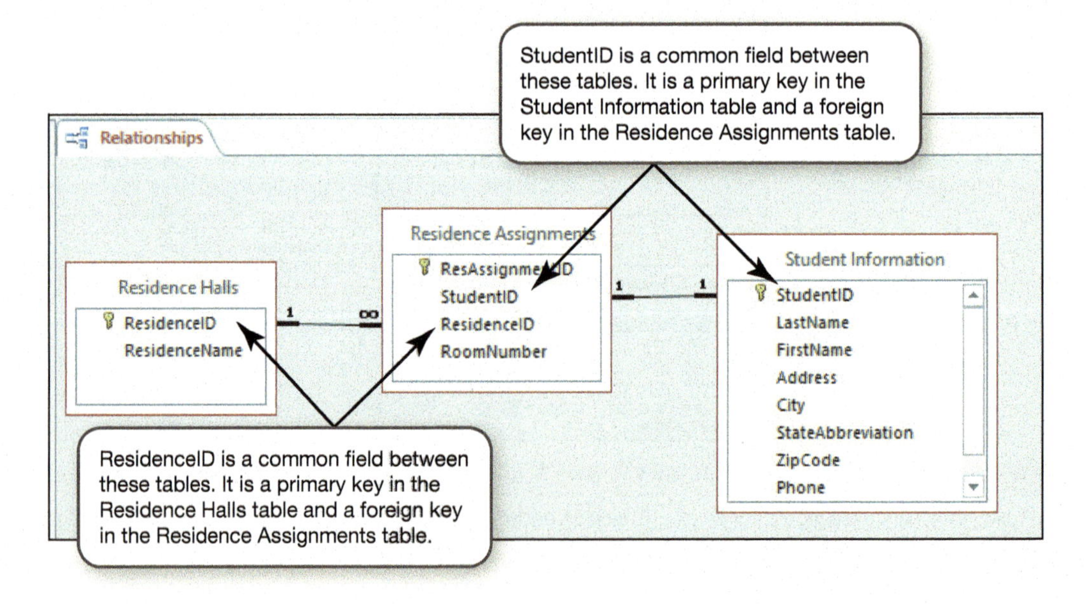

Figure 11.5 Tables are linked by a common field, creating a relationship. *(Courtesy of Microsoft Corp.)*

Bits&Bytes Normal Forms

A series of guidelines, known as *normal forms*, ensure that databases are structured efficiently and logically to avoid problems such as data redundancy and duplicity. Normal forms are numbered from 1 to 5 (1NF–5NF). First Normal Form (1NF) is the most basic level of normalization and 5NF is the most complex. At a minimum, databases should be structured to meet 1NF. 1NF sets the basic rules for an organized database and includes:

- Groups of related data must be placed in separate tables.
- Column names must be unique.
- Each row of data must be identified by a unique identifier (primary key).

2NF and 3NF must meet the requirements of the previous level and, in addition, 2NF sets rules to avoid duplicate data and 3NF ensures that all columns in a table relate to the primary key. Usually, care is taken by DBAs to ensure that databases are structured to meet the requirements through 3NF. 4NF is applied less often and 5NF is the most complex and applied only in special situations.

Is there a way to ensure that data is organized efficiently? The process to ensure that data is organized most efficiently is called normalization. Normalization removes issues such as data redundancy and inconsistency. For example, applying normalization rules to the data represented in Figure 11.4 would have the student information in a distinctly separate table and not repeated in the class registration and residence hall tables. Removing redundancy and inconsistency not only improves the integrity of the data but also improves the efficiency of the database.

Using Databases

In this Helpdesk, you'll play the role of a helpdesk staffer, fielding questions about databases, their benefits, their components, and how relational databases organize and manipulate data.

Object-Oriented Databases

Objective 11.4 *Describe features of object-oriented databases.*

When is a relational database not sufficient? Relational databases are great for text-based or structured (analytical) data (such as Bill or 345). However, some data, referred to as unstructured data, such as audio clips (such as MP3 files), video clips, pictures, and extremely large documents, are handled better in an object-oriented database. Data of this type is also known as a binary large object (BLOB) because it's encoded in binary form.

How does an object-oriented database organize data? An object-oriented database stores data in objects rather than in tables. Objects contain not only data but methods for processing or manipulating that data. This allows object-oriented databases to store more types of data than relational databases and to access that data faster.

Because businesses today need to store a greater variety of data, object-oriented databases are becoming more popular. For a business to use its data in an object-oriented database, the data needs to undergo a costly conversion process. However, the faster access and reusability of the database objects can provide advantages for large businesses.

Object-oriented databases also need to use a query language to access and manage data. A query language is a specially designed computer language used to manipulate data in or extract data from a database. Many object-oriented databases use Object Query Language (OQL), which is similar in many respects to Structured Query Language (SQL), a standard language used to construct queries to extract data from relational databases.

Multidimensional Databases

Objective 11.5 *Describe features of multidimensional databases.*

What is a multidimensional database? A multidimensional database stores data that can be analyzed from different perspectives, called *dimensions*. This distinguishes it from a relational database, which stores data in tables that have only two dimensions (fields and records). Multidimensional databases organize data in a cube format. Each data cube has a *measure attribute*, which is the main type of data that the cube is tracking. Other elements of the cube are known as *feature attributes*, which describe the measure attribute in some meaningful way.

For example, sales of automobiles (measure attribute) could be categorized by various dimensions such as region, automobile model, or sales month—all feature attributes (see Figure 11.7).

Bits&Bytes Use a Graph Database to Track Social Network Data

Think about all your Instagram, Facebook, or other social network connections. If you were to illustrate these connections, most likely they would look like a mind-map, in which the relationships of associated images, words, and other concepts flow from a central concept (see Figure 11.6). In essence, that is what a graph database is. Sure, there are databases that can capture the discrete pieces of personal data generated from social networks (gender, employer, marital status, education, etc.), but those databases are not sufficient to capture the "connectedness" of the relationships between the contacts. Graph databases store both the discrete data and the relationships. More important, gathering insights from the data is much faster with graph databases.

Graph databases have other uses beyond social networks. For example, recommendation engines can use graph databases to store relationships between categories such as customer interests and purchase history. The recommendations are then based on products or services other customers with similar interests or histories have purchased. Banks can also use graph databases to engage in fraud detection, for example by detecting suspicious relationships between a purchaser and an email address or credit card number that have been identified in a known fraud case. If you want to try your hand at using a graph database, Neo4j is a free and open source option.

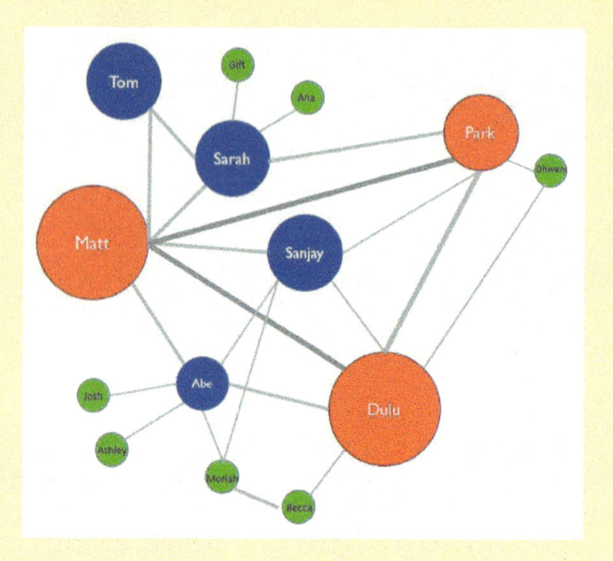

Figure 11.6 Graph databases are used to show relationships with connected data.

In addition, the database could be constructed to define different levels within a particular feature attribute, such as state and town within a region.

What are the advantages of multidimensional databases? The two main advantages of multidimensional databases are that:

- They can be customized to provide information to a variety of users based on their needs.
- They can process data much faster than pure relational databases can.

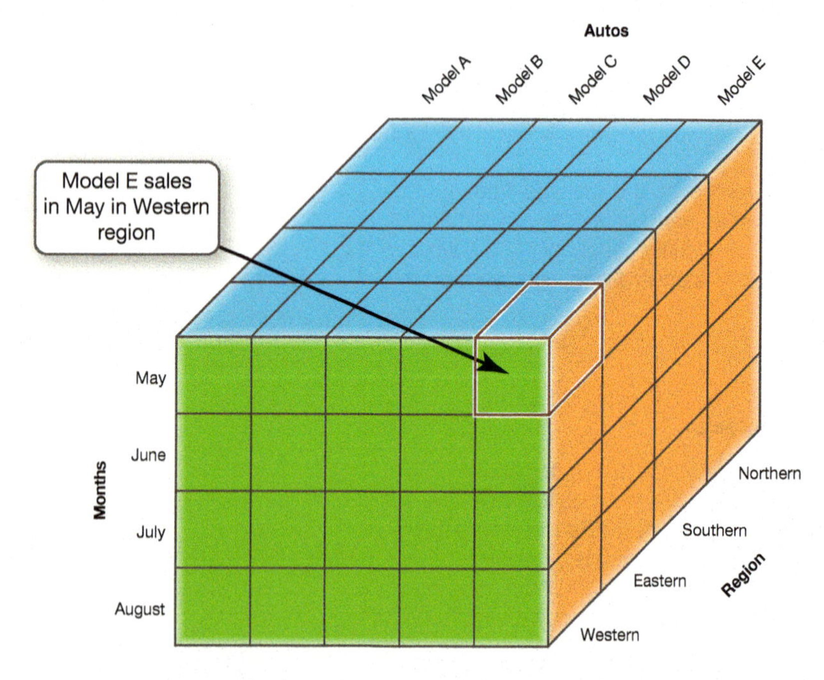

Figure 11.7 Multidimensional databases store data in at least three dimensions.

The need for processing speed is critical at any time, but especially when deploying a large database that's accessed through the Internet. Therefore, large databases such as the one used by eBay or Amazon are usually designed as multidimensional databases. Multidimensional databases are used in data warehouses and online analytical processing (OLAP) systems, which we discuss in Part 2 of this chapter.

NoSQL Databases

Objective 11.6 ***Describe how dynamic, web-created data is managed in a database.***

Massive amounts of unstructured data are generated from social media sharing and other online activities, online personal settings, photos, and videos. In addition, websites track usage metrics, customer location–based information, click streams, and more. Trying to store, process, and analyze these types of unstructured data led to the development of alternatives to the structural requirements of relational and object-oriented databases.

Relational databases are structured, storing related data in tables, columns, and rows, and object-oriented databases store data as objects. Instead, NoSQL (non-relational or not only SQL) databases group related information of all types into documents that are more like file folders. Documents hold everything—including a person's address, phone numbers, and photos as well as their social media posts and feedback and online shopping preferences—without the requirement of specific categorization like other databases.

NoSQL systems store and manage data so that the data can be accessed quickly and provide the database design and structure with greater flexibility. Many NoSQL databases were developed by big online companies like Google, Amazon, and Facebook so their massive amounts of online data could be stored and processed better.

NoSQL systems don't generally provide the same level of data consistency as SQL databases but offer speedier access and greater sizing capabilities. Because of these trade-offs, it's possible that companies would have the need for both types of databases to manage inherently different types of data.

Using Databases

Databases are created and managed using specially designed application software (such as Oracle Database or Microsoft Access) to capture and analyze data. Once a database is designed and tested to ensure that it works properly, you can begin to populate the database by creating individual records. After the database has been populated with data, users can employ certain methods to extract subsets of data from the database and then output the data in a meaningful and presentable format. In this section, we look at the components that make up a relational database and how these components function.

Relational Database Components and Functions

Objective 11.7 *Describe how relational databases organize and define data.*

What is needed to manage a database? A database management system (DBMS) is the software that allows a computer to perform database functions. The four main operations of a DBMS include:

1. Storing and defining data.
2. Viewing (or browsing), adding, deleting, and modifying data.
3. Querying (extracting) data.
4. Outputting data.

Storing and Defining Data

How is data stored and organized in a relational database? As described earlier, an advantage of many databases is that they provide an organized structure for data to make it more meaningful and useful. As shown in Figure 11.8, in a relational database, data is organized using fields, records, and tables:

1. **Fields.** Categories of data are called fields. Fields are displayed in columns. Each field is identified by a field name, which is a way of describing the information in the field. Figure 11.8 contains seven fields (Student ID, Last Name, First Name, Address, City, State, and Zip Code).
2. **Records.** A group of related fields is called a record. For example, in Figure 11.8, a student's ID, name, and complete address comprise a record.
3. **Tables (files).** A group of related records is called a table (or file). Tables are usually organized by a common subject. Figure 11.8 shows a table (called Student Information) that contains contact information for students.

How do I know what fields are needed in my database? Careful planning is required to identify each distinct category of data you need to capture. It's often easiest to begin with the anticipated output—that is, what fields will be required in reports. Then you can work backward to determine what other components of data are needed to support the output and begin to organize

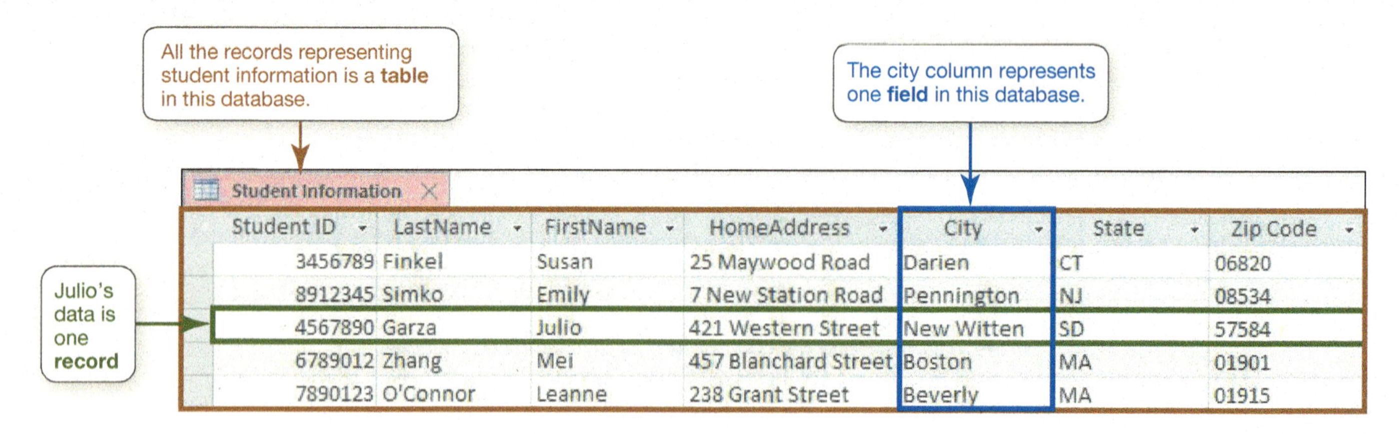

Figure 11.8 In a database, a category of information is stored in a *field*. A group of related fields is called a *record*, and a group of related records is called a *table*. *(Courtesy of Microsoft Corp.)*

the fields into distinct tables. Each field should describe a unique piece of data and should not combine two pieces of data. For example, first and last names are separate pieces of data, and you'd want to create a separate field for each rather than having one field for Name. Keeping data into the smallest units possible makes sorting and filtering data more efficient. For instance, suppose you want to send an e-mail to students that addresses all of them by their first name (such as Dear Geri). If Geri's first and last names are in the same field in the database, extracting just her first name for the salutation will be difficult. Similarly, if the name data is combined as first name last name (Geri Akman), it will be difficult to sort by last name (unless the data is inputted with the last name first).

Are there specific rules for establishing field names? The following are some guidelines in establishing field names:

- Field names must be unique within a table. For example, as shown in Figure 11.9, you can't have two fields labeled Name in one table; they must be different, such as FirstName and LastName.
- It's helpful to distinguish similarly named fields in different tables to avoid confusion if those fields appear together in other components of the database. For example, in a college database, there may be separate tables for Student and Faculty. Each table would have fields for FirstName and LastName for both students and faculty. Those fields may be used together in the same report

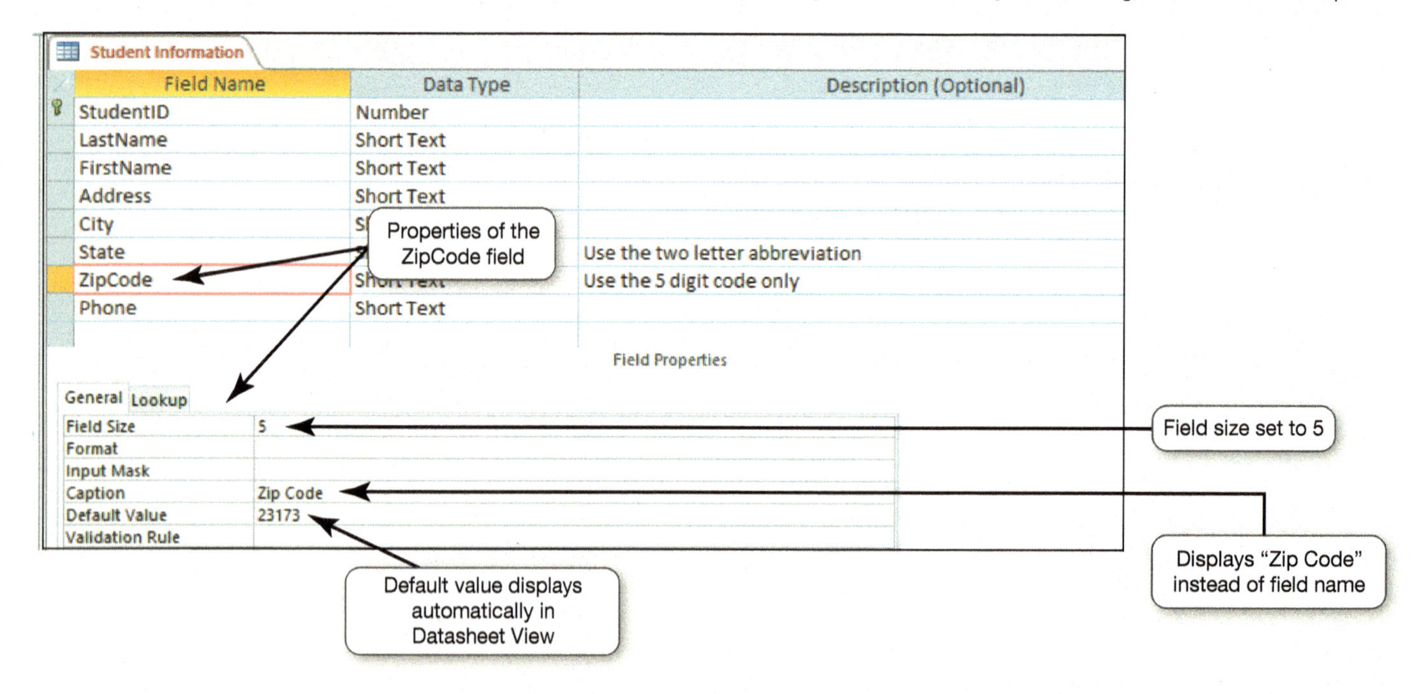

Figure 11.9 The Student Information table, displayed in Design View, shows field names, data types, and other field properties. *(Courtesy of Microsoft Corp.)*

that details a student's schedule. To avoid confusion, it would be better to use field names with some notation that identifies the table that the field is from, such as S_LastName and S_FirstName in the Student table and F_LastName and F_FirstName in the Faculty table.

- It is best practice not to use spaces in field names because they can cause errors when the objects are involved in programming tasks. Underscores or PascalCase notation can be used in lieu of spaces.

Can any type of data be entered in any field? When fields are created in the database, the user assigns each field a data type (or field type). The data type indicates which type of data can be stored in the field and prevents the wrong type of data from being entered. For example, you would not be able to enter an alphabetic character in a field with a number data type.

Table 11.1 lists the most common data types, with examples of each.

In addition to data type, are there other ways to define field data? In addition to specifying the data type for each field, you can also define field data with other properties, referred to as field properties. Common field properties include:

- Field size defines the maximum number of characters that a field can hold. It is important to ensure that the selected field length is long enough to enable storage of the longest possible piece of data that might be entered in the field. Otherwise, data may be truncated (a portion of it chopped off and lost) during data entry. If a field size is set to 5, it can hold a number that has up to five digits (from 1 to 99999) or text with five characters. For the ZipCode field in Figure 11.9, the field size property is set to 5 (five characters), allowing enough space for a five-digit zip code. If the zip code were to include the ZIP + 4 system, the field size would need to be 9 or 10 (allowing for the hyphen). The default field size is determined by the data type, and not all data types (such as Date/Time) require a field size property.
- Default value is the value the database automatically uses for the field unless a different value is entered. Default values are useful for field data that's frequently the same. For example, if most

Table 11.1 Common Data Types in Microsoft Access

Data Type	Used to Store	Examples
Short Text	Letters, symbols, or combinations of letters, symbols, and numbers (up to 255 characters). Short text also contains numbers that can't be used in calculations, such as phone numbers and zip codes.	Cecilia PSY 101 (610) 555-1212
Long Text	Long blocks of text (up to 65,535 characters).	Four score and seven years ago our fathers brought forth on this continent, a new nation, conceived in Liberty, and dedicated to the proposition that all men are created equal.
Number, Large Number, and Currency	Numeric data that can be used in calculations or that is represented as currency.	512 1.789 $1,230
Date/Time, Date/Time Extended	Dates and times in standard notation. Data may be used in calculations, such as calculating the number of days between today and your next birthday.	2/21/2018 10:48.01
Yes/No	Data that has one of two values	True/False
Calculated	Results of a calculation; often involves numeric or date/time data from other fields in the database.	Price * Quantity * .06
OLE Object	Pictures, charts, or files from another Windows-based application.	JPG file Word document
Hyperlink	Alphanumeric data stored as a hyperlink address to a web page, an e-mail address, or an existing file.	pearsonhighered.com techinactionname@email.com

students live in Delaware, then setting a default value for the State field to DE saves users from having to enter that value for each student.

- Caption enables you to display the field name in a more meaningful or readable manner. As described earlier, PascalCase notation or underscores are often used when creating field names. In addition, field names might include other notations to distinguish similar field names on different tables. In these cases, the field name used to create the table may not be the best to display on forms and reports. The Caption property in Microsoft Access enables you to display a field name that is more readable, such as Zip Code instead of ZipCode, as shown in Figure 11.9.

Using Primary Keys

Can fields have the same values in the same table? In most fields, it is possible to have the same value in multiple records. For example, two students can live in the same town or have the same last name. However, to keep records distinct, each record must have one field, called a primary key field, that has a value unique to that record. For example, as shown in Figure 11.10, in the Student Information table, the primary key field is Student ID. Notice that each Student ID value in the Student Information table is different. Establishing a primary key and ensuring that its values are unique makes it impossible to have duplicate records, thereby maintaining data integrity.

What makes a good primary key? The most important constraint is that the primary key does not allow duplicate data. For example, when you place an order with *Amazon.com*, your transaction gets a unique order number. This order number is a primary key in Amazon's database. It's essential to have a unique number for each order because it would be difficult to keep track of individual orders without one. Similarly, your student ID number is a unique number, most likely generated in some serial manner. Although some numbers, such as Social Security numbers or driver's license numbers, are unique to a person and may be used as primary keys for those specific databases, they do not make great primary keys for other uses. Social Security numbers and other personal identifiers usually aren't a good choice as primary keys for other databases because of the fear of identity theft if the data were misused.

Sound Byte MyLab IT

Creating and Querying an Access Database

In this Sound Byte, you learn how to create an Access database to catalog a collection of CDs and DVDs. You also learn how to create input forms, queries, and reports to simplify maintenance of your CD and DVD database.

Primary keys don't have to be numbers that already represent something. In many instances, a primary key value is some form of serial number that begins with the first record and increases serially as each new record is generated. Many database programs have a means of automatically generating a unique identifier. In Microsoft Access, the AutoNumber data type provides the means to generate a serial number as each new record is added.

How is the primary key used to establish relationships between tables? Recall that relational databases link tables by using a common field to establish a relationship between the tables. The common field in one table is that table's primary key, and the same field in the related table is called the foreign key (refer to Figure 11.5).

When the relationship is established, you have the option of enforcing referential integrity for each relationship. Because data changes continually in a database, referential integrity ensures that the data in related tables is synchronized and that a record isn't added or modified in one table and not in the other. For instance, going back to our earlier example, you wouldn't be able to enter a student's record in the Residence Assignments table with a StudentID number that doesn't exist in

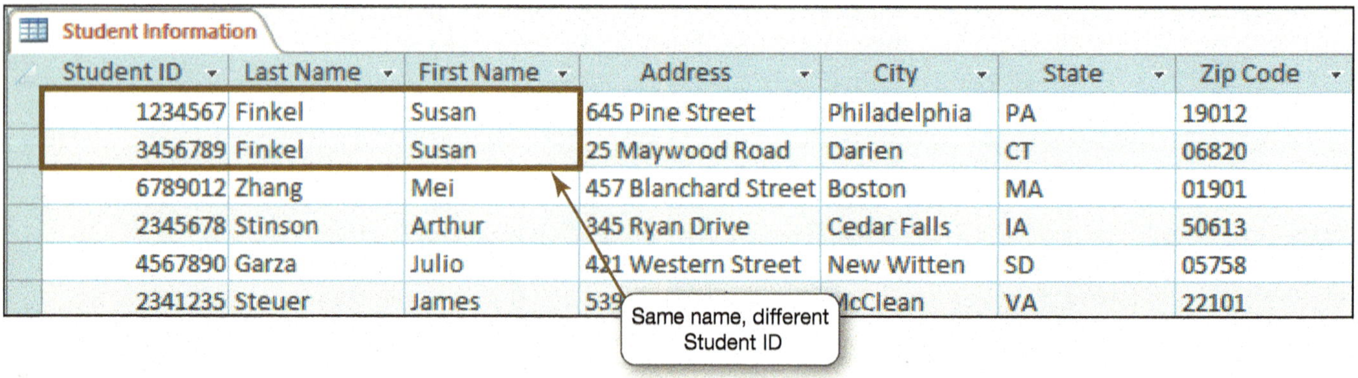

Student Information

Student ID	Last Name	First Name	Address	City	State	Zip Code
1234567	Finkel	Susan	645 Pine Street	Philadelphia	PA	19012
3456789	Finkel	Susan	25 Maywood Road	Darien	CT	06820
6789012	Zhang	Mei	457 Blanchard Street	Boston	MA	01901
2345678	Stinson	Arthur	345 Ryan Drive	Cedar Falls	IA	50613
4567890	Garza	Julio	421 Western Street	New Witten	SD	05758
2341235	Steuer	James	539 [illegible]	McClean	VA	22101

Same name, different Student ID

Figure 11.10 The StudentID field makes an ideal primary key field because even students with the same name won't have the same ID number. *(Courtesy of Microsoft Corp.)*

Bits&Bytes Music Streaming Services Use Databases

When you stream music on your smartphone, you usually don't think about accessing a database, but that's what you're doing when you subscribe to music services such as Tidal, Apple Music, and Spotify (see Figure 11.11). Spotify is a data-driven company, incorporating databases in almost everything it does. The company uses massive databases to keep track of and organize over 20 million songs, with thousands of new songs added every day. Moreover, Spotify uses databases to manage its millions of users, their playlists, and their payment and access preferences. Databases are also used to keep track of the monies owed and paid to rights holders. User data is analyzed to create recommended playlists tailored to each user's preferences, and sophisticated algorithms are used to manipulate the data so when songs are shuffled the same artist isn't played in back-to-back songs. Spotify's databases keep track of information such as who is listening to what music, their location, the time music is played, and which tracks are skipped. Spotify is enabling artists and others in the music industry to tap into these databases to gain insight into who is listening to what music, which provides valuable information for artists to improve revenues from touring and merchandising. Now, as you play a song recommendation or shuffle your playlist, you might think of the databases that are being used.

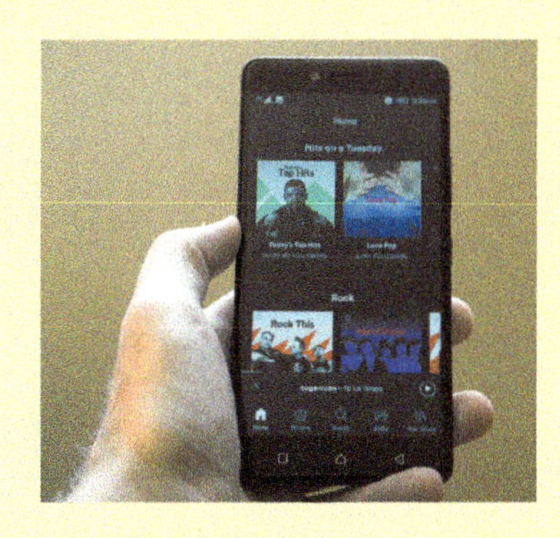

Figure 11.11 Spotify uses databases so you can create and share your playlists or listen to those created by your friends or others. *(Pe3k/Shutterstock)*

the Student Information table (the primary key table) if referential integrity is being enforced. You first would need to add the student to the Student Information table and then enter the student in the Residence Assignments table.

How can I keep track of everything in a database? Databases have many components, such as tables, reports, queries, and forms and fields. It's often difficult to keep track of all the components in a database. Therefore, it's useful, especially for large databases, to access the data dictionary (or the database schema). A database dictionary captures all the database components in a list. The data dictionary is like a map of the database; it not only defines the features of the database (tables, queries, reports, and forms) but also defines the relationships between the tables as well as the features or attributes of the fields in the database.

The attributes that define the data in the data dictionary are metadata: data that describes other data. Metadata is an integral part of the data dictionary and includes:

- The field name
- The *data type* (type of data in the field, such as text, numeric, or date/time)
- The description of the field (optional)
- Any properties (decimals, formatting, etc.) of that particular type of data
- The *field size* (the maximum expected length of data for each field)

Describing the data in this way helps to categorize it and sets parameters for entering valid data into the database (such as a 10-digit number in a phone number field).

Most database management systems keep the data dictionary hidden from users to prevent them from accidentally modifying the contents. In Microsoft Access, the data dictionary can be produced using the Database Documenter command found on the Database Tools tab.

Inputting and Managing Data

Objective 11.8 *Describe how data is inputted and managed in a database.*

How do I get data into the database? You can manually enter data directly into a database table, but that is often inefficient and subject to error. Alternatively, because a great deal of data already exists in some type of digital format, such as web-based data, spreadsheets, and data from other databases, most databases can import data from other files, which can save an enormous amount of keying and reduces the number of data-entry errors.

When importing data, it must exactly match the format of the database. Most databases usually apply filters to the data to determine whether it's in the correct format. Nonconforming data is flagged (either on-screen or in a report) so that you can modify the data to fit the necessary format.

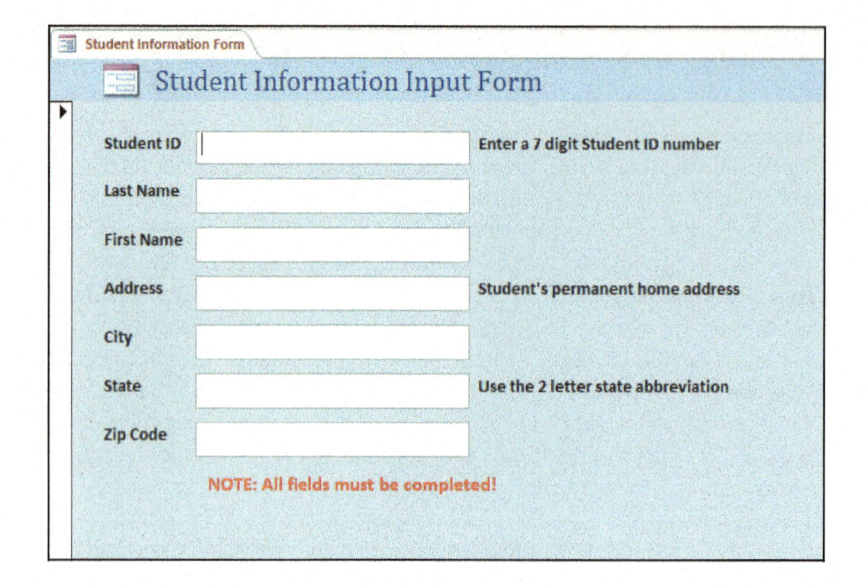

Figure 11.12 Input forms are used to enter data into tables. *(Courtesy of Microsoft Corp.)*

How can I make manual entry into a database more efficient? Input forms are used to control how new data is entered into a database. Figure 11.12 shows an example of an input form for the Student Information table. Each field has a label that indicates the data to be placed in the field. The data is inputted into the blank boxes. Notes have been added to the form to guide users. In addition to being used to input new data, input forms are used to make changes to existing data in the database. Because a form can be configured to display individual records, making changes to data by using a form ensures that the change is made to the correct record.

Data Validation

How can I ensure that only valid data is entered into a field? Recall that one of the main reasons to use a database is to ensure data integrity. Validation is the process of ensuring that data entered into a field meets specified guidelines so that it's accurate and reliable. A validation rule sets a standard for the data contained in a particular field and is specified in the field properties. Violations of validation rules usually result in an error message displayed on the screen with a suggested action so that the error can be addressed. Common validation rules include:

- **Range check.** A range check ensures that the data entered into the field falls within a certain range of values. For instance, you could set a field constraint (a property that must be satisfied for an entry to be accepted into the field) to restrict the rate of pay to fall within a certain range. Figure 11.13 shows a range check in the Field Properties pane for an Access database.

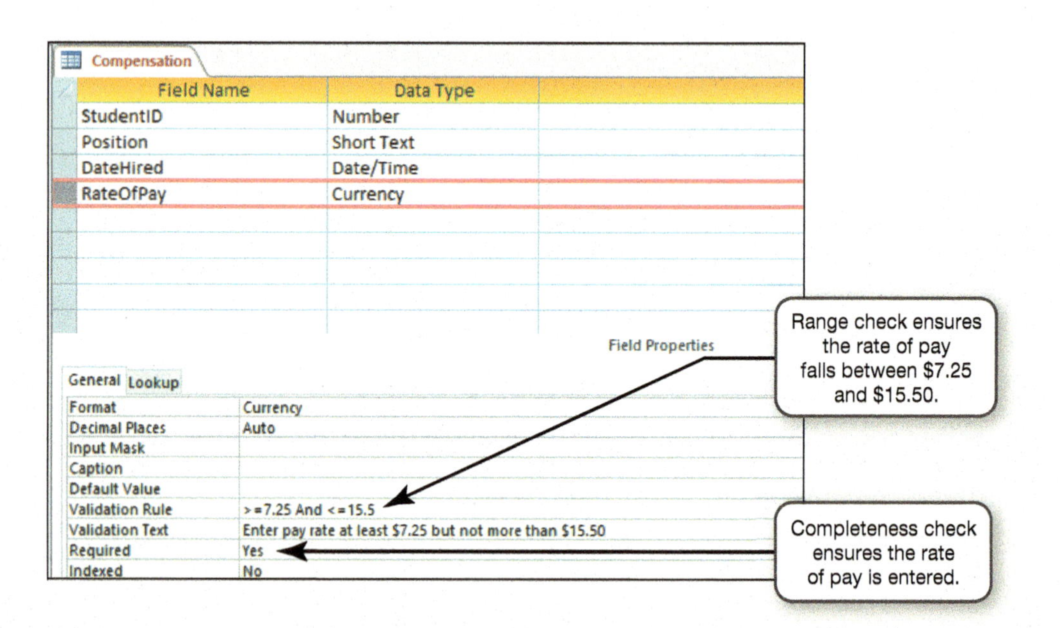

Figure 11.13 Validation rules help ensure that correct data is entered. *(Courtesy of Microsoft Corp.)*

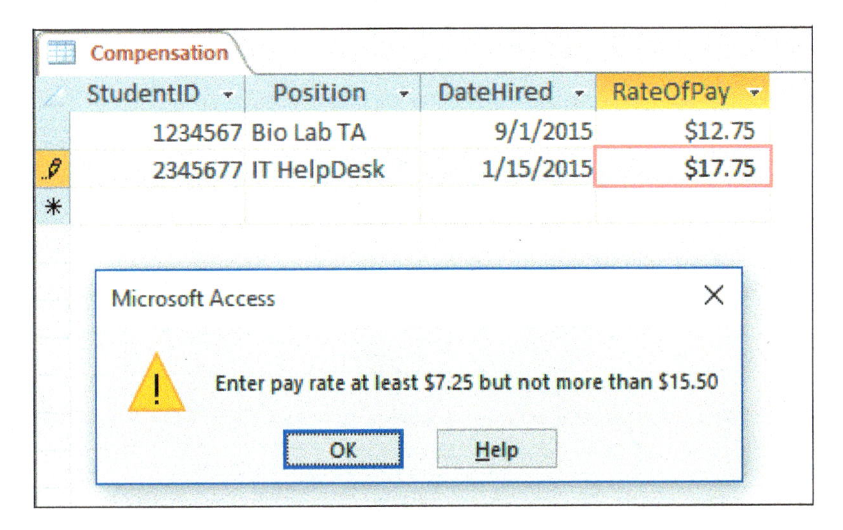

Figure 11.14 Access users see this pop-up when they enter a value that falls outside the stated range of acceptable values. The displayed message is generated from the text typed in the Validation Text property. *(Courtesy of Microsoft Corp.)*

A validation rule restricts the entries for a student's rate of pay for his or her campus job to amounts between $7.25 and $15.50. If users tried to enter a rate of pay less than $7.25 or greater than $15.50, they would be notified of a range error and the input would not be accepted, as shown in Figure 11.14.

- **Completeness.** A completeness check ensures that all required fields have been completed. For example, Figure 11.13 shows the Required property set to Yes, so when data is not entered in the field, as is the case in Figure 11.15, the database performs a completeness check, notices the blank RateOfPay field, and displays an error message to alert the user of the omission.
- **Consistency check.** A consistency check compares the values of data in two or more fields to see whether those values are reasonable. For example, your birth date and the date you enrolled in school are often in a college's database. It's not possible for you to have enrolled in college before you were born. Furthermore, most college students are at least 16 years old. Therefore, a consistency check on these fields might ensure that your birth date is at least 16 or more years before the date when you enrolled in college.
- **Alphabetic and numeric checks.** An alphabetic check confirms that only textual characters are entered in a field. A numeric check confirms that only numbers are entered in a field. With a numeric check in place, $J2.5n would not be accepted as the price of a product. Setting the data type will help ensure that the correct type of data is entered into a field and acts as an alphabetic and numeric check.

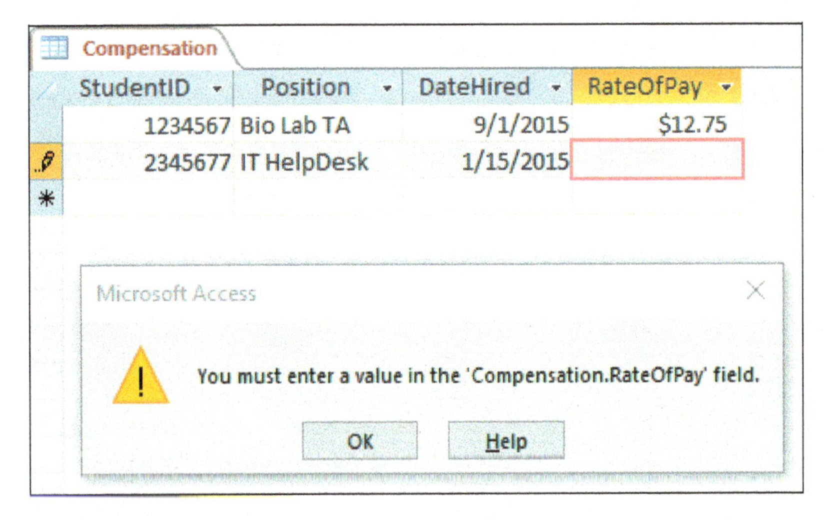

Figure 11.15 Access users see this pop-up when they do not enter a value in a field that has been indicated as required. *(Courtesy of Microsoft Corp.)*

Viewing and Sorting Data

How can I view data in a database? In many instances, you'll want to view the data only one record at a time and not display all the data in an entire table. Forms, such as the one in Figure 11.12, are used to display individual records, either alone or in conjunction with related data. Displaying the tables on-screen and browsing through all the data is an option with most databases and is the better format when you want to reorder or show only a subset of the data.

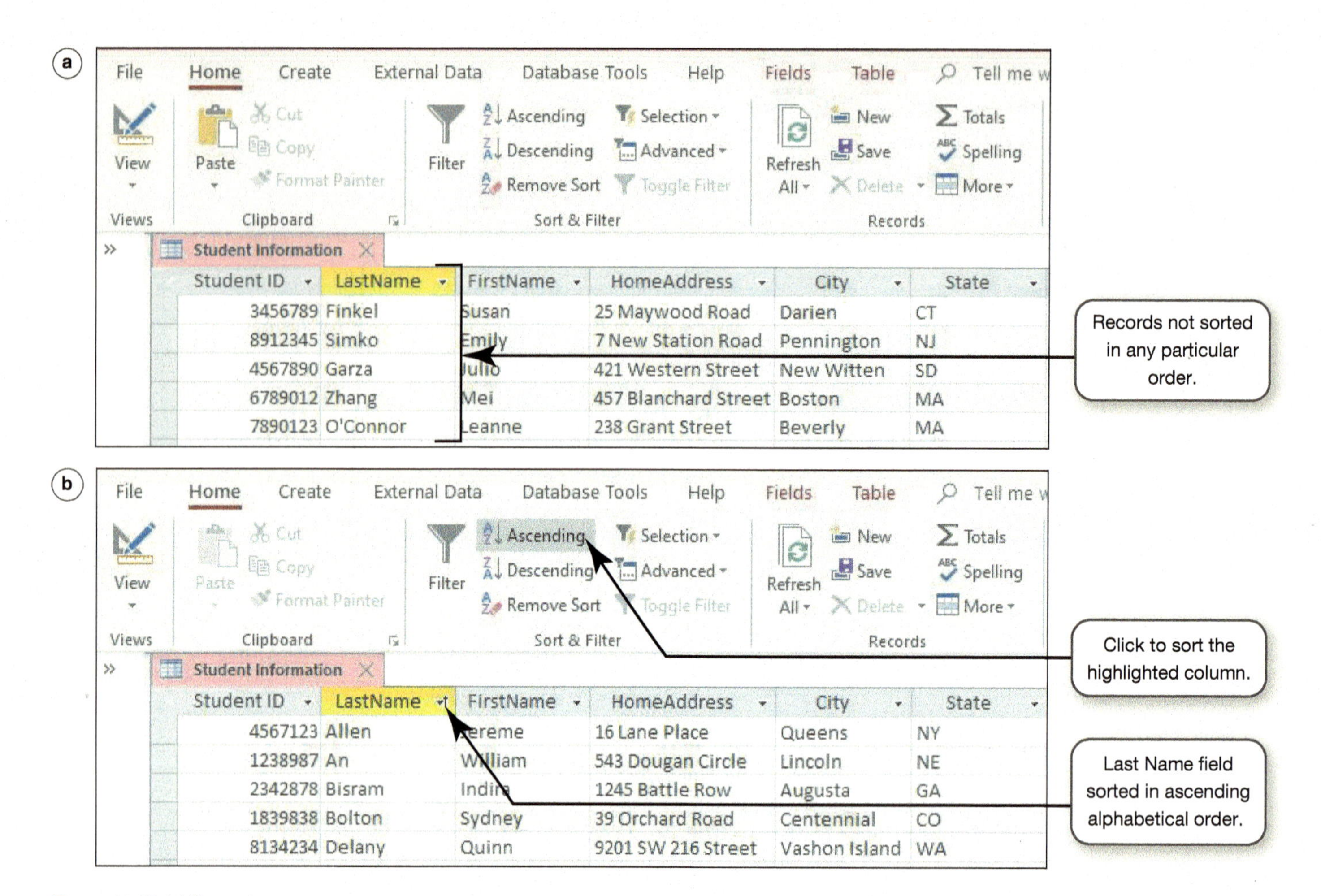

Figure 11.16 (a) Shown is an unsorted table. Notice that the Last Name column is selected (highlighted) for sorting. (b) Selecting the Ascending sorting option produces the sorted output with the records sorted in ascending alphabetical order by last name. *(Courtesy of Microsoft Corp.)*

How can I reorder records in a database? You can reorder data in ascending or descending order by using the sort feature. Figure 11.16a shows an Access database table in which the records were inputted in no particular order. By highlighting a column (in this case, Last Name) and then sorting the data in ascending order, the database displays the records in alphabetical order by last name, as shown in Figure 11.16b.

Extracting or Querying Data

What if I want to display only a certain subset of data in a database? One of the primary benefits of using a database is the ability to manipulate the data so it reveals only those records that meet specific criteria. There are two ways to display records that match certain criteria:

1. **Use a filter.** A filter temporarily displays records that match certain criteria. Filters are useful for quick analyses and are appropriate if you don't need to review the results at another time. In addition, filters can be applied only to fields from the same table, so if you need to combine fields from multiple tables to achieve your subset of data, queries are a better option.
2. **Create a query.** Like a filter, a database query is a way of retrieving a subset of data that matches certain criteria. A query can be used to extract data from one or multiple tables, and unlike a filter, queries can be saved and run again whenever necessary to obtain updated results. When you create a query, you choose the fields to be included in the results and specify the criteria for selecting records. For example, as shown in Figure 11.17a, to create a list of students living in the Boyer 1 dorm and their room numbers, you would pull data from the Student Information table (FirstName and LastName), Residence Halls table (ResidenceName), and Residence Assignments table (RoomNumber). Then you'd apply criteria to specify that you want only those records pertaining to a specific dorm, in this case Boyer 1. Figure 11.17b displays the results for this query.

How do I develop queries for my database? Modern database systems provide wizards or other mechanisms to guide you through the process of creating queries. Figure 11.17a shows an example of an Access query being created in Design View that allows you to select query elements

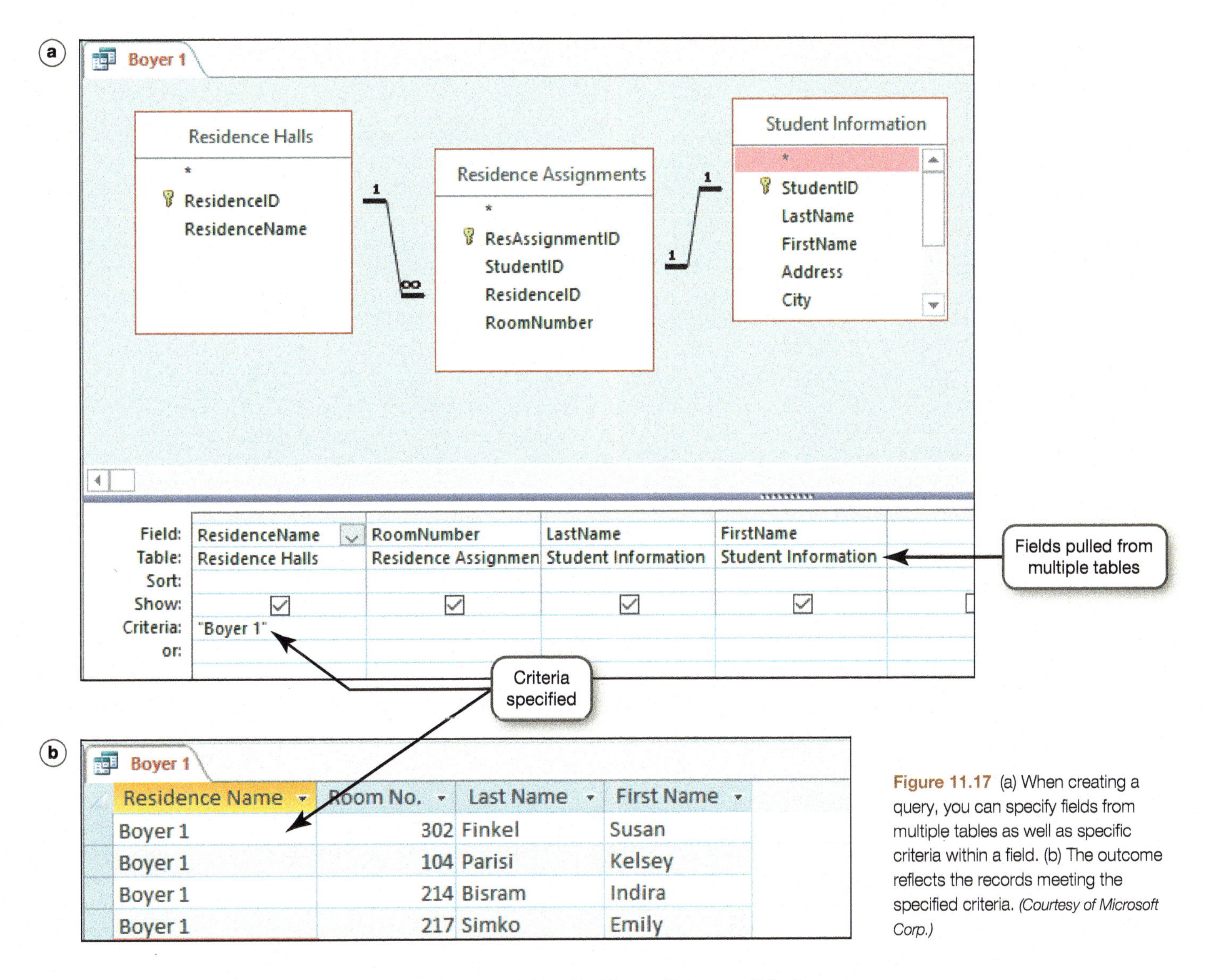

Figure 11.17 (a) When creating a query, you can specify fields from multiple tables as well as specific criteria within a field. (b) The outcome reflects the records meeting the specified criteria. *(Courtesy of Microsoft Corp.)*

easily. You can also create a query using a Simple Query Wizard in Microsoft Access. Whether you use the Simple Query Wizard or Design View, you're actually using SQL commands without realizing it. All modern DBMSs contain a query language that the software uses to retrieve and display records. A query language consists of its own vocabulary and sentence structure that you use to frame the requests. Query languages are similar to full-blown programming languages but are usually much easier to learn. The most popular query language used today is SQL.

Rather than relying on the features in Access to create queries, you may create your own SQL queries in Access, modify existing queries at the SQL language level, or view the SQL code that the wizard created. Figure 11.18 shows the SQL code that the query in Figure 11.17 created.

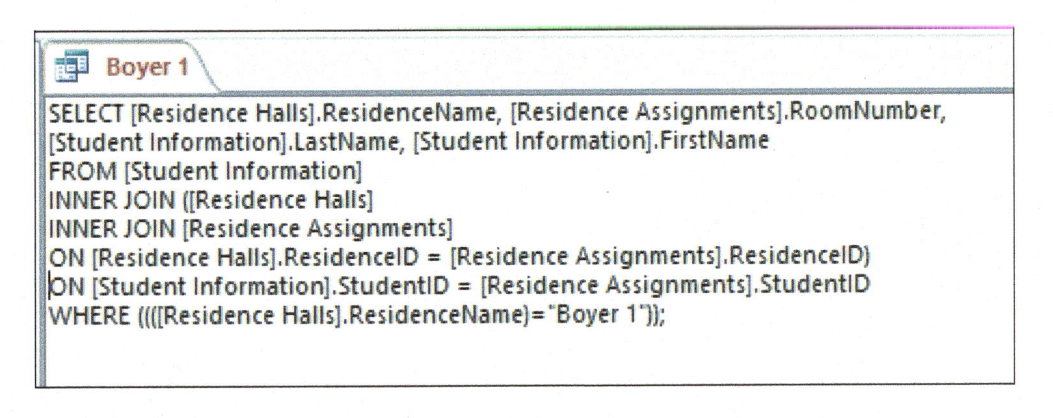

Figure 11.18 The SQL View window in Access shows the SQL code for the query in Figure 11.17. Although it is a relatively simple SELECT statement, it is much easier to create the query using the Access tools provided.

> To display SQL code for a query, open a query and select ***SQL View*** *from the* ***View*** *drop-down list on the* ***Home*** *tab of the Ribbon. (Courtesy of Microsoft Corp.)*

Dig Deeper Structured Query Language (SQL)

To extract records from a database, you use a query language. Almost all relational and object-oriented databases today use Structured Query Language, or SQL. Oracle, Microsoft SQL Server, Microsoft Access, IBM DB2, and MySQL are examples of popular databases that use SQL.

SQL uses relational algebra to extract data from databases. Relational algebra is the use of English-like expressions that have variables and operations, much like algebraic equations. Variables include table names, field names, or selection criteria for the data you wish to display. Operations include directions such as *select* (which enables you to pick variable names), *from* (which tells the database which table to use), and *where* (which enables you to specify selection criteria). The two most common queries used to extract data using relational algebra are select queries and join queries.

A select query displays a subset of data from a table (or tables) based on the criteria you specify. A typical select query has the following format:

```
SELECT (Field Name 1, Field Name 2, . . .)
FROM (Table Name)
WHERE (Selection Criteria)
```

The first line of the query contains variables for the field names you want to display. The FROM statement enables you to specify the table name from which the data will be retrieved. The last line (the WHERE statement) is used only when you wish to specify which records need to be displayed (such as all students with GPAs greater than 3.3). If you wish to display all the rows (records) in the table, then you don't use the WHERE statement (see Figure 11.19).

Suppose you want to create a phone list that includes all students from the Student Information table in Figure 11.19a. The SQL query you would send to the database would look like this:

```
SELECT (FirstName, LastName, CellPhone)
FROM (Student Information)
```

Figure 11.19b shows the output from this query.

What if you want a phone list that shows only students from Massachusetts? In that case, you would add a WHERE statement to the query, as follows:

```
SELECT (FirstName, LastName, CellPhone)
FROM (Student Information)
WHERE (State = MA)
```

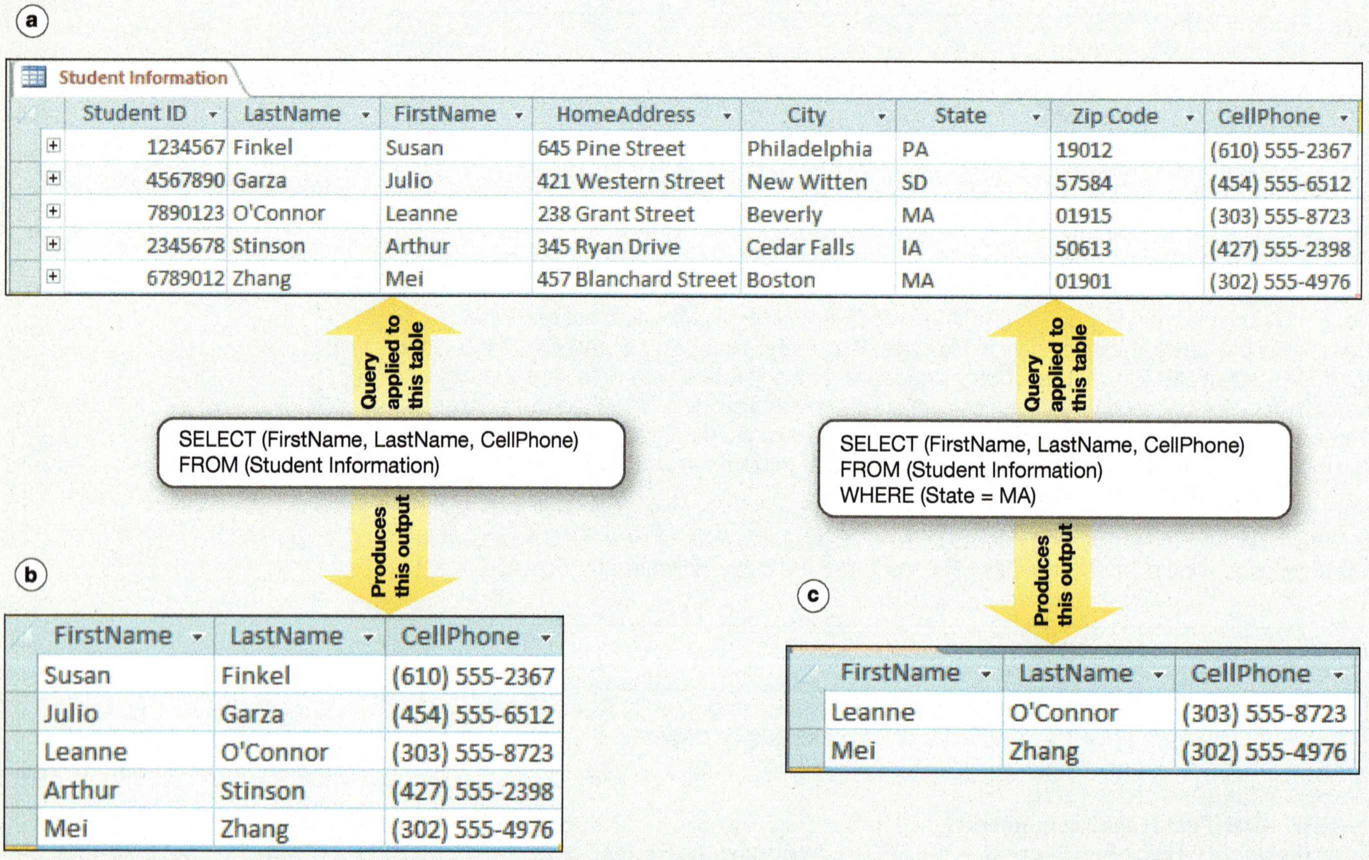

(a)

Student Information

Student ID	LastName	FirstName	HomeAddress	City	State	Zip Code	CellPhone
1234567	Finkel	Susan	645 Pine Street	Philadelphia	PA	19012	(610) 555-2367
4567890	Garza	Julio	421 Western Street	New Witten	SD	57584	(454) 555-6512
7890123	O'Connor	Leanne	238 Grant Street	Beverly	MA	01915	(303) 555-8723
2345678	Stinson	Arthur	345 Ryan Drive	Cedar Falls	IA	50613	(427) 555-2398
6789012	Zhang	Mei	457 Blanchard Street	Boston	MA	01901	(302) 555-4976

Query applied to this table

SELECT (FirstName, LastName, CellPhone)
FROM (Student Information)

Produces this output

(b)

FirstName	LastName	CellPhone
Susan	Finkel	(610) 555-2367
Julio	Garza	(454) 555-6512
Leanne	O'Connor	(303) 555-8723
Arthur	Stinson	(427) 555-2398
Mei	Zhang	(302) 555-4976

Query applied to this table

SELECT (FirstName, LastName, CellPhone)
FROM (Student Information)
WHERE (State = MA)

Produces this output

(c)

FirstName	LastName	CellPhone
Leanne	O'Connor	(303) 555-8723
Mei	Zhang	(302) 555-4976

Figure 11.19 When the query on the left is applied to the (a) Student Information table, it restricts the output to (b) only a phone list. The query on the right, which uses a WHERE statement, further restricts the phone list to (c) only students from Massachusetts. *(Courtesy of Microsoft Corp.)*

This query restricts the output to students who live in Massachusetts, as shown in Figure 11.19c. Notice that the State field in the Student Information table can be used by the query (in this case, as a limiting criterion) but that the contents of the State field aren't required to appear in the query results. This explains why the output shown in Figure 11.19c does not show the State field.

When you want to extract data that is in two or more tables, you use a join query. The query actually links (or joins) the two tables using the common field in both tables and extracts the relevant data from each. The format for a simple join query for two tables is as follows:

SELECT (Field Name 1, Field Name 2)
FROM (Table 1 Name, Table 2 Name)
WHERE (Table 1 Name.Common Field Name = Table 2 Name.
Common Field Name)
AND (Selection Criteria)

Notice how similar this is to a select query, although the FROM statement must now contain two table names. In a join query, the WHERE statement is split into two parts. In the first part (right after WHERE), the relation between the two tables is defined by identifying the common fields between the tables. The second part of the statement (after AND) is where the selection criteria are defined.

The AND means that both parts of the statement must be true for the query to produce results (i.e., the two related fields must exist and the selection criteria must be valid). Figure 11.20 illustrates using a join query for the Student Information and Roster Master tables to produce a class roster for students.

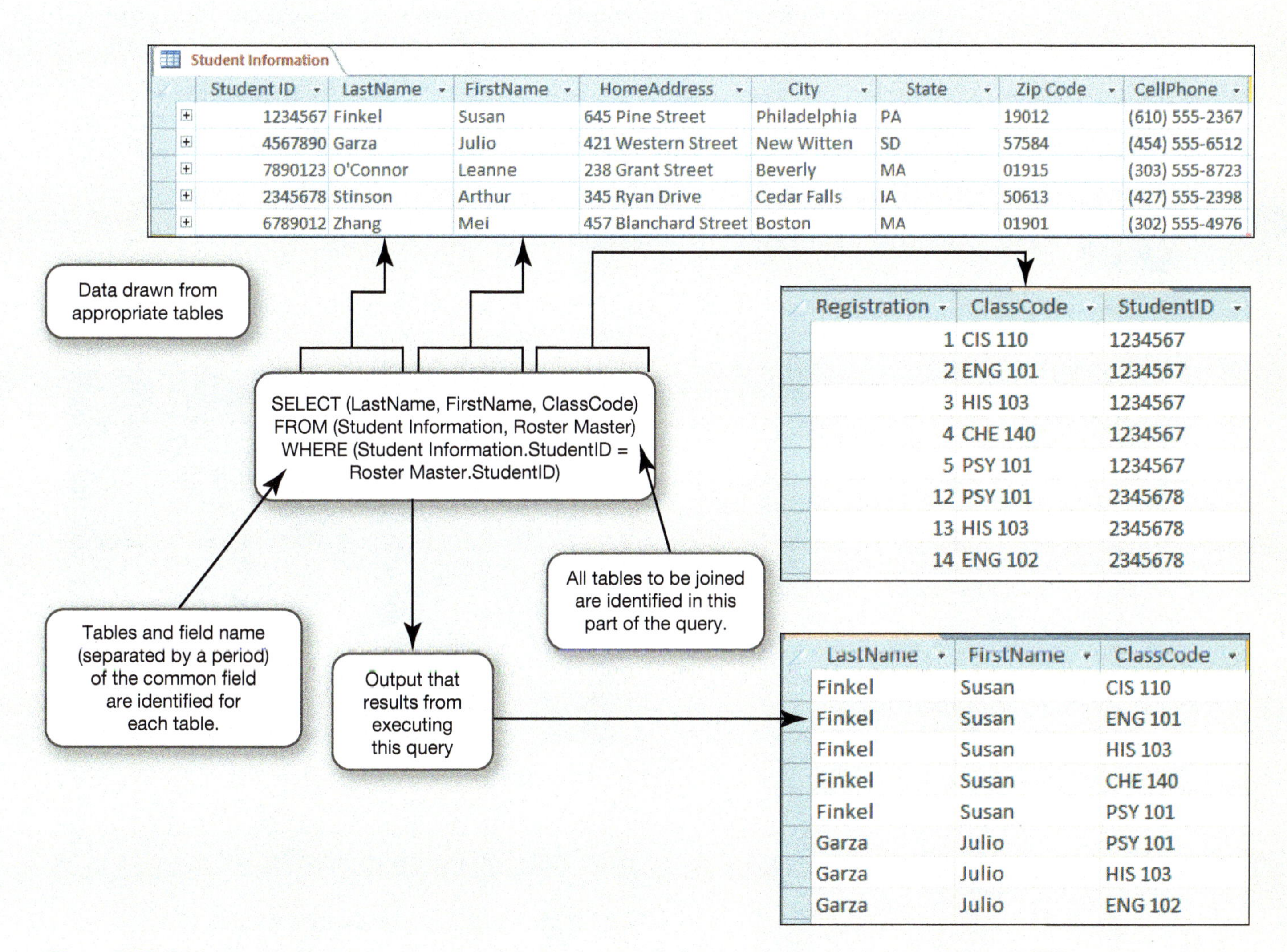

Figure 11.20 This join query will display a student roster for each student in the Student Information table. Notice that the WHERE statement creates the join by defining the common field (in this case, StudentID) in each table. *(Courtesy of Microsoft Corp.)*

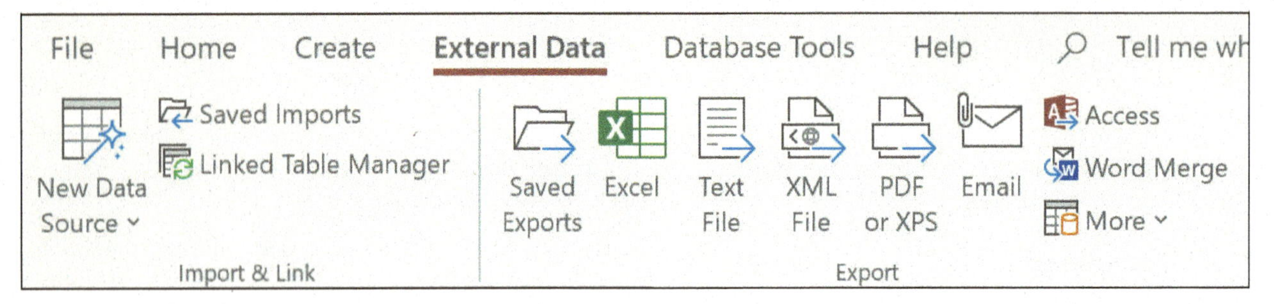

Figure 11.21 Data can be imported and exported between Access and other Microsoft Office applications as well as converted to PDF, XML, and other formats for use with other applications. *(Courtesy of Microsoft Corp.)*

Outputting Data

How do I present data from a database? Data by itself is not often meaningful. Information is derived by querying, sorting, and filtering the data, as we've been discussing. Although you can print out the data directly from tables or from the results of queries you create, the resulting format is usually not displayed in the most presentable manner. Therefore, it's best to create a report. When you create a report, you can adjust how the information is displayed, such as grouping like information and including aggregate information such as totals. Businesses routinely summarize the data in their databases and compile summary data reports. For instance, at the beginning of each semester, you receive an invoice of amounts owed based on the number of classes you have registered for, and at the end of each semester, your school generates a grade report for you.

Can I transfer data from a database to another software application? There are times when you need to export data from a database to another application, such as Microsoft Excel or Word, for further analysis or manipulation or to be included in a report or other document. Exporting data involves putting it into an electronic file in a format that another application can understand. For example, as shown in Figure 11.21, data can be imported and exported among Microsoft Access and Word and Excel as well as converted to and from other formats such as PDF and XML for use with other applications.

Bits&Bytes Data Dashboards: Useful Visualization Tools

Data is abundant in today's business world, but all that data is useless unless it's analyzed appropriately and provides insights that lead to better business solutions and decisions. Moreover, in a fast-paced and competitive business environment, having meaningful metrics in one place and updated in real time is powerful for any decision maker.

A *data dashboard* is a combination of company metrics, or key performance indicators (KPIs), and is often displayed publicly in real time so many employees can have access to the information at once. Data dashboards are important tools for good decision making at any level. They can be tailored to display department-specific metrics or incorporate data from a variety of corporate sources such as sales, customer relationships, website analytics, and accounting. Data dashboards can foster better collaboration between departments and promote better decision making for the entire corporation. In the past, data could only be visualized with charts that used static data. With cloud-based data and tools, however, dashboards can be updating live on screens around the office, providing a means to spot trends or pick up on issues before they become major problems.

(Rawpixel/123RF)

Before moving on to Part 2:

1. Watch Chapter Overview Video 11.1.
2. Then take the Check Your Understanding quiz.

Check Your Understanding // Review & Practice

For a quick review to see what you've learned so far, answer the following questions.

multiple choice

1. Two lists showing different values for the same data in different places of a database is an example of
   a. data redundancy.
   b. data duplication.
   c. incomplete data.
   d. data inconsistency.
2. NoSQL databases are best used with what kind of data?
   a. Data generated from social media sites
   b. Static data from flat databases
   c. Dynamic data from object oriented databases
   d. Data with more than two attributes
3. Which type of relationship should be used to connect a Password table to a User Account table?
   a. One-to-many relationship
   b. One-to-one relationship
   c. Many-to-many relationship
   d. One-to-none relationship
4. Which type of database organizes data by measure and feature attributes?
   a. Multidimensional database
   b. Relational database
   c. Object-oriented database
   d. All of the above
5. The field that has a unique entry for each record in a database table is called the
   a. foreign key.
   b. logical key.
   c. primary key.
   d. master field.

chew on this

(3Dalia/Shutterstock)

You've been asked to establish a new database for a local auto repair shop. The owner has given you the following data categories that she feels should be in the database. Organize the data categories into two tables, label each table with a name, and identify in each table the primary key field.

Customer ID	Car Make	Last Name
Address	Phone	Vehicle ID
Model Year	Model	City
State	First Name	License Plate

Go to **MyLab IT** to take an autograded version of the *Check Your Understanding* review and find all media resources for the chapter.

For the IT Simulation for this chapter, see MyLab IT.

Try This

Using Excel's Database Functions

Databases are great for organizing lots of data, but they can be difficult to understand and to learn how to use. If your data is flat, that is, if it doesn't need to be divided into multiple tables, you can use Excel's database functions to organize your data. In this Try This exercise, you learn about some of Excel's useful database functions: sorting one or multiple columns, applying filters, and applying data validation. All the database commands in Excel are on the Data tab. For more step-by-step instructions, watch the Try This video on MyLab IT.

Sorting in Alphabetical Order

You can sort one column of data in alphabetical or numerical order.

Step 1 Place your cursor anywhere in the column you want to sort.

Step 2 In the Sort & Filter group, click **Sort A to Z** or **Sort Z to A**, depending on whether you want to sort in ascending or descending order.

Applying Filters

When you want to isolate a subset of data, you should apply a filter.

Step 1 Click anywhere in the table and then click **Filter** in the Sort & Filter group. Filter arrows will display next to each column heading.

Step 2 Click the filter arrow beside the column heading for the column you want to filter.

Step 3 Clear the checkmark from Select All and then click the filter value(s) you'd like to apply. Click **OK**.

Step 4 Review the data to ensure that the filter has been applied correctly. (*Note:* To remove the filter, click **Filter** to deselect the option.)

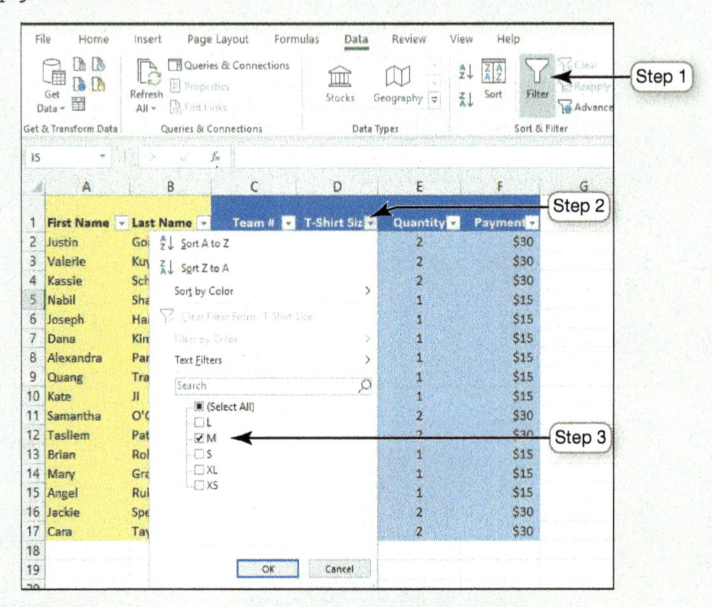

(Courtesy of Microsoft Corp.)

Applying Data Validation

You can apply rules to data by using the Data Validation tool to ensure that the proper data is being entered.

Step 1 Click **Data Validation** in the Data Tools group to open the Data Validation dialog box.

Step 2 On the Settings tab in the Data Validation dialog box, choose the type of data value to be allowed and the value limits in the Data and Maximum boxes. (*Note:* The criteria in the Data box and other limits will vary.)

Step 3 On the Error Alert tab, select the style of error message you want displayed (Stop, Warning, or Information) and then type the title of the error message and the error message text. Click **OK**.

Step 4 Test the data validation by entering an unacceptable value. When an unacceptable value is entered, an error message appears.

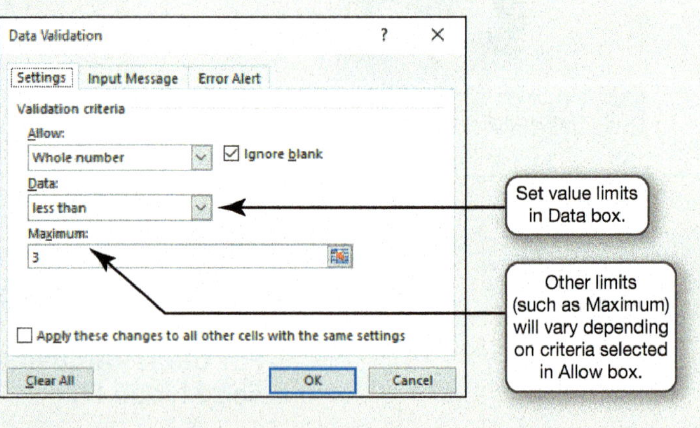

(Courtesy of Microsoft Corp.)

Make This

TOOL: App Inventor 2 or Thunkable

A Family Shopping List

In this exercise, you'll create an app that allows each family member to add items to a shopping list from his or her own phone.

This requires the TinyWebDB component in App Inventor, which allows you to store information on the Internet so it can be recalled by anyone using your application. Now your app can trade information between all its users!

Your app can store values on the Internet so they are accessible by everyone using your app.

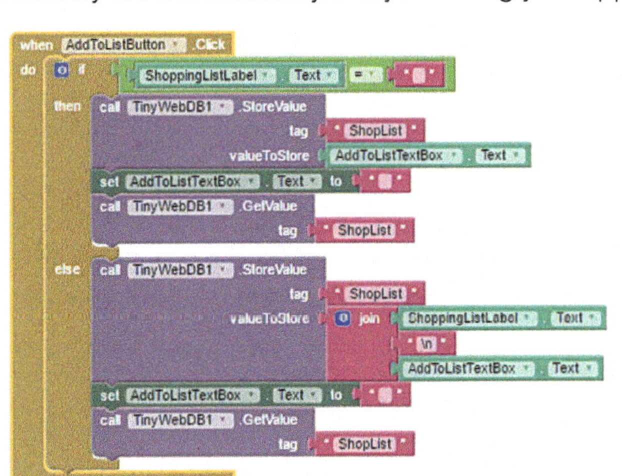

(Copyright MIT, used with permission.)

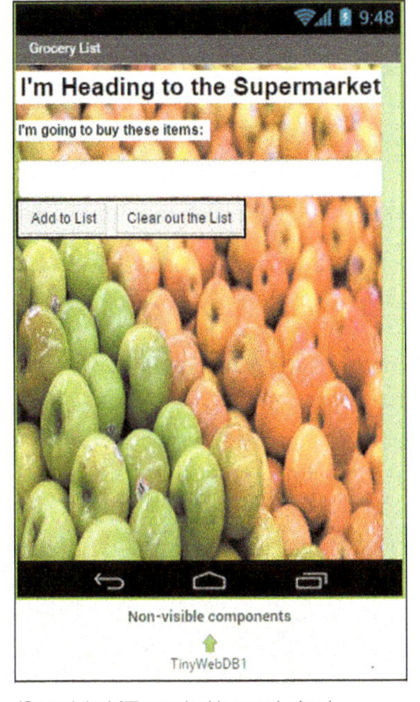

(Copyright MIT, used with permission.)

To create this app for iOS, go to *Thunkable.com*, a programming platform based on App Inventor.

For the instructions for this exercise, go to MyLab IT.

Part 2  For an overview of this part of the chapter, watch **Chapter Overview Video 11.2**.

How Businesses Use Databases

Learning Outcome 11.2 **You will be able to explain how businesses use data warehouses, data marts, and data mining to manage data and how business information systems and business intelligence are used to make business decisions.**

Large organizations accumulate massive amounts of data, both internal and external. By using data warehousing, data marts, and data mining, these organizations use databases to tap into a wealth of information, improving their internal operations as well as their customer relationships. In this section, we look at how they do just that.

Data Warehousing and Storage

At the simplest level, data is stored in a single database on a database server, and you retrieve the data as needed. This works fine for small databases and simple enterprises. Problems can arise when the organization gets much larger, requiring data to be stored in separate, department-specific databases. These databases weren't designed to be accessed together to retrieve and combine data from multiple sources, thus eliminating the benefits of accessing data from all databases. Large storage repositories called *data warehouses* and *data marts* are being used to help solve this problem.

Data Warehouses and Data Marts

Objective 11.9 *Explain what data warehouses and data marts are and how they are used.*

What is a data warehouse? A data warehouse is a large-scale collection of data that organizes and stores in one place all the data from an organization's multiple databases. Individual databases contain a wealth of information, but each database's information usually pertains to one area in the organization. Department-focused databases do not easily allow data to be shared and accessed by other departments or allow management to help make decisions at an organizational level. For instance, the order database at Amazon contains information about product orders, such as the buyer's name, address, and payment information and the product name or ID. However, the order database doesn't contain information on inventory levels of products, nor does it list suppliers from which out-of-stock products can be obtained. Data warehouses, therefore, consolidate information from various operational systems to present an enterprise-wide view of business operations that is helpful in decision making, revealing trends, and garnering a deeper understanding of how the business is doing as a whole.

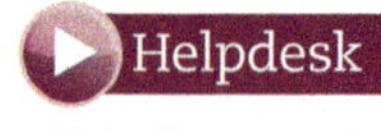

MyLab IT

How Businesses Use Databases

In this Helpdesk, you'll play the role of a helpdesk staffer, fielding questions about how businesses use databases.

Where does a data warehouse get its data? Data warehouses can get data from three sources:

1. **Internal sources.** Sales, billing, inventory, and customer databases all provide a wealth of information. Spreadsheets and other ad hoc analysis tools may contain data that can be loaded into the data warehouse.
2. **External sources.** Vendors and suppliers often provide data regarding product specifications, shipment methods and dates, billing information, and so on.
3. **Clickstream data.** Software used on company websites to capture information about each click that users make as they navigate through the site is referred to as clickstream data. Using clickstream data–capture tools, a company can determine which webpages users visit most often, how long users stay on each webpage, which sites directed users to the company site, and user demographics. Such data can provide valuable clues to what a company needs to improve on its website to stimulate sales.

Do data warehouses capture data from only one time period? Data warehouse data is time-variant data, meaning it doesn't all pertain to one period in time. The warehouse contains current values, such as amounts due from customers, as well as historical data. If you want to examine the buying habits of a certain type of customer, you need data about both current and prior purchases. Having time-variant data in the warehouse gives managers the ability to analyze the past, examine the present in light of historical data, and make projections about the future.

How does data get into a data warehouse? Although source databases might contain similar information (such as customer information or sales data), the format of the data is most likely

Bits&Bytes Data Warehouses Are Going to the Cloud

Data warehouses enable companies to compile and analyze big data. However, the costs of installing and maintaining the hardware and software required for data warehouses can often be quite high. Therefore, data warehouses are moving to the cloud. Amazon Redshift, for example, is a cloud-based data warehouse. Cloud-based platforms make it easier to increase the size of the data warehouse, which is important given the escalating amounts of data being collected. Redshift, for example, is designed to handle petabytes of information. Cloud-based data warehouses are becoming the standard to meet the quickly increasing demand for data storage and analytics and replacing the traditional on-premise platforms.

different in each database. Therefore, prior to going into a data warehouse, data must go through a formatting or cleansing process to enable data of different sources and types to be comingled for analysis. That process is referred to as ETL (extract, transform, and load). Data staging is an intermediate storage area used for data processing during the ETL process. The data staging area sits between the data source and the data warehouse. Data staging is vital because different data must be extracted and then reformatted to fit the data structure defined in the data warehouse's DBMS. In addition, sometimes the timing of data collection may vary by department. For example, sales data may be collected daily, whereas financial data may require monthly collection. In these cases, data staging is necessary to collect and store data before going into the data warehouse. Finally, the transformed and integrated data is moved to the data warehouse.

How is data in a data warehouse organized? Data in a data warehouse is organized by subject such as sales, operational data, inventory data, and external data instead of by department, which might have components of data from all those subjects. For example, a large retailer sells many types of electronics. Different departments in the retailer are responsible for each type of product and track the products they sell in different databases (a database each for TV, cell phone, and computer sales, for example), as shown in Figure 11.22.

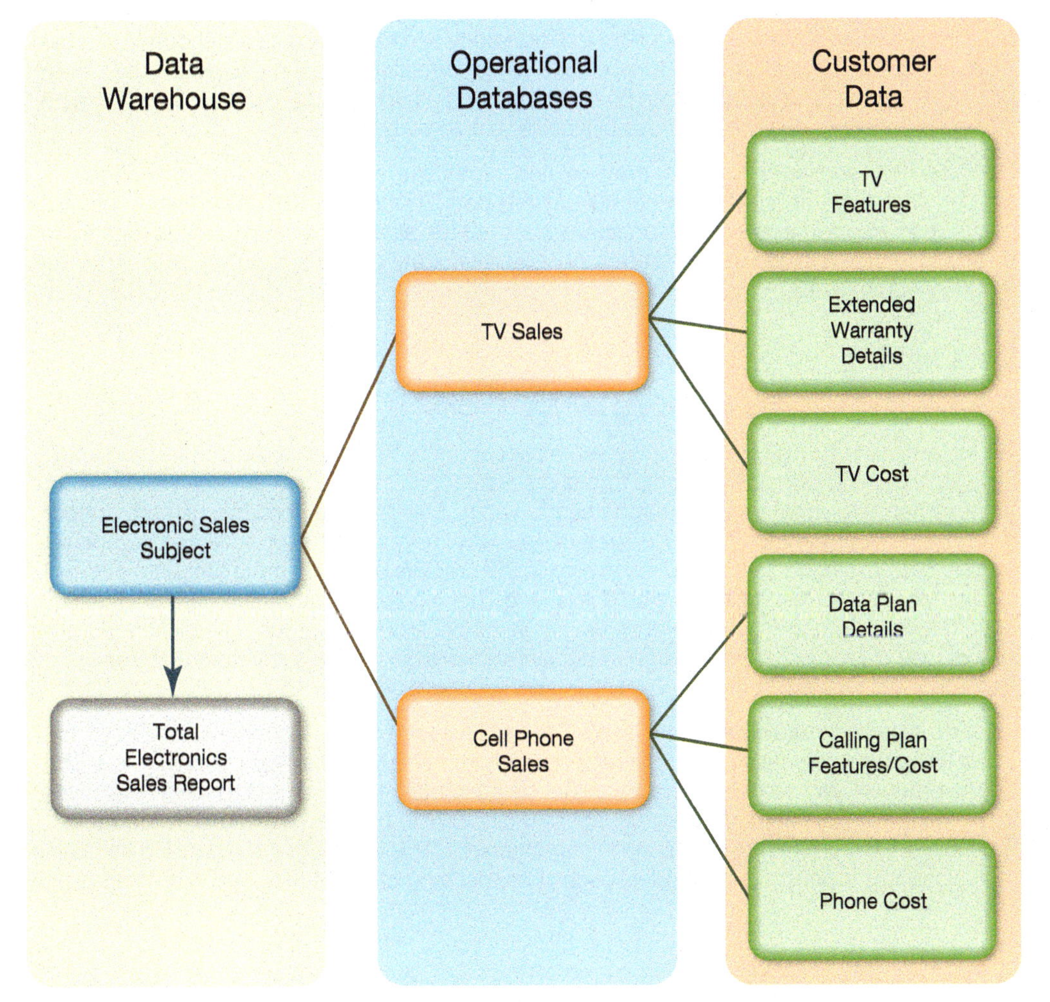

Figure 11.22 Data from individual databases is drawn together in a data warehouse. Managers can then produce comprehensive reports that would be difficult to create from individual databases.

These databases capture specific information about each type of electronics. The TV Sales database captures information such as inventory, wholesale and retail costs, extended warranty details (and costs), and perhaps even human resource data on the TV Sales department employees. All this information is pertinent to determining the ultimate value of TV sales. The Cell Phone Sales database captures data similar to TV but related to cell phones, and so forth with all the various operational databases.

However, data on purchasing, sales, and inventory on all the electronic devices is critical to the management of the electronics retailer, no matter which types of products are involved. Therefore, an electronics retailer's data warehouse might have a subject called Electronics Sales (see Figure 11.22) that would contain information about all electronic devices sold throughout the company (such as TVs, cell phones, computers, entertainment equipment). Although the Electronics Sales Subject is a database that contains information from the other databases the company maintains, all data in the Electronics Sales database specifically relates to electronics sales (as opposed to, say, appliances, which the retailer also carries).

From the Electronics Sales database, it's easy for managers to produce comprehensive reports, dashboards with key performance indicators, and other analytics that can contain information pertaining to all types of electronics sales.

How can data stored in the data warehouse be extracted and used? Managers can query the data warehouse in much the same way you would query an Access database. However, because there is more data in the data warehouse, special software is needed to perform such queries. Online analytical processing (OLAP) software provides standardized tools for viewing and manipulating data in a data warehouse. The key feature of OLAP tools is that they enable flexible views of the data, which the software user can easily change.

Data Marts

What if a smaller amount of data is needed? Small slices of the data warehouse, each called a data mart, are often created so that companies can analyze a related set of data that is grouped together and separated out from the main body of data in the data warehouse. Whereas data warehouses have an enterprise-wide depth, the information in data marts pertains to a single component of the business. Data can be extracted using powerful OLAP query tools, or it can be stored in specialized data marts for use by specific employee groups.

For instance, if you need accurate, sales-related information and you don't want to wade through customer service data, accounts payable data, and product shipping data to get it, a data mart that contains information relevant only to the sales department can be created to make the task of finding this data easier. An overview of the data-warehousing process is illustrated in Figure 11.23.

Data Mining

Objective 11.10 *Describe data mining and how it works.*

What is data mining? Just because you have captured data in an organized fashion and have stored it in a format that seems to make sense doesn't mean that an analysis of the data will automatically reveal everything you need to know. Trends can sometimes be hard to spot if the data isn't organized or analyzed in a certain way. To make data work harder, companies employ data-mining techniques. Data mining is the process by which great amounts of data are analyzed and investigated. The objective is to spot significant patterns or trends within the data that would otherwise not be obvious. For instance, by mining student enrollment data, a school may discover that there's been a consistent increase in new engineering-degree students who are women.

Why do businesses mine their data? The main reason businesses mine data is to understand their customers better. If a company can better understand the types of customers who buy its products and can learn what motivates its customers to do so, it can market effectively by concentrating its efforts on the populations most likely to buy. You may have noticed that products are frequently moved around in supermarkets. This is usually the result of data mining. With electronic scanning of bar codes, each customer's purchase is recorded in a database. By

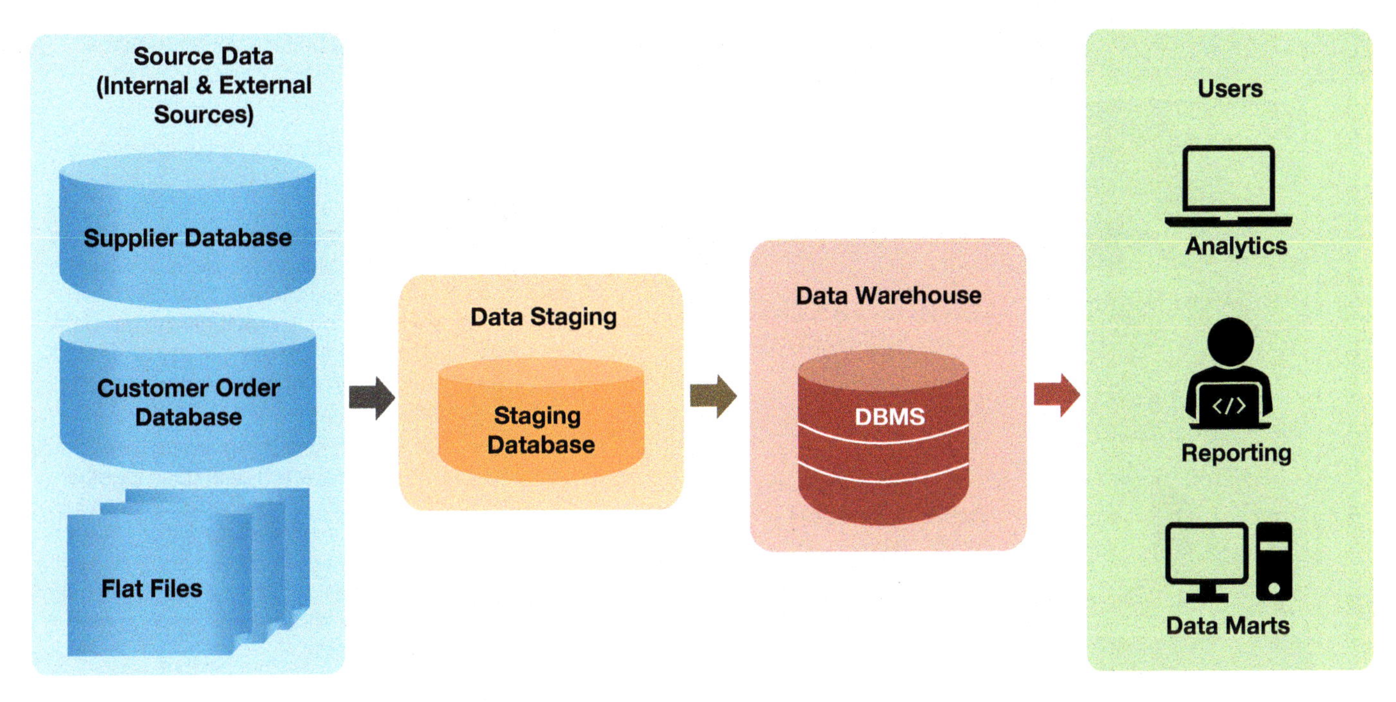

Figure 11.23 An Overview of the Data-Warehousing Process. *(Sirtravelalot/Shutterstock, Redav/Shutterstock)*

classifying the data and using analytical tools and methods, supermarket managers can determine which products people usually purchase with other products. The store then places these products close to each other so that shoppers can find them easily. For instance, if analysis shows that people often buy potato chips with soft drinks, it makes sense to place these items in the same aisle.

How do businesses mine their data? Data mining enables managers to sift through data in several ways. Each technique produces different information on which managers can then base their decisions.

Bits&Bytes Hadoop: How Big Data Is Being Managed

Hadoop is an open-source platform sponsored by the Apache Software Foundation that makes big, complex, and unstructured data easier to manage. It eliminates the constraints placed on how data is traditionally stored and processed by spreading out the data and using many processors that work in parallel. The unique aspects of Hadoop reside in how it stores files and processes data:

- **File storage.** Hadoop facilitates storing files bigger than what can be stored in a centralized database system. Instead of using one server that is common in a centralized database system, Hadoop breaks the data into chunks that it then spreads across to different servers, so it can store many files and very large files.
- **Data processing.** In addition, Hadoop uses a system called MapReduce for processing big data sets. It makes working with large data sets over a network much more efficient. Rather than moving the data to the software, as is done with traditional data processing systems, MapReduce moves the software to the data. MapReduce sends code to each of the dozens of servers storing the data, and then each server uses multiple processors to process its set of data. Results are then delivered back in a unified result set.

Hadoop is being used in a variety of industries and markets. For example, Hadoop can be used to conduct accurate portfolio evaluation and risk analysis in the financial industry as well as organize and manage retail data better to give customers a better response to their searches.

Table 11.2 Data-Mining Techniques

	**Anomaly Detection** Identify the outliers to provide meaningful insights into fixing problems or capturing new opportunities.
	**Association or Affinity Grouping** Determine which data goes together.
	**Classification** Define data classes to help spot trends and then apply the class definitions to all unclassified data.
	**Clustering** Organize data into similar subgroups, or clusters, without using predefined classes.
	**Estimation and Regression** Assign a value to data based on some criterion.
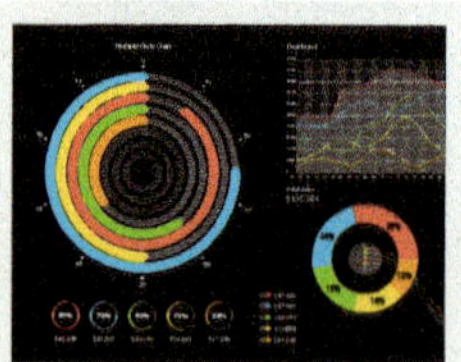	**Visualization** Use charts and dashboards to represent data visually so that managers can interpret it in new and different ways.

(Corund/Shutterstock; Nitr/Shutterstock; Esoxx/Shutterstock; Igor Kisselev/Shutterstock; Phipatbig/Shutterstock; Zmicier kavabata/Shutterstock)

Managers make their data meaningful through the following techniques described next and listed in Table 11.2.

- **Anomaly detection.** Not all data fits together or is as expected. There will be outliers. Outliers can occur because of errors or because of unusual circumstances. This process can discover when current data has changed from previously collected data or differs from expectations. One common use of anomaly detection is credit card fraud detection. Having access to such anomalies can provide meaningful insights into fixing problems or capturing new opportunities.
- **Association or affinity grouping.** When mining data, managers can determine which data goes together by applying affinity grouping or association rules to the data. For example, suppose analysis of a sales database indicates that two items are bought together 60% of the time. Based on this data, managers might decide that these items should be pictured on the same page of their website.
- **Classification.** Before mining, managers define data classes they think will be helpful in spotting trends. They then apply these class definitions to all unclassified data to prepare it for analysis. For example, good credit risk and bad credit risk are two data classes managers could establish to determine whether to grant car loans to applicants. Managers would then identify factors such as credit history and yearly income that they could use to classify applicants as good or bad credit risks.
- **Clustering.** Clustering involves organizing data into similar subgroups, or clusters. It's different from classification in that there are no predefined classes. The data-mining software makes the decision about what to group, and it's up to managers to determine whether the clusters are meaningful. For example, the data-mining software may identify clusters of customers with similar buying patterns. Further analysis of the clusters may reveal that certain socioeconomic groups have similar buying patterns.
- **Estimation or regression.** Regression or estimation predicts a value based on current data. Regression is often used with profit and sales data to predict future results based on historical outcomes. Or, in real estate, the value of a house could be predicted based on factors such as location, school district, number of rooms, and lot size.
- **Visualization.** Data is transformed into graphs and charts through which trends and relationships can often be identified better than with pure numeric data. In addition, key performance indicators are combined into meaningful dashboards for quick assessment and reference.

As we continue to generate data and businesses continue to accumulate data, the development of faster and bigger databases will be a necessity. You can expect to interact with more and more databases every year, even if you don't realize you're doing so.

Ethics in IT

Data, Data Everywhere—But Is It Protected?

Organizations are generating, acquiring, processing, and storing massive amounts of data every day. Unfortunately, protecting that data from being mishandled is difficult to do. Inadvertently exposing information to inappropriate or unauthorized individuals is known as a *data breach*. According to the Identity Theft Resource Center, thousands of data breaches occur each year. Although companies continue to strengthen data security measures, breaches continue to occur, with the heaviest impact falling on the business sector. It's highly likely that at some point your data will be involved in a breach, so it's good to know what your personal data is being used for and how it will be used.

What Can You Do?

Providing information and having it recorded in databases are part of our way of life now. Ask the following questions related to data you're providing:

- **For what purpose is the data being gathered?** When the clerk at the electronics superstore asks for your zip code, ask why he or she wants it.
- **Are the reasons for gathering the data legitimate or important to you?** You might not want to share your information for marketing purposes but might want to in order to initiate a product warranty. Similarly, disclosing personal information to your pharmacist may be important for receiving good care and therefore important to you. If you don't see the advantage, then ask more questions or don't reveal the information.
- **How will the collected information be protected?** Ask about data protection policies before you give information. Most websites provide access to their data protection policies when they ask you for information. If an organization does not have a data protection policy, be wary of giving them sensitive information. Data protection does not just refer to keeping data secure; it also means restricting access to the data to employees of the organization who need to use that data. A shipping clerk might need to see your address, for example, but does not need to see your credit card information.
- **Will the collected information be used for purposes other than the purpose for which it was originally collected?** This might be covered in a data protection policy. If it is not, ask about it. Will your information be sold to or shared with other companies? Will it be used for marketing other products to you?
- **Could the information asked for be used for identity theft?** Identity thieves usually need your Social Security number and your birth date to open credit card accounts in your name. Be especially wary when asked for this information, and make sure there is a legitimate need for it. Do you really need an e-mail from someone on your birthday, advertising a product? It really is not worth exposing your birth date to potential misuse.
- **Are organizations safeguarding your data?** Don't just consider new requests for information. Think about organizations, such as banks, that already have your information. Have they been in the news lately because of a major data breach? You might want to consider switching institutions if yours has a poor record of data security.

Think carefully before providing information and be vigilant about monitoring your data when you can. It may make the difference between invasion of privacy and peace of mind.

Using Databases to Make Business Decisions

Making intelligent decisions about developing new products, creating marketing strategies, and buying raw materials requires timely, accurate information. In this section, we look at the different types of database systems that are used to help managers make good business decisions.

Business Information Systems

Objective 11.11 *Describe the main types of business information systems and how they are used by business managers.*

What is an information system? An information system is a software-based solution used to gather and analyze data. It is made up of the database, application programs, and manual and machine procedures necessary to process data into meaningful information. For example, an information system is used to deliver up-to-the-minute sales data on shoes to the computer of Zappos' president. Databases are integral parts of information systems because they store the information that makes information systems functional.

All information systems perform similar functions, including acquiring data, processing that data into information, storing the data, and providing the user with a number of output options with which to make the information meaningful and useful (see Figure 11.24). There are several types of information systems. Each system responds to a particular need of an organization and is generally specific to the type of information required. Generally, the information can be categorized as operational, managerial, and strategic. Specifically, information systems include transaction-processing systems (operational), management information systems (managerial), decision support systems (senior managers), and business intelligence systems (executive managers). Businesses operate more efficiently by using one or more information systems. Finally, enterprise resource planning systems integrate data from department-specific information systems across the entire enterprise. This facilitates decision making at many managerial levels, improving efficiency.

Transaction-Processing Systems

What is a transaction-processing system? A transaction-processing system (TPS) is an operational-level system that keeps track of everyday business transactions or activities, such as order tracking, order processing, payroll, and cash management. For example, colleges have TPSs in place to track transactions that occur frequently, such as registering students for classes, accepting tuition payments, and printing course catalogs. When computers were introduced to the business world, they often were first put to work hosting TPSs.

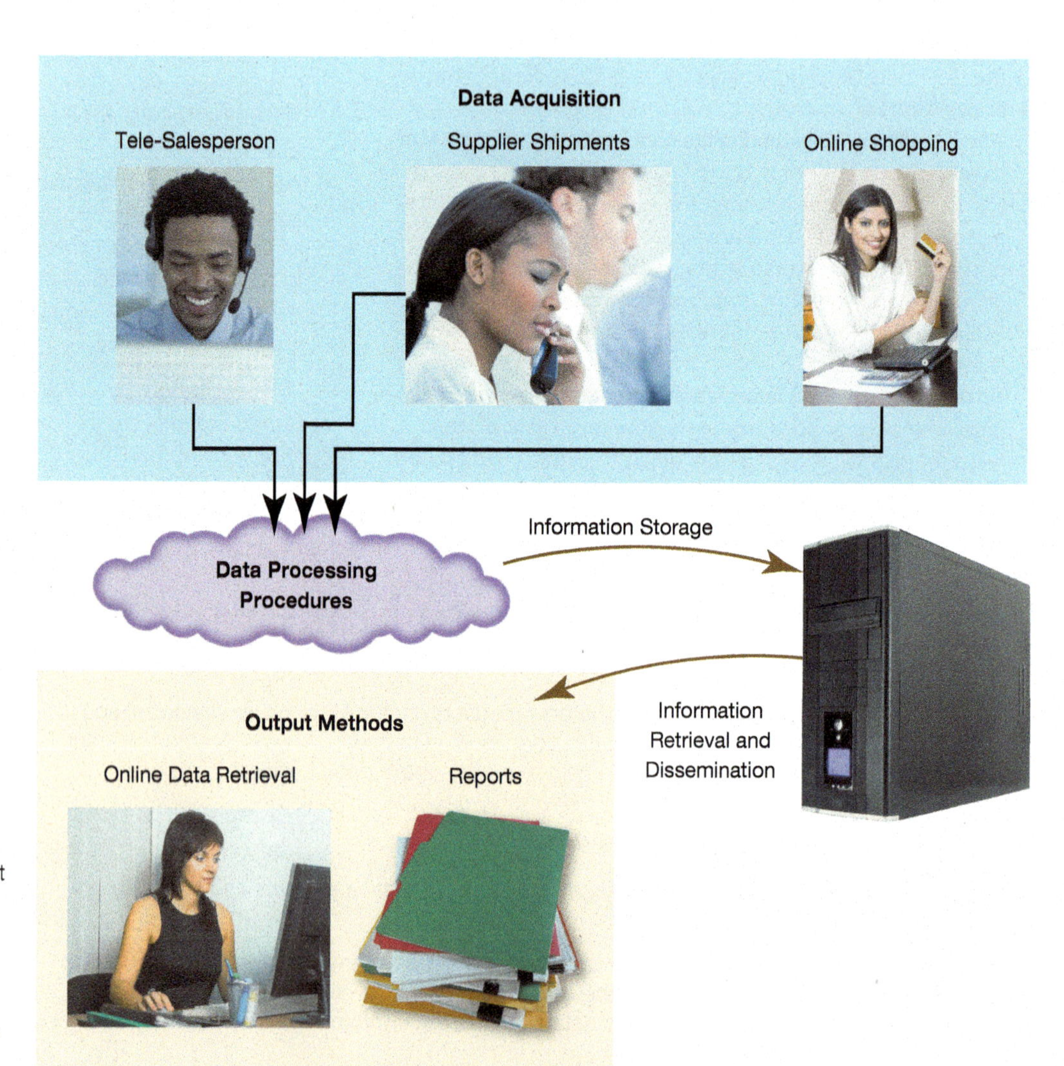

Figure 11.24 Information systems acquire data, process that data into information, store the data, and provide the user with a number of output options to make the information useful. *(Wavebreakmedia/Shutterstock; Wavebreakmedia/Shutterstock; Michaeljung/Shutterstock; Raisa Kanareva/Shutterstock; Ryan McVay/Photodisc/Getty Images; Zoomstudio/E+/Getty Images)*

How are transactions entered in a TPS? Transactions can be entered in real time or through batch processing:

- For most activities, processing and recording transactions in a TPS occur in real time. Real-time processing means that the database is updated while the transaction is taking place. For instance, when you register for classes online, the database immediately records your registration in the class to ensure that you have a spot. This online transaction processing (OLTP) ensures that the data in the TPS is current.
- Batch processing means that transaction data is accumulated until a certain point is reached and then several transactions are processed at once. Batch processing is appropriate for activities that aren't time sensitive. For example, you may not receive a bill for each charge you make at the bookstore. Instead, the college may collect your charges and batch them together into one monthly billing. It's more efficient to batch and process all requests periodically.

Various departments in an organization then access the TPSs to extract the information they need to process additional transactions, as shown in Figure 11.25.

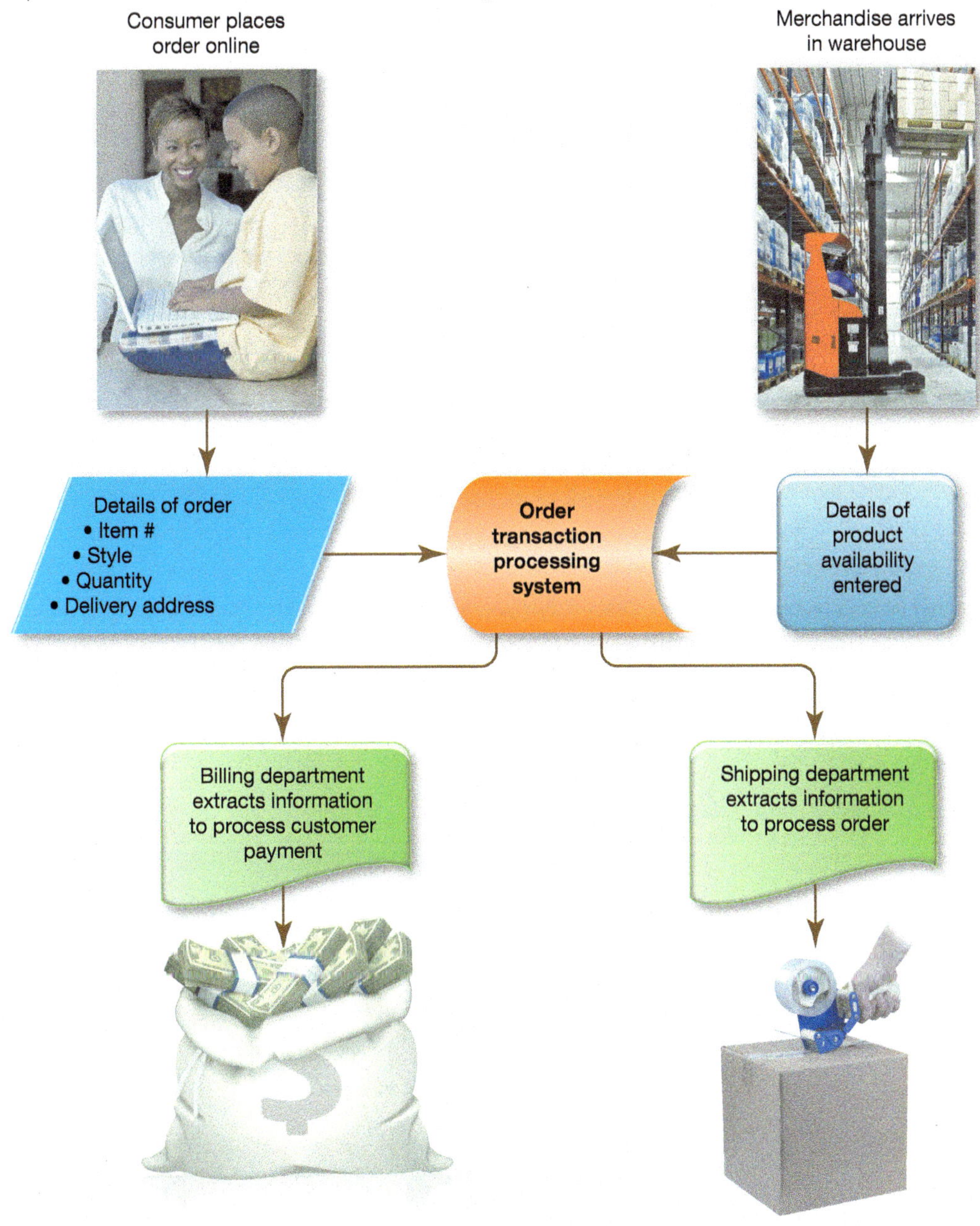

Figure 11.25 Transaction-processing systems help capture and track critical business information needed for successful completion of business transactions such as selling merchandise over the Internet. *(Sirtravelalot/Shutterstock; Penka Todorova Vitkova/Shutterstock; Natykach Nataliia/Shutterstock; Chuck Rausin/Shutterstock)*

How do transaction-processing systems maintain accuracy? To ensure that transactions are processed accurately, transaction processing systems must go through a series of checks for atomicity, consistency, isolation, and durability (the ACID test).

- *Atomicity:* All components of a transaction are treated as one or the transaction is not completed.
- *Consistency:* At the end of the transaction process, a new piece of data is completed (such as a successful sale or registration). If any failure occurs, the transaction is not completed.
- *Isolation:* While it's being processed, each transaction is treated separately from any other transaction.
- *Durability:* Completed transactions are saved by the TPS and cannot be undone, so that the data is always available.

Management Information Systems

What is a management information system? A management information system (MIS) provides timely and accurate information that enables managers to make critical business decisions. MISs were a direct outgrowth of TPSs. Managers quickly realized that the data contained in TPSs could be an extremely powerful tool only if the information could be organized and outputted in a useful form. Periodically, the MIS can create prescheduled reports, which managers use to make strategic and tactical decisions. MIS reports are also useful in conducting what-if analyses that offer potential results on given sets of variables.

What types of reports are generated by MISs? MISs generate three types of reports:

- A detail report provides a list of the transactions that occurred during a certain time period. For example, during registration at your school, the registrar might receive a detail report that lists the students who registered for classes each day. Figure 11.26a shows an example of a detail report on daily enrollment.
- A summary report provides a consolidated picture of detailed data. These reports usually include some calculations (totals) or visual displays of information (such as charts and graphs). Figure 11.26b shows an example of a summary report displaying total daily credits enrolled by division.
- An exception report shows conditions that are unusual or that need attention by system users. The registrar at your college may get an exception report when all sections of a course are full, indicating that it may be time to schedule additional sections. Figure 11.26c shows an example of such an exception report.

Decision Support Systems

What is a decision support system? A decision support system (DSS) is another type of business information system designed to help managers develop solutions for specific problems. With a DSS, managers can make queries of data for specific answers in an effort to evaluate the impact of

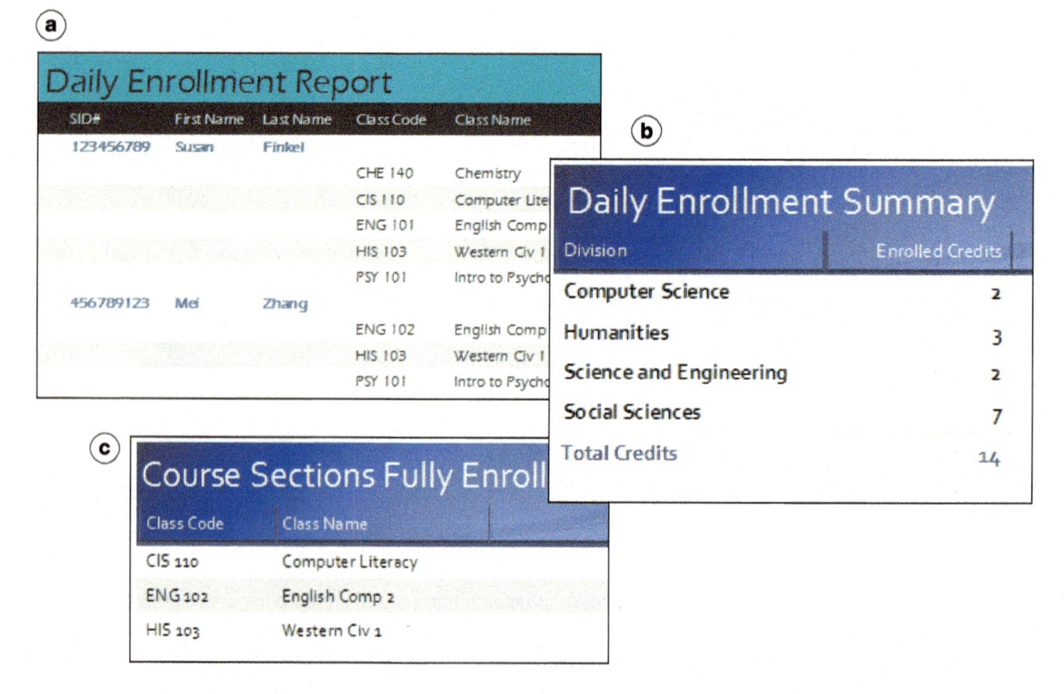

Figure 11.26 The three types of management information system reports are (a) detail reports, (b) summary reports, and (c) exception reports.

a decision before it is implemented. For example, a DSS for a marketing department might provide statistical information on customer attributes, such as income levels or buying patterns, that would assist managers in making advertising strategy decisions. A DSS not only uses data from databases and data warehouses but also enables users to add their own insights and experiences and apply them to the solution. As shown in Figure 11.27, many data-related systems work together to provide the user of the DSS with a broad base of information on which to base decisions.

How does a DSS get data? Internal and external data sources provide a stream of data that is integrated into the DSS for analysis. Internal data sources are maintained by the same company that operates the DSS. For example, internal TPSs can provide a wealth of statistical data about customers, ordering patterns, inventory levels, and so on. An external data source is any source not owned by the company that owns the DSS, such as customer demographic data purchased from third parties, mailing lists, or statistics compiled by the government.

What function does a model management system perform? A model management system is software that assists in building management models in a DSS. A management model is an analysis tool that, through the use of internal and external data, provides a view of a particular business situation for the purposes of decision making. Models can be built to describe any business situation, such as the classroom space requirements for next semester or a listing of alternative sales outlet locations. Model management systems typically contain financial and statistical analysis tools that are used to analyze the data provided by models or to create additional models.

Business Intelligence Systems

What is a business intelligence system? Business intelligence (BI) is the ability to improve business decision making with databases and other fact-based support systems. A business intelligence system is another form of business information system and is used primarily at the executive level. A business intelligence system is a set of technologies used to analyze and interpret large sets of data

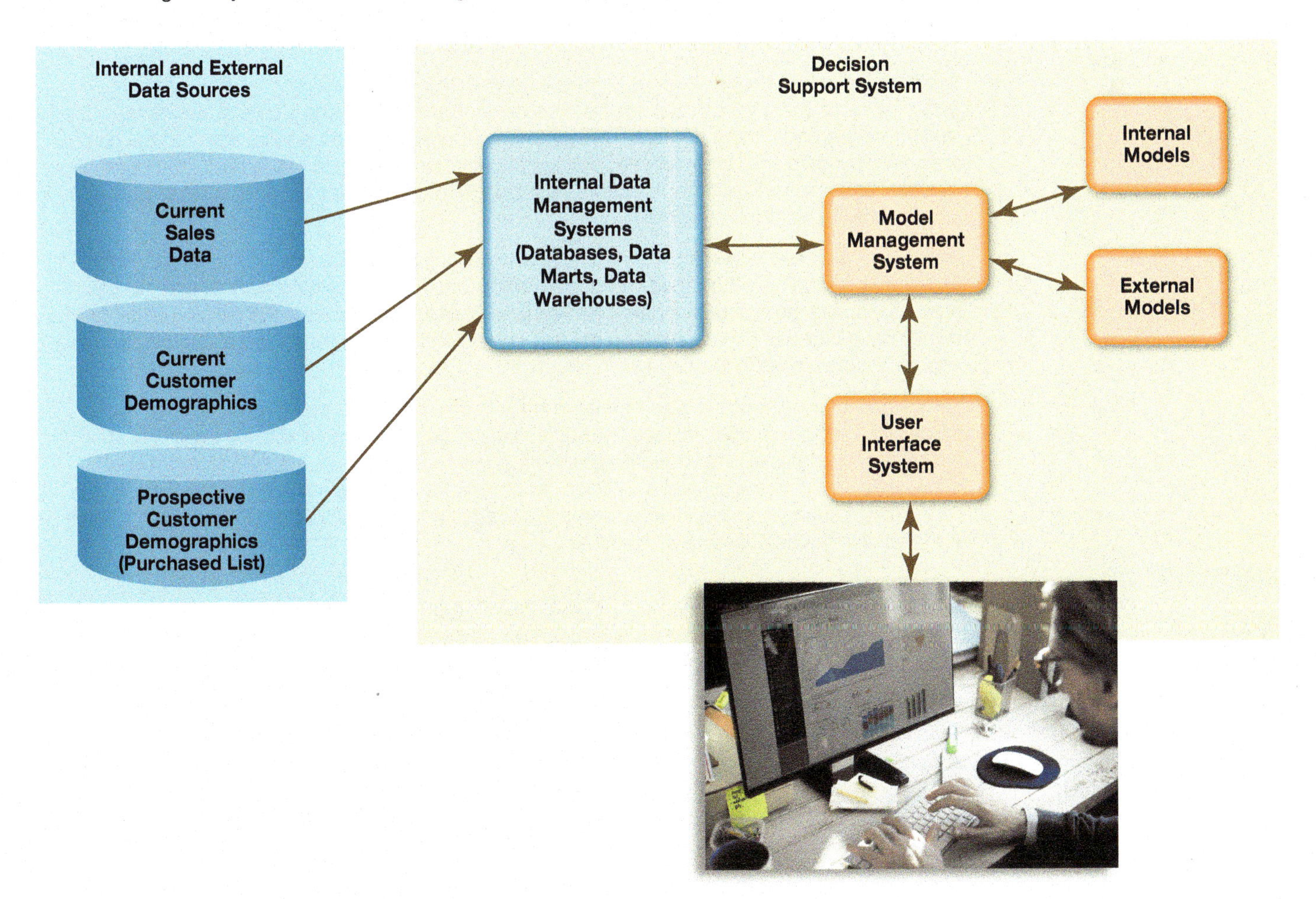

Figure 11.27 This figure shows the major components of a DSS. Through the user interface, models are analyzed and manipulated to provide information on which management decisions are based. *(Rawpixel/123RF)*

MyLab IT

Analyzing Data with Microsoft Power BI Suite

In this Sound Byte, you explore how the Microsoft Power BI Suite is used to analyze big data.

(often referred to as big data) to enable executives and senior managers to make informed decisions about how best to run a business. Often business intelligence systems use data gathered from data warehouses or data marts because they enable access to information gathered from multiple sources. Increased access to data and the ability to transform it into useful information for business analysis purposes can lead managers to identify new opportunities and implement effective business strategies.

Enterprise Resource Planning Systems

What does an enterprise resource planning system do? All businesses have data and information to manage, and large, complex organizations can benefit from managing that information with a central piece of software. An enterprise resource planning (ERP) system is a software system that accumulates all information relevant to running a business and makes it readily available to whoever needs it to make decisions. ERP systems use a common database to store and integrate information. This makes the information accessible across multiple areas of an enterprise.

Human resource functions (such as the management of hiring, firing, promotions, and benefits) and accounting functions (such as payroll) are often the first processes integrated into an ERP system. Having such information reside in one database makes the management and compensation of employees more streamlined. If manufacturing operations were also integrated into the ERP system, then the employees and payroll data could be used to determine the cost of running an assembly line or for scheduling workers to run the assembly line.

Knowledge-Based Systems and Artificial Intelligence

What is a knowledge-based system? A knowledge-based system provides intelligence that supplements the user's own intellect. It could be an expert system that tries to replicate the decision-making processes of human experts to solve specific problems. For example, an expert system might be designed to take the place of a physician in a remote location. A physician expert system would ask the patient about symptoms just as a live physician would, and the system would make a diagnosis based on the algorithms programmed into it.

Another type of knowledge-based system is a natural language processing (NLP) system. NLP systems enable users to communicate with computer systems by using a natural spoken or written language instead of using a computer programming language. Individuals just speak to the computer and it understands what they're saying, without users needing to use specific computer commands. The personal assistants in today's operating systems, such as Siri and Cortana, as well as Amazon's Alexa, are NLP systems.

All knowledge-based systems fall under the science of artificial intelligence. Artificial intelligence (AI) is the branch of computer science that deals with the attempt to create computers that think like humans. To date, no computers have been constructed that can replicate the thinking patterns of a human brain because scientists still do not fully understand how humans store and integrate knowledge and experiences to form human intelligence.

How does a knowledge-based system help in the decision-making process? Databases and the models provided by model management systems tend to be extremely analytical and mathematical in nature. If we relied solely on databases and models to make decisions, answers would be derived with a yes or no mentality, allowing no room for human thought. Fortunately, human users are involved in these systems, providing an opportunity to inject human judgment into the decision-making process.

Bits&Bytes Virtual Agents: Expert Systems Replace People on the Web

An offshoot of expert systems consists of the virtual agents that you encounter on the web or on your smartphone. Virtual agents are an animated, computer-generated form of artificial intelligence that is frequently used to interface with a database to answer customers' questions. For example, a typical virtual agent on the web features a picture, a name, and a box for entering questions. No virtual agent can be programmed to contain all the possible questions a human might ask. Therefore, the virtual agent software breaks the question down into key words and phrases, which it compares to a database containing question responses. It picks the most likely response and provides the answer. Customers provide feedback on whether the response adequately answered their question. Usually, if results from the virtual agent aren't satisfactory or if customers need more information, they can connect to a live customer service agent. Virtual agents, although not perfect, save companies money by answering common questions without using expensive human labor.

Trends in IT Mobile Business Intelligence

Business intelligence uses databases and other fact-based support systems to improve business decision making. Smartphones, tablets, and more universal access to wireless Internet have enabled business intelligence to go mobile. Mobile business intelligence systems and technologies give companies access to data at their fingertips and can interact in real time with customers and business partners. Such capabilities can improve service and productivity (see Figure 11.28).

Mobile business intelligence systems make data completely accessible all the time. Mobile delivery of business intelligence can be accomplished with a mobile web browser or a mobile app. Mobile business intelligence products integrate with business intelligence systems, such as knowledge-based and ERP systems, to deliver access to key performance indicators, business metrics, and dashboards on mobile devices. Systems such as Oracle's Business Intelligence Enterprise Edition and SAP's BusinessObjects Explorer have mobile business intelligence applications that integrate with expansive data sources. By using the capabilities of mobile business intelligence, a corporate executive can, for example, monitor on his or her mobile device a dashboard of key metrics of critical offshore manufacturing facilities while running a meeting thousands of miles away.

Figure 11.28 Mobile business intelligence applications on smartphones and other mobile devices enable managers to have access to data all the time. *(BrightSpace/Shutterstock)*

Mobile business intelligence solutions are enabling companies to improve business processes, increase employee productivity, and deliver better customer service by providing connections to real-time, on-demand access to business-critical information. In addition, mobile business intelligence is making advancements with collaboration and with using cloud computing, capturing and capitalizing on social data. However, as great as mobile business intelligence seems, there is room for improvement in many areas, especially security and privacy.

The knowledge-based system also provides an opportunity to introduce experience into the mix. Knowledge-based systems support the concept of fuzzy logic. Normal logic is highly rigid: If *x* happens, then *y* will happen. Fuzzy logic enables the interjection of experiential learning into the equation by considering probabilities. Whereas an algorithm in a database has to be specific, an algorithm in a knowledge-based system could state that if *x* happens, 70% of the time *y* will happen.

For instance, managers at Best Buy would find it extremely helpful if their DSS informed them that 45% of customers who bought an iPad also bought a Smart Cover for it. This could suggest that designing a discount program for Smart Covers bought with iPads might spur sales. Fuzzy logic enables a system to be more flexible and consider a wider range of possibilities than would conventional algorithmic thinking.

Before moving on to the Chapter Review:

1. Watch Chapter Overview Video 11.2.
2. Then take the Check Your Understanding quiz.

Check Your Understanding // Review & Practice

For a quick review to see what you've learned so far, answer the following questions.

multiple choice

1. Which best describes a data warehouse?
   a. A building that stores computer servers
   b. A collection of all the data from an organization's databases
   c. A software program that assists with file management
   d. A database that organizes data from multiple organizations
2. Before going into a data warehouse, data is held in an intermediate storage area called
   a. a transaction processing center.
   b. an online analytical processing area.
   c. a data staging area.
   d. a data scrubbing area.
3. In the process of data mining, organizing data into similar subgroups is known as
   a. clustering.
   b. data staging.
   c. affinity grouping.
   d. data mining.
4. A system that is designed to help keep track of everyday business activities is a(n)
   a. transaction-processing system.
   b. management information system.
   c. decision support system.
   d. enterprise resource planning system.
5. Natural language processing systems, artificial intelligence, and expert systems are all types of which larger type of database system?
   a. Business intelligence system
   b. Decision support system
   c. Knowledge-based system
   d. Voice management system

chew on this

(3Dalia/Shutterstock)

Machine learning is a form of artificial intelligence that enables software to learn by analyzing previous data patterns in an effort to improve future responses and tasks. One component of machine learning relies on neural networks in which the program simulates human thinking by trying to duplicate the biological connections (neurons) of the human brain. Making computers more human-like raises concerns. Discuss the prospects of artificial intelligence and machine learning creating computers with human-like capabilities in the future. Explain why this is something you are concerned about or not.

MyLab IT Go to **MyLab IT** to take an autograded version of the *Check Your Understanding* review and find all media resources for the chapter.

For the IT Simulation for this chapter, see MyLab IT.

11 Chapter Review

Summary

Part 1

Database Fundamentals

Learning Outcome 11.1 **You will be able to explain the basics of databases, including the most common types of databases and the functions and components of relational databases in particular.**

The Need for Databases

Objective 11.1 *Explain what a database is and why databases are useful.*

- Databases are electronic collections of related data that can be organized so that data is more easily accessed and manipulated. Properly designed databases cut down on data redundancy and duplicate data by ensuring that relevant data is recorded in only one place. This also helps eliminate data inconsistency, which comes from having different data about the same transaction recorded in different places. When databases are used, multiple users can share and access information at the same time.
- Some of the main advantages of using databases are that they can manage large amounts of data efficiently, enable information sharing, and promote data integrity. Data sharing is made possible because databases provide data centralization, a shared source that everyone can access. Data centralization also promotes better data integrity.

Database Types

Objective 11.2 *Describe features of flat databases.*

- A flat database, often represented as a list or simple table, is used to organize simple data. Tables created in Microsoft Word or in a Microsoft Excel spreadsheet serve as a flat database.
- Flat databases are subject to several types of problems, including data redundancy, data inconsistency, inappropriate data, and incomplete data.

Objective 11.3 *Describe features of relational databases.*

- A relational database operates by organizing data into various tables based on logical groupings and then creating a relationship between tables by linking through a common field.
- There are three main types of relationships. A one-to-many relationship has a record appearing only once in a table while also appearing many times in a related table. A one-to-one relationship is when there is only one instance of a record in a table and only one corresponding record in a related table. A many-to-many relationship has multiple instances of records in both the primary and related tables.

Objective 11.4 *Describe features of object-oriented databases.*

- Object-oriented databases store data in objects rather than in tables and provide methods for processing or manipulating the data. Object-oriented databases are best for handling unstructured data, such as media files and extremely large documents.

Objective 11.5 *Describe features of multidimensional databases.*

- Multidimensional databases store data that can be analyzed from different dimensions, or perspectives. Multidimensional databases organize data with three dimensions, giving it a cubelike format. In addition to the field and record dimensions of other databases, a measure attribute is also used.

Objective 11.6 *Describe how dynamic, web-created data is managed in a database.*

- NoSQL databases store the unstructured data generated from web-based applications such as data from social media sharing and other online activities, online personal settings, photos, and videos. In addition, NoSQL databases store website-usage metrics, location-based information, and clickstream data.
- NoSQL systems group related information into folder-like documents that hold a variety of data instead of the rigid row-and-column format of tables.

Using Databases

Objective 11.7 *Describe how relational databases organize and define data.*

- Database management systems (DBMSs) are specially designed applications (such as Microsoft Access) that interact with the user, other applications, and the database itself to capture and analyze data. The main operations of a DBMS are creating databases, entering data, viewing (or browsing) data, sorting (or indexing) data, extracting (or querying) data, and outputting data.
- Data is organized using fields, records, and tables. A category of information in a database is stored in a field. Fields are assigned a field name and a data type that indicates what type of data can be stored in the field. Common data types include short text, long text, numeric, calculated, date/time, yes/no, object linking and embedding (OLE) object, and hyperlink. A group of related fields is a record. A group of related records is a table or file. Field properties reflect data dictionary details. Some common field properties include field size, default value, and captions.
- To keep records distinct, each record must have one field that has a value unique to that record. This unique field is a primary key (or a key field).

Objective 11.8 *Describe how data is inputted and managed in a database.*

- Input forms are used to control how new data is entered in a database as well as how changes are made to data. Data validation helps ensure that only valid data is entered in a field. Common validation rules include a range check, a completeness check, a consistency check, and an alphabetic or numeric check.
- Queries and filters are used to display a subset of data. A query language is used to extract records from a database. Almost all relational databases today use Structured Query Language (SQL). However, most DBMSs include wizards that enable you to query the database without learning a query language. Reports are used to output data from a database, or data can be transferred to another software application for further distribution or modification.

Part 2

How Businesses Use Databases

Learning Outcome 11.2 **You will be able to explain how businesses use data warehouses, data marts, and data mining to manage data and how business information systems and business intelligence are used to make business decisions.**

Data Warehousing and Storage

Objective 11.9 *Explain what data warehouses and data marts are and how they are used.*

- A data warehouse is a large-scale collection of data that contains and organizes in one place all the relevant data for an organization. Data warehouses often contain information from multiple databases. Because it can be difficult to find information in a large data warehouse, small slices of the data warehouse, called data marts, are often created.
- Data usually goes through an extraction and transformation process to remove data from other databases and transform it so the collected data is formatted similarly. Then, data is stored in data staging until it is ready or needed to go to the data warehouse.
- The information in data marts pertains to a single department within the organization.
- Data warehouses and data marts consolidate information from a wide variety of sources to provide comprehensive pictures of operations or transactions within a business.

Objective 11.10 *Describe data mining and how it works.*

- Data mining is the process by which large amounts of data are analyzed to spot otherwise hidden trends. Through processes such as classification, estimation, affinity grouping, clustering, and description (visualization), data is organized so that it provides meaningful information that managers can use to identify business trends.

Using Databases to Make Business Decisions

Objective 11.11 *Describe the main types of business information systems and how they are used by business managers.*

- An information system is a software-based solution used to gather and analyze information. All information systems perform similar functions, including acquiring, processing, and storing data and providing the user with a means to output the results into meaningful and useful information.

- A transaction-processing system (TPS) is used to keep track of everyday business activities.
- A management information system (MIS) provides timely and accurate information that enables managers to make critical business decisions.
- A decision support system (DSS) is designed to help managers develop solutions for specific problems. A model management system is software that assists in building analysis tools for DSSs.
- Executive managers use business intelligence systems to analyze and interpret data to make informed decisions about how best to run a business.
- An enterprise resource planning (ERP) system is a large software system that gathers information from all parts of a business and integrates it to make it readily available for decision making.
- A knowledge-based system provides intelligence to decision making. There are several kinds of knowledge-based systems. An expert system tries to replicate the decision-making process of human experts to solve specific problems. A natural language processing (NLP) system uses natural spoken or written language rather than a computer programming language to communicate with a computer. Artificial intelligence (AI) attempts to create computers that think like humans.

Be sure to check out **MyLab IT** for additional materials to help you review and learn. And don't forget to watch the Chapter Overview Videos.

Key Terms

alphabetic check **431**
artificial intelligence (AI) **450**
batch processing **447**
binary large object (BLOB) **423**
business intelligence (BI) **449**
business intelligence system **449**
caption **428**
clickstream data **440**
completeness check **431**
consistency check **431**
data centralization **419**
data dictionary (database schema) **429**
data inconsistency **421**
data integrity **420**
data mart **442**
data mining **442**
data redundancy **421**
data staging **441**
data type (field type) **427**
data warehouse **440**
database **418**
database administrator (DBA) (database designer) **420**
database management system (DBMS) **425**
decision support system (DSS) **448**
default value **427**
detail report **448**
enterprise resource planning (ERP) system **450**
exception report **448**
expert system **450**
field **425**
field constraint **430**
field name **425**
field properties **427**
field size **427**
filter **432**
flat database **420**
foreign key **428**
fuzzy logic **451**
information system **445**
input forms **430**
join query **435**
knowledge-based system **450**
management information system (MIS) **448**
many-to-many relationship **422**
metadata **429**
model management system **449**
multidimensional database **423**
natural language processing (NLP) system **450**
normalization **423**
NoSQL database **425**
numeric check **431**
object-oriented database **423**
Object Query Language (OQL) **423**
one-to-many relationship **422**
one-to-one relationship **422**
online analytical processing (OLAP) **442**
online transaction processing (OLTP) **447**
primary key field **428**
query **432**
query language **423**
range check **430**
real-time processing **447**
record **425**
referential integrity **428**
relational algebra **434**
relational database **422**
relationship **422**
select query **434**
structured (analytical) data **423**
Structured Query Language (SQL) **423**
summary report **448**
table (file) **425**
time-variant data **440**
transaction-processing system (TPS) **446**
unstructured data **423**
validation **430**
validation rule **430**

Chapter Quiz // Assessment

For a quick review to see what you've learned, answer the following questions. Submit the quiz as requested by your instructor. If you're using **MyLab IT**, the quiz is also available there.

multiple choice

1. When is a flat database sufficient to use when organizing data?
   - **a.** When the data needs to be organized into two or more tables
   - **b.** When the data is simple and can be organized in one table
   - **c.** When more than one person needs to access the data
   - **d.** When the data includes video and audio
2. A customer information table linked to a password table in a business database reflects which type of relationship?
   - **a.** One-to-one
   - **b.** Many-to-one
   - **c.** Many-to-many
   - **d.** One-to-many
3. Which type of database is best for handling unstructured types of data such as audio and video files?
   - **a.** Relational
   - **b.** Multidimensional
   - **c.** NoSQL
   - **d.** Object-oriented
4. Which of the following is NOT considered metadata?
   - **a.** Data type
   - **b.** Database name
   - **c.** Field name
   - **d.** Field size
5. Data captured as users navigate through a website is called
   - **a.** staged data.
   - **b.** web-user data.
   - **c.** time-variant data.
   - **d.** clickstream data.
6. NoSQL databases are best used with what type of data?
   - **a.** Unstructured, dynamic, web-based data
   - **b.** Data represented in flat lists
   - **c.** Data with multiple attributes
   - **d.** Clickstream data
7. Which of the following is used to create a subset of data that matches certain criteria and that can be saved for future reference or use?
   - **a.** Relationship
   - **b.** Query
   - **c.** Request
   - **d.** Filter
8. Which of the following data-mining techniques is used to create charts and dashboards?
   - **a.** Classification
   - **b.** Affinity grouping
   - **c.** Clustering
   - **d.** Visualization
9. Unnecessary duplication of data in a database is referred to as
   - **a.** data redundancy.
   - **b.** normalized data.
   - **c.** validated data.
   - **d.** data inconsistency.
10. Which of the following is a method of inputting data into a transaction-processing system while the transaction is taking place?
    - **a.** Batch processing
    - **b.** Input processing
    - **c.** Real-time processing
    - **d.** Click data processing

true/false

_____ **1.** A field labeled PHONE that accepts data such as (610) 555-1212 would use the Number data type.

_____ **2.** ETL refers to extract, transform, and load.

_____ **3.** A small slice of a data warehouse is called a data mine.

_____ **4.** Data warehouses capture data only from one time period.

_____ **5.** A primary key is used to identify records uniquely.

What do you think now?

1. Most apps offer privacy settings for users, enabling you to determine how much and what types of information are shared or stored. Discuss how you've changed your privacy settings on your apps. If you haven't, explain why.
2. Locking your device is one of the first steps you should take to protect your data in the event of a lost or stolen device. Think about the password or passcode you have set on your smartphone. Do you think it's strong enough to keep your data safe in the event your phone is lost or stolen?

Team Time

Redesigning Social Media

problem

Social media sites are built around databases. Each site determines what data it requires from its account holders, and that data is displayed in specific areas in the site. But, perhaps you, as a user, would design a social media site differently.

task

Your class has volunteered to work as a focus group for a social media site to assess the usefulness of the information it gathers.

process

1. Divide the class into small groups.
2. Your group members should determine a social media site to review and examine their individual accounts on that site. What fields do you consider to be most useful? Which fields are unnecessary and could be eliminated? What fields are missing that you think would be useful?
3. Investigate your account settings and application settings. What changes would you make to these settings? Are options missing that you would find helpful?
4. Present your group's findings to the class. Compare your suggestions to those of other groups.
5. Prepare a list of recommendations for improvements to the current home page and settings pages. Clearly indicate how the proposed changes will benefit both users and the social media site developers (retaining users, being able to target advertising to users better, and so on).

conclusion

To remain competitive, social media sites need to consider the input of users to ensure that it delivers a product with features its customers want and need.

Ethics Project

Private Information on Public Databases

In this exercise, you research and then role-play a complicated ethical situation. The role you play might not match your own personal beliefs, but your research and use of logic will enable you to represent the view assigned. An arbitrator will watch and comment on both sides of the arguments, and, together, the team will agree on an ethical solution.

problem

A tremendous amount of data is accumulated about us in online databases. Much of this material is accessible in databases that are searchable by anyone with Internet access or by anyone willing to pay a small fee. Websites such as *spokeo.com* and *411.com* comb through publicly accessible databases, such as social networking sites, online phone books, and business and government websites, to compile information on individuals. Your online information may expose you to risks such as identify theft.

research areas to consider

- Electronic information privacy
- Electronic Privacy Information Center (*epic.org*)
- Protecting your online privacy
- Protecting yourself from data breaches

process

1. Divide the class into teams.
2. Research the areas cited and devise a scenario in which someone has complained about a website providing information that led to theft of his or her identity.
3. Team members should determine a character they want to represent in the role-playing event, such as a victim of identity theft, a website owner, or an arbitrator. Then each member should write a summary that provides background information for his or her character and that details the character's behaviors to set the stage for the role-playing event. Team members should then create an outline to use during the role-playing event.
4. Team members should present their case to the class or submit a PowerPoint presentation for review by the rest of the class, along with the summary and resolution they developed.

conclusion

As technology becomes ever more prevalent and integrated into our lives, ethical dilemmas will present themselves to an increasing extent. Being able to understand and evaluate both sides of the argument, while responding in a personally or socially ethical manner, will be an important skill.

Solve This with Access

MyLab IT Grader

At this time, an autograded version of this project is not available in **MyLab IT**.

College Database

You help in the registrar's office at your college. You are being asked to help update the college's database by modifying some of the tables and creating some queries and reports. You will use Microsoft Access to complete this exercise.

You will use the following skills as you complete this activity:	
• Create and Modify Data Table	• Modify Field Properties
• Import Excel Data	• Establish Table Relationships
• Create AND Query	• Create Multi-table Query

Instructions

1. Open the *TIA_Ch11_Start.accdb* file and save the database as **TIA_Ch11_LastFirst.accdb**.
2. Open the **Faculty table** in **Design View**. Change the Data Type of FullTime to **Yes/No**. Add a **primary key** to the FacultyID field. **Move** the DeptID field to follow the Title field.
3. In the Faculty table, modify the **Field Size** and **Caption properties** as follows:

Field Name	Field Size	Caption
FacultyFirst	10	First Name
FacultyLast	15	Last Name
DeptID	3	Dept ID
FullTime		Full Time Status

4. Add a new field with the name **HireDate** and the **Date/Time data type**. Save and close the Faculty table.
5. On the External Data tab, click **New Data Source**, click **From File**, and then select **Excel** in the Import & Link group. Browse to your student data files and select the *TIA_Ch11_Faculty.xlsx* file. In the *Specify how and where you want to store the data in the current database* section, click **Append a copy of the records to the table** and select the **Faculty table**. Accept all other default options in the wizard. Close the Get External Data - Excel Spreadsheet wizard and then open the Faculty table to ensure that the records were imported correctly. You should see 261 records. Close the Faculty table.
6. Open the **Students table** in **Datasheet view**. Adjust the columns so the column sizes best fit the data in each field. In **Design view** of the Students table, change the data type for the Email field to **Hyperlink**. Add a Default Value of **CT** for the State field. Save and close the Students table.
7. Create a form, using the **Form Wizard** and using all fields from the **Students** table. Select a **Columnar layout** and name the form **Students**. In the Students form, click in the **Student Last text box**. In the search box in the Navigation bar at the bottom of the Student form, type **Lopez**. Check that the StudentID is 444096 and then update the address field to **4024 Broad Street** to reflect the student's new address. Save and close the Students form.
8. On the Database Tools tab, click **Relationships**. Add the Faculty table and create a relationship between the Faculty table and the Courses table using the FacultyID field. **Enforce referential integrity**. Save and close the Relationships window.
9. Create a query from the Courses table, using **Query Design**. Display the CourseID, Building, Room, Days, Start, and End fields. Enter criteria to display records for courses meeting in the Park building on Tuesdays and Thursdays. You should see 78 records. Save the query as **Park TTH**. Run and then close the query.
   Hint: Check how Days is displayed in the table and use the same format for Days in the criteria.
10. Create a query, using **Query Design** that includes CourseID from the Roster table; Course Description from the Courses table; FacultyLast from the Faculty table; and StudentLast, StudentFirst, and Email from the Students table. Display only MGT110 IC courses. Save the query as **MGT110 IC Roster**. Run and then close the query. You should have four records.
11. Use the **Report Wizard** to create a report from the **MGT110 IC Roster query**. Include all fields except CourseID and FacultyLast. View the data by Courses; do not add additional grouping. **Sort** by StudentLast in ascending order. Accept the default layout options and save the report as **Rosters**.
12. Save and close the database and submit based on your instructor's directions.

Chapter

12 Behind the Scenes: Networking and Security in the Business World

For a chapter overview, watch the **Chapter Overview Videos**.

PART 1

Client/Server Networks and Topologies

Learning Outcome 12.1 **You will be able to describe common types of client/server networks, servers found on them, and network topologies used to construct them.**

Client/Server Network Basics 462

Objective 12.1 *List the advantages for businesses of installing a network.*

Objective 12.2 *Explain the differences between a client/server network and a peer-to-peer network.*

Objective 12.3 *Describe the common types of client/server networks as well as other networks businesses use.*

Servers and Network Topologies 468

Objective 12.4 *List the common types of servers found on client/server networks.*

Objective 12.5 *Describe the common types of network topologies and the advantages and disadvantages of each one.*

Helpdesk: Using Servers

Sound Byte: Network Topology and Navigation Devices

PART 2

Setting Up Business Networks

Learning Outcome 12.2 **You will be able to describe transmission media, network operating system software, and network navigation devices and explain major threats to network security and how to mitigate them.**

Transmission Media 480

Objective 12.6 *Describe the types of wired and wireless transmission media used in networks.*

Network Adapters and Navigation Devices 482

Objective 12.7 *Describe how network adapters help data move around a network.*

Objective 12.8 *Define MAC addresses, and explain how they are used to move data around a network.*

Objective 12.9 *List the various network navigation devices, and explain how they help route data through networks.*

Helpdesk: Transmission Media and Network Adapters

Network Operating Systems and Network Security 486

Objective 12.10 *Explain why network operating systems are necessary for networks to function.*

Objective 12.11 *List major security threats to networks, and explain how network administrators mitigate these threats.*

Sound Byte: A Day in the Life of a Network Technician

MyLab IT All media accompanying this chapter can be found here.

(Deepadesigns/Shutterstock; Sashkin/Shutterstock; Yuriy Kirsanov/123RF; Scanrail/123RF; Lukas Gojda/123RF)

What do you think?

We're starting to get comfortable with **facial recognition technology** being used to solve crimes and scan for terrorists. But how would you feel if artificial intelligence was trying to determine whether you and your friends were having fun yet?

Companies like Affectiva are already marketing **AI technology used to sense moods**. The potential applications are wide-ranging:

- Determining customer satisfaction levels when shopping or in check-out lines
- Seeing how you react when playing a game or using a website
- Predicting crimes more accurately

Is this just another attempt for **marketers to sell you more things?** Or does this harken back to George Orwell's *1984*, where Big Brother is constantly monitoring the citizens?

Which of the following statements do you agree with?

- *I would never feel comfortable with AI technology trying to guess my mood.*
- *I would feel comfortable with AI mood detecting if there was a reason for it and I knew I was being surveilled.*
- *We're under surveillance all the time—why should emotion sensing make a difference?*

See the end of the chapter for a follow-up question.

(Joshua Resnick/123RF)

Part 1

For an overview of this part of the chapter, watch **Chapter Overview Video 12.1**.

Client/Server Networks and Topologies

Learning Outcome 12.1 **You will be able to describe common types of client/server networks, servers found on them, and network topologies used to construct them.**

You learned about peer-to-peer networks and home networking in Chapter 7. In this chapter, you expand your knowledge about networks by delving into the client/server networks typically used in businesses.

Client/Server Network Basics

Recall that a *network* is a group of two or more computing devices (or nodes) that are configured to share information and resources such as printers, files, and databases. Businesses such as your school or an insurance company gain advantages from deploying networks, similar to the advantages gained with a home network.

Networking Advantages

Objective 12.1 *List the advantages for businesses of installing a network.*

What advantages do businesses gain from networks? Networked computers have many advantages over individual stand-alone computers (see Figure 12.1):

- **Networks enable expensive resources to be shared.** Networks enable people to share peripherals such as printers or resources such as an Internet connection. Without a network, each

Figure 12.1 Benefits of Business Networks

Enable resource sharing

- Expensive peripherals, such as printers, can be shared.
- Networks can share a single internet connection.

Facilitate knowledge sharing

- Data can be accessed by multiple people.

Enable software sharing

- Software can be delivered to client computers from a server.

Enhance communication

- Information sharing is more effective when employees are connected.

computer would have to be individually connected to a printer and to the Internet. Attaching the computers to a network avoids the cost of providing each computer with duplicate resources.

- **Networks facilitate knowledge sharing.** The databases you learned about in Chapter 11 become especially powerful when deployed on a network. Networked databases can serve the needs of many people at one time and increase the availability of data. Your college's databases are much more useful when all college employees can look up student records at the same time.
- **Networks enable software sharing.** Installing a new version of software on everyone's desktop in a business with 800 employees can be time-consuming. However, if the computers are networked, all employees can access the same copy of a program from the server. Although the business must still purchase a software license for each user, with a network it can avoid having to install the program on every computer.
- **Networks enable enhanced communication.** Social networking tools, e-mail, and instant messaging are powerful applications when deployed on a network, especially one that's connected to the Internet. Business colleagues can easily exchange information with each other and can share valuable data by transferring files to other users.

Are there disadvantages to using networks? Because business networks are often complex, additional personnel are usually required to maintain them. These people, called network administrators, have training in computer and peripheral maintenance and repair, networking design, and the installation of networking software.

Another disadvantage is that operating a network requires special equipment and software. However, most companies feel that the cost savings of peripheral sharing and the ability to give employees simultaneous access to information outweigh the costs associated with network administrators and equipment.

Comparing Client/Server and Peer-to-Peer Networks

Objective 12.2 *Explain the differences between a client/server network and a peer-to-peer network.*

Where do I find client/server networks? Aside from the smallest networks—such as peer-to-peer (P2P) networks, which are typically used in homes and small businesses—the majority of computer networks are based on the client/server model of computing. As you learned in Chapter 7, a client/server network (also called a server-based network) contains servers as well as client computers. A *server* is a computer that both stores and shares resources on a network, whereas a *client* is a computer that requests those resources.

Recall that in a P2P network, each node connected to the network communicates directly with every other node rather than using a server to exercise central control over the network. Many tasks that individual users must handle on a P2P network can be handled centrally at the server in a client/server network.

For instance, data files are normally stored on a server. Therefore, backups for all users on a business network can be performed by merely backing up all the files on that server. Security, too, can be exercised over the server instead of on each user's computer; this way, the server, not the individual user, coordinates data security. Therefore, client/server networks are said to be centralized; P2P networks, on the other hand, are decentralized.

Why do businesses use client/server networks? The main advantage of a client/server relationship is that it makes data flow more efficiently than in P2P networks. Servers can respond to requests from many clients at the same time. In addition, servers can be configured to perform specific tasks, such as handling e-mail or database requests, efficiently.

Why aren't P2P networks used more in business settings? P2P networks become difficult to administer when they're expanded beyond 10 users. Each computer may require updating if there are changes to the network, which isn't efficient with large numbers of computers. And as noted previously, security can't be implemented centrally on a P2P network but instead must be handled by each user.

Figure 12.2 shows a small client/server arrangement. The server in this figure provides printing and Internet-connection services for all the client computers connected to the network. The server is performing tasks that would need to be done by each of the client computers in a P2P network. This frees resources on the client computers so that they can more efficiently perform processor-intensive tasks such as viewing a video or accessing a database.

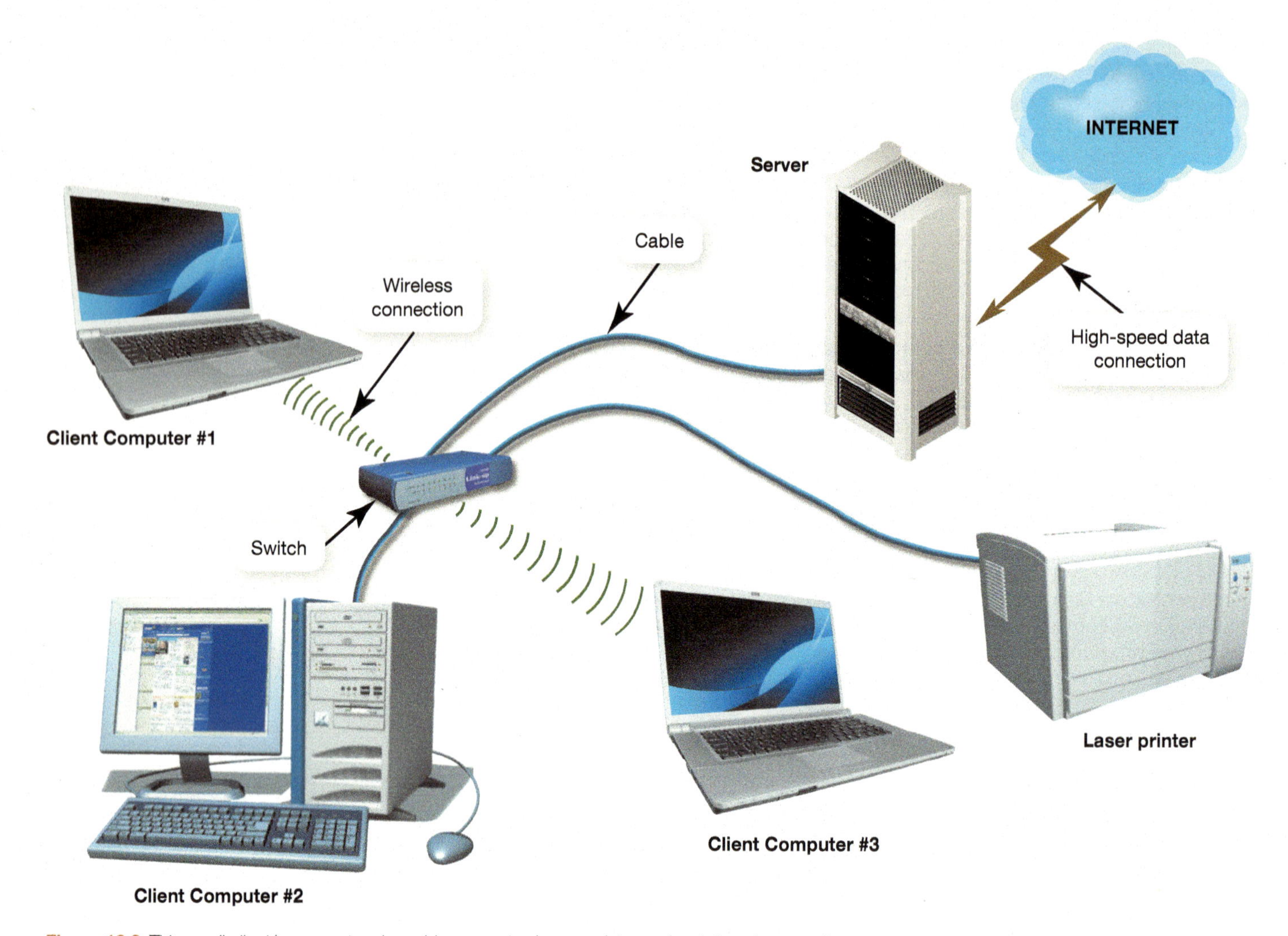

Figure 12.2 This small client/server network enables users to share a printer and an Internet connection. *(Vovan/Shutterstock)*

Besides having a server, what makes a client/server network different from a P2P network? Client/server networks also have increased scalability. Scalability means that more users can be added easily without affecting the performance of the other network nodes. Because servers handle the bulk of the printing, Internet access, and other tasks performed on the network, it's easy to accommodate more users by installing additional servers to help with the increased workload. Installing additional servers on a network is relatively simple and can usually be done with minimal disruption to existing users.

Types of Client/Server Networks

Objective 12.3 *Describe the common types of client/server networks as well as other networks businesses use.*

As you learned in Chapter 7, networks are generally classified according to their size and the distance between the physical parts of the network. Table 12.1 lists the five most common types of client/server network classifications.

What are the most common types of client/server networks encountered in businesses? The two types of client/server networks most commonly encountered in businesses are *local area networks* and *wide area networks*:

- Local area network (LAN): A LAN is generally a small group of computers and peripherals linked over a relatively small geographical area. The computer lab at your school or the network serving the floor of the office building where you work is probably a LAN.

Table 12.1 Classifications of Client/Server Networks

Network Type	Description	Where Used in Business
PAN (Personal Area Network)	Devices used by one person connected by wireless media	Usually by employees traveling on business
LAN (Local Area Network)	A network consisting of nodes covering a small geographical area	In small businesses or self-contained units of a large business (such as one or more floors of the same office building)
HAN (Home Area Network)	A type of small LAN installed in a home	Not usually deployed by businesses, except small home-based businesses
WAN (Wide Area Network)	Two or more LANs connected together, often over long distances	Connecting business LANs over long distances such as between branches in two cities
MAN (Metropolitan Area Network)	WANs constructed by municipalities to provide connectivity in a specific geographical area	Although not deployed by businesses, employees often use them while traveling

(Georgejmclittle/123RF; Viperagp/123RF; Cathy Yeulet/123RF; Yewkeo/123RF; Shark749/123RF)

- Wide area network (WAN): A WAN comprises large numbers of users over a wider physical area or separate LANs that are miles apart. Businesses often use WANs to connect two or more geographically distant locations. For example, a college might have a west and an east campus located in different towns. The LAN at the west campus is connected to the LAN at the east campus, forming one WAN connected either by dedicated telecommunications lines or by satellite links. Students on both campuses can share data and collaborate through the WAN.

What other sorts of networks do businesses use? An intranet is a private network set up by a business or an organization that's used exclusively by a select group of employees, customers, suppliers, volunteers, or supporters. It can facilitate information sharing, database access, group scheduling, videoconferencing, and other employee collaborations. An intranet isn't accessible by unauthorized individuals; a firewall and other security measures protect it from unauthorized access through the Internet.

An area of an intranet that only certain corporations or individuals can access is called an extranet. The owner of an extranet decides who will be permitted to access it. For example, a company's customers and suppliers may be permitted to access information on the company's extranet.

Extranets are useful for enabling electronic data interchange (EDI), which allows the exchange of large amounts of business data (such as orders for merchandise) in a standardized electronic format. Walmart has an extranet that allows its employees, vendors, and contractors to share information easily. Other uses of extranets include providing access to catalogs and inventory databases and sharing information among partners or industry trade groups.

How is information kept secure on intranets and extranets? Intranets and extranets often use virtual private networks to keep information secure. A virtual private network (VPN) uses the public Internet communications infrastructure to build a secure, private network among various locations. Although WANs can be set up using private leased communications lines, these lines are expensive and tend to increase in price as the distance between points increases. VPNs use special security technologies and protocols that enhance security, enabling data to traverse the Internet as securely as if it were on a private leased line. Installing and configuring a VPN requires special hardware such as VPN-optimized routers and firewalls. In addition, VPN software must be installed on users' computing devices. However, many VPN providers (such as ExpressVPN) are now offering VPN services to businesses and individuals. This alleviates the aggravation and hassle of installing and maintaining VPN equipment.

How do VPNs work? The main technology for achieving a VPN is called tunneling. In tunneling, data packets are placed inside other data packets. The format of these external data packets is encrypted and can be understood only by the sending and receiving hardware, which is known as a *tunnel interface*. The hardware is optimized to seek efficient routes of transmission through the Internet. This provides a high level of security and makes information much more difficult to intercept and decrypt. Using a VPN is the equivalent of hiring a limousine and an armed guard to drive a package (your data) through a private tunnel directly to the destination (see Figure 12.3).

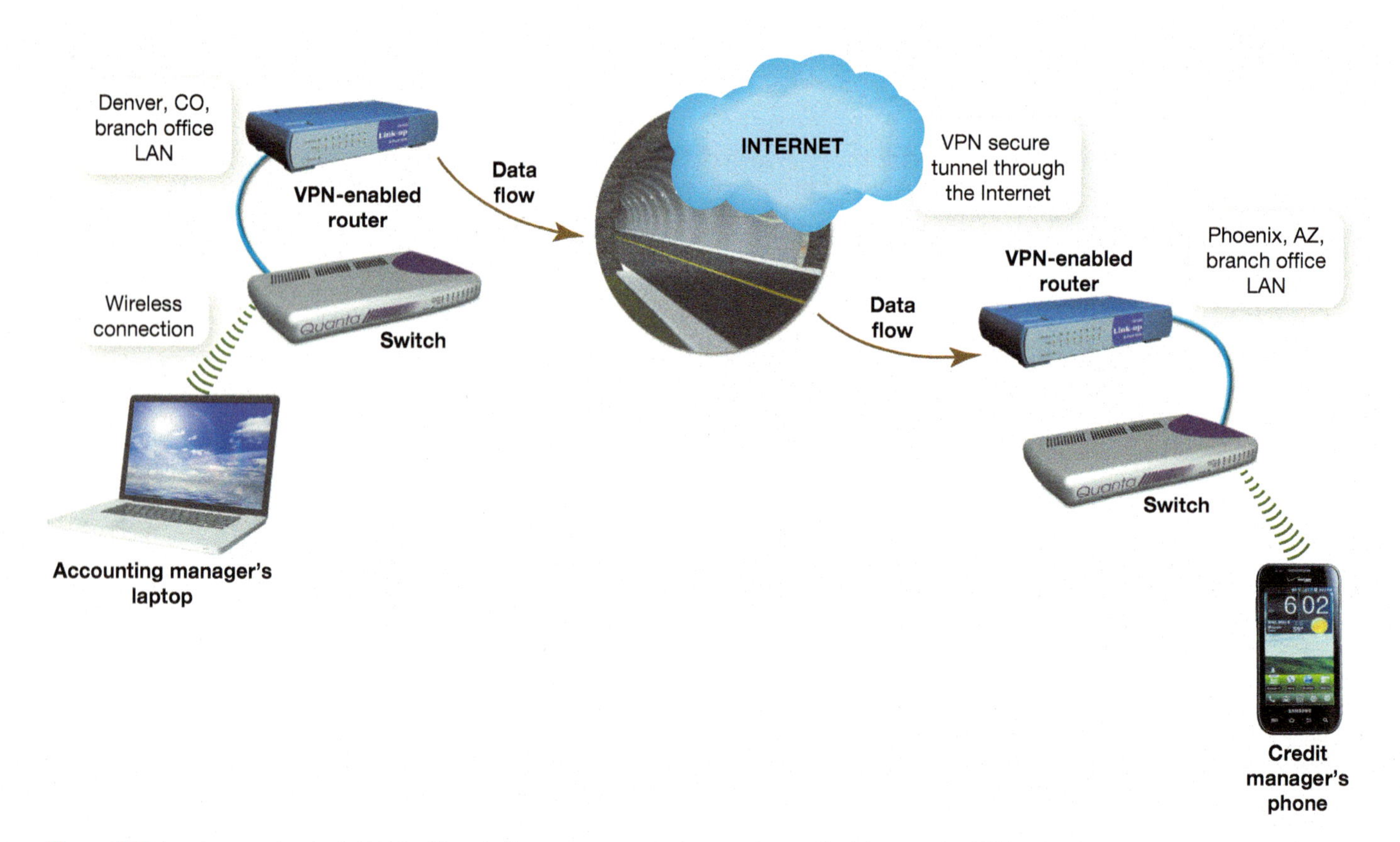

Figure 12.3 Local area networks (LANs) in different cities can communicate securely over the Internet using VPN technology. *(Vovan/Shutterstock)*

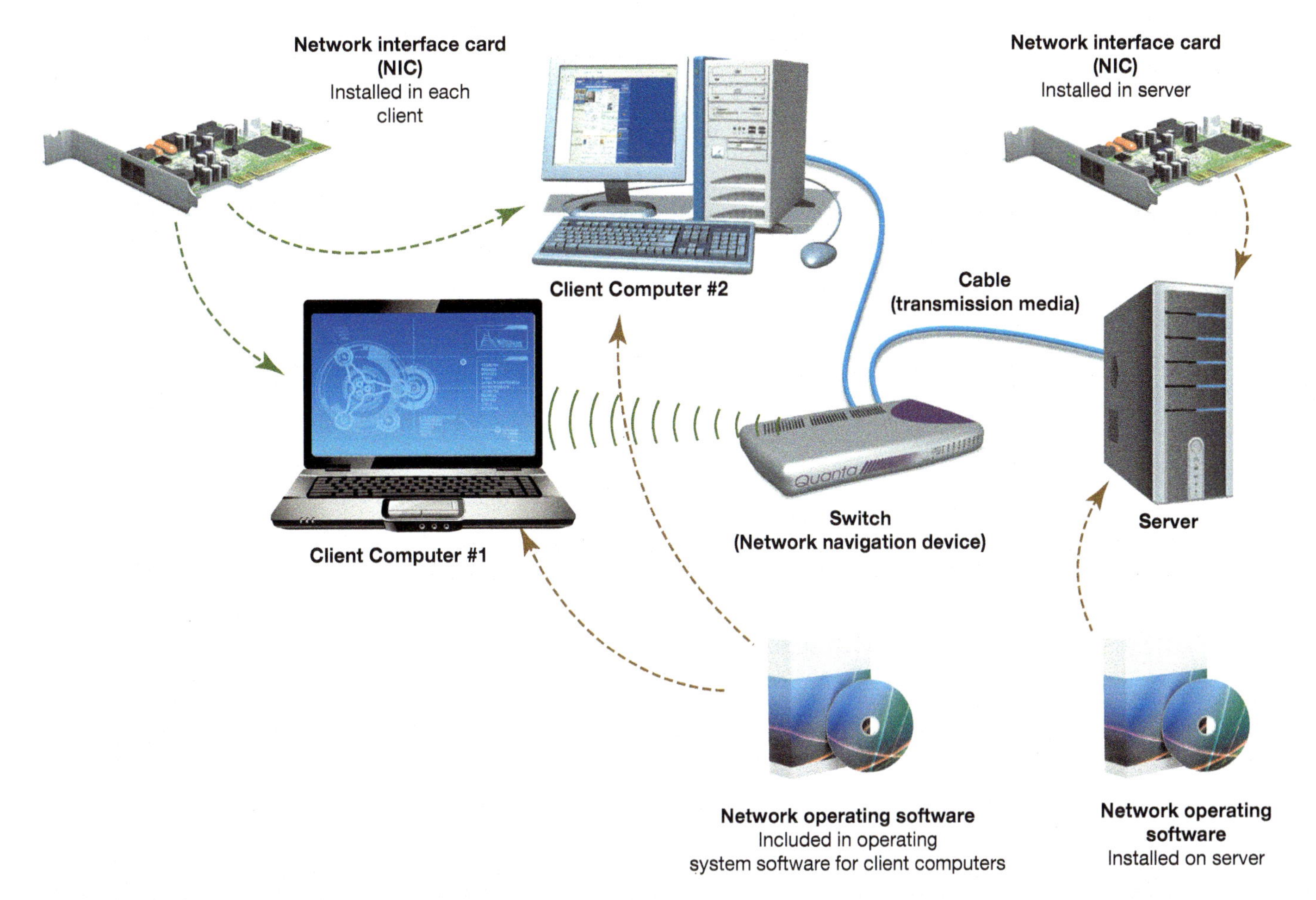

Figure 12.4 Basic Components of a Typical Client/Server Network *(Viktor Gmyria/123RF; Juffin/DigitalVision Vectors/Getty Images)*

What are the key components of a client/server network? The key components of a client/server network are:

- Servers
- Network topologies (the layout of the components)
- Transmission media
- Network adapters
- Network navigation devices
- A network operating system

Figure 12.4 shows the components of a simple client/server network. In the following sections, we explore each component in more detail.

Bits&Bytes Make Your Browser Protect You!

Data breaches are a regular occurrence today. However, studies by Google have shown that many users are ignoring the fact that passwords have been breached, and they continue to use these passwords on other sites.

To help combat this problem, Google developed a Chrome extension called Password Checkup. It is available from the Chrome Web Store for free.

Once installed, the extension checks your login credentials every time you log on to a website. Google maintains a database of billions of credentials from public data breaches that are known to be compromised. If your user ID password combination matches a breached combination, Chrome displays a pop-up that urges you to change the credentials. Password Checkup was designed jointly by Google personnel and cryptography experts at Stanford University to ensure that your credentials are not revealed to Google and hackers can't use it to obtain your passwords.

So install this Chrome extension today and let your browser help you remember to change your exposed credentials. You'll sleep better at night!

Servers and Network Topologies

Servers are the workhorses of the client/server network. They interface with many network users and assist them with a variety of tasks. The number and types of servers on a client/server network depend on the network's size and workload. Small networks (such as the one pictured in Figure 12.2 on page 464) would have just one server to handle all server functions. *Network topology* refers to the physical or logical arrangement of computers, transmission media (cables), and other network components. In the following sections, we explore the types of servers found in client/server networks and the typical topologies you may encounter.

Servers

Objective 12.4 *List the common types of servers found on client/server networks.*

What types of servers are found on larger client/server networks? A dedicated server is a server used to fulfill one specific function, such as handling e-mail. When more users are added to a network, dedicated servers are also added to reduce the load on the main server. Once dedicated servers are deployed, the original main server can become a dedicated server.

What functions do dedicated servers handle? Any task that's repetitive or demands a lot of time from a computer's processor (CPU) is a good candidate to relegate to a dedicated server. Common types of dedicated servers (shown in Table 12.2) are authentication servers, file servers, print servers, application servers, database servers, e-mail servers, communications servers, web servers, and cloud servers. Servers are connected to a client/server network so that all client computers that need to use their services can access them.

Table 12.2 Major Categories of Servers

Print Servers	Application Servers	Database Servers	Authentication Servers

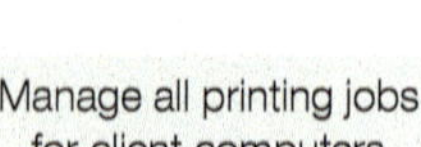 Manage all printing jobs for client computers	Serve as a repository for application software	Provide client computers with database access	Keep track of users logged on to the network
File Servers	**E-Mail Servers**	**Communication Servers**	**Web/Cloud Servers**
     			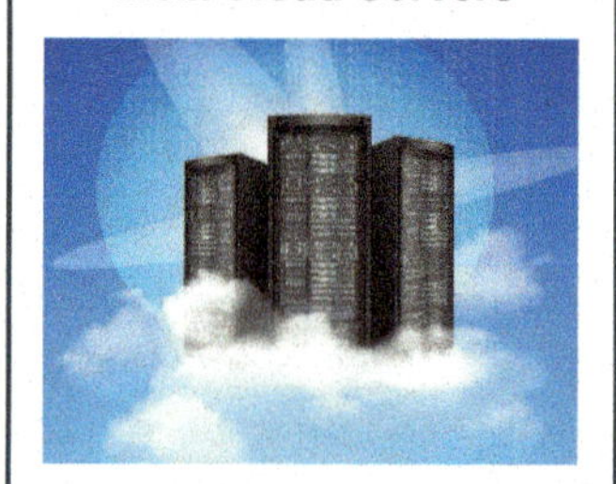
Store and manage files for network users	Process and deliver incoming and outgoing e-mail	Handle all communications between the network and other networks	Host websites to make them available through the Internet

(Maksym yemelyanov/123RF; Iaroslav Neliubov/123RF; Claudio Ventrella/123RF; Sashkin/Shutterstock; Agnieszka Murphy/123RF; Valentint/123RF; Galina Peshkova/123RF; Macrovector/123RF)

Authentication and File Servers

What are authentication and file servers? An authentication server keeps track of who is logging on to the network and which services on the network are available to each user. Authentication servers usually act as overseers for the network. They manage and coordinate the services provided by all other dedicated servers located on the network.

A file server stores and manages files for network users. On the network at your workplace or school, you may be provided with space on a file server to store files you create.

Print Servers

How does a print server function? Print servers manage all client-requested printing jobs for all printers on a network, which helps client computers complete more productive work by relieving them of printing duties. When you tell your computer to print a document, it passes off the task to the print server. This frees the CPU on your computer to do other jobs.

How does the printer know which documents to print? A print queue (or *print spooler*) is a software holding area for print jobs. When the print server receives a printing request from a client computer, it puts the job into a print queue on the print server. Normally, each printer on a network has its own uniquely named print queue. Jobs receive a number when they enter the queue and go to the printer in the order in which they were received. Thus, print servers organize print jobs into an orderly sequence to make printing more efficient on a shared printer.

Another useful aspect of print servers is that network administrators can set them to prioritize print jobs. Different users and types of print jobs can be assigned different priorities so that higher-priority jobs will be printed first. For instance, in a company in which documents are printed on demand for clients, you would want those print jobs to take precedence over routine employee correspondence.

Application Servers

What function does an application server perform? In many networks, all users run the same application software on their computers. In a network of thousands of personal computers, installing application software on each computer is time-consuming. An application server acts as a repository for application software.

When a client computer connects to the network and requests an application, the application server delivers the software to the client computer. Because the software doesn't reside on the client computer itself, this eases the task of installation and upgrading. The application needs to be installed or upgraded only on the application server, not on each client computer.

Database Servers

What does a database server do? A database server provides client computers with access to information stored in databases. Often, many people need to access a database at the same time. For example, multiple college advisors can serve students at the same time because the advisors all have access to the student information database. This is made possible because the database resides on a database server that each advisor's computer can access through the network. If the database were on a stand-alone computer instead of a network, only one advisor could use it at a time.

E-Mail Servers

How is e-mail handled on a large client/server network? The sole function of an e-mail server is to process and deliver incoming and outgoing e-mail. The volume of e-mail on a large network could quickly overwhelm a server that was attempting to handle other functions as well. On a network with an e-mail server, when you send or receive an e-mail, it goes through the e-mail server, which then handles the routing and delivery of your message.

Communications Servers

What types of communications does a communications server handle? A communications server handles all communications between the network and other networks, including managing Internet connectivity. All requests for information from the Internet and all messages being sent through the Internet pass through the communications server. Because Internet traffic is substantial at most organizations, the communications server has a heavy workload.

Using Servers

In this Helpdesk, you'll play the role of a helpdesk staffer, fielding questions about various types of servers and client/server software.

The communications server is often the only device on the network connected to the Internet. E-mail servers, web servers, and other devices needing to communicate with the Internet usually route all their traffic through the communications server. Providing a single point of contact with the outside world makes it easier to secure the network from hackers.

Web Servers and Cloud Servers

What function does a web server perform? A web server is used to host a website so that it will be available through the Internet. Web servers run specialized software such as Apache HTTP Server or Microsoft Internet Information Services (IIS) that enable them to host web pages. Not every large network has a web server; many businesses use a third-party web-hosting company to host their websites instead.

What is a cloud server? Servers no longer need to be physically located at a company's offices. Cloud servers are maintained by hosting companies, such as HostGator, and are connected to networks by the Internet. A company could choose to have any of the server types discussed previously hosted on cloud servers instead of maintaining them locally. Small businesses that don't have a large staff of computer professionals often choose to use cloud servers to save money.

Virtual Servers

Do you always need a dedicated server for each individual function your server needs to perform? Dedicated servers are mission critical when those servers handle processor-intensive operations. For instance, a major online retailer would have dedicated web servers handling e-commerce traffic so that customers could purchase items quickly and efficiently. Small businesses, however, don't necessarily require the computing power of a dedicated server for all their computing needs. In addition, today's powerful multi-core processors allow computers to process data more quickly and efficiently than single-core machines. Sometimes just dedicating a computer to one task, such as an authentication server, might not come close to using that server's computing potential fully.

In addition, computers produce a lot of heat and take up physical space. As a company grows, continually adding servers may exceed the available floor space and cooling capacity of a company's data center. And budgets may limit a company's ability to purchase additional dedicated servers when needed. For these reasons, *virtualization* was born.

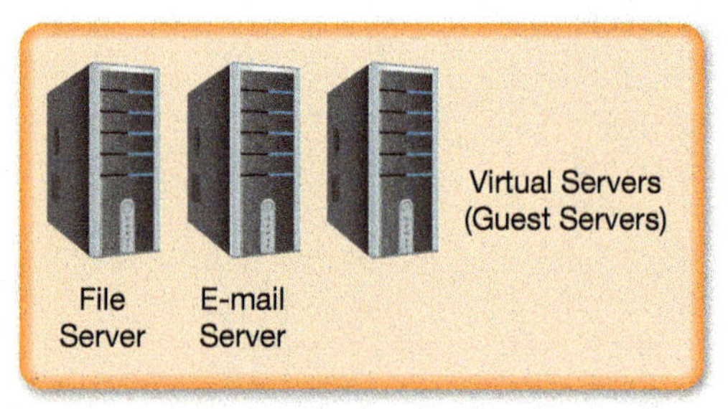

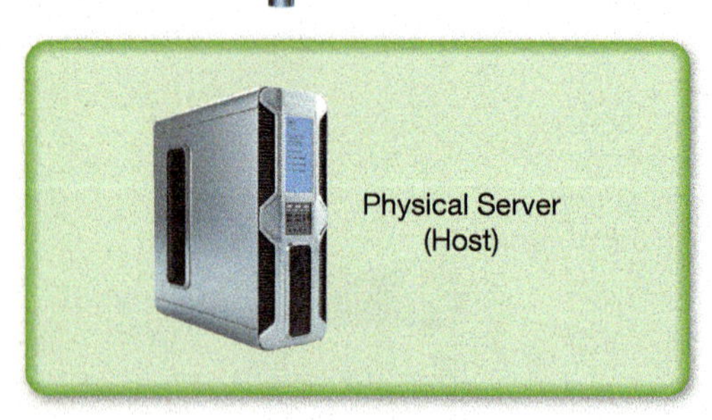

Figure 12.5 Virtualization software allows one physical server to appear to be multiple separate servers capable of handling multiple different processes. *(Sashkin/Shutterstock)*

What is virtualization? Virtualization involves using specialized software to make individual physical servers behave as though they are more than one physical device (see Figure 12.5). Each virtual server can operate as a separate device and can even run its own operating system (OS). Therefore, you could have a virtual e-mail server running Windows and a virtual web server running Linux on the same physical server.

How do you turn a regular server into a virtual server? Creating virtual servers requires running specialized virtualization software on a physical server. The physical server then becomes known as the *host*. Virtual servers running on a host machine are known as *guest servers*. The computing power of the CPU is split between the guests running on the host. Therefore, it wouldn't be practical to run dozens of guest environments on one host because the computing power would become diluted. However, it's very possible to run two or three guest servers on one multi-processor physical server if the applications running on the virtual servers aren't overly processor intensive. Workstation Player, Microsoft's Hyper-V Server, and XenServer are popular virtualization software packages.

Virtualization is a great way to provide backup servers at a low cost. Mission-critical servers, such as web servers, need to always be running and available to customers. If a web server for an e-commerce business fails, the company is unable to sell products until the server is repaired. With virtual servers, you could have a copy of your web server running virtually on two separate physical machines (such as servers 1 and 2). If server 1 breaks, you could immediately have the virtual copy of your web server on physical server 2 take over, and the company could still be processing orders and conducting business.

Trends in IT Protecting Privacy in Your Home Office

Many people are concerned about protecting their personal privacy when using the Internet. With the vast shift toward telecommuting, your boss is probably concerned about privacy in your home office as well. Because the major browsers all have a secure mode (Incognito for Chrome, for example), you merely need to start a secure browser session before surfing, right? Maybe not! In 2020, a class-action lawsuit was filed against Alphabet (the parent company of Google) alleging that Google was tracking users' browsing even when they were using an Incognito window. Naturally, this set off alarm bells among privacy advocates. Fortunately, you can exercise several easily implemented options to increase your privacy significantly:

Figure 12.6 Just because you're at a home office doesn't mean you shouldn't worry about your online privacy. *(Khakimullin Aleksandr/ Shutterstock)*

- *Use the Tor browser:* The most well-known of secure browsers is Tor (discussed in Chapter 8), which is highly touted by security experts. It was developed and is maintained by a nonprofit entity dedicated to protecting privacy. Unfortunately, although it may provide a more secure experience, it will make your browsing experience slower.
- *Use an open source browser:* Proprietary companies like Google and Microsoft might have a financial motive for tracking your browsing history, because they can sell that information for marketing purposes. However, open source browsers are developed by a community of developers without the usual profit motives. Therefore, using a browser such as Firefox (or an offshoot based on Firefox such as Waterfox) will be a more secure alternative.
- *Use plug-ins to make your browser safer:* You might be forced to use a certain browser by your employer or by the nature of the websites you need to access. (Some sites only support certain browsers.) No matter what browser you use, safety and security can always be improved. Plug-ins are software that work along with browsers to increase or change its functionality. HTTPS Everywhere is a popular plug-in that enforces SSL security wherever it's possible. It works with Chrome, Firefox, Opera, and Tor. Ad-blocking plug-ins are helpful too, because they shut down many types of tracking activity. The entities that maintain most browsers provide a directory of plug-ins compatible with the browser. Seek out and search the directory for your browser and investigate security plug-ins.
- *Use a VPN.* Virtual private networks (VPNs) greatly enhance your security when surfing the web. When using a VPN, your computer connects directly to a server at the VPN provider. Data flowing between your computer and the VPN server is encrypted. When you enter a URL in your browser, the request is forwarded to the VPN server, which then requests the information from the website's computer and forwards it to your computer. The website doesn't know your computer made the request...it is only aware of the VPN server. VPN providers like ExpressVPN and NordVPN offer cost-effective solutions for home office workers. Your employer might even pay for it! Many employers provide their employees with VPN access at no charge to the employees.

Enhancing your Internet security at home should contribute greatly to your peace of mind ... and make your boss feel better too.

Network Topologies

Objective 12.5 *Describe the common types of network topologies and the advantages and disadvantages of each one.*

Just as buildings have different floor plans depending on their uses, networks have different layouts according to their purpose. As noted earlier, network topology refers to the physical or logical arrangement of computers, transmission media (cables), and other network components. *Physical topology* refers to the layout of the real components of the network, whereas *logical topology* refers to the virtual connections among network nodes. Logical topologies are usually determined by network protocols instead of the physical layout of the network or the paths that electrical signals follow on the network.

What are network protocols? A protocol is a set of rules for exchanging communications. Although many people think that Ethernet is a type of network topology, it's actually a communications protocol. Therefore, an Ethernet network could be set up using almost any type of physical topology.

In this section, we explore the most common network topologies (bus, ring, and star) and discuss when each topology is used. The type of network topology used is important because it can affect a network's performance and scalability. Knowing how the basic topologies work—and knowing the strengths and weaknesses of each one—will help you understand why particular network topologies were chosen on the networks you use.

Bus Topology

What does a bus topology look like? In a bus (or linear bus) topology, all computers are connected in sequence on a single cable, as shown in Figure 12.7. This topology has largely become legacy technology because star topologies are more efficient on Ethernet networks, and a bus topology isn't designed to support wireless connections easily. However, bus topologies are still found in some manufacturing facilities where groups of computer-controlled machines are connected.

Each computer on the bus network can communicate directly with every other computer on the network. Data collisions, which happen when two computers send data at the same time and the sets of data collide somewhere in the connection media, are a problem on all networks. When data collides, it's often lost or damaged. A limitation of bus networks is that data collisions can occur fairly easily because a bus network is essentially composed of one main communication medium (a single cable).

Because two signals transmitted at the same time on a bus network may cause a data collision, an access method must be established to control which computer can use the transmission media at a certain time. Computers on a bus network behave like a group of people having a conversation. The computers listen to the network data traffic on the media. When no other computer is transmitting data (that is, when the conversation stops), the computer knows it's allowed to transmit data. This means taking turns talking prevents data collisions.

How does data get from point to point on a bus network? The data is broadcast throughout the network through the media to all devices connected to the network. The data is broken into small segments, each called a *packet*. Each packet contains the address of the computer or peripheral device to which it's being sent. Each computer or device connected to the network listens for data that contains its address. When it hears data addressed to it, it takes the data off the media and processes it.

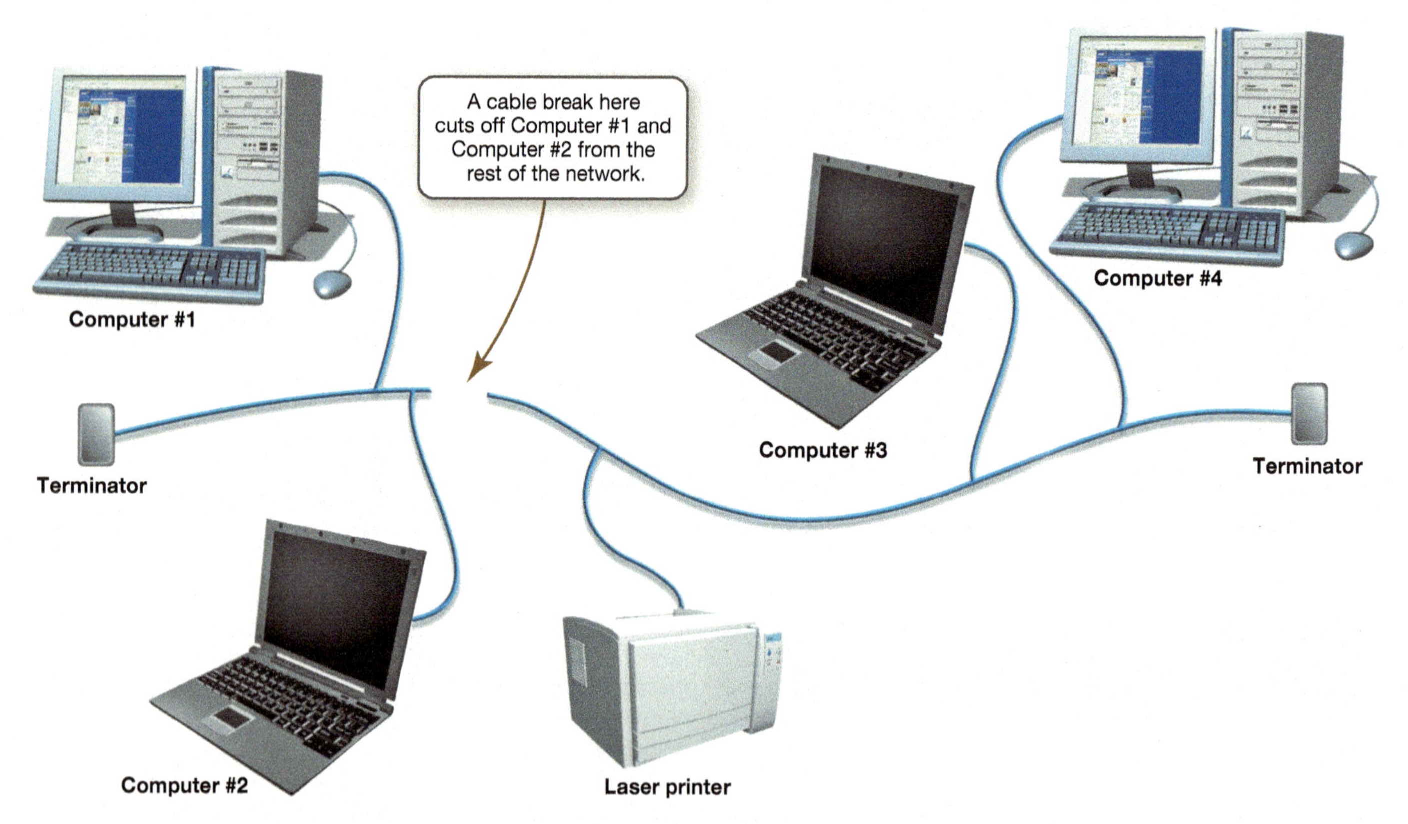

Figure 12.7 In a linear bus topology, all computers are connected in a sequence. *(Pearson)*

The devices (nodes) attached to a bus network do nothing to move data along the network. This makes a bus network a passive topology. The data travels the entire length of the medium and is received by all network devices. The ends of the cable in a bus network are capped off by terminators (as shown in Figure 12.7). A terminator is a device that absorbs a signal so that it's not reflected back onto parts of the network that have already received it.

What are the advantages and disadvantages of bus networks? The simplicity and low cost of bus network topology are its key advantages. As shown in Figure 12.7, the major disadvantage is that if there is a break in the cable, the bus network is effectively disrupted because some computers are cut off from others on the network. In addition, because only one computer can communicate at a time, adding many nodes to a bus network limits performance and causes delays in sending data.

Ring Topology

What does a ring topology look like? Not surprisingly, given its name, the computers and peripherals in a ring (or loop) topology are laid out in a configuration resembling a circle, as shown in Figure 12.8. Data flows around the circle from device to device in one direction only. Because data is passed using a special data packet called a token, this type of topology was once commonly called a *token-ring topology*.

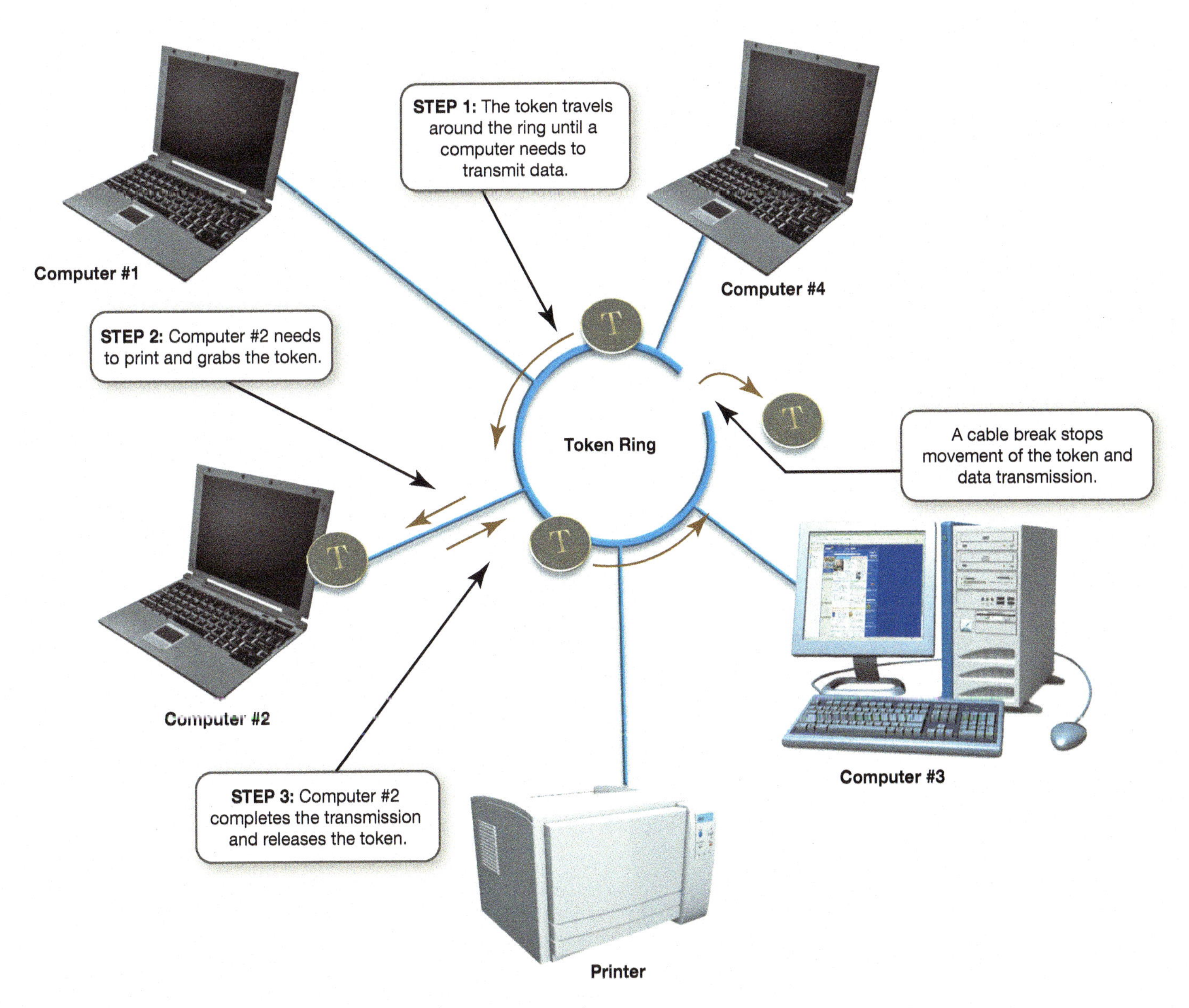

Figure 12.8 A ring topology provides a fair allocation of resources. *(Pearson)*

How does a token move data around a ring? A token is passed from computer to computer around the ring until it's grabbed by a computer that needs to transmit data. The computer holds on to the token until it has finished transmitting data. Only one computer on the ring can hold the token at a time, and usually only one token exists on each ring.

Network Topology and Navigation Devices

In this Sound Byte, you learn about common network topologies, the types of networks they are used with, and various network navigation devices.

If a node has data to send, such as a document that needs to go to the printer, it waits for the token to be passed to it. The node then takes the token out of circulation and sends the data to its destination. When the receiving node gets a complete transmission of the data (in this example, when the printer receives the document), it transmits an acknowledgment to the sending node. The sending node then generates a new token and starts it going around the ring again. This is called the token method and is the access method that ring networks use to avoid data collisions.

A ring topology is an active topology, which means that nodes participate in moving data through the network. Each node on the network is responsible for retransmitting the token or the data to the next node on the ring. Large ring networks can use multiple tokens to help move data faster.

What are the advantages and disadvantages of a ring topology? A ring topology provides a fairer allocation of network resources than does a bus topology. By using a token, a ring network enables all nodes on the network to have an equal chance to send data. One chatty node can't monopolize the network bandwidth as easily as in a bus topology because it must pass the token on after sending a batch of data. In addition, a ring topology's performance remains acceptable even with large numbers of users.

As shown in Figure 12.8, one disadvantage of a ring network is that if one computer fails, the entire network can come to a halt because the failed computer is unavailable to retransmit tokens and data. Another disadvantage is that problems in the ring can be hard for network administrators to find. It's easier to expand a ring topology than a bus topology, but adding a node to a ring causes the ring to cease to function while the node is being installed.

Star Topology

What is the layout for a star topology? A star topology is the most common client/server network topology because it offers the most flexibility for a low price. In a star topology, the nodes connect to a central communications device called a *switch* in a pattern resembling a star, as shown in Figure 12.9. The switch receives a signal from the sending node and retransmits it to the node on the network that needs to receive the signal. Each network node picks up only the transmissions addressed to it. Because the switch retransmits data signals, a star topology is an active topology. The only drawback is that if the switch fails, the network no longer functions. However, it's relatively easy to replace a switch.

How do computers on a star network avoid data collisions? Because most star networks are Ethernet networks, they use the method used on all Ethernet networks to avoid data collisions: CSMA/CD (short for *carrier sense multiple access with collision detection*). With CSMA/CD, a node connected to the network uses carrier sense (that is, it listens) to verify that no other nodes are currently transmitting data signals. If the node doesn't hear any other signals, it assumes that it's safe to transmit data. All devices on the network have the same right (that is, they have multiple access) to transmit data when they deem it safe. It's therefore possible for two devices to begin transmitting data signals at the same time. If this happens, the two signals collide.

What happens when the signals collide? As shown in Figure 12.10, when two nodes (#1 and #2) begin transmitting data signals at the same time (Step 1), signals collide, and a node on the network (#3) detects the collision. Node #3 then sends a special signal called a jam signal to all network nodes, alerting them that a collision has occurred (Step 2). The original nodes #1 and #2 then stop transmitting and wait a random amount of time before retransmitting their data signals (Step 3). The wait times need to be random; otherwise, both nodes would start retransmitting at the same time and another collision would occur.

What are the advantages and disadvantages of a star topology? The major advantages of a star network are that:

- The failure of one computer doesn't affect the rest of the network. This is extremely important in a large network, when having one disabled computer affect the operations of several hundred other computers would be unacceptable.
- It's easy to add nodes to star networks.
- Performance remains acceptable even with large numbers of nodes.

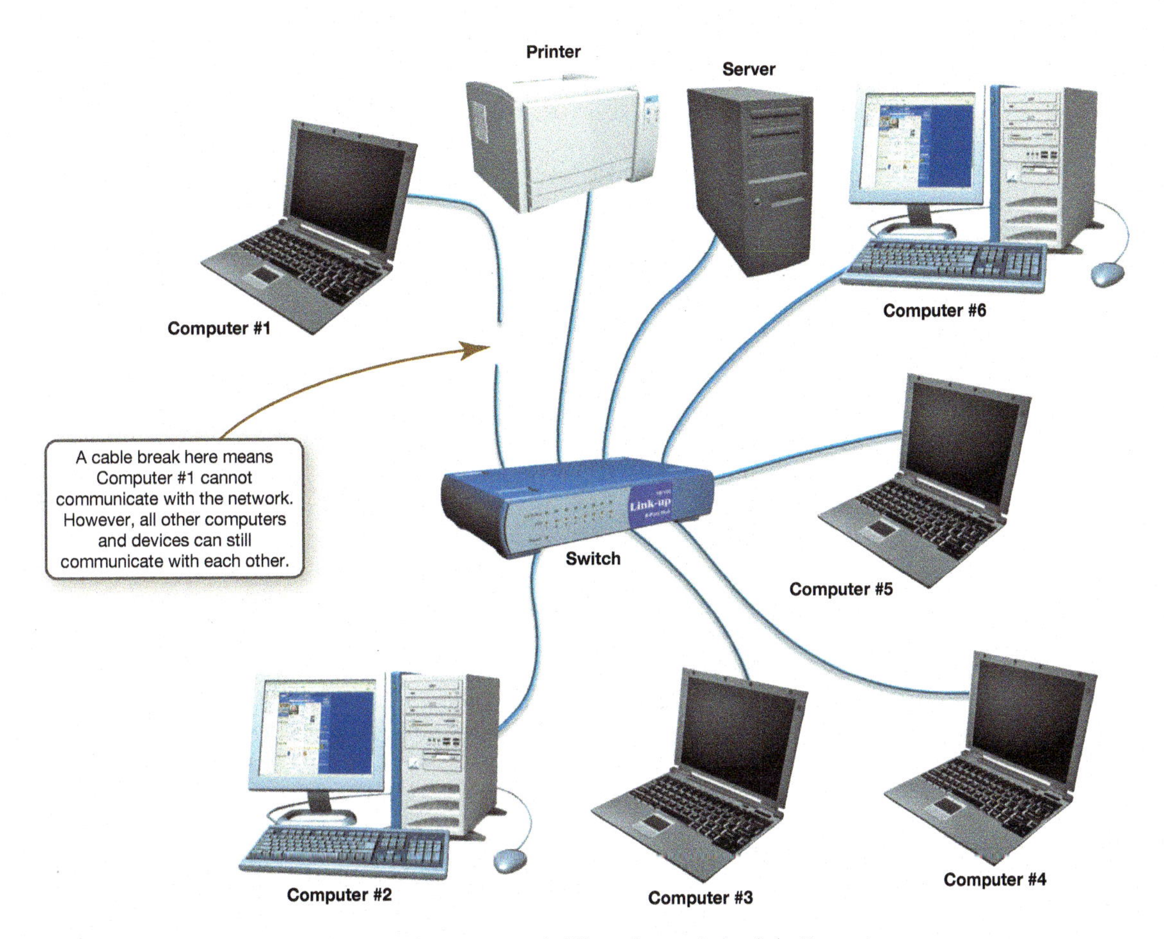

Figure 12.9 In a star topology, network nodes are connected through a central switch. *(Pearson)*

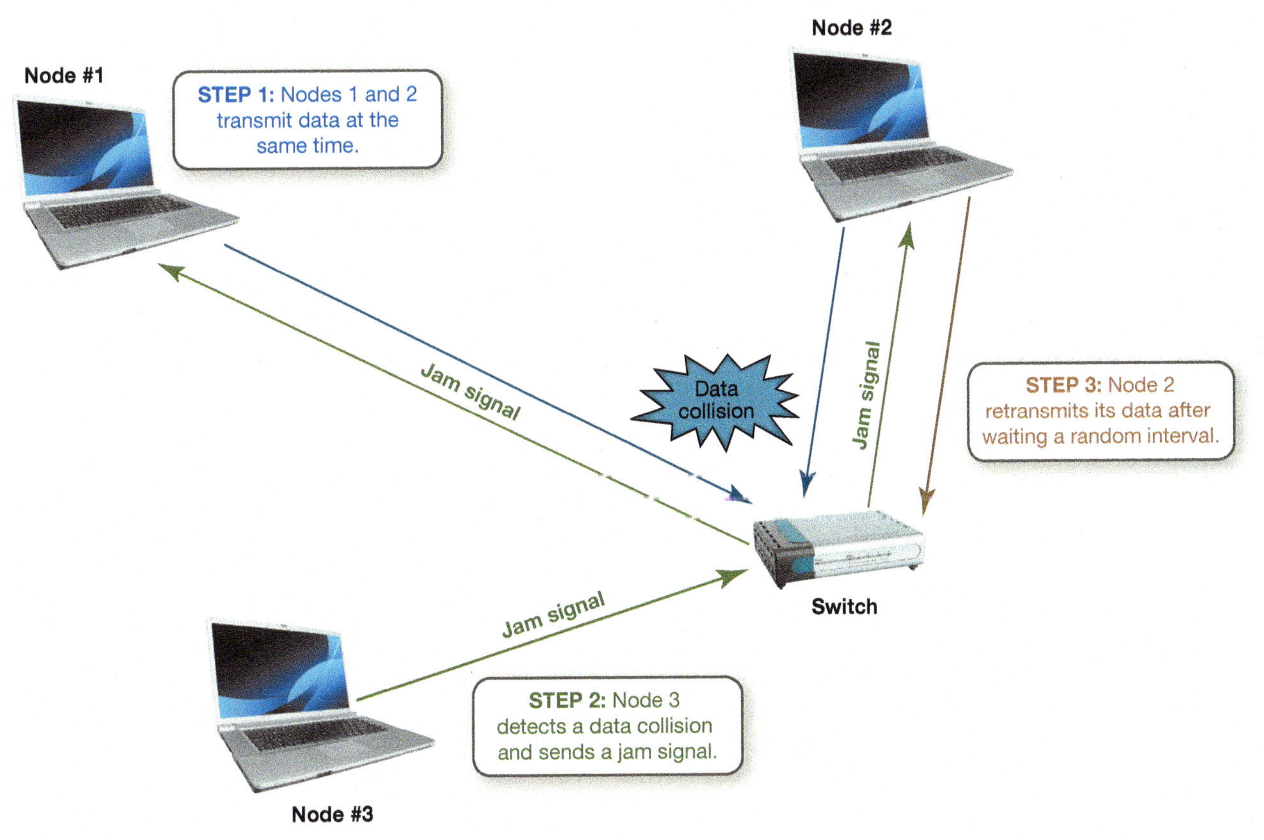

Figure 12.10 Jam signals are used to handle data collisions on an Ethernet network. *(Albo003/Shutterstock)*

- Centralizing communications through a switch makes troubleshooting and repairs easier. Technicians can usually pinpoint a communications problem just by examining the switch, as opposed to searching for a particular length of cable that has broken in a ring network.

The disadvantage of star networks used to be cost. Because of the complexity of the layout of star networks, they require more cable and used to be more expensive than bus or ring networks. However, the price of cable has fallen, and wireless nodes are replacing many wired nodes on networks, removing cost as a barrier in most cases.

Comparing Topologies

Which topology is the best one? Table 12.3 lists the advantages and disadvantages of bus, ring, and star topologies. Star topologies are the most common, mainly because large networks are constantly adding new users. The ability to add new users easily—by installing an additional switch—without affecting users already on the network is the deciding factor. Bus topologies have become all but extinct now that most home networks use a star topology. Ring topologies are still popular in certain businesses where fair allocation of network access is a major requirement of the network.

Can topologies be combined within a single network? Because each topology has its own unique advantages, topologies are often combined to construct business networks. Combining multiple topologies into one network is known as constructing a hybrid topology. For instance, fair allocation of resources may be critical for reservation clerks at an airline (thereby requiring a ring network), but the airline's purchasing department may require a star topology.

Table 12.3 Advantages and Disadvantages of Bus, Ring, and Star Topologies

Topology	Advantages	Disadvantages
Bus	• Uses a minimal amount of cable. • Installation is easy, reliable, and inexpensive.	• Breaks in the cable can disable the network. • Large numbers of users decrease performance because of high volumes of data traffic.
Ring	• Allocates access to the network fairly. • Performance remains acceptable even with many users.	• Adding/removing nodes disables the network. • Failure of one node can bring down the network. • Problems in data transmission can be difficult to find.
Star	• Failure of one node doesn't affect other nodes on the network. • Centralized design simplifies troubleshooting and repairs. • High scalability: Adding computers is easy. • Performance remains acceptable even with many users.	• Requires more cable (and possibly higher installation costs) than a bus or ring topology. • The switch is a single point of failure; if it fails, all computers connected to it are affected.

Before moving on to Part 2:

1. Watch Chapter Overview Video 12.1.
2. Then take the Check Your Understanding quiz.

Check Your Understanding // Review & Practice

For a quick review to see what you've learned so far, answer the following questions.

multiple choice

1. Which of the following is NOT an advantage of installing a network in a business?
   a. Enables peripheral sharing
   b. Decentralization of files and data
   c. Enables software sharing
   d. Centralization of files and data

2. Why aren't client/server networks usually installed in the home?
   a. Security is weaker on client/server networks than on P2P networks.
   b. Client/server networks aren't scalable enough for home networks.
   c. Client/server networks are too complex for most home networks.
   d. Client/server networks can't share peripherals.

3. A network consisting of nodes covering a wide geographic area is a
   a. PAN.
   b. WAN.
   c. HAN.
   d. LAN.

4. A(n) _____ server handles all data traffic between the network and other networks.
   a. authentication
   b. database
   c. communications
   d. application

5. A _____ is a set of rules for exchanging communications.
   a. blueprint
   b. domain
   c. topology
   d. protocol

chew on this

(3Dalia/Shutterstock)

Most schools have drafted acceptable-use policies for computers and Internet access to inform students and employees of the approved uses of college computing assets. Should college employees be allowed to use their computers and Internet access for personal tasks? If so, how much time per day is reasonable for employees to spend on personal tasks? Should student computer and Internet usage be monitored to ensure compliance with the personal use policies? Should the college inform students that they're being monitored?

Go to **MyLab IT** to take an autograded version of the *Check Your Understanding* review and find all media resources for the chapter.

For the IT Simulation for this chapter, see MyLab IT.

Try This

Sharing Printers on a Network Using Windows

You can share a printer with all the computers on your network, provided you have configured it for sharing in Windows. In this tutorial, you learn how to configure a printer for sharing. For more step-by-step instructions, watch the Try This video on MyLab IT.

Step 1 Ensure that the printer you wish to share is connected to a computer on your network and that the printer is powered on. Press the **Start Button** and select **Settings** > **Devices** > **Printers & scanners**. Choose the printer you wish to share and then select the **Manage** button.

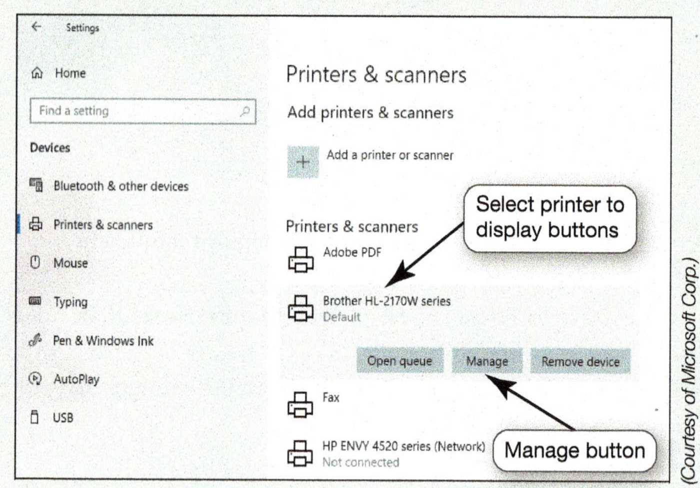

(Courtesy of Microsoft Corp.)

Step 2 On the next screen, click the **Printer properties** link and then select the Sharing tab in the Printer Properties dialog box. Select the check box beside **Share this printer** to enable sharing. You may change the **Share name** if you wish. Select **Apply** and then select **OK** to close the dialog box.

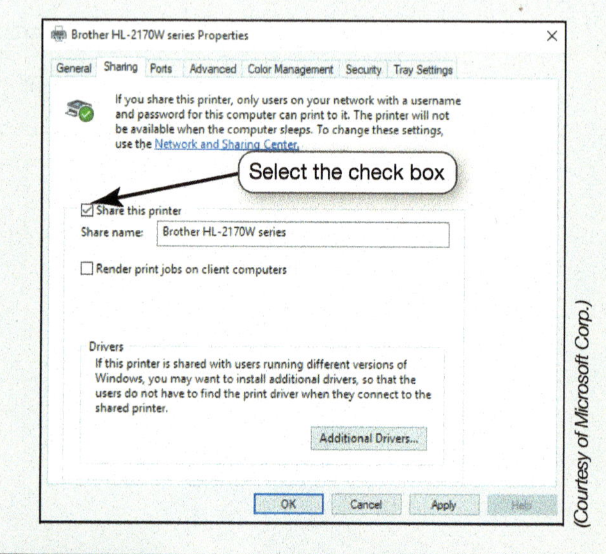

(Courtesy of Microsoft Corp.)

Step 3 On a different computer, Select the **Start** button and select **Settings** > **Devices** > **Printers & scanners**. Select **Add a printer or scanner**. Choose the printer you want from the list displayed, then select the **Add device** button. The printer is now accessible from this computer, provided the printer (and any computer it is attached to) is powered on.

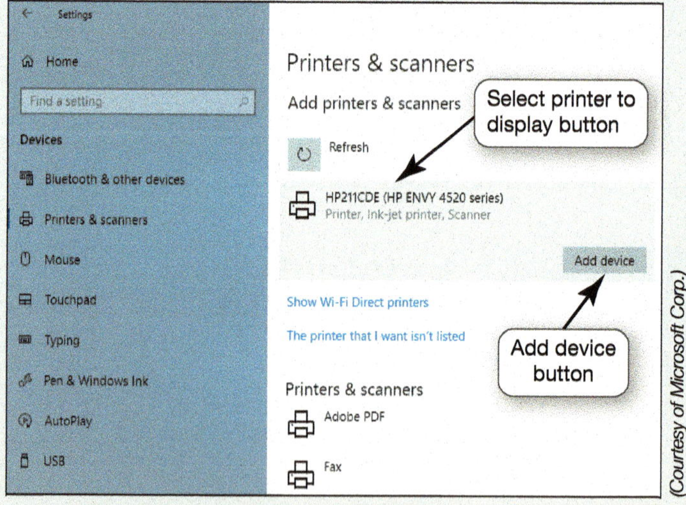

(Courtesy of Microsoft Corp.)

Make This  TOOL: App Inventor 2 or Thunkable

An App That Shares

Want to share with others something you have created in an app?

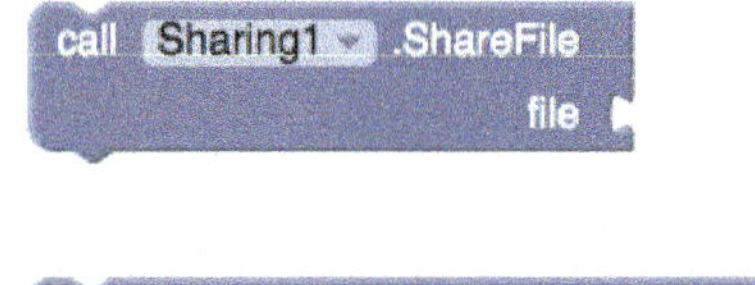

In this exercise, you continue your mobile app development by using the Sharing component of the Social drawer in App Inventor to share information.

(Copyright MIT, used with permission.)

This component will pull up a list of all the applications on your device that can share a file. Make your work part of a larger community by sharing it!

Use the Sharing component to fire up all the ways your users can exchange information.

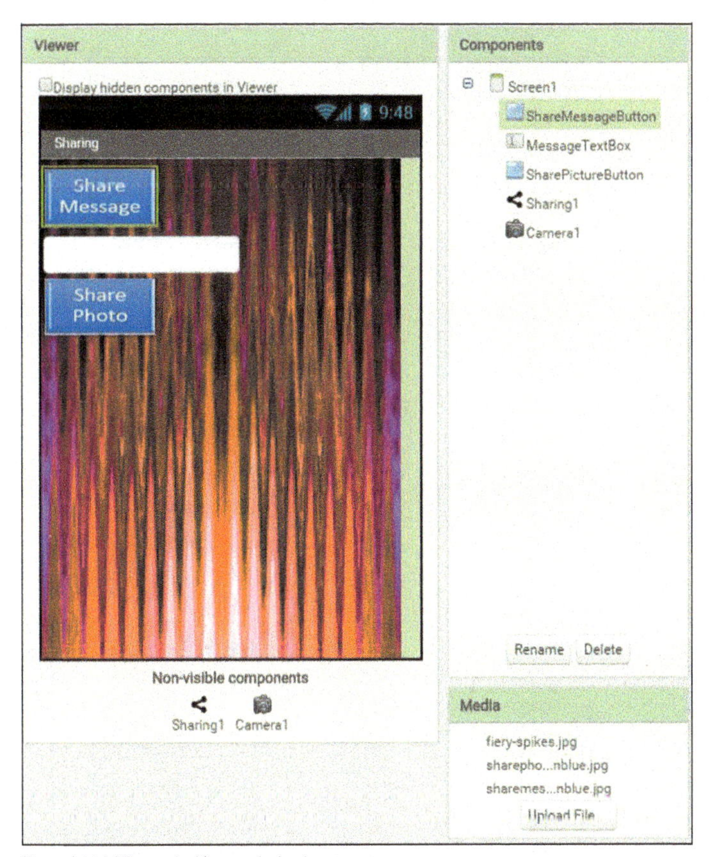

(Copyright MIT, used with permission.)

To create this app for iOS, go to *Thunkable.com*, a programming platform based on App Inventor.
For the instructions for this exercise, go to MyLab IT.

Part 2 For an overview of this part of the chapter, watch **Chapter Overview Video 12.2**.

Setting Up Business Networks

Learning Outcome 12.2 **You will be able to describe transmission media, network operating system software, and network navigation devices and explain major threats to network security and how to mitigate them.**

Setting up business networks is similar to configuring home networks. You need to consider the type of transmission media, ensure that all nodes have network adapters, and install the appropriate network communication devices.

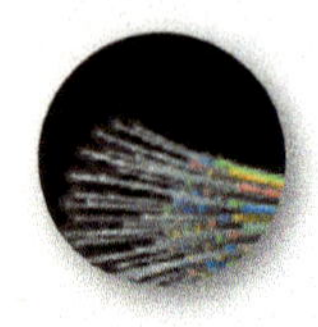

Transmission Media

You may have only wireless transmission media in your home network. But businesses have different needs than home users. Let's start by looking at the transmission media used in business networks.

Figure 12.11 Factors to Consider When Choosing Network Cable

Maximum run length
- How far a cable can run before data signal degrades
- Distance between nodes determines run length needed

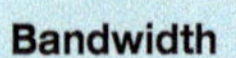

Bandwidth
- Amount of data transmitted across medium
- Measured in bits per second (bps)

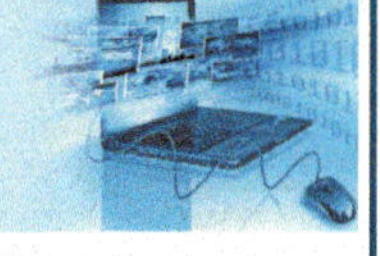

Flexibility (bend radius)
- How much a cable can be bent before it is damaged
- Lots of corners? Need cable with a high bend radius.

Cable cost
- Cost is different for each cable type
- Budget may limit choice of cable type

Installation cost
- Twisted-pair and coaxial cable are inexpensive to install (low $)
- Fiber-optic cable requires special training and equipment (high $)

Interference
- Twisted-pair most susceptible to interference
- Fiber-optic immune to interference

(Maridav/123RF; Toria/Shutterstock; Nicholas Piccillo/Shutterstock; Ion Popa/123RF; Auremar/123RF; Ermess/123RF)

Wired and Wireless Transmission Media

Objective 12.6 *Describe the types of wired and wireless transmission media used in networks.*

What exactly do we mean by transmission media? Transmission media, whether for wired or wireless communications technology, make up the physical system that data takes to flow between devices on the network. Without transmission media, network devices would be unable to communicate. Most corporate networks contain a combination of wired and wireless media.

Why are wired connections still used in business networks? Wired connections are popular in business networks because they generally provide higher throughput and better security than wireless connections. Desktop computers still provide more computing power for less money than laptops, which makes desktop computers popular choices for business networks. Because desktops aren't often moved around, they're usually connected to a network with a wired connection.

What are the important factors in choosing a cable type? For business networks, the three main cable types that are used are *twisted-pair*, *coaxial*, and *fiber-optic*. Although each cable type is different, the same six factors always need to be considered when choosing a cable type (see Figure 12.11): maximum run length, bandwidth, flexibility (bend radius), cable cost, installation cost, and interference.

What causes interference with data signals? Signals traveling down a cable are subject to two types of interference:

1. *Electromagnetic interference (EMI)*, which is caused when the cable is exposed to strong electromagnetic fields, can distort or degrade signals on the cable. Fluorescent lights and machinery with motors or transformers are the most common sources of EMI emissions.
2. Cable signals also can be disrupted by *radio frequency interference (RFI)*, which is usually caused by broadcast sources (TV and radio signals) located near the network.

Both coaxial cable and twisted-pair cable send electrical impulses down conductive material to transmit data signals, making them more subject to interference. Fiber-optic cable transmits data signals as pulses of light. Because EMI and RFI don't affect light waves, fiber-optic cable is virtually immune to interference.

In the sections that follow, we discuss the characteristics of each of the three major types of cable. We also discuss the use of wireless media as an alternative to cable.

Figure 12.12 Anatomy of (a) Unshielded Twisted-Pair (UTP) cable and (b) Shielded Twisted-Pair (STP) cable.

Twisted-Pair Cable

Why are the wires in twisted-pair cable twisted? Twisted-pair cable consists of pairs of copper wires twisted around each other and covered by a protective sheath (jacket). The twists are important because they cause the magnetic fields that form around the copper wires to intermingle, making them less susceptible to outside interference. The twists also reduce the amount of crosstalk interference (the tendency of signals on one wire to interfere with signals on a wire next to it).

If the twisted-pair cable contains a layer of foil shielding to reduce interference, it's called shielded twisted-pair (STP) cable. Most home networks use unshielded twisted-pair (UTP) cable that doesn't have the foil shielding. UTP cable is more susceptible to interference than STP cable. Figure 12.12 shows illustrations of both types of twisted-pair cable. Because of its lower price, UTP is used in business networks unless significant sources of interference must be overcome, such as in a production environment where machines create magnetic fields.

Coaxial Cable

Is coaxial cable still used in business networks? Although not as popular as it once was, coaxial cable is still used in some manufacturing facilities where machinery creates heavy electrical interference. Coaxial cable (as shown in Figure 12.13) consists of four main components:

1. A core (usually copper) is in the very center and is used for transmitting the signal.
2. A solid layer of nonconductive insulating material (usually a hard, thick plastic) surrounds the core.
3. A layer of braided metal shielding covers the insulation to reduce interference with signals traveling in the core.
4. An external jacket of lightweight plastic covers the internal cable components to protect them from damage.

Figure 12.13 Coaxial cable consists of four main components: a core, an insulated covering, a braided metal shielding, and a plastic jacket.

Fiber-Optic Cable

What does fiber-optic cable look like? As shown in Figure 12.14, fiber-optic cable is composed of:

- A glass (or plastic) fiber (or a bundle of fibers called a *core*) through which the data is transmitted.
- A protective layer of glass or plastic cladding is wrapped around the core to protect it.
- For additional protection, it has an outer jacket (sheath), which is often made of a durable material such as Kevlar (the substance used to make bulletproof vests).

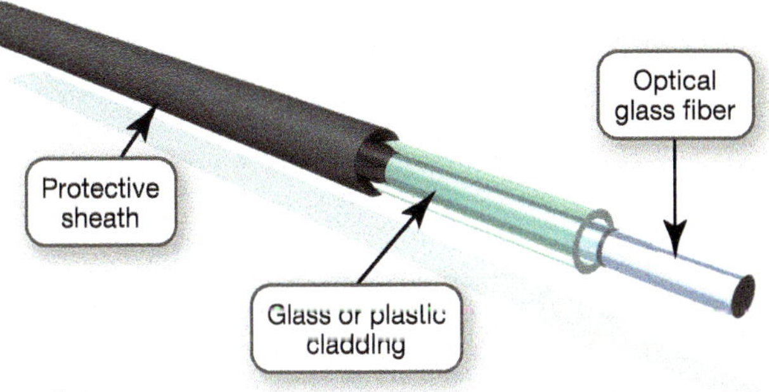

Figure 12.14 Fiber-optic cable is made up of a glass or plastic fiber (or a bundle of fibers), a glass or plastic cladding, and a protective sheath.

Data transmissions can pass through fiber-optic cable in only one direction. Therefore, at least two fibers (or cores) are contained in most fiber-optic cables to enable transmission of data in both directions.

Bits&Bytes Encryption: Not Just for Businesses Any Longer

Businesses have custody of tremendous amounts of data, much of it valuable proprietary information or sensitive customer details (such as credit card numbers). To keep hackers from absconding with useful information during a data breach, many companies turn to encryption software. Encryption software uses a series of algorithms to scramble information, rendering it unreadable until it's decrypted by another algorithm. The Advanced Encryption Standard (AES) is the norm and uses either 128- or 256-bit encryption key lengths, which are very hard to hack. Large companies use proprietary encryption software such as IBM Guardium or Trend Micro Endpoint Encryption, which can be expensive. But what if you feel you need a basic level of data security at home (see Figure 12.15)? Fortunately, some good free options are available.

Apple FileVault is built into the macOS operating system. You can enable FileVault and it will automatically encrypt all the data on your hard drive. Microsoft includes its Bitlocker encryption product but only in Windows 10 Pro or Windows 10 Enterprise. You can upgrade your Windows 10 home edition to Pro if you want to take advantage of Bitlocker. Both FileVault and Bitlocker enable you to use either 128- or 256-bit encryption. With either of these products enabled, no one without credentials could access the data on your device if it was lost or stolen. So why not protect the next great American novel you are writing ... the peace of mind is free!

Figure 12.15 Encryption isn't just for the workplace. You can encrypt the data on your personal devices too! *(LuckyStep/Shutterstock)*

Wireless Media Options

What wireless media options are there? Most business networks use the same Ethernet standards as home networks. Therefore, the wireless options for business networks are very similar to those available for home networks. *Wireless access points* are installed to provide coverage wherever employees will be working with portable devices, such as in conference rooms.

Comparing Transmission Media

Which medium is best for business networks? Network engineers specialize in the design and deployment of networks and are responsible for selecting the appropriate network topologies and media types. Their decision of which transmission medium a network will use is based on the topology selected, the length of the cable runs needed, the amount of interference present, and the need for wireless connectivity.

As noted earlier, most large networks use a mix of media types. For example, fiber-optic cable may be appropriate for the portion of a network that traverses the factory floor, where interference from magnetic fields is significant. However, UTP cable may work fine in a general office area. Wireless media may be required in areas where employees are likely to connect their portable computing devices or where it's impractical or expensive to run cable.

Network Adapters and Navigation Devices

As mentioned earlier, data flows through the network in packets. Data packets are like postal letters. They don't get to their destinations without some help. In this section, we explore the various conventions and devices that help speed data packets on their way through the network.

Network Adapters

Objective 12.7 *Describe how network adapters help data move around a network.*

What is a network adapter? As noted in Chapter 7, client and server computers and peripherals need an interface to connect with and communicate on the network. Network adapters are

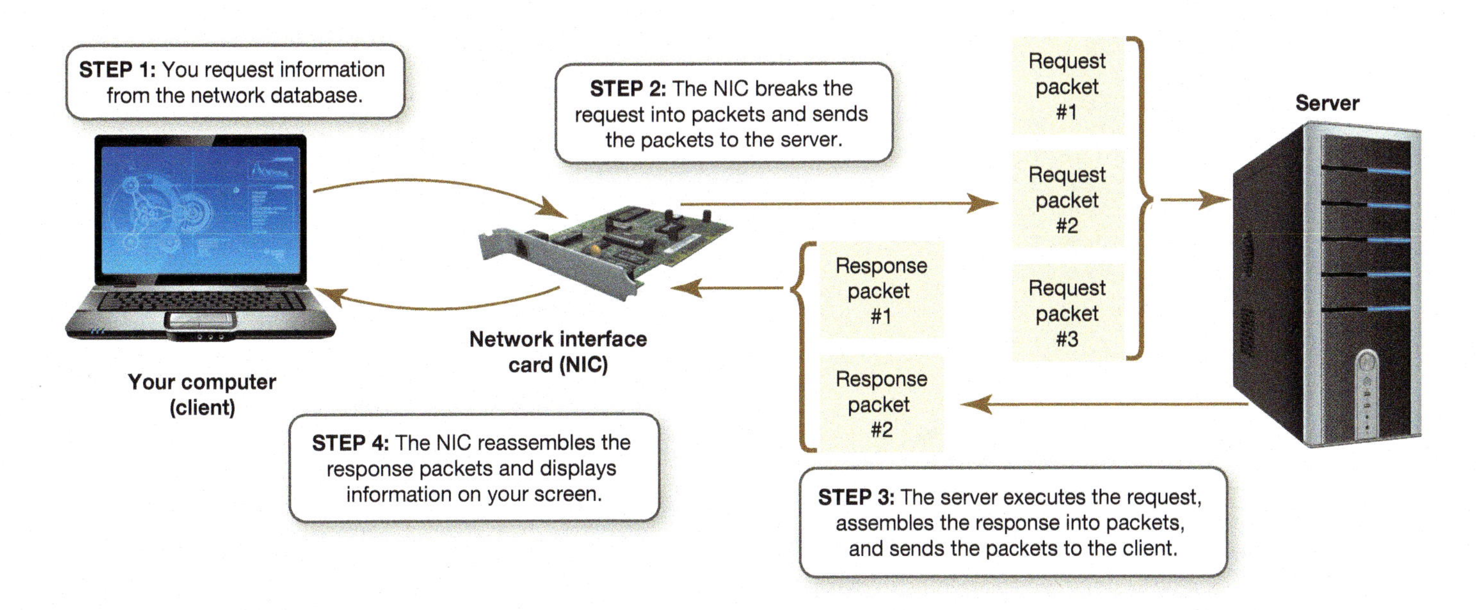

Figure 12.16 A network interface card is responsible for breaking down data into packets, preparing packets for transmission, receiving incoming data packets, and reconstructing them. *(Alan Evans; Juffin/DigitalVision Vectors/Getty Images)*

devices that perform specific tasks to enable nodes to communicate on a network. Network adapters are installed inside computers and peripherals. These adapters are referred to as *network interface cards (NICs)*. Most NICs are now integrated into the motherboard of the computing device.

What do network adapters do? Network adapters perform three critical functions:

1. *They generate high-powered signals to enable network transmissions.* Digital signals generated inside the computer are fairly low-powered and would not travel well on cable or wireless network media without network adapters. Network adapters convert the signals from inside the computer to higher-powered signals that have no trouble traversing the network media.
2. *They're responsible for breaking the data into packets and for transmitting and receiving data.* They're also responsible for receiving incoming data packets and, in accordance with networking protocols, reconstructing them, as shown in Figure 12.16.
3. *They act as gatekeepers for information flowing to and from the client computer.* Much like a security guard in a gated community, a network adapter is responsible for permitting or denying access to the client computer (the community) and controlling the flow of data (visitors).

You should note that the number of response packets won't always be the same as request packets. The number of packets depends on the volume of the data being sent. A simple response may have less data than a complex one.

Are there different types of network adapters? Although there are different types of network adapters, almost without exception Ethernet is the standard communications protocol used on most client/server networks. Therefore, the adapter cards that ship with computers today are Ethernet compliant.

Do wireless networks require network adapters? A computing device that connects to a network using wireless access needs to have a special network adapter card, called a wireless network interface card (wireless NIC), installed in it. Laptop computers and other portable computing devices contain wireless NICs.

To allow wireless connections, a network also must be fitted with devices called wireless access points. A wireless access point (WAP) gives wireless devices a sending and receiving connection point to the network.

Figure 12.17 shows an example of a typical small business network with a wireless access point. The access point is connected to the wired network through a conventional cable. When a laptop or other device with a wireless NIC is powered on or near a wireless access point, it establishes a

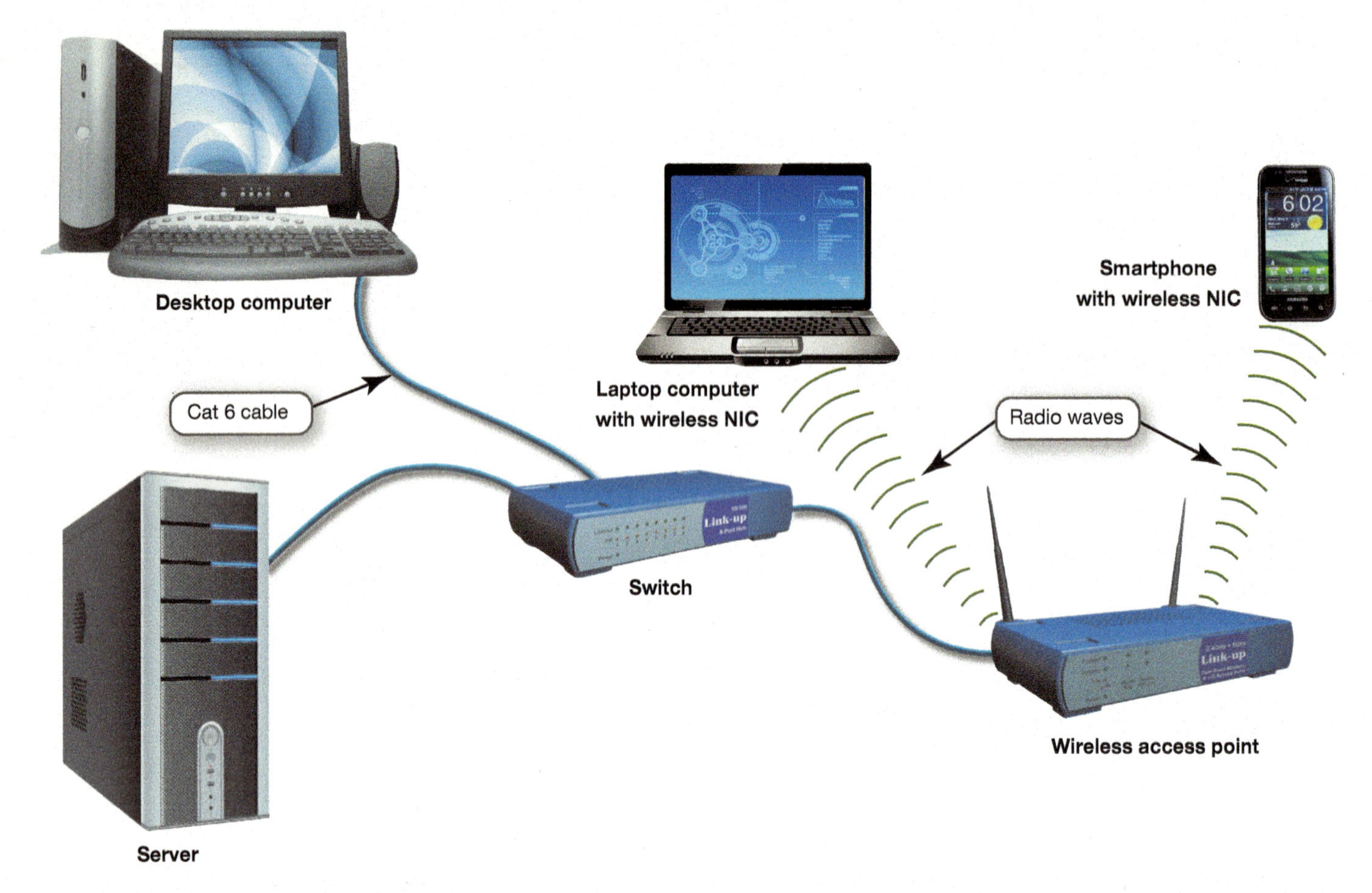

Figure 12.17 This small business network has an added wireless access point. *(Alan Evans; Nico_blue/iStock/Getty Images ;Juffin/DigitalVision Vectors/ Getty Images)*

connection with the access point, using radio waves. Many devices can communicate with the network through a single wireless access point.

Do network adapters require software? Special communications software called a device driver is installed on all client computers in the client/server network. Device drivers enable the network adapter to communicate with the server's OS and with the OS of the computer in which the adapter is installed.

MAC Addresses

Objective 12.8 *Define MAC addresses, and explain how they are used to move data around a network.*

How do network adapters know where to send data packets? Each network adapter has a physical address, like a serial number on an appliance. This address is called a media access control (MAC) address, and it's made up of six two-position characters, such as 01:40:87:44:79:A5. (Don't confuse this MAC with the Apple computers of the same name.) The first three sets of characters (in this case, 01:40:87) specify the manufacturer of the network adapter, and the second set of characters (in this case, 44:79:A5) makes up a unique address. Because all MAC addresses must be unique, an IEEE (Institute of Electrical and Electronics Engineers) committee is responsible for allocating blocks of numbers to network adapter manufacturers.

Are MAC addresses the same as IP addresses? MAC addresses and IP (Internet Protocol) addresses are not the same thing. A MAC address is used for identification purposes *internally* on a network. An IP address is the address *external* entities use to communicate with your network. Think of it this way: The postal carrier delivers a package (data packet) to your dorm building based on its street address (IP address). The dorm's mail clerk delivers the package to your room because it has your name on it (MAC address) and not that of your neighbor. Both pieces of information are necessary to ensure that the package (or data) reaches its destination.

How are data packets packaged for transmission? Data packets aren't necessarily sent alone. Sometimes groups of data packets are sent together in a package called a frame. A frame is a container that can hold multiple data packets. This is like placing several letters going to the same postal address in a big envelope. While the data packets are being assembled into frames, the network operating system (NOS) software assigns the appropriate MAC address to the frame. The NOS keeps track of all devices and their addresses on the network. Much like an envelope that's entrusted to the postal service, the frame is delivered to the MAC address that the NOS assigned to the frame.

What delivers the frames to the correct device on the network? In a small bus network, frames just bounce along the transmission medium until the correct client computer notices that the frame is addressed to it and pulls the signal off the medium. However, this is inefficient in a larger network. Therefore, many types of devices have been developed to deliver data to its destination efficiently. These devices are designed to route signals and exchange data with other networks.

Are MAC addresses useful for anything besides identifying a particular network device? On networks with wireless capabilities, MAC addresses can be used to enhance network security. Because each MAC address is unique, you can input a list of authorized MAC addresses in the router. If someone who is using an unauthorized network adapter attempts to connect to the network, he or she will be unable to make a connection.

Switches, Bridges, and Routers

Objective 12.9 *List the various network navigation devices, and explain how they help route data through networks.*

Which devices are used to route signals through a single network? Switches are used to send data on a specific route through the network. A switch makes decisions, based on the MAC address of the data, as to where the data is to be sent and rebroadcasts it to the appropriate network node. This improves network efficiency by helping ensure that each node receives only the data intended for it. Figure 12.18 shows a switch being used to rebroadcast a message.

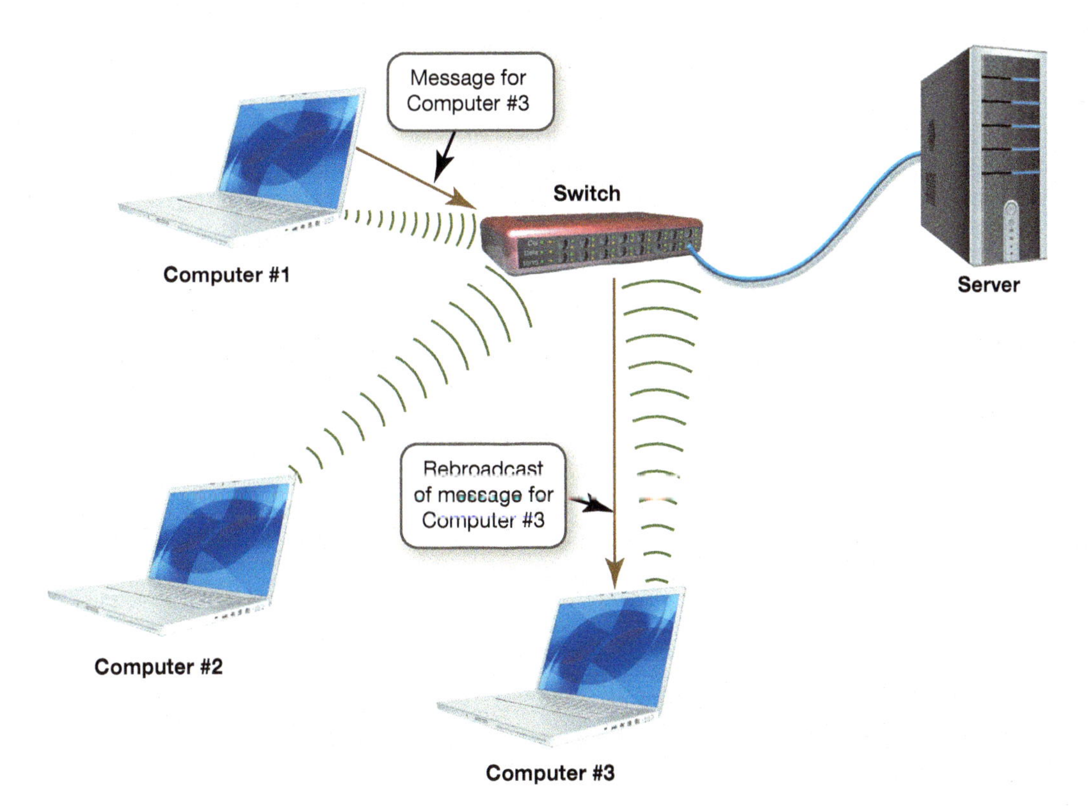

Figure 12.18 Switches rebroadcast messages—but only to the devices to which the messages are addressed.
(LdF/iStock/Getty Images; Juffin/DigitalVision Vectors/Getty Images)

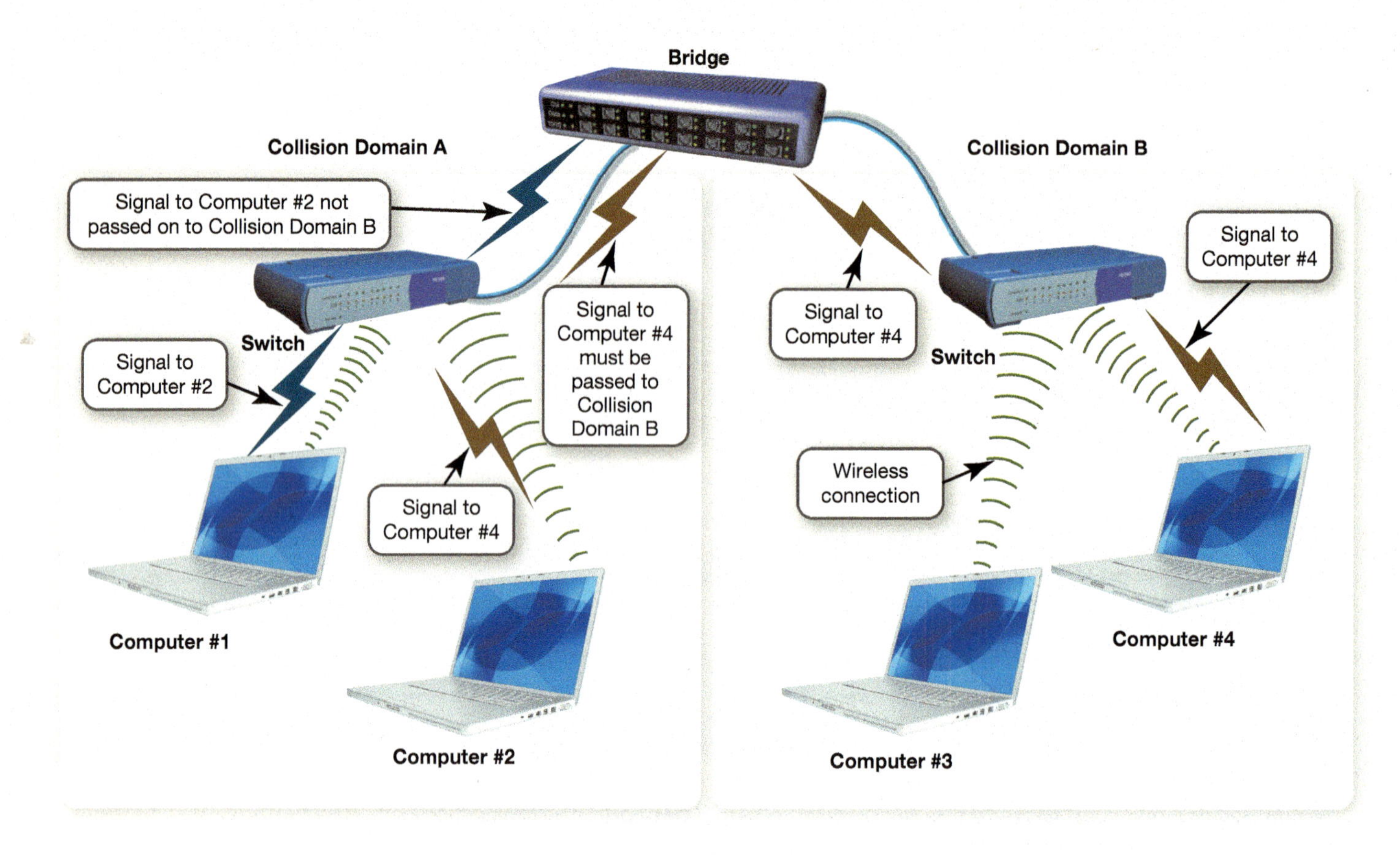

Figure 12.19 Bridges are devices used to send data between different network collision domains. *(LdF/iStock/Getty Images)*

Do all Ethernet networks need a switch? Switches are needed on Ethernet networks whether they are installed in a home or a business. Routers sold for home use have switches built in to them.

Are switches sufficient for moving data efficiently across all sizes of networks? When a corporate network grows, performance can decline because many devices compete for transmission time on the network media. To solve this problem, a network can be broken into multiple segments known as *collision domains*. A bridge is a device that's used to send data between different collision domains, depending on where the recipient device is located, as indicated in Figure 12.19. Signals received by the bridge from Collision Domain A are forwarded to Collision Domain B only if the destination computer is in that domain. (Note that most home networks contain only one segment and therefore do not require bridges.)

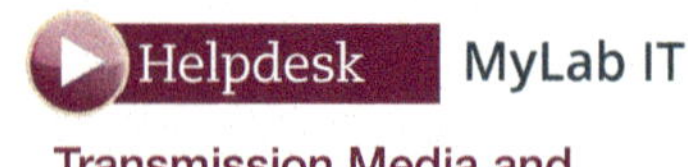

Transmission Media and Network Adapters

In this Helpdesk, you'll play the role of a helpdesk staffer, fielding questions about transmission media and network adapters.

What device does a network use to move data to another network? Whereas switches and bridges perform their functions within a single network, as you learned in Chapter 7, a router is designed to send information between two networks. To accomplish this, the router must look at higher-level network addresses (such as IP addresses), not MAC addresses. When the router notices data with an address that doesn't belong to a device on the network from which it originated, it sends the data to another network to which it is attached or out to the Internet.

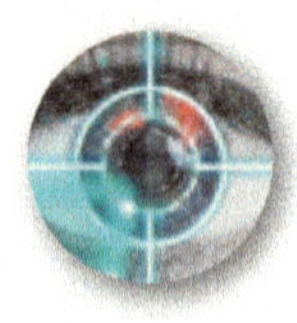

Network Operating Systems and Network Security

Networks need specialized software so that the devices attached to the networks can communicate with each other. Business networks need even more security than home networks to protect them from hackers. In this section, we explore the software that allows networks to function and discuss security measures that help keep networks safe from prying eyes.

Network Operating Systems

Objective 12.10 *Explain why network operating systems are necessary for networks to function.*

What is a network operating system? Merely using media to connect computers and peripherals does not create a client/server network. Special software known as a network operating system (NOS) needs to be installed on each client computer and server that's connected to the network to provide the services necessary for them to communicate. The NOS provides a set of common rules (a *protocol*) that controls communication among devices on the network. Modern operating systems, such as Windows and macOS, include NOS client software as part of the basic installation. However, large networks, such as those found in businesses, require more sophisticated NOS server software.

Why is NOS server software needed on large networks? OS software is designed to facilitate communication between the software and hardware components of your computer. NOS software is specifically designed to provide server services, network communications, management of network peripherals, and storage. Popular server software includes Windows Server, Red Hat Enterprise Linux, and SUSE Linux Enterprise Server. NOS software is installed on servers.

To provide network communications, the client computers must run a small part of the NOS in addition to their OS. Windows and macOS contain network communication capabilities; therefore, no other NOS software is usually installed on these client computers.

Do P2P networks need special NOS software? The software that P2P networks require is built into the Windows, Linux, and macOS operating systems. So, if you have a simple P2P network, there is no need to purchase specialized NOS software.

How does the NOS control network communications? Each NOS has its own proprietary communications language, file-management structure, and device-management structure. The NOS also sets and controls the protocols for all devices wishing to communicate on the network. Proprietary protocols from one vendor's NOS won't work with another vendor's NOS.

However, because the Internet uses an open protocol (called TCP/IP) for communications, almost all corporate networks use TCP/IP as their standard networking protocol regardless of the manufacturer of their NOS. All modern NOSs support TCP/IP.

Are NOS protocols standardized? The Institute of Electrical and Electronics Engineers has taken the lead in establishing worldwide networking protocols, including a standard of communications called the Open Systems Interconnection (OSI) reference model. The OSI model, which has been adopted throughout the computing world, provides the protocol guidelines for all modern networks. All modern NOS protocols are designed to interact in accordance with the standards set out in the OSI model. By following standardized protocols set forth by the OSI model, NOS software can communicate happily with the computers and peripherals attached to the network as well as with other networks.

The OSI model divides communications tasks into seven processes called *layers*. Each layer of an OSI network has a specific function and knows how to communicate with the layers above and below it. Figure 12.20 shows the layers of the OSI model and their functions.

This layering approach makes communications more efficient because specialized pieces of the NOS perform specific tasks. The layering approach is akin to assembly-line manufacturing. Producing thousands of cars per day would be difficult if one person had to build a car on his or her own. However, by splitting up the work of assembling a car into specialized tasks and assigning the tasks to people who perform them well, greater efficiency is achieved. By handling specialized tasks and communicating only with the layers above and below them, OSI layers make communications more efficient.

Can a network use two different NOSs? Many large corporate networks use several different NOSs at the same time. This is because different NOSs provide different features, some of which are more useful than others in certain situations. For instance, although the employees of a corporation may be using a Windows environment for their desktops and e-mail, the file servers and web servers may be running a Linux-based NOS.

Figure 12.20 Layers of the OSI Model and Their Functions

Application Layer

- Handles all interfaces between the application software and the network
- Translates user information into a format the presentation layer can understand

Presentation Layer

- Reformats data so that the session layer can understand it
- Compresses and encrypts data

Session Layer

- Sets up a virtual (not physical) connection between the sending and receiving devices
- Manages communications sessions

Transport Layer

- Creates packets and handles packet acknowledgment

Network Layer

- Determines where to send the packets on the network

Data Link Layer

- Assembles the data into frames, addresses them, and sends them to the physical layer for delivery

Physical Layer

- Transmits (delivers) data on the network so it can reach its intended address

Bits&Bytes Extending Smart Homes: Smart Yards Coming Soon!

Smart home devices are selling at a brisk pace. Currently, however, smart home devices rely on communications networks that tend to have limited range (like Wi-Fi and Bluetooth). Considering how much time residents spend in their yard (barbecues, parties, etc.), connectivity that reaches beyond the walls of your house would be optimal. Shouldn't you be able to move your Amazon Echo outside to your deck to enjoy music without losing connectivity with your Wi-Fi network?

Amazon has envisioned a solution. Called Sidewalk, it uses the 900 MHz spectrum, which has greater range and uses radio devices for connectivity that use less power than competing systems. Amazon Sidewalk devices will form a mesh network with Sidewalk devices installed in other homes to extend range even farther. Imagine your pet wearing a Sidewalk-enabled tracker when it strays beyond your yard. You could effectively track it through the neighborhood and find it quickly. So let's start thinking of ways to extend smart home functionality beyond the walls of the house and make yards smart too!

Dig Deeper Intent-Based Networking (IBN): Networks That Think for Themselves

Networks currently require a great deal of manual support and intervention from network administrators to set them up and keep them functioning. But with advances in fields of artificial intelligence, companies like Cisco are working on the next generation of networking software. Known as *intent-based networking (IBN)* the goal is to have networks become more autonomous in their functionality.

Intent-based networking is based on four basic functions:

- *Translation and validation:* The network administrators define high-level business policies, based on input from end users, about what the network needs to do. The IBN software then translates the policies into the network configuration necessary to achieve the policy goals. At this point, the software also confirms that the network is capable of being configured in the correct way or points out deficiencies that need to be corrected before moving forward.
- *Automated implementation:* The software takes charge of the network and configures it appropriately across the network hardware. This is achieved with no (or minimal) user input. The software has the intelligence and the authority to make appropriate configuration changes.
- *Network state monitoring:* Once the network is configured, the IBN software monitors the functioning of the network to ensure that it's operating properly.
- *Assurance and remediation:* The software interprets data collected during the monitoring phase and analyzes that data against the original business policies and goals to ensure that objectives are being achieved. The software also has the authority to take corrective action (such as redirecting the flow of network traffic) if the goals are not being met.

This doesn't mean network administrators will be replaced by software. IBN has the potential to relieve administrators of tedious setup and monitoring tasks so that they can spend more time interacting with users and ensuring that their needs are met.

Client/Server Network Security

Objective 12.11 *List major security threats to networks, and explain how network administrators mitigate these threats.*

What sources of security threats do all network administrators need to watch for? A major advantage that client/server networks have over P2P networks is that they offer a higher level of security. With client/server networks, users can be required to enter a user ID and a password to gain access to the network. The security can be centrally administered by network administrators, freeing individual users of the responsibility of maintaining their own data security, as they must do on a P2P network. Still, client/server networks are vulnerable to threats. These threats can be classified into three main groups:

1. *Human errors and mistakes.* Everyone makes mistakes. For example, a member of the computer support staff could mistakenly install an old database on top of the current one. Even physical accidents fall into this category; for example, someone could lose control of a car and drive it through the wall of the main data center.
2. *Malicious human activity.* Malicious actions can be perpetrated by current employees, former employees, or third parties. For example, a disgruntled employee could introduce a virus to the network or a hacker could break into the student database server to steal credit card records.
3. *Natural events and disasters.* Some events—such as floods and fires—are beyond human control. All can lead to the inadvertent destruction of data.

Because networks are vulnerable to such security threats, one of the network administrator's key functions is to keep network data secure. Let's explore the challenges network administrators face in keeping a client/server network secure.

Authentication

How do network administrators ensure that only authorized users access the network? Authentication is the process whereby users prove they have authorization to use a computer network. For example, correctly entering the user ID and password on your college network proves to the network that you have authorized access. The access is authorized because the ID was generated by a network administrator when you became an authorized user of the network.

However, authentication can also be achieved using biometric devices (discussed later in this chapter) and possessed objects. A possessed object is any object that users carry to identify themselves and that grants them access to a computer system or computer facility. Examples include identification badges, magnetic key cards, and smart keys (similar to flash drives).

Can hackers use my account to log on to the network? If a hacker knows your user ID and password, he or she can log on and impersonate you. Impersonation can also happen if you fail to log out of a terminal on a network and someone comes along and uses your account. Sometimes network user IDs are easy to figure out because they have a certain pattern, such as your last name and the first initial of your first name.

If a hacker can deduce your user ID, he or she might use a software program that tries millions of combinations of letters and numbers as your password in an attempt to access your account. Attempting to access an account by repeatedly trying different passwords is known as a brute-force attack. To prevent these attacks from succeeding, network administrators often configure accounts so that they'll disable themselves after a set number of login attempts using invalid passwords have been made.

Access Privileges

How can I gain access to everything on a network? The simple answer is that you can't! When your account is set up on a network, certain access privileges are granted to indicate which systems you're allowed to use. For example, on your college network, your access privileges probably include the ability to access the Internet. However, you weren't granted access to the grade-entering system because this would enable you to change your grades.

A Day in the Life of a Network Technician

In this Sound Byte, you learn firsthand about the exciting, fast-paced job of a computer technician. Interviews with network technicians and tours of networking facilities will provide you with a deeper appreciation for the complexities of the job.

How does restricting access privileges protect a network? Because network-access accounts are centrally administered on the authentication server, it's easy for the network administrator to set up accounts for new users and grant them access only to the systems and software they need. The centralized nature of the network-access creation and the ability to restrict access to certain areas of the network make a client/server network more secure than a P2P network.

Aside from improper access, how else do data theft and destruction occur? A major problem that portable devices such as flash drives pose is theft of data or intellectual property. Because these devices are so easy to conceal and have such large memory capacity, it's easy for a disgruntled employee to walk out the front door with stacks of valuable documents tucked in his or her pocket. Industrial espionage has never been easier—and no spy cameras are needed! Flash drives can also introduce viruses or other malicious programs to a network, either intentionally or unintentionally.

How should network administrators protect their networks from portable storage devices? To head off problems with portable storage devices, the following actions should be taken:

- Educate employees about the dangers posed by portable media devices.
- Create policies regulating the use of portable media devices in the workplace.
- Install security measures such as firewalls or antivirus software to protect all computers in the company. The firewalls should be able to prevent malicious programs from running, even if they're introduced to a computer by a flash drive.
- Limit and monitor the use of portable media devices. Microsoft networking software allows network administrators to shut off access to the USB ports on computers; this prevents employees from using flash drives and other USB devices for illegitimate purposes.
- Inform employees that their use of portable media devices is being monitored. This alone will often be enough to scare employees away from connecting untrusted devices to the network.

Other software products, such as DeviceLock DLP and Safend Data Protection Suite, can be deployed on a network to provide options such as detailed security policies. Such products can also monitor USB device connections and track which users have connected devices to the network (including devices other than flash drives).

Physical Protection Measures

What physical measures are used to protect a network? Restricting physical access to servers and other sensitive equipment is critical to protecting a network. Where are the servers that power your college network? They're most likely behind locked doors to which only authorized personnel have access. Likewise, the network's switches are securely tucked away in ceilings, walls, or closets, safe from anyone who might tamper with them to sabotage the network or breach its security.

As shown in Figure 12.21, many devices can be used to control access:

- An access card reader is a relatively cheap device that reads information from a magnetic strip on the back of a credit card–like access card, such as your student ID card. The card reader, which can control the lock on a door, is programmed to admit only authorized personnel to the area. Card readers are easily programmed by adding authorized ID card numbers.
- A biometric authentication device uses a unique characteristic of human biology to identify authorized users. Some devices read fingerprints or palm prints when you place your hand on a scanning pad. Other devices shine a laser beam into your eye and read the unique patterns of your retina. Facial-recognition systems store unique characteristics of an individual's face for later comparison and identification. When an authorized individual uses a device for the first time, his or her fingerprints, retinal patterns, or face patterns are scanned and stored in a database.

Financial institutions and retail stores are considering using such devices to attempt to eliminate the growing fraud problems of identity theft and counterfeiting of credit and debit cards. If fingerprint authorization were required at the supermarket to make a purchase, a thief who stole your wallet and attempted to use your credit card would be unsuccessful.

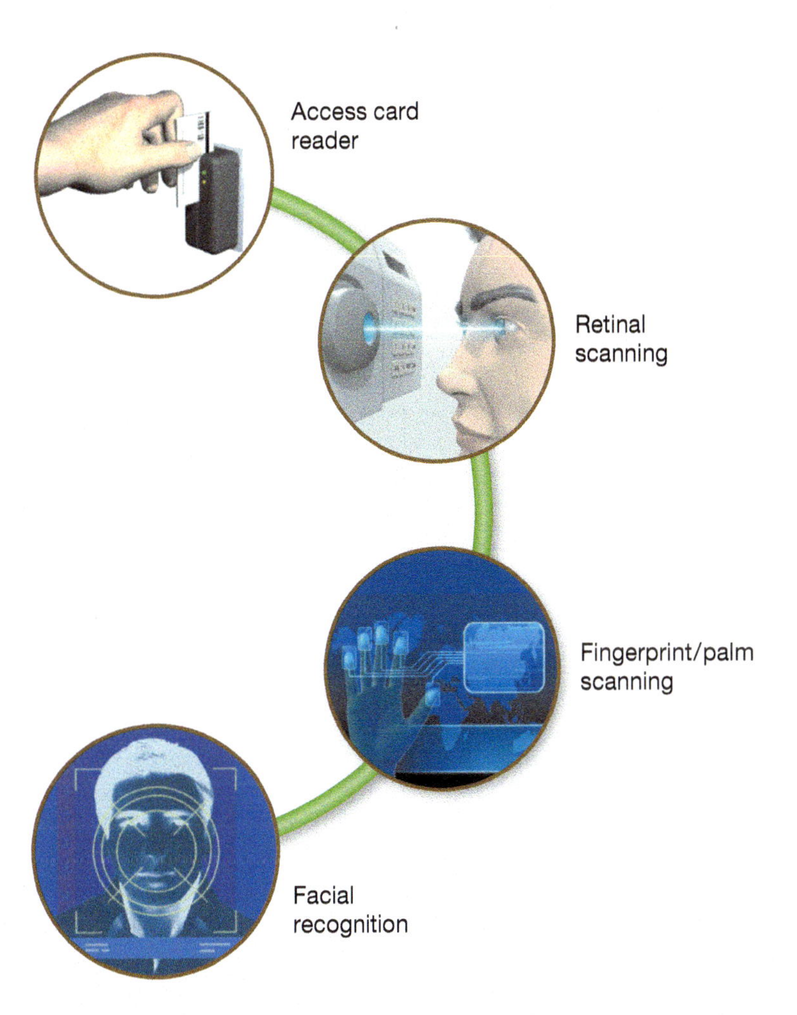

Figure 12.21 Access card readers provide a relatively inexpensive solution for areas needing low-level security. Biometric authentication devices, such as fingerprint/palm, retinal, and facial-recognition scanners, are now used to log on to personal computing devices. *(Pariwat Intrawut/123RF)*

What are the problems with biometric devices? The biometric devices currently on the market don't always function as intended. Facial-recognition and retinal-scanning systems can sometimes be fooled by pictures or videos of an authorized user. Researchers have fooled biometric fingerprint readers by using fingers made from modeling clay, using the fingers of cadavers, or having unauthorized persons breathe on the sensor, which makes the previous user's fingerprint visible.

Next-generation fingerprint readers will use specially designed algorithms that will detect moisture patterns on a person's fingers. Another approach may involve readers that detect an electrical current when a finger touches the reader, which is possible because the human body conducts electrical current. Future retinal readers may check whether a person blinks or if his or her pupils contract when a bright light shines on them.

What's the best security method for sensitive systems and locations? For high-level security, multi-factor authentication should be used. As you learned in Chapter 9, there are three independent authentication factors (Figure 12.22):

- Knowledge: something the user knows (password, PIN)
- Possession: something the user has (ATM card, mobile phone)
- Inherence: something only the user is (biometric characteristics, such as fingerprints or iris pattern)

Multi-factor authentication requires two of these three factors to be demonstrated before authorization is granted. Perhaps at the entrance to the server room, you would need to have your retina scanned (inherence factor) and key in a password (knowledge factor).

Figure 12.22 Multi-factor authentication depends on proving two of the three factors shown here. *(PAISAN HOMHUAN/123RF)*

Firewalls

Are Internet connections on client/server networks vulnerable to hackers? Just like a home network, any company's network that's connected to the Internet can attract hackers. For this reason, a well-defended business network, just like a well-defended home network, includes a firewall. Firewalls can be composed of software or hardware, and many sophisticated firewalls include both. Routers are often equipped to act as hardware firewalls.

Does a firewall on a client/server network work the same way as a personal firewall installed on a home network? Although a firewall on a business network may contain a few extra security options, making it even harder to breach than a personal firewall, the firewalls for business networks work on the same basic principles as a personal firewall. At a minimum, most firewalls work as packet screeners. Packet screening involves having an *external screening router* examining incoming data packets to ensure that they originated from or are authorized by valid users on the internal network. The firewall discards unauthorized or suspect packets before they reach the network.

Packet screening can also be configured for outgoing data to ensure that requests for information to the Internet are from legitimate users. This is done by an *internal screening router* and helps detect Trojan horse programs that may have been installed by hackers. Trojan horses often try to disguise where they're sending data from by using bogus IP addresses on the packets the programs send instead of using an authorized IP address belonging to the network.

With packet screening in place, packets going into and out of the network are checked to ensure that they're either from or addressed to a legitimate IP address on the network. If the addresses aren't valid addresses on the network, the firewall (or the screening routers) discards them.

What other security measures does the firewall on a client/server network use? To increase security even further, most large networks add a bastion host, which is a heavily secured server located on a special perimeter network between the company's secure internal network and the firewall.

To external computers, the bastion host gives the appearance of being the internal network server. Hackers can waste a lot of time and energy attacking the bastion host. However, even if a hacker breaches the bastion host server, the internal network isn't vulnerable because the bastion host isn't on the internal network. Moreover, during the time the hackers spend trying to penetrate the bastion host, network administrators can detect and thwart their attacks.

How do bastion hosts help protect systems from hackers? Bastion hosts are a type of honey pot. A honey pot is a computer system that's set up to attract unauthorized users by appearing to be a key part of a network or a system that contains something of great value.

Ethics in IT

How Should Companies Handle Data Breaches?

Data breaches occur all too frequently, and it seems we're always seeing reports in the media about yet another hacker obtaining supposedly private information. Many companies have sensitive data about you. But what is their ethical responsibility to protect your data? And what should a company do for its customers when a data breach occurs?

Data Confidentiality

The objective for any company that possesses sensitive data should be information assurance, which means ensuring that information systems are adequately secured against tampering. The five key attributes of secure information systems are:

1. *Availability:* The extent to which a data processing system can receive and process data. A high degree of availability is usually desirable.
2. *Integrity:* A quality that an information system has if the processing of information is logical and accurate and the data is protected against unauthorized modifications or destruction.
3. *Authentication:* Security measures designed to protect an information system against acceptance of a fraudulent transmission of data by establishing the validity of a data transmission or message or the identity of the sender.
4. *Confidentiality:* The assurance that information isn't disclosed to unauthorized persons, processes, or devices.
5. *Nonrepudiation:* A capability of security systems that guarantees that a message or data can be proven to have originated from a specific person and to have been processed by the recipient. The sender of the data receives a receipt for the data, and the receiver of the data gets proof of the sender's identity. The objective of nonrepudiation is to prevent either party from later denying having handled the data.

From your perspective, you are probably the most concerned about authentication and confidentiality. Companies that have your data shouldn't share it with unauthorized parties and should keep unauthorized people out of their databases. Unfortunately, you have very little control over authentication safeguards. That's the responsibility of the company that holds your data.

Authentication Failures

The data breaches that occur usually represent authentication failures—unauthorized people gaining access to data. Clearly, when this happens, a corporation has failed in its responsibility to protect its customers' data. No system is ever 100% safe against breaches, but most systems are adequately protected against all but the most determined hackers. But what ethical responsibilities does a company have to its customers after a data breach?

The first thing a company should do is admit to the problem and inform its customers. The Federal Trade Commission (FTC) recommends the following course of action for companies who experience data breaches:

1. Notify the appropriate law enforcement agencies.
2. Notify other affected businesses, such as banks and credit card companies. These companies can then monitor the affected accounts for fraudulent activity.
3. Notify the individuals affected. The company should explain what happened, what information was compromised, who to contact in the organization, and what customers should do to protect themselves (which will vary based on the information compromised).

How can you be aware of a data breach if the company that suffered it never tells you that your data was exposed? You should monitor websites such as the Privacy Rights Clearinghouse (*privacyrights.org/data-breach*) and Identity Theft Resource Center (https://www.idtheftcenter.org/data-breaches/), which track data breaches. This might be the fastest way to find out whether your information has been compromised so you can act. The FTC website on identity theft (*consumer.ftc.gov/topics/identity-theft*) provides good guidelines to follow when your data has been exposed.

Granting customers free identity-theft protection is one of the more common forms of compensation offered to placate customers after a data breach, but not every company does it. Some companies merely stop at notifying customers and offer nothing in terms of compensation for the inconvenience.

Although companies may behave ethically after a data breach, your best protection is still your own vigilance!

Data breaches are bound to happen...but what happens afterward?
(Sergey Nivens/123RF)

Bits&Bytes Logging on? Try Finger Vein Recognition

IT department personnel hate passwords because it is easy to forget, share, or guess them. Many hours are wasted yearly by helpdesk staffers on password resets. Biometrics has been making inroads on portable devices such as phones and tablets by allowing logging on with fingerprints or facial recognition. But another technology is poised to make an appearance: finger vein recognition.

Much like fingerprints, the vein patterns in your fingers don't change over time. Because veins are located inside your body, they are almost impossible to copy or fake. And authentication requires the finger be attached to a live body. Users would hold their hand up to the device's camera and if the finger vein patterns match, access to the device is granted. Already in use in Japan, this technology should be rolling out to laptops in the United States soon.

Bastion hosts are often configured as proxy servers. A proxy server acts as a go-between, connecting computers on the internal network with those on the external network (the Internet). All requests from the internal network for Internet services are directed through the proxy server. Similarly, all incoming requests from the Internet must pass through the proxy server. It's much easier for network administrators to maintain adequate security on one server than it is to ensure that security is maintained on hundreds or thousands of computers in a college network, for example.

Figure 12.23 shows a network secured by a firewall, a bastion host and proxy server, and a screening router. Now that you know a bit more about business networks, you should be able to navigate the network at your college or your place of employment comfortably and understand why certain security measures have been taken to protect network data.

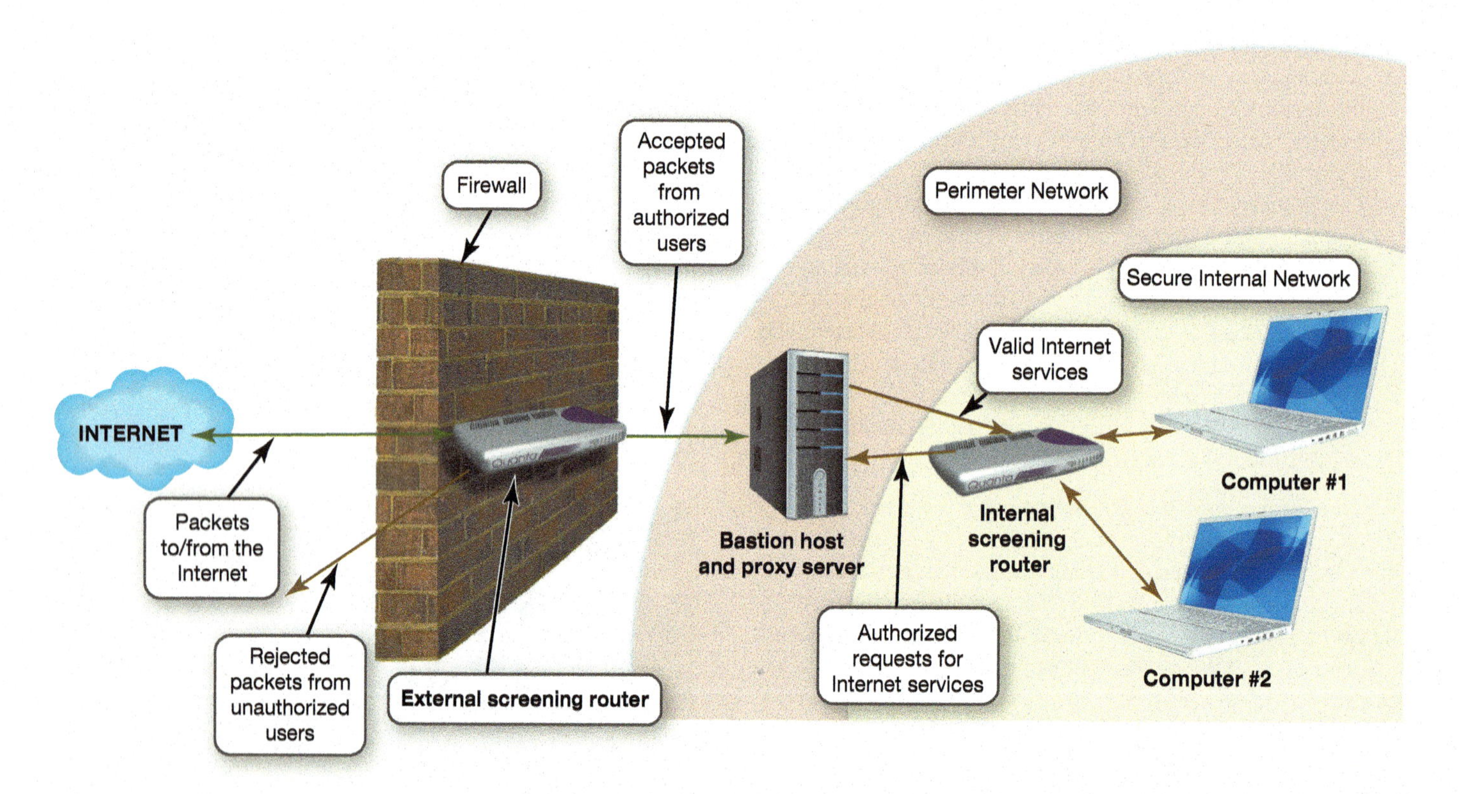

Figure 12.23 Business firewalls provide sophisticated protection from hackers.

Before moving on to the Chapter Review:

1. Watch Chapter Overview Video 12.2.
2. Then take the Check Your Understanding quiz.

Check Your Understanding // Review & Practice

For a quick review to see what you've learned so far, answer the following questions.

multiple choice

1. Fiber-optic cable most likely would be used in a business network when
   a. saving money is more important than speed.
   b. the installation budget is very limited.
   c. very short cable runs are required.
   d. electrical or magnetic interference is present.
2. Network adapters use ______ addresses to route data on an internal network.
   a. router
   b. authentication
   c. MAC
   d. data collision
3. On client/server networks, bridges
   a. transfer data between two networks.
   b. move data efficiently from node to node on the internal network.
   c. route data between two collision domains on a single network.
   d. are necessary only in networks using the ring topology.
4. NOS software is needed
   a. only if a communications server is deployed on a client/server network.
   b. only on the servers in a client/server network.
   c. on all computers in a client/server network.
   d. only when configuring a network in a star topology.
5. Which of the following is used to increase security on a client/server network?
   a. Bridge
   b. Bastion host
   c. Switch
   d. Terminator

chew on this

3Dalia/Shutterstock

Software tools for monitoring computer usage are available on the Internet, often for free. In most jurisdictions, it's legal for employers to install monitoring software on computers they provide to employees. However, it is illegal for the employees to do the same thing for purposes of monitoring computer usage of coworkers or bosses. Do you think this double standard is fair? What circumstances do you think would justify an employee monitoring other employees' computer usage?

MyLab IT Go to **MyLab IT** to take an autograded version of the *Check Your Understanding* review and find all media resources for the chapter.

For the IT Simulation for this chapter, see MyLab IT.

12 Chapter Review

Summary

Part 1
Client/Server Networks and Topologies

Learning Outcome 12.1 **You will be able to describe common types of client/server networks, servers found on them, and network topologies used to construct them.**

Client/Server Network Basics

Objective 12.1 *List the advantages for businesses of installing a network.*

- Networks enable expensive resources, such as printers, to be shared.
- Networks facilitate knowledge sharing, because multiple people can access information at the same time.
- Software can be shared from network servers, thereby reducing the costs of installation on each user's computer.
- Information sharing is more effective when employees are connected by networks.

Objective 12.2 *Explain the differences between a client/server network and a peer-to-peer network.*

- In a peer-to-peer network, each node connected to the network can communicate directly with every other node on the network. In a client/server network, a separate device (the server) exercises control over the network.
- Data flows more efficiently in client/server networks than in peer-to-peer networks.
- Client/server networks have increased scalability, so users can be added to the network easily.

Objective 12.3 *Describe the common types of client/server networks as well as other networks businesses use.*

- Local area networks (LANs) are small groups of computers (as few as two) and peripherals linked over a small geographical area.
- Wide area networks (WANs) comprise large numbers of users (or of separate LANs) that are miles apart and linked.
- An intranet is a private network set up by a business or an organization that's used exclusively by a select group of employees, customers, suppliers, volunteers, or supporters.
- Virtual private networks (VPNs) use the public Internet communications infrastructure to build a secure, private network among various locations.

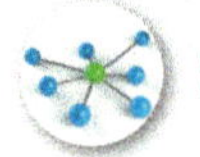

Servers and Network Topologies

Objective 12.4 *List the common types of servers found on client/server networks.*

- Dedicated servers are used on large networks to increase efficiency.
- Authentication servers control access to the network and ensure that only authorized users can log on.
- File servers provide storage and management of user files.
- Print servers manage and control all printing jobs initiated on a network.
- Application servers provide access to application software.
- Database servers store database files and provide access to users who need the information in the databases.
- E-mail servers control all incoming and outgoing e-mail traffic.
- Communications servers are used to control the flow of information from the internal network to outside networks.
- Web servers are used to host websites.
- Cloud servers provide storage and access to data on the Internet.

Objective 12.5 *Describe the common types of network topologies and the advantages and disadvantages of each one.*

- In a bus topology, all nodes are connected to a single linear cable. Installation is easy, reliable, and inexpensive. Breaks in the cable disable the network, and large numbers of users decrease performance.

- Ring topologies are made up of nodes arranged roughly in a circle. The data flows from node to node in a specific order. Access to the network is allocated fairly among nodes. Failure of one node can bring down the network.
- In a star topology, nodes are connected to a central communication device (a switch) and branch out like points of a star. Failure of one node doesn't affect other nodes and the network is highly scalable. Requires more cable to install, and network navigation devices are a single point of failure that can affect many nodes.

Part 2
Setting Up Business Networks

Learning Outcome 12.2 **You will be able to describe transmission media, network operating system software, and network navigation devices and explain major threats to network security and how to mitigate them.**

Transmission Media

Objective 12.6 *Describe the types of wired and wireless transmission media used in networks.*

- Transmission media make up the physical system that data takes to flow between devices on a network. Twisted-pair cable consists of pairs of wires twisted around each other to reduce interference. Coaxial cable is the same type of cable used by your cable TV company to run a signal into your house. Fiber optic cable uses bundles of glass or plastic fiber to send signals by using light waves. It provides the largest bandwidth but is expensive and difficult to install. Wireless media uses radio waves to send data between nodes on a network.

Network Adapters and Navigation Devices

Objective 12.7 *Describe how network adapters help data move around a network.*

- A network adapter provides three critical functions: (1) It converts low-power data signals generated by the computer into higher-powered signals that can traverse network media easily. (2) It breaks the data generated by the computer into packets and packages them for transmission across the network media. (3) It acts as a gatekeeper to control the flow of data to and from the computer.

Objective 12.8 *Define MAC addresses, and explain how they are used to move data around a network.*

- A media access control (MAC) address represents the physical address of a network adapter. It is made up of six, two-position characters, such as 01:40:87:44:79:A5. A frame is a container that can hold multiple data packets. While the data packets are being assembled into frames, the network operating system (NOS) software assigns the appropriate MAC address to the frame. The NOS keeps track of all devices and their addresses on the network.

Objective 12.9 *List the various network navigation devices, and explain how they help route data through networks.*

- Switches are devices that read the addresses of data packets and retransmit a signal to its destination instead of to every device connected to the switch. Bridges are devices used to send data between two segments (collision domains) of the same network. Routers are used to route data between two networks, such as between a business network and the Internet.

Network Operating Systems and Network Security

Objective 12.10 *Explain why network operating systems are necessary for networks to function.*

- Network operating system (NOS) software needs to be installed on each computer and server connected to a client/server network to provide the services necessary for the devices to communicate. The NOS provides a set of common rules (called a *protocol*) that controls communication among devices on the network.

Objective 12.11 *List major security threats to networks, and explain how network administrators mitigate these threats.*

- Human errors and mistakes, malicious human activity (such as unauthorized access), and natural events and disasters are major categories of security threats.
- Access to most networks requires authentication procedures, such as having users enter a user ID and password, to ensure that only authorized users access the network. The system administrator defines access privileges for users so that they can access only specific files.
- Network equipment is physically secured behind locked doors. Access to equipment is often restricted by biometric devices, which use unique physical characteristics of individuals for identification purposes. Multi-factor authentication provides a high level of security.
- Firewalls, packet screeners, and bastion hosts are employed to keep hackers from attacking networks successfully.
- Data backups help secure data against destruction by natural events and disasters (such as fire).

Be sure to check out **MyLab IT** for additional materials to help you review and learn. And don't forget to watch the Chapter Overview Videos.

Key Terms

access card reader **491**
access method **472**
active topology **474**
application server **469**
authentication **489**
authentication server **469**
bastion host **492**
biometric authentication device **491**
bridge **486**
brute-force attack **490**
bus (linear bus) topology **472**
centralized **463**
client/server network (server-based network) **463**
cloud server **470**
coaxial cable **481**
communications server **469**
CSMA/CD **474**
data collision **472**
database server **469**
decentralized **463**
dedicated server **468**
device driver **484**
electronic data interchange (EDI) **466**
e-mail server **469**
extranet **466**
fiber-optic cable **481**
file server **469**
frame **485**
honey pot **492**
hybrid topology **476**
information assurance **493**
intranet **465**
jam signal **474**
local area network (LAN) **464**
media access control (MAC) address **484**
multi-factor authentication **491**
network adapter **482**
network administrator **463**
network operating system (NOS) **487**
network topology **471**
Open Systems Interconnection (OSI) **487**
packet screening **492**
passive topology **473**
possessed object **490**
print queue **469**
print server **469**
protocol **471**
proxy server **494**
ring (loop) topology **473**
router **486**
scalability **464**
shielded twisted-pair (STP) cable **481**
star topology **474**
switch **485**
terminator **473**
token **473**
token method **474**
transmission media **480**
tunneling **466**
twisted-pair cable **481**
unshielded twisted-pair (UTP) cable **481**
virtual private network (VPN) **466**
virtualization **470**
web server **470**
wide area network (WAN) **465**
wireless access point (WAP) **483**
wireless network interface card (wireless NIC) **483**

Chapter Quiz // Assessment

For a quick review to see what you've learned, answer the following questions. Submit the quiz as requested by your instructor. If you are using **MyLab IT**, the quiz is also available there.

multiple choice

1. Which of the following is an advantage of installing a client/server network in a business?
- **a.** Centralization of network adapters
- **b.** Decentralization of peripherals
- **c.** Decentralization of files and data
- **d.** Sharing of peripherals

2. On client/server networks, more users can be added without affecting the performance of other nodes. This is known as network
- **a.** scalability.
- **b.** decentralization.
- **c.** topology.
- **d.** flexibility.

3. Which server would a client/server network include to allow users to share software?
   a. Cloud
   b. Communications
   c. Authentication
   d. Application
4. Which of the following is an area of an intranet that only certain corporations or individuals can access?
   a. Extranet
   b. VPN
   c. PAN
   d. HAN
5. Sometimes groups of data packets are sent together in a package called a
   a. protocol.
   b. bridge.
   c. MAC.
   d. frame.
6. Which of the following is NOT one of the factors used in multi-factor authentication?
   a. Inherence
   b. Identity
   c. Possession
   d. Knowledge
7. Which network cable type is least susceptible to signal interference?
   a. UTP
   b. STP
   c. Coaxial
   d. Fiber-optic
8. The process whereby users prove they have authorization to use a computer network is known as
   a. authentication.
   b. decentralization.
   c. proxy control.
   d. repudiation.
9. Which topology is most widely used today?
   a. Bus
   b. Star
   c. Ring
   d. VPN
10. Which is NOT a key component of a typical client/server network?
   a. Transmission media
   b. Network navigation devices
   c. Network interface card
   d. Bastion host

true/false

_____ 1. Switches are used to send data on a specific route through an internal network.

_____ 2. A database server is used to host websites on a client/server network.

_____ 3. Network adapters perform specific tasks to enable nodes to communicate on a network.

_____ 4. Most modern operating systems do not include any type of NOS functionality.

_____ 5. Ring topologies are the most widely used topology.

What do you think now?

(Joshua Resnick/123RF)

1. Assuming that emotion-sensing facial recognition software deployment might be inevitable, what practical applications do you think there are for this type of innovation? Are there any situations in which you think this technology should never be deployed?
2. How should deployment of this technology be regulated?

Team Time

Developing a Bring-Your-Own-Device Policy

problem

Employees, especially in developing economies, are often bringing their own computing device (BYOD) to work and using these devices to access company networks, databases, and the Internet. Although this can keep equipment costs down, BYOD can present unique challenges for data security.

task

Cool Tees, a local T-shirt company, has experienced a large rise in BYOD by its employees. It is concerned about the security of data that may be on the devices. The Cool Tees CEO has asked you to assist him in constructing a BYOD policy for his company.

process

1. Divide the class into small teams. Research BYOD security issues. Review BYOD policies that are in place for businesses and schools. Consider what types of devices (laptops, phones, tablets, etc.) might be used to access company data. Draft a BYOD policy for Cool Tees employees that sets forth the company's expectations with regard to personal devices.
2. Meet as a class to discuss each group's findings. Do you feel most employees are aware of the security issues related to BYOD? Try to obtain a consensus about which precautions are the most important ones for employees to take to protect the company's data.
3. Have each group prepare a multimedia presentation (PowerPoint, video, etc.) to educate employees about the security risks of BYOD.

conclusion

Having personal computing devices and bringing them to work is fast becoming the norm in the twenty-first century. However, the best defense against loss of control of company data is to understand the consequences of accessing data with and storing data on personal computing devices.

Ethics Project

Using Wireless Networks Without Permission

In this exercise, you research and then role-play a complicated ethical situation. The role you play might not match your own personal beliefs, but your research and use of logic will enable you to represent the view assigned. An arbitrator will watch and comment on both sides of the argument, and together the team will agree on an ethical solution.

problem

Piggybacking occurs when people use a wireless network without the owner's permission. Although piggybacking is illegal in many jurisdictions, it's often hard to detect. Piggybacking often happens inadvertently when people trying to connect to their own network accidentally connect to their neighbor's network. And because of the close proximity of many businesses, the potential for piggybacking of business wireless networks also exists. Sharing wireless connections between two entities (whether they are two households or two businesses) may violate the terms of service of the Internet service provider.

research areas to consider

- Detecting wireless piggybacking
- Piggybacking laws (legality of piggybacking)
- Securing wireless networks

process

- Divide the class into teams. Research the areas cited and devise a scenario in which the owner of a coffee shop has accused the proprietor of the sandwich shop next door of encouraging the sandwich shop's patrons to piggyback on the coffee shop's wireless network.
- Team members should write a summary that provides background information for their character—for example, coffee shop owner, sandwich shop owner, and arbitrator—and that details their character's behaviors to set the stage for the role-playing event. Team members should then create an outline to use during the role-playing event.
- Team members should present their case to the class or submit a PowerPoint presentation for review by the rest of the class, along with the summary and resolution they developed.

conclusion

As technology becomes ever more prevalent and integrated into our lives, ethical dilemmas will present themselves to an increasing extent. Being able to understand and evaluate both sides of the argument, while responding in a personally or socially ethical manner, will be important skills.

Solve This with Word

MyLab IT Grader

Cyber-Security Flyer and Mail Merge

You have been asked to modify a flyer that announces an upcoming cyber-security presentation to highlight National Cyber Security Awareness Month. In addition, you have been told to mail the flyer to some key community members. You will use Mail Merge and other features of Microsoft Word to complete this task.

You will use the following skills as you complete this activity:

- Apply Artistic Effect
- Insert Shapes
- Apply Page Borders
- Create Custom Bullets
- Use Mail Merge

Instructions

1. Open *TIA_Ch12_Start.docx* and save as **TIA_Ch12_LastFirst.docx**.
2. Scroll to page 2. Select the image at the top left of the page, click the **Picture Tools Format** tab, click **Artistic Effects** in the Adjust group, and then click **Photocopy effect**.
3. Click the **Insert tab**, click **Shapes** in the Illustrations group, and then click **Rectangle** from the Rectangles group. Place the cursor in the top left image and draw a rectangle. In the Size group, modify the size to be 2.2" high and 4.1" wide. Use the arrow keys on the keyboard to nudge the rectangle in any direction so that it is centered within the image.
4. Change the Shape Fill color of the rectangle to **Blue, Accent 5, Lighter 60%**. Type **October is**, press **Enter**, and then type **National Cyber Security Awareness Month**. Change the font to **Century Gothic** and change the font color to **Black, Text 1**. Change the font size of the first line to **14** and the remaining text to **28**. Click the **Paragraph dialog box launcher** and then change Spacing Before to **12 pt** and Line spacing At to **.85**.
5. With the text box still selected, click the **Drawing Tools Format tab**, click **Shape Effects** in the Shape Styles group, and then select **Soft Edges, 5 point**.
6. Select **Cyber Security Presentation** (upper right corner of page 2). Change the font to **Britannic Bold** and the font color to **Dark Red** in the Standard Colors group.
7. Select the first set of bullets, click the **Bullets arrow** in the Paragraph group, click **Define New Bullet**, and then click **Picture**. Click **Browse** next to From a File and select **TIA_Ch12_Image1** from your student data files location. Select the second set of bullets and repeat the process to define a new bullet, using **TIA_Ch12_Image2**.
8. Click the **Design tab** and then click **Page Borders** in the Page Background group. Click Box in the *Setting* section and then click the **triple line** in the *Style* section. Ensure that *Whole document* appears in the Apply to box.
9. On Page 1, select **National Cyber Security Awareness Month!** Click **WordArt** in the Text group on the Insert tab, select **Fill: Blue, Accent color 1; Shadow**. Click the **Text Fill arrow** and then click **Dark Red** in the Standard Colors group.
10. In the second paragraph, beginning with *You are invited*, **bold** *Springer Auditorium, October 22, from 12:00–2:30 PM* and change the font color to **Dark Red** in the Standard Colors group.
11. On the Mailings tab, click **Start Mail Merge** in the Start Mail Merge group and select **Letters**. Click **Select Recipients** and click **Use an Existing List**. Navigate to the student data files and select **TIA_Ch12_MailList**. Ensure that Sheet1 is selected and then click **OK**.
12. Click **Edit Recipient List** in the Start Mail Merge group, click the **City header** to sort the addresses by city, and then clear the two records from Elk Grove Village. Click **OK**.
13. Place the insertion point at the beginning of the first paragraph, beginning with *We are all living*. In the Write & Insert Fields group, click **Address Block**. Ensure that *Joshua Randall Jr.* is selected in the *Specify address elements* section and accept all other default settings. Click **OK**. Press **Enter**.
14. Place the insertion point at the beginning of the first paragraph, beginning with *We are all living*. In the Write & Insert Fields group, click **Greeting Line** and then, in the *Greeting line format* section, ensure that *Dear* appears in the first box. Select **Joshua** in the second box and ensure that a comma appears in the third box. Click **OK**. Press **Enter**.
15. Click **Preview Results** and click **Next Record** several times to advance through the letters.
16. Select the recipient and address lines. On the Home tab, click **No Spacing** in the Styles group. Click at the end of the zip code and press **Enter**.
17. Save the file and submit it, based on your instructor's directions.

Chapter

13 Behind the Scenes: How the Internet Works

For a chapter overview, watch the **Chapter Overview Videos.**

PART 1

Inner Workings of the Internet

Learning Outcome 13.1 **You will be able to explain how the Internet is managed and the details of how data is transmitted across the Internet.**

Internet Management and Networking 504

Objective 13.1 *Describe the management of the Internet.*

Objective 13.2 *Explain how the Internet's networking components interact.*

Objective 13.3 *List and describe the Internet protocols used for data transmission.*

Internet Identity 509

Objective 13.4 *Explain how each device connected to the Internet is assigned a unique address.*

Objective 13.5 *Discuss how a numeric IP address is changed into a readable name.*

Helpdesk: Understanding IP Addresses, Domain Names, and Protocols

Sound Byte: Creating Web Pages with Squarespace

PART 2

Coding and Communicating on the Internet

Learning Outcome 13.2 **You will be able to describe the web technologies used to develop web applications.**

Web Technologies 519

Objective 13.6 *Compare and contrast a variety of web development languages.*

Objective 13.7 *Compare and contrast server-side and client-side application software.*

Helpdesk: Keeping E-Mail Secure

Communications over the Internet 525

Objective 13.8 *Discuss the mechanisms for communicating via e-mail and instant messaging.*

Objective 13.9 *Explain how data encryption improves security.*

Sound Byte: Client-Side Web Page Development

MyLab IT All media accompanying this chapter can be found here

(Pasko Maksim/Shutterstock; Matthias Pahl/Shutterstock; Bowie15/123RF; ra2 studio/Shutterstock)

What do you think?

The amazing growth of the Internet over a relatively short period of years has left us with a **shortage** of trained programmers and designers. Many aspects of **web programming** offer opportunities: website design and development, but also development of the backend server code, security and testing, designing, and monitoring the user experience. These positions offer **high salaries**, good stability and growth, and **flexible** work conditions. What pathway would you need to take advantage of these career options?

What is the minimum training you think should be required to enter the field of web programming?

- *Two-year degree*
- *Bachelor's degree in Computer Science*
- *Bachelor's degree but not necessarily in Computer Science*
- *Self-taught (no degree necessary)*

See the end of the chapter for a follow-up question. For more information on careers in IT, see Appendix B.

(Clement Philippe/Arterra Picture Library/Alamy Stock Photo)

Part 1  For an overview of this part of the chapter, watch **Chapter Overview Video 13.1**.

Inner Workings of the Internet

Learning Outcome 13.1 **You will be able to explain how the Internet is managed and the details of how data is transmitted across the Internet.**

We use the Internet so routinely that it may not have occurred to you to ask some fundamental questions about how it's administered. In this part of the chapter, we discuss this and other topics.

Internet Management and Networking

To keep a massive network like the Internet functioning at peak efficiency, it must be governed and regulated. However, no single entity is in charge of the Internet. In addition, new uses for the Internet are created every day by a variety of individuals and companies.

Management

Objective 13.1 *Describe the management of the Internet.*

Who owns the Internet? The local networks that constitute the Internet are owned by different entities, including individuals, universities, government agencies, and private companies. Government entities such as the National Science Foundation (NSF), as well as many privately held companies, own pieces of the communications infrastructure (the high-speed data lines that transport data between networks) that makes the Internet work.

Does anyone manage the Internet? Several nonprofit organizations and user groups, each with a specialized purpose, are responsible for the Internet's management. Table 13.1 shows major organizations that play a role in the governance and development of the Internet.

Many of the functions these nonprofit groups handle were previously handled by US government contractors because the Internet developed out of a defense project. However, because the Internet serves the global community, assigning responsibilities to organizations with global membership is the best way to guarantee worldwide engagement in determining the direction of the Internet.

Who pays for the Internet? The National Science Foundation (NSF), which is a US government–funded agency, still pays for a large portion of the Internet's infrastructure and funds research and development for new technologies. The primary source of NSF funding is federal taxes. Other countries also pay for Internet infrastructure and development.

Table 13.1 Major Organizations in Internet Governance and Development

Organization	Purpose	Web Address
Internet Society	Professional membership society that provides leadership for the orderly growth and development of the Internet	internetsociety.org
Internet Engineering Task Force (IETF)	A subgroup of the Internet Society that researches new Internet technologies to improve its capabilities and keep the infrastructure functioning smoothly	ietf.org
Internet Architecture Board (IAB)	Technical advisory group to the Internet Society and an IETF committee; provides direction for the maintenance and development of Internet protocols	iab.org
Internet Corporation for Assigned Names and Numbers (ICANN)	Organization responsible for managing the Internet's domain name system and the allocation of IP addresses	icann.org
World Wide Web Consortium (W3C)	Consortium of organizations that sets standards and develops protocols for the web	w3.org

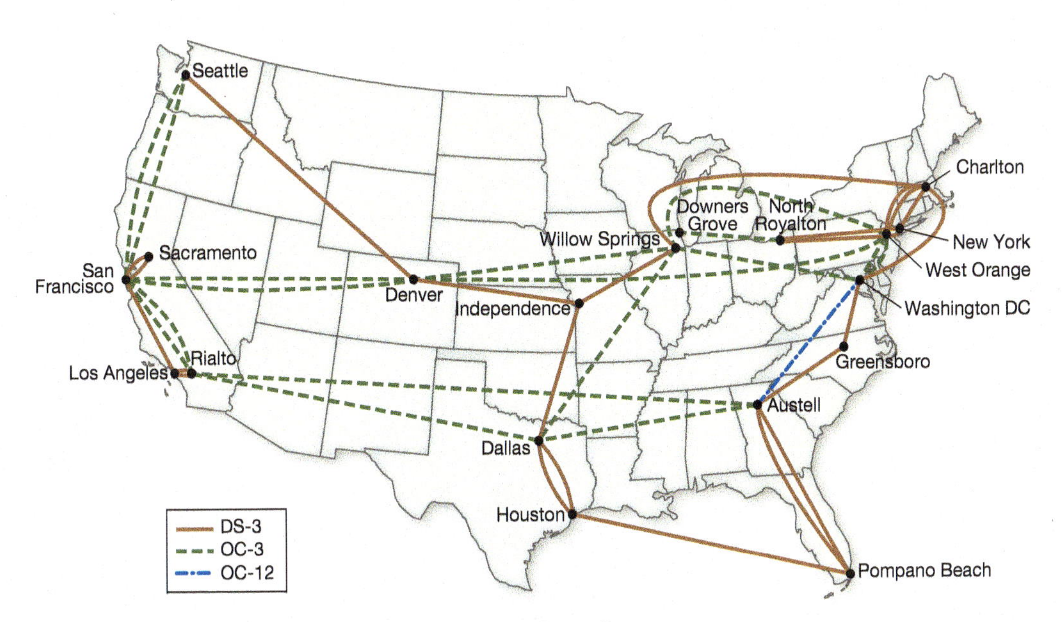

Figure 13.1 A Typical Internet Backbone Provided by a Single Company.

Networking Components

Objective 13.2 *Explain how the Internet's networking components interact.*

How are computers connected to the Internet? As a network of networks, the Internet is similar to the US highway system, where smaller roads feed into larger, faster highways. The main paths of the Internet, along which data travels the fastest, are known collectively as the Internet backbone. The Internet backbone is a collection of large national and international networks, most of which are owned by commercial, educational, or government organizations. These backbone providers, which are required to connect to other backbone providers, have the fastest high-speed connections. Figure 13.1 shows a typical Internet backbone provided by a single company. Large US companies that provide backbone connectivity include Verizon, AT&T, and Sprint.

How do the Internet service providers (ISPs) that form the Internet backbone communicate? A backbone is typically a high-speed fiber-optic line, designated as an optical carrier (OC) line. OC lines come in a range of speeds, from 0.052 Gbps for OC-1 to the fastest OC-3840, which runs at almost 200 Gbps. In Figure 13.1, you can see the optical carrier lines run across the country by one company in green (OC-3) and blue (OC-12).

Backbone ISPs initially connected with T lines. A T line carried digital data over twisted-pair wires. T-1 lines, which were the first to be used, transmitted data at a throughput rate of 1.544 Mbps. T-3 lines, which were developed later, transmitted data at 45 Mbps. Figure 13.1 shows the T-3 lines in orange. (In the figure, they are labeled as DS-3 lines.)

The bandwidth of the connections between ISPs and end users depends on the amount of data traffic required. Whereas your home might connect to the Internet with cable or fiber-optic lines, the volume of Internet traffic at your college probably requires it to use an OC line to move data to the school's ISP. Large companies usually must connect to their ISPs by using high-throughput OC lines.

How are ISPs connected to each other? ISPs like Verizon and Comcast are connected through an Internet exchange point (IXP). A typical IXP is made up of one or more network switches to which ISPs connect. As you'll recall from Chapter 7, switches are devices that send data on a specific route through a network. By connecting directly to each other through IXPs, networks can reduce their costs and improve the speed and efficiency with which data is exchanged.

How do individuals connect to an ISP? Individual Internet users enter an ISP through a point of presence (POP), which is a bank of modems, servers, routers, and switches (see Figure 13.2) through which many users can connect to an ISP simultaneously. ISPs maintain multiple POPs throughout the geographic area they serve.

What type of network model does the Internet use? The majority of Internet communications follow the client/server model of network communications. The clients are devices such as computers, tablets, and smartphones that use browsers to request services such as web pages.

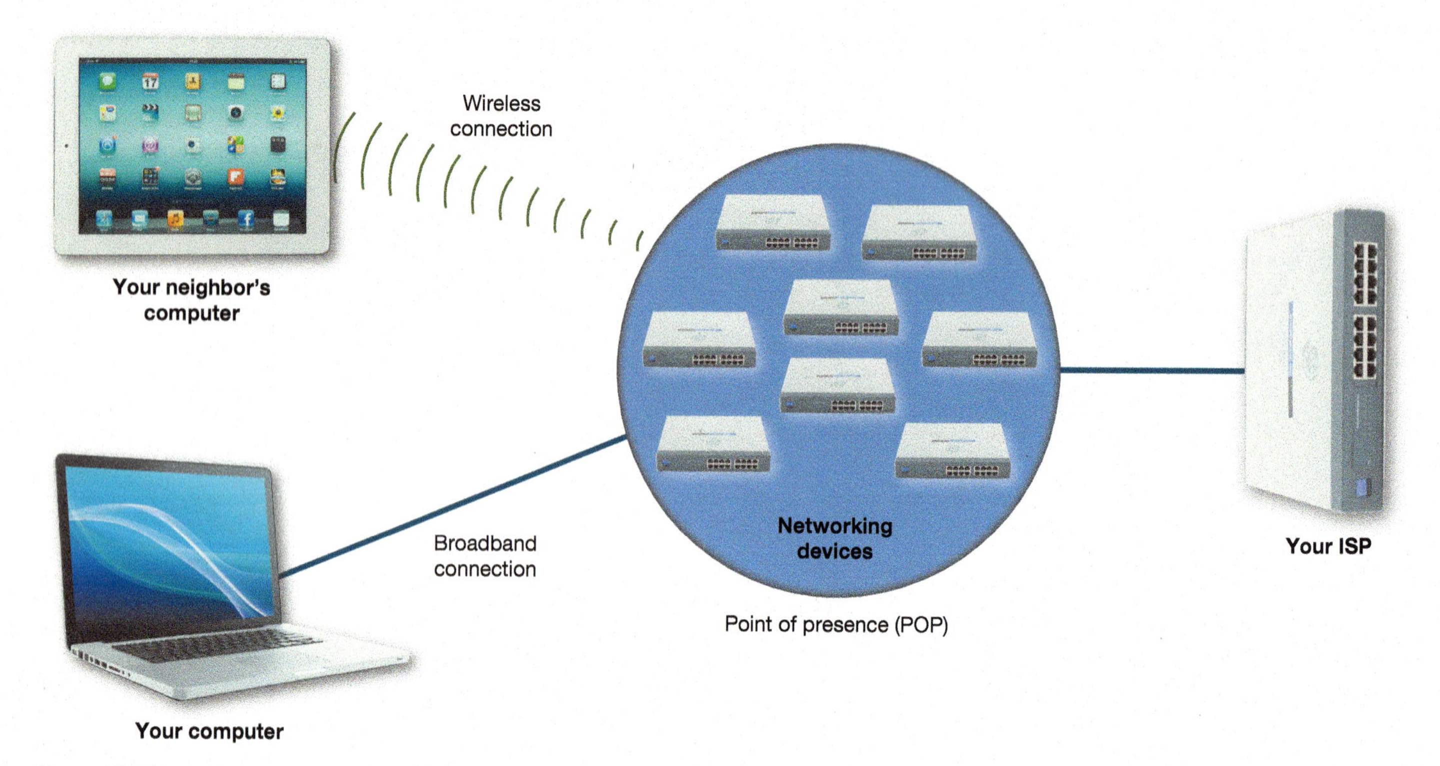

Figure 13.2 Home users connect to their ISPs through a single point of presence that can handle many simultaneous connections. *(Klaus Westermann/Agencja Fotograficzna Caro/Alamy Stock Photo, Ifong/Shutterstock; Robert Hyrons/Alamy Stock Photo, Sashkin/Shutterstock)*

Various types of servers from which clients can request services are deployed on the networks that make up the Internet:

- Web servers: Computers that run specialized operating systems, enabling them to host web pages and other information and provide requested information to clients.
- Commerce servers: Computers that host software that enables users to buy goods and services over the web. These servers generally use special security protocols to protect sensitive information, such as credit card numbers, from being intercepted.
- File servers: Computers that are deployed to provide remote storage space or to act as storehouses for files that users can download. Cloud service providers offering online storage services have a huge collection of file servers.

Data Transmission

Objective 13.3 *List and describe the Internet protocols used for data transmission.*

What is a protocol? Just like any other network, the Internet follows a standard set of rules to send information between computers. A computer protocol is the set of rules for exchanging electronic information. If the Internet is the information superhighway, then protocols are the rules of the road.

Bits&Bytes A Free Cloud-Based Server for You

Interested in trying out web application development but stop in your tracks when you think about setting up and running your own server? No longer! Head to AWS Cloud9 (aws.amazon.com/cloud9), where you can set up and run as many Ubuntu workspaces as you need. Each workspace is an easy-to-use Linux desktop that lives in the cloud and that you access through your browser. It is tightly integrated with Amazon Web Services, so working with their platforms is simplified. Use one workspace just to practice your skills on a Linux command line or use the languages Cloud9 supports, like Ruby, C++, or PHP, to start coding your own server in the cloud.

Why were Internet protocols developed? The idea of a protocol is that anyone can use it on his or her computer system and be able to communicate with any other computer using the same protocol. The most common Internet tasks—communicating, collaborating, creating content, seeking information, and shopping—are all executed the same way on any system following accepted Internet protocols.

When common communication protocols are followed, networks can communicate even if they have different topologies, transmission media, or operating systems. To accomplish the early goals of the Internet, protocols needed to be written and agreed on by users. Each protocol had to be an open system, meaning its design would be made public for access by any interested party. This was in direct opposition to the proprietary system (private system) model that was the norm at the time.

Were there problems developing an open system Internet protocol? Agreeing on common standards was relatively easy. The tough part was developing a new method of communication because the technology available in the 1960s—circuit switching—was inefficient for computer communication.

Circuit Switching

Why don't we use circuit switching to connect two computers? Circuit switching has been used since the early days of the telephone for establishing communication. In circuit switching, a dedicated connection is formed between two points (such as two people on phones), and the connection remains active for the duration of the transmission. This method of communication is important when communications must be received in the order in which they're sent, like phone conversations.

When applied to computers, however, circuit switching is inefficient. As a computer processor performs the operations necessary to complete a task, it transmits data in a group, or burst. The processor then begins working on its next task and ceases to communicate with output devices or other networks until it's ready to transmit data in the next burst. Circuit switching is inefficient for computers because either the circuit would have to remain open, and therefore unavailable to any other system, with long periods of inactivity, or it would have to be reestablished for each burst.

Packet Switching

If they don't use circuit switching, what do computers use to communicate? Packet switching is the communications methodology that makes computer communication efficient. Packet switching doesn't require a dedicated communications circuit to be maintained. With packet switching, data is broken into smaller chunks called packets (or data packets). The packets are sent over various routes at the same time. When the packets reach their destination, they're reassembled by the receiving computer. This technology resulted from one of the original goals of creating the Internet: If an Internet node is disabled or destroyed, the data can travel an alternate route to its destination.

What information does a packet contain? Packet contents vary, depending on the protocol being followed. At a minimum, all packets must contain the following:

1. An address to which the packet is being sent
2. The address from where the packet originates
3. The data that's being transmitted
4. Reassembly instructions, if the original data is split between multiple packets

Sending a packet is like sending a letter. Assume you're sending a large amount of information in written format from your home in Philadelphia to your aunt in San Diego. The information is too large to fit in one small envelope, so you mail three envelopes to your aunt. Each envelope includes your aunt's address, your return address, and the information being sent inside it. The pages of the letters in each envelope are numbered so that your aunt will know in which order to read them.

Each envelope may not find its way to San Diego by the same route. However, even if the letters are routed through different post offices, they'll all eventually arrive in your aunt's mailbox. Your aunt will then reassemble the message in the right order and read it. The process of sending a message through the Internet works in much the same way. This process is illustrated in Figure 13.3, which traces an e-mail message sent from a computer in Philadelphia to a computer in San Diego.

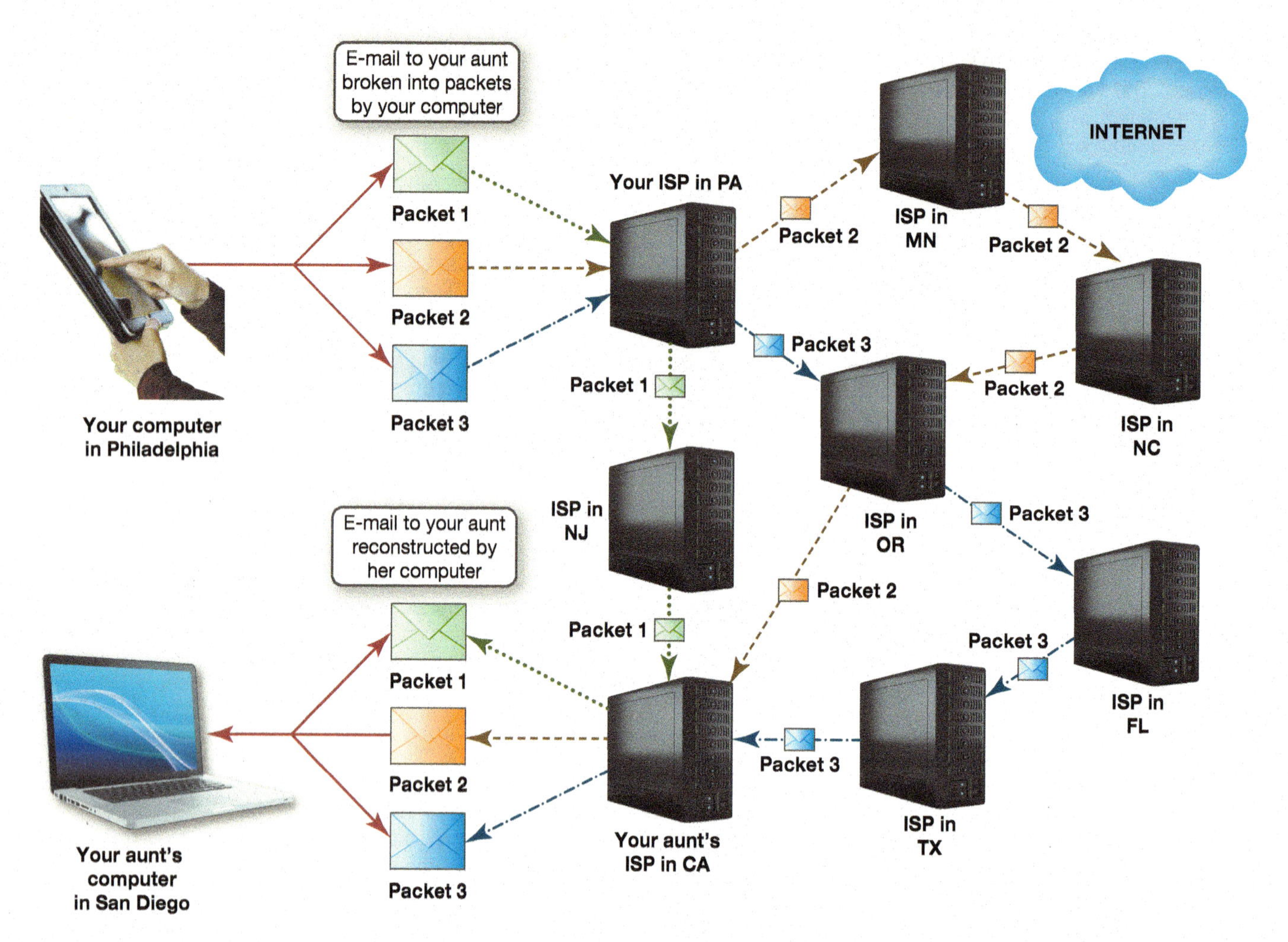

Figure 13.3 Each packet sent through the Internet can follow its own route to its final destination. Sequential numbering of packets ensures that they're reassembled in the correct order at their destination. *(Nikada/E+/Getty Images, Ifong/Shutterstock, Adventtr/E+/Getty Images)*

Bits&Bytes Packet Analysis

If you want to see packet traffic actually being sent back and forth, you can use a packet analysis tool. Wireshark is a very popular open source packet analyzer. It shows you the exact protocol being used, the source and destination of each packet, and other information, including length. The quickest way to understand the concepts around data communication over the Internet is to make the ideas visible; using a packet analyzer lets you see inside each packet sent your way.

Why do packets take different routes, and how do they decide which route to use? The routers that connect ISPs with each other monitor traffic and decide on the most efficient route for packets to take to their destination. The router works in the same way a police officer does while directing traffic. To ensure a smooth flow of traffic, police officers are deployed in areas of congestion, directing drivers along alternate routes to their destinations. The Try This exercise in this chapter uses the Windows utility *tracert* to show you the details on the exact route your request takes to the destination server.

TCP/IP

What protocol does the Internet use for transmitting data? Although many protocols are available on the Internet, the main suite of protocols used is TCP/IP. The suite is named after the original two protocols that were developed for the Internet: the Transmission Control Protocol (TCP) and the Internet Protocol (IP). Although most people think that the TCP/IP suite consists of only two protocols, it actually comprises many interrelated protocols, the most important of which are listed in Table 13.2.

Table 13.2 TCP/IP Protocol Suite—Main Protocols	
Internet Protocol (IP)	Sends data between computers on the Internet
Transmission Control Protocol (TCP)	Prepares data for transmission and provides for error checking and resending of lost data
User Datagram Protocol (UDP)	Prepares data for transmission; lacks resending capabilities
File Transfer Protocol (FTP)	Enables files to be downloaded to a computer or uploaded to other computers
Telnet	Enables users to log on to a remote computer and work on it as if sitting in front of it
Hypertext Transfer Protocol (HTTP) and HTTP Secure (HTTPS)	Transfers Hypertext Markup Language (HTML) data from servers to browsers; HTTPS is an encrypted protocol for secure transmissions
Simple Mail Transfer Protocol (SMTP)	Used for transmission of e-mail messages across the Internet
Dynamic Host Configuration Protocol (DHCP)	Shares a pool of IP addresses with hosts on a network on an as-needed basis
Real-time Transport Protocol (RTP)	Network protocol for delivering audio and video over IP

Which particular protocol actually sends the information? The IP is responsible for sending the information from one computer to another. The IP is like a postal worker who sends a letter (a packet of information) that was mailed (created by the sending computer) on to another post office (router), which in turn routes it to the addressee (the receiving computer). The postal worker never knows whether the recipient actually receives the letter. The only thing the postal worker knows is that the letter was handed off to an appropriate post office that will assist in completing the delivery of the letter.

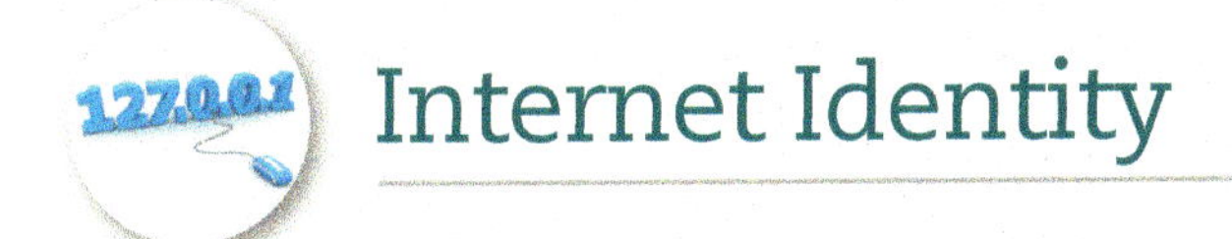

Internet Identity

Each computer, server, or device connected to the Internet is required to have a unique identification number, or IP address. However, because humans are better at remembering and working with words than with numbers, the numeric IP addresses were given more human, word-based addresses. Thus, domain names were born. In this section, we look at both IP addresses and domain names.

IP Addresses

Objective 13.4 *Explain how each device connected to the Internet is assigned a unique address.*

What is an IP address? You'll recall from Chapter 3 that an IP address is a unique identification number that defines each computer, service, or other device connected to the Internet. IP addresses fulfill the same function as street addresses. For example, to send a letter to Rosa Juarez's house in Metaville, Illinois, you have to know her address. Rosa might live at 456 Walnut Street, which isn't a unique address because many towns have a Walnut Street, but 456 Walnut Street, Metaville, IL 00798 is a unique address.

The numeric zip code is the unique postal identification for a specific geographical area. IP addresses must be registered with the Internet Corporation for Assigned Names and Numbers (ICANN) to ensure that they're unique and haven't been assigned to other users. ICANN is responsible for allocating IP addresses to network administrators.

What does an IP address look like? A typical IP address is expressed as follows:

197.169.73.63

An IP address in this form is called a dotted decimal number (or a dotted quad). The same IP address in binary form is as follows:

11000101.10101001.01001001.00111111

Bits&Bytes What's Your IP Address?

Curious about what your IP address is? Go to Google and type "what is my ip" (see Figure 13.4). Google will show your IP address at the top of the search results. If you want additional information, such as the geographical location of an IP address or even the telephone area code associated with it, visit *whatismyipaddress.com*.

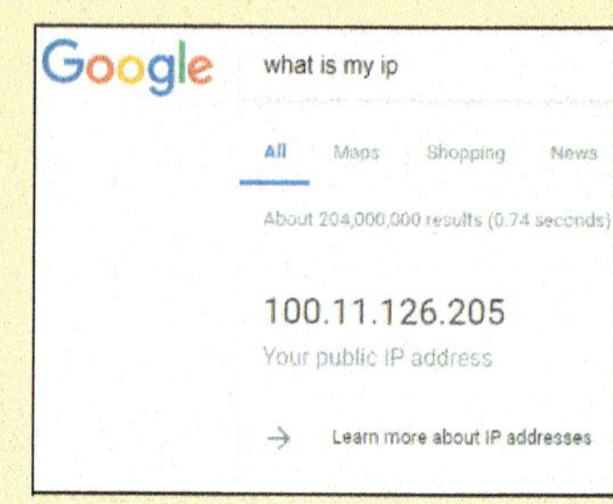

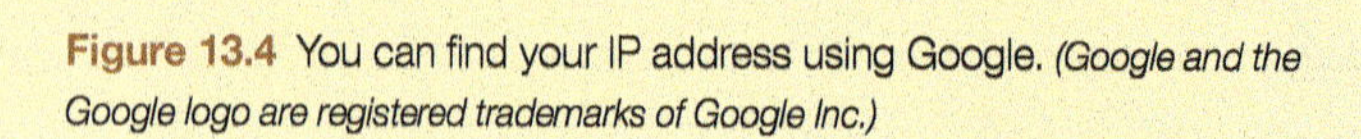
Figure 13.4 You can find your IP address using Google. *(Google and the Google logo are registered trademarks of Google Inc.)*

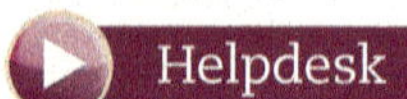
Helpdesk MyLab IT

Understanding IP Addresses, Domain Names, and Protocols

In this Helpdesk, you'll play the role of a helpdesk staffer, fielding questions about which data transmissions and protocols the Internet uses and why IP addresses and domain names are important for Internet communications.

Each of the four numbers in a dotted decimal number is referred to as an octet. This is because each number would have eight positions when shown in binary form. Because 32 positions are available for IP address values (four octets with eight positions each), IP addresses are considered 32-bit numbers. A position is filled by either a 1 or a 0, resulting in 256 (which is 2^8) possible values for each octet, from 0 to 255. The entire 32-bit address can represent 4,294,967,296 values (or 2^{32}), which is quite a few Internet addresses!

Will we ever run out of IP addresses? When the original IP addressing scheme, Internet Protocol version 4 (IPv4), was created in 1981, no one foresaw the explosive growth of the Internet in the 1990s. Four billion values for an address field seemed like enough to last forever. However, as the Internet grew, it quickly became apparent that we would run out of IP addresses.

Why do we need so many IP addresses? Not only laptops need an IP address. The Internet of Things (IoT) is a term for the explosive growth of Internet-enabled devices. Appliances, switches, light bulbs, and smoke detectors now send and receive messages over the Internet. To use the smart devices around us, we need each to connect to the Internet so its data can be stored or sent to cloud-based computers that can use it in sophisticated algorithms. That means each IoT device needs its own unique IP address.

How did Internet addresses change to allow more of them? Internet Protocol version 6 (IPv6) is an IP addressing scheme developed by the IETF to make IP addresses longer, thereby providing more available IP addresses. IPv6 uses 128-bit addressing instead of 32-bit addressing. An IPv6 address would have the following format:

XXXX:XXXX:XXXX:XXXX:XXXX:XXXX:XXXX:XXXX

Each digit X is a hexadecimal digit. Hexadecimal is a base-16 number system, and each hexadecimal digit is one of 16 possible values: 0–9 or A–F. Hex addressing provides a much larger field size, which will make available a much larger number of IP addresses (approximately 340 followed by

Bits&Bytes Amazing Applications of IoT

Many countries are facing an exploding demographic of elderly people. To help manage their safety and health, Internet of Things solutions are abounding (Figure 13.5). IoT devices are being introduced to help track those suffering dementia so they can be kept safe from wandering. Smart fabrics are being used to monitor respiration rate and heart rate and to track incidence of heart attacks. A small transmitter attached to the smart fabric can send data to cloud platforms so physicians can be alerted at the first sign of danger to a patient. Temperature and humidity are also monitored since the elderly are at a higher risk for heatstroke. IoT devices are helping monitor and control diabetes by helping the elderly keep track of food intake, exercise, and blood sugar levels. Data between appointments can be stored or directly transmitted to caretakers for monitoring and quick intervention.

Figure 13.5 Internet of Things devices have brought us many new solutions for managing people's safety and health. *(Wrightstudio/123RF)*

36 zeros). This should provide a virtually unlimited supply of IP addresses and will allow many kinds of non-PC devices, such as cell phones and home appliances, to join the Internet more easily in the future. All modern operating systems can handle both IPv4 and IPv6 addresses. Although the majority of routing on the Internet still takes place using IPv4 addresses, the conversion to IPv6 addressing should accelerate now that we're running out of IPv4 addresses.

Creating Web Pages with Squarespace

In this Sound Byte, you use Squarespace, an online web page development tool, to design a modern interactive web page with no need for coding or web development tools.

How does my device get an IP address? IP addresses are assigned either statically or dynamically:

- Static addressing means that the IP address for a computer never changes and is most likely assigned manually by a network administrator or an ISP.
- Dynamic addressing, in which your computer is assigned a temporary address from an available pool of IP addresses, is more common.

A connection to an ISP could use either method. If your ISP uses static addressing, then you were assigned an IP address when you applied for your service and had to configure your computer manually to use that address. More often, though, an ISP assigns a computer a dynamic IP address, as shown in Figure 13.6.

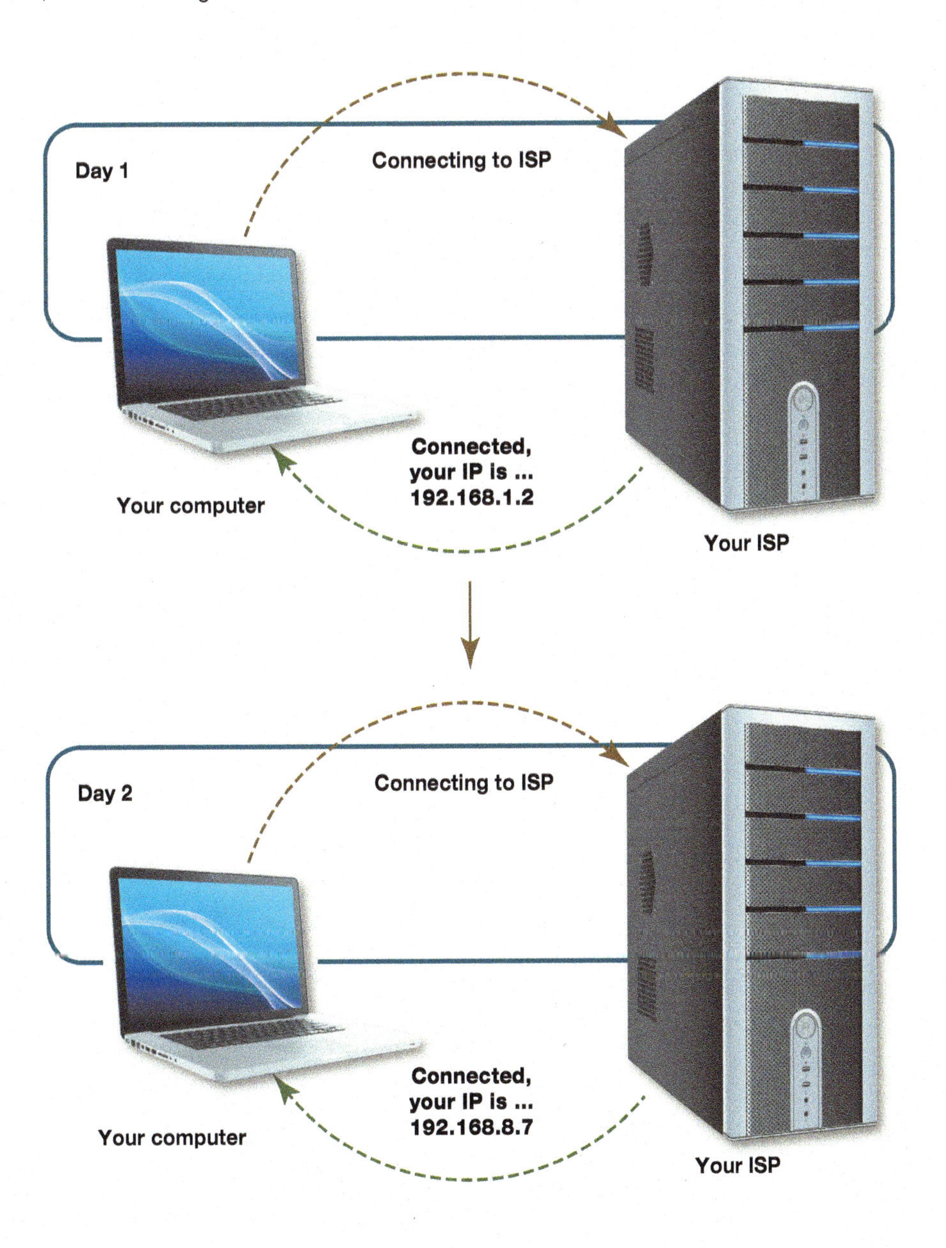

Figure 13.6 With dynamic IP addressing, your IP address changes every time you connect to the Internet. *(Ifong/Shutterstock; Juffin/DigitalVision Vectors/Getty Images)*

How exactly are dynamic addresses assigned? Dynamic addressing is normally handled by the Dynamic Host Configuration Protocol (DHCP), which belongs to the TCP/IP protocol suite. DHCP shares a pool of IP addresses with hosts on the network on an as-needed basis. ISPs don't need to maintain a pool of IP addresses for all their subscribers because not everyone is logged on to the Internet at one time. Thus, when a user logs on to an ISP's server, the DHCP server assigns that user an IP address for the duration of the session. Similarly, when you log on to your computer at work in the morning, DHCP assigns an IP address to your computer. These temporary IP addresses may or may not be the same from session to session.

What are the benefits of dynamic addressing? Although having a static address would seem to be convenient, dynamic addressing provides a more secure environment by keeping hackers out of computer systems. Imagine how hard it would be for burglars to find your home if you changed your address every day!

Dig Deeper Connection-Oriented Versus Connectionless Protocols

The Internet Protocol is responsible only for sending packets on their way. The packets are created by either the TCP or the User Datagram Protocol (UDP). You don't decide whether to use TCP or UDP. The choice of protocol is made for you by the developers of the computer programs you're using or by the other protocols that interact with your data packet (such as those listed in Table 13.2).

As explained earlier, data transmission between computers is highly efficient if connections don't need to be established (as in circuit switching). However, there are benefits to maintaining a connection, such as reduced data loss. The difference between TCP and UDP is that TCP is a *connection-oriented* protocol, whereas UDP is a *connectionless* protocol.

A connection-oriented protocol requires two computers to exchange control packets, thereby setting up the parameters of the data-exchange session, before sending packets that contain data. This process is referred to as handshaking. TCP uses a process called a three-way handshake to establish a connection, as shown in Figure 13.7.

1. Your computer establishes a connection to the ISP and announces it has e-mail to send.
2. The ISP server responds that it's ready to receive the e-mail.
3. Your computer then acknowledges the ready state of the server and begins to transmit the e-mail.

A connectionless protocol doesn't require any type of connection to be established or maintained between two computers exchanging information. Just like a letter that's mailed, the data packets are sent without notifying the receiving computer or receiving any acknowledgment that the data was received. UDP is the Internet's connectionless protocol.

Besides establishing a connection, TCP provides for reliable data transfer, ensuring that all the data packets are delivered to the receiver free from errors and in the correct order. TCP achieves reliable data transfer by using acknowledgments and providing for the retransmission of data, as shown in Figure 13.8.

Assume that two systems, X and Y, have established a connection. When Y receives a data packet that it can read from X, it sends back a positive acknowledgment (ACK). If X doesn't receive an ACK in an appropriate period of time, it resends the packet. If the packet is unreadable (damaged in transit), Y sends a negative acknowledgment (NAK) to X, indicating the packet wasn't received in understandable form. X then retransmits that packet. Acknowledgments ensure that the receiver has received a complete set of data packets. If a packet is unable to get through after being resent several times, the user is generally presented with an error message, indicating that the communications were unsuccessful.

You may wonder why you wouldn't always want to use a protocol that provides for reliable data transfer. On the Internet, speed is often more important than accuracy. For certain

Figure 13.7 Two computers establish communication by using a three-way handshake. *(Ifong/Shutterstock, Juffin/DigitalVision Vectors/Getty Images)*

Domain Names

Objective 13.5 *Discuss how a numeric IP address is changed into a readable name.*

I've been on the Internet, so why have I never seen IP addresses? A domain name is a name that takes the place of an IP address, making it easier for people to remember. For example, *google.com* is a domain name. The server where Google's main website is deployed has an IP address (such as 173.194.205.101), but it's much easier for you to remember to tell your browser to go to *google.com*.

How are domains organized? Domains are organized by level. The portion of the domain name after the dot is the top-level domain (TLD). The TLDs are standardized pools such as .com and .org established by ICANN (see Figure 13.9).

applications (such as e-mail), it's important for your message to be delivered completely and accurately. For streaming multimedia, it's not always important to have every frame delivered accurately because most streaming media formats provide for correction of errors caused by data loss. However, it's extremely important for streaming media to be delivered at a high rate of speed. Therefore, a protocol such as TCP, which uses handshakes and acknowledgments, would probably not be appropriate for transmitting a movie trailer over the Internet; in this instance, the Real-time Transport Protocol would be better.

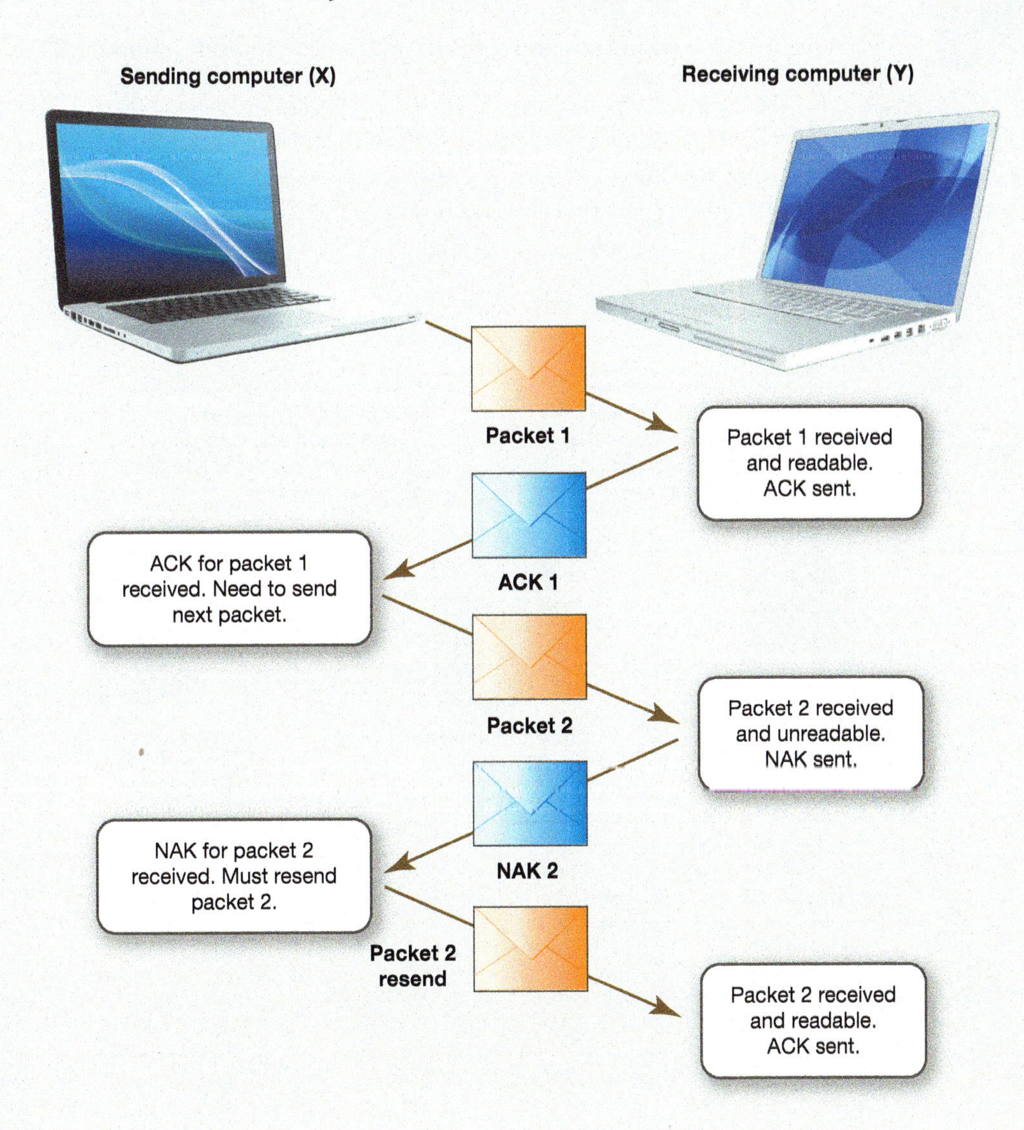

Figure 13.8 Positive (ACK) and negative (NAK) packet acknowledgments provide reliable data transfer. *(Ifong/Shutterstock, LdF/iStock/Getty Images)*

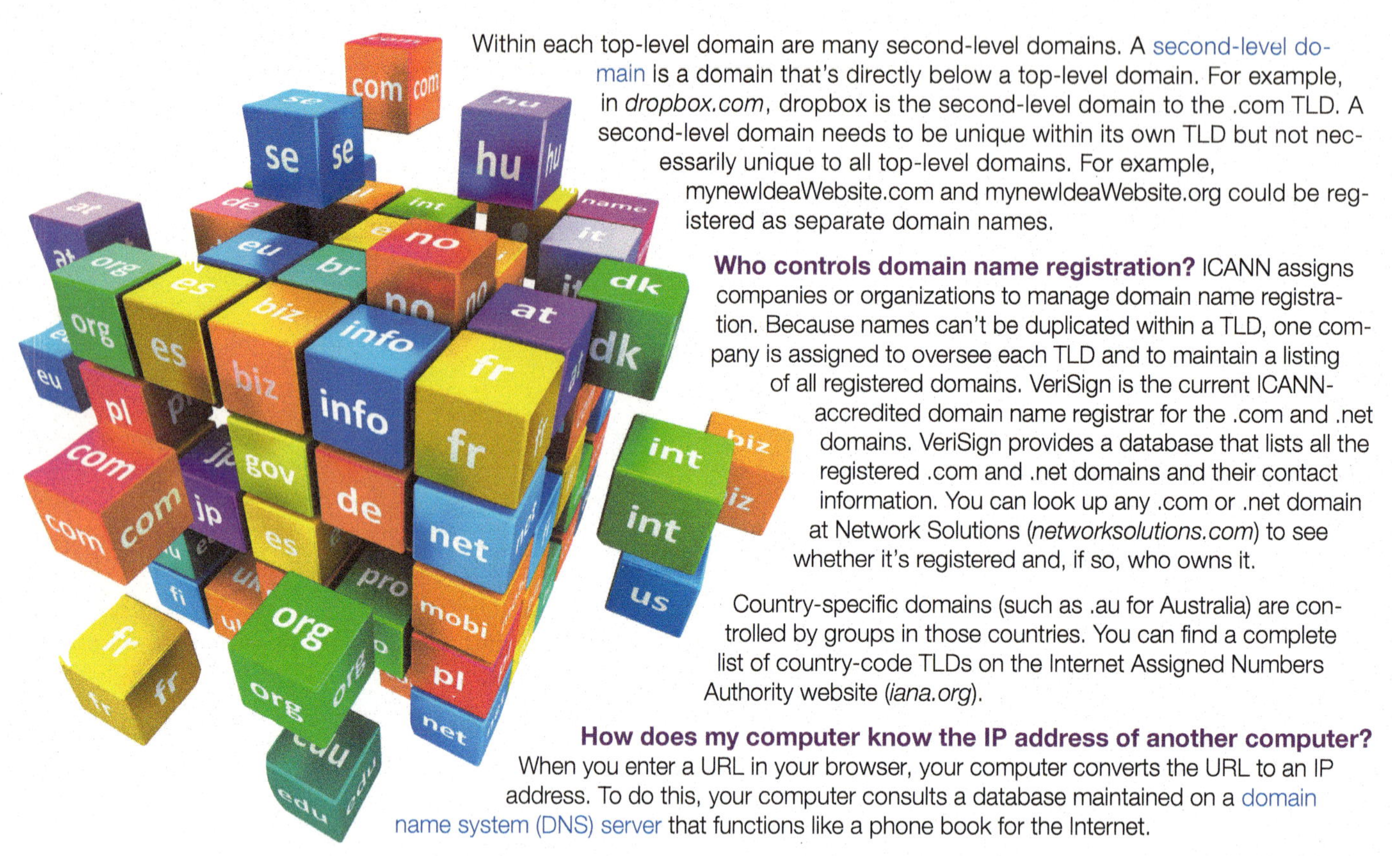

Within each top-level domain are many second-level domains. A second-level domain is a domain that's directly below a top-level domain. For example, in *dropbox.com*, dropbox is the second-level domain to the .com TLD. A second-level domain needs to be unique within its own TLD but not necessarily unique to all top-level domains. For example, mynewIdeaWebsite.com and mynewIdeaWebsite.org could be registered as separate domain names.

Who controls domain name registration? ICANN assigns companies or organizations to manage domain name registration. Because names can't be duplicated within a TLD, one company is assigned to oversee each TLD and to maintain a listing of all registered domains. VeriSign is the current ICANN-accredited domain name registrar for the .com and .net domains. VeriSign provides a database that lists all the registered .com and .net domains and their contact information. You can look up any .com or .net domain at Network Solutions (*networksolutions.com*) to see whether it's registered and, if so, who owns it.

Country-specific domains (such as .au for Australia) are controlled by groups in those countries. You can find a complete list of country-code TLDs on the Internet Assigned Numbers Authority website (*iana.org*).

How does my computer know the IP address of another computer? When you enter a URL in your browser, your computer converts the URL to an IP address. To do this, your computer consults a database maintained on a domain name system (DNS) server that functions like a phone book for the Internet.

Figure 13.9 The number of generic top-level domains has grown to include .museum and .job. ICANN is responsible for organizing TLD names. *(Scanrail/123RF)*

Your ISP's web server has a default DNS server (one that's convenient to contact) that it goes to when it needs to translate a URL to an IP address (illustrated in Figure 13.10).

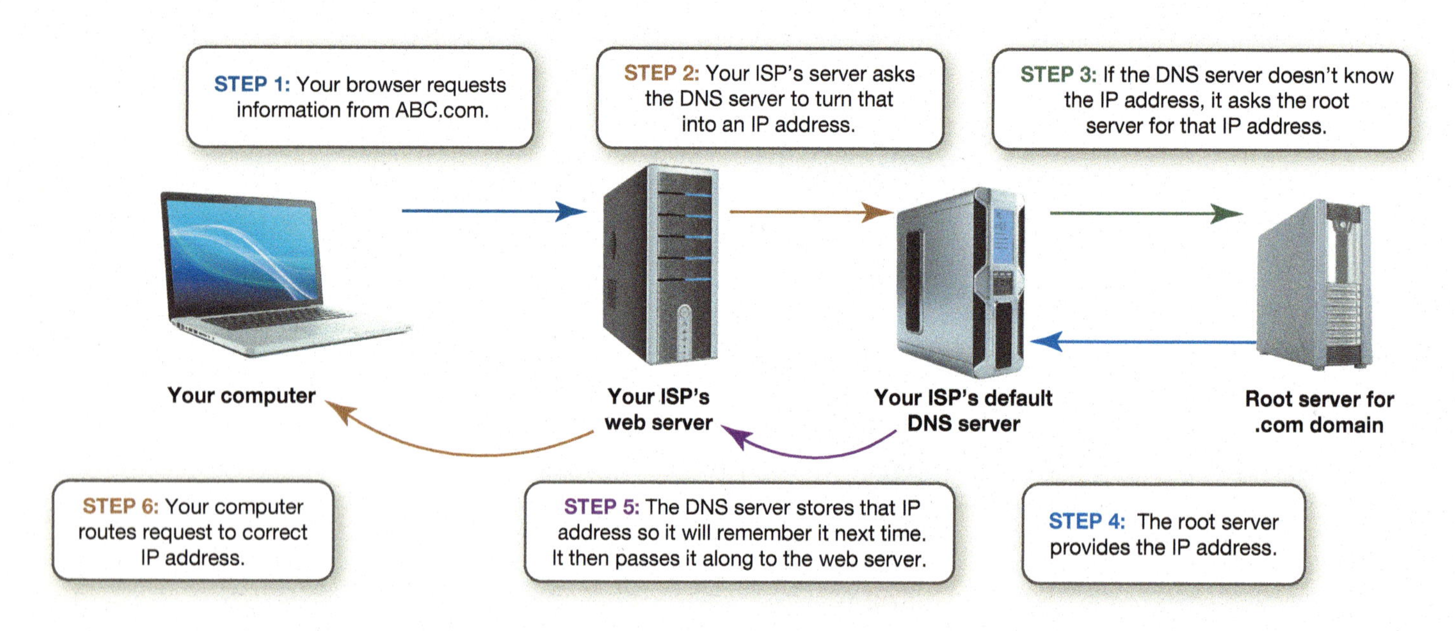

Figure 13.10 DNS Servers in Action. *(Ifong/Shutterstock, Juffin/DigitalVision Vectors/Getty Images, Sashkin/Shutterstock, Alex Mit/Shutterstock)*

It uses the following steps:

1. Your browser requests information from a website (for example, ABC.com).
2. Your ISP doesn't know the address of ABC.com, so your ISP must request the address from its own default DNS server.
3. The default DNS server doesn't know the IP address of ABC.com either, so it queries one of the 13 root DNS servers maintained throughout the Internet. Each root DNS server knows the location of all the DNS servers that contain the master listings for an entire top-level domain.
4. The root server provides the default DNS server with the appropriate IP address of ABC.com.
5. The default DNS server stores the correct IP address for ABC.com for future reference and returns it to your ISP's web server.
6. Your computer then routes its request to ABC.com and stores the IP address in cache for later use.

How can you make your browser translate IP addresses faster? The Internet cache is a section of your hard drive that stores information you may need again, such as IP addresses and frequently accessed web pages. The caching of domain name addresses also takes place in DNS servers. This helps speed up Internet access time because a DNS server doesn't have to query master DNS servers for TLDs constantly.

However, caches have limited storage space, so entries are held in the cache only for a fixed period of time and are then deleted. The time component associated with cache retention is called *time to live (TTL)*. Without caches, surfing the Internet would take a lot longer.

Remember you can't always tell whether your browser is loading the current version of a page from the web or a copy from your Internet cache. If a website contains time-sensitive information (such as a snow day alert on a college website), clicking the browser's Reload or Refresh button will ensure that the most current copy of the page loads into your browser.

Before moving on to Part 2:

1. **Watch Chapter Overview Video 13.1.**
2. **Then take the Check Your Understanding quiz.**

Check Your Understanding // Review & Practice

For a quick review to see what you've learned so far, answer the following questions.

multiple choice

1. The Internet is managed by
   a. the US government.
   b. no one.
   c. the National Science Foundation.
   d. several nonprofit organizations.
2. In the client/server model, clients are phones or computers and servers are computers
   a. with software allowing you to make purchases.
   b. hosting web pages.
   c. that are storehouses for files.
   d. any of the above
3. Packet switching sends information between two points
   a. by finding one route between the two points and sending the data.
   b. by breaking it into small chunks and sending the chunks across different routes.
   c. by encrypting the data and then sending it.
   d. differently for each company using it.
4. The rise of the Internet of Things has meant
   a. we need longer IP addresses than are used in IPv4.
   b. there are more machines with wireless network adapters.
   c. light bulbs can join your Wi-Fi network at home.
   d. all of the above
5. The main set of protocols used to move information across the Internet is
   a. DNS.
   b. TCP/IP.
   c. POP.
   d. XML.

chew on this

(3Dalia/Shutterstock)

Should there be one agency in charge of the Internet? How would business, government, and education be impacted if the Internet were suddenly no longer available? Does this warrant establishing a governing body to regulate and monitor the Internet?

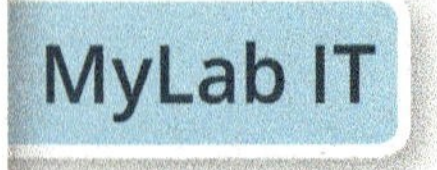

Go to **MyLab IT** to take an autograded version of the *Check Your Understanding* review and find all media resources for the chapter.

For the IT Simulation for this chapter, see MyLab IT.

Ping Me

Several programs that help you understand networking are part of your Windows operating system but aren't on your desktop. In this Try This, we use the **ping command** and the **tracert command** to get some detailed info about the Internet. For more step-by-step instructions, watch the Try This video.

Step 1 In Windows, in the search box, type **command**. Click **Command Prompt** from the search results.

Step 2 Type **ping google.com**, and press **Enter**. Ping is a utility that sends a packet of information to a specific computer and measures the time it takes to travel a round trip to the destination and back. You can use ping to test an IP address or a computer name. Ping will send a few packets and then report back to you the minimum, maximum, and average round-trip times in milliseconds.

Step 3 Now let's try another utility, tracert. Type **tracert google.com**, and press **Enter**. Tracert traces the route that the packet takes from your computer to the destination computer and back. Here, we see it starting off on a home system and jumping on to a computer that ISP Verizon operates in the Philadelphia area. The packet makes one more hop, on the Verizon system, jumps onto backbone network IAD8.ALTER.NET, and then begins to step down to local computers until it reaches Google. Tracert displays the total number of hops to reach the destination. Here, we see it took 10 hops to reach google.com. It also shows the round-trip time for a packet to travel to each hop. Tracert sends three packets to each point and reports the round-trip time for each one. Note that the specifics you see when you run tracert will be different because you are starting at a different location.

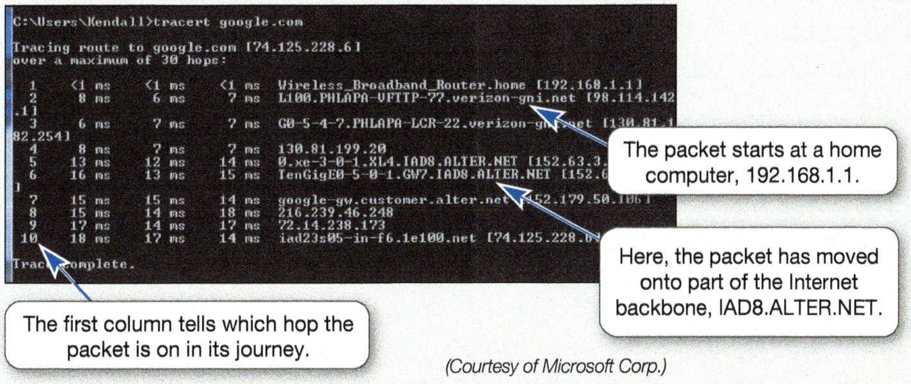

```
C:\Users\Kendall>tracert google.com

Tracing route to google.com [74.125.228.6]
over a maximum of 30 hops:

  1    <1 ms    <1 ms    <1 ms  Wireless_Broadband_Router.home [192.168.1.1]
  2     8 ms     6 ms     7 ms  L100.PHLAPA-VFTTP-77.verizon-gni.net [98.114.142
.1]
  3     6 ms     7 ms     7 ms  G0-5-4-7.PHLAPA-LCR-22.verizon-gni.net [130.81.1
82.254]
  4     8 ms     7 ms     7 ms  130.81.199.20
  5    13 ms    12 ms    14 ms  0.xe-3-0-1.XL4.IAD8.ALTER.NET [152.63.3.
  6    16 ms    13 ms    15 ms  TenGigE0-5-0-1.GW7.IAD8.ALTER.NET [152.6
]
  7    15 ms    15 ms    14 ms  google-gw.customer.alter.net [152.179.50.106]
  8    15 ms    14 ms    18 ms  216.239.46.248
  9    17 ms    14 ms    17 ms  72.14.238.173
 10    18 ms    17 ms    14 ms  iad23s05-in-f6.1e100.net [74.125.228.6]

Trace complete.
```

The packet starts at a home computer, 192.168.1.1.

Here, the packet has moved onto part of the Internet backbone, IAD8.ALTER.NET.

The first column tells which hop the packet is on in its journey.

(Courtesy of Microsoft Corp.)

Step 4 Now let's try to reach a location far away from home base. At the prompt, type **tracert www.cityoflondon.gov.uk** and press **Enter**. Here, a computer in Philadelphia is trying to reach a computer in London, **www.cityoflondon.gov.uk**. The packet moves to a Verizon computer in Philadelphia and, then travels to Washington, D.C. Next, there is a much longer trip time, 85 ms. This is probably as the packet moves across the Atlantic Ocean. The packet finally reaches **www.cityoflondon.gov.uk** [86.54.118.84]. (*Note:* You may see some timeouts along the way. You can exit the tracert command by pressing Ctrl + C.) By using ping and tracert, you can see exactly where a connection is failing.

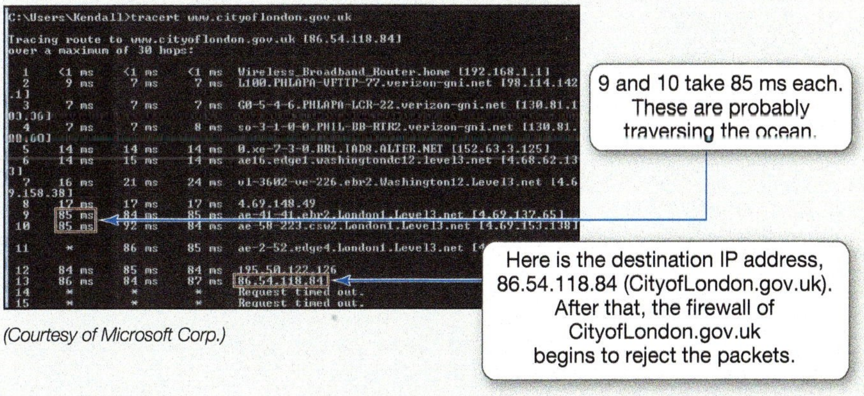

```
C:\Users\Kendall>tracert www.cityoflondon.gov.uk

Tracing route to www.cityoflondon.gov.uk [86.54.118.84]
over a maximum of 30 hops:

  1    <1 ms    <1 ms    <1 ms  Wireless_Broadband_Router.home [192.168.1.1]
  2     9 ms     7 ms     7 ms  L100.PHLAPA-VFTTP-77.verizon-gni.net [98.114.142
.1]
  3     7 ms     7 ms     7 ms  G0-5-4-6.PHLAPA-LCR-22.verizon-gni.net [130.81.1
03.36]
  4     7 ms     7 ms     8 ms  so-3-1-0-0.PHIL-BB-RTR2.verizon-gni.net [130.81.
00.60]
  5    14 ms    14 ms    14 ms  0.xe-7-3-0.BR1.IAD8.ALTER.NET [152.63.3.125]
  6    14 ms    15 ms    14 ms  ae16.edge1.washingtondc12.level3.net [4.68.62.13
3]
  7    16 ms    21 ms    24 ms  vl-3602-ve-226.ebr2.Washington12.Level3.net [4.6
9.158.38]
  8    17 ms    17 ms    17 ms  4.69.148.49
  9    85 ms    84 ms    85 ms  ae-41-41.ebr2.London1.Level3.net [4.69.137.65]
 10    85 ms    92 ms    84 ms  ae-58-223.csw2.London1.Level3.net [4.69.153.138]

 11     *       86 ms    85 ms  ae-2-52.edge4.London1.Level3.net [4

 12    84 ms    85 ms    84 ms  195.50.122.126
 13    86 ms    84 ms    87 ms  86.54.118.84
 14     *        *        *     Request timed out.
 15     *        *        *     Request timed out.
```

9 and 10 take 85 ms each. These are probably traversing the ocean.

Here is the destination IP address, 86.54.118.84 (CityofLondon.gov.uk). After that, the firewall of CityofLondon.gov.uk begins to reject the packets.

(Courtesy of Microsoft Corp.)

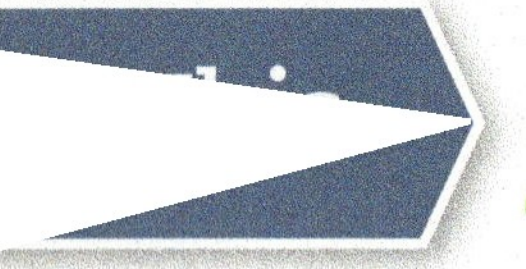

TOOL: App Inventor 2 or Thunkable

An Earthquake Detector

Want your apps to talk to other computers and grab current information from them? Web services allow you to do just that. Thousands of free web services are available to you, like one that shows current stock prices or one that tells the most recent earthquake activity.

In this exercise, you continue your mobile app development by using the Web component of the Connectivity drawer in App Inventor to have a conversation with a web service. You will send a request for the specific information you want and then process it with the **when Web1.GotText** event handler. Using web services, your app can talk with other computers and collect useful information.

(Copyright MIT, used with permission.)

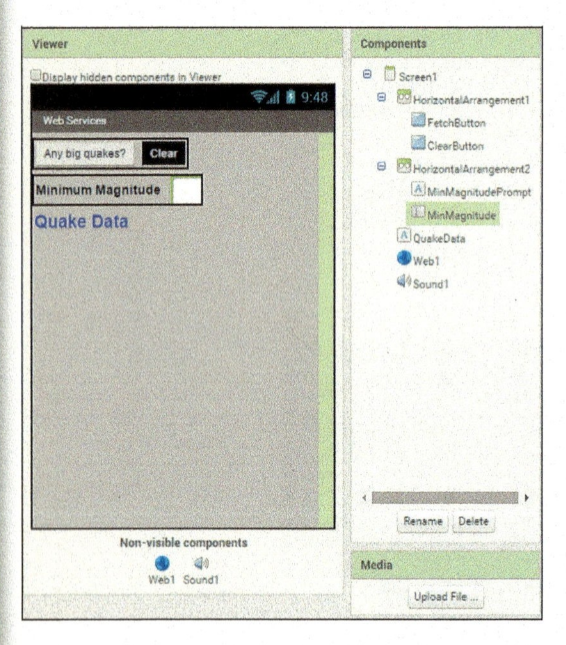

(Copyright MIT, used with permission.)

To create this app for iOS, go to *Thunkable.com*, a programming platform based on App Inventor.

For the instructions for this exercise, go to MyLab IT.

For an overview of this part of the chapter, watch **Chapter Overview Video 13.2**.

Coding and Communicating on the Internet

Learning Outcome 13.2 **You will be able to describe the web technologies used to develop web applications.**

The web uses special languages such as HTML and protocols such as HTTP to facilitate communication between computers that use different system and application software. Communication between any two points on the Internet is possible because of HTML and communication protocols. In this part of the chapter, we explore these topics.

Web Technologies

Let's examine in more detail the programming languages and communication protocols that support the Internet. We introduced some of these ideas in Chapter 10. Now we explore them in a bit more depth.

Web Development

Objective 13.6 *Compare and contrast a variety of web development languages.*

Websites require more than just simple formatting commands. They are interactive and display high-quality audio and video. This is all controlled with a variety of development tools and languages.

HTML

How are web pages formatted? Web pages are text documents formatted using HTML (Hypertext Markup Language). HTML isn't a programming language; rather, it's a set of rules for marking up blocks of text so that a browser knows how to display them.

Blocks of text in HTML documents are surrounded by pairs of HTML tags (such as <b> and </b>, which indicate bolding). HTML tags surround and define HTML content. Each pair of tags and the text between them are collectively referred to as an element. The browser interprets the elements, and appropriate effects are applied to the text. The following is an element from an HTML document:

```
<i>This should be italicized.</i>
```

The browser would display this element as:

This should be italicized.

The first tag, <i>, tells the browser that the text following it should be italicized. The ending tag </i> indicates that the browser should stop applying italics. Note that multiple tags can be combined in a single element:

```
<b><i>This should be bolded and italicized.</i></b>
```

The browser would display this element as:

This should be bolded and italicized.

Tags for creating hyperlinks appear as follows:

```
<a> href="www.pearsonhighered.com">Pearson Higher Education</a>
```

The code <a href="www.pearsonhighered.com"> defines the link's destination. The </a> tag is the anchor tag and creates a link to another resource on the web. Here, the link is to the www.pearsonhighered.com web page. The text between the open and close of the anchor tag, Pearson Higher Education, is the link label. The link label is the text (or image) that's displayed on the web page as clickable text for the hyperlink.

Bits&Bytes CodePen: An Editing Community for Web Designers

Web pages are collections of types of files. There is the HTML file that governs formatting and basic structure, the CSS file that refines the formatting, and often a set of JavaScript code that runs animations or other interactivity on the page (see Figure 13.11).

CodePen (*codepen.io*) is a web-based platform for working with all three kinds of files at once. As a community, developers post code for pages or effects they have created so you can use them to expand your skills.

Figure 13.11 Tools like CodePen make it easier to develop client-side web application code. *(Courtesy of CodePen Inc., Reprinted by permission.)*

How do I see the HTML coding of a web page? HTML documents are merely text documents with tags applied to them. If you want to look at the HTML coding behind any web page, just right-click and select View page source from the shortcut menu in most browsers, and the HTML code for that page will appear, as shown in Figure 13.12.

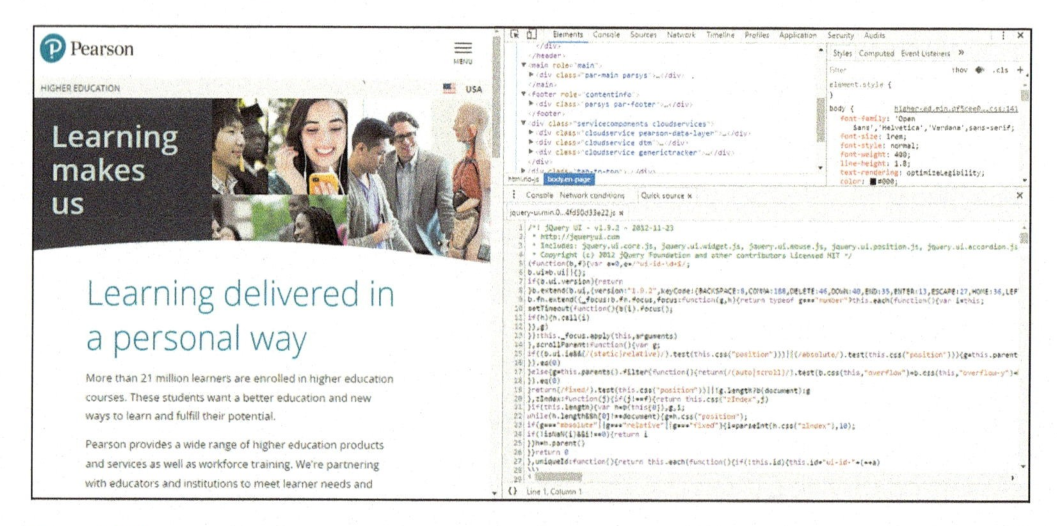

Figure 13.12 You can view the source code of a website by using built-in browser tools. *(Pearson Education, Inc.) > In Chrome, select More tools and then select Developer tools.*

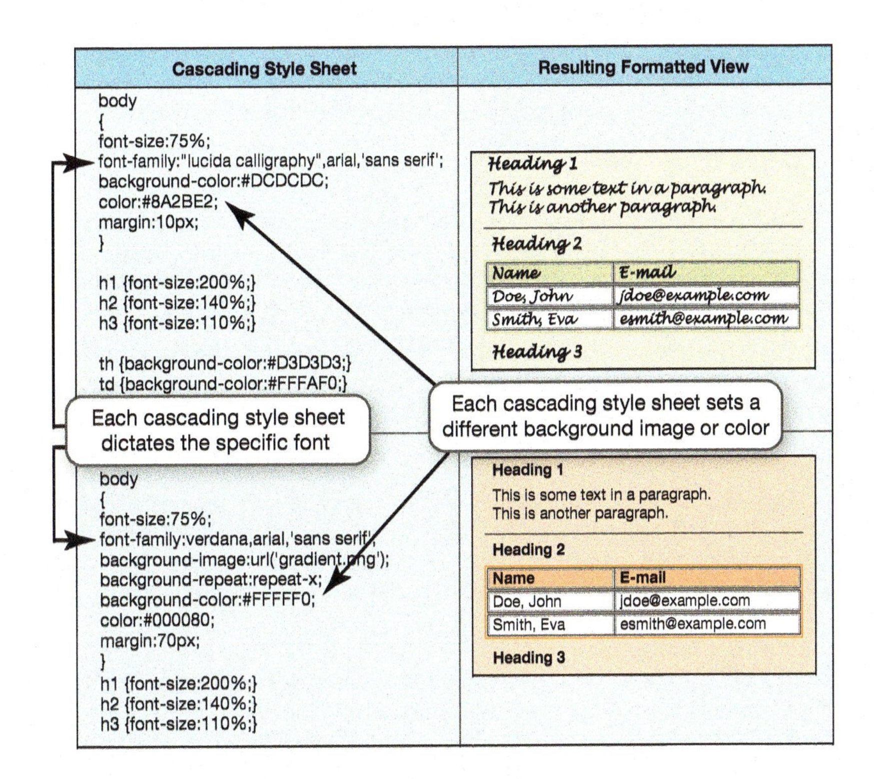

Figure 13.13 Cascading style sheets enable developers to create formatting templates. Just as all the pages of this book have a similar look and feel, one style sheet can control the formatting of many web pages.

What is the current HTML standard? HTML5 is the current version of the HTML standard. It modernizes a number of features in HTML, such as:

- Reducing the need for external plug-ins (like Flash)
- Supporting better error handling
- Supporting advanced animation and audio effects
- Making it easier to draw graphics

To see what is possible in a browser with HTML5, play with the drawing tool at *muro.deviantart.com*.

How can developers easily change the formatting of HTML elements? A cascading style sheet (CSS) is a list of rules that defines in one single location how to display HTML elements. Style rules enable a web developer to define a kind of formatting for each HTML element and apply it to all those elements on a set of web pages. Essentially, a template is created on which the formatting for many web pages within a site will be based. When a global change is necessary, the developer only needs to change the style on the style sheet, and all the elements in the web document will update automatically (see Figure 13.13).

For example, a web page has an heading tag, and all tags in the web page are formatted with a white background and an orange border. Before cascading style sheets, if you wanted to change the border color from orange to yellow, you had to change the border color of each tag. With cascading style sheets, the change from orange to yellow needs to happen only once on the style sheet; all the tags on the web pages then update to yellow without needing individual changes to be made.

Where does the cascading come in? In web documents, there are different layers of styles:

- External: stored in a separate file
- Embedded: stored inside the current HTML document
- Inline: stored within a single line inside the HTML document

Therefore, it's possible that different rules can be created for the same type of element located in different places (external, embedded, or inline). In other words, in an external style sheet, there might be a rule that defines the background color for all paragraphs as blue. In an embedded style sheet, a rule for the background color for all paragraphs might be white. In an inline style sheet, the background color might be light pink. Eventually, all these style sheets must be merged to form one style sheet for the document. This creates conflicts among rules. Therefore, each type of rule is understood to have a specific weight so that when the rules are collected and merged, the rule or style with a higher weight overrides the rule or style with a lower weight. This hierarchy of competing styles creates a cascade of styles ranked according to their assigned weights.

XML and JSON

How is XML different from HTML? The eXtensible Markup Language (XML) describes the content in terms of what data is being described rather than how it's to be displayed. Instead of being locked into standard tags and formats for data, users can build their own markup languages to accommodate particular data formats and needs.

For example, say you want to capture a credit card number. In HTML, the paragraph tags (<P> and </P>) are used to define text and numeric elements. Almost any text or graphic can fall between these tags and be treated as a paragraph. The HTML code could appear as:

```
<P>1234567890123456</P>
```

The browser will interpret the data contained within the <P> and </P> tags as a separate paragraph. However, the paragraph tags tell us nothing about the data contained within them. A better approach lies in creating tags that actually describe the data contained within them. Using XML, we could create a credit card number tag that's used exclusively for credit card data:

```
<credit_card_number>1234567890123456</credit_card_number>
```

What custom XML packages are available? XML has spawned quite a few custom packages for specific communities. A few examples are listed in Table 13.3. These packages illustrate the goal of XML—information exchange standards that can be easily constructed and customized to serve a growing variety of online applications.

Is XML the only format websites use to speak to each other? Another very popular format is used to transfer information between computers, named JSON. It stands for JavaScript Object Notation and is a data interchange standard that is easy for humans to read and write. JSON is faster and easier to use than XML. It consists of name and value pairs like:

```
"firstName": "Kendall"
```

HTTP

Which Internet protocol does a browser use to send requests? The Hypertext Transfer Protocol (HTTP) was created especially for the transfer of hypertext documents across the Internet. (Recall that hypertext documents are documents in which text is linked to other documents or media.)

How does a browser safeguard secure information? Hypertext Transfer Protocol Secure (HTTPS) ensures that data is sent securely over the web. HTTPS is actually an acronym that's the combination of HTTP and Secure Sockets Layer (SSL), a network security protocol. Transport Layer Security (TLS) is an updated extension of SSL. These protocols provide data integrity and security for transmissions over the Internet. Commerce servers use security protocols to protect sensitive information from interception by hackers.

Application Architecture

Objective 13.7 *Compare and contrast server-side and client-side application software.*

Web application software typically consists of two parts: the client-side software, which lives on the user's computer, and the server-side programs, which live on the server computer. Let's look more closely at both aspects of web application development.

Table 13.3 XML Custom Packages

XML Custom Package	Sample Tags
MathML	<times>, <power>, <divide>
X3D	<viewpoint>, <shape>, <scene>
MusicXML	<score>, <beat>, <measure>, <clef>

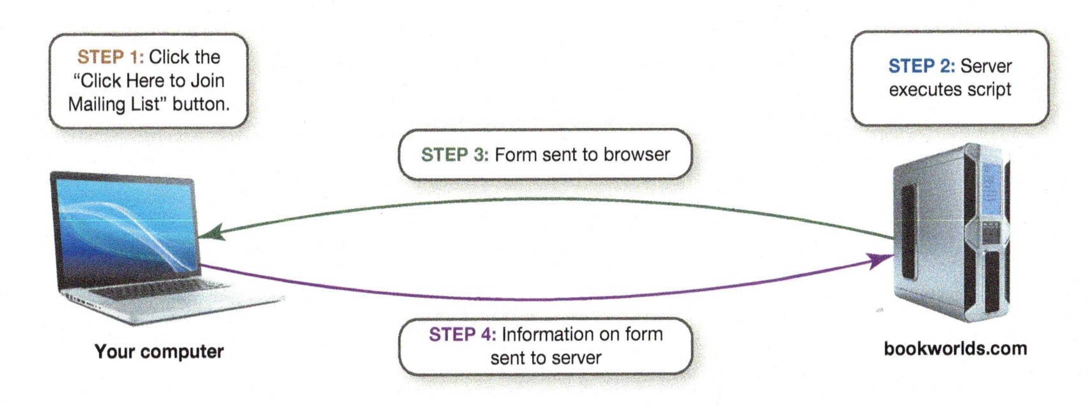

Figure 13.14 Information Flow When a Browser Action Executes a Program on a Web Server. *(Ifong/Shutterstock; Sashkin/Shutterstock)*

Client-Side Web Page Development

In this Sound Byte, we use a popular tool named CodePen to examine client-side web applications that use HTML, CSS, and JavaScript to create amazing results.

Server-Side Applications

What are server-side applications? The web is a client/server network. The type of program that runs on a web server rather than on your computer is referred to as a server-side program. Server-side program execution can require many communication sessions between the client and the server to achieve the goal, but server-side programs can perform very complex operations.

What are popular server-side interfaces? Several methods of designing programs are available that execute on the server and give a web page more sophisticated capabilities. PHP and .NET are very popular choices for coding server-side programs. A very popular choice for those beginning to explore programming server-side code is to use JavaScript, using the Node.js library.

How do the two programs communicate? Let's examine a specific example: to collect a user's answers in a form and then create and send out an e-mail message with that information.

Begin by looking at Figure 13.14.

Step 1. A button on bookworlds.com's website may say, "Click Here to Join Mailing List."

Step 2. Clicking the button may execute a script file, called "mailinglist.js," from the web server hosting the site.

Step 3. This file generates a form that's sent to your browser. The form includes fields for a name and e-mail address and a Submit button.

Step 4. After you fill in the fields and click the Submit button, the mailinglist.js program sends the information back to the server. The server then records the information in a database.

Client-Side Applications

What are client-side applications? A client-side program is a program that runs on the client computer and requires no interaction with a web server. Once a web server processes a web page and sends the page to the client computer that requested it, the receiving computer can't get new data from the server unless a new request is made. If interactivity is required on a web page, this exchange of data between the client and server can make the interactivity inefficient and slow. Often it's more efficient to run programs on your computer (the client). Therefore, client-side programs are created.

How is client-side programming done? There are two main approaches to client-side programming:

1. An HTML embedded scripting language, which tucks programming code directly into the HTML of a web page. The most popular embedded language is JavaScript.
2. An applet, which is a small application actually located on the server. When requested, the applet is downloaded to the client. It's there ready to run when needed, without additional data transfers between the client and server. The Java language is the most common language used to create applets for use in browsers.

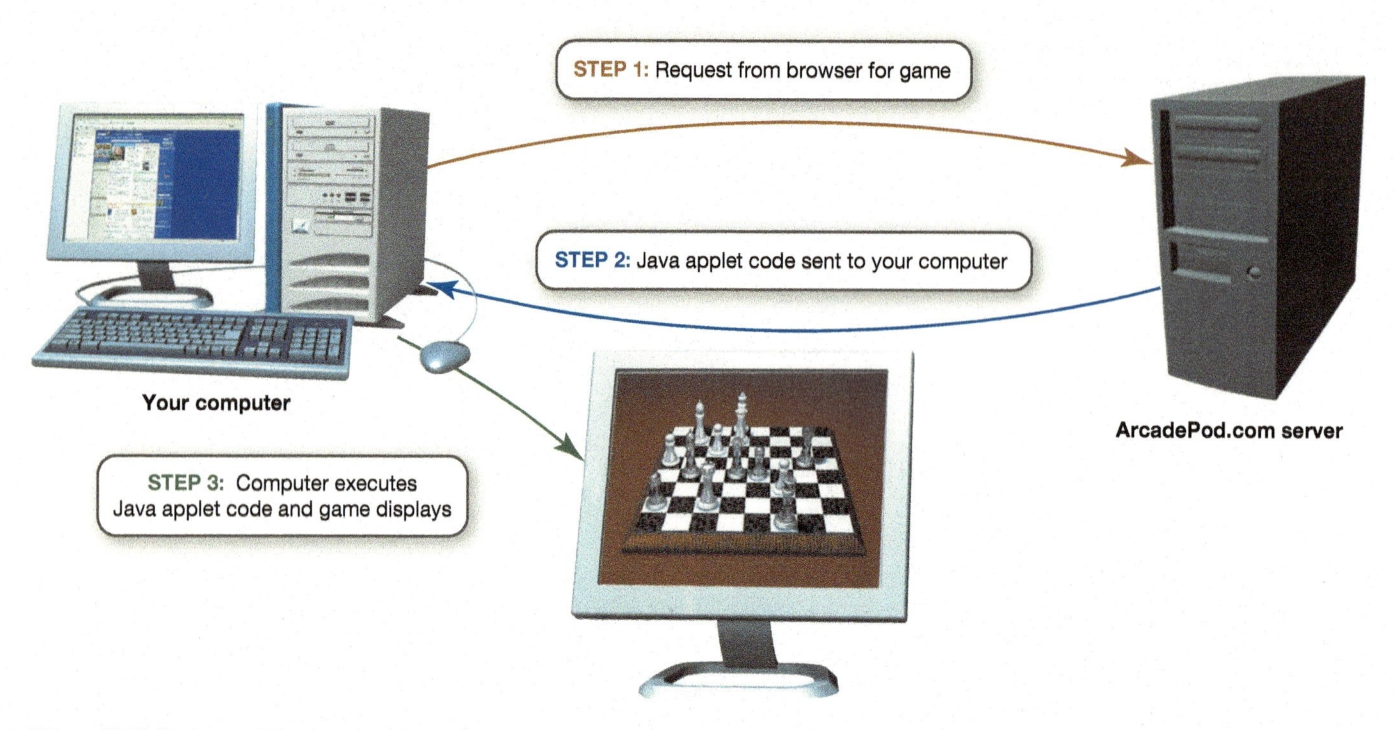

Figure 13.15 Deployment of a Java Applet on a Computer. *(Pearson Education, Inc.)*

Isn't there a delay as an applet downloads? You may experience some delay in functionality while waiting for a Java applet to download to your computer, but once the applet arrives, it can execute all its functions without further communication with the server. Let's look at an example, as shown in Figure 13.15.

Step 1. Your browser makes contact with a game on the ArcadePod (*arcadepod.com*) game site and makes your request to play a game.

Step 2. The web server returns the Java applet that contains all the code to run the game on your computer.

Step 3. Your computer executes the applet code, and the game runs on your computer.

What are some scripting technologies? Dynamic HTML (DHTML) is a combination of technologies—HTML, cascading style sheets, and JavaScript—that's used to create lively and interactive websites. DHTML technologies allow a web page to change (that is, to be dynamic) after it's been loaded. Such change generally occurs in response to user actions, such as when you click your mouse or mouse over objects on a page. DHTML also brings special effects to otherwise static web pages without requiring users to download and install plug-ins or other special software.

What is JavaScript? JavaScript is a commonly used scripting language for creating DHTML effects. JavaScript is often confused with the Java programming language because of the similarity in their names. However, although they share some common elements, the two languages are different.

With JavaScript, HTML documents can be made responsive to mouse clicks and typing. For example, JavaScript is often used to validate the information you input in a web form, for example, to make sure you filled in all required fields.

When JavaScript code is embedded in an HTML document, it's downloaded to the browser with the HTML page. All actions dictated by the embedded JavaScript commands are executed on the client computer. Without JavaScript and other scripting languages, web pages would be lifeless.

How does JavaScript control the components of a web page? Just as cascading style sheets organize and combine the attributes of objects on a web page, JavaScript uses the Document Object Model (DOM) to organize the objects and page elements. The Document Object

Bits&Bytes Editors for Web Development

You can type up an HTML web page in any text editor, even in as basic a tool as Notepad. But to get modern conveniences like smart auto-completion and color coding, try using an editor meant for web development. Brackets is a very popular open source choice. It supports a live preview so you can see the changes in the target web design instantly as you type changes in the HTML and CSS code. Because it's open source, new extensions are constantly being developed that do useful things like beautify the code or run a W3C validator. Check out Brackets (*brackets.io*) and see whether it helps you be more productive.

Model defines every item on a web page—including graphics, tables, and headers—as an object. Then with DOM, as with cascading style sheets, web developers can easily change the look and feel of these objects.

Where is web programming headed? Web pages can interact with servers at times other than when a page is being fetched. Web pages can have an ongoing exchange of information, communicating with the server and updating information on the page without requiring the user to do a page refresh or leave the page. AJAX is the acronym for Asynchronous JavaScript and XML, a group of technologies that facilitates the creation of this style of web applications. A huge number of web application frameworks are now available, like Ruby on Rails. These packages make it simpler to create an application that talks to a web server, queries a database, and then presents information in a nicely formatted template.

Communications over the Internet

In this section, we explore electronic communications media and how to keep your information exchanges efficient and secure.

Types of Internet Communication

Objective 13.8 *Discuss the mechanisms for communicating via e-mail and instant messaging.*
Two of the most common communication methods on the Internet are e-mail and instant messaging.

E-Mail

Who invented e-mail? In 1971, Ray Tomlinson, a computer engineer who worked on the development of the ARPANET, the precursor to the Internet, created e-mail. E-mail grew from a simple program that Tomlinson wrote to enable computer users to leave text messages for each other on a single machine. The logical extension of this was sending text messages between machines on the Internet. E-mail became the most popular application on ARPANET; by 1973, it accounted for 75% of all data traffic.

How does e-mail travel the Internet? Just like other kinds of data that flow along the Internet, e-mail has its own protocol. The Simple Mail Transfer Protocol (SMTP) is responsible for sending e-mail along the Internet to its destination. SMTP is part of the Internet Protocol suite. As in most other Internet applications, e-mail is a client/server application. On the way to its destination, your mail will pass through e-mail servers—specialized computers whose sole function is to store, process, and send e-mail.

Where are e-mail servers located? If your ISP provides you with an e-mail account, it runs an e-mail server that uses SMTP. For example, as shown in Figure 13.16, say you're sending an e-mail message to your friend Cheyenne. Cheyenne uses Verizon as her ISP. Therefore, your e-mail to her is addressed to Cheyenne@verizon.net.

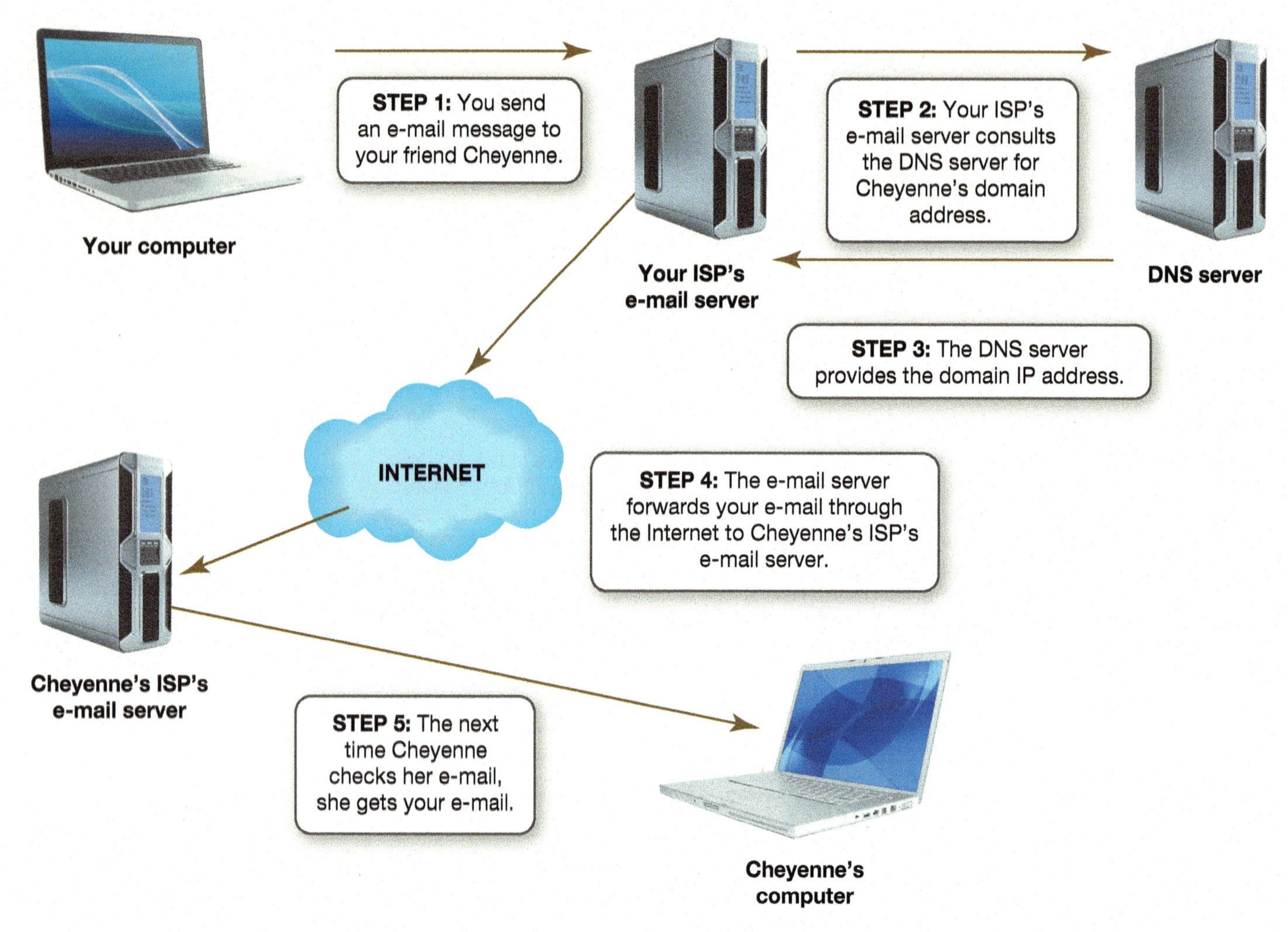

Figure 13.16 Deployment of a Java Applet on a Computer. *(Ifong/Shutterstock, Sashkin/Shutterstock, LdF/iStock/Getty Images)*

Let's follow that e-mail:

Step 1. When you send the e-mail message, your ISP's e-mail server receives it.

Step 2. The e-mail server reads the domain name (verizon.net) and communicates with a DNS server to determine the location of verizon.net.

Step 3. The DNS server turns the domain name into the associated IP address.

Step 4. The e-mail message is forwarded to verizon.net, and it arrives at an e-mail server maintained by Cheyenne's ISP.

Step 5. The e-mail is then stored on Cheyenne's ISP's e-mail server. The next time Cheyenne checks her e-mail, she'll receive your message.

How are we able to send files as attachments? SMTP was designed to handle text messages. When the need arose to send files by e-mail, the Multipurpose Internet Mail Extensions (MIME) specification was introduced. All e-mail client software now uses this protocol to attach files. E-mail is still sent as text, but the e-mail client using the MIME protocol now handles the encoding and decoding for users. When attaching a file, your e-mail client transparently encodes and decodes the file for transmission and receipt.

Bits&Bytes AI and Your Inbox

The use of artificial intelligence (AI) to automate our lives seems to have no end. Gmail now uses AI techniques to suggest reply messages for you as e-mails arrive. For example, the Smart Reply feature may suggest "Sounds great!" as a response to an e-mail it feels is making a suggestion. It was trained on a collection of over 500,000 e-mails.

Smart Compose goes one step further. As you are writing an e-mail, it suggests phrases that would complete your thought. Type "Take a look" and Smart Compose might suggest "and let me know what you think." You may have to enable Smart Compose as an experimental feature in Gmail if you are not seeing its suggested phrases (see Figure 13.17).

Other things Gmail can help with include suggesting e-mails you should reply to but have forgotten and automatically adding people to your Contacts list when you first send them an e-mail. You can also add security features to the e-mails you send that prevent readers from forwarding or printing them.

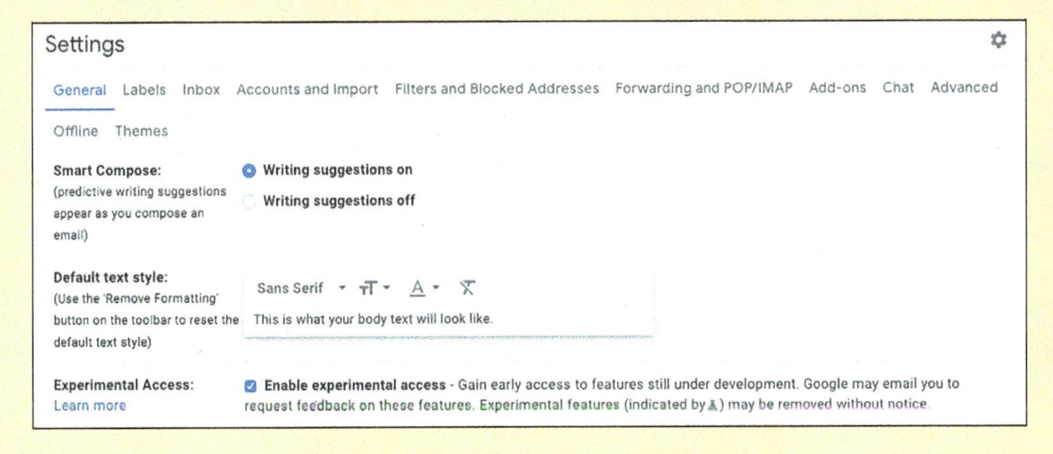

Figure 13.17 When enabled, Smart Compose suggests phrases that complete your thought. *(Courtesy of Google, Inc.)*

Instant Messaging

What do you need to run instant messaging? Instant messaging (IM) requires the use of a client program that connects to an IM service. WhatsApp and Facebook Messenger are two popular IM services in use today.

How does instant messaging work? The client software running on your device makes a connection with the chat server and provides it with connection information for your device (such as the IP address), as shown in Figure 13.18.

Because both your device and your friend's device have the connection information (the IP addresses) for each other, the server isn't involved in the chat session. Chatting takes place directly between the two devices over the Internet.

Is sending an IM secure? Most IM services do not use a high level of encryption for their messages—if they bother to use encryption at all. In addition to viruses, worms, and hacking threats, IM systems are vulnerable to eavesdropping, in which someone using a packet sniffer listens in on IM conversations. Although several measures are under way to increase the security of this method of real-time communication, major vulnerabilities still exist. Furthermore, employers can install monitoring software to record IM sessions. Therefore, it's not a good idea to send sensitive information by IM because it's susceptible to interception and possible misuse by hackers.

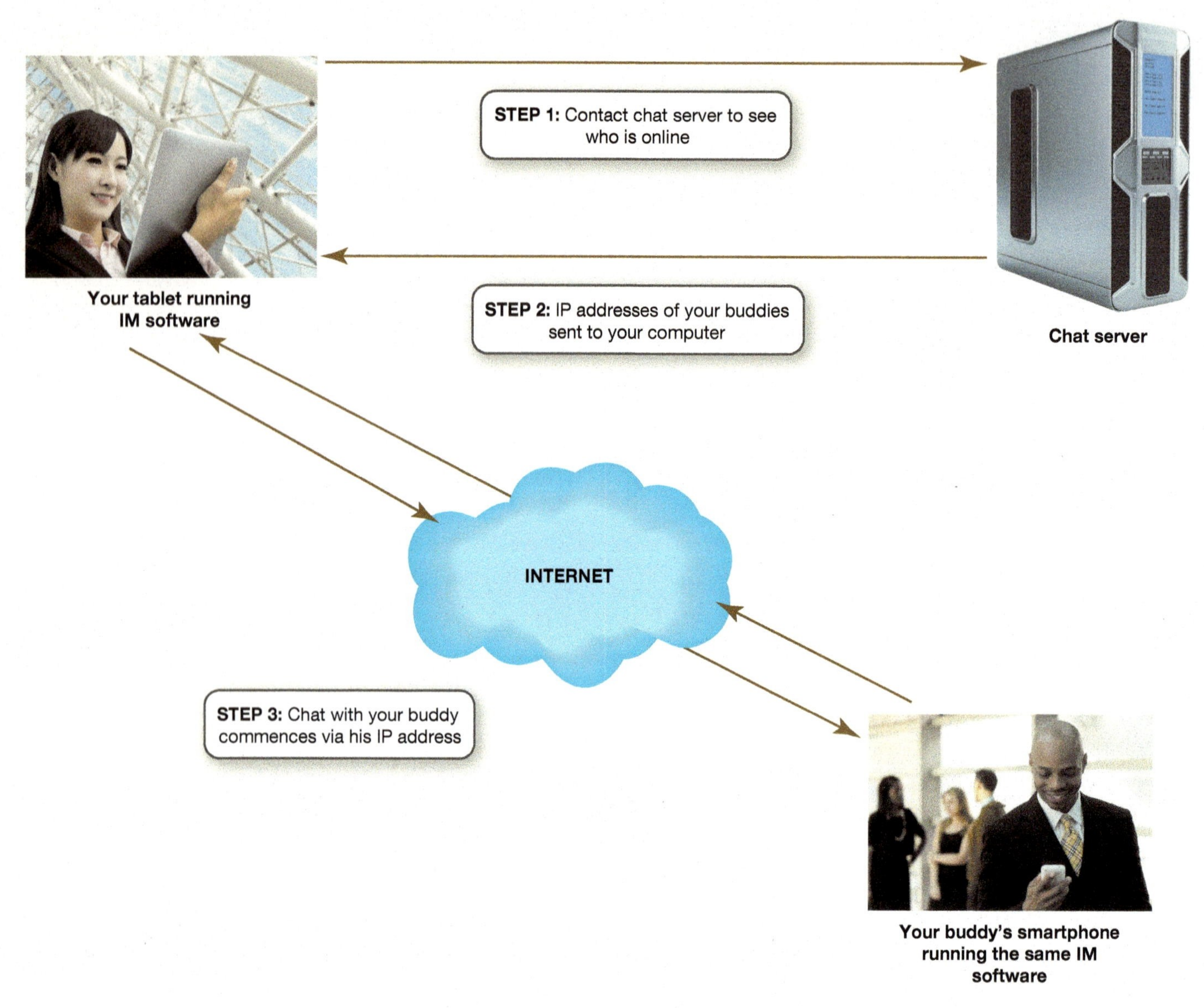

Figure 13.18 How an Instant Messaging Program Works. *(Tom Wang/Shutterstock; Sashkin/Shutterstock; Tyler Olson/Shutterstock)*

Encryption

Objective 13.9 *Explain how data encryption improves security.*

Can other people read my e-mail? E-mail is highly susceptible to being read by unintended parties because it's sent in plaintext. In addition, copies of your e-mail messages may exist, temporarily or permanently, on numerous servers as the messages make their way through the Internet. To protect your sensitive e-mail messages, encryption practices are available.

How do I encrypt my e-mail? Many e-mail services offer built-in encryption. ProtonMail (*protonmail.com*) and Tutanota *(tutanota.com)* offer free versions of their secure e-mail services. You can sign up on their websites and experiment with sending encrypted e-mail. You don't need to abandon your current e-mail accounts—just use your secure account when you require secure communications.

How does encryption work? Encryption refers to the process of coding your e-mail so that only the person with the key to the code (the intended recipient) can decipher the message. There are two basic types of encryption: private-key encryption and public-key encryption.

What is private-key encryption? In private-key encryption, only the two parties involved in sending the message have the code. This could be a simple shift code in which letters of the alphabet are shifted to a new position (see Figure 13.19).

Bits&Bytes Random Numbers: We Wouldn't Have Encryption Without Them!

E-mail encryption, SSL encryption, and just about anything we do to achieve privacy on the Internet requires random numbers. Encryption is accomplished using random number sequences, which are sequences of numbers in which no patterns can be recognized. Even for an e-commerce transaction that uses SSL encryption to encode your credit card number, as many as 368 bits of random data might be needed. Only 128 bits are needed for the encryption key, but other random data is needed to create authentication codes and to prevent replay attacks. Replay attacks occur when hackers attempt to copy packets traveling across the Internet and extract data (such as encryption codes) from them. The hackers then can reuse the data to gain access to networks or transactions.

So where do all these random numbers come from? Generating true random sequences is more difficult than it sounds. Most random number generators are really pseudorandom because they're all based on some sort of pattern to generate the numbers. However, the site *random.org* is dedicated to providing true random numbers for web applications such as encryption algorithms. The numbers are generated based on atmospheric noise, which is truly random. The noise is gleaned from radios not tuned to a particular station, so they're broadcasting static. Anyone can access the website and download random numbers to be used for encryption or other vital services such as lottery drawings. For more information, check out *random.org*.

For example, in a two-position right-shift code, the letter *a* becomes *c*, *b* becomes *d*, and so on. Alternatively, it could be a more complex substitution code (a = h, b = r, c = g, etc.). The main problem with private-key encryption is key security. If someone steals a copy of the code or is savvy about decoding, the code is broken.

What is public-key encryption? In public-key encryption, two keys, known as a key pair, are created. You use one key for coding and the other for decoding. The key for coding is generally distributed as a public key. You can place this key on your website, for instance. Anyone wishing to send you a message can then download your public key and code the message using your public key.

When you receive the message, you use your private key to decode it. You're the only one who ever possesses the private key, and therefore it's highly secure. The keys are generated in such a way that they can work only with each other. The private key is generated first. The public key is then generated using a complex mathematical formula, often using values from the private key. The computations are so complex that they're considered unbreakable. Both keys are necessary to decode a message. If one key is lost, the other key can't be used by itself.

A = C	N = P
B = D	O = Q
C = E	P = R
D = F	Q = S
E = G	R = T
F = H	S = U
G = I	T = V
H = J	U = W
I = K	V = X
J = L	W = Y
K = M	X = Z
L = N	Y = A
M = O	Z = B

The word **COMPUTER** using the two-position code at the left now becomes:

EQORWVGT

This is difficult to interpret without the code key at the left.

Figure 13.19 Writing the word *COMPUTER* using a two-position right-shift encryption code results in *EQORWVGT*.

Which type of encryption is used on the Internet? Public-key encryption is the most commonly used encryption on the Internet. The OpenPGP (Open Pretty Good Privacy) standard is implemented by a number of products, like Canary Mail for Mac and EverDesk for Windows. OpenPGP is free to use and available for download at *openpgp.org*. After obtaining the OpenPGP software, you can generate key pairs to provide a private key for you and a public key for the rest of the world.

What does a key look like? A key is a binary number. Keys vary in length, depending on how secure they need to be. A 12-bit key has 12 positions and might look like this:

100110101101

Longer keys are more secure because they have more values that are possible. A 12-bit key provides 4,096 possible values, whereas a 40-bit key allows for 1,099,511,627,776 possible values. The key and the message are run through a complex algorithm in the encryption program that converts the message into unrecognizable code. Each key turns the message into a different code.

Is a private key really secure? Because of the complexity of the algorithms used to generate key pairs, it's impossible to deduce the private key from the public key. However, that doesn't mean your coded message can't be cracked. A brute-force attack occurs when hackers try every possible key combination to decode a message. This type of attack can enable hackers to deduce the key and decode the message.

Do We Really Want Strong Encryption?

Encryption has fueled the enormous growth of electronic commerce. Because people feel safe in transmitting private data like credit card numbers across the Internet, online banking, shopping, and communication have exploded. But ardent shoppers are not the only people relying on encryption to keep their information hidden.

Criminals are also increasing their use of encryption. "Going dark" is the term for encrypting the phone and e-mail accounts that carry the details of their illegal transactions. For federal agencies, it becomes more and more difficult to gather the evidence needed for a conviction as easy-to-use encryption software spreads. The encryption standards are now so strong that there is no way to break into some encrypted messages. Apple announced with the release of iOS 8 that the default encryption its products support could not be broken by Apple. So if a government agency received a warrant for specific records, Apple could not possibly comply and hand over readable information.

This exact scenario occurred when the FBI began investigating a terrorist attack that took place in San Bernardino, California. The couple who carried out the attack were both killed in the subsequent shootout with police. The FBI wanted to access the iPhone the shooter used to search for additional background on the crime. Neither the FBI nor the National Security Agency (NSA) was able to enter the phone. Apple's security system meant that if anyone tried to enter the phone with the incorrect password more than 10 times, all data on the phone would be erased.

Concern about just such an event has pushed national security officials and law enforcement agencies to lobby for tech companies to undermine their own encryption schemes. But if the encryption were weakened, allowing some mechanism for it to be broken upon request, there would be no information that was completely secure. This idea of a backdoor or a key under the doormat, allowing every secure message to be broken, has been controversial for years, and the debate continues (see Figure 13.20).

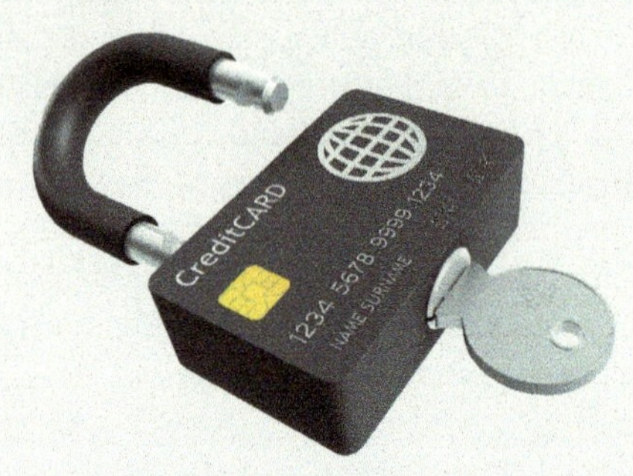

Figure 13.20 Should security experts be forced by law to leave a way to break their encryption systems? *(Johan2011/123RF)*

The FBI was granted a court order forcing Apple to create a new operating system that could be used to allow law enforcement to enter the phone. Apple refused. As tensions grew, it suddenly was announced that the Department of Justice had found a way into the phone. The *Washington Post* reported that the FBI had hired professional hackers to do the job.

The opposition, which includes many security experts as well as civil libertarians, points to the fact that this would be bad for commercial business. Deliberately making security products less secure may be useful to some government agencies but exposes all users to possible security leaks by hackers.

So do you want strong encryption? And just how strong?

What kind of key is considered safe? In the early 1990s, 40-bit keys were thought to be completely resistant to brute-force attacks and were the norm for encryption. However, in 1995, a French programmer used a unique algorithm of his own and 120 workstations simultaneously to attempt to break a 40-bit key. He succeeded in just 8 days. After this, 128-bit keys became the standard. However, using supercomputers, researchers have had some success cracking 128-bit encryption. Therefore, strong encryption now calls for 256-bit keys. It's believed that even with the most powerful computers in use today, it would take hundreds of billions of years to crack a 256-bit key.

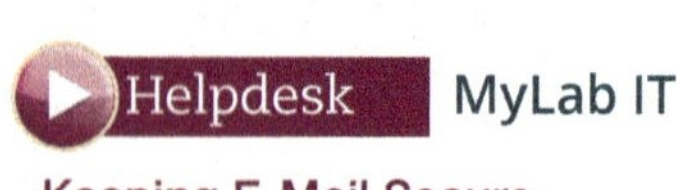

Keeping E-Mail Secure

In this Helpdesk, you'll play the role of a helpdesk staffer, fielding questions about how e-mail works and how messages are kept secure.

Do corporations use special encryption? Businesses often pay for encryption services that also provide other features such as confirmation of message delivery, message tracking, and overwriting of e-mail messages when they're deleted to ensure that copies don't exist. Companies such as Zix Corporation (*zixcorp.com*) provide these higher levels of service to businesses.

Trends in IT Cognitive Computing

We're used to thinking of computers as pure calculators, good at performing fixed tasks with specific instructions very quickly. But computers have begun to mimic how our minds work. This is referred to as *cognitive computing*.

Imagine a computer program that can read through your Facebook posts, your tweets, and the online essays in your blog and then "think" about them. Instead of just returning a word count or computing the writing level of your work, what if it could return the most dominant emotion of your writing—perhaps anger with a strong undercurrent of fear? IBM's Watson tone analyzer program does that now (try it out by searching for the Watson tone analyzer demo).

One of the hallmarks of cognitive computing is strong natural language processing. Cognitive computing combines natural language processing with advanced machine learning algorithms, pattern recognition, and data mining to be able to extract new insights from information faster and on a grander scale than we can as humans. One application uses a vast array of legal information to create an artificially intelligent lawyer, who can conduct legal research at blazing speeds. Another project is looking at creating spontaneous networks of idle cars, which have cameras, sensors, and computers onboard already, so they can communicate. Together these connected cars could produce information about where the nearest open parking space is, if there is a gas leak nearby, or even locate missing persons.

The applications are not just static either. They can learn over time from their past mistakes and improve their performance. Veterinarians are using IBM's artificial intelligence supercomputer Watson to digest hundreds of thousands of pages of medical journals, textbooks, and current research to assist in diagnosing animal illness (see Figure 13.21). As more information is added and more cases are presented, the software builds better techniques for diagnosis. Meanwhile, businesses are using cognitive computing to spot trends and coordinate feedback from a variety of sources to adjust their products. We may soon have to work hard to outthink our laptops.

Figure 13.21 IBM's Watson can communicate with natural language and deduce emotion, tone, and mood from visual and textual data. *(Caroline Seidel/Dpa picture alliance archive/Alamy Stock Photo)*

Before moving on to the Chapter Review:

1. Watch Chapter Overview Video 13.2.
2. Then take the Check Your Understanding quiz.

Check Your Understanding//Review & Practice

For a quick review to see what you've learned so far, answer the following questions.

multiple choice

1. JSON, JavaScript Object Notation, is useful because it
   a. allows web services to exchange information.
   b. uses OpenPGP to encrypt data.
   c. makes the Document Object Model usable in HTML.
   d. is hexadecimal codes that execute quickly on computers.
2. Cascading style sheets allow you to change quickly
   a. from server-side processing to client-side processing.
   b. the rules for applying formatting to an HTML document.
   c. to a more secure transport protocol.
   d. from HTML to XML.
3. Node.js and PHP are languages used for
   a. e-mail transfer protocol.
   b. client-side programming.
   c. formatting of HTML.
   d. server-side programming.
4. A server-side program
   a. lives on the user's computer.
   b. typically has information exchange with the client.
   c. requires the use of SMTP protocol.
   d. defines every item on a web page.
5. A key pair is a set of two keys you use to
   a. encrypt and decrypt an e-mail message.
   b. open doors into websites so you can pass through a firewall.
   c. execute a CSS script.
   d. attach a photo to an e-mail message.

chew on this

Where is web programming headed? It will be driven by the needs of users, so think about what kinds of things you need web pages to do for you that they currently cannot do. Should biometric identification replace passwords, for example?

MyLab IT

Go to **MyLab IT** to take an autograded version of the *Check Your Understanding* review and find all media resources for the chapter.

For the IT Simulation for this chapter, see MyLab IT.

13 Chapter Review

Summary

Part 1
Inner Workings of the Internet

Learning Outcome 13.1 **You will be able to explain how the Internet is managed and the details of how data is transmitted across the Internet.**

Internet Management and Networking

Objective 13.1 *Describe the management of the Internet.*

- Management of the Internet is carried out by several nonprofit organizations and user groups such as the Internet Society, the Internet Engineering Task Force (IETF), the Internet Architecture Board (IAB), the Internet Corporation for Assigned Names and Numbers (ICANN), and the World Wide Web Consortium (W3C). Each group has different responsibilities and tasks. Currently, the US government funds a majority of the Internet's costs. Other countries also pay for Internet infrastructure and development.

Objective 13.2 *Explain how the Internet's networking components interact.*

- The largest paths of the Internet, along which data travels the most efficiently and quickly, make up the Internet backbone. Homes and all but the largest businesses connect to the Internet through regional or local connections, which then connect to the Internet through the entities that make up the Internet backbone. The largest businesses, educational centers, and some government agencies make up the Internet backbone.
- Optical carrier (OC) lines are high throughput and are used to connect large corporations to their ISPs.
- Service providers use an Internet exchange point (IXP) to connect directly to each other, reducing cost and latency time.
- Individual users enter their ISP through a point of presence (POP), a bank of routers and switches through which many users can connect simultaneously.

Objective 13.3 *List and describe the Internet protocols used for data transmission.*

- Data is transmitted along the Internet using packet switching. Data is broken up into discrete units known as *packets*, which can take independent routes to the destination before being reassembled.
- A computer protocol is a set of rules for exchanging electronic information. Although many protocols are available on the Internet, the main suite of protocols used to move information over the Internet is TCP/IP. The suite is named after the original two protocols that were developed for the Internet: the Transmission Control Protocol (TCP) and the Internet Protocol (IP). TCP is responsible for preparing data for transmission, but IP actually sends data between computers on the Internet.

Internet Identity

Objective 13.4 *Explain how each device connected to the Internet is assigned a unique address.*

- An IP address is a unique number assigned to all computers connected to the Internet. The IP address is necessary so that packets of data can be sent to a particular location (computer) on the Internet.

Objective 13.5 *Discuss how a numeric IP address is changed into a readable name.*

- A domain name is merely a name that stands for a certain IP address and that makes it easier for people to remember it. DNS servers act as the phone books of the Internet. They enable your computer to find out the IP address of a domain by looking up its corresponding domain name.

Part 2

Coding and Communicating on the Internet

Learning Outcome 13.2 **You will be able to describe the web technologies used to develop web applications.**

Web Technologies

Objective 13.6 *Compare and contrast a variety of web development languages.*

- The Hypertext Transfer Protocol (HTTP) is the protocol used on the Internet to display web pages in your browser.
- The Hypertext Markup Language (HTML) is a set of rules for marking up blocks of text so that a browser knows how to display them. Most web pages are generated with at least some HTML code. Blocks of text in HTML documents are surrounded by a pair of tags (such as <b> and </b> to indicate bolding). These tags and the text between them are referred to as *elements*. The current HTML standard is HTML5, which offers much better support for media like audio and video.
- Because HTML wasn't designed for information exchange, the eXtensible Markup Language (XML) was created. Instead of locking users into standard tags and formats for data, XML enables users to create their own markup languages to accommodate particular data formats and needs. JSON is another standard supporting the exchange of information between web services.

Objective 13.7 *Compare and contrast server-side and client-side application software.*

- Software is developed both on the server side and on the client side to make web pages more responsive and interactive. Server-side programming includes Node.js and PHP. Client-side programs can run faster because they don't need to wait for information exchanges with the server. Embedded JavaScript and Java applets are two common approaches for writing client-side programs.

Communications over the Internet

Objective 13.8 *Discuss the mechanisms for communicating via e-mail and instant messaging.*

- Simple Mail Transfer Protocol (SMTP) is the protocol responsible for sending e-mail over the Internet. E-mail is a client/server application. E-mail passes through e-mail servers, whose functions are to store, process, and send e-mail to its ultimate destination. ISPs and portals such as Gmail maintain e-mail servers to provide e-mail functionality to their customers. Your ISP's e-mail server uses DNS servers to locate the IP addresses for the recipients of the e-mail you send.
- Instant messaging (IM) software communicates with a chat server to deliver communications quickly to your friends who are online. It's not secure and can be vulnerable to eavesdropping.

Objective 13.9 *Explain how data encryption improves security.*

- Encryption software, such as OpenPGP, is used to code messages so that they can be decoded only by authorized recipients.

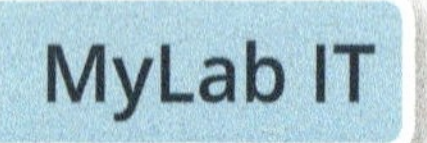

Be sure to check out **MyLab IT** for additional materials to help you review and learn. And don't forget to watch the Chapter Overview Videos.

Key Terms

AJAX **525**
applet **523**
cascading style sheet (CSS) **521**
circuit switching **507**
client/server model **505**
client-side program **523**
commerce server **506**
computer protocol **506**
connectionless protocol **512**
connection-oriented protocol **512**
Document Object Model (DOM) **524**
domain name system (DNS) server **514**
dotted decimal number (dotted quad) **509**
dynamic addressing **511**
Dynamic Host Configuration Protocol (DHCP) **512**
dynamic HTML (DHTML) **524**
element **519**
e-mail server **525**
encryption **528**
eXtensible Markup Language (XML) **522**
file server **506**
handshaking **512**
hexadecimal digit **510**
HTML embedded scripting language **523**
HTML tag **519**
Hypertext Transfer Protocol (HTTP) **522**
Hypertext Transfer Protocol Secure (HTTPS) **522**
Internet backbone **505**
Internet cache **515**
Internet Corporation for Assigned Names and Numbers (ICANN) **509**
Internet exchange point (IXP) **505**
Internet of Things (IoT) **510**
Internet Protocol (IP) **508**
Internet Protocol version 4 (IPv4) **510**
Internet Protocol version 6 (IPv6) **510**
IP address **509**
JavaScript **524**
JSON **522**
key pair **529**
Multipurpose Internet Mail Extensions (MIME) **526**
negative acknowledgment (NAK) **512**
octet **510**
open system **507**
OpenPGP **529**
optical carrier (OC) line **505**
packet (data packet) **507**
packet switching **507**
point of presence (POP) **505**
positive acknowledgment (ACK) **512**
private key **529**
private-key encryption **528**
proprietary system **507**
public key **529**
public-key encryption **529**
root DNS server **515**
second-level domain **514**
Secure Sockets Layer (SSL) **522**
server-side program **523**
Simple Mail Transfer Protocol (SMTP) **525**
static addressing **511**
T line **505**
TCP/IP **508**
three-way handshake **512**
Transmission Control Protocol (TCP) **508**
Transport Layer Security (TLS) **522**
User Datagram Protocol (UDP) **512**
web server **506**

Chapter Quiz // Assessment

For a quick review to see what you've learned, answer the following questions. Submit the quiz as requested by your instructor. If you are using **MyLab IT**, the quiz is also available there.

multiple choice

1. The Internet is primarily funded by
   a. your state income taxes.
   b. voluntary contributions.
   c. your federal income taxes.
   d. Google.
2. The collection of large national and international networks used to establish the Internet is called the
   a. Internet backbone.
   b. highway system.
   c. OC.
   d. T1.
3. The DNS server
   a. lets you look up an IP address from a domain name.
   b. lets you look up a domain name from an IP address.
   c. requires you to set up a specific computer in your home.
   d. is not often used any longer.
4. IPv6 uses 128 bits to assign a(n)
   a. data packet to each transmission.
   b. memory address to the CPU.
   c. destination Internet address to each e-mail.
   d. address to every device connected to the Internet.
5. HTTPS is different from HTTP because
   a. of how it transforms or reformats the data.
   b. it extracts data from source databases.
   c. it is a combination of HTTP and the Secure Sockets Layer (SSL).
   d. it stores data in the data warehouse.
6. A private key is important
   a. to reboot system hardware.
   b. to reset your public key if you lose a password.
   c. because it enables you to use MIME.
   d. to decode an encrypted message.
7. Encryption is used to
   a. shrink the amount of space it takes to store e-mail.
   b. code your e-mail so only one person can decipher it.
   c. transfer e-mail more quickly.
   d. make shopping on the Internet more available.
8. An applet is
   a. a program distributed by Apple Inc.
   b. a small application located on a server computer.
   c. a hardware device that encrypts e-mail messages.
   d. a programming language used to help create dynamic web pages.
9. SMTP is the set of rules used to
   a. shrink the amount of space it takes to store e-mail.
   b. send e-mail along the Internet to its destination.
   c. encrypt e-mail messages.
   d. make shopping on the Internet more available.
10. CSS stands for
   a. continuous signal source.
   b. coding smart source code.
   c. cascading style sheet.
   d. color swatch selection.

true/false

_____ **1.** The POP is a bank of switches that allows many users to connect to the ISP at the same time.
_____ **2.** Your Internet service provider has a default DNS server to translate a URL into an IP address.
_____ **3.** Instant messaging is secure because of the type of encryption used.
_____ **4.** MIME is the protocol that supports sending files as e-mail attachments.
_____ **5.** The term *dotted quad* means you have four dots in a row in your Java source code.

What do you think now?

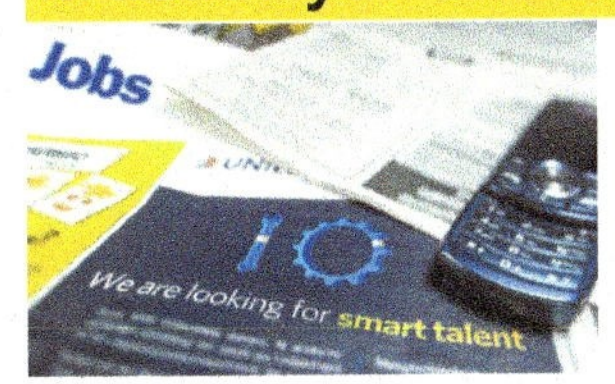

(Clement Philippe/Arterra Picture Library/Alamy Stock Photo)

Think about the breadth of knowledge that is needed for web development. What set of education, training, and skills would you now say you need for a career in web development?

Team Time

Using Crowdsourcing in a Business

background

Determining the right merchandise to sell is critical. Having a website that's designed to use crowdsourcing—either to gather information about products you carry, such as customer reviews, or to determine what merchandise you're going to stock—provides you with a competitive advantage. In this Team Time, you consider how to take advantage of crowdsourcing in running a business.

task

Three individuals starting new businesses have requested help in designing their websites and customer interaction programs. The businesses are (a) a coffee shop that offers live music, (b) a clothing company, and (c) a company that sells iPhone accessories. You need to help these businesses choose a crowdsourcing strategy.

process

Break the class into small teams. Each team should select one of the businesses described previously. Each team should prepare a report as follows:

1. Decide on the types of products the company will sell. For each category of product, decide what attributes you feel customers will comment on when giving input on products.
2. Decide whether the company will solicit feedback from potential customers on all products or will offer a core selection of products without soliciting feedback. Divide the products into two lists: core products and products subject to crowdsourcing.
3. Determine a social networking strategy for the company. How will customers be driven to the company website to participate in the crowdsourcing for incentives, discount coupons, contests, and such?
4. Develop a PowerPoint presentation to summarize your findings and present your proposed solution to the class.

conclusion

E-commerce websites are relatively easy to deploy, but risk exists when financing and stocking products. Crowdsourcing feedback can help businesses minimize the risk of stocking unwanted products.

Ethics Project

Privacy at School

In this exercise, you research and then role-play a complicated ethical situation. The role you play might not match your own personal beliefs; regardless, your research and use of logic will enable you to represent the view assigned. An arbitrator will watch and comment on both sides of the arguments, and together the team will agree on an ethical solution.

problem

Parents are especially concerned about protecting their children's privacy rights. School administrators are quickly learning that they need to craft policies that set appropriate boundaries and guidelines for monitoring students. Should school districts be allowed to monitor students? Should school districts be required to inform students and parents of any monitoring that takes place?

research areas to consider

- Lower Merion school district in Pennsylvania using webcams to monitor students in their homes
- Schools using the Geo Listening service to monitor students' social media posts
- The Family Educational Rights and Privacy Act (FERPA)

process

1. Divide the class into teams. Research the areas cited and devise a scenario in which a school district has been monitoring students without their knowledge and has potentially violated their privacy.
2. Team members should write a summary that provides background information for their character—for example, student, parent, school official, or arbitrator—and details their character's behaviors to set the stage for the role-playing event. Then team members should create an outline to use during the role-playing event.
3. Team members should present their case to the class or submit a PowerPoint presentation for review by the rest of the class, along with the summary and resolution they developed.

conclusion

As technology becomes ever more prevalent and integrated into our lives, ethical dilemmas will present themselves to an increasing extent. Being able to understand and evaluate both sides of the argument, while responding in a personally or socially ethical manner, will be an important skill.

Solve This with Notepad

MyLab IT Grader

Creating an HTML Document

You work part time for a small computer repair and education center. As part of your job, you need to be able to edit HTML files. Your boss has started a brief document about the basics of HTML that she'll be using for a training class. She would like you to add the HTML tags so she can put it on the website. You'll be using Notepad to complete this exercise.

You will use the following skills as you complete this activity:

- Apply HTML Tags
- View an HTML Doc in a Browser

Instructions

1. In Windows, type **Notepad** in the Search box on the taskbar. Click **Notepad Desktop App**.
2. Click File, click Open, and then navigate to your student data files. Click *TIA_Ch13_Start.txt*.
3. Press **Ctrl + Home** on your keyboard to place the cursor at the top of the document.
4. Type **<!DOCTYPE html>**. Press **Enter**. Type **<html>**. Press **Enter**. Type. Press **Enter**. These tags define the document as an HTML document.
5. Place the cursor at the beginning of the first paragraph that starts with *HTML stands for* and type **What is HTML?** Press **Enter**.
6. Place the cursor at the beginning of the second paragraph, which starts with *You can easily create*, and type **Creating an HTML Document**. Press **Enter**.
7. At the beginning of the third paragraph, type **Formatting Tags**, press **Enter**, and at the beginning of the fourth paragraph, type **Viewing an HTML Page**. Press **Enter**.
8. Place the cursor at the beginning of the first header, *HTML Beginning Tutorial*. Add a header start tag by typing <h1>. Press **End** and then type the end tag </h1>.
9. Place the cursor at the beginning of the second header, *What is HTML?* Add a third-level header start tag by typing <h3>. Press **End** and then add an end tag by typing </h3>.
10. Add third-level header tags for the other three headers by repeating step 9 for each.
11. Place the cursor at the beginning of the first paragraph, which begins with *HTML stands for*. Add paragraph tags to the beginning and end of the paragraph by typing <p> at the start of the paragraph and </p> at the end of the paragraph. Repeat for the remaining three paragraphs.
12. Press **Ctrl** + **Home**, click **Edit** on the Menu bar, and use **Find** to locate the first occurrence of the word *tags*. With the cursor at the beginning of *tags*, add a bold formatting start tag by typing <b>. Place the cursor at the end of the word *tags* and add a bold formatting end tag by typing </b>.
13. In the same paragraph, add start and end bold tags to the phrases *Start tags* and *End tags*.
14. Place the cursor between the 3 and the > in the first </h3> tag. Press **Space** and then type **style="color:blue"**.
15. Repeat step 14 for the remaining three headers.
16. Place the cursor before the paragraph end tag at the end of the third paragraph and type **For more information on creating HTML pages work through this <a href="**http://www.w3schools.com/html/**"> HTML Tutorial</a>**
17. Press **Ctrl** + **End** and then type </body>. Press **Enter**.
18. Type </html>.
19. Click **File**, click **Save As**, and then type **TIA_Ch13_LastFirst.html**. (Make sure you type **.html**). Navigate to where you are saving your student files and click **Save**. Close Notepad.
20. Open File Explorer. Navigate to where you saved *TIA_Ch13_LastFirst.html*. Double-click the file to open it in your default browser. Check to ensure that all headings have been formatted correctly, that bolding formats have been applied to select words and phrases, and all paragraphs start and end appropriately. Click the hyperlink to ensure that it works.
21. Close the browser and submit *TIA_Ch13_LastFirst.html* for grading.

Appendix

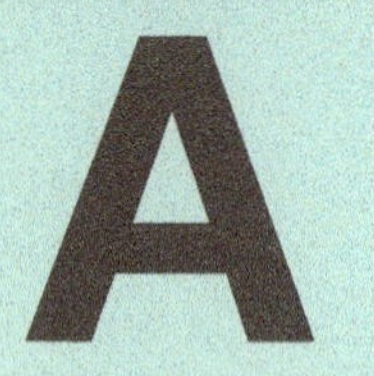

The History of the Personal Computer

For an overview of how the PC got its start, watch the **History of the PC video**.

Learning Outcome A.1 You will be able to describe the history of personal computer hardware and software development.

Have you ever wondered how big the first personal computer was or how much the first laptop weighed? Computers are such an integral part of our lives that we don't often stop to think about how far they've come or where they got their start. Computers have evolved from expensive, huge machines that only corporations could own to small, powerful devices that almost anyone can have. In this appendix, we look at the history of the personal computer.

Early Personal Computers

Our journey through the history of the personal computer starts in 1975. At that time, most people weren't familiar with the mainframes and supercomputers that large corporations and the government owned. With price tags exceeding the cost of buildings, and with few (if any) practical home uses, these monster machines weren't appealing to or attainable by average Americans.

The Altair 8800

1975

- Only 256 bytes of memory
- No keyboard or monitor
- Switches on front used to enter data in machine code (1s and 0s)
- Lights on front indicated results of a program

Apple I

1976

- Had to provide your own keyboard and monitor
- First unit produced was used in a high school math class

Apple II

June 1977

- Featured color monitor, sound, and game paddles
- 4 KB of RAM
- Operating system stored in ROM
- Optional floppy disk to load programs (mostly games)

Commodore PET

October 1977

- Featured in *Popular Science* magazine
- Popular in business settings in Europe

Figure A.1 Timeline of Early Personal Computer Development. *(B Christopher/Alamy Stock Photo; Tony Avelar/EPA/Shutterstock; Science & Society Picture Library/Getty Images; Jerry Mason/Science Source; INTERFOTO/History/Alamy Stock Photo; INTERFOTO/History/Alamy Stock Photo; Science & Society Picture Library/Getty Images)*

The First Personal Computer: The Altair

Objective A.1 *Describe the earliest personal computer ever designed.*

In 1975, the first personal computer, the Altair 8800 (see Figure A.1), was born. At around $400 for a do-it-yourself kit or around $600 for a fully assembled unit (about $2,900 in today's dollars), the price was reasonable enough that computer fanatics could finally own their own computers.

The Altair was primitive by today's standards—no keyboard, no monitor, and completely not user friendly. Despite its limitations, computer enthusiasts flocked to the machine. Many who bought it had been taught to program, but until that point, they had access only to big, clumsy computers at their jobs. With the Altair, they could create their own programs at home. Within three months, 4,000 orders were placed for the machine.

The release of the Altair marked the start of the personal computer boom. In fact, two men whose names you might be familiar with were among the first Altair owners: Bill Gates and Paul Allen, who were so enamored by the minicomputer that they wrote a compiling program (a program that translates user commands into commands the computer can understand) for it. The two later convinced the Altair's developer to buy their program, marking the start of a company called Microsoft. We get to that story later. First, let's see what their rivals were up to.

The Apple I and II

Objective A.2 *Describe the distinguishing features of the Apple I and Apple II.*

Around the time the Altair was released, Steve Wozniak, an employee at Hewlett-Packard, was dabbling with his own computer design. Steve Jobs, who was working for computer game manufacturer Atari at the time, liked Wozniak's prototypes and made a few suggestions. Together, the two built a personal computer, the Apple I, in Wozniak's garage (see Figure A.1) and formed the Apple Computer Company in 1976.

TRS-80

- Introduced by Radio Shack
- Monochrome display
- 4 KB of RAM
- Circuitry hidden under keyboard
- Wildly popular with consumers—sold 10,000 units in the first month

November 1977

IBM PC (5150)

- Marketed to businesses and consumers
- 64 KB to 256 KB of RAM
- Floppy disk drives optional
- Hard disks not supported in early models

August 1981

April 1981

Osborne

- First "portable" computer
- Weighed 24.5 pounds
- 5-inch screen
- 64 kilobytes of RAM
- Two floppy disk drives
- Preinstalled with spreadsheet and word processing software

Bits&Bytes **Why Is It Called "Apple"?**

Steve Jobs wanted Apple Computer to be the perfect computer company. Having recently worked at an apple orchard, Jobs thought of the apple as the perfect fruit because it was high in nutrients, came in a nice package, and was not easily damaged. Thus, he and Wozniak decided to name their computer company Apple.

(Frommenwiler Peter/Prisma by Dukas Presseagentur GmbH/Alamy Stock Photo)

One year later, in 1977, the Apple II was born (see Figure A.1). The Apple II included a color monitor, sound, and game paddles. Priced around $1,300 (about $5,600 in today's dollars), one of its biggest innovations was that the operating system was stored in read-only memory (ROM). Previously, the operating system had to be rewritten every time the computer was turned on. The friendly features of the Apple II operating system encouraged less technically oriented computer enthusiasts to write their own programs.

An instant success, the Apple II would eventually include a spreadsheet program, a word processor, and desktop publishing software. These programs gave personal computers like the Apple functions beyond gaming and special programming and led to their popularity.

Enter the Competition

Around the time Apple was experiencing success with its computers, a number of competitors entered the market. The largest among them were Commodore, Radio Shack, Osborne, and IBM. The Commodore PET (see Figure A.1) was aimed at the business market and did well in Europe, whereas Radio Shack's TRS-80 (see Figure A.1) was clearly aimed at the US consumer market. Just one month after its release in 1977, the TRS-80 had sold about 10,000 units.

The Osborne: The Birth of Portable Computing

Objective A.3 *Describe the first portable computer.*

The Osborne Computer Corporation introduced the first portable computer, the Osborne, in 1981 (see Figure A.1). Although portable, the computer weighed 24.5 pounds. It featured a minuscule five-inch screen and carried a price tag of $1,795 (about $5,200 today). The Osborne was an overnight success, and its sales quickly reached 10,000 units per month. However, despite the computer's initial popularity, the Osborne Computer Corporation filed for bankruptcy two years later.

IBM PCs

Objective A.4 *Describe the development of the IBM PC.*

Until 1980, IBM primarily made mainframe computers, which it sold to large corporations, and hadn't taken the personal computer seriously. In 1981, however, IBM released its first personal computer, the IBM PC (see Figure A.1). Because many companies were familiar with IBM mainframes, they adopted the IBM PC. The term *PC* soon became the term used to describe all personal computers.

IBM marketed its PC through retail outlets such as Sears to reach home users, and it quickly dominated that market. In January 1983, *Time* magazine, playing on its annual person of the year issue, named the computer "1982 Machine of the Year."

Bits&Bytes Continuing Changes

Between 1975 and the early 1980s, huge advancements were made in personal computer technology so that by the early 1990s, when access to the Internet became readily available to more people, PCs had begun to be a common (and necessary) home device. Technology advancements have continued with devices becoming smaller yet more powerful and faster. Table A.1 lists some devices that were common at the turn of the 21st century but have since been replaced by more advanced technology. How many of these devices do you remember?

Table A.1 Devices That Have Become Obsolete Since 2000

Obsolete Device	Use	Replaced by
Floppy disk	External file storage with maximum capacity of 1.44 MB	Online storage and flash drives
Zip disk	External file storage with maximum capacity of 100 MB–750 MB	Online storage and flash drives
CRT monitor	Output display device	LCD, LED, and OLED displays
Trackball mouse and Rollerball mouse	Input device	Optical mouse
Trackpoint	Input device for laptops	Touchpad
PDA	Small mobile device for storing and accessing digital information including contacts, calendars, modified productivity software, songs, games, and photos	Smartphones
Dot matrix printer	Output device that used tiny hammer-like keys to strike paper through an inked ribbon	Inkjet and laser printers
Serial port	Connect printers and mice to computer	USB ports and wireless
Parallel port	Connect printers to computer	USB ports and Wi-Fi printers
PS/2 port	Connect mice and keyboards to computer	USB ports and wireless mice and keyboards
Encyclopedia on CD	Comprehensive reference source on CD	Wikipedia
Camcorder	Portable video recording device	Smartphones
Sony Walkman	Portable music device	iPod and smartphones
Cassette tape and player	Music storage and player	Online streaming services
VCR (video cassette recorder)	Video recorder/player	DVR, DVD, Blu-ray players

(Claudio Divizia/Shutterstock; AjayTvm/Shutterstock; Lipskiy/Shutterstock; Enterphoto/Shutterstock; Lukas Wroblewski/Shutterstock; J.Hurley/Shutterstock; Zern Liew/Shutterstock; Ngaga/Shutterstock; John Voos/The Independent/Shutterstock; 3DMAVR/Shutterstock; AlexLMX/Shutterstock; JOKE777/Shutterstock; Jiri Hera/Shutterstock; Jiri Hera/Shutterstock)

Other Important Advancements

It wasn't just personal computer hardware that was changing. At the same time, advances in programming languages and operating systems and the influx of application software were leading to more useful and powerful machines.

The Importance of BASIC

Objective A.5 *Explain why BASIC was an important step in revolutionizing the software industry.*

The software industry began in the 1950s with programming languages such as FORTRAN, ALGOL, and COBOL. These languages were used mainly by businesses to create financial, statistical, and engineering programs. However, in 1964, John Kemeny and Thomas Kurtz designed the Beginner's All-Purpose Symbolic Instruction Code (BASIC) at Dartmouth College and revolutionized the software industry. BASIC was a language that beginning programming students could easily learn. It thus became enormously popular—and the key language of the PC. In fact, Bill Gates and Paul Allen used BASIC to write their program for the Altair, which led to the creation of Microsoft, a company that produced (and continues to produce) computer software.

The Advent of Operating Systems

Objective A.6 *Explain why the development of the operating system was an important step in PC development.*

Because data on the earliest personal computers was stored on audiocassettes, many programs weren't saved or reused. This meant that programs had to be rewritten whenever they were needed. In 1978, Steve Wozniak designed a 5.25-inch floppy disk drive so that programs could be saved easily and operating systems developed.

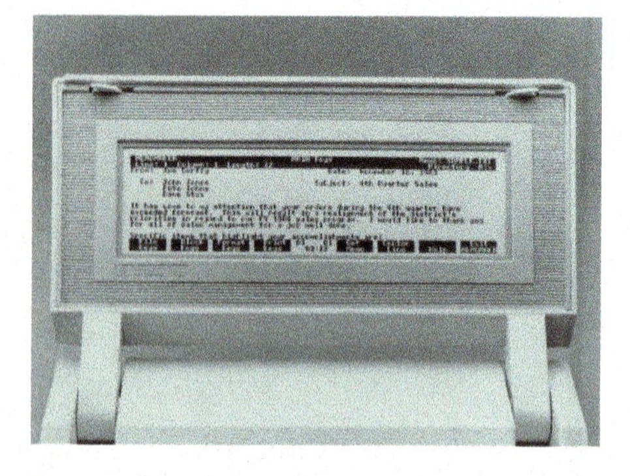

Figure A.2 This 1983 Hewlett-Packard computer used an early version of the MS-DOS operating system as well as the Lotus 1-2-3 spreadsheet program. *(Everett Collection/SuperStock)*

Operating systems are written to coordinate with the specific processor chip that controls the computer. At that time, Apple computers ran on a Motorola chip, whereas PCs (IBMs and so on) ran on an Intel chip. The Disk Operating System (DOS), developed by Wozniak (in conjunction with Paul Laughton of Shepardson Microsystems) and introduced in 1977, was the OS that controlled the first Apple computers. The Control Program for Microcomputers (CP/M), developed by Gary Kildall, was the OS designed for the Intel 8080 chip (the processor for PCs).

In 1980, when IBM was entering the personal computer market, it approached Bill Gates at Microsoft to write an OS program for the IBM PC. Gates recommended that IBM investigate the CP/M OS, but IBM couldn't arrange a meeting with the founder, Gary Kildall. Microsoft reconsidered the opportunity to write an OS program and developed MS-DOS for IBM computers. Eventually, virtually all PCs running on the Intel chip used MS-DOS as their OS (see Figure A.2). Microsoft's reign as one of the dominant players in the personal computer landscape had begun.

The Software Application Explosion: VisiCalc and Beyond

Objective A.7 *List early application software that was developed for the PC.*

Because the floppy disk was a convenient way to distribute software, its inclusion in personal computers set off an application software explosion. In 1978, Harvard Business School student Dan Bricklin recognized the potential for a personal computer spreadsheet program. He and his friend Bob Frankston created the program VisiCalc, which became an instant success. Finally, ordinary home users could see the benefit of owning a personal computer. More than 100,000 copies of VisiCalc were sold in its first year.

After VisiCalc, other electronic spreadsheet programs entered the market. Lotus 1-2-3 came on the market in 1983, and Microsoft Excel entered the scene in 1985. These products became so popular that they eventually put VisiCalc out of business.

Meanwhile, word processing software was also gaining a foothold in the industry. Until then, there were separate, dedicated word processing machines; personal computers, it was believed, were

Table A.2 Application Software Development

Year	Application
1978	**VisiCalc:** First electronic spreadsheet application **WordStar:** First word processing application
1980	**WordPerfect:** Best DOS-based word processor; was eventually sold to Novell and later acquired by Corel
1983	**Lotus 1-2-3:** Added integrated charting, plotting, and database capabilities to spreadsheet software **Word for MS-DOS:** Introduced in the pages of *PC World* magazine on the first magazine-inserted demo disk
1985	**Microsoft Excel:** One of the first spreadsheets to use a graphical user interface **PageMaker:** The first desktop publishing software

for computation and data management only. However, once WordStar, the first word processing application, became available for personal computers in 1979, word processing became another important use for the personal computer. Competitors such as Word for MS-DOS (the precursor to Microsoft Word) and WordPerfect soon entered the market. Table A.2 lists some of the important dates in application software development.

The Graphical User Interface and the Internet Boom

Other important advancements related to personal computers came in the form of the graphical user interface and the Internet.

Xerox and Apple's Lisa and Macintosh

Objective A.8 *Describe the features of the Lisa and Macintosh computers and their predecessor, the Xerox Alto.*

A graphical user interface (GUI), which uses icons to represent programs and actions, allows users to interact with the computer more easily. Until that time, users had to use complicated command- or menu-driven interfaces. Apple was the first company to take full commercial advantage of the GUI, but the GUI was not created by a computer company.

In 1972, a few years before Apple launched its first personal computer, photocopier manufacturer Xerox was designing a personal computer of its own. Named the Alto (see Figure A.3), the computer included a word processor, based on the What You See Is What You Get (WYSIWYG) principle, that incorporated a file management system with directories and folders. It also had a mouse and could connect to a network. None of the other personal computers of the time had these features. For a variety of reasons, Xerox never sold the Alto commercially. Several years later, it developed the Star Office System, which was based on the Alto. Despite its convenient features, the Star never became popular because no one was willing to pay the $17,000 asking price.

Figure A.3 The Alto was the first computer to use a GUI, and it provided the basis for the GUI that Apple used. However, because of marketing problems, the Alto never was sold. *(Josie Lepe/MCT/Newscom)*

Xerox's ideas were ahead of their time, but many of the ideas present in the Alto and Star would soon catch on. In 1983, Apple introduced the Lisa—the first successful personal computer brought to market that used a GUI (see Figure A.4). Legend has it that Jobs had seen the Alto during a visit to Xerox in 1979 and was influenced by its GUI. He therefore incorporated a similar user interface in the Lisa, providing features such as windows, drop-down menus, icons, a file system with folders and files, and a point-and-click device called a *mouse*. The only problem with the Lisa was its price. At $9,995 (about $26,100 in today's dollars), few buyers were willing to take the plunge.

One year later, in 1984, Apple introduced the Macintosh, shown in Figure A.5. The Macintosh was everything the Lisa was and then some, at about a third of the cost. The Macintosh was also the first personal computer to use 3.5-inch floppy disks with a hard cover, which were smaller and sturdier than the previous 5.25-inch floppies.

Figure A.4 The Lisa was the first computer to introduce a GUI to the market. Priced too high, it never gained the popularity it deserved. *(SSPL/Getty Images)*

Figure A.5 The Macintosh became one of Apple's best-selling computers, incorporating a GUI along with other innovations such as the 3.5-inch floppy disk drive. *(Interfoto/Alamy Stock Photo)*

The Internet Boom

Objective A.9 *List the first successful web browsers.*

The GUI made it easier for users to work on the computer. The Internet provided another reason for people to buy computers. Now people could conduct research and communicate in a new way. In 1993, the web browser Mosaic was introduced. This browser allowed users to view multimedia on the web, causing Internet traffic to increase by nearly 350%.

Meanwhile, companies discovered the Internet as a means to do business, and computer sales took off. IBM-compatible PCs became the computer system of choice when, in 1995, Microsoft introduced Internet Explorer, a browser that integrated web functionality into Microsoft Office applications, and Windows 95, the first Microsoft OS designed to be principally a GUI OS.

About one year earlier, in 1994, a team of developers launched the Netscape Navigator web browser, which soon became a predominant player in browser software. However, pressures from Microsoft became too strong, and in 1998, Netscape announced it would no longer charge for the product and would make the code available to the public.

Making the Personal Computer Possible: Early Computers

Billions of personal computers have been sold over the past four decades. However, the computer is a compilation of parts, each of which is the result of individual inventions. Let's look at some early machines that helped create the personal computer we know today.

The Pascalene Calculator and the Jacquard Loom

Objective A.10 *Discuss the features of the Pascalene calculator and the Jacquard loom that inspired modern computer elements.*

From the earliest days of humankind, we have been looking for a more systematic way to count and calculate. Thus, the evolution of counting machines led to the development of the computer we know today. The Pascalene (also known as Pascal's calculator) was the first accurate mechanical calculator. This machine, created by French mathematician Blaise Pascal in 1642, used revolutions of gears, like odometers in cars, to count by tens. The Pascalene could be used to add, subtract, multiply, and divide. The basic design of the Pascalene was so sound that it lived on in mechanical calculators for more than 300 years.

Nearly 200 years later, Joseph Jacquard revolutionized the fabric industry by creating a machine that automated the weaving of complex patterns. Although not a counting or calculating machine,

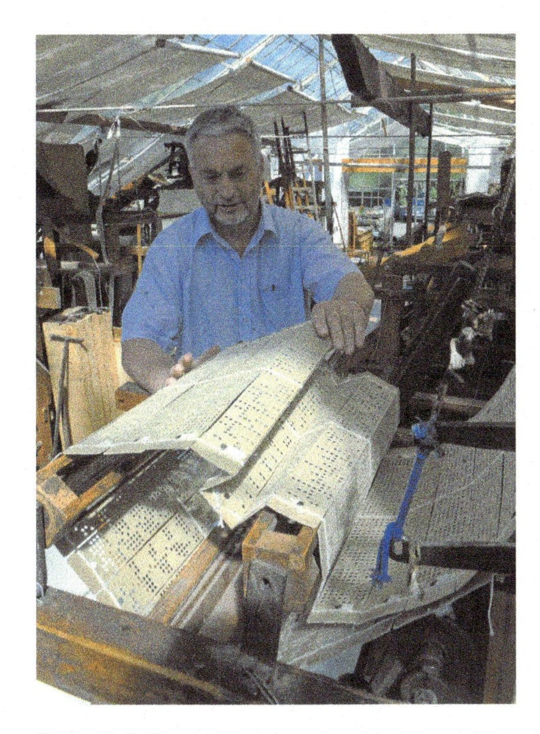

Figure A.6 The Jacquard loom used holes punched in stiff cards to make complex designs. This technique would later be used in punch cards that controlled the input and output of data in computers. *(Janek Skarzynski/AFP/Getty Images)*

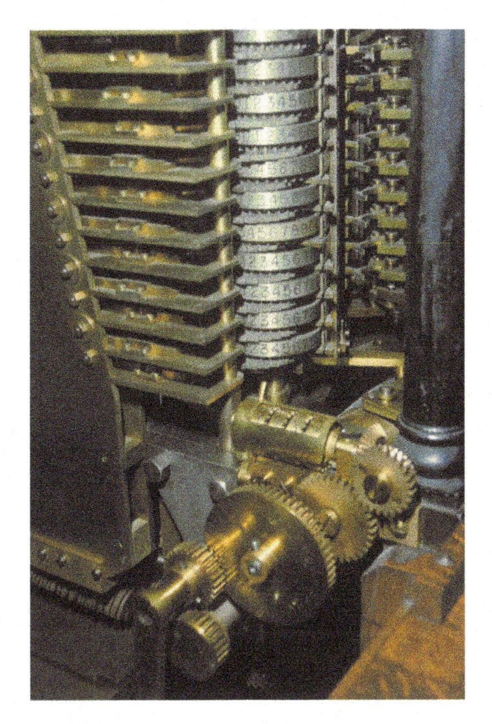

Figure A.7 The Analytical Engine, designed by Charles Babbage, was never fully developed but included components similar to those found in today's computers. *(Chris Howes/Wild Places Photography/Alamy Stock Photo)*

the Jacquard loom (shown in Figure A.6) was significant because it relied on stiff cards with punched holes to automate the weaving process. Much later, this punch-card process would be adopted as a means for computers to record and read data.

Babbage's Engines and the Hollerith Tabulating Machine

Objective A.11 *Discuss the contributions of Babbage's engines and the Hollerith Tabulating Machine to modern computing.*

In 1834, Charles Babbage designed the first automatic calculator, called the Analytical Engine (see Figure A.7). The machine was actually based on another machine, called the Difference Engine, which was a huge steam-powered mechanical calculator that Babbage designed to print astronomical tables. Although the Analytical Engine was never developed, Babbage's detailed drawings and descriptions of the machine include components similar to those found in today's computers, including the store (akin to RAM) and the mill (a central processing unit) as well as input and output devices. This invention gave Charles Babbage the title of the father of computing.

Meanwhile, Ada Lovelace, who was the daughter of poet Lord Byron and a student of mathematics (which was unusual for women of the time), was fascinated with Babbage's engines. She translated an Italian paper on Babbage's machine and, at the request of Babbage, added her own extensive notes. Her efforts are thought to be the best description of Babbage's engines.

In 1890, Herman Hollerith, while working for the US Census Bureau, was the first to apply Jacquard's punch-card concept to computing with his Hollerith Tabulating Machine. Until that time, census data had been tabulated manually in a long, laborious process. Hollerith's tabulating machine automatically read data that had been punched onto small punch cards, speeding up the tabulation process. Hollerith's machine became so successful that he left the Census Bureau in 1896 to start the Tabulating Machine Company. His company later changed its name to International Business Machines, or IBM.

Figure A.8 The Atanasoff–Berry Computer laid the design groundwork for many computers to come. *(Frederick News Post/AP Images)*

The Z1, the Atanasoff–Berry Computer, and the Harvard Mark I

Objective A.12 *Discuss the features of the Z1, the Atanasoff–Berry Computer, and the Mark I.*

German inventor Konrad Zuse is credited with a number of computing inventions. His first, in 1936, was a mechanical calculator called the Z1. The Z1 is thought to be the first computer to include certain features integral to today's systems, such as a control unit and separate memory functions.

In 1939, John Atanasoff, a professor at Iowa State University, and his student Clifford Berry built the first electrically powered digital computer, called the Atanasoff–Berry Computer (ABC), shown in Figure A.8. The computer was the first to use vacuum tubes, instead of the mechanical switches used in older computers, to store data. Although revolutionary at the time, the machine weighed 700 pounds, contained a mile of wire, and took about 15 seconds for each calculation. (In comparison, today's personal computers can perform billions of calculations in 15 seconds.) Most important, the ABC was the first computer to use the binary system and to have memory that repowered itself upon booting. The design of the ABC would be central to that of future computers.

From the late 1930s to the early 1950s, Howard Aiken and Grace Hopper designed the Mark series of computers used by the US Navy for ballistic and gunnery calculations. Aiken, an electrical engineer and physicist, designed the computer, and Hopper did the programming. The Harvard Mark I, finished in 1944, could add, subtract, multiply, and divide.

However, many believe Hopper's greatest contribution to computing was the invention of the compiler—a program that translates English-language instructions into computer language. The team was also responsible for a common computer-related expression. Hopper was the first to debug a computer when she removed a moth that had flown into the Harvard Mark I and had caused the computer to break down (see Figure A.9). After that, problems that caused a computer not to run were called *bugs*.

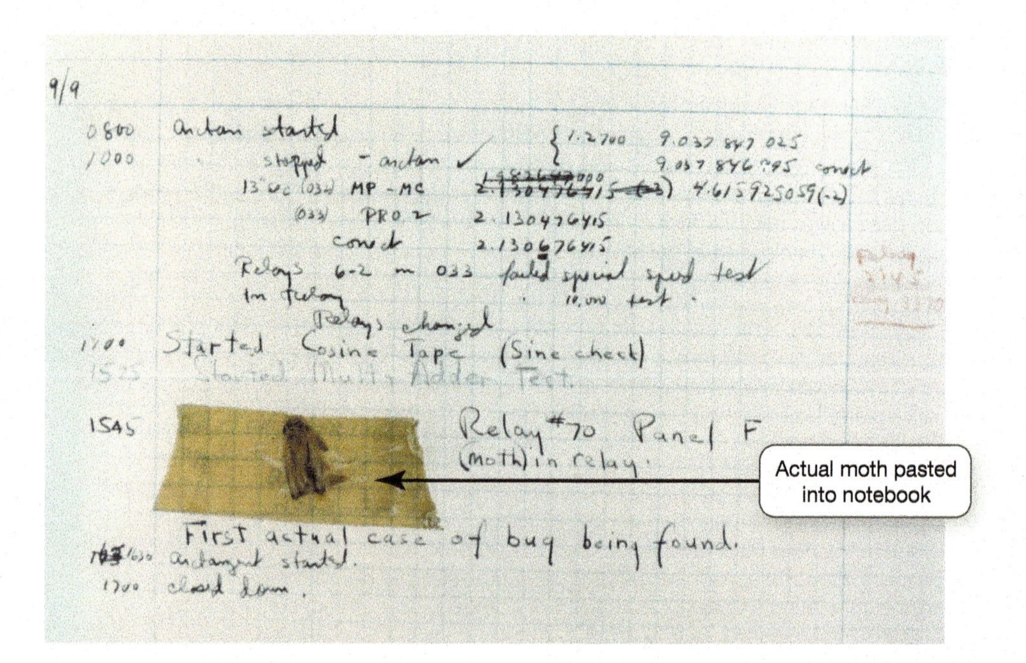

Figure A.9 Grace Hopper coined the term computer bug when a moth flew into the Harvard Mark I, causing it to break down. *(Everett Historical/Shutterstock)*

Figure A.10 The ENIAC took up an entire room and required several people to manipulate it. *(Interfoto/History/ Alamy Stock Photo)*

The Turing Machine, the ENIAC, and the UNIVAC

Objective A.13 *Discuss the major features of the Turing Machine, the ENIAC, and the UNIVAC.*

Meanwhile, in 1936, British mathematician Alan Turing created an abstract computer model that could perform logical operations. The Turing Machine was not a real machine but rather a hypothetical model that mathematically defined a mechanical procedure (or algorithm). In addition, Turing's concept described a process by which the machine could read, write, or erase symbols written on squares of an infinite paper tape. This concept of an infinite tape that could be read, written to, and erased was the precursor to today's RAM.

The Electronic Numerical Integrator and Computer (ENIAC), shown in Figure A.10, was another US government–sponsored machine developed to calculate the settings used for weapons. Created by John W. Mauchly and J. Presper Eckert at the University of Pennsylvania, it was put into operation in 1944. Although the ENIAC is generally thought of as the first successful high-speed electronic digital computer, it was big and clumsy. The ENIAC used nearly 18,000 vacuum tubes and filled approximately 1,800 square feet of floor space. Although inconvenient, the ENIAC served its purpose and remained in use until 1955.

Figure A.11 UNIVACs were the first computers to use magnetic tape for data storage. *(Underwood Archives/ UIG/Shutterstock)*

The Universal Automatic Computer, or UNIVAC, was the first commercially successful electronic digital computer. Completed in 1951, the UNIVAC operated on magnetic tape (see Figure A.11), setting it apart from its competitors, which ran on punch cards. The UNIVAC gained notoriety when, in a 1951 publicity stunt, it was used to predict the outcome of the Stevenson–Eisenhower presidential race. After analyzing only 5% of the popular vote, the UNIVAC correctly identified Dwight D. Eisenhower as the victor. After that, UNIVAC soon became a household word. The UNIVAC and computers like it were considered first-generation computers and were the last to use vacuum tubes.

Figure A.12 Transistors were one-tenth the size of vacuum tubes, were faster, and produced much less heat. *(Boris Sosnovyy/Shutterstock)*

Transistors and Beyond

Objective A.14 *Describe the milestones that led to each generation of computers.*

Only one year after the ENIAC was completed, scientists at the Bell Telephone Laboratories in New Jersey invented the transistor (see Figure A.12). The transistor replaced the bulky vacuum tubes of earlier computers and was smaller and more powerful than tubes. It was used in almost everything, from radios to phones. Computers that used transistors were referred to as second-generation computers. Still, transistors were limited in how small they could be made.

A few years later, in 1958, Jack Kilby, while working at Texas Instruments, invented the world's first integrated circuit—a small chip capable of containing thousands of transistors. This consolidation in design enabled computers to become smaller and lighter. The computers in this early integrated-circuit generation were considered third-generation computers.

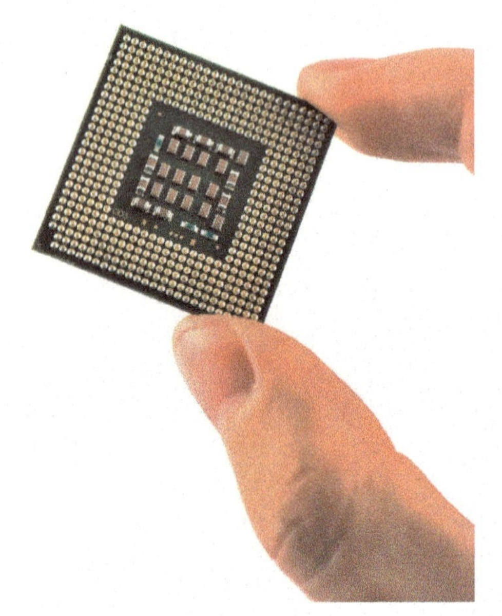

Figure A.13 Today's microprocessors can contain billions of transistors. *(Tudor Voinea/Shutterstock)*

Other innovations in the computer industry further refined the computer's speed, accuracy, and efficiency. However, none was as significant as the 1971 introduction by the Intel Corporation of the microprocessor chip—a small chip containing millions of transistors (see Figure A.13). The microprocessor functions as the CPU, or brains, of the computer. Computers that use a microprocessor chip are called fourth-generation computers. Over time, Intel and Motorola became the leading manufacturers of microprocessors. Today, the Intel Core i9 is one of Intel's most powerful processors.

As you can see, personal computers have come a long way since the Altair and have a number of inventions and people to thank for their amazing popularity. What will the future bring?

Check Your Understanding // Review & Practice

For a quick review of what you've learned, answer the following questions.

multiple choice

1. What was the name of the first web browser?
   a. Internet Explorer
   b. Netscape
   c. Mosaic
   d. Firefox

2. Which was the first electronic spreadsheet program?
   a. Excel
   b. Numbers
   c. VisiCalc
   d. Lotus 1-2-3

3. Which invention replaced transistors in computers?
   a. The integrated circuit
   b. Magnetic tape
   c. The microprocessor chip
   d. The vacuum tube

4. Which computer was touted as the first portable personal computer?
   a. Lisa
   b. Commodore PET
   c. Altair
   d. Osborne

5. Who was responsible for creating a weaving machine that used punch cards?
   a. Jacquard
   b. Pascal
   c. Babbage
   d. Hollerith

6. Which computer had the first bug?
   a. ENIAC
   b. Apple I
   c. UNIVAC
   d. Harvard Mark I

7. Which early programming language was developed for beginning programming students?
   a. BASIC
   b. FORTRAN
   c. COBOL
   d. PASCAL

8. Which was the first personal computer sold as a kit?
   a. Apple I
   b. Altair 8800
   c. TRS-80
   d. Commodore PET

9. What was Grace Hopper well known for?
   a. Creating the first portable computer
   b. Developing magnetic storage
   c. Removing a moth from a computer
   d. Developing the BASIC programming language

10. Which early personal computer was the first to use 3.5-inch floppy disks for external storage?
    a. Lisa
    b. Apple I
    c. Apple II
    d. Macintosh

Appendix

B Careers in IT

Learning Outcome B.1 **Describe the various categories of IT jobs available, explain why IT jobs are in demand, and discuss various ways to prepare for IT employment.**

It's hard to imagine an occupation in which computers aren't used in some fashion. Even such previously low-tech industries as waste disposal and fast food use computers to manage inventories and order commodities. In this appendix, we explore various information technology (IT) career paths open to you.

Rewards of Working in Information Technology

Objective B.1 *List the reasons IT fields are attractive to students pursuing bachelor's degrees.*

There are many great reasons to work in the exciting, ever-changing field of IT. In this section, we explore some reasons the IT field is so attractive to graduates looking for entry-level positions.

(Vlastas/Shutterstock)

IT Workers Are in Demand

If you're investigating a career with computers, the first question you may have is, "Will I be able to get a job?" Consider the following:

- According to projections by the US Department of Labor's Bureau of Labor Statistics, computer-related jobs are expected to be among the fastest-growing occupations through 2028 (see Table B.1).
- In 2020, *U.S. News & World Report* published a list of the 100 best careers to consider, based on employment opportunity, salary, work–life balance, and job security. Software developer leads the list at number one, and computer systems analyst, information security analyst, computer systems administrator, database administrator, web developer, computer network architect, operations research analyst, computer support specialist, and IT manager were also on the list.
- According to the National Association of Colleges and Employers (NACE) Winter 2020 Salary Survey, the average starting salary for computer science graduates was projected to be $67,411 and was the second highest salary on the list, just below engineering majors.

The number of students pursuing computer science degrees has also been increasing over the past several years. Fields of specialization such as game development, information security, and mobile app development are very popular. Yet, shortages of computing professionals in the United States are still projected over the next 5–10 years. What does all this mean? In terms of job outlook, now is a perfect time to consider an IT career.

IT Jobs Pay Well

As you can see from Table B.1, median salaries in IT careers are robust. But what exactly affects your salary in an IT position? Your skill set and your experience level are obvious answers, but the size of an employer and its geographical location are also factors. Large companies tend to pay more, so if you're pursuing a high salary, set your sights on a large corporation. Remember that making a lot of money isn't everything—be sure to consider other quality-of-life issues, such as job satisfaction. Of course, choosing a computer career isn't a guarantee you'll get a high-paying job. Just as in any other profession, you'll need appropriate training and on-the-job experience to earn a high salary.

So how much can you expect to start out earning? Although starting salaries for some IT positions, such as computer support specialists, are more on the modest side ($45,000), starting salaries for students with bachelor's degrees in IT are fairly robust.

Table B.1 High-Growth IT Jobs

Occupation	Median Pay ($)	10-Year Growth Rate (%)	10-Year Total New Jobs
Information security analysts	99,730	32	35,500
Software developers	105,590	21	284,100
Computer and information research scientists	118,370	16	5,200
Web developers	73,760	13	20,900
Computer support specialists	54,760	10	83,100
Database administrators	93,750	9	10,500
Computer systems analysts	90,920	9	56,000
Computer network architects	112,690	5	8,400
Network and computer systems administrators	83,510	5	18,200

Source: Excerpted from *Occupational Outlook Handbook, 2020 Edition,* US Bureau of Labor Statistics.

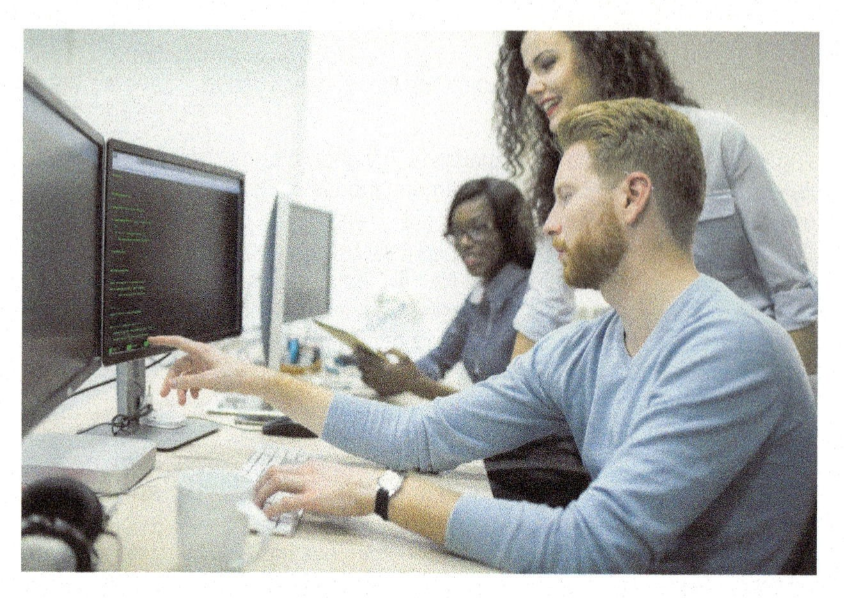

Figure B.1 Although many companies outsource IT jobs to third-party companies, this does not mean the jobs necessarily go offshore. *(NDAB Creativity/Shutterstock)*

To obtain the most accurate information, you should research salaries yourself in the geographical area where you expect to work. *Salary.com* provides a free salary wizard to help you determine how much IT professionals in your area are making compared with national averages. You can add information such as your degree and the size of the company to your search selection to fine-tune the figures further. Hundreds of IT job titles are listed in *Salary.com*, so you can tailor your search to the specific job, location, and industry in which you're interested.

IT Jobs Are Not Going Offshore

In the global economy in which we now operate, job outlook includes the risk of jobs being outsourced, possibly to other countries. Outsourcing is a process whereby a business hires a third-party firm to provide business services (such as customer-support call centers) that were previously handled by in-house employees (see Figure B.1). Offshoring occurs when the outsourcing firm is located (or uses employees) outside the United States.

India, China, and Romania and other former Eastern Bloc countries are major players in providing outsourcing services for US companies. The big lure of outsourcing and offshoring is cost savings: Considering that the standard of living and salaries are much lower in many countries than they are in the United States, offshoring is an attractive option for many US employers.

However, outsourcing and offshoring don't always deliver the cost savings that CEOs envision. Demand for personnel overseas has led to increased costs (primarily due to wage increases) for service providers in foreign markets. Furthermore, other, less tangible factors can outweigh the cost savings from outsourcing. Communication problems can arise between internal and external employees, for example, and cultural differences between the home country and the country doing the offshoring can result in software code that needs extensive rework by in-house employees. Data can also be less secure in an external environment or during the transfer between the company and an external vendor. Although outsourcing and offshoring won't be going away, companies are approaching these staffing alternatives with more caution and are looking more to US companies to provide resources.

So, many IT jobs are staying in the United States. According to *InformationWeek* magazine, most jobs in the following three categories (see Table B.2) will stay put:

1. **Customer interaction:** jobs that require direct input from customers or that involve systems with which customers interface daily
2. **Enablers:** jobs that involve accomplishing key business projects, often requiring technical skills beyond the realm of IT and good people skills
3. **Infrastructure:** jobs that are fundamental to moving and storing the information that US-based employees need to do their jobs

Table B.2 Jobs That Will Likely Remain Onshore

Customer Interaction	Enablers	Infrastructure Jobs
Web application developers	Business process analysts	Network security
Web interface designers	Application developers (when customer interaction is critical)	Network installation technicians
Database and data warehouse designers/developers	Project managers (for systems with users who are located predominantly in the United States)	Network administrators (engineers)
Customer relationship management (CRM) analysts		Wireless infrastructure managers and technicians
Enterprise resource planning (ERP) implementation specialists		Disaster recovery planners and responders

In addition, jobs that require specific knowledge of the US marketplace and culture, such as social media managers, are also very likely not to be offshored.

Women Are in High Demand in IT Departments

Currently, women make up about 26 percent of the IT workforce (per the National Center for Women and Information Technology). This presents a huge opportunity for women who have IT skills because many IT departments are actively seeking to diversify their workforces. In addition, although a salary gender gap (the difference between what men and women earn for performing the same job) exists in IT careers, it's smaller than in most other professions.

You Have a Choice of Working Location

In this case, location refers to the setting in which you work. IT jobs can be office based, field based, project based, or home based. Because not every situation is perfect for every individual, you can look for a job that suits your tastes and requirements. Table B.3 summarizes the major job types and their locations.

You Constantly Meet New Challenges

In IT, the playing field is always changing. New software and hardware are constantly being developed. You'll need to work hard to keep your skills up to date. You'll spend a lot of time in training and self-study, trying to learn new systems and techniques. Many individuals thrive in this type of environment because it keeps their jobs from becoming dull or routine.

You Work in Teams

When students are asked to describe their ideal jobs, many describe jobs that involve working in teams. Despite what some people think, IT professionals are not locked in lightless cubicles, basking in the glow of their monitors and working alone on projects. Most IT jobs require constant interaction with other workers, usually in team settings. People skills are highly prized by IT departments. If you have good leadership and team-building skills, you'll have the opportunity to exercise them in an IT job.

You Don't Need to Be a Mathematical Genius

Certain IT careers such as programming involve a fair bit of math. But even if you're not mathematically inclined, you can explore many other IT careers. IT employers also value such attributes as creativity, marketing, and artistic style, especially in jobs that involve working on the Internet or with social media.

Table B.3 Where Do You Want to Work?

Type of Job	Location and Hours	Special Considerations
Office-Based 	Report for work to the same location each day and interact with the same people on a regular basis; requires regular hours of attendance (such as 9:00 a.m. to 5:00 p.m.)	May require working beyond normal working hours; may also require workers to be on call 24/7
Field-Based 	Travel from place to place as needed and perform short-term jobs at each location	Involves a great deal of travel and the ability to work independently
Project-Based 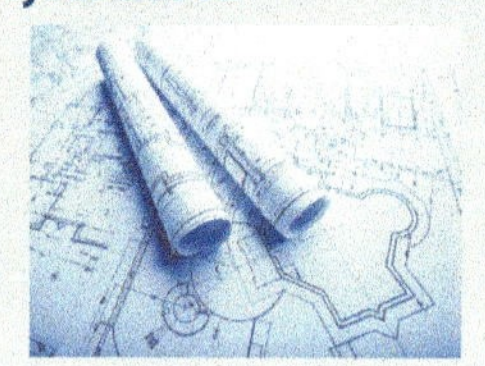	Work at client sites on specific projects for extended periods of time (weeks or months)	Can be especially attractive to individuals who like workplace situations that vary on a regular basis
Home-Based (Telecommuting) 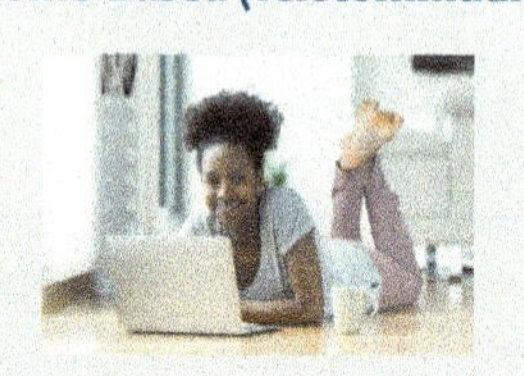	Work from home	Involves very little day-to-day supervision and requires an individual who is self-disciplined

(Rawpixel.com/Shutterstock; Muellek Josef/Shutterstock; Maksym Dykha/Shutterstock; Samuel Borges Photography/Shutterstock)

IT Skills Are Transferable

Most computing skills are transferable from industry to industry. A networking job in the clothing manufacturing industry uses the same primary skill set as a networking job for a supermarket chain. Therefore, if something disastrous happens to the industry you're in, you should be able to switch to another industry without having to learn an entirely new skill set. Combining business courses with IT courses will also make you more marketable when changing jobs. For example, as an accounting major, if you minor in IT, employers may be more willing to hire you because working in accounting today means constantly interfacing with management information systems and manipulating data.

Challenges of IT Careers

Objective B.2 *Explain the various challenges of IT careers.*

Although there are many positive aspects of IT careers, there can be some challenges. The discussions that follow aren't meant to discourage you from pursuing an IT career but merely to make you aware of exactly what challenges you might face in an IT department.

Stress

Most IT jobs are hectic (see Figure B.2). Whereas the average American works 42 hours a week, a survey by *InformationWeek* revealed that the average IT staff person works 45 hours a week and is

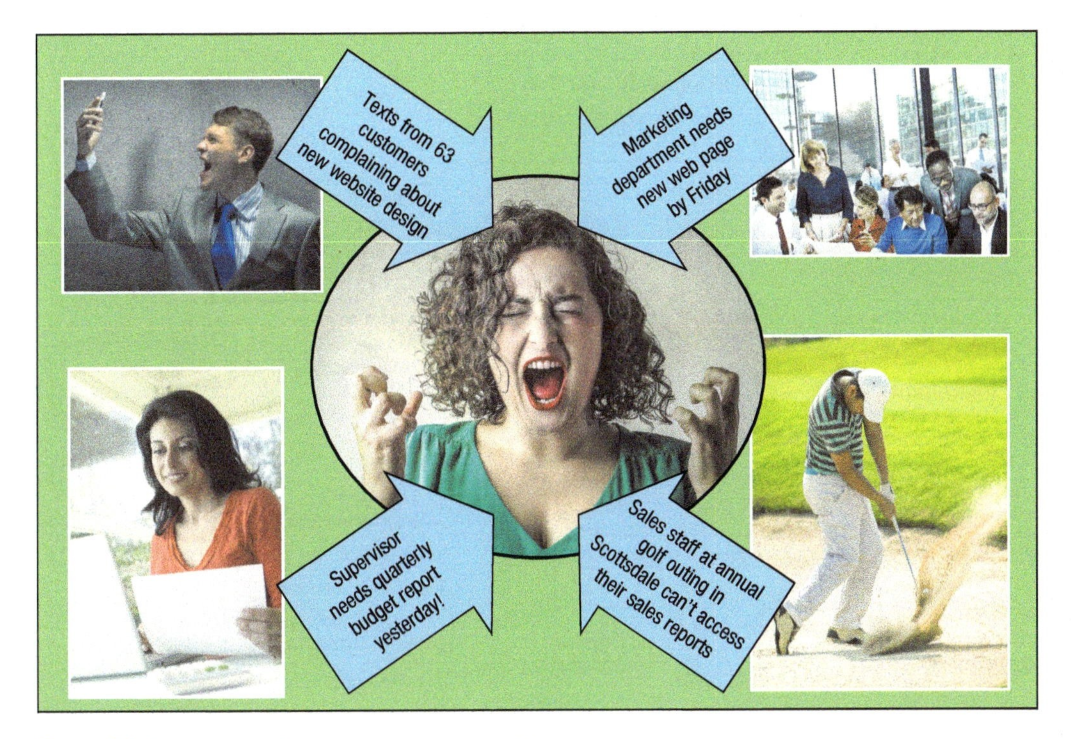

Figure B.2 Stress comes from multiple directions in IT jobs.

(Khakimullin Aleksandr/Shutterstock; Shutterstock; Rawpixel.com/Shutterstock; Torwaistudio/Shutterstock; Ollyy/Shutterstock)

on call for another 24 hours. On-call time (hours an employee must be available to work in the event of a problem) has been increasing because most IT systems require 24/7 availability.

Women Are in the Minority

A majority of IT jobs are filled by men, so some women view IT departments as *Dilbert*-like microcosms of antisocial geeks and don't feel as though they would fit in. Unfortunately, it is true that in most IT departments, women are underrepresented. As companies begin to push to address this imbalance, new opportunities are emerging for women with technical skill sets.

Lifelong Learning Is Required

Although the constantly changing nature of IT can alleviate boredom, keeping up with the changes can also cause some stress. You'll need to take training courses, do self-study, and perhaps take additional college courses, such as getting a graduate degree, to keep up with the vast shifts in technology.

Choosing Your Realm of IT

Objective B.3 *List and describe the various IT careers for which you can train.*

Figure B.3 shows an organizational chart for a modern IT department at a large corporation that should help you understand the variety of careers available and how they interrelate. The chief information officer (CIO) has overall responsibility for the development, implementation, and maintenance of information systems and their infrastructure. Usually, the CIO reports to the chief operating officer (COO).

The responsibilities below the CIO are generally grouped into two units:

1. Development and integration (responsible for the development of systems and websites)
2. Technical services (responsible for the day-to-day operations of the company's information infrastructure and network, including all deployed hardware and software)

Figure B.3 This chart represents a typical structure for an IT department at a large corporation.

In large organizations, responsibilities are distinct, and jobs are defined more narrowly. In medium-sized organizations, there can be overlap between position responsibilities. At a small company, you might be the network administrator, database administrator, computer support technician, and social media manager all at the same time. Let's look at the typical jobs found in each department.

Working in Development and Integration

Two distinct paths exist in this division: web development and systems development. Because everything involves the web today, there's often a great deal of overlap between these paths.

Web Development. When most people think of web development careers, they usually equate them with being a webmaster. Webmasters used to be individuals who were solely responsible for all aspects of a company's website. However, today's webmasters are usually supervisors with responsibility for certain aspects of web development. At smaller companies, they may also be responsible for tasks that other individuals in a web development group usually do, such as:

- Web content creators generate the words and images that appear on the web. Journalists, other writers, editors, and marketing personnel prepare an enormous amount of web content, whereas video producers, graphic designers, and animators create web-based multimedia. Web content creators have a thorough understanding of their own fields as well as of HTML, PHP, and JavaScript.
- Interface designers work with graphic designers and animators to create a look and feel for the site and make it easy to navigate.
- Web programmers build web pages to deploy the materials that the content creators develop. They wield software tools such as Adobe Dreamweaver and InDesign to develop the web pages. They also create links to databases, using products such as Oracle and SQL Server to keep information flowing between users and web pages. They must possess a solid understanding of client- and server-side web languages (HTML, XML, Java, JavaScript, ASP, PHP, Silverlight, and Perl) and of development environments such as the Microsoft .NET Framework.
- Customer interaction technicians provide feedback to a website's customers. Major job responsibilities include answering e-mail, sending requested information, funneling questions to appropriate personnel (technical support, sales, and so on), and providing suggestions to web programmers for site improvements. Extensive customer service training is essential to work effectively in this area.
- Social media directors are responsible for directing the strategy of the company on all social media sites where the company maintains a presence. Often, while supervising customer interaction technicians, social media directors make sure that customers have a quality experience while interacting with company employees and customers on sites such as Facebook, Twitter, and Yelp. Responding to comments left on such sites, developing promotional strategies, and designing functionality of the company's social media sites are common job responsibilities.

As you can see in Figure B.4, many kinds of people can work on the same website. Web programming jobs often require a four-year college degree in computer science, whereas graphic designers are often hired with two-year art degrees.

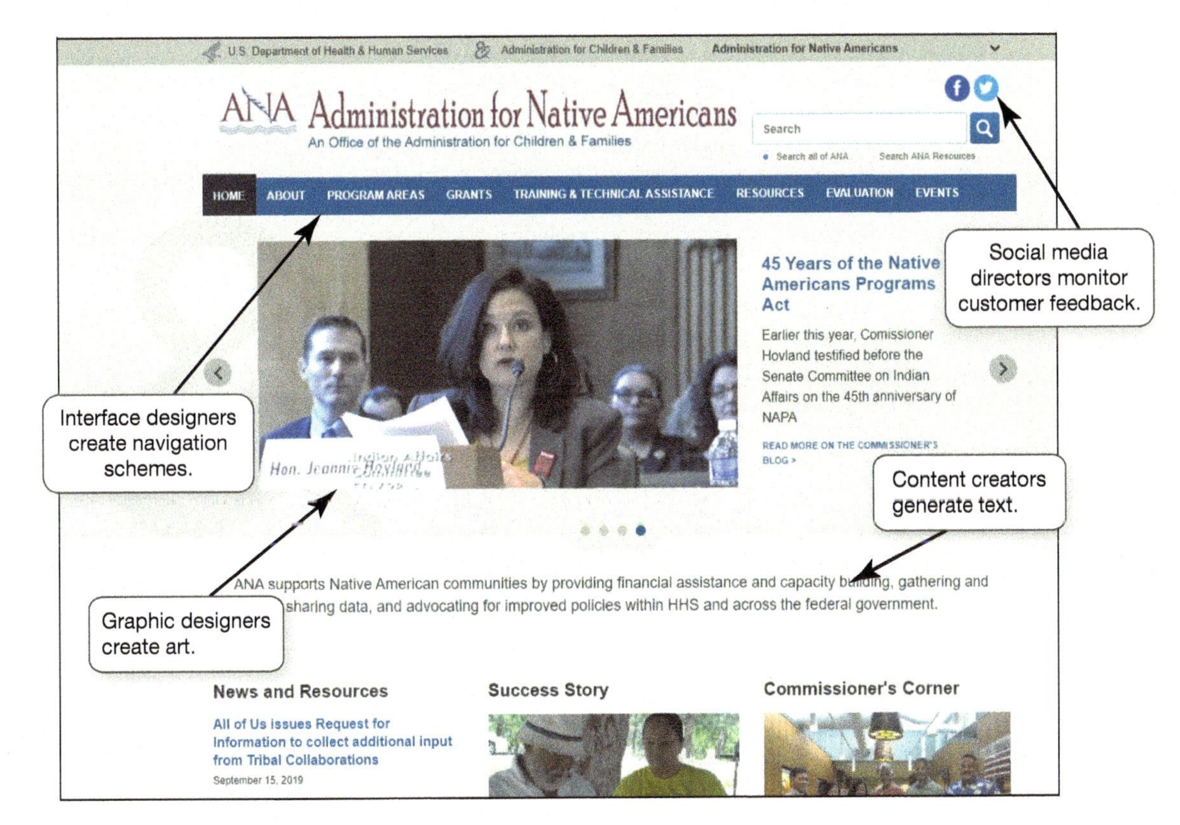

Figure B.4 It takes a team to create and maintain a website.

(United States Department of Agriculture)

Bits&Bytes Matching a Career to Your Skills

Trends in technology and business often give clues to up-and-coming IT jobs. Consider these trends when considering your educational choices:

Drone aircraft. Drones are making their way from the military to businesses and households. Companies that make or use drones will need drone programmers.

Cyber spying. Recent revelations of widespread cyber spying and espionage point to a continued demand for IT security professionals.

The Internet of Things. More and more devices are featuring Internet connectivity every year. Companies will need designers for user interfaces and software applications to drive connected devices.

Driverless cars. As vehicles become ever more sophisticated, so do their IT systems. Developers of artificial intelligence systems for vehicles and business equipment will be highly sought after.

Systems Development. Ask most people what systems developers do and they'll answer, "Programming." However, programming is only one aspect of systems development. Because large projects involve many people, there are many job opportunities in systems development, most of which require four-year college degrees in computer science or management information systems:

- Systems analysts gather information from end users about problems and existing information systems. They document systems and propose solutions to problems. Having good people skills is essential to success as a systems analyst. In addition, systems analysts work with programmers during the development phase to design appropriate programs to solve the problem at hand. Therefore, many organizations insist on hiring systems analysts who have solid business backgrounds and programming experience (at least at a basic level).
- Programmers attend meetings to document user needs, and they work closely with systems analysts during the design phase of program development. Programmers need excellent written communication skills because they often generate detailed systems documentation for end-user training purposes. Because programming languages are mathematically based, it is essential for programmers to have strong math skills and an ability to think logically.
- Project managers manage the overall systems development process: assigning staff, budgeting, reporting to management, coaching team members, and ensuring that deadlines are met. Project managers need excellent time management skills because they're often pulled in several directions at once. They usually have prior experience as programmers or systems analysts. Many project managers obtain master's degrees to supplement their undergraduate degrees.

In addition to these key players, the following people are also involved in the systems development process:

- Technical writers generate systems documentation for end users and for programmers who may make modifications to the system in the future.
- Network administrators help the programmers and analysts design compatible systems, because many systems are required to run in certain environments (UNIX or Windows, for instance) and must work in conjunction with other programs.
- Database developers design and build databases to support the software systems being developed.

Large development projects may have all of these team members on the project. Smaller projects may require an overlap of positions, such as a programmer also acting as a systems analyst. As shown in Figure B.5, team members work together to build a system.

Working in Technical Services

Technical services jobs are vital to keeping IT systems running. The people in these jobs install and maintain the infrastructure behind the IT systems and work with end users to make sure they can interact with the systems effectively. There are two major categories of technical services careers: information systems and support services. Note that these also are the least likely IT jobs to be outsourced because hands-on work with equipment and users is required on a regular basis.

- **Programmers build the software that solves problems.**

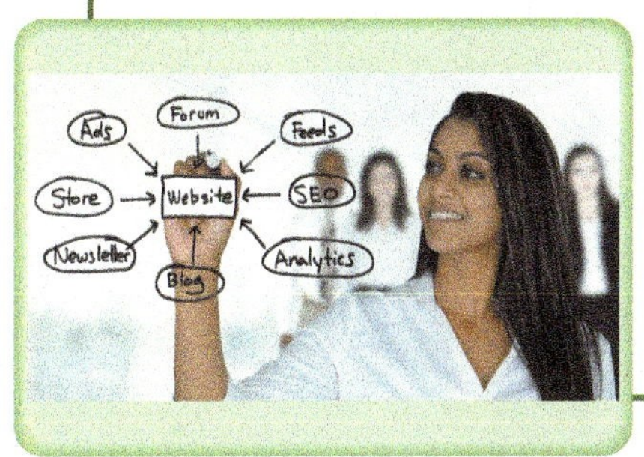

- **Systems analysts document systems and propose solutions.**

- **Project managers supervise, organize, and coach team members.**

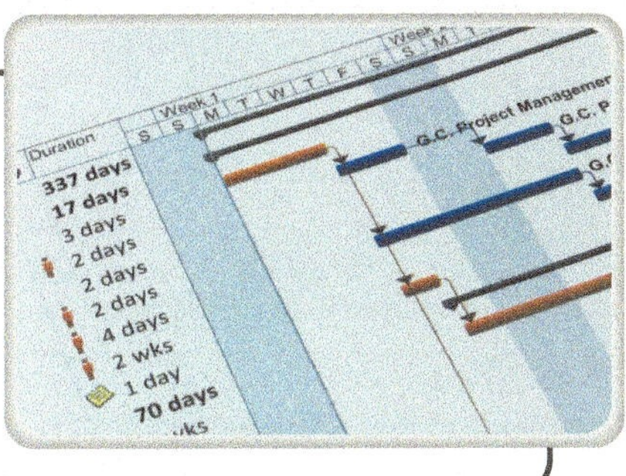

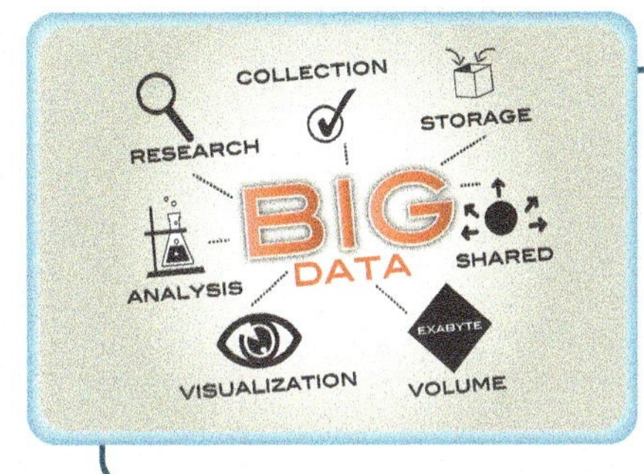

- **Database developers ensure data is accessible when it's needed.**

Figure B.5 There are many important team members in systems development.

(Dmitry Nikolaev/Shutterstock; Rob Marmion/Shutterstock; Paul Barnwell/Shutterstock; Trueffelpix/Shutterstock)

Information Systems. The information systems department keeps the networks and telecommunications up and running at all times. Within the department, you'll find a variety of positions.

- Network administrators (sometimes called *network engineers*) are involved in every stage of network planning and deployment (see Figure B.6). They decide what equipment to buy and what media to use, determine the network's topology, and help install the network (by supervising contractors or doing it themselves). Network administrators plan disaster-recovery strategies (such as what to do if a fire destroys the server room). When equipment and cables break, network administrators must fix the problem. They also obtain and install updates to network software and evaluate new equipment to determine whether the network should be upgraded. In addition, they monitor the network's performance and often develop policies regarding network usage, security measures, and hardware and software standards.
- Database administrators (DBAs) install and configure database servers and ensure that the servers provide an adequate level of access to all users.
- Web server administrators install, configure, and maintain web servers and ensure that the company maintains Internet connectivity at all times.
- Telecommunications technicians oversee the communications infrastructure, including training employees to use telecommunications equipment. They are often on call 24 hours a day.

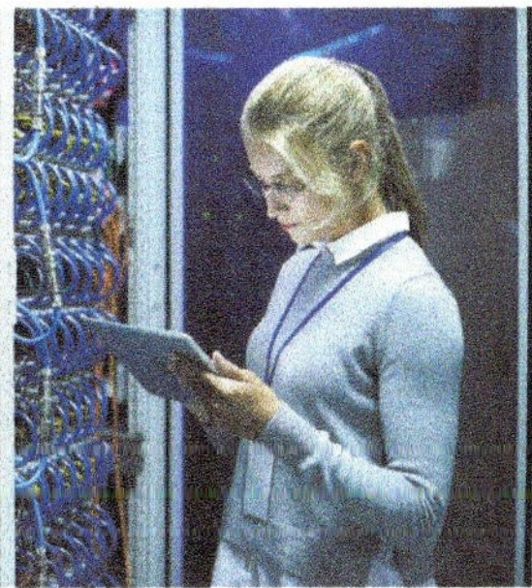

Figure B.6 At smaller companies, network administrators may be fixing a user's computer in the morning, installing and configuring a new network operating system in the afternoon, and troubleshooting a server problem (shown here) in the evening.

(SeventyFour/Shutterstock)

Support Services. As a member of the support services team, you interface with users (external customers or employees) and troubleshoot their computer problems. Support service positions include:

- Helpdesk analysts staff the phones, respond to Internet live chats, or respond to e-mail and solve problems for customers or employees, either remotely or in person. Often, helpdesk personnel are called on to train users on the latest software and hardware.
- Computer support technicians go to a user's location and fix software and hardware problems. They also often have to chase down and repair faults in the network infrastructure.

As important as these people are, they may receive a great deal of abuse by angry users whose computers are not working. When working in support services, you need to be patient and not be overly sensitive to insults!

Trends in IT — Get in the Game: Careers in Game Development

The video gaming industry in the United States has surpassed the earning power of the Hollywood movie industry. In 2019, the US video game industry brought in sales of $14.6 billion worth of video game software, hardware, and accessories, whereas Hollywood took in just over $11.4 billion. Although some aspects of game development, such as scenery design and certain aspects of programming, are being sent offshore, the majority of game development requires a creative team whose members need to work in close proximity. Therefore, it's anticipated that many game development jobs will stay in the United States.

Consoles such as the Xbox and the PlayStation generate demand for large-scale games. The popularity of mobile devices such as smartphones and tablets is driving demand for lower-end, casual game applications. Casual games are games that can be played relatively quickly or occasionally, such as massively multiplayer online (MMO) strategy games like Clash of Clans. Demand for family-friendly games without violence, sex, and profanity is on the rise (see Figure B.7). With all this demand, there are many opportunities for careers in game development.

Game development jobs are usually split along two paths: designers and programmers.

- Game designers tend to be artistic and are responsible for creating 2D and 3D art, game interfaces, video sequences, special effects, game levels, and scenarios. Game designers must master software packages such as Autodesk 3ds Max, Autodesk Maya, NewTek LightWave, Adobe Photoshop, and Adobe Flash.
- Game programmers are responsible for coding the scenarios developed by the designers. Programmers use languages and toolsets such as Objective-C, Unity, Apple Xcode, C, C++, Assembly, and Java to build the game and ensure that it plays accurately.

Aside from programmers and designers, play testers and quality-assurance professionals play the games with the intent of breaking them or discovering bugs within the game interfaces or worlds. Play testing is an essential part of the game development process because it assists designers in determining which aspects of the game are most intriguing to players and which parts of the game need to be repaired or enhanced.

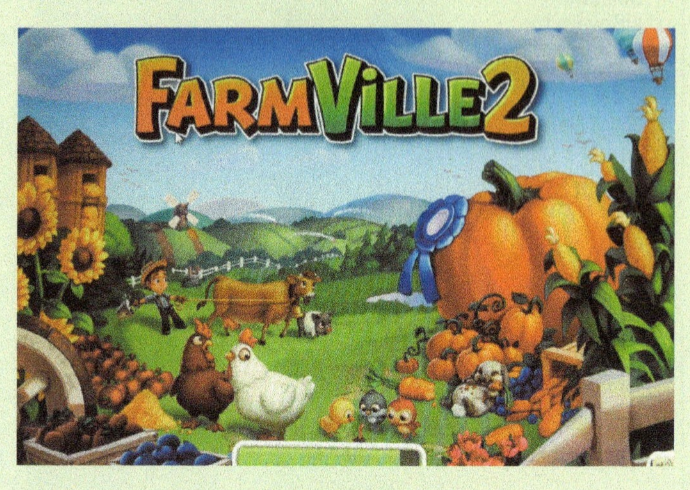

Figure B.7 FarmVille2, in which you manage a virtual farm, is an example of a popular casual game available for mobile platforms.

(Jake Charles/Alamy Stock Photo)

No matter what job you may pursue in the realm of gaming, you'll need to have a two- or four-year college degree. If you're interested in gaming, look for a school with a solid animation or 3D art program or a computer game–programming curriculum. Programming requires a strong background in math and physics to enable you to program environments realistically that mimic the real world. Proficiency with math (especially geometry) also helps with design careers.

For more information on gaming careers, check out the International Game Developers Association site (*igda.org*) and the Game Career Guide (*gamecareerguide.com*).

Technical services jobs often require a two-year college degree or training at a trade school or technical institute. At smaller companies, job duties tend to overlap between the helpdesk and technician jobs. These jobs are in demand and require staffing in local markets, so they are staying onshore. You can't repair a computer's power supply if you are located in another country!

Preparing for a Job in IT

Objective B.4 *Explain the various ways you can prepare for a career in IT.*

A job in IT requires a robust skill set and formal training and preparation. Most employers today have an entry-level requirement of a college degree, a technical institute diploma, appropriate professional certifications, experience in the field, or a combination of these. How can you prepare for a job in IT?

1. **Get educated.** Two and four-year colleges and universities normally offer three degrees to prepare students for IT careers: computer science, management information systems, and information technology (although titles vary). Alternatives to colleges and universities are privately licensed technical (or trade) schools. Generally, these programs focus on building skill sets rapidly and qualifying for a job in a specific field. The main advantage of technical

Bits&Bytes **Certifications: How Important Are They?**

Employees with certifications generally earn more than employees who aren't certified. However, most employers don't view a certification as a substitute for a college degree or a trade school program. You should think of certifications as an extra edge beyond your formal education that will make you more attractive to employers. To ensure that you're pursuing the right certifications, ask employers which certifications they respect, or explore online job sites to see which certifications are listed as desirable or required.

schools is that their programs usually take less time to complete than college degrees. However, to have a realistic chance of employment in IT fields other than networking or web development, you should attend a degree-granting college or university.

2. **Investigate professional certifications.** Certifications attempt to provide a consistent method of measuring skill levels in specific areas of IT. Hundreds of IT certifications are available, most of which you get by passing a written exam. Software and hardware vendors (such as Microsoft and Cisco) and professional organizations (such as the Computing Technology Industry Association) often establish certification standards. Visit *microsoft.com, cisco.com, comptia.org*, and *https://education.oracle.com/certification* for more information on certifications.
3. **Get experience.** In addition to education, employers want you to have experience, even for entry-level jobs. While you're still completing your education, consider getting an internship or part-time job in your field of study. Many schools will help you find internships and allow you to earn credit toward your degree through internships.
4. **Do research.** Find out as much as you can about the company and the industry it's in before going on an interview. Start with the company's website, and then expand your search to business and trade publications such as *Businessweek* and *Inc.* magazines.

Getting Started in an IT Career

Objective B.5 *Explain the various ways you can find a career in IT.*

Training for a career is not useful unless you can find a job at the end of your training. Here are some tips on getting a job:

1. **Use the career resources at your school.** Many employers recruit at schools, and most schools maintain a placement office to help students find jobs. Employees in the placement office can help you with résumé preparation and interviewing skills and can provide you with leads for internships and jobs.
2. **Develop relationships with your instructors.** Many college instructors still work in or previously worked in the IT industry. They can often provide you with valuable advice and industry contacts.
3. **Start networking.** Many jobs are never advertised but instead are filled by word of mouth. Seek out contacts in your field and discuss job prospects with them. Find out what skills you need, and ask them to recommend others in the industry with whom you can speak. Professional organizations such as the Association for Computing Machinery (ACM) offer one way to network. These organizations often have chapters on college campuses and offer reduced membership rates for students. Table B.4 lists major professional organizations you should consider investigating.

   If you're a woman and are thinking about pursuing an IT career, many resources and groups cater to female IT professionals and students (see Table B.5). The oldest and best-known organization is the Association for Women in Computing, founded in 1978.
4. **Use online career tools.** Sites such as *LinkedIn.com* help you build a social network of business contacts and locate people working for companies in which you are interested. *Glassdoor.com* features company reviews posted anonymously by employees and job seekers to provide you with a transparent view of the culture of individual workplaces.
5. **Check corporate websites for jobs.** Many corporate websites list current job opportunities. For example, Google provides searchable job listings by geographical location. Check the sites of companies in which you are interested and then do a search on the sites for job openings or, if provided, click their employment links.

Table B.4 Professional Organizations

Organization Name	Purpose	Website
Association for Computing Machinery (ACM)	Oldest scientific computing society; maintains a strong focus on programming and systems development	*acm.org*
Association for Information Systems (AIS)	Organization of professionals who work in academia and specialize in information systems	*aisnet.org*
CompTIA IT Pro and Student Membership	Heavy focus on IT education and development of seminars and learning materials	*comptia.org*
Institute of Electrical and Electronics Engineers (IEEE)	Provides leadership and sets engineering standards for all types of network computing devices and protocols	*ieee.org*
Information Systems Security Association (ISSA)	Not-for-profit, international organization of information security professionals and practitioners	*issa.org*

Table B.5 Resources for Women in IT

Organization Name	Purpose	Website
Anita Borg Institute	Organization whose aim is to ensure "a future where the teams that imagine and build technology mirror the people and societies for whom they build it"	*anitab.org*
Association for Women in Computing (AWC)	A not-for-profit organization dedicated to promoting the advancement of women in computing professions	*awc-hq.org*
The Center for Women in Technology (CWIT)	An organization dedicated to providing global leadership in achieving women's full participation in all aspects of IT	*cwit.umbc.edu*
Women in Technology International (WITI)	A global trade association for tech-savvy, professional women	*witi.com*

6. **Visit online employment sites.** Most of these sites allow you to store your résumé online, and many sites allow employers to browse résumés to find qualified employees. Begin looking at job postings on these sites early in your education because these job postings detail the skill sets employers require. Focusing on coursework that will provide you with desirable skill sets will make you more marketable. Table B.6 lists employment sites as well as sites that offer other career resources.

The outlook for IT jobs should continue to be positive in the future. We wish you luck with your education and job search.

Table B.6 Resources for IT Employment

CareerBuilder.com	*Indeed.com*
ComputerJobs.com	*Jobs.Gamasutra.com*
ComputerWork.com	*JustTechJobs.com*
Crunchboard.com	*LinkedIn.com*
Dice.com	*Monster.com*
Glassdoor.com	*TechCareers.com*
Hired.com	*TheLadders.com*
Idealist.org	*WhiteTruffle.com*

Check Your Understanding // Review & Practice

For a quick review of what you've learned, answer the following questions.

multiple choice

1. The individuals responsible for generating images for websites are referred to as
   a. network administrators.
   b. web programmers.
   c. graphic designers.
   d. interface designers.
2. What type of job involves traveling as needed and performing short-term jobs at each location?
   a. Home based
   b. Project based
   c. Office based
   d. Field based
3. Which position is NOT typically a part of the web development department?
   a. Graphic designer
   b. Interface designer
   c. Social media director
   d. Computer support technician
4. Outsourcing is thought to be an attractive option for many companies because of
   a. the emphasis on employee training.
   b. increased data security.
   c. the cost savings that can be realized.
   d. decreased travel and entertainment costs.
5. Which of the following IT positions is responsible for directing a company's strategy on sites such as Facebook, Twitter, and Yelp?
   a. Project manager
   b. Social media director
   c. Customer interaction technician
   d. Social web analyst
6. Which of the following statements about IT careers is TRUE?
   a. Most IT jobs require little interaction with other people.
   b. Women who have IT skills have very few opportunities for securing IT employment.
   c. Many IT jobs are staying in the United States.
   d. IT salaries are declining.
7. Which task is typically performed by a network administrator?
   a. Developing network usage policies
   b. Going to a user's location and fixing software and hardware problems
   c. Web programming
   d. Generating systems documentation for end users
8. Which of the following is NOT a challenge in an IT career?
   a. Stress often accompanies IT jobs.
   b. You usually work alone.
   c. Women are in the minority.
   d. Lifelong learning is the norm.

true/false

_____ **1.** IT certifications are considered important by most employers.
_____ **2.** Online employment sites cannot provide guidance about coursework you should take.

Glossary

2-in-1 PC A laptop computer that can convert into a tablet-like device.

3D printer Used to print three-dimensional models.

3D sound card An expansion card that enables a computer to produce omnidirectional or three-dimensional sounds.

4G Stands for fourth generation. The current mobile communication standard with faster data transfer rates than 3G.

5G Stands for fifth generation. The latest mobile communication standard being rolled out as a mobile data service and also as a home Internet service. 5G will have faster data transfer rates than 4G.

A

acceptable use policies Guidelines regarding usage of computer systems.

access card reader A device that reads information from a magnetic strip on the back of a credit card–like access card.

access method A method to control which computer is allowed to use transmission media at a certain time.

access point Device that is connected by a cable to the main router that amplifies a wireless signal to otherwise unreachable locations.

access time The time it takes a storage device to locate its stored data.

accounting software An application program that helps business owners manage their finances more efficiently by providing tools for tracking accounting transactions such as sales, accounts receivable, inventory purchases, and accounts payable.

Active Server Pages (ASP) Programming language used to build websites with interactive capabilities; adapts an HTML page to the user's selections.

active topology A type of network design where the nodes participate in moving data through the network.

adware A program that downloads on your computer when a user installs a freeware program, game, or utility. Generally, adware enables sponsored advertisements to appear in a section of a browser window or as a pop-up ad.

affective computing A type of computing that relates to emotion or that deliberately tries to influence emotion.

aggregator A software program that finds and retrieves the latest update of web material (usually podcasts) according to your specifications.

AJAX (Asynchronous JavaScript and XML) A collection of technologies that allow the creation of web applications that can update information on a page without requiring the user to refresh or leave the page.

algorithm A set of specific, sequential steps that describe exactly what the computer program must do to complete the required work.

all-in-one computer A desktop system unit that houses the computer's processor, memory, and monitor in a single unit.

all-in-one printer A device that combines the functions of a printer, scanner, copier, and fax.

alphabetic check In a database, confirms that only textual characters are entered in a field.

American Standard Code for Information Interchange (ASCII) The American National Standards Institute's (ANSI, pronounced "AN-see") standard code, pronounced "AS-key," to represent each letter or character as an 8-bit (or 1-byte) binary code.

amoral behavior When a person has no sense of right and wrong and no interest in the moral consequences of his or her actions.

analog Waves that illustrate the loudness of a sound or the brightness of the colors in an image at a given moment in time.

Android A mobile operating system designed by Google and used by Samsung, Google, and other manufacturers.

antivirus software Software specifically designed to detect viruses and protect a computer and files from harm.

app creation software Programming environments that can produce apps that run on various mobile devices.

applet A small application located on a server; when requested, the applet is downloaded to the client.

application programming interface (API) A set of software routines that allows one software system to work with another.

application server A server that acts as a repository for application software.

application software The set of programs on a computer that helps a user carry out tasks such as word processing, sending e-mail, balancing a budget, creating presentations, editing photos, taking an online course, and playing games.

architecture neutral A feature of Java whereby code needs to be compiled only once, after which the code can be run on many different CPUs.

arithmetic logic unit (ALU) The part of the CPU that performs arithmetical and logic calculations.

artificial intelligence (AI) The branch of computer science that deals with the attempt to create computers that think like humans.

artificial neural networks (ANNs) Computer systems constructed based on the structure of the human brain using loosely connected artificial neurons.

aspect ratio The width-to-height proportion of a monitor.

assembly language A computer language that allows programmers to write programs using a set of short, English-like commands that speak directly to the CPU and that give the programmer direct control of hardware resources.

assistive (adaptive) technology Any device, software feature, or app that is designed to improve the functional capabilities for individuals with disabilities.

audio-editing software Programs that perform basic editing tasks on audio files such as cutting dead air space from the beginning or end of a song or cutting a portion from the middle.

audio MIDI interface Interface technology that allows a user to connect musical instruments and microphones to his or her computer.

augmented reality A combination of our normal sense of the objects around us while displaying an overlay of information.

authentication The process of identifying a computer user, based on a login or username and password. The computer system determines whether the computer user is authorized and what level of access is to be granted on the network.

authentication server A server that keeps track of who's logging on to the network and which services on the network are available to each user.

B

backdoor program A program that enables a hacker to take complete control of a computer without the legitimate user's knowledge or permission.

backup A copy of a computer file that can be used to replace the original if it's lost or damaged.

backward compatibility The accommodation of current devices being able to use previously issued software standards in addition to the current standards.

base class In object-oriented analysis, the original class.

base-10 number system (decimal notation) The system you use to represent all of the numeric values you use each day. It's called *base 10* because it uses 10 digits—0 through 9—to represent any value.

basic input/output system (BIOS) A program that manages the data between a computer's operating system and all the input and output devices attached to the computer; also responsible for loading the operating system (OS) from its permanent location on the hard drive to random access memory (RAM).

bastion host A heavily secured server located on a special perimeter network between the company's secure internal network and its firewall.

batch processing The accumulation of transaction data until a certain point is reached, at which time several transactions are processed at once.

beta version A version of the software that's still under development. Many beta versions are available for a limited trial period and are used to help the developers correct any errors before they launch the software on the market.

Big Data Very large data sets that are analyzed to reveal patterns, trends, and associations.

binary decision A type of decision point in an algorithm that can be answered in one of only two ways: yes (true) or no (false).

binary digit (bit) A digit that corresponds to the on and off states of a computer's switches. A bit contains a value of either 0 or 1.

binary language The language computers use to process data into information, consisting of only the values 0 and 1.

binary large object (BLOB) Audio clips, video clips, pictures, and extremely large documents.

binary number system (base-2 number system) A method of representing a number in powers of 2.

biometric authentication device A device that uses some unique characteristic of human biology to identify authorized users.

bit rot (data degradation) Data loss on hard drives due to the passage of time and wearing out materials that store data.

black-hat hacker A hacker who uses his or her knowledge to destroy information or for illegal gain.

blockchain A massive public online ledger that keeps track of transactions made in a digital currency.

blog (weblog) A personal log or journal posted on the web; short for *web log*.

Bluetooth A type of wireless technology that uses radio waves to transmit data over short distances often used to connect peripherals such as printers and keyboards to computers or headsets to cell phones.

Blu-ray disc (BD) A method of optical storage for digital data, developed for storing high-definition media. It has the largest storage capacity of all optical storage options.

Bookmarks Features in some browsers that place markers of websites' Uniform Resource Locators (URLs) in an easily retrievable list.

Boolean operators Words used to refine logical searches. For Internet searches, the words AND, NOT, and OR describe the relationships between keywords in the search.

boot process The process for loading the operating system (OS) into random access memory (RAM) when the computer is turned on.

boot-sector virus A virus that replicates itself into the master boot record of a flash drive or hard drive.

bot accounts Automated programs retweeting news stories and quotes.

botnet A large group of software applications (called *robots* or *bots*) that run without user intervention on a large number of computers.

breadcrumb trail A navigation aid that shows users the path they have taken to get to a web page or where the page is located within the website; it usually appears at the top of a page.

bridge A device that is used to send data between different collision domains in a network, depending on where the recipient device is located.

broadband A high-speed Internet connection such as cable, satellite, or digital subscriber line (DSL).

brute-force attack An attempt to access an account by repeatedly trying different passwords.

bus (linear bus) topology A system of networking connections in which all devices are connected in sequence on a single cable.

business intelligence (BI) The ability to improve business decision making with databases and other fact-based support systems.

business intelligence system Systems used to analyze and interpret data in order to enable managers to make informed decisions about how best to run a business.

business-to-business (B2B) E-commerce transactions between businesses.

business-to-consumer (B2C) E-commerce transactions between businesses and consumers.

byte Eight binary digits (bits).

C

C A programming language originally developed for system programmers.

C++ A programming language; takes C to an object-oriented level.

C# A Microsoft programming language developed to compete with Java.

cable Internet A broadband service that transmits data over coaxial cables.

cache memory Small blocks of memory, located directly on and next to the central processing unit (CPU) chip, that act as holding places for recently or frequently used instructions or data that the CPU accesses the most.

caption A property in a database that enables the field name to appear in a more meaningful or readable manner.

cascading style sheet (CSS) A list of rules that defines in one single location how to display HTML elements.

catfishing A type of Internet harassment by which some individual scams others into a false romantic relationship.

Cat 5e cable A UTP cable commonly found in wired Ethernet networks, designed for 100 Mbps-wired Ethernet networks.

Cat 6 cable A UTP cable that provides more than 1 Gbps of throughput.

Cat 6a cable A UTP cable that is designed for ultrafast 10GbE networks.

Cat 7 cable A UTP cable that is designed for 10GbE networks and offers greater throughput than Cat 6a cable.

central processing unit (CPU or processor) The part of the system unit of a computer that is responsible for data processing; it is the largest and most important chip in the computer. The CPU controls all the functions performed by the computer's other components and processes all the commands issued to it by software instructions.

centralized A characteristic of client/server networks where the server, not the individual nodes, provide services such as data security.

charge-coupled device (CCD) arrays Electronic sensors used to capture images on digital cameras.

Chromebook Any laptop or tablet running the Chrome OS as its operating system.

circuit switching Where a dedicated connection is formed between two points (such as two people on telephones) and the connection remains active for the duration of the transmission.

class A category of input identified in object-oriented analysis; classes are defined by information and actions.

clickstream data Software used on company websites to capture information about each click that users make as they navigate through the site.

client A computer that requests information from a server in a client/server network (such as your computer when you are connected to the Internet).

client/server model A model of network communications where a client device such as a computer, tablet, or smartphone uses browsers to request services from networks that make up the Internet.

client/server network A type of network that uses servers to deliver services to computers that are requesting them (clients).

client-side program A program that runs on the client computer and that requires no interaction with a web server.

clock cycle Steady beats or ticks of the system clock.

clock speed The steady and constant pace at which a computer goes through machine cycles, measured in hertz (Hz).

cloud computing The process of storing data, files, and applications on the web, which allows access to and manipulation of these files and applications from any Internet-connected device.

cloud-ready printer Printers that connect directly to the Internet and register themselves with Google Cloud Print.

cloud server A server that is maintained by a hosting company and that is connected to networks via the Internet.

cloud storage A service that keeps files on the Internet (in the cloud) rather than storing files solely on a local device.

cluster The smallest increment in which data is stored on hard disks; hard disks are divided into *tracks*, then *wedges*, then *sectors*, and then *clusters*.

coaxial cable A single copper wire surrounded by layers of plastic insulation, metal sheathing, and a plastic jacket; used mainly in cable television and cable Internet service.

code editing The step of programming in which a programmer types the code to be executed.

codec A rule, implemented in either software or hardware, which squeezes a given amount of audio and video information into less space.

coding Translating an algorithm into a programming language.

cognitive surplus The combination of leisure time and the tools to be creative.

cold boot The process of starting a computer from a powered-down or off state.

collaborative consumption Joining together as a group to use a specific product more efficiently.

command-driven interface Interface between user and computer in which the user enters commands to communicate with the computer system.

comment A note left by a programmer in the program code to explain the purpose of a section of code, to indicate the date the program was written, or to include other important information about the code so that other programmers can more easily understand and update it.

commerce server Computers that host software that enables users to buy goods and services over the web.

communications server A server that handles all communications between the network and other networks, including managing Internet connectivity.

compilation The process by which code is converted into machine language—the language the CPU can understand.

compiler A program that understands both the syntax of the programming language and the exact structure of the CPU and its machine language.

completeness check In a database, ensures that all required fields have been completed.

computer A data-processing device that gathers, processes, outputs, and stores data and information.

computer-aided design (CAD) A 3D modeling program used to create automated designs, technical drawings, and model visualizations.

computer literacy Being familiar enough with computers that a user knows how to use them and understands their capabilities and limitations.

computer protocol A set of rules for exchanging electronic information.

computer vision The ability of a computing device to interpret visual information the way humans do.

connection-oriented protocol A protocol for exchanging information that requires two computers to exchange control packets, thereby setting up the parameters of the data-exchange session, before sending packets that contain data.

connectionless protocol A protocol that a host computer can use to send data over the network without establishing a direct connection with any specific recipient computer.

connectivity port A port that enables a computing device to be connected to other devices or systems such as networks, modems, and the Internet.

consistency check In a database, compares the values of data in two or more fields to see if those values are reasonable.

consumer-to-consumer (C2C) E-commerce transactions between consumers through online sites such as eBay.

control structure General term used for a keyword in a programming language that allows the programmer to direct the flow of the program based on a decision.

control unit Unit of the CPU that manages the switches inside the CPU.

cookie A small text file that some websites automatically store on a client computer's hard drive when a user visits the site.

copyleft A simplified licensing scheme that enables copyright holders to grant certain rights to a work while retaining other rights.

copyright A creator's exclusive rights to use a work of intellecutal property, such as making copies of it or selling it.

copyright infringement Any violation of someone's rights under copyright laws, such as reproducting his or her work without permission.

core A complete processing section from a central processing unit, embedded into one physical chip.

course management software A program that provides traditional classroom tools, such as calendars and grade books, over the Internet, as well as areas for students to exchange ideas and information in chat rooms, discussion forums, and e-mail.

CPU benchmarks Measurements used to compare performance between processors.

CPU usage The percentage of time the central processing unit (CPU) is working.

CPU usage graph Records central processing unit (CPU) usage and displays it visually.

crisis-mapping tool A tool that collects information from e-mails, text messages, blog posts, and Twitter tweets and maps them, making the information publicly available instantly.

crowdfunding Asking for small donations from a large number of people, often using the Internet; a style of generating capital to start a business through social media.

crowdsourcing Obtaining information or input for a project (such as development of a product) by obtaining the opinions of many different people.

cryptocurrency (virtual, cyber, or digital currency) A currency specifically created for the purpose of making payments. Bitcoin is the most well-known cryptocurrency.

CSMA/CD The method used on Ethernet networks to avoid data collisions; short for *carrier sense multiple access with collision detection*. A node connected to the network uses carrier sense to verify that no other nodes are currently transmitting data signals.

cursor An onscreen icon (often shown by a vertical bar or an arrow) that helps the user keep track of exactly what is active on the display screen

custom installation The process of installing only those features of a software program that a user wants on the hard drive.

cyber harassment Using technology to harass or intimidate another individual.

cybercrime Any criminal action perpetrated primarily through the use of a computer.

cyberloafing Doing anything with a computer that's unrelated to a job (such as playing video games) while one's supposed to be working. Also called *cyberslacking*.

D

dark web A very well-concealed part of the deep web where many illegal and illicit activities take place.

data Numbers, words, pictures, or sounds that represent facts, figures, or ideas; the raw input that users have at the start of a job.

data breach When sensitive or confidential information is copied, transmitted, or viewed by an individual who is not authorized to handle the data.

data centralization When data is maintained in only one file.

data collision When two computers send data at the same time and the sets of data collide somewhere in the transmission media.

data dictionary (database schema) A map of a database that defines the features of the fields in the database.

data file A file that contains stored data.

data-flow diagram Diagrams that trace all data in an information system from the point at which data enters the system to its final resting place (storage or output).

data inconsistency Data for a record being different in two different lists.

data integrity When the data in a database is accurate and reliable.

data mart Small slices of a data warehouse grouped together and separated from the main body of data in the data warehouse so that related sets of data can be analyzed.

data mining The process by which great amounts of data are analyzed and investigated. The objective is to spot significant patterns or trends within the data that would otherwise not be obvious.

data plan A connectivity plan or text messaging plan in which data charges are separate from cell phone calling charges and are provided at rates different from those for voice calls.

data redundancy Duplicated data in a list.

data staging An intermediate storage area used for data processing during the extract, transform, and load process; between the data source and the data warehouse.

data transfer rate (bandwidth) The maximum speed at which data can be transmitted between two nodes on a network; measured in megabits per second (Mbps) or gigabits per second (Gbps).

data type (field type) (1) Describes the kind of data being stored at the memory location; each programming language has its own data types (although there is some degree of overlap). (2) In a database, indicates what type of data can be stored in a field and prevents the wrong type of data from being entered into the field.

data warehouse A large-scale collection of data that contains and organizes in one place all the data from an organization's multiple databases.

database A digital collection of related data that can be stored, sorted, organized, and queried.

database administrator (database designer) An information technology professional responsible for designing, constructing, and maintaining databases.

database management system (DBMS) Specially designed application software used to create and manage databases.

database server A server that provides client computers with access to information stored in databases.

database software An electronic filing system best used for larger and more complicated groups of data that require more than one table and the ability to group, sort, and retrieve data and generate reports.

debugger A tool in an integrated development environment that helps programmers analyze a program as it runs.

debugging The process of running a program over and over to find and repair errors and to make sure the program behaves in the way it should.

decentralized A characteristic of peer-to-peer networks where the individual nodes provide services such as data security.

decision point A place where a program must choose from a list of actions based on the value of a certain input.

decision support system (DSS) A type of business intelligence system designed to help managers develop solutions for specific problems.

dedicated server A server used to fulfill one specific function, such as handling e-mail.

deep learning (DL) AI systems capable of learning from their mistakes (just as humans do). DL systems therefore improve their accuracy going forward.

deep web Any part of the Internet that you can't find by using a search engine, such as areas protected by a password or a paywall.

deepfake Digital videos or images that are manipulated using artificial intelligence to create images and sounds that appear to be real.

default value The value a database automatically uses for a field unless the user enters another value.

denial-of-service (DoS) attack An attack that occurs when legitimate users are denied access to a computer system because a hacker is repeatedly making requests of that computer system that tie up its resources.

derived class In object-oriented analysis, the modified class.

desktop The primary working area of Windows 10.

desktop client A software program that runs on a computer and is used to send and receive e-mail through an Internet service provider's server.

desktop computer A computer that is intended for use at a single location. A desktop computer consists of a case that houses the main components of the computer, plus peripheral devices.

desktop publishing (DTP) software Programs for incorporating and arranging graphics and text to produce creative documents.

detail report A report generated by a management information system that provides a list of the transactions that occurred during a certain time period.

device driver Software that facilitates the communication between a device and its operating system.

digital audio workstation software (DAW) Software used to record and edit music.

digital convergence The use of a single unifying device to handle media, Internet, entertainment, and telephony needs.

digital divide The discrepancy between those who have access to the opportunities and knowledge that computers and the Internet offer and those who do not.

digital forensics A discipline which analyzes computer devices with specific techniques to gather potential legal evidence.

digital manipulation Altering media so that the images captured are changed from the way they were originally seen by the human eye.

digital rights management (DRM) A system of access control that allows only limited use of material that has been legally purchased.

digital video-editing software A program for editing digital video.

digital wallet A software-based storage device that stores the information needed to complete cryptocurrency transactions.

digitized Information measured and converted (from analog data) to a stream of numerical values.

directory A hierarchical structure that includes files, folders, and drives used to create a more organized and efficient computer.

Disk Cleanup A Windows utility that removes unnecessary files from the hard drive.

disk defragmentation The process of regrouping related pieces of files on the hard drive, enabling faster retrieval of the data.

DisplayPort A digital display interface used to connect computing devices to display devices. Has backward compatibility with HDMI and DVI.

display screen (monitor) A common output device that displays text, graphics, and video as soft copies (copies that can be seen only on screen).

distributed denial-of-service (DDoS) attack An automated attack that's launched from more than one zombie computer at the same time.

distributions (distros) Linux download packages.

Document Object Model (DOM) Used by JavaScript to organize objects and page elements.

documentation Description of the technical details of the software, how the code works, and how the user interacts with the program; in addition, all the necessary user documentation that will be distributed to the program's users.

domain name A part of a Uniform Resource Locator (URL). Domain names consist of two parts: the site's host and a suffix that indicates the type of organization (e.g., popsci.com, where *popsci* is the domain name and *com* is the suffix).

domain name system (DNS) server A server that maintains a database of domain names and converts domain names to Internet protocol addresses.

dotted decimal number (dotted quad) The form of an Internet protocol address, where sets of numerals are separated by decimals (i.e., 197.169.73.63).

drawing software (illustration software) Programs for creating or editing two-dimensional, line-based drawings.

drive-by download The use of malicious software to attack a computer by downloading harmful programs onto a computer, without the user's knowledge, while they are surfing a website.

DSL (digital subscriber line) A type of connection that uses telephone lines to connect to the Internet and that allows both phone and data transmissions to share the same line.

dynamic addressing A way of assigning Internet protocol addresses where a computer is assigned a temporary address from an available pool of addresses.

dynamic decision making The ability of a web page to decide how to display itself based on the choices the reader makes.

Dynamic Host Configuration Protocol (DHCP) A protocol for assigning dynamic Internet protocol addresses.

dynamic HTML (DHTML) A combination of technologies—HTML, cascading style sheets, and JavaScript—used to create lively and interactive websites.

E

e-commerce (electronic commerce) The process of conducting business online for purposes ranging from fund-raising to advertising to selling products.

editor A special tool in an integrated development environment (IDE) that helps programmers as they enter code.

educational fair use Extension of fair use guidelines that grants additional exceptions to copyright infringement for teachers and students when the material is used for educational purposes.

electronic data interchange (EDI) The exchange of large amounts of data in a standardized electronic format.

electronic ink (E Ink) A very crisp, sharp grayscale representation of text achieved by using millions of microcapsules with white and black particles in a clear fluid.

electronic text (e-text) Textual information stored as digital information so that it can be stored, manipulated, and transmitted by electronic devices.

element In HTML, a pair of tags and the text between them.

e-mail Internet-based communication in which senders and recipients correspond.

e-mail server A server whose function is to process and deliver incoming and outgoing e-mail.

e-mail virus A virus transmitted by e-mail that often uses the address book in the victim's e-mail system to distribute itself.

embedded computer A specially designed computer chip that resides inside another device, such as a car. These self-contained computer devices have their own programming and typically neither receive input from users nor interact with other systems.

embodied agents Robots that look and act like human beings.

encryption The process of coding e-mail so that only the person with the key to the code (the intended recipient) can decode the message.

encryption virus A malicious program that searches for common data files and compresses them into a file using a complex encryption key, thereby rendering the files unusable.

End User License Agreement (EULA) An agreement between the user and the software developer that must be accepted before installing the software on a computer.

enterprise resource planning (ERP) system A business intelligence system that accumulates in a central location all information relevant to running a business and makes it readily available to whoever needs it to make decisions.

e-reader A device that can display e-text and that has supporting tools, like note taking, bookmarks, and integrated dictionaries.

ergonomics How a user sets up his or her computer and other equipment to minimize risk of injury or discomfort.

error handling The part of a problem statement where programmers describe what the program should do if the input data is invalid or just gibberish.

Ethernet network A network that uses the Ethernet protocol as the means (or standard) by which the nodes on the network communicate.

Ethernet port A port that transfers data at speeds of up to 10 Gbps; used to connect a computer to a modem or to a network.

ethics The study of the general nature of morals and the choices individuals make.

event The result of an action—such as a keystroke, mouse click, or signal to the printer—in the respective device (keyboard, mouse, or printer) to which the operating system responds.

exception report A report generated by a management information system that shows conditions that are unusual or that need attention by system users.

executable program The binary sequence that instructs the CPU to run the programmer's code.

expansion cards (or adapter cards) A circuit board with specific functions that augment the computer's basic functions and provide connections to other devices; examples include sound cards and video cards.

expert system A system that tries to replicate the decision-making processes of human experts in order to solve specific problems.

exploit kits A software toolkit used to take advantage of security weaknesses found in apps or operating systems, usually to deploy malware.

extended reality (XR) Full set of ways reality can be manipulated with digital intervention, whether augmented reality (AR), new virtual realities (VR), or a combination of real world and digitally created reality (mixed reality).

eXtensible Markup Language (XML) A markup language that enables designers to define their own data-based tags, making it much easier for a website to transfer the key information on its page to another site; it defines what data is being described rather than how it's to be displayed.

extension (file type) In a file name, the three letters that follow the user-supplied file name after the dot (.); the extension identifies what kind of family of files the file belongs to, or which application should be used to read the file.

extranet An area of an intranet that only certain entities or individuals can access; the owner decides who will be permitted to access it.

F

fair use Guidelines allowing people to use portions of a copyrighted work for specific purposes without receiving prior permission from the copyright holder.

fake news Stories that are invented rather than being factual and that are intended to spread misinformation.

Favorites A feature in Microsoft Internet Explorer that places a marker of a website's Uniform Resource Locator (URL) in an easily retrievable list in the browser's toolbar. (called *Bookmarks* in some browsers.)

fiber-optic cable A cable that transmits data via light waves along glass or plastic fibers.

fiber-optic service Internet access that is enabled by transmitting data at the speed of light through glass or plastic fibers.

field The component of a database in which the database stores each category of information.

field constraint A property that must be satisfied for an entry to be accepted into the field.

field name A way of describing the information in the field.

field properties In an Access database table, the field name, data type, and other data elements.

field size Defines the maximum number of characters that a field can hold in a database.

fifth-generation language (5GL) A computer language in which a problem is presented as a series of facts or constraints instead of as a specific algorithm; the system of facts can then be queried; considered the most "natural" of languages.

file A collection of related pieces of information stored together for easy reference.

file compression utility A program that takes out redundancies in a file in order to reduce the file size.

File Explorer The main tool for finding, viewing, and managing the contents of your computer by showing the location and contents of every drive, folder, and file; called Windows Explorer prior to Windows 8.

File History A Windows utility that automatically creates a duplicate of your libraries, desktop, contacts, and favorites and copies it to another storage device, such as an external hard drive.

file management The process by which humans or computer software provide organizational structure to a computer's contents.

file name The first part of the label applied to a file; generally the name a user assigns to the file when saving it.

file path The exact location of a file, starting with the drive in which the file is located and including all folders, subfolders (if any), the file name, and the extension (e.g., C:\Users\username\Documents\Illustrations\EBronte.jpg).

file server A server that stores and manages files for network users or that acts as a storehouse for files that users can download.

File Transfer Protocol (FTP) A protocol used to upload and download files from one computer to another over the Internet.

filter In a database, temporarily displays records that match certain criteria.

financial planning software Programs for managing finances, such as Intuit's Quicken, which include electronic checkbook registers and automatic bill-payment tools.

firewall A software program or hardware device designed to prevent unauthorized access to computers or networks.

firmware System software that controls hardware devices.

first-generation language (1GL) The machine language of a CPU; the sequence of bits that the CPU understands.

flash drive (jump drive, USB drive, or thumb drive) A drive that plugs into a universal serial bus (USB) port on a computer and that stores data digitally.

flash memory card A form of portable storage; this removable memory card is often used in digital cameras, smartphones, video cameras, and printers.

flat database A type of database that is often represented as a list or simple table.

flatbed scanner Used to create a digital image from a tangible image (like a paper photo).

flowchart Visual diagram of a process, including the decisions that need to be made along the way.

folder A collection of files stored on a computer.

For In Visual Basic, programmers use the keyword For to implement a loop; after the keyword For, an input or output item is given a starting value, and then the statements in the body of the loop are executed.

foreign key In a relational database, the common field between tables that's not the primary key.

fourth-generation language (4GL) A computer language type that includes database query languages and report generators.

frame Groups of data packets that are sent together in a package.

freeware Any copyrighted software that can be used for free.

full backup A file backup that encompasses all files on a computer regardless of whether they have changed since the last backup.

full installation The process of installing all the files and programs from the distribution media to the computer's hard drive.

fuzzy logic Enables the interjection of experiential learning into knowledge-based systems by allowing the consideration of probabilities.

G

general availability (GA) The point in the release cycle, where, after release to manufacturers, software is available for purchase by the public.

gigabit Ethernet (GbE) The most commonly used wired Ethernet standard deployed in devices designed for home networks; provides bandwidth of up to 1 Gbps.

gigabyte (GB) About a billion bytes.

Google Chrome OS A web-based OS.

graphical user interface (GUI) Unlike the command- and menu-driven interfaces used in earlier software, GUIs display graphics and use the point-and-click technology of the mouse and cursor, making them much more user friendly.

graphics double data rate 6 (GDDR6) A standard of video memory.

graphics processing unit (GPU) A specialized logic chip that's dedicated to quickly displaying and calculating visual data such as shadows, textures, and luminosity.

gray-hat hacker A cross between black and white—a hacker who will often illegally break into systems merely to flaunt his or her expertise to the administrator of the system he or she penetrated or to attempt to sell his or her services in repairing security breaches.

green computing (green IT) A movement that encourages environmentally sustainable computing (or IT).

H

hacker Anyone who unlawfully breaks into a computer system (whether an individual computer or a network).

hacktivism Using computers and computer networks in a subversive way to promote an agenda.

handshaking In a connection-oriented protocol, the process of exchanging control packets before exchanging data packets.

hard drive The computer's nonvolatile, primary storage device for permanent storage of software and data.

hardware Any part of a computer or computer system you can physically touch.

head crash Impact of the read/write head against the magnetic platter of the hard drive; often results in data loss.

hexadecimal digit A digit with 16 possible values: 0–9 and A–F.

hexadecimal notation A base-16 number system, meaning it uses 16 digits to represent numbers instead of the 10 digits used in base 10 or the two digits used in base 2. The 16 digits it uses are the 10 numeric digits, 0 to 9, plus six extra symbols: A, B, C, D, E, and F.

hibernate A power-management mode that saves the current state of the current system to the computer's hard drive.

high definition A standard of digital TV signal that guarantees a specific level of resolution and a specific *aspect ratio*, which is the rectangular shape of the image.

high-definition multimedia interface (HDMI) port A compact audio–video interface standard that carries both high-definition video and uncompressed digital audio.

hoax Anything designed to deceive another person either as a practical joke or for financial gain.

home area network (HAN) A network located in a home that's used to connect all of its digital devices.

home network server A device designed to store media, share media across the network, and back up files on computers connected to a home network.

honey pot A computer system that's set up to attract unauthorized users by appearing to be a key part of a network or a system that contains something of great value.

host The portion of a domain name that identifies who maintains a given website. For example, berkeley.edu is the domain name for the University of California at Berkeley, which maintains that site.

HTML embedded scripting language A programming language that tucks programming code directly within the HTML of a web page; the most popular example is JavaScript.

HTML tag Tags that surround and define HTML content (such as <b> and</b>, which indicate bolding).

hybrid topology Combining multiple topologies on one network.

hyperlink A type of specially coded text that, when clicked, enables a user to jump from one location, or web page, to another within a website or to another website altogether.

Hypertext Markup Language (HTML) A series of tags that define how elements on a website should be displayed in a browser.

Hypertext Transfer Protocol (HTTP) The protocol that allows files to be transferred from a web server so that you can see them on your computer by using a browser.

Hypertext Transfer Protocol Secure (HTTPS) The Internet protocol that ensures data is sent securely over the web.

hyperthreading A technology that permits quicker processing of information by enabling a new set of instructions to start executing before the previous set has finished.

I

identity theft The process by which someone uses personal information about someone else (such as the victim's name, address, and Social Security number) to assume the victim's identity for the purpose of defrauding another.

if else In C++, a binary decision in the code where the program can follow one of two paths: If the decision is made one way, the program follows one path; if made the other way (else), the program follows another path.

image backup (system backup) A copy of an entire computer system, created for restoration purposes.

image-editing software Programs for editing photographs and other images.

incremental backup (partial backup) A type of backup that only backs up files that have changed since the last time files were backed up.

information Data that has been organized or presented in a meaningful fashion; the result, or output that users require at the end of a job.

information assurance Ensuring that information systems contain accurate information and are adequately secured against tampering.

information system A system that includes data, people, procedures, hardware, and software that help in planning and decision making; a software-based solution used to gather and analyze information.

information technology (IT) The set of techniques used in processing and retrieving information.

inheritance In object-oriented analysis, the ability of a new class to automatically pick up all the data and methods of an existing class and then extend and customize those to fit its specific needs.

initial value A beginning point.

inkjet printer A nonimpact printer that sprays tiny drops of ink onto paper.

inoculation A process used by antivirus software; compares old and current qualities of files to detect viral activity.

input device A hardware device used to enter, or input, data (text, images, and sounds) and instructions (user responses and commands) into a computer.

input forms Used to control how new data is entered into a shared database.

instant messaging (IM) A program that enables users to communicate online in real time with others who are also online.

instruction set The collection of commands that a specific CPU can execute. Each CPU has its own unique instruction set.

intangible personal property Property that can't be touched—or potentially even seen—yet it still has value.

integrated circuits (or chips) Tiny regions of semiconductor material that support a huge number of transistors.

integrated development environment (IDE) A developmental tool that helps programmers write and test their programs; one IDE can be configured to support many different languages.

intellectual property Refers to products derived from the mind, such as works of art and literature, inventions, and software code.

intelligence The ability to acquire and apply knowledge and skills.

intelligent personal assistant Software designed to perform tasks or services for individuals.

interactive white board An input/output device that projects the computer's display onto a surface, which allows users to interface with the board and computing device.

Internet A network of networks that's the largest network in the world, connecting billions of computers globally.

Internet backbone The main pathway of high-speed communications lines over which all Internet traffic flows.

Internet cache A section of the hard drive that stores information that may be needed again, such as Internet protocol addresses and frequently accessed web pages.

Internet Corporation for Assigned Names and Numbers (ICANN) The organization that registers Internet protocol addresses to ensure they're unique and haven't been assigned to other users.

Internet exchange point (IXP) A way of connecting Internet service providers (ISPs) that's made up of one or more network switches to which the ISPs connect.

Internet of Things (IoT) The interconnection of uniquely identifiable embedded computing devices that transfer data over a network without requiring human-to-human or human-to-computer interaction.

Internet protocol (IP) One of the original two protocols that were developed for the Internet (the other was TCP).

Internet protocol (IP) address The means by which all computers connected to the Internet identify each other. It consists of a unique set of four numbers separated by dots, such as 123.45.178.91.

Internet protocol version 4 (IPv4) The original Internet protocol addressing scheme, created in 1981.

Internet protocol version 6 (IPv6) An Internet protocol addressing scheme that makes IP addresses longer, thereby providing more available addresses.

Internet service provider (ISP) A company that specializes in providing Internet access. ISPs may be specialized providers, like Juno, or companies that provide other services in addition to Internet access (such as phone and cable television).

interpreter For a programming language, translates the source code into an intermediate form, line by line; each line is then executed as it's translated.

interrupt A signal that tells the operating system that it's in need of immediate attention.

interrupt handler A special numerical code that prioritizes requests from various devices. These requests then are placed in the interrupt table in the computer's primary memory.

intranet A private network set up by a business or an organization that's used exclusively by a select group of employees, customers, suppliers, volunteers, or supporters.

iOS The operating system for Apple mobile devices like the iPhone and iPad.

IP address *See Internet protocol address.*

J

jam signal A special signal sent to network nodes alerting them that a data collision has occurred.

Java An object-oriented programming language that has a large set of existing classes.

Java applet A small Java-based program.

JavaScript An Internet protocol (IP) addressing scheme that makes IP addresses longer, thereby providing more available addresses.

Javascript Object Notation (JSON) A syntax for exchanging information between computers.

JavaServer Pages (JSP) Programming language used to build websites with interactive capabilities; adapts the HTML page to the user's selections.

join query A query used to extract data that's in two or more tables in a database.

K

kernel The essential component of the operating system that's responsible for managing the processor and all other components of the computer system. Because it stays in random access memory (RAM) the entire time the computer is powered on, the kernel is called *memory resident*.

key pair The two keys used in public-key encryption.

keystroke logger (keylogger) A type of spyware program that monitors keystrokes with the intent of stealing passwords, login IDs, or credit card information.

keyword (1) A specific word a user wishes to query (or look for) in an Internet search. (2) A specific word that has a predefined meaning in a particular programming language.

kilobyte (KB) A unit of computer storage equal to approximately 1,000 bytes.

knowledge-based system A business intelligence system that provides intelligence that supplements the user's own intellect and makes the decision support system more effective.

knowledge representation Encoding information about the world into formats that an AI system can understand.

L

laptop (or notebook) computer A portable computer with a keyboard, a monitor, and other devices integrated into a single compact case.

large-format printer A printer that prints on oversized paper. Often used for creating banners and signs.

laser printer A nonimpact printer known for quick and quiet production and high-quality printouts.

latency (rotational delay) The process that occurs after the read/write head of the hard drive locates the correct track and then waits for the correct sector to spin to the read/write head.

light-emitting diode (LED) An energy-efficient technology used in monitors. It may result in better color accuracy and thinner panels than traditional LCD monitors.

linked data Data that is formally defined and that can be expressed in relationships.

Linux An open-source operating system based on UNIX. Because of the stable nature of this operating system, it's often used on web servers.

liquid crystal display (LCD) The technology used in flat-panel computer monitors.

local area network (LAN) A network in which the nodes are located within a small geographic area.

locally installed software Software that is installed on your device's internal storage.

logic bomb A computer virus that runs when a certain set of conditions is met, such as when a program is launched a specific number of times.

logical error An error in a program that produces unintended or undesired output, where the syntax is correct but some other human error has occurred.

logical port A virtual communications gateway or path that enables a computer to organize requests for information (such as web page downloads and e-mail routing) from other networks or computers.

logical port blocking A condition in which a firewall is configured to ignore all incoming packets that request access to a certain port so that no unwanted requests will get through to the computer.

loop A type of decision point in an algorithm. In a loop, a question is asked, and if the answer is *yes*, a set of actions is performed. Once the set of actions has finished, the question is asked again, creating a loop. If the answer to the question is *no*, the algorithm breaks free of the loop and moves on to the first step that follows the loop.

M

machine cycle The series of steps a central processing unit goes through when it performs a program instruction.

machine language The sequence of bits that the CPU understands.

machine learning A form of artificial intelligence that provides computers with the ability to learn without being explicitly programmed.

macOS The first commercially available operating system to incorporate a graphical user interface (GUI) with user-friendly point-and-click technology.

macro A small program that groups a series of commands to run as a single command.

macro virus A virus that's distributed by hiding it inside a macro.

mainframe A large, expensive computer that supports hundreds or thousands of users simultaneously and executes many different programs at the same time.

malware Software that's intended to render a system temporarily or permanently useless or to penetrate a computer system completely for purposes of information gathering. Examples include spyware, viruses, worms, and Trojan horses.

management information system (MIS) A type of business intelligence system that provides timely and accurate information that enables managers to make critical business decisions.

many-to-many relationship In a relational database, records in one table are related to multiple records in a second table, and vice versa.

master boot record A small program that runs whenever a computer boots up.

media access control (MAC) address The physical address, similar to a serial number, of each network adapter.

megabyte (MB) A unit of computer storage equal to approximately 1 million bytes.

memory module (memory card) A small circuit board that holds a series of random access memory (RAM) chips.

menu-driven interface A user interface in which the user chooses a command from menus displayed on the screen.

metadata Data that describes other data.

metasearch engine A search engine, such as Dogpile, that searches other search engines rather than individual websites.

method The process of how a program converts inputs into the correct outputs.

metropolitan area network (MAN) A wide area network (WAN) that links users in a specific geographic area (such as within a city or county).

microblog Social media in which users post short text with usually frequent updates.

microprocessors The chips that contain a central processing unit (CPU, or processor).

Microsoft account Registered user profile with specific user id and password to log into Windows account from any machine and access familiar desktop and applications.

mobile broadband Connection to the Internet through the same cellular network that cell phones use to get 3G or 4G Internet access.

mobile commerce (m-commerce) Conducting commercial transactions online through a smartphone, tablet, or other mobile device.

mobile hotspot Devices that enable you to connect more than one device to the Internet; they require access to a data plan. Most smartphones have this capability built-in.

mobile operating system An operating system that includes many features of personal computer operating systems but are modified to be more functional on handheld devices.

model management system Software that assists in building management models in decision support systems; an analysis tool that, through the use of internal and external data, provides a view of a particular business situation for the purposes of decision making.

model release A release that usually grants a photographer the right to use an image of the model (or subject of the photo) commercially.

modem A device that connects a network to the Internet.

Moore's Law A prediction, named after Gordon Moore, the co-founder of Intel; states that the number of transistors on a central processing unit chip will double every two years.

motherboard A special circuit board in the system unit that contains the central processing unit, the memory (RAM) chips, and the slots available for expansion cards; all of the other boards (video cards, sound cards, and so on) connect to it to receive power and to communicate.

mouse A hardware device used to enter user responses and commands into a computer.

multidimensional database A database that stores data from different perspectives, called *dimensions*, and organizes data in cube format.

multi-factor authentication A process that requires two of the three assigned factors be demonstrated before authentication is granted.

multimedia software Programs that include image-, video-, and audio-editing software, animation software, and other specialty software required to produce computer games, animations, and movies.

multi-partite virus Literally meaning "multi-part" virus; a type of computer virus that attempts to infect computers using more than one method.

Multipurpose Internet Mail Extensions (MIME) Specification for sending files as attachments to e-mail.

multi-task The ability of an operating system to perform more than one process at a time.

multi-user operating system (network operating system) An operating system that enables more than one user to access the computer system at one time by efficiently juggling all the requests from multiple users.

N

natural language processing A feature of AI systems that allows the system to understand written and spoken words and to interact with humans using language.

natural language processing (NLP) system A knowledge-based business intelligence system that enables users to communicate with computer systems using a natural spoken or written language instead of a computer programming language.

negative acknowledgment (NAK) In data exchange, the communication sent from one computer or system to another stating that it did not receive a data packet in readable form.

network A group of two or more computers (or nodes) that are configured to share information and resources such as printers, files, and databases.

network adapter A device that enables the computer (or peripheral) to communicate with the network using a common data communication language, or protocol.

network address translation (NAT) A process that firewalls use to assign internal Internet protocol addresses on a network.

network administration Involves tasks such as (1) installing new computers and devices, (2) monitoring the network to ensure it's performing efficiently, (3) updating and installing new software on the network, and (4) configuring, or setting up, proper security for a network.

network administrator Person who maintains networks for a business or organization.

network architecture The design of a computer network; includes both physical and logical design.

network-attached storage (NAS) device A specialized computing device designed to store and manage network data.

network navigation device A device on a network such as a router or switch that moves data signals around the network.

network operating system (NOS) Software that handles requests for information, Internet access, and the use of peripherals for network nodes, providing the services necessary for the computers on the network to communicate.

network-ready device A device (such as a printer or an external hard drive) that can be attached directly to a network instead of needing to attach to a computer on the network.

network topology The physical or logical arrangement of computers, transmission media (cable), and other network components.

Next In Visual Basic, programmers use the keyword Next to implement a loop; when the Next command is run, the program returns to the For statement and increments the value of the input or output item by 1 and then runs a test cycle.

node A device connected to a network such as a computer, a peripheral (such as a printer), or a communications device (such as a modem).

nonvolatile storage Permanent storage, as in read-only memory (ROM).

normalization A process to ensure data is organized most efficiently in a database.

NoSQL database A nonrelational database developed to store unstructured data generated primarily from Web 2.0 applications such as clickstream data, social media data, and location-based information.

number system An organized plan for representing a number.

numeric check In a database, confirms that only numbers are entered in a field.

O

object An example of a class in object-oriented analysis.

object-oriented analysis A type of analysis in which programmers first identify all the categories of inputs that are part of the problem the program is meant to solve.

object-oriented database A database that stores data in objects rather than in tables.

Object Query Language (OQL) A query language used by many object-oriented databases.

Objective C The programming language most often used to program applications to run under OS X.

octet A reference to each of the four numbers in a dotted decimal number Internet protocol address, so called because each number would have eight numerals in binary form.

one-to-many relationship In a relational database, a record appears only once in one table while having the capability of appearing many times in a related table.

one-to-one relationship In a relational database, each record in a table has only one corresponding record in a related table.

online analytical processing (OLAP) Software that provides standardized tools for viewing and manipulating data in a data warehouse.

online reputation Information available about you in cyberspace that can influence people's opinions of you in the real world.

online transaction processing (OLTP) The real-time processing of database transactions online.

OpenPGP A free public-key standard for encryption available for download on the Internet.

open-source software Program code made publicly available for free; it can be copied, distributed, or changed without the stringent copyright protections of proprietary software products.

open system A system having the characteristic of being public for access by any interested party; as opposed to a *proprietary system*.

Open Systems Interconnection (OSI) A networking protocol established by the Institute of Electrical and Electronics Engineers (IEEE) that provides guidelines for modern networks.

operating system (OS) The system software that controls the way in which a computer system functions, including the management of hardware, peripherals, and software.

operator A coding symbol that represents a fundamental action of the programming language.

organic light-emitting diode (OLED) displays Displays that use organic compounds to produce light when exposed to an electric current. Unlike LCDs, OLEDs do not require a backlight to function and therefore draw less power and have a much thinner display, sometimes as thin as 3 mm.

output device A device that sends processed data and information out of a computer in the form of text, pictures (graphics), sounds (audio), or video.

overclocking Running the central processing unit at a speed faster than the manufacturer recommends.

P

packet (data packet) A small segment of data that's bundled for sending over transmission media. Each packet contains the address of the computer or peripheral device to which it's being sent.

packet analyzer (sniffer) A computer hardware device or software program designed to detect and record digital information being transmitted over a network.

packet filtering A process in which firewalls are configured so that they filter out packets sent to specific logical ports.

packet screening Having a screening router examine data packets to ensure they originated from or are authorized by valid users on a network.

packet switching A communications methodology that makes computer communication efficient by breaking up data into smaller chunks called *packets*.

paging The process of swapping data or instructions that have been placed in the swap file for later use back into active random access memory (RAM). The contents of the hard drive's swap file then become less active data or instructions.

paper mills Websites that sell prewritten or custom-written research papers to students.

parallel processing A large network of computers, with each computer working on a portion of the same problem simultaneously

passive topology In a network, a type of topology where the nodes do nothing to move data along the network.

patents Grant inventors the right to stop others from manufacturing, using, or selling (including importing) their inventions for a period of 20 years from the date the patent is filed.

path (subdirectory) The information after the slash that indicates a particular file or path (or subdirectory) within the website.

path separator The backslash mark (\) used by Microsoft Windows and DOS in file names. Mac files use a colon (:), and UNIX and Linux use the forward slash (/) as the path separator.

peer-to-peer (P2P) network A network in which each node connected to the network can communicate directly with every other node on the network.

peer-to-peer (P2P) sharing The process of users transferring files between computers.

peripheral device A device such as a monitor, printer, or keyboard that connects to the system unit through a data port.

persistence of information The tendency for pictures, videos, and narratives to continue to exist in cyberspace even after you delete them.

personal area network (PAN) A network used for communication among devices close to one person, such as smartphones, laptops, and tablets, using wireless technologies such as Bluetooth.

personal ethics The set of formal or informal ethical principles that an individual uses to guide their ethical decisions.

personal firewall A firewall specifically designed for home networks.

personal information manager (PIM) software Programs such as Microsoft Outlook or Lotus Organizer that strive to replace the various management tools found on a traditional desk such as a calendar, address book, notepad, and to-do lists.

personal property Things of value that you own (such as a car or laptop).

petabyte 10^{15} bytes of digital information.

pharming Planting malicious code on a computer that alters the browser's ability to find web addresses and that directs users to bogus websites.

phishing The process of sending e-mail messages to lure Internet users into revealing personal information such as credit card or Social Security numbers or other sensitive information that could lead to identity theft.

PHP (PHP: Hypertext Preprocessor) Programming language used to build websites with interactive capabilities; adapts the HTML page to the user's selections.

physical memory The amount of random access memory (RAM) that's installed in a computer.

piggybacking The process of connecting to a wireless network without the permission of the owner of the network.

pinning The process through which you choose which applications are visible on the Windows Start screen.

pipelining A technique that allows the CPU to work on more than one instruction (or stage of processing) at the same time, thereby boosting CPU performance.

pixel A single point that creates the images on a computer monitor. Pixels are illuminated by an electron beam that passes rapidly back and forth across the back of the screen so that the pixels appear to glow continuously.

plagiarism The act of copying text or ideas from someone else and claiming them as your own.

platform The combination of a computer's operating system and processor. The two most common platform types are the PC and the Apple.

platter A thin, round, metallic storage plate stacked onto the hard drive spindle.

Plug and Play (PnP) The technology that enables the operating system, once it is booted up, to recognize automatically any new peripherals and to configure them to work with the system.

podcast A clip of audio or video content that's broadcast over the Internet using compressed audio or video files in formats such as MP3.

point of presence (POP) A bank of modems, servers, routers, and switches through which Internet users connect to an Internet service provider.

polymorphic virus A virus that changes its virus signature (the binary pattern that makes the virus identifiable) every time it infects a new file. This makes it more difficult for antivirus programs to detect the virus.

port An interface through which external devices are connected to the computer.

portability The capability to move a completed solution easily from one type of computer to another.

positive acknowledgment (ACK) In data exchange, the confirmation sent from one computer or system to another saying that the computer has received a data packet that it can read.

positive psychology A field that attempts to discover the causes of happiness rather than to address the treatment of mental dysfunctions.

possessed object Any object that users carry to identify themselves and that grants them access to a computer system or facility.

power supply Regulates the wall voltage to the voltages required by computer chips; it's housed inside the system unit.

power-on self-test (POST) The first job the basic input/output system (BIOS) performs, ensuring that essential peripheral devices are attached and operational. This process consists of a test on the video card and video memory, a BIOS identification process (during which the BIOS version, manufacturer, and data are displayed on the monitor), and a memory test to ensure memory chips are working properly.

preemptive multi-tasking When the operating system processes the task assigned a higher priority before processing a task that has been assigned a lower priority.

presentation software An application program for creating dynamic slide shows, such as Microsoft PowerPoint or Apple Keynote.

pretexting The act of creating an invented scenario (the pretext) to convince someone to divulge information.

Pretty Good Privacy (PGP) A public-key package for encryption available for download on the Internet.

primary key field In a database, a field that has a value unique to a record.

print queue A software holding area for print jobs; also called a *print spooler*.

print server A server that manages all client-requested printing jobs for all printers on a network.

private key The key for decoding retained as private in public-key encryption.

private-key encryption A type of encryption where only the two parties involved in sending the message have the code.

problem statement The starting point of programming work; a clear description of what tasks the computer program must accomplish and how the program will execute those tasks and respond to unusual situations.

processing Manipulating or organizing data into information.

productivity software Programs that enable a user to perform various tasks generally required in home, school, and business. Examples include word processing, spreadsheet, presentation, personal information management, and database programs.

program A series of instructions to be followed by a computer to accomplish a task.

program development life cycle (PDLC) The process of performing a programming project, which consists of five stages: describing the problem, making a plan, coding, debugging, and testing and documentation.

program file Files that are used in the running of software programs and that do not store data.

program specification A clear statement of the goals and objectives of a project.

programming The process of translating a task into a series of commands a computer will use to perform that task.

programming language A kind of "code" for the set of instructions the CPU knows how to perform.

proprietary (commercial) software Custom software application that's owned and controlled by the company that created it.

proprietary system A system having the characteristic of being closed to public access (private), as opposed to an *open system*.

protocol A set of rules for conducting communications.

prototype A small model of a program built at the beginning of a large project.

proxy server A server that acts as a go-between, connecting computers on an internal network with those on external networks (like the Internet).

pseudocode A text-based approach to documenting an algorithm.

public domain Copyrightable works whose copyright has expired, or that never had copyright protection.

public key The key for coding distributed to the public in public-key encryption.

public-key encryption A type of encryption where two keys, known as a *key pair*, are created. One key is used for coding and the other for decoding. The key for coding is distributed as a public key, whereas the private key is retained for decoding.

Q

quarantining The placement (by antivirus software) of a computer virus in a secure area on the hard drive so that it won't spread infection to other files.

query In a database, a way of retrieving information that defines a particular subset of data; can be used to extract data from one or more tables.

query language A specially designed computer language used to manipulate data in or extract data from a database.

R

RaaS (Ransomware as a Service) Allows less technically skilled criminals to still conduct ransomware attacks by using the code of a master developer.

RAID 0 The strategy of running two hard drives in one system, cutting in half the time it takes to write a file.

RAID 1 The strategy of mirroring all the data written on one hard drive to a second hard drive, providing an instant backup of all data.

random access memory (RAM) The computer's temporary storage space or short-term memory. It's located in a set of chips on the system unit's motherboard, and its capacity is measured in megabytes or gigabytes.

range check Ensures that the data entered into the field of a database falls within a certain range of values.

ransomware Type of malware that attacks a system by encrypting critical files and then demands payment so that files can be restored.

rapid application development (RAD) An alternative program-development method; instead of developing detailed system documents before they produce the system, developers first create a prototype, then generate system documents as they use and remodel the product.

read-only memory (ROM) A set of memory chips, located on the motherboard, that stores data and instructions that cannot be changed or erased; it holds all the instructions the computer needs to start up.

read/write head The mechanism that retrieves (reads) and records (writes) the magnetic data to and from a data disk.

real property Immoveable property such as land or a home (often called *real estate*).

real-time operating system (RTOS) A program with a specific purpose that must guarantee certain response times for particular computing tasks or else the machine's application is useless. Real-time operating systems are found in many types of robotic equipment.

real-time processing The processing of database transactions in which the database is updated while the transaction is taking place.

Really Simple Syndication (RSS) An XML–based format that allows frequent updates of content on the web.

recommendation engines AI systems that help people discover things they may like but are unlikely to discover on their own.

record A group of related fields in a database.

recovery drive A drive that contains all the information needed to reinstall your operating system if it should become corrupted. Often the manufacturer will have placed a utility on your system to create this.

Recycle Bin A folder on a Windows desktop in which deleted files from the hard drive are held until permanently purged from the system.

redundant array of independent disks (RAID) A set of strategies for using more than one hard drive in a computer system.

referential integrity In a relational database, a condition where, for each value in the foreign key table, there's a corresponding value in the primary key table.

registers Special memory storage areas built into the CPU, which are the most expensive, fastest memory in your computer.

registry A portion of the hard drive containing all the different configurations (settings) used by the Windows operating system as well as by other applications.

relational algebra The use of English-like expressions that have variables and operations, much like algebraic equations, to extract data from databases using Structured Query Language.

relational database A database type that organizes data into related tables based on logical groupings.

relationship In a relational database, a link between tables that defines how the data is related.

release to manufacturers (RTM) The point in the release cycle, where, after beta testing, a manufacturer makes changes to the software and releases it to other manufacturers, for installation on new machines, for example.

Reset this PC A utility program in Windows 10 that attempts to diagnose and fix errors in Windows system files that are causing a computer to behave improperly.

resolution The clarity or sharpness of an image, which is controlled by the number of pixels displayed on the screen.

restore point A type of backup that doesn't affect your personal data files but saves all the apps, updates, drivers, and information needed to restore your computer system to the exact way it's configured at that time. That way, if something does go awry, you can return your system to the way it was before you started.

reusability In object-oriented analysis, the ability to reuse existing classes from one project for another project.

ring (loop) topology A type of network topology where computers and peripherals are laid out in a configuration resembling a circle.

root directory The top level of the filing structure in a computer system. In Windows computers, the root directory of the hard drive is represented as C:\.

root DNS server A domain name system (DNS) server that contains the master listings for an entire top-level domain.

rootkit Programs that allow hackers to gain access to your computer and take almost complete control of it without your knowledge. These programs are designed to subvert normal logon procedures to a computer and to hide their operations from normal detection methods.

router A device that routes packets of data between two or more networks.

rules-based systems Software that asks questions and responds based on preprogrammed algorithms. Early attempts at expert systems were rules based.

runtime error An error in a program that occurs when a programmer accidentally writes code that divides by zero, a mathematical error.

S

sampling rate The number of times per second a signal is measured and converted to a digital value. Sampling rates are measured in kilobits per second.

satellite Internet A way to connect to the Internet using a small satellite dish, which is placed outside the home and is connected to a computer with coaxial cable. The satellite company then sends the data to a satellite orbiting Earth. The satellite, in turn, sends the data back to the satellite dish and to the computer.

scalability A characteristic of client/server networks where more users can be easily added without affecting the performance of other network nodes.

scareware A type of malware that's downloaded onto your computer and that tries to convince you that your computer is infected with a virus or other type of malware.

scope creep An ever-changing set of requests from clients for additional features as they wait longer and longer to see a working prototype.

script A list of commands (mini-programs or macros) that can be executed on a computer without user interaction.

scripting language A simple programming language that's limited to performing a set of specialized tasks.

search engine A set of programs that searches the web for specific words (or keywords) you wish to query (or look for) and that then returns a list of the websites on which those keywords are found.

search engine optimization (SEO) Designing a website to ensure ranking near the top of search engine results.

second-generation language (2GL) A computer language that allows programmers to write programs using a set of short, English-like commands that speak directly to the central processing unit and that give the programmer direct control of hardware resources; also called *assembly language*.

second-level domain A domain that's directly below a top-level domain.

sector A section of a hard drive platter, wedge-shaped from the center of the platter to the edge.

Secure Sockets Layer (SSL) A network security protocol that provides for the encryption of data transmitted using the Internet. The current versions of all major web browsers support SSL.

seek time The time it takes for the hard drive's read/write heads to move over the surface of the disk to the correct track.

select query In Structured Query Language, displays a subset of data from a table (or tables) based on the criteria specified.

semantic web (Web 3.0) An evolving extension of the web in which information is defined in such a way as to make it more easily readable by computers.

sensor Any device that detects or measures something. Computing devices, especially smartphones, contain many sensors.

server A computer that provides resources to other computers on a network.

server-side program A type of program that runs on a web server rather than on a computer.

service mark Essentially the same as a trademark, but it applies to a service as opposed to a product.

service pack A software update.

service set identifier (SSID) A network name that wireless routers use to identify themselves.

shielded twisted-pair (STP) cable Twisted-pair cable that contains a layer of foil to reduce interference.

Simple Mail Transfer Protocol (SMTP) The protocol responsible for sending e-mail along the Internet to its destination; part of the Internet Protocol suite.

simulation programs Software, often used for training purposes, that allows the user to experience or control an event as if it's reality.

Sleep mode A low-power mode for computing devices that saves electric power consumption and saves the last-used settings. When the device is "woken up," work is resumed more quickly than when cold-booting the device.

smart home A dwelling in which devices and appliances are automated or controlled by apps.

smartphone Phone with features of a computer including a wide assortment of apps, media players, high-quality cameras, and web connectivity.

social bookmarking Assigning a keyword or term to a web resource such as a web page, digital image, or video.

social commerce A subset of e-commerce that uses social networks to assist in marketing and purchasing products.

social engineering Any technique that uses social skills to generate human interaction for the purpose of enticing individuals to reveal sensitive information.

social media Websites or apps that allow users to create and share content and/or participate in social networking with others.

social networking A means by which people use the Internet to communicate and share information among their immediate friends and to meet and connect with others through common interests, experiences, and friends.

software The set of computer programs or instructions that tells the computer what to do and that enables it to perform different tasks.

software license An agreement between the user and the software developer that must be accepted before installing the software on a computer.

software piracy Violating a software license agreement by copying an application onto more computers than the license agreement permits.

Software as a Service (SaaS) Software that's delivered on demand over the Internet.

solid-state drive (SSD) A storage device that uses the same kind of memory that flash drives use but that can reach data in only a tenth of the time a flash drive requires.

solid-state hybrid drive (SSHD) A drive that is a combination of both a mechanical hard drive and an SSD in a single device.

sound card An expansion card that attaches to (or is integrated into) the motherboard inside the system unit and that enables the computer to produce sounds; also provides a connection for the speakers and microphone.

source code The instructions programmers write in a higher-level language.

spam Unwanted or junk e-mail.

spam filter An option you can select in your e-mail account that places known or suspected spam messages into a folder other than your inbox.

spear phishing A targeted phishing attack that sends e-mails to people known to be customers of a company. Such attacks have a much greater chance of successfully getting individuals to reveal sensitive data.

spooler A program that helps coordinate all print jobs being sent to the printer at the same time.

spreadsheet software An application program such as Microsoft Excel or Lotus 1-2-3 that enables a user to do calculations and numerical analyses easily.

spyware An unwanted piggyback program that downloads with the software you want to install from the Internet and then runs in the background of your system.

star topology The most widely deployed client/server network topology, where the nodes connect to a central communications device called a *switch* in a pattern resembling a star.

Start menu A feature in Windows 10 that provides access to all applications in one convenient screen.

statement A sentence in a code.

static addressing A way of assigning Internet protocol addresses where the address for a computer never changes and is most likely assigned manually by a network administrator or an Internet service provider.

stealth virus A virus that temporarily erases its code from the files where it resides and hides in the active memory of the computer.

structured (analytical) data Data such as "Bill" or "345," as opposed to *unstructured data*.

Structured Query Language (SQL) A database programming language used to construct queries to extract data from relational databases; one example of a fourth-generation language.

stylus A pen-shaped device used to tap or write on touch-sensitive screens.

subscription software A model whereby the user pays a fee to use the software. The software is downloaded and installed locally but is routinely updated by connection to the manufacturer's server.

summary report A report generated by a management information system that provides a consolidated picture of detailed data; these reports usually include some calculation or visual displays of information.

supercomputer A specially designed computer that can perform complex calculations extremely rapidly; used in situations in which complex models requiring intensive mathematical calculations are needed (such as weather forecasting or atomic energy research).

SuperFetch A memory-management technique used by Windows. Monitors the applications you use the most and preloads them into your system memory so that they'll be ready to go.

supervised learning Training an AI system using a huge number of examples.

surge protector A device that protects computers and other electronic devices from power surges.

surround sound A type of audio processing that makes the listener experience sound as if it were coming from all directions.

surround-sound system A system of speakers set up in such a way that it surrounds an entire area (and the people in it) with sound.

swap file (page file) A temporary storage area on the hard drive where the operating system "swaps out" or moves the data or instructions from random access memory (RAM) that haven't recently been used. This process takes place when more RAM space is needed.

Swift A programming language introduced by Apple for developing for iOS and macOS.

switch A device for transmitting data on a network that makes decisions, based on the media access control address of the data, as to where the data is to be sent.

syntax An agreed-on set of rules defining how a language must be structured.

syntax error A violation of the strict set of rules that define the programming language.

system clock This internal clock is actually a special crystal that acts like a metronome, keeping a steady beat and controlling when the CPU moves to the next stage of processing.

system development life cycle (SDLC) A process used to develop information systems; it consists of the following six steps: problem and opportunity identification, analysis, design, development, testing and installation, and maintenance and evaluation.

system evaluation The process of looking at a computer's subsystems, what they do, and how they perform to determine whether the computer system has the right hardware components to do what the user ultimately wants it to do.

system files The main files of an operating system.

system requirements The set of minimum storage, memory capacity, and processing standards recommended by the software manufacturer to ensure proper operation of a software application.

System Restore A utility in Windows that restores system settings to a specific previous date when everything was working properly.

system restore point In Windows, a snapshot of your entire system's settings used for restoring your system to a prior point in time.

system software The set of programs that enables a computer's hardware devices and application software to work together; it includes the operating system and utility programs.

system unit The metal or plastic case that holds all the physical parts of the computer together, including the computer's processor (its brains), its memory, and the many circuit boards that help the computer function.

T

T line A communications line that carries digital data over twisted-pair wires.

table (file) A group of related records in a database.

tablet computer A mobile computer, such as the Apple iPad, integrated into a flat multi-touch-sensitive screen. It uses an onscreen virtual keyboard.

tag Electronically identifying someone in a photo or post.

tangible personal property Property that has substance like your smartphone or car.

Task Manager A Windows utility that shows programs currently running and permits you to exit unresponsive programs when you click End Task.

taskbar In later versions of Windows operating systems, a feature that displays open and favorite applications for easy access.

tax preparation software An application program, such as Intuit's TurboTax or H&R Block's At Home, for preparing state and federal taxes. Each program offers a complete set of tax forms and instructions as well as expert advice on how to complete each form.

TCP/IP The main suite of protocols used for transmitting data over the Internet; short for Transmission Control Protocol (TCP) and Internet Protocol (IP).

template A form included in many productivity applications that provides the basic structure for a particular kind of document, spreadsheet, or presentation.

terabyte (TB) 1,099,511,627,776 bytes or 2^{40} bytes.

terminator A device that absorbs a signal so that it's not reflected back onto parts of the network that have already received it.

terms of use Legal documents (or disclaimers) that delineate how copyrighted material can be used.

test condition A check to see whether the loop in an algorithm is completed.

testing plan The part of the problem statement that lists specific input numbers the programmers would typically expect the user to enter; the plan then lists the precise output values that a perfect program would return for those input values.

texting Using the Short Message Service (SMS) to send short messages between mobile devices.

third-generation language (3GL) A computer language that uses symbols and commands to help programmers tell the computer what to do, making 3GL languages easier for humans to read and remember; examples of 3GL languages include BASIC, FORTRAN, COBOL, C/C++, and JAVA.

thrashing A condition of excessive paging in which the operating system becomes sluggish.

three-way handshake In Transmission Control Protocol, the process used to establish a connection between two computers before exchanging data. The steps in a three-way handshake are as follows: One computer establishes a connection to the Internet service provider (ISP) and announces it has e-mail to send, the ISP server responds that it's ready to receive, and the computer acknowledges the ready state of the server and begins to transmit the e-mail.

throughput The actual speed of data transfer that's achieved. It's usually less than the data transfer rate.

Thunderbolt port A high-speed input/output port.

time bomb A virus that's triggered by the passage of time or on a certain date.

time-variant data Data that doesn't pertain to one time period.

token A special data packet used to pass data on a network.

token method The access method used by ring networks to avoid data collisions.

top-down design A systematic approach in which a problem is broken up into a series of high-level tasks.

top-level domain The suffix, often of three letters (such as .com or .edu), in the domain name that indicates the kind of organization the host is.

Tor (The Onion Router) browser An app used to access the dark web, which allows you to locate hidden websites in the .onion domain.

touch pad (trackpad) A small, touch-sensitive surface at the base of a laptop keyboard that's used to direct the cursor.

touch screen A type of monitor (or display in a smartphone or tablet computer) that accepts input from a user touching the screen.

track A concentric circle that serves as a storage area on a hard drive platter.

trademark A word, phrase, symbol, or design—or a combination of these—that uniquely identifies and differentiates the goods (or services) of one party from those of another.

transaction-processing system (TPS) A type of business intelligence system for keeping track of everyday business activities.

Transmission Control Protocol (TCP) One of the original two protocols developed for the Internet.

transmission media The radio waves or the physical system (cable) that transports data on a network.

Transport Layer Security (TLS) An updated extension of the *secure sockets layer*.

Trojan horse A computer program that appears to be something useful or desirable (such as a game or a screen saver), but at the same time does something malicious in the background without the user's knowledge.

trolling The act of posting inflammatory remarks online for the sheer pleasure of soliciting an angry or negative response.

tunneling The main technology for achieving a virtual private network; the placement of data packets inside other data packets.

Turing test A simple test to distinguish between a human and a computer system. Named after Alan Turing, a computing pioneer.

twisted-pair cable Cable made of copper wires that are twisted around each other and are surrounded by a plastic jacket (such as traditional home phone wire).

U

ultrabook A full-featured but lightweight laptop computer that features a low-power processor and a solid-state drive; it tries to reduce its size and weight to extend battery life without sacrificing performance.

unethical behavior Not conforming to a set of approved standards of behavior—cheating on an exam, for example

unicode An encoding scheme that uses 16 bits instead of 8 bits. Unicode can represent nearly 1,115,000 code points and currently assigns more than 128,000 unique character symbols.

Uniform Resource Locator (URL) A website's unique address; an example is microsoft.com.

universal serial bus (USB) port A port that can connect a wide variety of peripheral devices to the computer, including keyboards, printers, mice, smartphones, external hard drives, flash drives, and digital cameras.

UNIX An operating system originally conceived in 1969 by Ken Thompson and Dennis Ritchie of AT&T's Bell Labs. In 1974, the UNIX code was rewritten in the standard programming language C. Today there are various commercial versions of UNIX.

unshielded twisted-pair (UTP) cable The most popular transmission media option for Ethernet networks. UTP cable is composed of four pairs of wires that are twisted around each other to reduce electrical interference.

unsupervised learning When an AI system can look at data on its own and build rules for deciding what it is seeing.

unstructured data Audio clips, video clips, pictures, and extremely large documents. Also called *binary large object (BLOB)*.

urban legends When hoaxes become so well known that they're incorporated into society as true events even though they're false.

User Datagram Protocol (UDP) An Internet protocol that creates data packets across the Internet.

user interface Part of the operating system that enables individuals to interact with the computer.

utility program A small program that performs many of the general housekeeping tasks for the computer, such as system maintenance and file compression.

V

validation The process of ensuring that data entered into a field of a database meets specified guidelines.

validation rule In a database, generally defined as part of the data dictionary and specified in the field properties for each field; violations result in an error message and a suggested action.

variable Each input and output item the program manipulates.

variable declaration Tells the operating system that the program needs to allocate storage space in RAM.

VB.NET An object-oriented programming language based in the Microsoft .NET framework.

VBScript A subset of Visual Basic, used to introduce dynamic decision making into web pages.

vertical market software Software that's developed for and customized to a specific industry's needs (such as a wood inventory system for a sawmill) as opposed to software that's useful across a range of industries (such as word processing software).

video card (video adapter) An expansion card that's installed inside a system unit to translate binary data (the *1*s and *0*s the computer uses) into the images viewed on the monitor.

video blog A personal online journal that uses video as the primary content in addition to text, images, and audio.

video memory Random access memory that's included as part of a video card.

virtual desktops A Windows 10 feature that allows you to organize groups of windows into different displays.

virtual memory The space on the hard drive where the operating system stores data if there isn't enough random access memory to hold all of the programs you're currently trying to run.

virtual private network (VPN) A network that uses the public Internet communications infrastructure to build a secure, private network among various locations.

virtual reality (VR) An artificial environment that is immersive and interactive.

virtualization Using specialized software to make individual physical servers behave as though they are more than one physical device.

virus A computer program that attaches itself to another computer program (known as the host program) and attempts to spread itself to other computers when files are exchanged.

virus signature A portion of the virus code that's unique to a particular computer virus and that makes it identifiable by antivirus software.

Visual Basic (VB) A programming language used to build a wide range of Windows applications.

visual programming languages Programming language that uses graphical blocks to represent control elements and variables.

voice recognition software Software that allows you to control your computing devices by speaking into the microphone instead of using a keyboard or mouse.

VoIP (voice over Internet protocol) A technology that facilitates making telephone calls across the Internet instead of using conventional telephone lines.

volatile storage Temporary storage, such as in random access memory. When the power is turned off, the data in volatile storage is cleared out.

W

warm boot The process of restarting a computing device while it's powered on.

Web 2.0 Tools and web-based services that emphasize online collaboration and sharing among users.

web authoring software Programs you can use to design interactive web pages without knowing any HyperText Markup Language (HTML) code.

web-based e-mail A type of e-mail system that's managed by a web browser and that allows access to e-mail from the web.

web browser (browser) Software installed on a computer system that allows individuals to locate, view, and navigate the web.

web server A computer running a specialized operating system that enables it to host web pages (and other information) and to provide requested web pages to clients.

web service Part of the Microsoft .NET Framework; programs that a website uses to make information available to other websites.

webcam A small camera that sits on top of a computer monitor or that's built into a computing device and is usually used to transfer live video.

webcast The (usually live) broadcast of audio or video content over the Internet.

white-hat hacker (ethical hacker) A hacker who breaks into systems just for the challenge of it (and who doesn't wish to steal or wreak havoc on the systems). Such hackers tout themselves as experts who are performing a needed service for society by helping companies realize the vulnerabilities that exist in their systems.

whole-house surge protector A surge protector that's installed on (or near) the breaker panel of a home and that protects all electronic devices in the home from power surges.

Wi-Fi The 802.11 standard for wireless data transmissions established by the Institute of Electrical and Electronics Engineers (IEEE).

wide area network (WAN) A network made up of local area networks (LANs) connected over long distances.

Windows Microsoft's operating system that incorporates a user-friendly, visual interface.

wireless access point (WAP) Gives wireless devices a sending and receiving connection point to the network.

wireless Internet service provider (wireless ISP) An ISP that provides service to wireless devices such as smartphones.

wireless network interface card (wireless NIC) A special network adapter card that allows a computing device to connect to a network using wireless media.

wireless range extender A device that amplifies your wireless signal to get it out to parts of your home that are experiencing poor connectivity.

wizard A step-by-step guide that walks a user through the necessary steps to complete a complicated task.

word processing software Programs used to create and edit written documents such as papers, letters, and résumés.

World Wide Web (WWW or the web) The part of the Internet used the most. What distinguishes the web from the rest of the Internet are (1) its use of common communication protocols (such as Transmission Control Protocol/Internet Protocol, or TCP/IP) and special languages (such as the HyperText Markup Language, or HTML) that enable different computers to talk to each other and display information in compatible formats and (2) its use of special links (called *hyperlinks*) that enable users to jump from one place to another on the web.

worm A program that attempts to travel between systems through network connections to spread infections. Worms can run independently of host file execution and are active in spreading themselves.

Z

zombie A computer that is controlled by a hacker who uses it to launch attacks on other computer systems.

Index

B

C

D

H

I

J

K

L

N

O

T

U

V

W

X

Y

Z